Food futures:

ethics, science and culture

Food futures:

ethics, science and culture

EurSafe 2016

Porto, Portugal

29 September – 1 October 2016

edited by:

I. Anna S. Olsson

Sofia M. Araújo

M. Fátima Vieira

Wageningen Academic
Publishers

EAN: 9789086862887
e-EAN: 9789086868346
ISBN: 978-90-8686-288-7
e-ISBN: 978-90-8686-834-6
DOI: 10.3921/978-90-8686-834-6

First published, 2016

Acknowledgements

This book and the associated conference are the result of collaborative work. The editors and the other members of the organizing committee (Júlio Borlido Santos, Luísa Neto, Paulo Vaz-Pires, Pedro Graça, Pedro Moreira, Rui Nunes) are grateful to the reviewers listed below for their review of and constructive input to the extended abstracts, to the EurSAFE board for their confidence and enthusiastic support during the conference preparation and to our host institutions at the University of Porto for highly professional support with all aspects of logistics and design. But above all, this book exists thanks to the authors who have contributed the nearly 100 papers presented here. The Luso-American Foundation and the Fundação para a Ciência e a Tecnologia have supported the conference financially, for which we are grateful. Thank you all!

Reviewers

S. Aerts
S.M.A. Araújo
F.H. Arion
V. Beusmann
J. Borlido Santos
B. Bovenkerk
J.V.J. de Tavernier
D.E. Dumitras
L. Escajedo San-Epifanio
N.H. Franco
P.M. Galvão
M. Gjerris
P. Graça
H.G. Grimm
S. Gunnarsson
K. Hagen
J. Hultgren
M.H. Huth
A. Inza
J.I.M. Jitea
M. Kaiser
L. Landeweerd
J.L. Lassen
M. Magalhães Sant'ana
F.L.B. Meijboom
S.M. Meisch
K. Millar
I.C. Muresan
M.L.N. Neto
R.M.L. Nunes
I.A.S. Olsson
F. Pirscher
C.B. Pocol
T. Potthast
H. Röcklinsberg
P. Sandin
S. Stirn
O. Varga
P.M.R. Vaz-Pires
F.V. Vieira
M.A.G. Von Keyserlingk
A. Wallenbeck

Table of contents

Section 5. Animal perspectives

Section 6. Corporate social responsibility and public procurement

Section 7. Food and literature

Section 8. Aquatic species

Section 12. Food policy

Section 13. Animal ethics in practice

Section 14. Food in communities

Section 15. Consumers and animal products

Section 16. Food in a local context

Section 17. Individuals and food choices

Section 18. Economy

Section 19. Animals in research

Section 20. Genetic modifications

Section 21. Food and environmental ethics

Section 22. Responsible research and innovation

Posters

Workshop

Preface

Food is at the centre of human existence. We eat every day, not only to satisfy our physical needs but also as part of cultural and social interaction. Food choices and markets shape the agricultural landscape and the cities we live in. Whereas what we choose to eat and feed our family is part of who we are, a growing number of actors compete to influence our food habits, through marketing strategies and nutritional advice. And ethical considerations are coupled with every choice over food – whether related to production, distribution, consumption, food waste, policy in general, marketing or advice. EurSAFE – European Society for Agricultural and Food Ethics provides a context for analysis and discussion of these topics among academic scholars as well as other practitioners in the field.

Given the variety of implications the 'food problem' entails, the construction of an inclusive society must redirect the concerns about food in the present to the imagination of future alternatives. 2016 marks the 500th anniversary of perhaps the most emblematic European literary work on such imaginations: Thomas More's Utopia. Celebrating this anniversary in the context of the conference is timely since utopian thinking will be instrumental in the search for innovative solutions to the food challenges. Utopian thinking forces us, first of all, to reflect on what we would consider to be an ideal society and to set goals – even knowing that they will never be totally reached. Second, utopian thinking is informed by the awareness that societies work as systems, and that if we change one aspect, all the other aspects will have to be changed as well. Third, utopian thinking forces us to reflect on alternative forms of organization, either by recycling solutions of the past and putting them into a new context, or by devising new ones. Thinking about food through holistic and prospective utopian thinking, aiming at the construction of an inclusive society where a variety of aspects are considered – the nature of families and communities; gender and class relations; rural/urban relations; the health and well-being of humans, other animals and ecosystems; the political system; decision-making, amongst others – was the challenge we set for the 2016 EurSAFE conference. The book you now have in your hand presents the responses of more than 80 authors to this challenge.

Keynote contributions

1. Food studies and utopia: why they need each other

Lyman Tower Sargent
University of Missouri-St. Louis, 1 University Blvd, St. Louis, MO 63121, USA; lyman.sargent@umsl.edu

Abstract

I have characterized utopianism as having 'three faces,' utopian literature, intentional communities, once known as utopian experiments, and utopian social thought. Food plays a central role in the first two, and I contend that the third, through F.L. Polak's idea of the importance of a 'positive image of the future,' Albert Camus's argument for a 'relative utopia,' and John Rawls's argument for a 'realistic utopia,' can help us understand the role that utopianism, and due to its centrality in utopianism, the role of food in a positive social theory. And if utopianism is to have such a role, it must depend on accurate information and that obviously means food studies. Utopianism is the only approach that takes all of life as its purview, and where food is concerned, this means everything from production through disposal including what is produced, how it is processed, procured, and prepared, how and by whom it is served, and how what is not consumed and whatever by-products are produced in the process are dealt with. All these issues are treated in much of utopian literature. And intentional communities, many of which act as real-time experiments regarding how every one of these steps can be improved in order to improve the quality of what is consumed and the health of the consumer, increase food security and sustainability, and protect the environment from further damage.

Keywords: utopian literature, intentional communities, Albert Camus, F.L. Polak, sustainability, food security

Utopianism

The title makes it obvious that I am making an argument, an argument that I think is quite simple and straightforward. But I have found that the word utopia tends to produce images and expectations that turn people off because they equate utopia with perfection, which, since we are dealing with human beings, is much too high a standard. So, the first step in my argument is to say that we need to get rid of the idea of perfection. Perhaps the closest one can come to perfection is the ever elusive goal of perfectibility.

The word utopia simply means no place or nowhere, but when Thomas More coined the word, he played with the word eu-topia or good place. As a result, the word utopia has come to mean a non-existent good place. And utopianism, which I call social dreaming, has three facets, utopian literature, intentional communities, which were once called utopian experiments and are now sometimes called practical utopias, and a somewhat amorphous category, utopian social theory.

Within utopian literature, while a few authors of utopias have called the utopias they wrote as describing perfect societies, most have not. More's 'Utopia' is certainly not perfect and neither are any of the best-known utopias. Some of the problem arises because a utopia is frequently presented as a photograph taken at a moment in time with some indication of what had happened before to bring the utopia into being but with little about the future. But most utopias, like More's, are open to change and improvement, which means that the current society is not thought to be perfect, and the words to use to characterize the societies presented in utopias are better and good, not perfect.

These words also more accurately reflect the fact that an important function of the projection of a better society is critique, and many utopias contain significant elements of satire both directly and by

EurSafe 2016 – DOI 10.3921/978-90-8686-834-6_1,

implication in showing the better future. Critique is also a function of an aspect of utopian literature that I shall only touch on, the dystopia or bad place that frequently extrapolates tendencies in the present that, unless checked, will produce a much worse future. Thus the dystopia is a warning, something of a jeremiad saying shape-up or else. The utopia says that changes in the present can bring about a much better future; the dystopia says not changing the present will result in a much worse future.

Intentional communities

In intentional communities people choose to experiment with their own lives in an attempt to create a society in line with a vision of what constitutes a better or good society. Many are short-lived but most last longer than the average small business, which in the U.S. is three years, some, all religious, are hundreds of years old, a fair number have passed their century, and many of those established in the third quarter of the twentieth century are fifty to seventy-five years old. Quite simply there are more such communities in existence throughout the world today than in any time in the past. As a result, they provide an immense but largely untouched source of experiments related to food. Of course, many, but not all, would simply tell researcher to go away and leave them alone.

There is a subset of such communities, ecovillages, that are particularly important for my argument because they are working to create and maintain ecologically sustainable communities, and they are experimenting with ways of producing food that come from food studies and are developing knowledge likely to be useful to food studies. Obviously such communities need accurate information in order to function, and, an aspect of utopias that is often missed is that they reflect their authors understanding of the reality of the world they live in. Of course, sometimes that reality is misunderstood and, as a result, quite a few utopias come up with simplistic answers to complex problems, but so do many contemporary politicians and pundits, My point, quite simply, is that for a utopia to be useful it has to be based on accurate information. Albert Camus, whose thought is again resonating with current issues, argued that what he called the 'absolute utopia' is impossible but that we can strive for the 'relative utopia', and John Rawls, although discussing issues in liberal philosophy, argued for what he called a 'realistic utopia', and it is the approach suggested by these words that I argue for.

But, why is utopianism important? In 1955 F.L. Polak, a Dutch sociologist, published 'The Image of the Future' in which he argued that our images of the future effect the actual future And he goes on to say that without such dreams civilization will come to an end, which certainly says that utopianism is not merely important but crucial. Polak emphasizes the West, which I find too limiting, but the fact that today's dominant image of the future is dystopian should deeply worry us. I do not follow Polak as far as he wants to go, but I also do not dismiss him entirely. My argument is that positive change requires a belief that it is possible and some idea of what a positive change would look like so that we can craft a way to move in that direction.

Food and utopia

The first element of what I have called the basic human utopia is a full stomach (the others are adequate clothing and shelter). Food is central to human life; therefore, it is central to utopianism, which I have characterized as social dreaming. Utopias say here are the serious problems that we should be aware of, here are a possible solutions to these problems, and here is the better life that we and our descendants can have if we can find the will to make changes. One thing we know with certainty about children, let alone teenagers, is that they have to be fed, and food has been a central focus of utopias from well before the word utopia was coined. And in both utopian literature and intentional communities, food can be both a way of bringing people together and the basis for disputes.

My approach to the issue of food begins with production and ends with disposal, and utopias, because they deal with all of life rather than just one part, are well positioned to reflect this approach. And since this is the 500th anniversary of the publication of Thomas More's 'Utopia', let me use the Utopia as an example of the role food plays in utopian literature. More discusses food production, including both large scale farming and small scale gardening, and distribution and consumption, including the arrangement of common meals. He implies something about preparation and service. He has nothing on disposal. Some 300 years later Edward Bellamy published 'Looking Backward', the most widely read utopia until well into the 20th century. In Looking Backward, Bellamy does not discuss production, an omission he corrects at some length in 'Equality', the sequel to Looking Backward, published in 1897. In Looking Backward, Bellamy has a lengthy section on preparation, service, and consumption. He has nothing to say about disposal. Thus, what are historically the two best-known utopias have much to say about food but do not cover all the basic issues. They discuss what seemed to their authors to be the most important issues facing their societies at the time of writing, scarcity at the time of More and radically unequal distribution at the time of Bellamy, both issues that still face our world today.

There is one theme that runs throughout the discussion of food in utopianism from production through processing, preparation, and disposal and that is, unsurprisingly, health. Specifics change over time, but producing healthy food and ensuring that it is still healthy by the time it reaches the table and that what reaches the table is right for those eating it are central concerns. Also, there is considerable interest in what happens to the leftovers and to human waste.

2. 'Just food': contemporary challenges for richer countries

E.A. Dowler
Department of Sociology, University of Warwick, Coventry, CV4 7AL, United Kingdom; e.dowler@warwick.ac.uk

Abstract

Social justice is often omitted from debates about food system crises and the urgent need to address sustainability. This is particularly true in relation to how all households can eat decently and sufficiently in times of general economic austerity. Although those who are poor in the majority world are of course more seriously affected, there is growing evidence that longstanding problems in household level food security for those on low incomes in a rich country are also worsening. There is little overt national policy attention to these issues, although some devolved administrations (in the UK) and local authorities are exploring different responses. The consequences of what might be termed 'food poverty' for social wellbeing and nutritional health can be severe, but are hidden and individually embodied rather than monitored and addressed by society. Public policy has been slow to respond, leaving local people and both longstanding and new networks to step in with charitable food and support. The implications for accountability, advocacy and the social contract – indeed, for food justice, may be profound. Visions of what a more 'just' food system might look like, not least in terms of progressive implementation of the right to food, and the development of sustainable, creative food systems, are growing but not yet fully articulated in the major public debates, nor supported by serious policy initiatives. The paper outlines and briefly explores some of these problematics and possibilities, drawing particularly, but not exclusively, on UK experiences.

Keywords: food poverty, food insecurity, food policy, food justice, UK

Introduction and context

It seems fitting to explore contemporary challenges of injustices in relation to the food system at a meeting to celebrate More's 'Utopia', wherein the link between a then contemporary source of social unrest – 'theft' – is discussed by his early correspondent as not so much evidence of moral degeneracy in the general population, as an act of desperation by those who have lost their means of production through appropriation by the rich. The increasingly widespread land enclosures at the time removed ordinary people's access to grazing rights and common land, forcing them into emerging urban industries or penury. There seem obvious parallels with contemporary loss in entitlement to decent, secure and reasonably paid work under growing neo-liberal regimes, and concomitant loss of food entitlements through purchase, retaining own production, or exchange, and increasing resort to charitable food as normative practice is the experience of rising numbers across Europe.

That this should be the case is largely unremarked or addressed in policy circles in richer countries. (For instance, both the implications for the food system, and how the UK would ensure that those who already struggle to afford their food needs could thrive, and not be even worse off, was hardly mentioned in public debates running up to the recent referendum on remaining in the European Union, despite attempts to raise such important issues; Anon, 2016; Lang and Shoen, 2016; Matthews, 2016.) Of course, those in the majority world face more substantial levels of malnutrition and food insecurity at national, local and household levels; nevertheless, growing numbers cannot reliably sustain their household food needs in rich countries such as the USA, Canada, UK and other parts of Europe (see Caraher and Cavacchia, 2014; Dowler and Lambie-Mumford, 2015a; Riches and Silvasti, 2014, for

EurSafe 2016 – DOI 10.3921/978-90-8686-834-6_2, © Wageningen Academic Publishers 2016

recent international review). While not a new phenomenon this is arguably a critical one, attracting some social and academic attention and deserving of more considered examination.

In a brief comment I cannot provide detailed account of factors behind the growth in numbers for different richer countries; the broad case is that, on top of ongoing socio-economic conditions and social policies which do not sufficiently address them, a combination of the economic crisis in 2008-2009, coupled with rising food prices globally, created circumstances in which increasing numbers find themselves struggling to provision their households with sufficient food to meet nutritional and social requirements, especially where housing and other essential costs have to take priority. Both nutritional and social requirements are best served by a diverse and varied diet, which in itself can be more expensive than the parsimonious least-cost diets beloved of modellers and policy makers. Food is of course more than a bundle of nutrients, and fulfils a complex and interwoven set of human needs (Max Neef, cited in Dowler and Tansey, 2003), and those unable to meet them face physiological and mental health consequences, as well as emotional, psychological pain; attention to rising global cardio-vascular disease, cancer, diabetes and obesity signals some of these costs (e.g. Beaglehole *et al.*, 2011). In addition, the growing awareness that climate change, loss of biodiversity, soil and water degradation, and population increase are having severe, and sometimes irreversible, effects on the food system, adds to the urgency with which sustainable food systems are being sought. Essential here are multi/inter-disciplinary research efforts where biological and social sciences work together to enable wellbeing for all sentient beings, living in healthy, just societies, within environmental limits.

Response to growing household food insecurity

In common in different countries is the continual inability to purchase or otherwise obtain enough appropriate food for a healthy life, with constant anxiety that such failure will recur (broadly, the condition of household food insecurity) which often manifests in hunger (c.f. Riches and Silvasti, 2014, for discussion of these and other terminologies such as food poverty). People increasingly turn to charitable sources for food, both those which offer informal, local help and those which are outlets of larger scale, more systematic operations. Many of the latter draw on food within industrialised systems which would otherwise be destined for landfill or other disposal; redirection of that constituted as 'waste' to feed 'hungry people' is often deemed a social and economic 'win: win'. All such responses might seem reasonable, indeed noble, in meeting fellow citizens' needs, but they pose significant issues of efficacy, accountability, appropriateness and simple justice (see Dowler and Lambie-Mumford, 2015b; Loopstra and Tarasuk, 2012; Loopstra *et al.*, 2015). Charitable responses can distract attention from critical structural issues and by locating problems in the arena of individual efficiency and competence contribute to institutionalising a blame as well as an aid mentality. By contrast, and certainly in the UK, many seeking charitable help are in fact struggling under punitive social security regimes, and/or in increasingly casualised and inadequately paid work (Caplan, 2016; Garrett *et al.*, 2016; Garthwaite, 2016; Maslan *et al.*, 2013; Perry *et al.*, 2014). The increasing evidence that this is the case is ignored, and charitable systems are growing in strength, thus enabling those in policy and the food sectors who should be addressing fundamental issues to ignore their responsibilities.

Immediate hunger can be assuaged by emergency help, but the conditions underpinning household food insecurity require systematic and structural response. Societies such as the UK too readily demonise those who are poor (and certainly the print and TV media have been happy to do this over those using food banks); the state continually asserts that paid work is the solution, paying insufficient attention to its current inadequacy for many. People need work which is both sufficiently secure and offering wages that can meet household needs, or, for those unable to work or without recourse to a pension, access to sufficient money from a social security system which offers support with dignity. In addition, for food security people need to be able to reach shops selling appropriate, healthy and sustainably produced food at affordable prices, and to have the skills and confidence to use the produce they purchase. The prevailing food culture should be healthy and fair, and not one which promotes food that undermines physical and social health. Furthermore, a sustainable food system is one where producers, processors, retail workers and all who work in the food and catering sectors have decent working conditions too (Food Ethics Council, 2010, 2013).

Food poverty/household food insecurity/hunger are not due to individual failings or happenstance: they are the outcomes of public and private sector policy decisions, whose results have been increasingly normalised as acceptable in civilised societies. 'Food' is located in private, personal, domestic 'choice' – one of the mantras of contemporary neo-liberalism – but in reality those in economic and social deprivation have little choice at all. When these issues are raised, the answer is usually that addressing the fundamentals would cost society too much: the reality is that it already does cost society a great deal, but these costs are borne by those with least responsibility and resources. Thus, these food issues illustrate deep rooted injustice: inequitable differences (rather than everyone being the same in terms of culture, taste, skills or outcomes) which are unfair. They could be addressed but at some level, society chooses to tolerate the unfairness.

Many structural issues have to be addressed at central government level (e.g. policies towards social security, access to and the cost of housing, levels of a national minimum wage) but there are policy activities at local levels with potential to improve household level food security. The sustainable city network (no date) and grassroot or alternative systems can make some difference (Caraher and Dowler, 2014). However, particularly pertinent in the UK context are attempts to involve devolved nation governments in addressing in a sustainable, fair food system (e.g. Johnstone *et al.*, 2016; Morgan, 2016; Scottish Food Coalition, 2016), which include rights based approaches (Tait, 2015). Some now argue the latter are essential in that they impose obligation on the state and other society actors, and introduce accountability and a normative basis for responses (Dowler and O'Connor, 2012; Riches and Silvasti, 2014).

Those living in poorer households the world over bear many of the consequences of global climate change, rising food prices and the economic crisis, for which they have had little if any responsibility, embodying the consequences in stress, anxiety, despair; poorer health, reduced well-being and shorter lives. Food justice is long overdue.

References

Anonymous (2016). The European Union, us, our food ... where next? *and* The stakes of the EU Referendum for our food system. Blog posts on Nourish Scotland website. EU Referendum. Available at: http://tinyurl.com/hx56wda. Accessed 27 July 2016.

Beaglehole, R., Bonita, R., Horton, R. (and 42 others) (2011). Priority actions for the non-communicable disease crisis. Lancet 377: 1438-1447.

Caplan, P. (2016). Big society or broken society?: Food banks in the UK. Anthropology Today 32: 5-9.

Caraher, M. and Cavacchi, A. (2014). Old crises on new plates or old plates for a new crises? Food banks and food insecurity. British Food Journal 116. DOI: http://dx.doi.org/10.1108/BFJ-08-2014-0285.

Caraher, M. and Dowler, E. (2014). Food for Poorer People: Conventional and 'Alternative' Transgressions? Aldershot: Ashgate, Surrey, UK. pp. 227-246.

Cooper, N., Purcell, S. and Jackson, R. (2014). Below the Breadline: The Relentless Rise of Food Poverty in Britain. CAP-OXFAM-Trussell Trust Available at: http://tinyurl.com/hkesqdt. Accessed 27 July 2016.

Dowler, E. and Tansey, G. (2003). Food and poverty. In: Mosley, P and Dowler, E. (eds) Poverty and Social Exclusion in North and South London: Routledge, Abingdon-on-Thames, UK. pp 189-207.

Dowler, E. and O'Connor, D. (2012). Rights-based approaches to addressing food poverty and food insecurity in Ireland and UK. Social Science and Medicine 74: 44-51.
Dowler, E. and Lambie-Mumford, H. (2015a). (eds) Hunger, Food Poverty and Social Policy in Austerity. Special themed issue Social Policy and Society 14: 417-508.
Dowler, E. and Lambie-Mumford, H. (2015b). How can households eat in austerity? Challenges for social policy in the UK. Social Policy and Society 14: 417-428.
Food Ethics Council (2010). Food justice: the report of the Food and Fairness Inquiry. Brighton, Food Ethics Council. Available at: http://tinyurl.com/h6lvkqq. Accessed 29 July 2016.
Food Ethics Council (2013). Beyond Business as Usual. Brighton: Food Ethics Council. Available at: http://tinyurl.com/jbuxcvc. Accessed 29 July 2016.
Garratt, E., Spencer, A. and Ogden, C. (2016). #stillhungry. Who is hungry, for how long, and why? Research Report by West Cheshire Foodbank, The University of Oxford, The University of Chester, The Trussell Trust, Cheshire West Citizens Advice Bureau UK. Available at: http://tinyurl.com/hty529k. Accessed 29 July 2016.
Garthwaite, K. (2016). Hunger pains. Life inside foodbank Britain. Policy Press, Bristol, UK. 176 pp.
Johnstone, M. and the Scottish Working Group on Food Poverty (2016). Dignity: Ending Hunger Together in Scotland. The Report of the Independent Working Group on Food Poverty. Edinburgh: The Scottish Government. Available at: http://tinyurl.com/jhuo65e. Accessed 20 July 2016.
Lang, T. and Schoen, V. (2016). Food, the UK and the EU: Brexit or Bremain? Briefing Paper. Food Research Collaboration. Available at: http://tinyurl.com/jbxgm96. Accessed 27 July 2016.
Loopstra, R. and Tarasuk, V. (2012). The Relationship between Food Banks and Household Food Insecurity among Low-Income Toronto Families. Canadian Public Policy 38: 497-514.
Loopstra, R., Reeves, A., Taylor-Robinson, D., Barr, B., McKee, M., and Stuckler, D. (2015). Austerity, sanctions, and the rise of food banks in the UK. BMJ 350:h1775.
Maslen, C., Raffle, A., Marriott, S., and Smith, N. (2013). Food Poverty: What Does The Evidence Tell Us? Bristol: Bristol City Council. Available at: http://tinyurl.com/hfj5my4. Accessed 27 July 2016.
Matthews, A. (2016). Implications of Brexit for UK Food and Agriculture. Presentation to AHDB. Outlook Conference Feb 2016. Available at: http://tinyurl.com/hjolsrb. Accessed 27 July 2016.
Morgan, K. (2016). Good Food for all. The Senedd Papers #3. Cardiff: Institute of Welsh Affairs. Available at: http://tinyurl.com/gs5sg4z. Accessed 27 July 2016.
Perry, J., Williams, M., Sefton, T. and Haddad, M. (2014). Emergency Use Only: understanding and reducing the use of food banks in the UK. London, CPAG, Church of England, Oxfam GB and The Trussell Trust. Available at: http://tinyurl.com/okdzl4m. Accessed 27 July 2016.
Riches, G. and Silvasti, T. (eds) (2014). First World Hunger Revisited, Basingstoke, PalgraveMacmillan, London, UK, 248 pp.
Riches, G. and Silvasti, T. (2014). Hunger in the Rich World: Food Aid and Right to Food Perspectives. Ch1 in: Riches, G. and Silvasti, T. (eds) First World Hunger Revisited, Basingstoke, PalgraveMacmillan, London, UK, pp 1-14.
Scottish Food Coalition. (2016). Plenty: food, farming and health in a new Scotland. Scottish Food Coalition c/o Nourish, Edinburgh. Available at: http://tinyurl.com/zcxkbuo. Accessed 20 July 2016.
Sustainable Food Cities Initiative (no date): Available at: http://sustainablefoodcities.org. Accessed 27 July 2016.
Tait, C. (2015). Hungry for Change: A Fabian Policy Report. London, Fabian Society. Available at: http://tinyurl.com/zkm6u2h. Accessed 12 Jan 2016.

3. Safe food for whom – consumers, animals and the environment in the European food safety regulations

A. Aragão
Faculty of Law and researcher at the Legal Institute, Research Group 'Risk, transparency and litigation', University of Coimbra, Pátio da Universidade, 3004-528 Coimbra, Portugal; aaragao@fd.uc.pt

Abstract

At first sight, European Union food safety law seems to display a grossly anthropocentric approach. But placing it in the context of environmental law and reading it in the light of the ethical principles enshrined in European Union Treaties and regulations, will reveal a totally different picture. In Europe, safe food is much more than merely consumers not getting sick when eating or drinking. Applying the life-cycle approach underlying food safety law in Europe – safety from the farm to the fork – a certain food product cannot receive the 'safe food' label whenever its production caused unnecessary and unacceptable damages to animal life or to the environment.

Keywords: food safety, animal welfare, anthropocentrism, ecocentrism, European law

Introduction

At first sight, European Union food safety law seems to display a grossly anthropocentric approach. According to the framework Regulation, safe food is a substance or product that does not kill or cause an 'adverse health effect'[1] when it is 'ingested by humans'[2].

Consumer protection and food safety go hand in hand to ensure 'a high level of protection of human health and consumers' interest in relation to food'[3]. Besides, the phrase 'high level of protection of human life and human health' is repeated to exhaustion (12 times in 24 pages) throughout the Regulation.

However, it is not so difficult to contradict the apparent anthropocentricity of EU food safety law. In fact, the Regulation cannot be read by itself, isolated from the whole European legal system. As a consequence, putting this seemingly anthropocentric food safety rule in the broader context of European environmental law, and taking into account some general principles of the Treaty, a totally different picture will be revealed.

General principles

In the provisions of the Treaty on the Functioning of the European Union which have general application[4], the so called 'integration principle' stands as the key strategy towards sustainable development: 'Environmental protection requirements must be integrated into the definition and implementation

[1] Definition of 'risk' in article 3/9 of Regulation 178/2002 of 28 January 2002, the main European regulation on food security, laying down the general principles and requirements of food law, establishing the European Food Safety Authority and laying down procedures in matters of food safety.

[2] Definition of 'food' in article 2 of the same Regulation.

[3] Article 1 of the food security Regulation 178/2002 of 28 January 2002 on 'aim and scope'.

[4] Articles 7 to 17 of the Treaty on the Functioning of the European Union.

I. Anna S. Olsson, Sofia M. Araújo and M. Fátima Vieira (eds.) ***Food futures: ethics, science and culture***
EurSafe 2016 – DOI 10.3921/978-90-8686-834-6_3, © Wageningen Academic Publishers 2016

of the Union's policies and activities, in particular with a view to promoting sustainable development'[5]. It is well-known that the requirements concerning protection of the environment in Europe demand aiming at nothing less than 'a high level of protection'[6].

Of all the European Union's areas of intervention food safety is indubitably one of those that are intrinsically associated with environmental protection. First of all, there is the risk of health impacts arising from pollution entering the food chain. The cases of soybean contaminated with an enteric pathogen (*Escherichia coli* from waste water)[7], of mercury in fish products[8], or of polycyclic aromatic hydrocarbons[9] in pressed hemp seed oil, are just a few examples[10].

Secondly, since large part of the typical European diet is composed of animal products, the biological act of ingesting food – as well as the cultural practice of preparing and eating it[11] – is also about animals and animal welfare.

Indeed, animal welfare became a core European goal[12] since 2009[13]: 'in formulating and implementing the Union's agriculture, fisheries, transport, internal market, research and technological development and space policies, the Union and the Member States shall, since animals are sentient beings, pay full regard to the welfare requirements of animals, while respecting the legislative or administrative provisions and customs of the Member States relating in particular to religious rites, cultural traditions and regional heritage'[14].

---

[5] Article 11 of the Treaty on the Functioning of the European Union.

[6] Article 191/2 of the Treaty on the Functioning of the European Union: 'Union policy on the environment shall aim at a high level of protection taking into account the diversity of situations in the various regions of the Union. It shall be based on the precautionary principle and on the principles that preventive action should be taken, that environmental damage should as a priority be rectified at source and that the polluter should pay'. Also article 37 of the Charter of fundamental rights of the European Union proclaims 'a high level of environmental protection and the improvement of the quality of the environment must be integrated into the policies of the Union and ensured in accordance with the principle of sustainable development'.

[7] For more information see the scientific report of EFSA 'Shiga toxin-producing *E. coli* (STEC) O104:H4 2011 outbreaks in Europe: Taking Stock' available at http://tinyurl.com/z3xh8ra.

[8] See the information note by EFSA available at http://tinyurl.com/hdu292a.

[9] http://tinyurl.com/gpc6xpv.

[10] Other examples are nitrates (Directive 91/676 of 12 December 1991 concerning the protection of waters against pollution caused by nitrates from agricultural sources), fertilizers (Regulation 2003/2003 of 13 October 2003 relating to fertilizers), or sewage sludge (Directive 86/278 of 12 June 1986 on the protection of the environment, and in particular of the soil, when sewage sludge is used in agriculture).

[11] Recognized by UNESCO as intangible heritage in 2013, the Mediterranean diet 'involves a set of skills, knowledge, rituals, symbols and traditions concerning crops, harvesting, fishing, animal husbandry, conservation, processing, cooking, and particularly the sharing and consumption of food. Eating together is the foundation of the cultural identity and continuity of communities throughout the Mediterranean basin. It is a moment of social exchange and communication, an affirmation and renewal of family, group or community identity. The Mediterranean diet emphasizes values of hospitality, neighbourliness, intercultural dialogue and creativity, and a way of life guided by respect for diversity' (http://tinyurl.com/heoucpp) (this and all the following sites were accessed in May 2016).

[12] Animal welfare is also a citizen's concern, as proven by the presentation of a Citizens' Initiative in 2012 for the improvement of the welfare of dairy cows (http://tinyurl.com/zbcreo9).

[13] Article 13 of the Treaty on the Functioning of the European Union was introduced by the Lisbon Treaty but the topic of animal welfare has been around in primary European Union Law since 1999, as a Protocol attached to the Amsterdam Treaty.

[14] Article 13 of the Treaty on the Functioning of the European Union.

It is worth highlighting the sharp justification for the protection of animals explicitly included in the text of the Treaty: '... since animals are sentient beings ...'. The primary reason for animal protection is the fact that animals have feelings. Inasmuch as they are caused by humans, negative feelings such as pain, distress or suffering[15] shall be avoided.

Besides, there are parallel legal requirements[16] for the hygiene of food[17] and feed[18]. One might think that rules on feed hygiene are justified on purely egoistic reasons: if the animal will be transformed into food it is advisable to produce it according to good hygienic practices. This narrow interpretation is contradicted by the obvious fact that the Regulation applies not only to animals for human consumption (cattle, pigs, sheep, poultry, etc.) or to food producing animals (eggs, milk) but also to ... pets!

The same regime is applicable to feeding stuffs intended for particular nutritional purposes. Despite recognizing that 'pet food is not part of the human food chain and has no environmental impact on arable land' the European Regulation considers that 'specific provisions for additives in pet food are appropriate'[19].

This is enough to confirm the non-instrumental features of animal protection in Europe. Animals must be protected for their own sake and not merely because they will be transformed into (safe) human food. In other words: animal welfare arguments prevail over food chain arguments[20]. Which leads us to the question: safe food for whom?

Safe food for whom?

A correct interpretation of the European Law will show that, in Europe, safe food is much more than merely consumers not getting sick when eating or drinking. Three additional questions will help us to understand the different angles of the wider concept of integrated food safety:

1. Can an animal product be considered safe food if it was obtained through a particularly painful or cruel way of killing animals?
2. Can a foodstuff product be considered safe food if its production caused the accidental death of other living beings?
3. Can a food product be considered safe food if it is likely to pose a serious risk to the environment?

Applying the life-cycle approach underlying food safety law in Europe – safety from the farm to the fork – a certain food product cannot receive the 'safe food' label whenever its production caused unnecessary and unacceptable damages to animal life or to the environment. In the three following chapters we will address these problems one by one.

---

[15] All mentioned in Regulation 1099/2009 of 24 September 2009 on the protection of animals at the time of killing.

[16] In what concerns the use of genetically modified ingredients, it is also one single regulation to deal with genetically modified food and feed (Regulation 1829/2003 of 22 September 2003).

[17] Regulation 852/2004 of 29 April 2004 on food hygiene.

[18] Regulation 183/2005 of 12 January 2005 on feed hygiene.

[19] Preamble of the Regulation 1831/2003 of 22 September 2003. Article 7/5 is also clear: 'These rules should, where appropriate, differentiate between requirements for feed additives in respect of food-producing animals and requirements in respect of other animals, in particular pets'.

[20] Food chain arguments explain the Regulation 999/2001 of 22 May 2001 laying down rules for the prevention, control and eradication of certain transmissible spongiform encephalopathies.

The oxymoron of 'humane killing'

What do cows and pigs, hare and seals, lobsters and bluefin tuna, have in common? They are all animals used for human consumption. And what are the differences among them? While cows and pigs are domestic animals raised in farms to be killed for food, the others are wild animals living in natural environments, until they are captured and killed. Worse, the latter are often victims of barbaric uses practiced by humans to turn them into human food.

In legal terms the difference is that cows, pigs and cattle in general are protected by law. A European Regulation on the protection of animals at the time of killing determines that 'animals shall be spared any avoidable pain, distress or suffering during their killing and related operations'[21]. A European Directive on the protection of animals kept for farming purposes[22] establishes rules on freedom of movement[23], buildings and accommodation[24], and feed, water and other substances[25], among several other requirements.

Hares, seals as well as beavers, badgers and other wild animals are only protected as far as the used killing means are insidious and inhumane. Since 1991, a Council Regulation prohibits the use of leghold traps in the EU and the introduction into the Community of pelts and manufactured goods of certain wild animal species originating in countries which did not yet ban trapping methods not meeting international humane trapping standards[26].

In 2009, the Parliament and the Council adopted a Regulation on trade in seal products 'in response to concerns of citizens and consumers about the animal welfare aspects of the killing and skinning of seals and the possible presence on the market of products obtained from animals killed and skinned in a way

---

[21] It's article 3 of the same Regulation on the protection of animals at the time of killing.

[22] Directive 98/58 of 20 July 1998.

[23] Annex 7 of Regulation 'The freedom of movement of an animal, having regard to its species and in accordance with established experience and scientific knowledge, must not be restricted in such a way as to cause it unnecessary suffering or injury. Where an animal is continuously or regularly tethered or confined, it must be given the space appropriate to its physiological and ethological needs in accordance with established experience and scientific knowledge'.

[24] Annex 8 of the same Regulation: '... Materials to be used for the construction of accommodation, and in particular for the construction of pens an equipment with which the animals may come into contact, must not be harmful to the animals and must be capable of being thoroughly cleaned and disinfected. ... Accommodation and fittings for securing animals shall be constructed and maintained so that there are no sharp edges or protrusions likely to cause injury to the animals. ... Air circulation, dust levels, temperature, relative air humidity and gas concentrations must be kept within limits which are not harmful to the animals. ... Animals kept in buildings must not be kept either in permanent darkness or without an appropriate period of rest from artificial lighting. Where the natural light available is insufficient to meet the physiological and ethological needs of the animals, appropriate artificial lighting must be provided'.

[25] Annex 14 of the mentioned Regulation: '... Animals must be fed a wholesome diet which is appropriate to their age and species and which is fed to them in sufficient quantity to maintain them in good health and satisfy their nutritional needs. No animal shall be provided with food or liquid in a manner, nor shall such food or liquid contain any substance, which may cause unnecessary suffering or injury. ... All animals must have access to feed at intervals appropriate to their physiological needs. ... All animals must have access to a suitable water supply or be able to satisfy their fluid intake needs by other means. ... Feeding and watering equipment must be designed, constructed and placed so that contamination of food and water and the harmful effects of competition between the animals are minimised. ... No other substance, with the exception of those given for therapeutic, or prophylactic purposes or for the purposes of zootechnical treatment ..., must be administered to an animal unless it has been demonstrated by scientific studies of animal welfare or established experience that the effect of that substance is not detrimental to the health or welfare of the animal'.

[26] Regulation 3254/91 of 4 November 1991 completed by Decision 98/596 of 14 October 1998. This Regulation applies mainly to animals used for their fur but also for obtaining other products such as meat.

that causes pain, distress, fear and other forms of suffering'[27]. The exception is that of 'the placing on the market of seal products shall be allowed only where the seal products result from hunts traditionally conducted by Inuit and other indigenous communities and contribute to their subsistence'[28].

For a large spectrum of other wild animals, Natura 2000 Directive forbids the use of perfidious hunting techniques such as blind or mutilated animals as live decoys, tape recorders, poisons or non-selective traps[29].

Finally, in what concerns lobsters, bluefin tuna and wild caught fish they are not protected at all – not even the method for killing them is regulated. For example, the main method for killing fish remains asphyxia on ice, even though scientists have pronounced this method not to be humane for fish[30]. Unfortunately these are not single cases. Several other animals are in the bottom of the 'caste-like' system suffering legally accepted discriminations in the moment of being killed.

Despite being a contradiction in terms, the concept of 'humane killing' at least reveals some concern with animal treatment.

Indirect killing beyond 'dolphin safe'

Leaving vegetarianism aside, and assuming that a meat diet is ethically acceptable provided that the animal was submitted to 'humane killing', there are still ethical questions which must be addressed by law. It is the case of unintended loss of animal lives as a side-effect of food production. This is particularly critical in fishing or hunting when using non-selective techniques. The dolphin safe logo, originated in the United States and pushed by consumer concerns about tuna fishing boats capturing dolphins accidentally is the best known example of indirect killing, concentrated on a charismatic mammal species.

Indirect killing of dolphins have been a recurrent hot topic discussed once again in the World Trade Organization in 2015[31]. Similar cases concerning other species[32] have been taken to the World Trade Organization framed as 'non-technical barriers to trade' regulated by the Agreement on Technical Barriers to Trade[33]. The state of the discussion in the World Trade Organization confirms the relevance of the legal consequences of commercial measures adopted for ethical reasons.

In Europe, the concern goes beyond animal species having widespread popular appeal. One of the strongest concerns of the European Regulation on the common fisheries policy and conservation of marine biological resources is promoting 'selective fishing' – 'fishing methods or fishing gears that target and capture organisms by size or species during the fishing operation, allowing non-target specimens

[27] Paragraph 5 of the Preamble of Regulation 1007/2009 of 16 September 2009.

[28] Article 3/1.

[29] Annex VI of Directive 92/43 of 21 May 1992 on the conservation of natural habitats and of wild fauna and flora, amended in 2006 and 2013.

[30] The EFSA Journal (2004), 45, 1-29, Welfare aspects of the main systems of stunning and killing the main commercial species of animals. p 22 (http://tinyurl.com/gwhufra). By contrast, for fish produced in aquaculture there is even support for promoting animal health and welfare (see articles 47, 48 and 56 of the. Regulation 508/2014 of the European Parliament and of the Council of 15 May 2014 on the European Maritime and Fisheries Fund).

[31] Dispute DS381 between Mexico and the United States on Measures concerning the importation, marketing and sale of tuna and tuna products (http://tinyurl.com/jmb3v5o).

[32] Shrimps and turtles in the 'shrimp-turtle' case (http://tinyurl.com/js378a8).

[33] Agreement on technical barriers to trade (http://tinyurl.com/ju4dq5z).

to be avoided or released unharmed'[34] – and eliminating fish 'discards' – 'catches that are returned to the sea'[35].

In what concerns hunting, the European Directive on the Natura 2000 network, mentioned above, bans the use of non-selective capturing and killing methods such as explosives, non-selective nets or traps, and semi-automatic or automatic weapons[36] for cynegetic species.

Even more subtle, and yet more important from an ecosystem services' point of view[37], is the protection of other animals, lower in the food chain, such as insects and most especially, bees and other pollinators. These fundamental species, crucial for the correct functioning of both agricultural ecosystems and natural habitats suffer great anthropogenic disturbances. Besides chemical pollution from phytosanitary products, electromagnetic radiation and invasive species are the main sources of disturbance.

This is the rationale behind the Directive for sustainable use of pesticides. The aim is 'reducing the risks and impacts of pesticide use on human health and the environment and promoting the use of integrated pest management[38] and of alternative approaches or techniques such as non-chemical alternatives to pesticides'[39]. At the same time, European Union supports agri-environmental measures to ensure the integration of environmental concerns into the Common Agricultural Policy[40] and produces scientific expertise on the subject such as the European Food Safety Authority 'Guidance document on the risk assessment of plant protection products on bees'[41] adopted in 2014.

Unacceptable environmental damages

Food production can cause serious and extensive environmental damages. Even knowing that nourishment is a basic necessity both of humans and animals, not all the environmental damages associated with it are acceptable. This is the reason why an environmental impact assessment must be performed before

[34] Article 4/12 of Regulation 1380/2013 of 11 December 2013.

[35] Article 4/10 of the same Regulation.

[36] It's the same Annex VI of Directive 92/43 of 21 May 1992 on the conservation of natural habitats and of wild fauna and flora, amended in 2006 and 2013.

[37] The Millennium Ecosystem Assessment is a worldwide report prepared to respond to the United Nations Secretary-General Kofi Annan challenge in the Millennium Summit in 2000. Prepared by more than 1,360 experts it had the objective of assessing the consequences of ecosystem change for human well-being. The Millennium Ecosystem Assessment provides a state-of-the-art scientific appraisal of the condition and trends in the world's ecosystems and the services they provide (clean water, food, forest products, flood control, and natural resources) and the options to restore, conserve or enhance the sustainable use of ecosystems. (http://tinyurl.com/nh3tzov).

[38] According to article 3/6 of the same Directive integrated pest management 'means careful consideration of all available plant protection methods and subsequent integration of appropriate measures that discourage the development of populations of harmful organisms and keep the use of plant protection products and other forms of intervention to levels that are economically and ecologically justified and reduce or minimise risks to human health and the environment. 'Integrated pest management' emphasises the growth of a healthy crop with the least possible disruption to agro-ecosystems and encourages natural pest control mechanisms'.

[39] Article 1 of Directive 2009/128 of 21 October 2009.

[40] Regulation 1305/2013 of 17 December 2013 on support for rural development by the European Agricultural Fund for Rural Development.

[41] The report was published on 04 July 2014 and is available at: http://tinyurl.com/htvqlcb.

the administrative authorization of several food producing activities, such as the ones mentioned in the Directive on the assessment of the effects of certain public and private projects on the environment[42].

The long list of environmental impacting activities[43] includes livestock and fish farming, intensive agriculture, large transformations of soil and waterbodies and industrial transformation of food. This is also the reason why an integrated prevention and pollution control (IPPC) of certain food producing activities is mandatory to ensure that 'all the appropriate preventive measures are taken against pollution', 'the best available techniques are applied', 'no significant pollution is caused', 'the necessary measures are taken to prevent accidents and limit their consequences' and 'the necessary measures are taken upon definitive cessation of activities to avoid any risk of pollution and return the site of operation to the satisfactory state'[44].

Under the Directive on IPPC, activities submitted to an integrated prevention and pollution control procedure include slaughterhouses, large installations for the production of food or feed from animal or vegetal raw materials (having a capacity of several tones per day), large installations for disposal or recycling of animal carcasses or animal waste and finally, intensive rearing of poultry or pigs[45].

In practice, what does this mean?

It means food producing activities are not innocuous and can cause extensive environmental harm. Therefore, an environmentally food production undertaking must ensure that all the relevant environmental impacts[46] have to be either prevented, or minimized or, as a last resort, offset[47].

Besides, applying the best available techniques[48] is a condition *sine qua non* for the authorization of such activities.

---

[42] Directive 2011/92 of 13 December 2011 amended by Directive 2014/52 of 16 April 2014 (to be transposed by Member States until 16 May 2017).

[43] Annex I activities: Installations for the intensive rearing of poultry or pigs with more than: (a) 85 000 places for broilers, 60 000 places for hens; (b) 3 000 places for production pigs (over 30 kg); or (c) 900 places for sows.
Annex II activities: 1. Agriculture, silviculture and aquaculture: (a) projects for the restructuring of rural land holdings; (b) projects for the use of uncultivated land or semi-natural areas for intensive agricultural purposes; (c) water management projects for agriculture, including irrigation and land drainage projects; (d) initial afforestation and deforestation for the purposes of conversion to another type of land use; (e) intensive livestock installations (projects not included in Annex I); (f) intensive fish farming; (g) reclamation of land from the sea. Extractive industry (iii) drilling for water supplies; 7. Food industry (a) manufacture of vegetable and animal oils and fats; (b) packing and canning of animal and vegetable products; (c) manufacture of dairy products; (d) brewing and malting; (e) confectionery and syrup manufacture; (f) installations for the slaughter of animals; (g) industrial starch manufacturing installations; (h) fish-meal and fish-oil factories; (i) sugar factories 10. Infrastructure projects (l) groundwater abstraction and artificial groundwater recharge schemes not included in Annex I.

[44] General principles governing the basic obligations of the operator set out in article 11 a, b, c, g, and h of Directive 2010/75 of 24 November 2010 on industrial emissions.

[45] Categories of activities listed and described in detail and with quantitative thresholds, in annex I (6.4, 6.5., 6.6.) of Directive 2010/75 of 24 November 2010.

[46] On human beings, on fauna and flora, on soil, water, air, climate and the landscape; on material assets and the cultural heritage; on the interaction between the previous factors (article 3 a-d).

[47] Annex IV 6. of Directive 2011/92 of 13 December 2011: 'a description of the measures envisaged to prevent, reduce and where possible offset any significant adverse effects on the environment'.

[48] The best available techniques reference document for the relevant food producing activities was adopted in May 2005 (eippcb.jrc.ec.europa.eu/reference/sa.html).

In extreme cases, Member States are allowed to refuse granting any authorization to food producing activities grounded on important environmental reasons. It's the case of the agricultural production of genetically modified food and feed. This activity, dramatically baptized in the European Union as 'deliberate release into the environment of genetically modified organisms' raises deep concerns among citizens[49] and has even led to the 'condemnation' of Member States reluctant to accept GMOs[50]. Since 2015 the European Union shifted into a more flexible position accepting the 'possibility for the Member States to restrict or prohibit the cultivation of genetically modified organisms (GMOs) in their territory'[51] based on the following grounds: environmental policy objectives; town and country planning; land use; socioeconomic impacts; avoidance of GMO presence in other products; agricultural policy objectives; public policy[52].

Moving towards sustainable development

Placed in the context of environmental law and in the light of the ethical principles enshrined in European Union Treaties and Regulations, integrated food safety law assumes a rather more ecocentric character. The three major trends in integrated food safety law in Europe are:

1. fighting unnecessary cruelty against animals;
2. contesting institutionalized speciesism; and
3. aiming at a high level of environmental protection.

To conclude, integrated food safety law is pushing European Law into a new era of interspecies-environmental-justice and moving Europe a step further towards sustainable development.

---

[49] Special Eurobarometer number 354 on food-related risks shows a clear picture concerning the risks associated with 'genetically modified organisms found in food and drink: 'the worry levels range from just under half of the sample in Ireland (46%), Malta, Sweden and the United Kingdom (all 48%) to over 80% in Greece and Lithuania (both 81%). Austria is the only Member State where 'GMO' is ranked in that country as the most serious worry (at 67%, the same level of concern as that for pesticides). Worry on this issue has increased significantly in 10 Member States since 2005, with Lithuania (81%; +18) showing the largest shift, followed by Portugal (67%; +13). The United Kingdom (48%; -6) is once more the only country that shows a marked decline in the level of worry since 2005' (page 30) (http://tinyurl.com/jb5ojdo).

[50] Joined Cases T-366/03 and T-235/04 (decided on 5 October 2005), and appeal C-439/05 P and C-454/05 P (decided on 13 September 2007), all against Austria (http://tinyurl.com/hfnl52w).

[51] Directive 2015/412 of 11 March 2015.

[52] Article 26 b) 3.

Section 1. Animal ethics

4. Bernard E. Rollin, 'gestalt shift' and animal ethics

J.L. Harfeld
Center for Applied Philosophy, Aalborg University, Kroghstraede 3, 9220 Aalborg, Denmark; jlh@learning.aau.dk

Abstract

Bernard E. Rollin is especially known for his work on the animal welfare concept telos – the essence of a given animal. This paper explores another and lesser known, albeit very central concept in the work of Rollin: the 'gestalt shift'. This is the idea that prior to the possibility of a valid ethical theory concerning actions in relation to animals there is a demand for an epistemological and metaphysical change of view. Rollin argues that we ought to understand the lives upon whom our actions have such profound. In effect, before any moral demands aimed at actions we are morally obliged to develop a certain awareness of animals that can enable a shift of our world-view and let us realize hitherto unseen aspects of our relationship with non-human beings. We do not observe something new as such, but by shifting gestalt, we shift our perspective on the same data. The first step is through sympathetic observations. This is the phenomenological approach of experiencing non-human animals around us either directly or through the help of others. An understanding of animals that is fully relevant to ethics must not only be accurate about animal physiology and behaviour, but must also involve sympathetic observations of the animal's life. This includes an *a priori* granting of a certain status or, at least at the outset, conceding that there is someone there. Only from sympathetic observations can an understanding meaningful to ethics arise – an empathetic understanding. Such empathetic understanding is the precursor to any possible gestalt shift in our world-view. Empathetic understanding means that there is an understanding beyond, but including, scientific facts. It is an understanding that is necessary for both moral actions and ethical theory in relation to animals. The concept of the gestalt shift has been an underlying subject in the work of Rollin throughout the last 35 years and in this paper, a number of aspects and underlying arguments are explored and brought together. Furthermore, the paper points to the problem that Rollin's idea of gestalt shift does not address the systems of thought and language which pervades society's current gestalt – in particular the idea of market thinking.

Keywords: sympathetic observations, empathetic understanding, moral progress

Situating Rollin in animal ethics

Two of the most important modern animal ethicists are arguably Peter Singer (Animal Liberation, 1976) and Tom Regan (The Case for Animal Rights, 1983). Together, Singer and Regan have come to outline the main theoretical diversity in contemporary animal ethics. Singer's approach is unmistakably utilitarian, with a focus on suffering and interests, and an extensive criticism of inequality and speciesism in traditional ethics. Regan defends a rights-based ethics on the basis of the inherent value of (some) animals – a theory with clear roots in the notion of dignity in the deontological tradition of Kant. Both philosophers work within a framework of normative ethics which is clearly aimed at providing clear answers to the question: 'how ought one act towards animal?'. In this sense, they are specific in their action guidance. This is in contrast to much of the work of Bernard E. Rollin who in his very first article on animal ethics Rollin writes:

> It is worth emphasizing in conclusion that we have not attempted to set up an ethical theory which will direct our behaviour towards animals ... I have simply tried to show that animals must be seen as object of moral concern in themselves, not in a derivative way. (Rollin, 1978)

EurSafe 2016 – DOI 10.3921/978-90-8686-834-6_4, © Wageningen Academic Publishers 2016

This has remained the focus for Rollin's work on animal ethics. He has worked extensively on exposing and analysing the many layers of arguments around animal welfare and the moral status of animals. Rather than pick a side (Singer's or Regan's or a third) Rollin addresses the issues of animal ethics through a pragmatic (common sense) position between rights and utilitarianism. He first adopts a teleological welfare view based on the animals' unique characteristic speciesness (telos), but rejects all-out utilitarianism on the grounds that certain interests and rights arise from these characteristics which cannot be readily dismissed by a greater sum of happiness.

> ... utilitarianism ... is only part of our social ethic. It must always be balanced by Kantian respect for individuals. This is, in essence, widely known as 'common sense' to people not involved in scientific ideology (Rollin, 2006: 85)

Rollin is especially known for his work on an animal welfare concept that includes the aforementioned concept of telos – the nature or essence of a given animal. Telos is not merely how an animal is at any given moment, but encompasses the extent of what it, due to its particular animalness, could be. Neither is the concept to be understood as simply a matter of the animal's physical and mental properties, but additionally as the animal's distinctive and active being itself. It is the positive or negative fulfilment of this telos that must necessarily be the foundation for any action guiding animal ethics. Rollin, however, does nowhere develop a system or a theory of principles to point us to how exactly such an action guiding ethics.

Shifting our stance

Instead of attempting to answer the question: 'how ought one act?', Rollin focuses on the two questions: 'why ought one act?' and 'how does one become capable of acting (ethically)?'. The question of 'why' is a matter of what is commonly known as 'moral status'. Rollin's views are here in line with the ideas of Singer and Regan in that he argues that, all other things considered, we have no good reasons for denying animals' moral status since there are no relevant differences between animals and humans on this matter. Contrary to Singer and Regan, Rollin does, in some of his works, differentiate between different kinds of moral status and, inspired by his long time colleague Holmes Rolston III, opens up for the possibility of a morals status for non-sentient entities. The main question for this paper, however, is not moral status itself, but rather how we come to recognize moral status and how we are enabled to act upon such recognition.

Before any ethical theory of action – indeed before the possibility of any ethical theory – we must first establish an understanding of the world. This world-view or this metaphysical position (Rollin, 1981) is necessary in order to understand what and who any given ethical theory is about. In this respect such an understanding could be seen as an ethical demand of how we view the world, or more correctly a demand to view the world correctly in some respect. This metaphysical position has been at the core of Rollin's animal ethics for almost four decades. Although Rollin has developed an elaborate theory of animal ethics with an emphasis on animal welfare (flourishing according to a specific telos), it is his ideas of pre-ethics theory that sets him apart from people like Singer and Regan. This has resulted in the formulation of a range of arguments in favour of changing our understanding of animals, a philosophical move which he calls 'a moral and metaphysical Gestalt shift' (*ibid.*) or just 'gestalt shift' (Rollin, 1992: 79) in our relationship with the natural world in general and animals in particular. Borrowing and modifying this concept from the broader tradition of gestalt perceptual psychology (*ibid.*: 79-80), Rollin uses it to argue that before dealing with what we ought to specifically do in our dealings with others, 'we are morally bound to understand the lives upon whom our actions have profound and considerable effect' (Rollin, 1992: 95). Before we relegate any time and effort towards action guiding ethical arguments we must 'escape the stranglehold in which the conceptual scheme of human separateness from nature has held us since antiquity' (Rollin, 1981). This epistemological escape is itself an ethical demand and could be described as a 'moral awareness' (Rollin, 1992: 78) of animals that will shift our world-view and let us realize hither though unseen aspects of our relationship with the rest of creation. We do not observe something new but this gestalt shift is a shift 'in perspective on the same data' (*ibid.*: 80) and thus an ontological clarification. Our historical ethical failure towards animals is not merely the harm we have brought upon them but more fundamentally the preceding lack of understanding them. One of the reasons for this lack is an irrational fear of anthropomorphism. Traditionally the scientific community has been dismissive of any arguments based upon or related to anthropomorphism – i.e. the notion that we can meaningfully use a great number of human emotional and mental concepts on certain animals. There is distinct and discernible inspiration from logical behaviourism and similar veins of thought when scientists reject that there can be a valuable debate about the unobservable mental states – the qualia – of animals. And this is not merely a trend steeped in the positivism of the early 20th Century but is still a heavy influence among prominent zoologists (Dawkins, 2012: 19). For Rollin, however, the full embrace of anthropomorphism is psychologically necessary to achieving the gestalt shift. This is not to say that we can meaningfully and appropriately utilize any mental states and thoughts, that we know from ourselves as humans, to animals. To know an animal is also to know the limits of the animal. To fully embrace should here be understood as presupposing the existence of an Other when dealing with animals, which, Rollin argues, is the only way of accessing the further understandings that can underpin moral considerations.

Our first step towards a new and a more proper epistemology of animals is through what Rollin calls 'sympathetic observations' (Rollin, 1992: 92). This is to be understood as a phenomenological approach in which one must experience non-human animals around us either directly or through the help of others (literature or media). Inspired by the animal phenomenologist Ralph Acampora, Rollin describes such sympathetic observations as a 'primordial experience of commonality between us and them in the shared struggle to survive, nourish ourselves, avoid being hurt, and all the other tasks entailed in sentient life' (Rollin, 2007). This is not the same as accumulating facts about animals. Although facts about animals and their physiology and behaviour are necessary in order to establish what is good for them, an understanding of them which is fully relevant to ethics must involve 'sympathetic observations of the animal's life and activities' (Rollin, 1992: 92) where the notion of 'sympathetic' means the a priory granting of a certain status. One is not sympathetic in dealings with or observations of chairs. This is due to the fact that we do not in our daily dealings with and observations of chairs judge it as having to do with someone. There is no one. To observe sympathetically is to acknowledge, at least at the outset, that there is someone there. Only from sympathetic observations can an understanding meaningful to ethics arise – an 'empathetic understanding' (*ibid.*: 94). Such empathetic understanding is the precursor to any possible gestalt shift in our world-view – or specifically our view of animals. By empathetic understanding is meant that it is an understanding beyond, but including, scientific facts. It is an understanding which includes a willingness to see and identify with the other someone.

The notion of the gestalt shift is Rollin's way of arguing that an ontological concept of moral status is necessary but that a certain epistemological stance is necessary for the ontological question of moral status to arise. We can, as on any epistemological question, be wrong on certain ways and in certain instances. It can, for example, be difficult to say where a line should be drawn as to which beings in the world I can meaningfully be said to have empathetic understanding. This would, however, under the given circumstances be a matter of erring on the side of caution.

Critical remarks

Rollin is an optimist when it comes to the question of moral progress concerning animals. This might be accurate to a certain extent. For example, more and more laws are implemented throughout especially western societies that focus on animal welfare and are steeped in public sentiment for animals. At least in America and Europe it seems as if the prevailing winds are in favour of a social ethic which more and more readily includes at least some animals. This, Rollin might argue, is among other things the result of many people experiencing a gestalt shift concerning animals and becoming capable of thinking ethically about them. What Rollin does very little to explain is why we are no further than we presently are. It seems as if all the necessary conditions for gestalt shifts are available to any modern European or American. What Rollin is lacking here is an analysis of the structures which inhibit our ways of thinking about animals. In a way he does approach this in his criticism of 'scientific ideology' (Rollin, 2006: 11) and the way that such narrow scientific thinking inhibits the escape of a certain frame of mind an allowing for concepts necessary for moral thinking. He fails, however, in pinpointing one of the main ideologies which currently systematically inhibit the kinds of thinking which is necessary for his gestalt shift: market thinking. Michael Sandel points to this when he argues that markets are not the neutral mechanisms of allocation which we are led to believe (Sandel, 2012: 9). Indeed the very buying and selling of things itself promotes and conveys certain kinds of thinking about those things – a certain gestalt towards them. The overwhelming power and pervasiveness of market thinking in relation to animals physically forces animals as units into production systems while the language of economic ideology objectifies them as columns on spreadsheets. Thus market thinking disengages citizens from the sympathetic experience of animals which must necessarily underlie a gestalt shift.

References

Dawkins, M.S. (2012). Why animals matter: animal consciousness, animal welfare, and human well-being. Oxford University Press, New York, NY, USA,

Regan, T. (1983). The case for animal rights. Routledge & Kegan Paul, London, UK.

Rollin, B.E. (1978). Of beasts and men: the scope of moral concern. The modern schoolman LV/March: 241-260.

Rollin, B.E. (1981). The metaphysics of anthropocentrism: a review of Paul Ehrenfeld's 'The arrogance of humanism' and Mary Midgley's 'Beast and man'. International Journal for the Study of Animal Problems 2: 75-80.

Rollin, B.E. (1992). Animal rights and human morality. Prometheus Books, Amherst, New York, NY, USA.

Rollin, B.E. (2006). Science and ethics. Cambridge University Press, New York, NY, USA.

Rollin, B.E. (2007). Corporal compassion: animal ethics and philosophy of body. Anthrozoös 20: 203-204.

Sandel, M.J. (2012). What money can't buy: the moral limits of markets. Farrar, Straus and Giroux, New York, NY, USA.

Singer, P. (1976). Animal liberation: a new ethics for our treatment of animals. Duckworth, London, UK.

5. How should death be taken into account in welfare assessments?

K.K. Jensen
Department of Food and Resource Economics, University of Copenhagen, Rolighedsvej 25, 1958 Frederiksberg C., Denmark; kkj@ifro.ku.dk

Abstract

It appears to be a widespread view among animal welfare researchers that death is not a welfare issue. This paper demonstrates that this view is based on the mistaken assumption that welfare assessment is absolute, which moreover is coupled with the assumption that 'welfare' means 'welfare at a time'. It also demonstrates that to exclude the welfare issues of being deprived of life from the ethical assessment of killing distorts the ethical considerations. In order to assess the welfare issues of death, it is necessary to structure welfare assessment as comparisons of possible whole lives of the animals.

Keywords: death, value of whole lives, comparative welfare

Introduction

It appears to be a widespread view among animal welfare researchers that death is not a welfare issue. 'The animal welfare issue is what happens before death, including how they are treated during last part of their lives, often the pre-slaughter period and then the method by which they are killed' says Donald Broom (2011: 126). 'Being dead' is not a problem, John Webster (1995: 15) adds. This view does not imply that killing an animal could not be an ethical issue; but whatever the ethical issue is, it cannot stem from the animal being worse off dead.

This paper aims to demonstrate that this view is based on a mistaken assumption of the structure of welfare assessment, which moreover is coupled with the assumption that 'welfare' means 'welfare at a time'. It also aims to demonstrate that to exclude the welfare issues of being deprived of life from the ethical assessment of killing distorts the ethical considerations.

Welfare assessment: comparative or absolute

Many welfare researchers appear to reason like this: welfare is about what matters to the animals. What matters to the animals is the quality of their mental states: what they experience in terms of positive or negative states. Hence, these researchers appear to be what philosophers call hedonists, although this is a term rarely used in the animal welfare literature. Hedonism is the view that how good a life is for an individual is determined entirely by the pleasures and pains it enjoys in its life (Broome, 1993: 172).

From this position, it is often assumed that, in order for something to be good or bad for an animal, it has to be experienced as good or bad; and in order for this to be possible, the animal has to be alive and conscious at the time. For instance, in a report about piglet mortality in Denmark, the Faculty of Agricultural Research (DJF) writes: 'Generally, it is a precondition for the presence of suffering and reduced animal welfare that an individual is alive and sufficiently conscious to experience sensorial input and interpret them as unpleasant' (DJF, 2010: 66, translated from Danish by the author).

Although Broom does not commit himself to hedonism, he seems to accept the assumption that in order for a state to be good or bad for an individual, the individual has to be alive at the time. For him,

EurSafe 2016 – DOI 10.3921/978-90-8686-834-6_5, © Wageningen Academic Publishers 2016

the notion of welfare does not apply to the state of being dead. Webster only partly commits himself to hedonism; so when he says that being dead is not a problem, he could simply agree with Broom; but he could also add the claim that there is no experience of this state. Hence, even from the non-hedonistic perspective of Broom, there is an assumption to the effect that, in order for an event to be good or bad for an individual, this event has to manifest itself in some way in the life of the individual.

But an event can be bad for an individual even though there is no experience of it as bad (or even though it does not manifest itself in some other way). Suppose a farmer between two batches of pigs makes a change in management which deteriorates the condition of animals. The new batch will not experience the change; it will not manifest itself as a change in the lives of these pigs. However, they have less good experiences compared with the pigs in the preceding batch. This is bad for them because they are worse off than they would have been, had the change not taken place.

Suppose a 24 year old person unknowingly and unexpectedly dies in her sleep. Her death does not show up as a bad experience or otherwise manifests itself as a bad thing in her life. Still, it is a bad thing for her to die, because she is deprived of what would otherwise likely have been a long and good life. It had been better for her to live a longer life enjoying the good things in life rather than the short life she end up living and the good things it contained.

I assume that Broom, Webster and other welfare researchers would agree that the longer life is better for the person than the shorter one. Still, they would presumably reject it was bad for the individual that she died. But the badness of her death is nothing else but the longer life being better than the shorter one. It is not a claim about the state of being dead.

Why is this simple point overlooked? According to John Broome (1993: 167), it is often overlooked that welfare is always comparative. The underlying comparison depends on the context. When we say about a particular individual alive now that it has a good life, we may for instance mean that it is better for the individual that it 'should continue to live rather than die now'. Possible lives for an individual could be ranked from the worst to the best.

To talk about welfare without relying on a comparison, Broome suggests, implicitly assumes a notion of absolute (non-comparative) goodness or badness: welfare is measurable on a scale with an absolute zero. From this perspective, all good and bad features of a life have to manifest themselves in that life. For instance, if death is bad, it is because it manifests itself as a particular bad experience. But what could be the absolute zero of welfare? Of course, we could choose some standard life as our zero; but then we still compare other lives with this standard. I shall not go deeper into a discussion of possibilities of giving meaning to an absolute zero of welfare (but see the discussion in Broome, 1993). I shall just point out that, as far as I am concerned, an absolute zero has never been identified in animal welfare research. Nevertheless, if the comparative nature of welfare assessments is overlooked, there is no room for the notion that an unconscious painless death can be bad for an individual.

Welfare at a time

However, there is another reason why death as a welfare issue is overlooked. In animal welfare research, 'welfare' almost always means 'welfare at a time'. Broom (2011: 122) is explicit about this. Elsewhere it is often only implicit. In Welfare Quality®, it appears implicit; the various measures cover varying slices of time, but always some period of time, never a whole life. It is explicit, however, that an assessment in Welfare Quality® only covers a part of an animal's life, e.g. rearing, producing or slaughtering (e.g. Welfare Quality, 2009: 9).

If the focus is only on 'welfare at a time', longevity never enters into the picture. It is simply not asked if it is better for an animal to live a long life rather than a short life. It follows that the badness of death as being deprived of future life cannot be assessed. Death only pops up, if some animals die within the studied period of time.

Thus, the measure 'motality' in Welfare Quality® is defined as 'the 'uncontrolled' death of animals as well as cases of euthanasia and emergency slaughter' (Welfare Quality, 2009; 33). It is treated on line with diseases as 'coughing, nasal discharge, ocular discharge, hampered respiration, diarrhoea, bloated rumen' (Welfare Quality, 2009: 22). 2% dead animals on a farm over a year (i.e. uncontrolled deaths, euthanasia or emergency slaughter) get the same weight as 4% animals (observed at a time) with nasal discharge. This weighting is coherent with the assumptions uncovered above: death is like a disease in that it may be painful for the individual, and like a disease, this pain stretches over a specific period of time; there is nothing else bad about death to be assessed within Welfare Quality®.

I conclude that death is a welfare issue when the individual is deprived of good experiences or other good things in life; but that animal welfare research is structured in a way where this issue cannot arise. The only way to make it visible is though comparing the welfare of whole possible lives.

Some authors (Kasperbauer and Sandøe, 2016; Yeates, 2010) who acknowledge that an animal can be harmed by being deprived of a longer life, claim that whether or not death is considered a welfare issue depends on the how 'welfare' is defined. I hope to have demonstrated that this is not the case. It is rather a matter of how welfare measurement is structured (absolute vs comparative, at a time vs whole life).

I should mention an influential philosophical critique of the view on the harm of death I have sketched here. Jeff McMahan says of this view that it has implications that are 'profoundly counterintuitive. For example, it implies that, if other things are equal, the killing of a foetus or infant is more seriously wrong than the killing of an older child or an adult' (McMahan, 2002: 192).

In the context of this paper, McMahan could add that it, presumably similarly counterintuitively, implies a stillborn pig, other things equal, suffers a greater harm than a pig slaughtered at the normal age, compared to lives they could have lived, had they not died or been slaughtered, respectively. Actually, following the assumptions of the DJF report quoted above, and relying on the evidence suggesting that piglets only gain consciousness when they start to breathe after birth (Mellor and Diesch, 2006), we reach the conclusion that a stillborn piglet suffers no welfare loss.

McMahan relies heavily on what he describes as intuitions about the badness of death. His ambition is therefore to find a theory which coheres better with these intuitions. Thus, he suggests 'the time-relative interest account of the wrongness of killing', according to which the harm by killing (which explains its wrongness) is the frustration of 'the victim's time-relative interest in continuing to live' (McMahan, 2002: 194). Roughly, an individual's time-relative interest in continuing to live at any given time is determined by its psychological connectedness to its future life, i.e. how much it matters for the individual at the time to continue living.

However, McMahan's account brings us back to the assumption that death only harms an individual to the extent that the loss of future life matters psychologically to the individual at the time of dying. This simply ignores Broome's point that an individual can be harmed by being deprived of future life, regardless of what it matters to it at the time of dying. McMahan thus reinstalls death as an event which in an absolute sense must have psychological significance when it happens.

On the other hand, McMahan acknowledges that a person can be harmed by being deprived of life, and the loss of welfare he or she suffers overall is measured by the loss of life-time welfare. The only reason for him to lastly claim that this is not what matters appears to be the intuitions already mentioned. This put heavy weight on these intuitions in his argument. But it is not clear what these intuitions exactly are about. It seems possible that they could be intuitions about the surviving persons' assessment of their experienced loss of the death of another individual. My experience of loss in the event of another's death is likely to be influenced by my psychological connectedness to the lost life of this individual. Even if they are intuitions about the badness of death from the perspective of the dying individual at the time of dying, this does not show that this is what matters when assessing the badness of death. So the question is whether McMahan has any independent argument for the claim that life-time welfare is not what matters? But this is a complicated discussion relating to issues about personal identity which I cannot do full justice to in this context.

The distortion of the ethical assessment of killing

To exclude any welfare issue from the ethical assessment of killing animals is to distort this assessment. It makes it possible to say: we have come to the assessment that it is justified to kill animals for human food, and provided the animals have a good life until slaughter and they are killed painlessly, there is no welfare issue involved in this, when instead they should have added: we acknowledge that we thereby bring animals to live a life that is less good than it could have been, had we not killed them at an early age. Strangely, the authors mentioned above who acknowledge death as a welfare issue do not discuss the killing of ordinary production animals.

One may of course ask what the relevant comparison is for production animals. It could be claimed that it is never on the table to let the animals live longer; the only feasible alternatives are either they live a short, productive life or they live no life at all, and the former might be better than the latter. This may well be true, but it does not justify the claim that there is no welfare issued involved in killing the animals at an early age. We are able to bring them to life and *ex hypothesi* make this life good. We are also able to let them continue this life after the normal time of killing, but this is what we decide not to do. By this decision, we deprive them of a longer life. To say that we would never want to make another decision can hardly justify that we make it; it is rather to imply that we do not care about possible alternatives.

Most animal welfare legislation is based on the assumptions that it is ethically acceptable to kill animals for meat production or kill them after their otherwise productive life, and that there is no welfare issue involved in the short lives of these animals. Most animal welfare research is based on the same assumptions. But if there is no welfare issue involved in killing production animals at an early age, and the killing is otherwise considered ethically acceptable, then how can there be any problem with the unintended deaths that result from the production technology, such as piglet mortality or the killing of day old male offspring of laying hens or dairy cattle? After all, they are the result of the choice of the most productive technology, and animal production may not be feasible without the using most efficient technology.

The only welfare problem in this regard which is visible under the commonly accepted assumptions is if these animals experience suffering or pain during the process of dying. As we saw in the case of piglet mortality, this appears to exclude any problem with stillborn piglets. This conclusion is not explicitly drawn in the DJF (2010) report commissioned by the Danish Government to address the problems of piglet mortality as response to a public uproar. However, the report does conclude: 'The main proportion of piglets that die after birth are likely to be exposed to serious suffering in terms of either pain, hunger, fear or stress lasting from a few minutes up till 12 hours (p. 23) ... The high proportion of dead piglets is therefore an ethical as well as a welfare problem, since most of the piglets that die after birth apparently die suffering'. (p. 50 translated from Danish by the author).

The report had the mandate to look at the increased mortality resulting from breeding for larger litters. Therefore, it also looked at technologies to reduce the mortality. However, if there is a welfare problem and an ethical problem with piglets that suffer during death, then this problem must be present for any positive level of piglet mortality. Apparently, some level of piglet mortality is unavoidable under the current intensive production systems. But it was never on the agenda to question the ethical acceptability of pig production as such, only to look for possibilities to reduce the present very high mortality rate. Hence, if some mortality is unavoidable under conditions otherwise considered ethically acceptable, this level of mortality seems also to be considered acceptable.

Again, this implication of the mandate is not made explicit. One may speculate whether the mandate and the report based upon it really address the ethical problems of piglet mortality as they are perceived by the general public. But once you allow for the thought that the unintended loss of life in animal production is a general welfare issue, you will have to admit that a comparable welfare issue is present also for the production animals killed at the expected time in their life.

Conclusions

In order to assess the welfare issues raised by killing animals, it is necessary to structure welfare assessment as comparisons of possible whole lives of the animals.

Acknowledgements

I have benefitted from discussions with many people, including Emma Baxter, Sandra Edwards, Vivi Aarestrup Moustsen, Kenny Rutherford, Peter Sandøe, Flemming Thorup and Simon Turner. I should also like to thank an anonymous referee for valuable comments.

References

Broom, D.M. (2011). A history of animal welfare science. Acta Biotheoretica 59: 121-137.

Broome, J. (1993). Goodness is reducible to betterness: the evil of death is the value of life. In: Koslowski, P. and Shionoya, Y. (eds.) The good and the economical: ethical choices in economics and management. Springer-Verlag, Berlin, Germany, pp. 70-84. Quoted from the re-issue in: Broome, J. (1999). Ethics out of economics. Cambridge University Press, Cambridge, UK, pp. 162-173.

Det Jordbrugsvidenskabelige Fakultet (DJF) (2010). Pattegrisedødelighed i DK. Muligheder for reduktion af pattegrisedødeligheden i Danmark. [The Faculty of Agricultural research. Piglet mortality in DK. Possibities for reduction of piglet mortality in Denmark]. DJF rapport, Husdyrbrug 86. Det Jordbrugsvidenskabelige Fakultet, 77 pp.

Kasperbauer, T.J. and Sandøe, P. (2016). Killing as a welfare issue. In: Visak, T. and Garner, R. (eds.) The ethics of killing animals. Oxford University Press, Oxford, UK, pp. 17-31.

McMahan, J. (2002). The ethics of killing. Problems at the margins of life. Oxford University Press, Oxford, UK, 540 pp.

Mellor, D.J. and Diesch, T.J. (2006). Onset of sentience: the potential for suffering in fetal and newborn farm animals. Applied Animal Behaviour Science 100: 48-57.

Webster, J. (1994). Animal welfare: a cool eye towards Eden. Blackwell Publishing, Oxford, UK, 284 pp.

Welfare Quality (2009). Welfare Quality® assessment protocol for cattle. Welfare Quality® Consortium, Lelystad, the Netherlands.

Yeates, J.W. (2010). Death is a welfare issue. Journal of Agricultural and Environmental Ethics 23: 229-241.

6. What is the problem of replaceability?

R. Miguel
LanCog, Centro de Filosofia, Universidade de Lisboa, Centro de Filosofia, Faculdade de Letras, Alameda da Universidade, 1600-214 Lisboa, Portugal; ricardomiguel@campus.ul.pt

Abstract

Singer's much-discussed replaceability argument states that non-self-conscious animals may be killed and replaced by new animals that will lead equally valuable lives. If sound, this argument can be used to justify the cycle of raising and killing animals for food. Thus, many have argued that Singer's theory, and utilitarianism in general, while committed to this argument, offers inadequate protection to animals. However, some utilitarians reject the argument and Singer himself was rather tentative in preventing its additional application to self-conscious beings. This tension, within utilitarianism, about how to best deal with the argument is the core of what I call 'the problem of replaceability'. My main goal here is to provide a precise description of this problem. Firstly, I distinguish between the general question of replacement permissibility and the specific question of whether (a certain version of) utilitarianism implies the replaceability argument. Focusing on the specific question I set the problem apart from other known objections to utilitarianism, like the value receptacles objection and some replacement-like difficulties. I also compare two versions of the replaceability argument and advance a better one. Finally, I point out how different interpretations of the argument affect the way out of the problem. I hope this understanding of the problem offers helpful insight to possible solutions and to further investigate its importance to animal food production and animals' moral status.

Keywords: animal ethics, moral status, replacement, utilitarianism, value receptacles

Introduction

Unlike persons, some beings have no prospective desires and, roughly speaking, live in the moment. If their lives are valuable because of their good experiences, it seems wrong to kill them, for this will cut short the amount of such experiences. But what if we also produce other good experiences? This question concerns the permissibility of killing non-person non-human animals (henceforth, 'animals') provided that new animals are bred to compensate the killed ones. If this idea of compensation is sound, it justifies a whole lot of human actions that imply killing animals, the most significant of which being animal food production. Here is how this could be done:

Although meat-eaters are responsible for the death of the animal they eat and for the loss of pleasure experienced by that animal they are also responsible for the creation of more animals, since if no one ate meat there would be no more animals bred for fattening. The loss meat-eaters inflict on one animal is thus balanced, on the total view, by the benefit they confer on the next. We may call this the 'replaceability' argument (Singer, 1979: 149).

At this point I need to clarify what being 'total' means for a utilitarian view. Utilitarianism says rightness depends on bringing about the most value. But value is spread across individuals – according to hedonism, it is in their pleasurable experiences. This, then, begs the question of how to aggregate value in order to compare states of the world. The total view says that the value of a state of the world is the sheer sum – the total – of each individual's net pleasure in that state.

The replaceability argument (RA) generated a hot debate but it has been focused on arguing that Singer's theory and, by association, utilitarianism in general, offers inadequate protection to animals. The point

EurSafe 2016 – DOI 10.3921/978-90-8686-834-6_6, © Wageningen Academic Publishers 2016

is that although the RA does not justify factory farms or any kind of food production which implies animal pain, it seems to justify 'humane' versions of animal food production where animals lead happy lives and are killed painlessly. On the other hand, efforts to dismiss the RA without leaving utilitarian ground have not been duly appreciated. And given that Singer's defence of the irreplaceability of self-conscious beings was rather tentative (changing throughout the three editions of 'Practical Ethics'), this kind of efforts do not seem misplaced.[1] This tension, within utilitarianism, about how to best deal with the RA is the core of what I call 'the problem of replaceability'. Others have been using this phrase in connection with the RA but it is often unclear what is essential to the argument and what distinct difficulties it raises. My main goal here is to provide a precise description of this problem. In doing so I hope to offer helpful insight to possible solutions and to further investigate its importance to animal food production and animals' moral status.

Distinctions

Consider these two questions about replaceability: (i) is it permissible to replace an animal?; (ii) does utilitarianism (or some version thereof) imply the RA? Question (i) does not pose an interesting problem. In general, answers are easily derived by the various normative theories, the exception being utilitarianism. That is why question (ii) is more interesting. Its answers depend on the specific utilitarian theory and it is even arguable whether the total view is committed with the RA as it seems to be. In addition, question (ii) is somewhat puzzling because utilitarianism attributes moral importance to animals but yet the question presupposes that it is doubtful that their lives are morally protected. The second and more specific question, then, is the one that gives rise to the problem of replaceability. Ultimately, a thorough investigation of this problem – out of the scope of this paper – has to examine whether or not utilitarians can coherently defend animals' lives, at least in a stronger way than the RA allegedly allows. Consequently, it has to assess how utilitarians could make such a defence – what other commitments would they need if the price to do so were to abandon the total view?

Turning our attention to question (ii) some other distinctions are in place. Although related, the value receptacles objection is distinct from the RA. This objection is connected to the following passage:

> The total version of utilitarianism regards sentient beings as valuable only insofar as they make possible the existence of intrinsically valuable experiences like pleasure. It is as if sentient beings are receptacles of something valuable and it does not matter if a receptacle gets broken, so long as there is another receptacle to which the contents can be transferred without any getting spilt. (Singer, 1979: 149)

According to Chappell (2015), this objection can take three forms, each one associated with the following undesirable consequences of the underlying theory: (1) death is bad only because it causes fewer future good; (2) what matters is utility rather than individuals' interests as such; and (3) individuals' interests are fungible means to the aggregate good. However, none of these consequences is immediately about replaceability in the sense of the RA. In this sense, replaceability goes beyond that account of the badness of death and also beyond what Chappell (2015: 325) calls 'utility fundamentalism'. These may look like implicit necessary conditions to run the RA but it is defensible that they are not.

To see this, first regarding (1), start by assuming some version of preference utilitarianism, where value/disvalue resides in preference satisfaction/frustration. Then it could be said (about some individuals) that 'besides preventing the creation of future goods, death is also positively disvaluable insofar as it

[1] There is also the worry that if the argument is sound concerning some animals, it is sound concerning humans with comparable characteristics or even 'normal' humans (see Pluhar, 1982).

involves the interruption and thwarting of important life plans, projects, and goals' (Chappell, 2015:, 323). Even so, it is unclear that this move blocks the RA, for it can be argued that the lost value caused by death can still be compensated by other new lives with sufficiently valuable plans (see Uniacke, 2002: 215-216). Arguing that (2) is not necessary to the RA is easier. Assume that actions are fundamentally right because they are good for individuals and not merely because they promote net utility. Now suppose that animals are the only nutritious resource available for a group of humans. In such a context replacing animals might be right because it is good for individuals – in this case, the humans – and not merely because it promotes net utility. The issue is that actions being right in virtue of accordance with individuals' interests advances nothing regarding cases where interests conflict. Thus, at least in such contexts, it is possible to have both rightness in the former sense and replaceability.

What is specific to the RA is that replacing implies killing and then creating to compensate the killing. In a poorly informative sense, individuals are said to be replaceable in virtue of (3) above, that is, because the aggregate good is assumed to have priority over all their interests and the interest to live is one among them. Still, this means that the RA translates into a more specific worry than the third form of the value receptacles objection. Accordingly, to evaluate the RA there is no need to consider whether or not individuals' interests are fungible. It will be sufficient to examine if the aggregate good may override the interest to live. To use the metaphor at hand, being a mere receptacle of some, but not all, value is compatible with irreplaceability.

Another distinction worth mentioning concerns life-boat situations. It is common to think of individuals as replaceable in such situations. For example, if women and children have priority and someone must go overboard, then men will be replaceable (assuming no priorities among them). I claim that this 'replaceability' is troublesome for other theories as well: individuals may even have intrinsic value but it will still be reasonable to accept priorities among them, i.e. that it is better that some go overboard before others. Moreover, this 'replaceability' involves no compensation for death. Here, then, talk about 'replacement' is only an imprecise way of saying that some individuals are equally valuable to a certain effect.

The best expression of the replaceability argument

Since the RA is central to our problem and since it has different presentations in the literature, it is important to consider the most plausible one. To my knowledge, the first detailed version of this 'argument' appeared only in Miller (1982: 'Editorial', 1). It states that it

> is permissible, *ceteris paribus*, to use an animal and to kill it (for food or research or anything else) provided that the following conditions are met: (a) the life of the animal is on balance a life worth living; (b) the animal otherwise would have no life at all (would not exist); and (c) the animal will be replaced, at or after death, by another animal for whom conditions a. and b. hold.

A more recent version (adapted from Višak, 2013: 50) says that it is permissible to kill an animal if the following conditions are satisfied:

> (d) the future value of the animal's life would have been positive; (e) his death does not cause uncompensated disvalue; (f) the animal will be replaced, at or after death, by another animal, whose life value is at least as positive as the future value of the killed animal's life would have been, and which would not otherwise exist.

Comparing with the first version a. was improved by d., b. was abandoned (as was the permissibility talk of using animals), c. was improved by f. and e. is a new condition.

The first improvement is that d. makes the argument stronger and more accurate. Demanding that the killed animal has a positive lifetime value is irrelevant and unnecessarily limits the argument's application. What really matters is the value the killing prevents and this is that value that spans from the moment the animal is killed to the moment of its possible natural death were it not killed for replacement. Let this be the future value of the animal's life and its actual value the one going from birth to its actual death (for replacement). Then, whenever the future value is positive, utilitarianism requires compensation regardless of the lifetime value being negative (actual value + future value <0). In other words, sometimes maximizing value is a matter of reducing disvalue.

The abandonment of condition b. also makes the RA more accurate by setting it aside from what is known as the 'logic of the larder argument' (LLA). In a nutshell, the LLA states that breeding animals that lead pleasant lives is better for them than not existing at all and, therefore, farmers are justified to use and kill them. The 'logic' here seems to be that of animals owing farmers a favour for their lives. However similar these arguments may be, the LLA does not involve compensation for death. Another difference is that the LLA implies that it is possible to benefit an animal by creating it while the RA is silent about this claim (albeit compatible). Also, contrarily to the RA, the LLA is hardly seen as utilitarian (but see Višak, 2013: 129-133). As noted by Višak (2013: 48-49), setting these two arguments apart also explains why this version of the RA is not about using and killing animals but only about killing them.

The second improvement has two parts. The first is that c. reiterates a. and b. for the new animal and we have just seen that a. was improved by d. and also why b. should be abandoned. The second part is that c. produces a non-utilitarian argument. To be utilitarian the act of replacement must be maximizing and, therefore, the lost value must be compensated. But c. only demands that the new animal has a life worth living, which means that it allows replacing animals even when the lost value gets no compensation. For instance, it would allow so when the value of the new animal's life, although positive, is still less than the future value of the killed animal's life.

Finally, the addition of condition e. is, I think, redundant. The motivation to include e. is to account for negative side-effects, e.g. causing pain to other animals or even humans, and this is not explicitly stated in condition f. But it need not be so. It is in the spirit of the argument that the new life compensates whatever disvalue is caused by killing an animal and not just the future value of the killed animal's life. If I am right, the RA may have a simpler and clearer expression with a slight change in condition f. that already takes e. into account. However, condition d. is slightly confusing since, *ceteris paribus*, utilitarianism allows killing an animal whose future value is negative. Of course the RA is not about these cases and yet, precisely for this reason, we should make clear that the conclusion concerns only that subset of animals satisfying condition d. Furthermore, strictly speaking, these two expressions of the RA present no argument but only a conditional whose antecedent is the conjunction of the three conditions and whose consequent is the permissibility of (using and) killing an animal. So, if the objective is to argue for replaceability, the permissibility sentence has to figure in the conclusion of an argument. But we can easily get this by adding the premiss that such and such conditions are fulfilled. Only then we have an argument.[2] So here is my candidate for the best expression of the RA:

[2] Although the RA is usually presented as a conditional, I suppose this is its intended form. This common imprecision in the literature perhaps happens because the antecedent seems to be easy to satisfy.

4. If killing animals whose future life would have a positive value will lead to the reation of other animals which would not exist otherwise and whose lives will have at least the same value as the one lost with the killed animals, then such killing is permissible.
5. Killing *this* animal whose future life would have a positive value will lead to the creation of another animal which would not exist otherwise and whose life will have at least the same value as the one lost with the killed animal.

Therefore, it is permissible to kill *this* animal.

The quest for a solution

To tackle the problem it is essential to analyse how different versions of utilitarianism have been dealing with the RA. Very briefly, Total-Hedonism embraces it, Total-Preferentism restricts it to non-self-conscious beings and Previous Existence is able to block it.[3] So, one needs to grasp what features of each theory are doing the relevant work either to imply replaceability or to prevent it. It looks plausible that theories with person-affecting restrictions (containing a personal account of value) may have tools to block the RA.[4] Nonetheless, those restrictions also create more general worries as discussed in and after Parfit's (1984) influential work. The problem, thus, involves understanding which solution has the most acceptable cost. Hence Singer's recent claim that 'there is no consistent solution to the problems we are discussing that does not involve biting at least one bullet' (Višak and Garner, 2016: 235).

In my view there is still an important ambiguity in the RA that affects the way out of the problem. The point rests on two readings of the 'leading to' relation between the killed animal and the new animal. I propose these two readings: (iii) easy replaceability – there is no questioning about the dependence between the animals killed and their replacements; and (iv) hard replaceability – such dependence is questioned because it is a matter of moral agency. Arguably, only with (iv) the RA remains utilitarian because (iii) ignores relevant information to evaluate the replacement, namely, whether it would be possible to create a new animal without killing any. Therefore, utilitarianism requires asking if killing an animal and creating another is maximizing. All other things being equal, it is not. Two worth-living lives are better than only one. This, of course, conflicts with the common intuition of procreation freedom, but since premiss 5., under this reading, is only true in unlikely circumstances, animal replaceability becomes a much smaller worry than critics like to present it. Even if animals are not irreplaceable, raising and killing them, even humanely, would still be wrong in most scenarios.

Conclusions

The problem of replaceability is a specific concern about the best utilitarian(ish) way to deal with the RA. The most significant practical implication of the RA is that it justifies the 'humane' farming of animals for food. Theoretically, the RA raises a different, more specific objection than the value receptacles objection. To block the RA there is no need to consider if individuals' interests are fungible in general – the focus should be on their interest to live. Also, life-boat situations do not involve replaceability in the same sense as the RA does. The distinctiveness of the RA comes from linking the morality of killing and that of procreation to attain a state of value compensation. Comparing with other presentations advanced in the literature I believe I offered the best expression of the RA. Finding a solution to the problem involves

[3] For Previous Existence morally worth beings need to exist independently of the action under evaluation. In this way there is no value in the new animal's life to compensate the killing (for a defence see Višak, 2013).

[4] At first sight Roberts' (2011) variabilism presents a solution because it gives no weight to the loss of value involved in the new animal's life. However, it implies a counterintuitive asymmetry between losses and gains: losses only matter at a world where the individual who incurs that loss does or will exist but gains do not matter likewise (see Roberts, 2011: 365). Applied to the RA this is just an *ad hoc* denial that the new animal's life has moral weight.

evaluating how different theories have been dealing with the RA (which bullets are more palatable?). I suggested that a utilitarian reading of the RA allows the conclusion that, all other things being equal, replacing is not maximizing. The badness of death is central to this problem but giving attention to the goodness of new lives seems just as important and may mitigate the problem.

Acknowledgements

This work was done with the support of FCT studentship SFRH/BD/107907/2015, cofinanced by POPH/FSE. I thank my colleagues of LanCog and Arg. Cl. and also two anonymous reviewers for helpful comments, corrections and suggestions to previous versions of this work.

References

Chappell, R. (2015). Value receptacles. Noûs 49: 322-332.
Miller, H. (1982). Ethics and Animals 3: 1.
Parfit, D. (1984). Reasons and persons. Clarendon Press, Oxford, UK, 543 pp.
Pluhar, E. (1982). On replaceability. Ethics and Animals 3: 96-105.
Roberts, M. (2011). An asymmetry in the ethics of procreation. Philosophy Compass 6: 765-776.
Singer, P. (1979). Killing humans and killing animals. Inquiry 22: 145-156.
Singer, P. (2011). Practical ethics. 3rd edition. Cambridge University Press, Cambridge, UK, 356 pp.
Uniacke, S. (2002). A critique of the preference utilitarian objection to killing people. Australasian Journal of Philosophy 80: 209-217.
Višak, T. (2013). Killing happy animals: explorations in utilitarian ethics. Palgrave Macmillan, Basingstoke, UK, 188 pp.
Višak, T. and Garner, R. (2016). The ethics of killing animals. Oxford University Press, Oxford, UK, 272 pp.

Section 2. Farm animal welfare and ethics

7. Understanding Swedish dairy farmers' view on breeding goals – ethical aspects of longevity

H. Röcklinsberg[1*], *C. Gamborg*[2], *M. Gjerris*[2], *L. Rydhmer*[3], *E. Tjärnström*[1] *and A. Wallenbeck*[1,3]
[1]*Dept of Animal Environment and Health, Swedish University of Agricultural Sciences, Box 7068, 75007 Uppsala, SE;* [2]*Dept of Food and Resource Economics, University of Copenhagen, Rolighedsvej 25, 1958 Frederiksberg C, DK;* [3]*Dept of Animal Breeding and Genetics, Swedish University of Agricultural Sciences, Box 7023, 75007 Uppsala, Sweden; helena.rocklinsberg@slu.se*

Abstract

What values underlie farmers' choice of breeding goals? Typical dairy breeding goals cover production traits (e.g. yield, feed conversion, and lactation curve) and functional traits (e.g. leg health, fertility and calving ability). One central and multifunctional trait correlated to all these traits is longevity. Even so, dairy cows in Sweden are culled on average after only 2.4 lactations (about 4.5 years of age) i.e. before the housing and management of the heifer has paid off. Reasons for culling given in the literature include infertility, mastitis, disease or leg and claw problems, but also problems related to resources like milking system or grazing and to management practices like herd size or pasture management. Farmers' practice varies depending on their values or attitudes, farm size and breed, but no previous study has mapped their values underlying choice of breeding goals or values related to longevity. A first part of the study mapped dairy farmers' ranking of breeding traits via a web questionnaire. Here we present and discuss the second part: results from follow up qualitative focus groups interviews aimed at getting a deeper understanding of how farmers reason about longevity in relation to ethical issues concerning welfare and non-welfare issues (e.g. death, age at death or life span). We notice two interesting discrepancies between the breeding goal of longevity and (a) the farming practice of culling at an age of 4.5 years; and(b) the lack of robustness in dairy cows. Leaving genetic concerns (b) aside we will now map factors leading to farmers culling a healthy cow even before the investment has been secured. Results show that although culling is emotionally difficult, curiosity in breeding and hope for even better heifers keep up recruitment levels and hence culling. Further, cow integrity is mentioned with regard to handling but not to culling. The results are discussed in the light of Farm Animal Welfare Council's concept 'life worth living' and ethical aspects of lifespan.

Keywords: culling, integrity, life span, sustainable farming

Introduction

The aim of farm animal breeding is to change the genetics of the animals in a population over time. The direction of this genetic change is determined by the breeding goals set up by breeders, e.g. the breeding companies and farmers involved in breeding. In breeding of production animals, the breeding objectives have traditionally focused on improvement of traits related to productivity e.g. milk production, growth rate and feed conversion. However, there are unfavourable genetic correlations between many of these production traits and important functional traits, such as fertility and disease resistance. This means that breeding for e.g. improved milk production without taking fertility into account would reduce the fertility of the cows in the population. Moreover, the negative effects of increased productivity have become increasingly apparent; and as further understanding of these effects is developed, questions about what is ethically acceptable in farm animal breeding are raised (Olsson *et al.*, 2006). Thus it is crucial in all animal breeding to improve not only traits increasing profit in a short term perspective but also to consider traits for a long term sustainable production, e.g. traits related to environmental load or animal welfare should be included in the breeding goal (Mark, 2004; Hansen Axelsson, 2013). In the 1960s,

EurSafe 2016 – DOI 10.3921/978-90-8686-834-6_7, © Wageningen Academic Publishers 2016

dairy breeders in the Scandinavian countries were pioneers in including not only production traits, but also functional traits such as fertility and mastitis resistance in the breeding objective (Philipsson and Lindhe, 2003), and there is a common genetic evaluation of the Nordic dairy breeds.

One caveat of farm animal breeding is that the environment where the breeding animals are evaluated should be the same as the environment where the production takes place. There may, however, be differences in production environments between herds that call for alternative breeding goals. One topical example is organic and conventional production systems. Although animal housing and management at organic farms to some extent differ from that on conventional farms (and hence animals are subject to different conditions and challenges), the same breeds are used, i.e. animals bred for high production in conventional farms. It may be wise to use the same type animals in both production systems, but at present we do not have enough knowledge to conclude that. The development of long term sustainable breeding goals, for any production system, needs to consider the requirements of different stakeholders in the food chain, e.g. farmers, retailers and consumers (Gamborg and Sandøe, 2005). The discussion on breeding for organic farming has increased in recent years; but has mainly taken place within academia and breeding companies and seldom included farmers. (See http://www.lowinputbreeds.org for an exemption, Niggli *et al.*, 2007)

Longevity is included in the Nordic breeding goals and there has been a pronounced genetic improvement for longevity in cows the last decades (Carlén and Åkesson, 2013), even so cows in dairy production are seldom living a biologically long life (Växa, 2014). Instead, cows are culled; i.e. sent to slaughter, euthanized or die without human involvement (Alvåsen, 2014) more than 15 years before they are biologically 'old'. This has been seen as unsustainable both from a financial perspective (Carlén and Åkesson, 2013) and from an environmental perspective (Hansen Axelsson, 2013). In Sweden, a cow reaches an economic break even when the costs of rearing the recruitment heifer until she starts producing milk is paid off by her own milk production, which is between 45 and 60 months of age depending on factors such as age at first calving and production level (Carlén and Åkesson, 2013) or on average at 2.4 lactations (Växa, 2014). Reasons for culling may be related to the actual animal such as infertility, mastitis, disease, leg and claw problems or trauma, to resources like milking system or lack of grazing pasture, or to management practices like herd size or pasture management. Farmers practice in this area vary, not only with personality but also with farm size and breed (Alvåsen, 2014; Carlén o Eriksson, 2014). Decisions are made based on a combination of these factors, and also reflects a wider set of values related to the farmer's production philosophy. Farm animal breeding thus raises a number of ethical issues ranging over all steps of the breeding process – from choice of breeding goal to health and welfare of the 'product', i.e. the offspring (Sandøe *et al.*, 1999).

Swedish farmers' views on breeding have been examined through web questionnaires (Ahlman *et al.*, 2014), including the ethical aspects (Wallenbeck *et al.*, 2013). In this paper we will focus on the underlying values and views expressed by dairy farmers participating in focus groups interviews to examine the gap between the breeding goal of increasing longevity and farming practice of culling at an age of 4-5 years. The key questions are: (a) what factors do conventional and organic farmers regard relevant for the decision on culling; (b) how do they reason regarding longevity from animal welfare or other point of view; and (c) do their reasoning reflect other factors than those mentioned in literature?

Methods

In this study we rely primarily on results documenting Swedish conventional and organic dairy farmers' views on longevity gathered through two focus group discussions. Questions asked were based on results from a web questionnaire study (Ahlman *et al.*, 2014; Wallenbeck *et al.*, 2013). Farmers could indicate interest in focus groups in the questionnaire but remained anonymous, and others were recruited through contacts with organisations for farm advisors. We strived for even distribution between organic and conventional farmers in the groups, as well as geographical diversity as conditions for production differs somewhat with location. However, it should be noted that results from focus groups are not intended to be representative, but to gain deeper insight into certain key issues (Grimen and Ingstad, 2007). The recruitment procedure was difficult due to an economic crisis for many farmers, which could indicate that the participants have a higher interest in breeding issues compared to the average dairy farmer. In total 10 conventional and 6 organic farmers participated. Each focus group discussion lasted for about 2+2 hours, with an informal lunch in between. Participants were informed about the background and aim of the project, promised confidentiality and all discussions were tape recorded and minutes taken during the discussion. As to the inter-subjectivity in interpretation the same facilitator (ML) led all groups assisted by one or two persons (HR and AW), and two researchers (HR and ET) analysed the material thematically along a cluster method independently of each other (Braun and Clarke, 2006; Graneheim and Lundman, 2004).

Results

Four key themes emerged from the analysis of the focus group discussions: (1) the relation between healthy animals and long life; (2) breeding goals in relation to farm situation and housing; (3) breeding goals in relation to farmers' own role in culling and (4) perception of the human-animal relationship.

Relation between healthy animals and long life

The most common problems leading to culling expressed by the participants were mastitis, udder shape and functionality, leg health, body size and not fitting with the housing system, as well as temperament. Problems with walking and manageability were specifically mentioned as key reasons for culling: 'there is often something that is causing trouble which makes it feel worthwhile sending the cow to slaughter rather than keeping her'; 'It is sometimes just easier to call the slaughter truck'. Further, several farmers saw a connection between high yield and short lifespan due to higher risk of low health, and expressed that they would like to keep cows longer if their health status was better financially rewarding also in short term. They further missed the perspective of 'lifetime production' and that 'It is great to be able to choose and have a surplus that you can sell or slaughter'. 10 years are by several participants considered as an optimal lifespan for a dairy cow, but not within today's practice in general, and hence even less keeping them for a natural life span of about 20 years.

Breeding goals in relation to farm situation and housing

Several participants stressed that the main breeding company in Scandinavia has a too narrow range of bulls, and that advisors recommend the same breed to everyone. Farmers saw economic reasons govern recommendations rather than concern about a sustainable cow bred to fit a specific housing system. Current breeding of dairy cows was perceived as not focusing enough on 'on-farm functionality', on evaluating longevity, and not always considering 'the needs of the cow', but at the same time several participants noted that, it is 'difficult to go back'. Farmers also showed curiosity to see what breeding would yield, having expectations that e.g. the new heifer will be better, but argued that a feedback system was missing as well as evaluation in terms of e.g. looking at average lifespan of offspring/cows from a certain bull. Without a proper evaluation of progress it is difficult to choose between different traits. In general farmers expressed that breeding moves very quickly but that the effect of breeding on their own cows is often not noticeable until after five years or so. Moreover, breeding companies have a very homogenous stock which farmers experience cannot accommodate their sometimes differing needs. At the same time farmers expressed a wish for more homogenous cows to fit their housing and production system. Particularly for loose-housing more conformity is needed e.g. in terms of body size, udder shape and temperament.

Breeding goals in relation to farmers' own role in culling

According to the participants, few farmers are interested in or have time to take much interest in breeding. Because of this, farmers are dependent on breeding advisors and experience too often to be advised to replace 'problem cows' with new heifers. The farmers also expressed they had become used to a high culling rate but that they were not really comfortable with this. For example, they said, it is particularly difficult choosing between longevity and other traits such as yield: 'Maybe we would make as much money if the cows instead lived longer'. A clear majority stated that productivity is important (or the most important thing), although measured in different ways, but total cost of a cow per lifetime or yield per lifetime was seldom used. When economic margins are small such long-term calculations appear to be overlooked. 'We are bad at measuring lifetime yield for the whole herd, we don't really talk about that'.

Perception of human-animal-relation

A majority of farmers stated that they like their cows but also that there are limitations to the human-animal relationship, and that in the end decisions on culling were based on economic reasoning. Several participants stressed that is was important to keep a 'necessary distance' to their cows, and that it was not possible (or advisable) to be 'too empathic'. Another farmer viewed the relationship with the cows as mutually beneficial: we 'take care of each other'.

Many of the participants said they valued and cared more for older cows compared to younger ones, but that at the same time that these older ones were seen as 'way behind genetically'. Again, breeding is a fast-paced business, and about 30% of cows in a herd are generally replaced every year. Some farmers in the focus groups considered this to be too much and gave 20-25% as a more reasonable proportion. The farmers expressed that it could be difficult to send an older cow to slaughter that had 'been around for long' and they had thus formed some sort of bond, but at the same time it did not feel good to slaughter too young cows either.

Discussion

The above mentioned discrepancy found within dairy farming between the breeding goal of longevity and the low average age of cows at culling seems to have a parallel discrepancy in farmers expressing a wish to consider their cows on long-term basis and their actual decisions based on a one year-basis. Both discrepancies show that longevity is regarded to be ethically important, but in practice dairy cows are culled at an age of 4.5 years.

As efficient food production has been a prominent goal for many years, and still is, first motivated by the scarcity of food during World War II and since driven by market mechanisms, this has strongly influenced choice of breeding traits. However, today some parts of the world produce more animal based food than the population consumes, whereas other suffers under food insecurity, which has led to a discussion of redistribution of resources and revision of production forms. This, in combination with the effects of animal production on climate change and current discussion on 'sustainable intensification' calls for a discussion of choice of breeding goals based on an ethical perspective.

Many of the ethical issues related to breeding are related to the fundamental moral assessment that animal welfare matters. From this perspective moral dimensions are inherent to all human handling of animals as the level of welfare is affected either through resources available or management offered. Recently limits of the so called Five freedoms (Webster, 1994) both in terms of an achievable ideal state and as guarantee of farm animal welfare have been highlighted and the need to acknowledge a wider range of animal experiences, including positive ones, as fundamental for animal existence (Mellor, 2016). The preconditions for perceived welfare also depends on the breed whereby animal welfare is to be considered for both breeding procedures and goals. Hence, it should be noted that in adopting a certain breeding scheme the question is not only how existing animals should be treated but more one of which animals, or what kinds of animals, should exist (Gamborg and Sandøe, 2008), and hence also what is a desirable aim with regard to life span and quality of life Albeit difficult to define 'quality of life' in general for farm animal species, it has been used in e.g. policy documents (Farm Animal Welfare Council (FAWC), 2009), and it is argued to be intuitively understandable and possess prescriptive force (Yeates, 2011). The ambition is not to replace other welfare concepts, 'but is an holistic idea of an animal's welfare over its whole life, based on more fundamental concepts' (Yeates, 2011: 397) and is suitable for reflection on breeding for longevity, or upon life span in itself.

Hence, not only welfare is in focus from an ethical point of view. There are issues inherent in breeding that have no effect on an animal's welfare, but still involves ethical dimensions. Death, age at death, or life span is such an issue (Yeates *et al.*, 2011). Intuitively many persons think it matters how old an animal is when slaughtered for food arguing for instance that it is morally worse to kill young animals than old. Also this issue includes a wide range of perspectives. From an animal rights perspective one can argue, that not only housing of animals for human purposes is ethically wrong, but also that the animal has a right to an as long life as possible. Its life situation however has to be taken into account in terms of life quality, i.e. a balance between longevity and suffering. Only in extreme forms of animal rights views an animal should be kept alive independently of whether it suffers from caused pain or invalidity in order not to violate its right to live (e.g. at some 'animal rescue centres').

From another ethical perspective one can argue that killing young animals is a prima facie evil which may, however, be righteously overruled by other purposes such as efficient broiler production or culling offspring that was born as enrichment for the parent animals at zoos. Somewhere in between lies the view that animals, in parallel to humans, have intrinsic value or integrity, calling for respect for each individual, which in case of e.g. severe pain includes a right to be painlessly euthanized (Francione, 2008). Longevity is thus an ethically laden concept entailing a range of connotations relevant for the ethical discourse on breeding goals.

References

Ahlman, T., Ljung, M., Rydhmer, L., Röcklinsberg, H., Strandberg, E. and Wallenbeck, A. (2004). Differences in preferences for breeding traits between organic and conventional dairy producers in Sweden. Livestock Science 162: 5-14.

Alvåsen, K. (2014). On-farm cow mortality in Swedish dairy herds. Diss. (summary). Sveriges lantbruksuniv, Uppsala. Acta Universitatis Agriculturae Sueciae 21: 1652-6880.

Braun, V. and Clarke, V. (2006). Using thematic analysis in psychology. Qualitative Research in Psychology 3: 77-101.

Carlén and Åkesson (2013). Kornas livslängd – hur påverkas gårdens resultat? Djurhälso- och Utfodringskonferensen. Växa.

Farm Animal Welfare Council (FAWC) (2009). Farm animal welfare in Great Britain: past, present and future. FAWC, London, UK.

Francione, G.L. (2008). Animals as persons: essays on the abolition of animal exploitation, Columbia University Press, New York, NY, USA.

Gamborg, C. and Sandøe, P. (2008). Sustainability in farm animal breeding: a review. Livestock Production Science 92: 221-231.

Graneheim, U.H. and Lundman, B. (2004). Qualitative content analysis in nursing research: concepts, procedures and measures to achieve trustworthiness. Nurse Education Today 24: 105-112.

Grimen, H. and Ingstad, B. (2007). Qualitative research. In: Laake, P., Benestad, H.B. and Olsen, B.R. (eds.) Research methodology in the medical and biological sciences. London Academic Press, London, UK, pp. 281-310.

Hansen Axelsson, H. (2013). Breeding for sustainable milk production. Acta Universitatis agriculturae Sueciae. PhD thesis, Swedish University of Agricultural Sciences, Uppsala, Sweden.

Mark, T. (2004). Applied genetic evaluations for production and functional traits in dairy cattle. Journal of Dairy Science 87: 2641-2652.

Mellor, D. (2016). Updating animal welfare thinking: moving beyond the 'five freedoms' towards 'a life worth living'. Animals 6(21): 1-20.

Niggli, U., Leifert, C., Alföldi, T., Lück, L. and Willer, H. (2007). Improving sustainability in organic and low input food production systems. Proceedings of the 3rd International Congress of the European Integrated Project Quality Low Input Food (QLIF). March 20-23, 2007. Ökologie and Landbau 141: 32-33. University of Hohenheim, Hohenheim, Germany.

Olsson, A., Gamborg, C. and Sandøe, P. (2006). Taking ethics into account in farm animal breeding: what can the breeding companies achieve? Journal of Agricultural and Environmental Ethics 19: 37-46.

Philipsson, J. and Lindhe, B. (2003). Experiences of including reproduction and health traits in Scandinavian dairy cattle breeding programmes. Livestock Production Science 83: 99-112.

Sandøe, P. and Gamborg, C. (2008). Animal breeding and biotechnology. In: Sandøe, P. and Christiansen, S.B (eds.) Ethics of animal use. Blackwell, Oxford, UK, pp. 137-152.

Sandøe, P., Nielsen, B.L., Christensen, L.G. and Sorensen, P. (1999). Staying good while playing God – the ethics of breeding farm animals. Animal Welfare 8: 313-328.

Statens Jordbruksverk (2015). Organic animal production 2014. JO 27 SM 1501. Available at: http://tinyurl.com/jgarvue.

Wallenbeck, A., Ahlman, T., Rydhmer, L. and Röcklinsberg, H. (2013). Farmers' views on the impact of breeding traits on profitability, animal welfare and environment. In: Röcklinsberg, H. and Sandin, P. (eds.) The ethics of consumption. The citizen, the market, and the law. Wageningen Academic Publishers, Wageningen, the Netherlands, pp. 215-220.

Yeates, J. (2011). Is 'a life worth living' a concept worth having? Animal Welfare 20: 397-406.

Yeates, J., Röcklinsberg, H. and Gjerris, M. (2011). Is welfare all that matters? A discussion of what should be included in policymaking regarding animals. Animal Welfare 20: 423-432.

8. For the sake of production. How agricultural colleges shape students' views on animal welfare

J. Lassen[1], P. Sandøe[1,2] and I. Anneberg[3]*
[1]Department of Food and Resource Economics, University of Copenhagen, Rolighedsvej 25, 1958 Frederiksberg C, Denmark; [2]Department of Large Animal Sciences, University of Copenhagen, Grønnegårdsvej 8, 1870 Frederiksberg C, Denmark; [3]Department of Animal Science, Aarhus University, Blichers Allé 20, 8830 Tjele, Denmark; jlas@ifro.ku.dk

Abstract

The aim of this paper is to document and discuss the way future farmers' views on animal welfare are formed and shaped during their studies at agricultural colleges. The paper is based on qualitative interviews with first year students and more advanced students in Denmark. It is shown that there are only minor differences between the groups in the way good animal welfare is justified as both groups mainly view good animal welfare as a means of obtaining profitable production. In contrast, views which justify good animal welfare on the basis of the animals' wellbeing or their right to have a natural life, which are known to be of importance to the public, were either absent or considered of minor importance.

Keywords: interviews, perceptions, education

Introduction

The level of farm animals' welfare is closely related to farm economy, buildings, technology, regulation and other material and structural factors on and around the farm. However, it has been demonstrated that farmers' attitudes also influence the way they and their employees treat the animals and, thus, also the overall welfare of the animals (Hemsworth *et al.*, 1994). Therefore, the aim of this study is to analyse how future farmers' perceptions are moulded during their education at agricultural colleges.

Today there are no specific courses on animal welfare at Danish agricultural colleges, although the topic must be included in the curriculum when relevant. It is fair to assume that the years spent at college has a decisive influence on the students' perception of animal welfare. Despite this, few if any studies have examined the role agricultural colleges play in forming students' perceptions.

The following research question was formulated to guide the research process: how does students' perception of animal welfare change during their studies at agricultural colleges?

Method

The paper draws on a qualitative study of students at three of the 17 agricultural colleges in Denmark. The colleges were chosen to ensure diversity regarding size and uptake of students from rural as well as urban areas.

The study comprised interviews and observations at the three colleges. The aim of the observations was to obtain a contextual understanding of students' perception of animal welfare. Observations were carried out during sessions in the classes, in the stables, while on excursions to farms and during students during daily life at the college.

EurSafe 2016 – DOI 10.3921/978-90-8686-834-6_8, © Wageningen Academic Publishers 2016

Since the aim was to study possible changes in perceptions of animal welfare, two sets of interviews were carried out. First, 11 individual interviews with first year students were conducted. Due to practical reasons these students were interviewed 2-3 months after starting college. Common to this group was the fact that they were all enrolled on basic training and had not yet chosen which subjects to specialize in. The level of practical experience from farms and with animals varied among the 11 students: some had grown up on farms and/or had work experience from farms, while others were from urban areas and had no or very little experience with farm animals.

Second, three focus group interviews (n=6-10) with older students were carried out. The participants had all been studying for 3.5 years and were in the final phase of their studies. They were all specialized in the production of either pigs or cattle. All focus groups included boys and girls as well as students with rural and urban backgrounds. Unlike some of the first year students, these had extensive work experience from their vocational training on farms.

All students were recruited by snowballing and were identified and approached by teachers at the colleges who followed detailed recruitment criteria.

All interviews followed an interview guide. For the individual interviews this included sections on the students' relations with animals and their reasons for choosing to study at an agricultural college. This was followed by sections on the students' relation to agriculture and their perceptions of animal production with a particular focus on the welfare of pigs and cattle. Finally, the guide included a ranking exercise concerning the so-called five freedoms. The guide for the focus groups included sections inviting the students to rank animals and farm animals and discuss issues about animal welfare in relation to the ranking. This was followed by a section where the interviewees were asked to identify and discuss aspects in the production of pigs and cattle they found problematic for animal welfare, and a discussion of the five freedoms. Finally, the students were asked to recall good and bad experiences with animals from their vocational training and the interview finished with a discussion of the questions whether and how their views of animals had changed during their studies.

All interviews were recorded and subsequently transcribed verbatim. The subsequent analysis included a thematic coding followed by a meaning-condensation where the essence was extracted from the coded sections. Finally, patterns in the condensations were identified and described.

Results

The observations showed that animal welfare plays an important role at the colleges. Thus the welfare of animals was addressed by teachers as well as students during the different courses and was often the subject of heated debate. When trying to define animal welfare, 3 points of view seem to prevail: some see animal welfare as the absence of maltreatment and cruelty; some interpret it more holistically to include care, environment, space, etc., while others see it as additional concerns after care has been taken of basic requirements regarding feed, water, shelter, etc.

When elaborating on these more general points of view and arguing for the need for animal welfare, the students referred to three distinct sets of justifications with claims being backed by concerns about production and economy; by concerns about the animal itself, or by concerns related to the students themselves.

A matter of production and economy

By far the most dominant point of view regarding good animal welfare among both groups of students was that it was important for financial reasons in terms of productivity. Such a view included arguments for animal welfare in terms of providing sufficient feed, water, space, rest as well as ensuring the absence of stress and the maintenance of good health. The common justification for this was the close relationship between welfare, yields and farm economy. A lack of high welfare standards was, thus, considered to reduce yields, while high welfare standards increased yields. Therefore, the sufficient level of welfare was determined by the farm's economy, as argued by these final year students:

> Alfred: 'Imagine a production system [where the dairy cows are let out]. Watch the animals and you will see that they fall when they are let out to graze; they fall in yield. And if it's raining, then they will stay inside the stables anyway. And if it's too warm, they don't like lying in the open field either.'
>
> Andy: 'Right, it has to be profitable, it has to be.'

It is important to note that the students were searching for a balance between animal welfare and profits, which some students – in particular the older ones with experience from vocational training – described using real life situations. In the following excerpt, Alfred is arguing why some of the animals at the farm where he was trained should have been put down, not for the sake of the animal, but because of the costs:

> [The boss says:] 'Let's try to leave it for the time being, if it can't ... until tomorrow. Then all of a sudden, a month has passed and she has been walking around with a bad leg. And the farmer, he knows that if there is an inspection, well, then they will give a huge fine, and I will hardly be able to pay it ... But [the farmers] do it nevertheless!'

In this perspective, the animal is viewed instrumentally as an 'animal at work' comparable with other means of production. This does not mean that all animals are viewed in this way; hence there is a sharp distinction between pets and production animals as noted by this older student who has been working with mink: 'I am just looking after the animals. It is not that I'm someone who says "cute, cute, cute" ... they are production animals that must work. This is a business, not a hobby; we are not meant to run around and pet them ...'

Not just an animal

Although viewing animal welfare in the context of production was by far the most common perspective, it was sometimes challenged by students who were opposed to the inherent instrumental logic. Here the argument was that animals cannot be reduced to things, they are beings and, therefore, have rights, e.g. freedom from suffering, sufficient space and room to display natural behaviour, etc. In the following, this position is expressed by a first year student, Maria, who nonetheless recognizes that there is a dilemma between the two views and a need to strike a balance: 'I know it may be hard, that it shouldn't just be a production animal in a sense, that "now I don't need to look after it because it's just a thing out there and it will soon turn into money ..."'

The most common position in defence of the assertion that we should care for animals because of the animal itself maintains that farm animals have an essential need for and right to have plenty of room and an absence of suffering. Such a view was most common among the older students, who often referred to their vocational training at farms, and can be seen as a reaction to concrete experiences where the students felt that the instrumental view prevailed. Others such as the older student, Anton, in the following expressed it in more general terms, justifying extended care with reference to the animals (here pigs), not the costs: 'Really, if there are no places to hide; if there are ... fights and scuffles and ..., then they are bruised and stressed. They must have places to hide ...'

Interviewees with another, less commonly held view, observed a difference between the farms and a natural situation, where animals are free to satisfy their needs and express normal behaviour, which is expressed in the following by the first year student, Lasse, who draws a parallel between how he as a human reacts and the animals:

> The most important is that they have something to do ... You need to have something to do, otherwise you will end up doing something else, perhaps something stupid. So they need toys, they do, and something to play with otherwise ... And then they should have the opportunity to ... or freedom to express normal behaviour. And there has to be enough room for them to be themselves ... so you can see that they are happy and satisfied.

For my own sake

Holders of a third position, which was also less common than the instrumental view, argued for the relevance of welfare with reference to the students themselves. For some caring for the animals and a high level of animal welfare was something that made them happy. As expressed in the following by the first year student, Anton, who feels good, when the animals have good conditions: 'I worked at a farm where there were many hospital pens. More than recommended. I found that very nice, that there was always space and more space than needed.'

Other students – mostly younger – stressed the importance of a personal relationship with the animals. In the following quote, Lasse explains why he prefers pigs to cattle: 'It's the contact, that you can touch them. That's what I like. It's not the same with cattle, the only time you touch them is when milking, and then it's only the teats. There is no human-animal contact.' Among the older students, the importance of direct contact was more often based on concrete experiences, and they often noted that the close relation both made them feel good, but also had a positive function in that it made the animals more manageable.

Discussion

Although there were some differences between the younger and older students, the most striking result is that they tended to justify animal welfare in a similar way. The prevailing argument in favour of good animal welfare is that it is a means of achieving a profitable farm. Other perspectives justifying care for the animals by referring to the animal itself or the student were also identified, although they were less common.

The major difference between the two groups seems to be the amount of experience drawn upon when backing up their positions. Older students with extensive experience from their periods of vocational training, more often referred to actual experiences they had had which, at times, were very unpleasant, while younger students tended to back up their arguments with reference to how they personally believe animals are and how they should be treated.

The similarity between the two groups may have a methodological explanation as the young students had already been at the college for 2-3 months at the time of the interviews and, thus, may already have been socialized into viewing animals as a means of production. It may, however, also be explained by the fact that many students already have these values when enrolled. Students from rural areas may have been raised with these values; and students with a more urban childhood may before entering the colleges, have realized that farming is balance between ideals and economy, thus not having a romanticized image of agricultural production.

In any case, it is striking that the farmers-to-be share an instrumental approach to animal welfare which is very much in contrast to the views of the general public who typically justify animal welfare with reference to the animal itself and emphasize freedom, space and naturalness (Lassen *et al.*, 2006; Vanhonacker *et al.*, 2008). The students were aware of this difference, but showed little respect for or understanding of the public's view.

We are not suggesting that the students should adopt the position of the public, but rather that a societal dialogue about the use of farm animals would benefit from the ability of future farmers to understand the public's stance. Here the agricultural colleges have an important role to play in presenting students with alternative viewpoints regarding animal welfare.

References

Hemsworth, P.H., Coleman, G.J. and Barnett, J.L. (1994). Improving the attitude and behaviour of stockpersons towards pigs and the consequences on the behaviour and reproductive performance of commercial pigs. Applied Animal Behaviour Science 39: 349-362.

Lassen, J., Sandøe, P. and Forkman, B. (2006). Happy pigs are dirty! Conflicting perspectives on animal welfare. Livestock Science 103: 221-230.

Vanhonacker, F., Verbeke, W., Van Poucke, E. and Tuyttens, F.A.M. (2008). Do citizens and farmers interpret the concept of farm animal welfare differently? Livestock Science 116: 126-136.

9. The lesser of two evils? The killing of day-old male chicks in the Dutch egg sector

B. Gremmen[*] and V. Blok
Philosophy Group, Wageningen University and Research centre, Hollandseweg 1, 6706 KN, Wageningen, the Netherlands; bart.gremmen@wur.nl

Abstract

The practice of killing day-old chicks in the Dutch egg sector is a recurrent subject of societal debate. Preventing the killing of young animals and *in ovo* sex determination are the two main alternatives for this problem available. An online questionnaire was held to ask the opinion of the Dutch public about these alternatives. The results show that no alternative will be fully accepted, or accepted by more than half of Dutch society. However, the survey does provide an insight to what people think is important for their choice: food safety and a good treatment of animals. Irrespective of the alternative chosen, these values should be safeguarded and communicated clearly.

Keywords: laying hens, societal debate, GM

Introduction

The practice of killing day-old chicks in the Dutch egg sector is a recurrent subject of societal debate, even though most people are unaware that all male chicks of layer breeds are killed immediately after hatch. However, when hearing about it, most people react disapprovingly (Leenstra *et al.*, 2011). Based on societal preferences and technical, and practical perspectives, two main approaches have been proposed as possible solutions for killing day-old chicks. The first approach aims at preventing the killing of very young animals, by keeping male chicks alive for a longer period of time, but an important disadvantage is that it is less efficient with regard to input of resources and waste output. The second approach aims at *in ovo* sex determination, in which different techniques are developed to avoid the killing of very young animals by preventing the animals to be born. For example, researchers of Wageningen University and Research have proposed a genetic modification technique which uses a gene from a jellyfish to turn male eggs green under certain light conditions, as it might facilitate sex determination before incubation. Other techniques aim at taking a sample from the hatching egg in order to find differences between male and female eggs, and the use of spectrometry. Both alternative main approaches solve the problem of killing day-old chicks but they also raise a lot of new animal welfare related dilemmas, like genetically modified (GM) eggs or eggs containing embryos that cannot be consumed in the Europe. In order to identify the preferences of consumers regarding these alternative directions and the new dilemmas they raise, an internet questionnaire was developed. In this paper, the results of this survey will be analysed, both in order to identify key preferences of society regarding alternative directions to solve the problem of killing day-old chicks, and their development over time (scientific contribution), and to find out what direction in research and innovation should deserve more –political and societal – support (societal contribution).

Material and methods

A framework was developed in which the problem of killing day-old chicks and the two mentioned alternative directions were analysed (Bruijnis *et al.*, 2015). The ethical matrix (Mepham, 2000) was used to structure the arguments of the current situation and two alternatives (killing day-old chicks, dual use of chickens and *in ovo* sex determination). This method includes the reflective equilibrium method, i.e. the identification and evaluation of moral intuitions, principles and facts (cf. Bolt *et al.*, 2005). The

 I. Anna S. Olsson, Sofia M. Araújo and M. Fátima Vieira (eds.) ***Food futures: ethics, science and culture***
EurSafe 2016 – DOI 10.3921/978-90-8686-834-6_9,

ethical matrix also enables showing the various perspectives of the different stakeholders with regard to the arguments used. This ethical framework was used as a basis for the questions in the questionnaire (see the ethical matrix in Table 1).

The questionnaire included 32 questions and 3 blocks of text to provide the respondents with some essential information. The online survey was held from the 16th of October until the beginning of December 2014. The questionnaire was filled out by 2,219 respondents; however, not all of these were useful for further analysis. Respondents who skipped more than 9 questions were removed. The same was done with respondents that did not live in the Netherlands. This resulted in a dataset with 1,022 respondents. The data was analysed using Statistical Package for the Social Sciences (SPSS, version 22).

Results

A first result of the questionnaire is that 55% of the respondents are aware of the killing of day-old chicks. Figure 1 shows that about 30% of the respondents think it is a good practice or do not have problems with killing day-old chicks. Almost half of the people disagree with killing day-old chicks, and would rather search for alternatives or indicate it has to stop right away.

When asked about preference for how to deal with day-old chicks (Figure 2), about 30% of the respondents want to keep the current situation as it is. When given a restricted choice, more people

Table 1. Ethical arguments about killing day-old chicks and its alternatives.

Respect for:	Wellbeing	Autonomy	Justice
Society (consumers of eggs)	• acceptability • food safety	• consumer choice	• affordability • availability
Egg sector	• maintain profitability (good income) • adequate labour conditions	• freedom of management	• fair treatment in trade and law
Day-old chicks	• animal welfare	• integrity • naturalness	• intrinsic value
Environment	• protection of the biota	• maintenance biodiversity	• sustainability

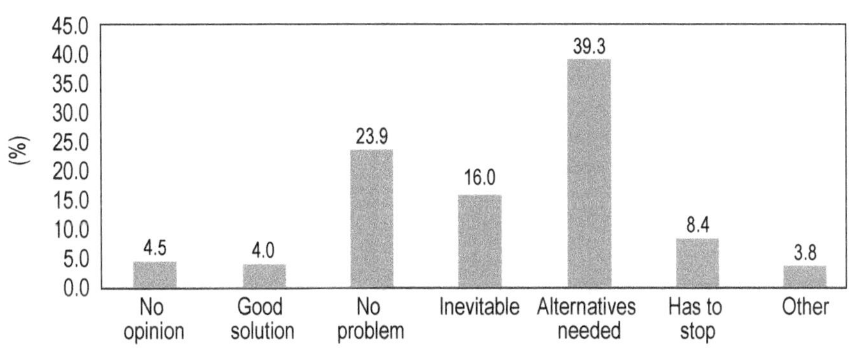

Figure 1. Opinion about killing day-old chicks (n=1,022).

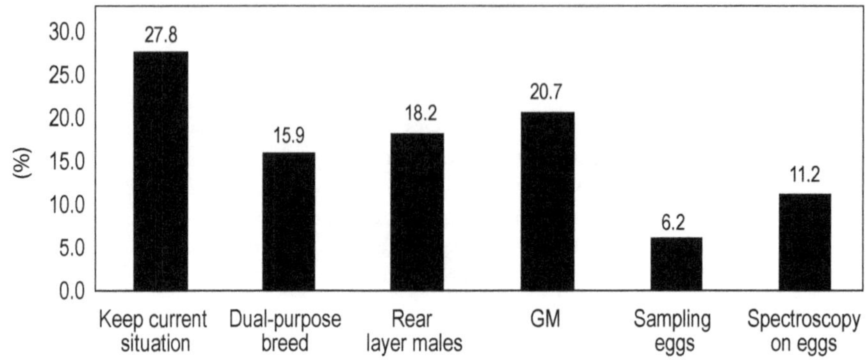

Figure 2. Preference with regard to dealing with day-old chicks (n=1,022).

choose for killing day-old chicks (59%) than for the use of GM (where no chicks or embryos are killed). Table 2 presents different arguments about how to deal with day-old chicks.

Discussion

The answer to our main research question, 'What is the preferred way for dealing with day-old male chicks in the egg sector?', is that there is not one alternative that will be fully accepted (or accepted by more than half) by society. However, the survey does give insight to what people think is important for their choice: food safety and a good treatment of animals are most important to people. Irrespective of the alternative, these values should be safeguarded and communicated clearly.

Our results show that people consider the GM alternative, part of the *in ovo* sex determination alternative, as the best of the alternatives. The opinions about GM are more extreme than for the other alternatives. Furthermore, people express their reservations about the GM alternative by using words like 'unless' or 'if', etc. The relative high concern about GM is also shown by the high scores for the different arguments about GM. In 2014 this lead also to a refusal of the Dutch parliament to start the proof-of-principle research of the theoretical GM-option of Wageningen University and Research. The State Secretary decided to financially support a new start-up company in Leiden which samples hatching eggs. However, it is interesting to see that our results show that this alternative receives the lowest number of 'votes'

Table 2. Different arguments about how to deal with day-old chicks.

Argument	Mean rank
Environment (use of resources and waste output)	3.60
Availability	3.71
Affordability	3.38
Food safety	4.99
Market position Dutch egg sector	2.93
Animal health and welfare	4.71
Respect for animals	4.69

(6.2%) while the GM option received 20.7% of the votes. Although the aspect of determining sex before incubation could make this technique interesting, sampling of hatching eggs or destroying hatching eggs with embryos is found problematic as well.

Our first sub-question, 'How do people in Dutch society perceive the killing of day-old male chicks in the egg sector?', is answered relatively positive because 75% of the respondents support the Dutch poultry sector. This means that there is clear favourable stance in defence of current practices of killing day-old male chicks. This answer seems to contradict the low values scored for income and welfare of farmers. The scores for Dutch poultry farmers are relatively low compared to arguments such as food safety and availability and animal health and welfare, but the majority wants to protect the sector at the same time. This shows that all arguments from the matrix are important, but that arguments relating to people's health and to the health and welfare of animals are relatively more important.

Conclusions

How to deal with day-old chicks in the egg sector responsibly? The data from our questionnaire show that there is no consensus among the respondents about how to deal with cockerels in the egg sector. The arguments that are most important for people relate to themselves and to the animals. Most important arguments are food safety, health and welfare of and respect for animals. As we described in (Bruijnis *et al.*, 2015), for this issue one could speak of a moral lock-in. Although we tried to incorporate the steps of responsible innovation, namely anticipate intended and potentially unintended future impacts of innovations, reflect on the purposes, motivation and potential impact of their innovations, deliberate with multiple stakeholders and respond to societal needs through participatory and anticipatory governance (Owen *et al.*, 2013), not one solution is considered to be ideal. We think that this is caused by the fact that the emergence of this practise is rooted in the development of a highly efficient production system, driven by other values than the values momentarily at stake. However, the process in addressing the question of how to deal with handling day-old chicks may be difficult (see also Franco *et al.*, 2014). Although the sector is not unwilling, it cannot make real changes. As long as the current practice seems to be the lesser of multiple evils, it might be better not to change our dealing with day-old chicks as long as there is not an alternative that is accepted more broadly. Advances in (the mentioned) techniques or maybe more extreme alternatives (e.g. eat plant based proteins) are needed to unlock the moral problem of killing day-old chicks.

References

Bolt, L.L.E., Verweij, M.F. and Van Delden, J.J.M. (2005). Ethiek in praktijk. Van Gorcum B.V., Assen, the Netherlands.

Bruijnis, M.R.N., Blok, V., Stassen, E.N. and Gremmen, H.G.J. (2015). Moral 'lock-in' in responsible innovation: the ethical and social aspects of killing day-old chicks and its alternatives. Journal of Agricultural and Environmental Ethics 28: 939-960.

Franco, N.H., Magalhães-Sant'Ana, M. and Olsson, I.A.S. (2014). Welfare and quantity of life. dilemmas in animal welfare. CABI, Wallingford, UK.

Leenstra, F., Munnichs, G., Beekman, V., Van den Heuvel-Vromans, E., Aramyan, L. and Woelders, H. (2011). Killing day-old chicks? Public opinion regarding potential alternatives. Animal Welfare 20: 37-45.

Mepham, B. (2000). A framework for the ethical analysis of novel foods: the ethical matrix. Journal of Agricultural and Environmental Ethics 12: 165-176.

Owen, R., Stilgoe, J., Macnaghten, P., Gorman, M., Fisher, E. and Guston, D. (2013). A framework for responsible innovation. In: Owen, R., Bessant, J. and Heintz, M. (eds.) Responsible innovation. John Wiley and Sons, Ltd. New York, NY, USA.

10. Does the opinion of dairy farmers about grazing change with increasing societal pressure?

C.C. de Lauwere[1], A.C.G. Beldman[1], J.W. Reijs[1], G.J. Doornewaard[1], A.C. Hoes[1] and A.P. Philipsen[2]*
[1]LEI Wageningen UR, P.O. Box 35, 6700 AA Wageningen, the Netherlands; [2]Livestock Research, Wageningen UR, P.O. Box 338, 6700 AH Wageningen, the Netherlands; carolien.delauwere@wur.nl

Abstract

Two studies were performed in the Netherlands in 2010 and 2015 to get insight into dairy farmers' attitudes to pasturing throughout the years. The preliminary results show that the intention of dairy farmers to keep on pasturing their cows is high for farmers who apply pasturing already. Farmers who do not apply pasturing have a low intention to start pasturing. Not surprisingly, farmers who apply pasturing emphasize advantages of pasturing and farmers who do not apply pasturing emphasize disadvantages. Whether farmers apply pasturing or not, to make dairy farms more sustainable they perceived reduction of antibiotics use, increasing longevity of cows and maintenance of pasturing as most important goals. However, farmers who applied pasturing seem to emphasize the maintenance of pasturing more. In addition to the three goals mentioned above, farmers who applied pasturing emphasized improving biodiversity more and farmers who did not apply pasturing lowering the emissions of greenhouse gasses and increasing nitrate efficiency. Farmers who do not apply pasturing then seem to emphasize other aspects of sustainability compared to colleagues who apply pasturing.

Keywords: pasturing, dairy farmers, intention, attitude, social norm, sustainability goals

Introduction

Pasturing of dairy cows is an important societal issue in the Netherlands. The representatives of the 'Sustainable Dairy Chain' (SDC)[1] thus make great efforts to maintain the current number of farmers who apply pasturing in the Netherlands. This is in contrast to the trend of expansion – also caused by milk quota abolition in 2015 – which causes an increasing number of farmers to keep their cows indoors all year round; especially at big farms[2] (De Lauwere *et al.*, 2015a; Reijs *et al.*, 2015). Nevertheless, SDC hopes that their efforts to emphasize the importance of pasturing for the sustainable image of dairy farming in the Netherlands, together with the increasing societal pressure, will positively influence dairy farmers' opinions about pasturing. This paper discusses the findings of two different studies concerning Dutch dairy farmers' attitudes to pasturing throughout the years, and includes dairy farmers who apply pasturing and farmers who do not apply pasturing.

Constructs of the Theory of Planned Behaviour were used to understand the farmers' behaviour with regard to pasturing. This theory assumes that a person's intention to perform a certain behaviour – in this case pasturing – depends on his/her attitude to the behaviour, the opinion about the behaviour of other people who are important to the person (social norm) and the extent to which the person perceives that he/she is capable of performing the behaviour (perceived behavioural control; e.g. – does a farmer belief that he/she has enough knowledge about pasture management to apply pasturing and/

[1] The Sustainable Dairy Chain is an initiative of the Dutch Dairy Association (NZO) and the Dutch Federation for Agriculture and Horticulture (LTO-Nederland), in which the dairy industry and its representatives strive to make the Dutch dairy sector world leader in sustainability; see for example De Lauwere *et al.* (2015b) for a short description.

[2] According to Reijs *et al.* (2015) the number of dairy farms that applied pasturing decreased from 81.2% in 2012 to 77.8% in 2014.

I. Anna S. Olsson, Sofia M. Araújo and M. Fátima Vieira (eds.) ***Food futures: ethics, science and culture***
EurSafe 2016 – DOI 10.3921/978-90-8686-834-6_10, © Wageningen Academic Publishers 2016

or does he/she have enough land to apply pasturing?) (Ajzen, 1991, 2002). This theory has proven to be useful to understand consumers' behaviour and has also been successfully applied in the agricultural domain to understand farmers' decisions (Breukers *et al.*, 2012; Colémont and Van den Broucke, 2008; De Lauwere *et al.*, 2012).

Materials and methods

The opinions of dairy farmers about pasturing were studied in 2010 and 2015 by means of two comparable surveys in which the same questions were asked:

- In 2010, a survey was conducted (SURV2010) among 127 dairy farmers, of whom 35 applied more than 12 hours pasturing per day during the grazing season, 35 applied less than 12 hours pasturing per day and 57 kept the cows indoors all year round.
- In 2015, a survey was conducted (SURV2015) among 159 dairy farmers of whom 33 applied more than 12 hours pasturing per day during the grazing season, 80 applied less than 12 hours pasturing per day and 39 did not apply pasturing (and it was unknown whether 7 remaining dairy farmers applied pasturing or not)[3].

The farmers' attitudes towards grazing were the central topic in both surveys. Most questions could be answered using a 7-point scale ranging from totally disagree to totally agree, very unlikely to very likely or not at all to very much. In addition, in 2015, a question was added about the farmers perceptions with regard to the most important sustainability goals.

The interviewed farmers in 2010 had on average 90.6±56.3 dairy cows and the interviewed farmers in 2015 had 124.0±95.6 dairy cows. According to Dutch FADN data, the national average was 82 dairy cows in 2010 and 102.5 dairy cows in 2015 (www.agrimatie.nl). This indicates that the farmers who participated in the survey had bigger farms than the national average.

Data were analysed by means of a univariate ANOVA analysis to find out whether differences existed in 2010 and 2015 between the groups of dairy farmers who applied pasturing or not.

Results

Intentions of dairy farmers to apply pasturing

The intentions of dairy farmers to apply pasturing were assessed in the same way in 2010 and 2015. The results show that the intention of dairy farmers to continue pasturing their cows is high for those farmers who already apply pasturing and that the intention of dairy farmers to start pasturing their cows is low for those farmers who do not apply pasturing. The figures of SURV2010 and SURV2015 are comparable (Figure 1).

Attitudes and behavioural beliefs of dairy farmers towards pasturing

In both 2010 and 2015, farmers who applied pasturing were more positive about it than farmers who did not apply pasturing (Table 1). Assessing the farmers' behavioural beliefs with regard to pasturing (indirect measurement of attitude according to the theory of planned behaviour (Ajzen, 2002)), showed that farmers who apply pasturing scored significantly higher on items concerning advantages of pasturing than farmers who did not apply pasturing. They believe that pasturing is beneficial for the welfare and

[3] When the study in 2010 was performed, it was not intended to repeat this study in 2015. For reasons of anonymity, it was not possible to approach the farmers who had participated in 2010 again in 2015.

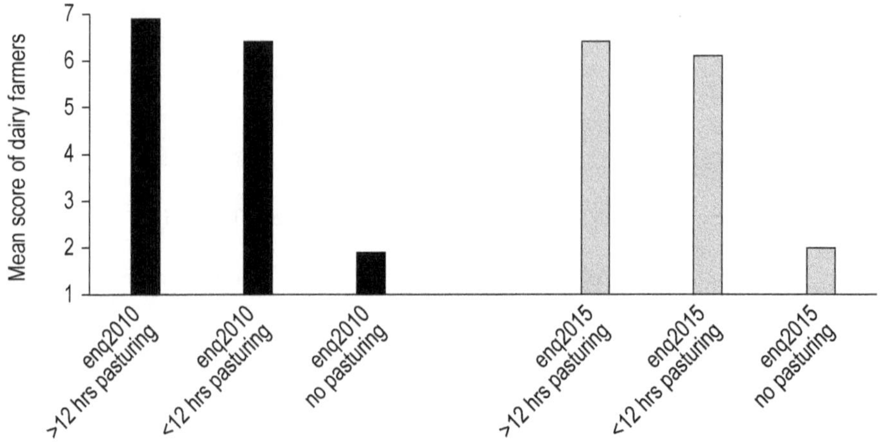

Figure 1. Intention of interviewed dairy farmers to continue or start pasturing. Mean scores of interviewed dairy farmers are given on the item 'I am planning to continue pasturing' for farmers who apply pasturing, and on the item 'I am planning to start pasturing' for farmers who do not apply pasturing; measurements at a 7 point scale ranging from totally disagree to totally agree.

health of the dairy cows and that it leads to fewer claw problems. Farmers who did not apply pasturing had significant higher scores on items concerning disadvantages of pasturing, for example because it decreases farm results, because it leads to more problems with the food composition and because it hampers cows from visiting the milking parlour or milking robot. These results were comparable in 2010 and 2015 (Table 1).

Social norms

In both 2010 and 2015, farmers who did not apply pasturing seemed to be less sensitive for social pressure of neighbours, advisors, and peers than farmers who applied pasturing. Using a 7-point scale ranging from 'not at all' to 'very much', they scored significantly lower on the questions 'to what extent would you be prepared to apply pasturing if this was recommended by your family, – your neighbour, – your veterinarian, – other dairy farmers, – the farmers' organisation and – the government than farmers who applied pasturing. In addition, in both 2010 and 2015, farmers who did not apply pasturing had significantly lower scores on the social norm related items 'dairy farmers whose opinions are important to me apply pasturing' and 'most people who are important to me think that I should apply pasturing' (Table 1).

Prioritizing sustainability goals

In 2015, farmers were asked to prioritize the goals of SDC to make dairy farms more sustainable. The results revealed that both farmers applying pasturing or not, perceived reduction of antibiotics use, increasing the longevity of cows and maintenance of pasturing as the most important goals. However, farmers who applied pasturing seemed to put more

emphasis on the maintenance of pasturing than the farmers who did not apply pasturing. Farmers who applied pasturing also mentioned improving biodiversity quite often as being important and farmers

Table 1. Attitude, behavioural beliefs and social norms of dairy farmers who applied pasturing or not in 2010 and 2015 (mean scores of farmers on a 7-point scale1 and standard deviation between brackets)[1,2,3]

	SURV2010				SURV2015			
	Pasturing >12 h	Pasturing <12 h	No pasturing	F	Pasturing >12 h	Pasturing <12 h	No pasturing	F
Attitude: for my farm applying pasturing is...								
Very disadvantageous – very advantageous	6.2 (1.1)[c]	5.3 (1.3)[b]	2.6 (1.3)[a]	105.2***	6.2 (0.9)[c]	5.4 (1.3)[b]	1.9 (0.9)[a]	149.9***
Very bad – very good	6.3 (0.9)[c]	5.6 (1.2)[b]	3.1 (1.3)[a]	98.8***	6.3 (0.7)[c]	5.6 (1.1)[b]	2.5 (1.2)[a]	123.3***
Behavioural beliefs:								
Pasturing leads to fewer claw problems	6.0 (1.3)[b]	6.1 (1.2)[b]	5.1 (1.9)[a]	6.7**	6.3 (1.0)[c]	5.4 (1.5)[b]	4.0 (1.8)[a]	22.7***
Pasturing is beneficial for animal welfare	6.7 (0.8)[b]	6.2 (1.3)[b]	5.1 (1.7)[a]	14.0***	6.7 (0.8)[c]	6.0 (1.2)[b]	3.8 (2.0)[a]	46.1***
Pasturing is beneficial for animal health	6.5 (1.1)[b]	6.2 (1.4)[b]	5.2 (1.7)[a]	9.3***	6.4 (1.0)[b]	5.9 (1.2)[b]	3.8 (1.9)[a]	39.0***
Pasturing decreases farm results	2.6 (1.8)[a]	2.7 (1.8)[a]	5.0 (1.7)[b]	27.2***	2.6 (1.6)[a]	3.2 (1.4)[a]	5.2 (1.7)[b]	28.8***
Pasturing leads to more problems with the food composition	4.9 (1.9)[a]	5.1 (1.7)[a]	6.5 (0.9)[b]	17.4***	4.5 (1.8)[a]	5.0 (1.7)[ab]	5.7 (1.5)[b]	4.6*
Pasturing hampers the cows from visiting the milking parlour/ milking robot	2.9 (2.1)[a]	2.5 (1.9)[a]	5.3 (2.2)[b]	23.3***	3.7 (2.0)[a]	3.7 (1.9)[a]	4.6 (2.1)[b]	3.2*
Social norm								
Dairy farmers whose opinions are important to me apply pasturing	4.8 (1.7)[b]	4.3 (2.1)[b]	3.0 (1.7)[a]	11.6***	4.4 (1.6)[b]	4.3 (1.6)[b]	3.1 (1.6)[a]	7.7*
Most people who are important to me think that I should apply pasturing	5.1 (2.1)[c]	5.4 (1.4)[b]	4.8 (1.2)[a]	10.4***	5.2 (1.5)[b]	4.0 (1.6)[b]	2.5 (1.2)[a]	37.3***

[1] Scale for intention ranges from totally disagree-totally agree; scale for behavioural beliefs ranges from very unlikely – very likely and scale for social norm ranges from absolutely not true – absolutely true.

[2] Here we refer to at least 6 hours pasturing per day during at least 120 days per year.

[3] Three years in 2010 [abc] different characters in a row mean a significant difference: *P<0.05; **P<0.01; ***P<0.001.

who did not apply pasturing mentioned lowering emissions of greenhouse gasses and increasing nitrogen efficiency quite often (Table 2).

Size of farms

In both 2010 and 2015, farmers who did not apply pasturing had more dairy cows than farmers who applied pasturing. In 2010, the interviewed farmers who applied pasturing more than 12 hours per day had on average 64.3±30.1 cows, farmers who applied pasturing 12 hours per day or less had on average 80.9±25.6 cows and farmers who did not apply pasturing had on average 113±74.2 cows. In 2015, these numbers were: 92.8±74.0, 110.1±72.5 and 186.6±130.9 respectively.

Table 2. Prioritization of the Sustainable Dairy Chain goals by the interviewed dairy farmers in 2012 and 2015 (percentage of farmers).

	pasturing	no pasturing
N	70	26
Lowering emissions of greenhouse gasses	12.9	23.1
Less use of gas and electricity	14.3	15.4
Production and use of sustainable energy	8.6	19.2
Reduction of the use of antibiotics	54.3	50.0
Lengthening the lifespan of the cows	70.0	61.5
More integral sustainable housing	8.6	11.5
Maintenance of pasturing	74.3	53.9
Sustainable soy and palm kernel flakes	8.6	0.0
Lowering phosphate in animal manure	17.1	15.4
Increasing nitrate efficiency	7.1	34.6
Improving biodiversity	24.3	15.4

Discussion and conclusions

Despite societal pressure and pressure of SDC, Dutch dairy farmers do not seem to have changed their attitudes towards pasturing between 2010 and 2015. Farmers who apply pasturing have a strong intention to continue this and farmers who do not apply pasturing have a weak intention to start pasturing. This is not very surprising, as the results also showed that farmers who did not apply pasturing seemed to be less sensitive to social pressure than farmers who applied pasturing. Comparable results were found by Beedell and Rehman (2000) with regard to conservation behaviour of farmers and by Fielding *et al.* (2008) with regard to riparian zone management: in both studies, less environmentally aware farmers felt less social pressure.

Farmers who applied pasturing emphasized the advantages of pasturing while farmers who did not apply pasturing emphasized disadvantages of pasturing. Comparable results were found by De Rooij *et al.* (2010) with regard to animal-friendly housing. This may be related to the Cognitive Dissonance Theory of Festinger that people tend to emphasize on negative aspects of behaviour which is commonly perceived as doing the right thing (in this case pasturing) if they do not perform this behaviour themselves (Cooper, 2012). These results indicate that it may be difficult to convince farmers who do not apply pasturing to start pasturing their cows. The fact that farmers who do not apply pasturing have more cows may contribute to this. Milk quota abolition in 2015 also has led to bigger farms. This interferes with the efforts of SDC to maintain the current number of farmers who apply pasturing in the Netherlands.

However, farmers who do not apply pasturing seem to emphasize other SDC goals than their colleagues who apply pasturing. Comparable results have been found by De Rooij *et al.* (2010): farmers with different ethical positions – in short growth oriented or oriented towards society – emphasized different aspects of sustainability (internal- or external social sustainability, economical sustainability or ecological sustainability) and contributed to sustainable farming in their own way.

References

Ajzen, I. (1991). The theory of planned behavior. Organizational Behavior and Human Decision Processes 50: 179-211.

Ajzen, I. (2002). Perceived behavioural control, self-efficacy, locus of control and the theory of planned behaviour. Journal of Applied Psychology 32: 665-683.

Beedell, J.D.C. and Rehman, T. (2000). Using social-psychology models to understand farmers' conservation behaviour. Journal of Rural Studies 16: 117-127.

Breukers, A., Asseldonk, M.A.P.M., Van Bremmer, J. and Beekman, V. (2012). Understanding growers' decisions to manage invasive pathogens at the farm level. Phytopathology 102: 609-619.

Colémont, A. and Van den Broucke, S. (2008). Measuring determinants of occupational health related behavior in Flemish farmers: an application of the theory of planned behaviour. Journal of Safety Research 39: 55-64.

Cooper, J. (2012). Cognitive dissonance theory. In: Lange, P.A.M., Kruglanski, A.W. and Higgings, T. (eds.) Handbook of theories of social psychology. Saga Publications Ltd., London, UK, pp. 377-397.

De Lauwere, C.C., Beldman, A.C.G., Reijs, J.W., Doornewaard, G.J., Van den Ham, A., Hoes, A.C. and Philipsen, A.P. (2015b). Towards a sustainable dairy chain in the Netherlands – the opinion of dairy farmers and their advisors. In: Dumitras, D.E., Jitea, I. M. and Aerts, S. (eds.) Know your food. Food ethics and innovation. Wageningen Academic Publishers, Wageningen, the Netherlands, pp. 120-140.

De Lauwere, C.C., Van Asseldonk, M.A.P.M., Van 't Riet, J.P., De Hoop, J.G. and Ten Pierick, E. (2012). Understanding farmers' decisions with regard to animal welfare: the case of changing to group housing for pregnant sows. Livestock Science 143: 151-161.

De Lauwere, C.C., Van den Ham, A., Reijs, J.W., Beldman, A.C.G., Doornewaard, G.J., Hoes, A.C. and Philipsen, B. (2015a). Adviseurs over verduurzaming in de zuivelketen. Report LEI 2015-002, Wageningen UR, Den Haag/Wageningen, the Netherlands.

Fielding, K.S., Terry, D.J., Masser, B.M. and Hogg, M.A. (2008). Integrating social identity theory and the theory of planned behaviour to explain decisions to engage in sustainable agricultural practices. British Journal of Social Psychology 47: 23-48.

Reijs, J.W., Doornewaard, G.J., Jager, J.H. and Beldman, A.C.G. (2015). Sectorrapportage duurzame zuivelketen. Prestaties 2014 in perspectief. Report LEI 2015-126, Wageningen UR, Den Haag/Wageningen, the Netherlands.

Rooij, S.J.G., De Lauwere, C.C. and Van der Ploeg, J.D. (2010). Entrapped in group solidarity? Animal welfare, the ethical positions of farmers and the difficult search for alternatives. Journal of Environmental Policy and Planning 12: 341-361.

11. A more sustainable pig production without tail biting and tail docking

M.L.V. Larsen[1], H.M.-L. Andersen[2] and L.J. Pedersen[1]*
[1]Department of Animal Science, Aarhus University, Blichers Allé 20, 8830 Tjele, Denmark; [2]Department of Agroecology, Aarhus University, Blichers Allé 20, 8830 Tjele, Denmark; mona@anis.au.dk

Abstract

Tail biting is a worldwide problem in the modern pig production. However, the predominant preventive measure – tail docking – also has several negative consequences. The question is: which of the two is the greater evil? This is not easy to answer and obviously, a more sustainable solution would be to avoid both. Two ongoing projects focus on exactly this. The project 'IntactTails' aims to develop a method to predict tail damage outbreaks in slaughter pigs, thereby giving the farmer the possibility to initiate preventive measures in a timely and efficient manner. One milestone of the project 'FareWellDock' is to investigate the relative effectiveness of straw provision and tail docking to reduce the risk of a tail damage outbreak. One study has been designed to support both projects. This study is ongoing and planned to include 112 pens of slaughter pigs divided between four batches. So far, three of these batches have been completed with a total of 80 slaughter pig pens of which there was a tail damage outbreak in 42 pens. Pens had pigs with either docked or undocked tails, and were either provided with 150 g straw per pig per day or not provided with straw at all. Preliminary results from project 'FareWellDock' will be presented. The risk of a tail damage outbreak was analysed using the Cox Proportional-Hazards Regression for survival data in R including pens without tail damage outbreaks as right-censored data (to compute hazard rate ratios (HRR) for a tail damage outbreak). It was found that both ending tail docking (HRR=3.71, 95% CI [1.24-11.09]) and no provision of straw (HRR=3.52, 95% CI [1.20-10.33]) increased the risk of a tail damage outbreak, and the risk was additively increased when the same pen had both undocked pigs and no straw provided (HRR=9.09, 95% CI [3.17-26.03]). However, no difference in risk of tail damage was seen between tail docking and straw provision (P=0.91; HRR=1.05, 95% CI [0.45-2.45]). This can lead to the interpretation, that tail docking and straw provision are equally protective measures in preventing tail damage outbreaks. The question is, therefore, no longer which of the two is the greater evil, but instead whether we can defend the use of tail docking, when we know of an equally effective measure that even brings other positive qualities with it? Future studies should focus on investigating whether smaller amounts of straw can reduce the risk of a tail damage outbreak to the same degree, and on developing production systems that make it possible to provide straw in an easy and efficient manner.

Keywords: animal welfare, animal behaviour, slaughter pigs, straw provision

The problem of tail biting

The modern pig production has been challenged by an increasing demand for improved welfare by consumers and policy makers. One such challenge is the tendency of slaughter pigs for biting the tail of other pigs resulting in serious tail damage. Tail biting is a worldwide phenomenon and, serious tail lesions are found on approximately 3% of all pigs slaughtered (EFSA, 2007, based on abattoir data). Considering that over a billion pigs are slaughtered annually, this corresponds to more than 10 million pigs with serious tail damage every year. Furthermore, abattoir data tends to underestimate the incidence of tail damage due to tail biting. Tail biting is a sign of underlying welfare problems and in itself, it reduces the welfare of both the pigs and the farmer. Reported negative consequences of tail biting include experienced stress and pain in the pig, spread of infection, reduced performance, and carcass

I. Anna S. Olsson, Sofia M. Araújo and M. Fátima Vieira (eds.) ***Food futures: ethics, science and culture***
EurSafe 2016 – DOI 10.3921/978-90-8686-834-6_11, © Wageningen Academic Publishers 2016

condemnation at slaughter (EFSA, 2007). There can be no doubt that tail biting should be prevented whenever possible.

The dilemma of tail biting and tail docking

The predominant method used to prevent tail biting is tail docking where a part of the pig's tail is cut off in a young age. This is an effective procedure, but it does not entirely eliminate tail damage caused by tail biting. Furthermore, tail docking has its own negative consequences. First, it does not solve the underlying welfare problem behind tail biting. Secondly, Simonsen *et al.* (1991) demonstrated that peripheral nerves could be traced to the tip of the tail and it has been shown, using both behavioural and physiological indicators, that the procedure leads to acute pain and stress in the pig (Nannoni *et al.*, 2014). Thirdly, Herskin *et al.* (2015) demonstrated the formation of neuromas in the tail stump which may result in chronic pain. It can be argued that tail docking should only be performed when it is absolutely necessary. This is supported by the EU legislation stating:

> Neither tail-docking nor ... must be carried out routinely but only where there is evidence that injuries to ... tails have occurred. Before carrying out these procedures, other measures shall be taken to prevent tail-biting ... For this reason inadequate environmental conditions or management systems must be changed.
>
> (COUNCIL DIRECTIVE 2008/120/EC of 18 December 2008 laying down minimum standards for the protection of pigs [2009] OJ L47/5)

Few countries have already banned tail docking (e.g. Finland, Lithuania, Norway, Sweden and Switzerland) which shows that it is possible to raise pigs without. However, it is not common knowledge how these farmers solve the problem of tail biting without tail docking. In most member states, farmers still perform tail docking routinely. Reasons for this could be that farmers do not see another solution to the problem of tail biting or simply that farmers view tail docking as a small price to pay to prevent tail biting. But is it possible to evaluate which of the two – tail biting or tail docking – is the greater evil? Is it worse to risk a slightly higher incidence of the more painful condition (tail biting) or to ensure that all pigs experience some level of pain (tail docking)? This dilemma has been further elaborated by Valros and Heinonen (2015). Obviously, a more sustainable solution would be to both decrease tail biting and eliminate tail docking at the same time, and two ongoing projects focus on exactly that.

'IntactTails' and 'FareWellDock'

Both projects have the overall goal to decrease the need for tail docking by decreasing the risk of tail damage caused by tail biting. Project 'IntactTails' (granted by the Green Development and Demonstration Programme under the Ministry of Food, Agriculture and Fisheries, Denmark) is a Danish project which aims to develop a method to predict tail damage outbreaks in slaughter pigs making it possible to apply preventive measures timely and efficiently. This will increase the farmer's control of the tail biting problem which may make him more willing to raise pigs with undocked tails. Based on a review by Larsen *et al.* (2016), the project focuses on behavioural changes prior to a tail damage outbreak including tail posture of pigs, activity level, and manipulative, feeding and drinking behaviour.

Project 'FareWellDock' (granted by ERA-net ANIHWA) is an EU project conducted across several member states. One milestone of the project is to investigate the relative effectiveness of enrichment materials and tail docking to reduce the risk of a tail damage outbreak. The European Commission has recently published new recommendations for the applications of the directive resulting in a staff working document (European Union, 2016). This document focus on environmental enrichment as a measure to reduce tail damage caused by tail biting, and highlights straw as one of the few materials that is optimal and can be used alone. Furthermore, straw has been shown to have other positive qualities and might solve the underlying welfare problem behind tail biting, also making the conditions better for the tail biter. In the following, the focus will be on the relative effectiveness of straw provision and tail docking to reduce the risk of a tail damage outbreak.

Presentation of methods used

One study has been designed to support both projects. This study is performed on an experimental farm at Aarhus University, Denmark. Only slaughter pigs from approximately 30 kg till slaughter are included and these pigs are housed under conditions similar to conventional production settings for slaughter pigs. The pigs are bought from a commercial pig farm right after weaning, and from approximately 7 to 30 kg, all pens are daily provided with straw. The study is ongoing and planned to include 112 pens of slaughter pigs divided between four batches. So far, three of these batches have been completed with a total of 80 slaughter pig pens. During this study, tail damage for each individual pig was registered three times per week through the entire slaughter pig period, and a tail damage outbreak at pen level was registered when at least one pig in the pen had a bleeding tail wound. After the first tail damage outbreak, tail damage was no longer recorded in that particular pen; therefore, per definition, only one tail damage outbreak could be registered per pen. Slaughter pig pens were randomly assigned to three different conditions: (1) TAIL: a pen either had pigs with undocked or docked tails; (2) STRAW: a pen was either not provided with straw or was provided with 150 g of straw per pig per day; (3) STOCK: a pen either had a high (18 pigs, 0.73 m^2/pig) or low stocking density (11 pigs, 1.19 m^2/pig). Furthermore, the study was performed all year round. In the following, preliminary results on the milestone of project 'FareWellDock' will be presented.

Preliminary results

Of the 80 pens included in the analysis, we recorded a tail damage outbreak in 42. The risk of a tail damage outbreak was analysed using the Cox Proportional-Hazards Regression for survival data in R (Therneau, 2014). The survival object was based on whether the single pen had a tail damage outbreak and when it occurred during the slaughter pig period. If the pen did not have a tail damage outbreak, it was treated as right-censored data and thereby, still included in the analysis. Two different analyses were performed. The first included the covariates TAIL, STRAW, STOCK, batch number, average pen start weight and the interaction between TAIL and STRAW to investigate whether the effect of tail docking and straw provision interacts. The second included the covariates STOCK, batch number, average pen start weight and a four level variable constructed from a combination of TAIL and STRAW to investigate the relative risk of a tail damage outbreak between the four levels: (1) D×S: docked tails and straw provided; (2) D×NS: docked tails and no straw provided; (3) UD×S: undocked tails and straw provided; and (4) UD×NS: undocked tails and no straw provided.

The lowest incidence of tail damage outbreaks was seen in pens with docked pigs provided with straw, while the highest incidence was seen in pens with undocked pigs not provided with straw. The two other levels (D×NS and UD×S) showed intermediate similar incidences (see Table 1 and Figure 1). In the first analysis, no interaction between TAIL and STRAW (P=0.48) was found, indicating that tail docking and straw provision have independent protective effects. A protective effect was seen of tail docking (P<0.001) and straw provision (P<0.01), while no protective effect was found when decreasing the stocking density (P=0.12); although the results numerically indicated a higher risk of tail damage outbreaks in pens with the higher stocking density (hazard rate ratio (HRR)=1.64, 95% CI [0.88-3.05]).

Results from the second analysis can be seen in Table 2. Because the lowest incidence of tail damage outbreaks was seen in pens with docked pigs provided with straw, this level was chosen as the reference.

Table 1. Relative comparison of incidence of tail damage outbreaks for four treatments.

Treatment[1]	No. of pens	No. of tail damage outbreaks	Incidence (95% CI)
D×S	22	5	22.7 (3.10-38.4)
D×NS	22	12	54.5 (28.2-71.2)
UD×S	18	10	55.6 (25.5-73.5)
UD×NS	18	15	83.3 (53.2-94.1)

[1] D×S=docked tails, straw provided; D×NS=docked tails, no straw provided; UD×S=undocked tails, straw provided; UD×NS=undocked tails, no straw provided.

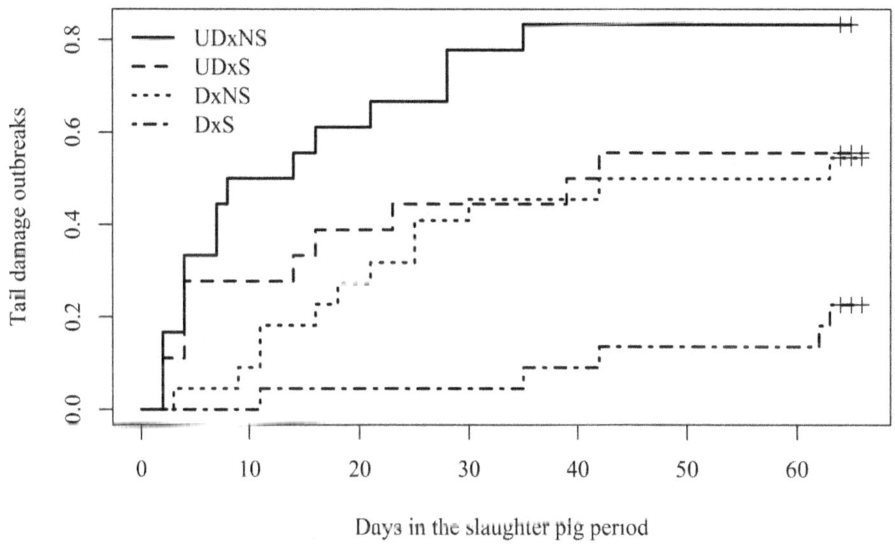

Figure 1. Cumulative incidence plot of tail damage outbreaks for four treatments (UD×NS: undocked tails, no straw provided; UD×S: undocked tails, straw provided; D×NS: docked tails, no straw provided; D×S: docked tails, straw provided).

If straw was not provided, the risk of a tail damage outbreak was more than three-fold higher (D×NS), and the same was seen when the pigs were not tail docked (UD×S). However, no difference was seen between not providing straw and not tail docking (P=0.91; HRR=1.05, 95% CI [0.45-2.45]). The risk of a tail damage outbreak was additively increased and more than nine-fold higher when the same pen had both undocked pigs and no straw provided (UD×NS).

What's next?

From the results of this study, the obvious conclusion is that both tail docking and straw provision were effective in reducing the risk of a tail damage outbreak. Furthermore, combining the two was even better. On the other hand, straw provision reduced the risk of a tail damage outbreak as effectively as tail docking under the current experimental conditions. This raises the question: can we defend the use of a measure

Table 2. Relative comparison of hazard rate ratios (HRR) for four treatments.

Treatment[1]	HRR (95% CI)	*P*-value
D×S	1.00 –	–
D×NS	3.52 (1.20-10.33)	0.02
UD×S	3.71 (1.24-11.09)	0.02
UD×NS	9.09 (3.17-26.03)	<0.0001

[1] D×S=docked tails, straw provided; D×NS=docked tails, no straw provided; UD×S=undocked tails, straw provided; UD×NS=undocked tails, no straw provided.

such as tail docking that we know causes pain to the pigs, when we know of an equally effective measure that even brings other positive qualities with it? To our knowledge, the relative effect of tail docking and other preventive measures have not previously been investigated, which may explain farmers' reluctant behaviour towards quitting tail docking. With this new knowledge, farmers may be more willing to avoid tail docking as other less invasive measures are available to avoid tail damage outbreaks. However, farmers could argue that the amount of straw used in this study is too much for their current production system. Therefore, the research community still need to investigate if smaller amounts of straw can reduce the risk of a tail damage outbreak with the same efficiency. There is also a need to investigate new production systems that can handle this or even larger amounts of straw provision. Currently, the Danish project 'StrawWell' (granted by the Green Development and Demonstration Programme under the Ministry of Food, Agriculture and Fisheries, Denmark) aims to develop a slurry system and flooring design that can handle straw in larger amounts as well as equipment for automated in-transport of straw for slaughter pigs. Hopefully, this development can aid to assure that new buildings are designed to handle larger amounts of straw. Thus, in new production buildings an alternative to tail docking may exist. The choice will then no longer be between two evils, because now there is a proven positive alternative for one of them, which we are obliged to incorporate in the modern pig production.

References

EFSA (2007). Scientific report on the risks associated with tail biting in pigs and possible means to reduce the need for tail docking considering the different housing and husbandry systems (question no. EFSA-Q-2006-013). Annex to the EFSA Journal 1-13.

Herskin, M.S., Thodberg, K. and Jensen, H.E. (2015). Effects of tail docking and docking length on neuroanatomical changes in healed tail tips of pigs. Animal 9: 677-681.

Larsen, M.L.V., Andersen, H.M.-L. and Pedersen, L.J. (2016). Can tail damage outbreaks in the pig be predicted by behavioural change? Veterinary Journal 209: 50-56.

Nannoni, E., Valsami, T., Sardi, L. and Martelli, G. (2014). Tail docking in pigs: a review on its short- and long-term consequences and effectiveness in preventing tail biting. Italian Journal of Animal Science 13.

Simonsen, H.B., Klinken, L. and Bindseil, E. (1991). Histopathology of intact and docked pigtails. British Veterinary Journal 147: 407-412.

Therneau, T. (2014). A package for survival analysis. In: S.R. package version 2.37-7.

Valros, A. and Heinonen, M. (2015). Save the pig tail. Porcine Health Management 1: 1-7.

12. Outdoor housing systems for Bísaro pig breed with a hoop barn: some effects on welfare

J.P. Araújo[1], I. Amorim[2,3], J. Santos Silva[4], P. Pires[5] and J. Cerqueira[2]*
[1]Mountain Research Centre, ESA-IPVC, Portugal; [2]Agrarian High School of Ponte de Lima, Polytechnic Institute of Viana do Castelo, 4990-706, Ponte de Lima, Portugal; [3]Institute of Biomedical Sciences Abel Salazar (ICBAS), University of Porto; [4]Ministry of Agriculture and Rural Development (Portugal), Guimarães, Portugal. [5]Technology and Management High School, Polytechnic Institute of Viana do Castelo, Portugal; pedropi@esa.ipvc.pt

Abstract

A growing interest has been shown in outdoor swine production systems due to the lower initial investment cost (facilities, buildings and equipment's). Simultaneously, concerns with animal welfare and awareness of niche marketing opportunities have increased the interest in the production of free-range animals. The Bísaro pig production in the northern region of Portugal has mainly adopted the traditional system rather than alternative outdoor systems. Facilities and accommodation factors influence both the environment and animal welfare and growth performance, making this issue a key point for efficient management, quality and food safety. The aim of the ongoing study presented in this paper is to test and to demonstrate an alternative building system for fattening Bísaro pigs (hoop barn with free access to open air) respecting good production practices and animal welfare. A total of 30 pigs of the Portuguese breed Bísaro, aged 3 months old, were equally distributed in number and gender in three batches, placed on a hoop barn with dimensions of 3 m^2/animal indoor and 200 m^2/animal outdoor (free access). Each animal has 30 cm of manger available and each batch has three fresh watering points. All the animals have been bimonthly evaluated according to the protocol assessment of Welfare Quality available for pigs, regarding the following parameters: (1) comfort around resting (manure on the body); (2) good health (absence/presence of injuries); (3) appropriate behaviour (fear of humans). Remarkably, to date almost the total of the animals regardless the batch in which they are included, presented no significant alterations in the first two measures exhibiting excellent scores values. Minor differences have been noticed in the animal's interaction with humans, intra-and inter-batches.

Keywords: pig, Bísaro breed, alternative production system, hoop barn, animal welfare

Introduction

Animal and environmental care, health, product safety and consumer acceptance are important factors for the assessment of pig housing (Von-Borell *et al.*, 2011). Alternative housing systems such as outdoor housing and application of environmental enrichment have gained interest (Millet *et al.*, 2005), and generally improve welfare by providing the opportunity to express species-specific and natural behaviour and engaging interaction with conspecifics (social behaviour). However, by permitting a wider range of behaviour, this kind of facilities might lead to other welfare problems mainly related to health (Barnett *et al.*, 1990) One alternative swine production system is deep-bedded hoop barns or hoops. Previous studies showed that pigs' growth performance is similar in hoops or confinement (Honeyman and Harmon, 2003).

The Bísaro breed is well adapted to the northern Iberian climate and environment. It was the most important swine breed of the north of Portugal until the middle of the 20th century. This breed was discovered on the brink of extinction in the beginning of 1990 and it has been preserved in small family farms where the animals are self-sufficient (Santos Silva and Tirapicos Nunes, 2013). Since

EurSafe 2016 – DOI 10.3921/978-90-8686-834-6_12, © Wageningen Academic Publishers 2016

1995, a conservation program has been developed. Currently, there are approximately 3,962 (year 2014) registered Bísaro sows in the herd book (ANCSUB, 2016), which are raised in traditional family farms (50%) and kept in semi-intensive outdoors systems or semi-extensively sufficient (Santos Silva and Tirapicos Nunes, 2013), on farms with an average of 30 sows (ANCSUB, 2016). The high prolificacy of this breed, the facility in raising the piglets and the quality of the meat provided are the main reasons for Bísaro breed conservation. The gentle character of these animals and their great ability to adapt to the traditional operating systems are also relevant factors. Bísaro meat allow the production of a great diversity of regional products, which in the current context of specific quality products, still have a market (niche) to explore displaying an important economic value sufficient (Santos Silva and Tirapicos Nunes, 2013).

The Bísaro pig production in the northern region of Portugal has mainly adopted the traditional system rather than alternative outdoor systems The main weaknesses of this system is related with the very small dimensions of the pig farms and with the buildings and animal facilities that are rudimentary, poorly dimensioned, which is problematic in terms of productivity, animal welfare and food safety sufficient (Santos Silva and Tirapicos Nunes, 2013). The accommodations used in the traditional system are very different and must be adapted (location, dimensions and materials) to the needs of the animals, according to physiological, geographical and environmental conditions of each region. Thus, a growing interest has been shown in outdoor swine production systems due to the lower initial investment cost (facilities, buildings and equipment's).

This work aims to demonstrate an alternative (sustainable) building and accommodation system for fattening Bísaro pigs (hoop barn with free access to open air) respecting good production practices and animal welfare.

Material and methods

A total of 30 pigs of the Portuguese Bísaro breed, aged 3 months old, arrived in November 2015 to Agrarian High School of the IPVC and were equally divided in number and gender in three batches, placed in a hoop barn with dimensions of 3 m^2/animal indoor and 200 m^2/animal outdoor (free access) (Figure 1). Each animal has 30 cm of manger available and each batch has three fresh watering points.

Figure 1. Hoop barn with outdoor access.

For the welfare assessment reported here, all the animals were bimonthly evaluated by two observers regarding 3 of the 12 parameters stated in the protocol of Welfare Quality® Assessment for pigs (Welfare Quality, 2009) (Table 1). Human animal relationship was assessed by entering the pens, walking around the group slowly until returning to the starting point and waiting for 30 s before again walking around slowly in the opposite direction. The animals' response to the presence of the assessor was registered.

Results and discussion

Concern with animal welfare and awareness of niche marketing opportunities have increased the interest in free range and other alternative production systems with outdoor access. In this study we report initial results of an animal-based welfare assessment of pigs kept in a hoop barn with free outdoor access. The welfare assessment method follows the approach defined in the Welfare Quality project, giving preference for animal-based measures (Czycholl *et al.*, 2015). Following this approach, the amount of manure on the body is used as an indicator of comfort during rest (Otten *et al.*, 2013; Temple *et al.*, 2011a). For this parameter, the three batches of pigs in the present study presented similar results with the majority of pigs showing signs of good welfare quality and achieving a null score (100%) (Figure 2). Differences in this measure has previously been found between pigs housed in a conventional indoor system in two different farms, where differences were attributed to higher temperatures at the farm with the highest score (dirtier pigs) (Renggaman *et al.*, 2015). In our study, all the batches were subjected to the same physical conditions of space and temperature, and the animals could choose freely between being inside the barn or in the outside area.

Health is an important aspect of animal welfare and must be appropriately considered (Broom, 2006). In terms of absence of injuries, all the animals were classified with score 0. Up to date, with exception of a unique successfully treated parasitic outbreak caused by *Ascaris suum*, no signs of disease were detected in any of the batches.

Table 1. Criteria and scoring scale for the evaluation of housing, health and appropriate behaviour.[1]

Main principles	Welfare criteria	Measures/ animal-based measures	Score	Description
Good housing	comfort around resting	manure on the body	0	Less than 20% of the body surface is soiled
			1	More than 20% but less than50% of the body surface is soiled
			2	Over 50% of the body surface is soiled
Good health	absence of injuries	wounds on body	0	If all regions of the body have up to 4 lesions
			1	When 5 to10 lesions are observed on up to 5 zones of the animal or one zone has from 11 to 15 lesions
			2	When more than 10 lesions are observed on at least 2 zones of the body, or if any zone has more than15 lesions
Appropriate behaviour	good human-animal relationship	human animal relationship	0	Up to 60% of the pigs show a panic response
			2	More than 60% of the animals show panic responses

[1] Each animal was evaluated only from one side since it is reported that the amount of lesions and dirtiness scored on the right and left side of the animal do not differ statistically (Courboulay and Foubert, 2007).

Figure 2. Manure on the body at different evaluation time points.

The human-animal relationship is very important and influences both animal production and welfare (Waiblinger *et al.*, 2006). Fear of humans is a direct reflection of how the pigs are handled. Other factors may also influence such as age, genetic background and space allowance (Temple *et al.*, 2011b); however in the present study these factors were constant. Our results clearly contrast with others (Renggaman *et al.*, 2015) that reported panic responses in pigs in a conventional indoor system in animals with different growing ages and space allowances. In a study of intensively and extensively housed Iberian pigs, the proportion of panic responses was higher in the latter (Temple *et al.*, 2011a). A possible explanation may be that in extensive systems, animals have fewer human contacts often experience human contact primarily in disturbing situations (Turner and Dwyer, 2007), resulting in animals being more fearful to the human presence. Fear is considered as a major welfare problem and the sudden, intense or prolonged elicitation of fear can seriously influence product quality (Waiblinger *et al.*, 2006). The quality of stockmanship is the main determinant of a good or bad human-animal relationship (Waiblinger and Spoolder, 2007). Remarkably, in the present study all the animals were scored with 0 regarding human animal interaction and more specifically, about 70% of the total animals showed no panic attitude towards the observer but instead approached with immense curiosity. These pigs received a daily food supplementation provided manually by the same technician. Pigs associate a rewarding experience of feeding with the handler and this will result in being less fearful of the handler and also of other humans (Hemsworth *et al.*, 1996). Animals that routinely receive frequent human contact are more able to ignore the stimulus person (Waiblinger *et al.*, 2006) while animals that receive less direct contact, such as extensively reared pigs, are more likely to show a clear response to the assessor presence. Most of the animals of the three batches show positive social behaviours such as sniffing, nosing, liking or even following the assessor. Interestingly, when the usual assessor was replaced by a person who was unfamiliar to the pigs, only 50% of the animals in each batch approached without any hesitation. The degree of curiosity of the animals can also affect the animals 'responses to people (De Passillé and Rushen, 2005).

The measures reported in this study correspond only to a part of a complete welfare assessment, but they all indicate high levels of animal welfare in this production system. This suggests that the system can be a viable solution that will enable sustainable development of high quality Bísaro products (regional, traditional and gourmet).

Acknowledgements

This project has received funding from the European Union's Horizon 2020 research and innovation programme under grant agreement No 634476 (Project acronym: TREASURE). The content of this paper reflects only the author's view and the European Union Agency is not responsible for any use that may be made of the information it contains.

References

ANCSUB (2016). Bísaro herdbook data. ANCSUB – Associação Nacional de Criadores de Suínos de Raça Bísara. Available at: http://tinyurl.com/gmhtvgg.

Barnett, J.L. and Hemsworth, P.H. (1990). The validity of physiological and behavioral measures of animal-welfare. Applied Animal Behaviour Science 25: 177-187.

Broom, D.M. (2006). Behaviour and welfare in relation to pathology. Applied Animal Behaviour Science 97: 73-83.

Courboulay, V. and Foubert, C. (2007). Testing different methods to evaluate pig welfare on farm. Animal Welfare 16: 193-196.

Czycholl, I., Büttner, K., Bailage, E.G. and Krieter, J. (2015). Review of the assessment of animal welfare with special emphasis on the 'Welfare Quality® animal welfare assessment protocol for growing pigs'. Archive Animal Breeding 58: 237-249.

De Passillé, A.M. and Rushen, J. (2005). Can we measure human-animal interactions in on-farm animal welfare assessment? Some unresolved issues. Applied Animal Behaviour Science 92: 193-209.

Hemsworth, P.H., Verge, J. and Coleman, G.J. (1996). Conditioned approach – avoidance responses to humans: the ability of pigs to associate feeding and aversive social experiences in the presence of humans with humans. Applied Animal Behaviour Science 50: 71-82.

Honeyman, M.S. and Harmon, J.D. (2003). Performance of finishing pigs in hoop structures and confinement during summer and winter. Journal of Animal Science 81: 1663-1670.

Millet, S., Moons, C.P.H., Van Oeckel, M.J. and Janssens, G.P.J. (2005). Welfare, performance and meat quality of fattening pigs in alternative housing and management systems: a review. Journal of the Science of Food and Agriculture 85: 709-719.

Otten, D., Annas, E. and Van den Weghe, H. (2013). The application of animal welfare standards in intensive production systems using the assessment protocols of welfare quality: fattening pig husbandry in Northwest Germany. International Journal of Livestock Production 4: 49-59.

Renggaman, A., Choi, H.L., Sudiarto, S.I.A., Alasaarela, L. and Nam, O.S. (2015). Development of pig welfare assessment protocol integrating animal-, environment-, and management-based measures. Journal of Animal Science and Technology 57: 1-11.

Santos Silva, J. and Tirapicos Nunes, J.L. (2013). Inventory and characterization of traditional mediterranean pig production systems. advantages and constraints towards its development. Acta Agriculturae Slovenica, Suppl. 4: 61-67.

Temple, D., Dalmau, A., Ruiz de la Torre, J.L., Manteca, X. and Velarde, A. (2011b). Application of the Welfare Quality® protocol to assess growing pigs kept under intensive conditions in Spain. Journal of Veterinary Behaviour: Clinical Applications and Research 6: 138-149.

Temple, D., Manteca, X., Velarde, A. and Dalmau, A. (2011a). Assessment of animal welfare through behavioural parameters in Iberian pigs in intensive and extensive conditions. Applied Animal Behaviour Science 131: 29-39.

Turner, S.P. and Dwyer, C.M. (2007). Welfare assessment in extensive animal production systems: challenges and opportunities. Animal Welfare 16: 189-192.

Von-Borell, E., Bockisch, F.-J., Buscher, W., Hoy, S., Krieter, J., Muller, C., Parvizi, N., Richter, T., Rudovsky, A., Sundrum, A. and Van den Weghe, H. (2001). Critical control points for on-farm assessment of pig housing. Livestock Production Science 72: 177-184.

Waiblinger, S. and Spoolder, H.A.M. (2007). Quality of stockmanship. In: Velarde, A. and Geers, R. (eds.) On farm monitoring of pig welfare. Wageningen Academic Publishers, Wageningen, the Netherlands, pp. 156-166.

Waiblinger, S., Boivin, X., Pedersen, V., Tosi, M.-V., Janczak, A.M., Visser, E.K. and Jones, R.B. (2006). Assessing the human-animal relationship in farmed species: a critical review. Applied Animal Behaviour Science 101: 185-242.

Welfare Quality (2009). Welfare Quality® applied to growing and finishing pigs. In: Dalmau, A., Velarde, A., Scott, K., Edwards, S., Veissier, I., Keeling, L. and Butterworth, A. (eds.) Welfare Quality® assessment protocol for pigs. Welfare Quality® consortium, Lelystad, the Netherlands.

13. Feeding of colostrum – animal welfare and ethical aspect of different feeding regimes

H. Röcklinsberg[1], C. Berg[2], L. Lidfors[2], A. Johansson[1] and C.E. Hernandez[3]*
[1]Dept. of Animal Environment and Health, Swedish University of Agricultural Sciences, P.O. Box 7068, 75007 Uppsala, Sweden; [2]Dept. of Animal Environment and Health, Swedish University of Agricultural Sciences, P.O. Box 234, 532 23 Skara;[3]Dept. of Animal Nutrition and Management, Swedish University of Agricultural Sciences, P.O. Box 7024, 75007 Uppsala, Sweden; helena.rocklinsberg@slu.se

Abstract

At birth, calves do not have sufficient immunity to fight disease. They rely on the transfer of passive immunity via ingestion of maternal immunoglobulins (Ig) present in colostrum and a sufficient intake shortly after birth is important. Traditionally calves are left to suckle the dam, but dairy farmers frequently bottle or bucket feed, and more recently some have started using oesophageal tube (OT) feeding to improve transfer of immunity. To date, most studies have focused on the effects of OT feeding on transfer of passive immunity, i.e. health outcomes and mortality rates. However, there are no studies evaluating how the calves react to the intubation procedure (behaviourally and physiologically) and whether or not this invasive procedure may compromise the welfare of the calves. This project aims to address routine use of OT feeding with regards to health and welfare using a multidisciplinary approach, and to investigate the public's perception of OT feeding of colostrum of dairy calves as a management routine. Practices in dairy farming raise a number of welfare and ethical issues. A pilot interview study with farmers, veterinarians and advisors (10 persons in total) indicate different perceptions of separation and of OT feeding. For some OT feeding of healthy calves is beyond imagination, referring to it as unnatural, whereas others advocate it, referring to perceived improved calf health. However, all informants rank OT feeding lower than suckling and bottle feeding both from a calf welfare point of view and from 'a consumer perspective'. Independently of farmers' own view, results show that they are worried about how the public may perceive routine OT feeding. Based on this revealed awareness of value differences we reflect on different perspectives on animal welfare concluding that transparency and trust are crucial aspects in any ethical elaboration on calf feeding routines.

Keywords: animal health, natural behaviour, oesophageal tube (OT) feeding, suckling, consumer/citizen

Introduction

A prerequisite for a sustainable dairy production is a high level of animal welfare, both from the producers' and consumers' perspective. The concept of sustainable production includes both economical, ecological and ethical aspects. Successful dairy calf management is the base for an efficient and sustainable dairy production. As for all mammals the first day of life is considered to be of vital importance for the development of the young to a grown up creature. Calves' immune system is not developed at birth and they are dependent on maternal immunoglobulins (Ig). This is present in the colostrum, which is traditionally supplied by allowing the calf to suckle the dam, but also bottle or bucket feeding is frequently used. More recently oesophageal tube (OT) colostrum feeding is used as a standard procedure, to improve transfer of immunity and reduce work load. The effects of routine OT feeding on the welfare of newborn dairy calves has however not been studied. In the project 'Effect of colostrum feeding routines on calf performance and welfare' we address routine use of OT feeding with regards to health and welfare using a multidisciplinary approach. These concerns raise the question of whether or not intubating a healthy calf with good vitality is justifiable given the invasiveness of the procedure. A

EurSafe 2016 – DOI 10.3921/978-90-8686-834-6_13, © Wageningen Academic Publishers 2016

holistic perspective based on systematically collected data on the OT routine is missing, where attention is paid to the effect on Ig-absorption, endocrine response triggered by feed intake, calf behaviour and welfare, growth rate, time spent on feeding and whether there is an acceptance among farmers and the public. The aim of our project is to evaluate the effect of OT feeding healthy calves on calf reaction to the intubation procedure (behaviourally and physiologically) and also the effects on calf health and welfare. Furthermore, we aim at investigating the public's perception of OT feeding of healthy dairy calves as a management routine. The present paper looks into some ethical aspects of OT feeding, but start with a general overview of the role of colostrum and feeding routines and a presentation of results from a pilot interview study.

The role of colostrum for newborn calves

Newborn cattle calves are not supplied with Ig via *in utero* transmission and therefore depend on successful passive transfer of maternal Ig from colostrum after birth. The ability to absorb Ig through the intestinal epithelium decreases linearly with time from birth up to 24 h *post-partum*, where after the gut epithelium is considered closed (Stott *et al.*, 1979; Weaver *et al.*, 2000). The timing of feeding the colostrum is hence critical for calf survival, and it might be difficult to reach an adequate passive transfer of Ig within the critical timeframe at night time (Chigerwe *et al.*, 2008). Early and adequate intake of colostrum may have long term positive effects beyond the survival of the newborn calf. Effects such as improved growth rate, reduced age at first calving and improved future lactation production are examples (Godden, 2008). The importance of colostrum feeding is much more than just a supplier of Ig as it stimulates intestinal growth and development, and long lasting effects on nutritional status, metabolism and endocrine system has been reported, where both amounts and time of colostrum intake affect metabolic and endocrine traits in calves (Hammon *et al.*, 2012; Blum *et al.*, 2006).

Colostrum feeding routines

Technical and practical aspects of colostrum feeding, such as the volume of colostrum fed to the neonate, time of first feeding in relation to birth, and method of feeding colostrum have been thoroughly investigated (Weaver *et al.*, 2000; Godden *et al.*, 2009). Suckling the mother, feeding colostrum using a nipple bottle or bucket feeding, and providing colostrum through an oesophageal tube are examples on how the colostrum is fed to calves on commercial farms.

During recent years, routine feeding of calves using OT has become increasingly common and even recommended in some countries (Besser *et al.*, 1991; Chigerwe *et al.*, 2008).

Whereas using an OT to feed weak or sick calves with a decreased ability to suckle is a well-accepted and uncontroversial practice, the routine OT feeding of healthy and strong calves requires careful consideration of its pros and cons (Persson Waller *et al.*, 2013). Examples of arguments supporting routine feeding using the OT technique are that the calf is ensured a sufficient amount of colostrum in early life. Farms routinely using OT colostrum feeding have a reduced failure of transfer of passive immunity to the calves when compared to bottle feeding or suckling (10.8, 19.3 and 61.4% of calves with <10 mg/ml serum IgG1 at 48 h of age, respectively; Besser *et al.*, 1991). Another argument in favour of routine use of OT feeding is that the farmer saves time since no time is needed to teach the calf to drink from a bucket or nipple bottle during the first meal.

Examples on drawbacks reported are that the calves' natural need to suckle will not be satisfied, and that the equivalent time will have to be spent on teaching the calf to drink at the subsequent meal instead. Feeding large volumes of milk may result in distension of abomasum and fore-stomachs with decreased absorption of colostral components as an effect (Sakai *et al.*, 2012). Furthermore, the oesophageal groove reflex, which directs the milk/colostrum directly to the abomasum (Leek, 1993) is not triggered when feeding the calves colostrum through an OT (Lateur-Rowet and Breukink, 1983). Therefore, the colostrum is rather deposited into the rumen or reticulum than in the abomasum delaying the uptake of Ig compared to when the calves are fed from bucket or teat (Lateur-Rowet and Breukink, 1983). In addition, OT is an invasive procedure that requires trained personnel to ensure correct restraining and intubation technique to avoid calf injury and ensure the tube is inserted into the oesophagus and not the trachea, which would result in colostrum being deposited directly into the lungs. In addition, the restraint and intubation itself could be stressful for the calf.

Materials and methods

This study is part of an ongoing project conducted at the Swedish Livestock Research Centre Lövsta, which belongs to the Swedish University of Agricultural Sciences. The aim of the project is to evaluate the effects of different colostrum feeding routines (tube feeding vs suckling the dam vs oesophageal tube feeding) on health, growth rate, Ig-status, behaviour and feed intake of the calves. In addition, the perception of the farmers, veterinarians, advisors and the general public about oesophageal tube feeding will be investigated using a combination of interviews and questionnaires. The present study relates to the public's perception part of the project and The following presentation is limited to a small interview study of 10 informants (8 farmers at 6 different farms, 2 advisories and 2 veterinarians) part of a master thesis in agronomy, and we also mention a parallel questionnaire among 30 veterinarians and advisers, but, based on this, can make no claims of representativity (Johansson, 2016). Rather, we were interested in the informants reflections. From such in-depth interviews 'saturation' may be achieved after 10-15 interviews, (no additional information provided). We used a questionnaire scheme with open questions for all three groups, with a section of questions relevant for the respective profession, but also offered time for free reflections. Questions posed considered their view on calves, calf advisory, colostrum and colostrum feeding. During interviews the advisory and veterinarians were asked to rank the feeding techniques (suckling on dam, bottle and OT feeding) according to four perspectives: animal welfare, animal health, farmer management and consumer perspectives. Animal welfare is considered a broader concept (incl. e.g. animal behaviour and feelings or 'mode') than animal health, and we were interested in whether informants regarded the practice of OT feeding differently as related to these different issues.

Results

In the interview study we were interested in how farmers, veterinarians and advisers regard feeding of colostrum in general, and especially how they consider the three methods suckling the dam, bottle feeding and OT feeding to healthy calves. All of the interviewed consider colostrum as a most important part of the calf rearing. The advisers and veterinarians however added that routines for colostrum feeding at the farms were not satisfying and argued some farmers do not have functioning routines for colostrum feeding. The farmer who used OT to all calves and the farmer who used it to about 1/3 of the calves regard it to be a good way to ensure that the calves get their colostrum in the right time. Other farmers thought that OT was a waste of time and some thought that it was cruel to the calf. Advisers and veterinarians in the study regarded OT to be a good complement, but that the calf should get the chance to suckle (dam or bottle) before OT should be used. Another clear result was that all interviewees thought that the person using OT feeding, depending on whether it is a healthy or a sick calf, should have been duly informed about the technique and its risks to the calf's health before feeding calves themselves.

The two veterinarians and the two advisers were also asked to rank the feeding regimes. From a health perspective all four ranked bottle feeding first, OT next and suckling the dam last, giving the rationale of farmer's control of amount and quality of colostrum. From a welfare perspective two ranked suckling the dam first and two ranked bottle feeding first, three ranked OT last and one ranked suckling last. Further, on the question what they thought consumers would think about using OT feeding to healthy calves on a routine basis most of them answered that consumers would think it is unacceptable and dissociate from it. They did not think it would help if consumers got an explanation about why OT is used and the advantages farmers see with it. The farmer who used OT to all his calves, however believed that if consumers got to know that the calf survival is better with OT, they would accept it. All the informants thought that consumers would accept OT feeding to sick and weak calves (Johansson, 2016).

Ethical considerations

Dairy production rests on the idea of separating cow and calf in order to take the milk for human consumption. Above the basic issue of keeping animals it involves two ethical issues: separation of cow- and calf and offering the calf milk of lower quality than the dam's milk. Traditionally this is ethically justified by referring to humans being more important, having a higher moral status than (other) animals, and hence having the right to take what they need from the animal kingdom, albeit sometimes it is clearly argued that this has to be done with respect or caution. Animal ethics literature has shown that this view is subject to much criticism, and need not be repeated here, as the focus of this paper is not the argumentative fundament for potential animal use. Rather the issue of ethical values like respect or caution is relevant, as different perceptions and values are shown to be linked to different levels of acceptance of dairy farming (Boogard *et al.*, 2011a; Boogard *et al.*, 2011b). There is a discrepancy between what farmers and industry on the one hand, and consumers or citizens on the other consider ethically acceptable (Bergstra *et al.*, 2015).

Results from the small interview study, also seen in the small questionnaire study, show that informants' answer on what is good from an animal welfare point of view totally overlaps with what they consider to be the consumers' ranking (Johansson, 2016). From both perspectives suckling the dam is ranked highest. We did not offer any definition of animal welfare, but they were asked to discriminate it from animal health, and since no questions were posed regarding this we can expect informants were aware of differences between these concepts. The most common definitions of animal welfare distinguishes between focus on health, on natural behaviour, or on subjective experience (Fraser, 1999), and interviews show focus on health among veterinarians and those farmers who use OT feeding on their own calves, whereas farmers critical of OT argue for the importance of suckling as a natural behaviour. This may explain why farmers in the study were worried about how the public may perceive routine OT feeding, independently of their own view of OT. Hence, these farmers were aware that consumers in general regard allowing for natural behaviour as ethically preferable (see e.g. Borkfelt *et al.*, 2015), although their own understanding of welfare might lend towards health, which implies an (implicit) awareness of value dimensions in welfare descriptions.

Another possible reason for discrepancy in understanding animal welfare may be found in whether one has an animal perspective or a human perspective, the latter called 'anthropocentric welfarism' by agro-economists meaning 'that only human utility counts and that animal welfare is a sub-set of man's welfare ... In human-centered preferences, people can be willing to pay to improve animal welfare quality, directly to the extent that their own utility is affected by such improvements, or indirectly through altruistic concern for other people's behaviour or opportunities or for the animals *per se*.' (Hansson and Lagerkvist, 2011). This latter version seems close to an animal perspective, but the former clearly focuses on human utility in supporting animal welfare. According to Hansson and Lagerkvist, a farmer's idea of moral obligation is expressed in measures taken in relation to animal welfare and care, and farmers may also adopt their practices in order to avoid negative evaluation by others (Hansson and Lagerkvist, 2011). A possible consequence would then be that farmers themselves positive to OT would restrain from it as standard feeding regime if there was a risk of negative reactions from consumers/citizens. As one farmer in another interview study put it, 'modern dairy farming practices are possible only thanks to consumers/citizens being ignorant about it' (Simonsson, 2016). Thus consumer/citizen trust in farming practices is perceived crucial by the farmers in this study, but is here combined with reluctance to inform them about feeding practices. Albeit a small number of informants, it is interesting to note this contradicts the more common view of the importance to build trust on transparency.

Conclusions

Both farmers and consumers/citizens recognize the discrepancy between an ethical ideal and what is practical reality, albeit from different points of departure. On the one hand it seems crucial to farmers to know about consumers'/citizens' views on animal welfare and their potentially different understanding of what is respectful enough, i.e. ethically acceptable, which probably matters in the case of calf rearing and colostrum feeding regimes, too. But on the other hand, at least some farmers in this study, seemed not to be interested in informing consumers/citizens if they are not willing to adopt criticism based on values they do not share. Hence, above welfare and health, transparency and trust are crucial aspects in any discussion on calf feeding routines.

Acknowledgements

The authors would like to thank Prof. Kerstin Svennersten-Sjaunja for initiating this project, receiving the funding and for valuable comments to the paper. Formas has funded this project and the Centre of Excellence for Farm Animal Welfare Research where Berg, Hernandez and Lidfors and Röcklinsberg are affiliated.

References

Besser, T.E., Gay, C.C. and Pritchett, L. (1991). Comparison of the three methods of feeding colostrum to dairy calves. Journal of the American Veterinary Medical Association 198: 419-422.

Bergstra, T.J., Gremmen, H.G.J. and Stassen, E.N. (2015). Moral values and attitudes toward Dutch sow husbandry. Journal of Agricultural and Environmental Ethics 28: 375-401.

Blum J.W. (2006). Nutritional physiology of neonatal calves. Journal of Animal Physiology and Animal Nutrition 90: 1-11.

Boogaard, B.K., Bock, B.B., Oosting, S.J., Wiskerke, J.S.C. and Van der Zijpp, A.J. (2011a). Social acceptance of dairy farming: The ambivalence between the two faces of modernity. Journal of Agricultural and Environmental Ethics 24: 259-282.

Boogaard, B.K., Oosting, S.J., Bock, B.B. and Wiskerke, J.S. (2011b). The sociocultural sustainability of livestock farming: an inquiry into social perceptions of dairy farming. Animal 5(9): 1458-1466.

Borkfelt, S., Kondrup, S., Röcklinsberg, H., Bjørkdahl, K. and Gjerris, M. (2015). Closer to nature? A critical discussion of the marketing of 'ethical' animal products. Journal of Agricultural and Environmental Ethics 28(6): 1053-1073.

Chigerwe, M., Tyler, J.W., Schultz, L.G., Middleton, J.R., Steevens, B.J. and Spain, J.N. (2008). Effect of colostrum administration by use of oroesophageal intubation on serum IgG concentrations in Holstein bull calves. American Journal of Veterinary Research 69: 1158-1163.

Fraser, D. (1999). Animal ethics and animal welfare science: Bridging the two cultures. Applied Animal Behaviour Science 65: 171-189.

Godden, S. (2008). Colostrum management for dairy calves. Veterinary Clinics: Food Animal Practice 24: 19-39.

Hammon, H.M., Steinhoff-Wagner, J., Schönhusen, U., Metges, C.C. and Blum, J.W. (2012). Energy metabolism in the newborn farm animal with emphasis on the calf: endocrine changes and responses to milk-born and systemic hormones. Domestic Animal Endocrinology 43: 171-185.

Hansson, H. and Lagerkvist, C.J. (2012). Measuring farmers' attitudes to animal welfare and health, British Food Journal 114(6): 840-852.

Johansson, A. (2016). Synen på kalvar, råmjölkstillförsel och kalvrådgivning hos några svenska lantbrukare, rådgivare och veterinärer. Sveriges Lantbruksuniversitet. Institutionen för Husdjurens miljö och hälsa.

Lateur-Rowet, H.J.M. and Breukink, H.J. (1983). The failure of the oesophageal reflex, when fluids are given with an oesophageal feeder to newborn and young calves. Veterinary Quarterly 5: 68-74.

Leek, B.F. (1993). Digestion in the ruminant stomach. In: Stevenson, M.J. and Reece, W.O. (eds.) Duke's physiology of domestic animals. Ithaca, New York, NY, USA 415 pp.

Lupoli, B., Johansson, B., Uvnäs-Moberg, K. and Svennersten-Sjaunja, K. (2001). Effect of suckling on the release of oxytocin, prolactin, cortisol, gstrin, cholecystokinin, somatostatin and insulin in dairy cows and their calves. Journal of Dairy Research 68: 175-187.

Persson Waller, K. (2013). Sondmatning av råmjölk till mjölkraskalvar – för- och nackdelar. Svensk Veterinär Tidning 2: 31-34.

Sakai, R.R. (2012). Effect of single oroesophageal feeding 3L versus 4 L of colostrum on absorption of coleostral IgG in Holstein bull calves. Livestock Science 148: 297-299.

Simonsson, H. (2016). Den moderna mjölkproduktionens etiska problem. En kvalitativ studie baserad på djupintervjuer med 8 nyckelpersoner. Master thesis, SLU, Uppsala, Sweden.

Stott, G.H., Marx, D.B., Menefee, B.E. and Nightengale, G.T. (1979). Colostral immunoglobulin transfer in calves I. period of absorption. Journal of Dairy Science 62: 1632-1638.

Weaver, D.M., Tyler, J.W., VanMetre, D.C., Hostetler, D.E. and Barrington, G.M. (2000). Passive transfer of colostral immunoglobulins in calves. Journal of Veterinary Internal Medicine 14: 569-577.

14. Increase sustainability of organic pig production with more vital piglets

S-L.A. Schild[*], *L. Rangstrup-Christensen and L.J. Pedersen*
Department of Animal Science, Aarhus University, Blichers Allé 20, 8830 Tjele, Denmark; sarah-lina.schild@anis.au.dk

Abstract

About every third piglet dies before weaning within the Danish organic pig production, making the organic piglet mortality higher, than what is seen in the conventional indoor production. The high piglet mortality is in conflict with both the idea of increased animal welfare and the organic principles. As most of the piglets die within the first days after farrowing, improving the thermal conditions around the time of farrowing is important, since the thermal environment in many cases may be a significant contributor to piglet death also for piglets that are categorized as crushed or starved to death. This paper describes a subproject under the larger Danish VIPiglets project with the overall aim to reduce piglet mortality in Danish organic pig production. The subproject described in this paper has three overall aims concerning (1) ways to improve the thermal environment around the time of farrowing; (2) how the temperature and humidity inside the farrowing hut relate to piglet mortality; and (3) how temperature affects farrowing duration and birth interval. Studies concerning aim two and three are still ongoing and so no results are presented. In the study concerning aim one the average total litter size was 17 piglets and total piglet mortality was 31%. The study showed that providing sows with access to poplar trees (a shaded area) significantly affected paddock use compared to sows without a shaded area (control sows), by increasing the use of the paddock before farrowing and day zero and one *post partum* (pp) ($F_{4,218}$=2.91, P<0.05) and the paddock use particularly increased at hot temperatures. Furthermore the sows with access to poplar were lying more when outside compared to controls (29 against 18% respectively; $F_{1,70}$=15.13, P<0.0005). In general the sows were seen lying more in the paddock at high temperature compared to low temperature particularly before farrowing, on day zero and one pp and again after day 11 pp.

Keywords: piglet mortality, heat stress, lactating sows, animal welfare, behaviour

Introduction

In organic pig production the animals must have access to an outdoor area and in some countries e.g. Denmark and the UK sows must be kept on pasture. One may therefore argue that organic production provides better animal welfare than conventional production since it gives animals more opportunity to perform natural behaviour. The organic production in general is based on four ethical principles; namely the principle of health, the principle of ecology, the principle of fairness and the principle of care (IFOAM, 2016). These four principles all stress the importance of sustainability within the organic production system.

Piglet mortality has been increasing since 1999 where a mortality of 18% was recorded (Kongsted and Larsen, 1999), in 2007 the total mortality had increased to 33% (Sørensen and Pedersen, 2015) and so the first part of the ongoing Danish VIPiglets project included quantifying the present level of mortality rate in the organic production. Preliminary data show a total mortality comparable to that recorded in 2007 (average 29%, range 18 to 38%) (Rangstrup-Christensen, 2015).

With a piglet mortality around 33% the mortality in organic pig production has increased to a level, which is higher than in the indoor production, and this high mortality conflicts with both the idea of

I. Anna S. Olsson, Sofia M. Araújo and M. Fátima Vieira (eds.) ***Food futures: ethics, science and culture***
EurSafe 2016 – DOI 10.3921/978-90-8686-834-6_14, © Wageningen Academic Publishers 2016

increased animal welfare and the organic principles. One way to reduce the mortality may be through the use of genotypes selected for low mortality under outdoor production conditions where farrowing assistance, aid for small under-weight piglets, use of foster sows, etc. are not as easily applied as in indoor conditions. Another way is to improve the environmental condition around the time of farrowing to sustain sow health and maternal behaviour as well as piglet viability. The current subproject deals with the latter issue.

Since animals housed outdoors are more exposed to fluctuations in the environmental temperature and both sows and piglets have a narrow thermal comfort zone, particularly around the time of farrowing, improving the thermal environment around farrowing could be one way to improve piglet survival. The thermal comfort zone of a lactating sow ranges from 12 to 22 °C (Black *et al.*, 1993). When sows are exposed to higher temperatures there is a risk of heat stress. Heat stress is known to be related to farrowing problems likely due to the inhibitory effect of stress on oxytocin (Lawrence *et al.*, 1997). Heat stress may therefore lead to an increased farrowing duration, longer inter birth intervals between piglets (Oliviero *et al.*, 2008), an increased number of sick sows and may cause reduced feed consumption and reduced milk production (Mullan *et al.*, 1992; Prunier *et al.*, 1997; Quiniou and Noblet, 1999). The behaviour of the sows is expected to change during heat stress – with an increased amount of risky behaviours (e.g. posture changes and resting away from the nest/hut) putting particularly small under-weight piglets at a high risk of crushing. Contrary to the sows piglets are more likely to experience cold stress during winter than heat stress during summer as the piglets thermal comfort zone lies around 30 to 35 °C (Black *et al.*, 1993; Mount, 1959). Cold stress is negatively related to survival of the piglets (Tuscherere *et al.*, 2000) and particularly small under-weight piglets are at risk. The thermal environment may therefore in many cases contribute to piglet death also for piglets that are categorized as crushed or starved to death. And so the aims of the current subproject of the VIPiglets project is to identify (1) ways to improve the thermal environment around the time of farrowing by providing access to a shaded area; (2) how the temperature and humidity inside the farrowing hut relate to piglet mortality; (3) how temperature affects farrowing duration and birth interval. The aims are addressed through three studies.

Methods

In all three studies the sows and gilts used are Landrace × Yorkshire crossbred and their piglets are Landrace × Yorkshire × Duroc crosses. The sows are kept on pasture all year and during farrowing and lactation in individual paddocks with access to an insulated A frame farrowing hut. If the temperature exceeds 15 °C in the shade a wallow is created as is required by law (Anonymous, 2015a). Within the first week of life the male piglets are castrated.

From farrowing and until day ten *post partum* (pp) piglets are prevented from leaving the hut. Piglets are weaned at seven weeks of age.

In all studies environmental recordings are done by use of data loggers (Hygrochron™ iButton (DS1923), Stockholm, Sweden). The loggers collect temperature and humidity inside the farrowing huts and outside on the field every tenth minute from approximately a week before expected farrowing until two weeks after expected farrowing.

Besides environmental recordings, in all studies, the herds record production data e.g. farrowing date and litter results.

Study 1

Study 1 focused on improving the thermal environment around the time of farrowing.

Although studies have shown less temperature fluctuations inside insulated farrowing huts compared to uninsulated huts (Edwards *et al.*, 1995; Randolph *et al.*. 2005), hut temperature during summer often exceeds the thermal comfort zone of lactating sows. The farrowing hut is usually the only shaded area in a standard farrowing paddock. Thus, as sows mainly regulate temperature through behaviour, thermal conditions may be improved by providing access to an alternative shaded area.

The aim of the current study was to investigate how sow behaviour and signs of hyperthermia are affected by giving sows access to poplar trees. We hypothesized that access to poplar would increase the sows use of the paddock and that paddock use would increase at increasing temperature.

This was an experimental field trial study with a parallel group design conducted from July to August 2015 in a private organic pig producing herd. Seventy-two lactating sows (median parity 4; range 1-8) and their litters were included. Fifty-eight test sows were housed in individual farrowing paddocks and allowed access to 8 × 4 rows of poplar trees. Control sows (n=14) were placed in standard individual farrowing paddocks without shade.

Sows were monitored using focal scan sampling on a daily basis from three days before until eight days after expected farrowing and again at approximate day 14 and 28 after farrowing.

Data were analysed using the proc mixed procedure in SAS (SAS 9.3, SAS Institute Inc., Cary, NC, USA).

The average total litter size was 17 piglets (range 5 to 23) and total piglet mortality until weaning (at day 49 pp) was 31%. The results showed that sows, which had access to poplar used the paddock more before farrowing and on day zero and one *post partum* (59.3 and 19%) (P<0.05) compared to controls (42 and 6.4%, respectively). When outside the sows used the poplar more before farrowing and after day 11 pp compared to the days in between (P<0.0001). Furthermore, when outside the huts, sows with access to poplar were observed lying more (29%) compared to controls (18%) (P<0.0005). How much the sows with access to poplar used the paddock depended on the hut temperature, thus the sows used the paddock more at high compared to low temperature (at 25 against 18 °C: 45 against 30% of time), for control sows there were no change in paddock use (at 25 against 18 °C: 30 against 30% of time). In general, sows were seen lying more in the paddock at high versus low temperature, particularly before farrowing, on day 0 and 1 pp and again after day 11 pp (P<0.01).

The use of the wallow depended on both treatment and day in lactation (P<0.005).

Thus sows with access to poplar used the wallow more (16%) compared to controls (5%) in the days before farrowing whereas controls used it more after day 11 in lactation (18 compared to 25% for poplar and control, respectively).

In conclusion access to poplar increased the sows use of the paddock particularly during hot periods, indicating that access to poplar does provide sows with a better opportunity for thermoregulation. Also as the use of the poplar depended on day in lactation and was increased before farrowing and after the piglets were able to leave the farrowing hut. Hence the results suggest that access to poplar will not keep the sows away from the huts (and their piglets) while piglets are restricted to the hut. In general there was a trend for all sows to lie down more when outside the hut at high temperature.

Since access to poplar also increased the use of the wallow before farrowing it is suggested that even though the sows have access to an alternative shaded area, they still need access to a wallow in order to thermoregulate properly.

Study 2

Study 2 aims at investigating how temperature and humidity inside the farrowing hut relate to piglet mortality. There are several indications of increased piglet mortality both during summer (Randolph *et al.*, 2005) and winter (Berger *et al.*, 1997; Edwards *et al.*, 1995). However, better documentation for the relation between hut climate and piglet mortality is required in order to clarify whether sows and piglets experience hyperthermia during summer and/or cold stress during winter. We hypothesize that when the temperature inside the hut, on day zero until day seven pp, exceeds 25 °C or falls below 15 °C piglet mortality will increase.

Study 2 was initiated at four Danish private organic pig producing herds in July 2015 and data collection will run for a one year period. Environmental recordings and production data are collected. Data loggers have been placed inside 30 standard A frame farrowing huts at each herd and recordings of about 720 farrowings are expected. As the study is still running no results are available. However, preliminary data from the temperature recordings made in the summer 2015 will be briefly mentioned. Despite the summer (June, July, August) of 2015 being the coolest Danish summer in three years, with an average temperature of only 15.2 °C (Anonymous, 2015b) temperatures above 35.5 °C and hours with an average temperature above 30 °C were recorded inside farrowing huts in all four herds. As this temperature exceeds the thermal comfort zone of lactating sows by around 10 °C it was possible that the sows were experiencing hyperthermia, despite the low average temperature of the summer months.

Study 3

In study 3 the aim was to investigate how temperature affects farrowing duration and birth interval. Heat stress can lead to an increased farrowing duration and longer inter birth intervals between piglets (Oliviero *et al.*, 2008), both of which are associated with an increased risk of piglet mortality (Baxter *et al.*, 2009; Panzardi *et al.*, 2013; Pedersen *et al.*, 2011). We therefore investigate the hypothesis that longer farrowing durations and increased inter birth intervals will be seen when hut temperature exceeds 25 °C and more stillborn and crushed piglets will be seen in litters born at >25 °C.

Study 3 was initiated November 2015 at a private free range herd and will continue for a one year period. Besides data loggers (recording temperature and humidity inside the hut), video cameras (IPC-HDBW4100EP-0360B; Dahua Technology, Irvine, CA, USA) have been placed inside 20 farrowing huts. As the study has just been initiated no results are presented.

Perspectives

By identifying at what temperatures and humidity peaks in piglet mortality occur it will be possible to identify when the thermal discomfort of the animals becomes so extreme that it is reflected in the piglet mortality. Furthermore, piglet mortality may be due to either stillbirth or death of live born piglets. These mortalities may increase at different environmental conditions meaning that the type of management initiatives needed and the time in lactation where they need to be implemented differ across the year. E.g. we expect that during summer the proportion of sows experiencing farrowing problems, and thus also the proportion of stillborn (and weak born) piglets, will increase. Hence (if this is the case), during summer, management initiatives should be implemented prior to farrowing and aim at reducing heat stress in the sows. Contrary we hypothesize that a larger proportion of live born piglets die during winter and so management routines during winter should be implemented after the birth of the piglets and aim at reducing cold stress among piglets.

The principle of fairness: 'insists that animals should be provided with the conditions and opportunities of life that accord with their physiology, natural behaviour and well-being' (IFOAM, 2016). Even though some aspects of animal welfare are increased in organic pig production, due to the animals' opportunity to display a broad spectrum of natural behaviours such as rooting and nest building, other aspects around the time of farrowing may be jeopardized in today's organic pig production. This is particularly reflected in the high piglet mortality.

In the large litters born by the high prolific sow lines used in both the Danish conventional and outdoor production, a large number of underweight (less viable) piglets are seen. These piglets are difficult to manage in an outdoor production system, where farrowing assistance, aid for small under-weight piglets, etc. are difficult to apply, and so the survival chances of underweight piglets are slim. One way to reduce early piglet mortality may therefore be through the use of genotypes selected for low piglet mortality under outdoor production conditions, e.g. sows that give birth to smaller litters consisting of larger sized piglets. Another way to reduce piglet mortality could be to improve the environmental conditions around the time of farrowing to sustain sow health and maternal behaviour as well as piglet viability. This would be in consensus with the organic principle of fairness as it would provide conditions that to a larger extent meet the animals' needs with regards to physiology and well-being.

References

Anonymous (2015a). LBK nr 1148 af 12/09/2015: Ministerial order concerning outdoor housing of pigs [Bekendtgørelse af lov om udendørs hold af svin].

Anonymous (2015b). The weather in Denmark – summer 2015 [Vejret i Danmark – sommer 2015]. Available at: http://tinyurl.com/hf5s5lo.

Baxter, E.M., Jarvis, S., Sherwood, L., Robson, S.K., Ormandy, E., Farish, M., Smurthwaite, K.M., Roehe, R., Lawrence, A.B. and Edwards, S.A. (2009). Indicators of piglet survival in an outdoor farrowing system. Livestock Science 124: 266-276.

Berger, F., Dagorn, J., Le Denmat, M., Quillient, J.P., Vaudelet, J.C. and Signoret, J.P. (1997). Perinatal losses in outdoor pig breeding. A survey of factors influencing piglet mortality. Annales de Zootechnie 46: 321-329.

Black, J.L, Mullan, B.P., Lorschy, M.L. and Giles, L.R. (1993). Lactation in the sow during heat stress. Livestock Production Science 35: 153-170.

Edwards, S.A., Riddoch, I. and Fordyce, C. (1995). Effect of outdoor farrowing hut insulation on piglet mortality and growth. Farm Building Progress 117: 33-35.

IFOAM Organics International Principles of organic agriculture – health, ecology, fairness and care. Available at: http://tinyurl.com/z5zasjr.

Kongsted, A.G. and Larsen, V.A. (1999). Piglet mortality in free range pig producing herds [Pattegrisedødelighed i frilandssohold]. Available at: http://tinyurl.com/jtglbxj.

Lawrence, A.B., McLean, K.A., Jarvis, S., Gilbert, C.L. and Petherick, J.C. (1997). Stress and parturition in the pig. Reproduction in Domestic Animals 32: 231-236.

Mount, L.E.(1959). The metabolic rate of the new-born pig in relation to environmental temperature and to age. The Journal of Physiology 147: 333-345.

Mullan, B.P., Brown, W. and Kerr, M. (1992). The response of the lactating sow to ambient temperature. Proceedings of the Nutrition Society of Australia 17: 215.

Oliviero, C., Heinonen, M., Valros, A., Hälli, O. and Peltoniemi, O. (2008). Effect of the environment on the physiology of the sow during late pregnancy, farrowing and early lactation. Animal Reproduction Science 105: 365-377.

Panzardi, A., Bernardi, M.L., Mellagi, A.P., Bierhals, T., Bortolozzo, F.P., Wentz, I. (2013). Newborn piglet traits associated with survival and growth performance until weaning. Preventive Veterinary Medicine 11: 206-213.

Pedersen, L.J., Berg, P., Jorgensen, G. and Andersen, I.L. (2011). Neonatal piglet traits of importance for survival in crates and indoor pens. Journal of Animal Science 89: 1207-1218.

Prunier, A., De Bragança, M.M. and Le Dividich, J. (1997). Influence of high ambient temperature on performance of reproductive sows. Livestock Production Science 52: 123-133.

Quiniou, N. and Noblet, J. (1999). Influence of high ambient temperatures on performance of multiparous lactating sows. Journal of Animal Science 77: 2124-2134.

Randolph, C.E., O'Gorman, A.J., Potter, R.A., Jones, P.H. and Miller, B.G. (2005). Effects of insulation on the temperature within farrowing huts and the weaning weights of piglets reared on a commercial outdoor pig unit. Veterinary Record 157: 800-805.

Rangstrup-Christensen, L. (2015). Oral Presentation at the organic congress (Økologikongres) of November 25, 2015. Vingsted, Denmark.

Sørensen, J.T. and Pedersen, L.J. (2013). Status, causes and challenges in relation to increased mortality among organic piglets [Status, årsager og udfordringer i forhold til løsning af forhøjet dødelighed hos økologiske pattegrise.] DCA Rapport nr 021.

Tuchscherer, M., Puppe, B., Tuchscherer, A. and Tiemann, U. (2000). Early identification of neonates at risk: traits of newborn piglets with respect to survival. Theriogenology 54: 371-388.

Section 3. Food waste

15. Combating food waste in Portugal: a case study of a civil society initiative

D. Lorena[1*] *and I. Pires*[2]
DARiACORDAR – Associação para a Recuperação do Desperdício, LABS LISBOA Rua Adriano Correia de Oliveira, 4ª 1600-312 Lisboa, Portugal; CICS.NOVA, Faculdade de Ciências Sociais e Humanas-UNL Av Berna 26C 1069-061 Lisboa; diogo.lorena@zerodesperdicio.pt

Abstract

Food waste is a transversal phenomenon that affects, although in different ways, all countries. A Food and Agriculture Organization (FAO) study has shown that there are about 1.3 billion tons of food in good condition for consumption wasted each year in the world. It is thus a global problem. However, the solutions must be local and must result from the mobilization of civil society through its awareness of this serious problem and desirably contributing to behaviour change while promoting sustainable development. The DARiACORDAR is a civil society non-profit association that aims at eliminating and recovering waste. It started its activity with the development, in partnership with ASAE (Portuguese authority of food safety) of procedures for recovering cooked food. In April 2012 it founded with other partners the Zero Waste Movement, creating a network for collecting surplus food with emphasis on cooked food. The Zero Waste initiative works through a system of governance based on partnerships among the various stakeholders, including the municipality, donors and food aid workers/volunteers. One of its advantages is that it can be replicated easily relying on the experience gained in the creation of the first networks. This network recovered by 2015, more than 1,200 tons of food equivalent to about 2.5 million meals with an estimated value of more than 6 million euros. We propose to introduce the association, its governance model and the innovative way it has contributed to the reduction of food waste in Portugal. We also aim to share the results of the pilot projects that are underway in some municipalities of the metropolitan area of Lisbon (AML), with particular emphasis on the recovery of cooked meals in different sectors of activity and its distribution to needed families with the collaboration of the municipal social service networks.

Keywords: redistribution

Introduction

At the global scale some 1.3 billion tons of food fit for human consumption is lost or wasted throughout the food supply chain (FSC) every year (FAO, 2011). Food waste (FW) is a global problem affecting both developed and underdeveloped countries although for different reasons and at different moments of the food supply chain (Godfray *et al.*, 2010; Parfitt *et al.*, 2010). In developed countries FW occurs mainly at the end of the FSC, namely distribution, restaurants, canteens and at the households (Evans, 2011; Falasconi, *et al.*, 2015; Guyomard *et al.*, 2012; William *et al.*, 2011). At the same time about 805 million people were estimated to be chronically undernourished (FAO, 2014). In the European Union total FW will have risen to approximately 126 million tons (a 40% increase) by 2020. To act against FW and to promote a sustainable food production and consumption preventive actions or measures are discussed and are being included in several European Union (EU) policies.

In Portugal an exploratory study on FW concluded that circa one million tons of food were wasted every year along the FSC (Batista *et al.*, 2012), with families and food services located on schools, universities and social institutions as well as in restaurants identified as the main responsible (Liz Martins *et al.*, 2014; Pires, 2013; Vaz, 2013).

For ethical but also for economic and environmental reasons FW should be reduced and ultimately banned. In the meantime, redistribution of food surpluses could be a temporary solution. To collect and distribute food among local non-profit, social solidarity institutions the first food bank in Portugal was established in Lisbon in 1992 and ever since many food banks have been opened. Today, the Portuguese Federation of Food Banks represents a network of 17 food banks across the country. However, for safety reasons cooked food is not collected.

In July of 2010 António Costa Pereira launched a public petition entitled 'Desperdício Alimentar' with the intent of altering European Regulation no. 178/2002, then perceived as the main roadblock to the legal recovery of cooked FW. This petition eventually gathered more than 70,000 signatures and galvanized a small group of citizens that came to form the association DARiACORDAR in 2011 with the aim of ending waste in all its forms.

This group was able to establish a collaboration with the Portuguese national food authority, the Autoridade de Segurança Alimentar e Económica (ASAE). From this common work came the demystification of the European ban on the recovery of food and the development of technical procedures by ASAE that, if followed, could ensure the legal and safe recovery of food surplus of this nature. Once prepared, such procedures were made publicly available with English translations available at the European Commission and FAO platforms, thus opening the way for several initiatives that came to be in the following years.

In Portugal just like many other countries, there are few studies of food redistribution (Hanssen *et al.*, 2014), despite the challenges that this activity face through practical difficulties and complex stakeholder relations (Alexander and Smaje, 2008). Thus, after ensuring that the process was legal, it became necessary to investigate whether, and demonstrate that, the proposed procedures were feasible. With that in mind a pilot network was established in 2012 with 4 municipalities of the greater Lisbon area – Lisbon, Loures, Cascais and Sintra. The aim is to quantify the flows of redistributed food from donators to organisations who will distribute them, food recovery composition analysis and how donators and receiver institutions are collaborating to identify possible bottlenecks and what can be done to improve the efficiency of food recovery and distribution.

Methodology

The pilot network, with the name Zero Waste Program, was developed with a concern of minimizing the generation of (new) waste streams, focusing on using (developing) resources already present in the field instead of building new ones (infrastructure).

Receiving entities recover food surplus at the donor's site at a previously agreed time. The donations are packed in food grade containers which are then placed inside isothermal boxes as to avoid temperature fluctuations. The transportation time from the pickup to delivery is as short as possible which is achieved through a proximity consideration during the donor/receiving entities matching process.

At arrival, donations are processed as soon as possible, with the highest priority given to those that need refrigeration. All donations are weighed and labelled according to provenience and registered according to 4 food types defined in the Zero Waste program: Uncooked foodstuff, Soup, Main Course and Complementary foods (e.g. bread, cakes, fruit, yogurts). The monthly aggregated values are reported in kg through an on-line form discriminated by donor.

For reporting purposes it is often necessary to convert the mass of recovered food in equivalent meals. The methodology followed for the conversion is as follow:

- For uncooked foodstuff only 80% of the recovered mass was considered, as to account for waste generated during the cooking process. For the reminiscent mass it was considered that each 0.500 kg would originate one full meal.
- For cooked food it was considered that a full meal was composed of a 0.350 kg portion of main course, a 0.200 kg portion of soup and a 0.100 kg portion of complementary foods.
- When it is not possible to group the recovered quantities into a complete meal a 0.500 kg portion of soup/main course/complementary food is converted into a meal and is signalled as being from that single source.
- The €2.5 per meal per person subsidy value attributed by the state to social canteens was used as a reference to calculate the economic value of each recovered meal.

Data used in the paper was collected from the reports receiving entities sent to Zero Waste Program.

Some preliminary data on food redistribution in Lisbon metropolitan area

In 2015 the pilot network in the Lisbon Metropolitan Area had 46 receiving institutions and 115 donors distributed through the 4 municipalities (Figure 1) with Lisbon taking the lead in the number of donors institutions. This difference may at least partly be due to the greater involvement of the local government at the parish level at the Lisbon municipality. Further data and studies are necessary to shed light on this matter.

Donors have been grouped in different categories according to theirs sector of activities. An official nomenclature as the European Union Economic Activity Classification System was not considered at this point, as often the food surplus generating activity of the donor is not related with its main activity. The selected categories were restaurants, distribution (e.g. supermarkets, hypermarkets), other private entities, public entities (like the Bank of Portugal or the National Mint), schools and hospitals (e.g. public and private) (Table 1).

The distribution sector, despite representing 33% of the total number of donors is responsible for the largest share of donations, donating mostly complementary foods and food reaching the sale by date limit that could not be recovered by traditional supply chains with longer product placement times, as can be seen in Table 2 that describes the percentage of each food type recovered per sector. They contributed with 78% of total donations during this period.

On the other hand restaurants, that represents 43% of the total number of donors account for no more than 4% of the donations. This lack of donations seems to relate not only to the smaller scale of this types

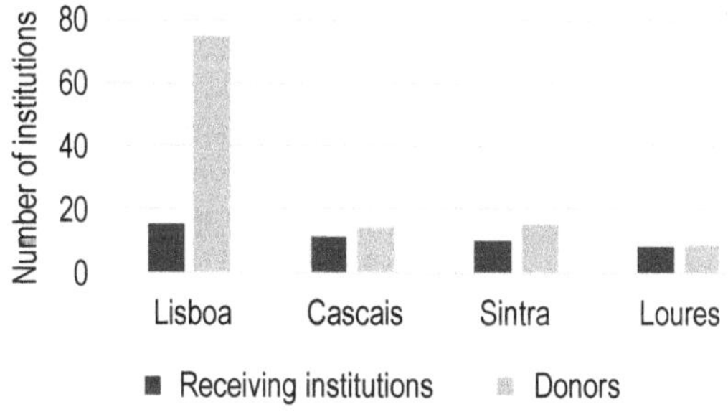

Figure 1. Receiving institutions and donors per municipality in 2015.

Table 1. Percentage of donors and donations per sector of activity in 2015.

	% of donors	% of donations
Restaurants	43%	4%
Distribution	33%	78%
Other private entities	8%	6%
Public entities	5%	4%
Hotels	5%	3%
Schools	3%	3%
Hospitals	3%	2%

Table 2. Percentage of food type recovered per sector.

	% of complementary foods	% of main course	% of soup	% of uncooked foodstuff
All donations	20	17	9	54
Distribution	36	11	4	49
Schools	9	56	33	2
Hospitals	4	54	42	0
Public entities	3	36	55	6
Other private entities	20	25	53	2
Hotels	19	55	26	0
Restaurants	49	24	22	5

of business but also with the predominance of pull supply chain strategies where most of the meals are made to order allowing for a better control of waste generation.

Comparing the average value of monthly donations for the different sectors (Figure 2) it is possible once more to identify the distribution and restaurants sectors as clear outliers of a central cluster where the other sectors remain.

The radical differences observed in the volume and nature of the donations of donors from different sectors reinforce the need for a proper matching of the receiving entity needs and recovery capacity with the donors' surplus availability and, for high volume donors, to manage multiple recipients simultaneously.

Although it is important to maintain a large diversity of donors, among other reasons because they assure a denser network and being closer to the receivers can contribute to maintain short routes. It is also important to include large distributors as they have a disproportional effect on the total recovered amount.

As explained before, the amounts of food donated by large supermarkets shows that uncooked food and complimentary foods are the large majority of the recovered mass while main courses and soup, although increasing during this period, remain less significant (Figure 3). Between 2013 and January 2016 uncooked food represented 54% and complimentary foods other 20% of food surplus donated.

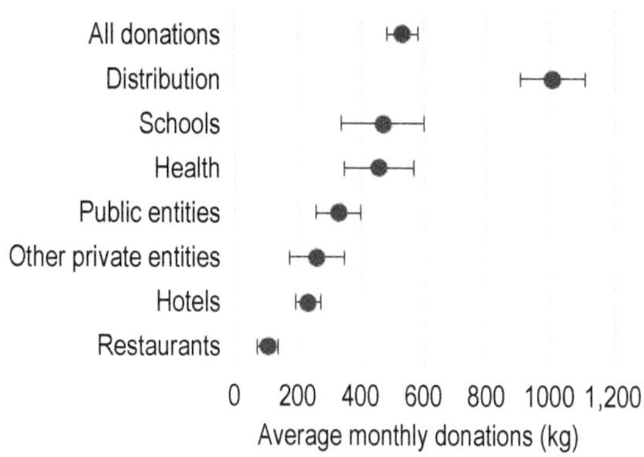

Figure 2. Average monthly donations in kg and respective 95% confidence intervals.

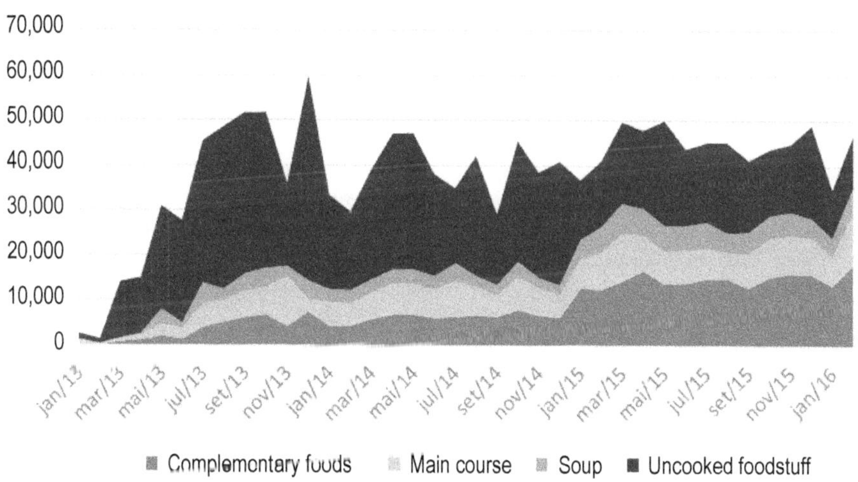

Figure 3. Quantities of Food Surplus recovered, by type, as reported by the donor

Conclusions

In Portugal, a wide amount of FW previously considered unrecoverable is now available for use as food aid to families in a situation of food insecurity and as a food resource for solidarity institutions.

Experience with the network currently in place demonstrated that it is possible to recover several tons of edible food by capacitating the institutions that are already present in the field, i.e. without involving huge investments. In a period of three years over 1,200 tons were recovered with nearly 500 tons in 2015 alone. This volume was achieved with a total of 115 donors which averaged 530 kg of donations a month per donor.

The model is based on using and developing existing infrastructure and can be replicated without the need of large investments.

As of this point the most important areas of investment to develop a sustainable network are those connected with training and implementation of health and safety measures as well as with the systematic collection and treatment of data for a better analysis of the network to maximize its efficiency.

Data already available indicates the necessity to look pass initial preconceptions of who the donor entities might be, as sectors which are less often mentioned, such as health care institutions, make significant contributions.

References

Alexander, C. and Smaje, C. (2008). Surplus retail food redistribution: an analysis of a third sector model. Resources, Conservation and Recycling 52: 1290-1298.

Baptista, P., Campos, I., Pires, I. and Vaz, S. (2012). Do Campo ao Garfo, Desperdício Alimentar em Portugal. Cestras, Lisbonne, Portugal.

FAO, IFAD and WFP (2014). The state of food insecurity in the world strengthening the enabling environment for food security and nutrition. Rome, Italy.

Hanssen, O., Ekegren, P., Gram-Hanssen, I., Korpel, P., Langevad-Clifforth, N., Skov-Olsen, K., Silvennoinen, K., Stare, M., Stenmarck, Å. and Svanes, E. (2014). Food redistribution in the Nordic region, experiences and results from a pilot study. Nordic Council of Ministers.

Falasconi, L., Vittuari, M., Politano, A. and Segrè, A. (2015). Food waste in school catering: an Italian case study. Sustainability 7: 14745-14760.

Martins, M.L., Cunha, L., Rodrigues, S. and Rocha, A. (2014). Determination of plate waste in primary school lunches by weighing and visual estimation methods: a validation study. Waste Management 34(8): 1362-1368.

Pires, I. and Vaz, S., (2013). Food waste: an exploratory study in Portugal. In: Medeiros, R. and Swatuk, L. (eds.) Sustainable development practice: advancing evidence based solutions for the post 2015 agenda. Proceedings of the 2013 International Conference on Sustainable Development Practice. The Global Association of Master's in Development Practice. New York, NY, USA, pp. 172179.

Evans, D. (2011). Sociology beyond the throwaway society: ordinary domestic practice and a sociological approach to household food waste. Sociology 46: 41-56.

Food and Agriculture Organization of the United Nations (FAO) (2011). Global food losses and food waste. FAO, Rome, Italy.

Godfray, H.C.J., Crute, I.R., Haddad, L., Lawrence, D., Muir, J.F., Nisbett, N., Pretty, J., Robinson, S., Toulmin, C. and Whiteley, R. (2010). The future of the global food system. Philosophical Transactions of the Royal Society B 365: 2769-2777.

Guyomard, H., Darcy-Vrillon, B., Esnouf, C., Marin, M., Russel, M. and Guillou, M. (2012). Eating patterns and food systems: critical knowledge requirements for policy design and implementation. Agriculture & Food Security 1: 13.

Hodges, R.J., Buzby, J.C. and Bennett, B. (2011). Postharvest losses and waste in developed and less developed countries: opportunities to improve resource use. Journal of Agricultural Science 149: 37-45.

Jones, T. (2004). The value of food loss in the American household. Bureau of Applied Research in Anthropology. Tilia Corporation, San Francisco, CA, USA.

Lundqvist, J., Fraiture, C. and Molden, D. (2008). Saving water: from field to fork – curbing losses and wastage in the food chain. SIWI policy brief. Stockholm International Water Institute, Huddinge, Sweden.

Parfitt, J., Barthel, M. and Macnaughton, S. (2010). Food waste within food supply chains: quantification and potential for change to 2050. Philosophical Transactions of the Royal Society B. 365: 3065-3081.

Williams, H., Wikstrom, F., Otterbring, T., Lofgren, M. and Gustafsson, A. (2012). Reasons for household food waste with special attention to packaging. Journal of Cleaner Production 24: 141-148.

16. Feeding cities sustainably: the contribution of a 'zero-foodwaste-city' to sustainable development goal 2, 'zero hunger'

L. Bellina
Leuphana University, Lüneburg, Scharnhorststr. 1, C11.006, 21335 Lüneburg, Germany; bellina@leuphana.de

Abstract

Lüneburg in northern Germany is one of 51 cities participating in the contest 'City of the Future 2030+' (https://www.wettbewerb-zukunftsstadt.de), for which the German government sponsors projects that bring together city government, citizens, and university to develop evidence-based visions for their sustainable city in 2030. Leuphana University Lüneburg is committed to research-based learning for sustainability and gave first semester students the opportunity to develop research projects within the larger question: how can the sustainable development goals (SDG) be implemented locally? Can citizen-'ownership' of those goals foster transformations that connect the local and the global in tangible, effective ways? Between October 2015 and March 2016, 25 courses with a total of about 100 student research teams set out to explore possibilities. The paper proposes that hunger and unsustainable agriculture – the objects of SDG2 – are intimately linked with food waste: the high demand for foodstuffs in wealthy (mostly Global North) cities drives global markets and prices, as well as crop choices in other countries, both of which affect food security of the poor. Each discarded food item also wastes large amounts of limited resources: from soil and water over inputs like fertilizers to human labour. And artificially low prices make sustainable livelihoods of farmers increasingly difficult, while making wastefulness even easier for those who have plentiful access to 'cheap' food. Therefore, ending food waste through various strategies that also foster a new valuation of food in its local-global context is an important contribution wealthy cities can make to realizing SDG 2: zero hunger. The 30 students of the 2030+ course I taught envisioned their city without food waste and set out, in six projects with different local actors, to explore specific opportunities for better practices. The paper reports on the systemic approach taken to combat food waste as a city, with the projects' results as examples within the emerging actors network. The shared vision is a whole-system change in which each location of the city's food system – from production via trade to consumption and waste management, from policy to education and science – functions in ways that support the others in reducing food waste.

Keywords: food justice, local-global food systems, social responsibility

Connecting the dots: food waste, hunger, and food justice

The right to food has been recognized since 1948 in the Universal Declaration of Human Rights, and was specified in the 1966 International Covenant on Economic, Social, and Cultural Rights as the right to sufficient quantity as well as quality of food, its cultural appropriateness, and access to such foods free of discrimination. Yet about a billion people still suffer from hunger, while inordinate amounts of food are wasted (Kreutzberger, 2011). In industrialized countries, this amounts to 222 million tons per year, up to 100 kg of thrown-away food per person (*ibid.*). The situation is absurd – and developing the capacity to change it requires an understanding of the links between food waste and the structures that produce hunger. Looking through the lens of the ecological, economic, and social aspects of sustainability in conjunction with a food justice framework helps to assess what we actually throw away.

Food has high ecological value: the production of each piece requires large amounts of resources like soil, water, energy, and land/space; they are limited and often overtaxed (FAO, 2013). The current

EurSafe 2016 – DOI 10.3921/978-90-8686-834-6_16, © Wageningen Academic Publishers 2016

globalized food system also has a particular resource flow: much of these resources are used in poor countries and then imported to wealthy countries in 'virtual' form (Chapagain *et al.*, 2008). Economic pressure, often connected to development debt, forces countries in the Global South to produce export crops rather than food for their own populations. Problems of land grabbing and destruction of entire ecosystems abound (Weltagrarbericht, 2015). In addition, most food is produced within a problematic agricultural system that not only uses soil and water but deteriorates their quality, is responsible for much of the loss of bio- and seed diversity, requires high inputs of fossil fuels, and produces significant amounts of climate gases (Weis, 2010). Food waste creates an artificially high demand on the products of this ecologically unsustainable system – about 1.3 billion tons of food per year are produced and never consumed (FAO, 2015) – and therefore directly drives the current decline of resources and ecosystem functions. This is not only an ecological disaster – throwing food away means throwing the basic ingredients of food security away. While many movements work towards a fundamental shift in agricultural systems, reducing waste would already reduce pressure on resources, especially where they are most needed for local food production.

Hunger is not only a matter of available quantity of food but a matter of its distribution, and therefore a matter of justice. Food justice can be defined as 'that the benefits and risks of where, what, and how food is grown and produced, transported and distributed, and accessed and eaten are shared fairly' (Gottlieb *et al.*, 2012: 6). The 'cheap' food that is so easily thrown away, however, is based not only on agricultural subsidies, but also on systematic exploitation of labour. As large scale industrial farming systems increased, millions of farmers were displaced from their land, turning from producers into labour, and therefore into a cost-factor for agribusiness (Bello, 2010). As (largely migrant) farm workers, they work for low pay under hazardous conditions, often subjected to human rights violations (Gottlieb *et al.*, 2012). Wasting food means we quite literally throw away the work and even lives of those who grew our food. Ironically, the poverty and bad living conditions of farm workers limit access to sufficient food, so that those who produce food are now one of the severely food insecure populations (*ibid.*). The production of 'cheap' food then, actually produces hunger – whereas a sustainable agriculture should produce sustainable livelihoods, a primary means to achieve food security (De Schutter, 2011). But even given such livelihoods, treating food as a commodity on the global market rather than as a human right causes price hikes (as in 2008) that can make it impossible for the poor to afford basic nutrition. This calls for a social re-valuation of food: thinking beyond ecological aspects to consider the impact of food production on workers everywhere, as well as on (others') food security. Not wasting food saves money, which can be put towards choosing food produced more sustainably, both ecologically and socially. However, this approach cannot simply be individualized in the sense of 'voting with your fork', since organic and fairly produced foods are accessible mostly to people with middle class and up incomes (Viertel, 2011).

Therefore, cities are called to act on the issue of food waste because they can engage multi-pronged strategies beyond the individual consumer: setting structural incentives to reduce waste and redistribute surplus foods, engaging knowledge production and education, but also providing access to means of food production to vulnerable communities, and supporting local, fair and sustainable agriculture models that provide affordable food, to name just a few. The projects described in the next section engage several such avenues.

Lüneburg – envisioning a zero-foodwaste-city

Systems-approach

We defined food waste as 'food appropriate for human consumption being discarded', as opposed to food loss, which refers to food deterioration along the production chain, or inevitable food waste, meaning inedible food parts (FAO, 2013: 8). To develop a vision of a zero-food-waste city, we began with assessing local nodes in the food system, defined as locations that engage with food in a particular way (Figure 1). They can be summarized as locations that directly engage food, and locations that influence food practices. Direct engagement includes producing (growers/farmers, bakers, etc.), selling (stores, markets, etc.), serving (restaurants, schools, etc.), redistributing (charity organizations), and consuming (households/individuals). Each of these locations faces particular challenges in regards to reducing food waste. Locations of influence include politics (city government, -services, etc.), science, education, citizen organizations, and public space.

As the city is nested within the larger food system and its legal/political structures, those nodes are nested within the city. The city has a 'hinge'-function between the larger and the local system: it can utilize its political freedoms to create a local policy-, incentive-, and funding-framework that encourages the nodes to employ better practices. Within the nodes, there are organizations, businesses, and people, many of whom already want to reduce food waste. Student projects connected to already interested actors to research possible strategies. We prioritized strategies according to their effects on sustainability and food justice: (1)Avoid generating food surplus that leads to waste by structurally changing incentives; (2) Fully use all edible food, including re-distribution of all unavoidable surplus; (3) Reuse spoiled/inedible leftover food for composting and animal feed.

Such prioritisation reveals conflicts among strategies, for example: #1, aimed at the necessary change of patterns of overproduction and -consumption that drive an ecologically and socially unsustainable food system, will reduce surplus that can be redistributed to poor people. Therefore, charitable redistribution of surplus (part of #2) is not a truly sustainable solution to local food security. #3 shows that reusing food waste should never become business that creates demand for something we seek to reduce (for

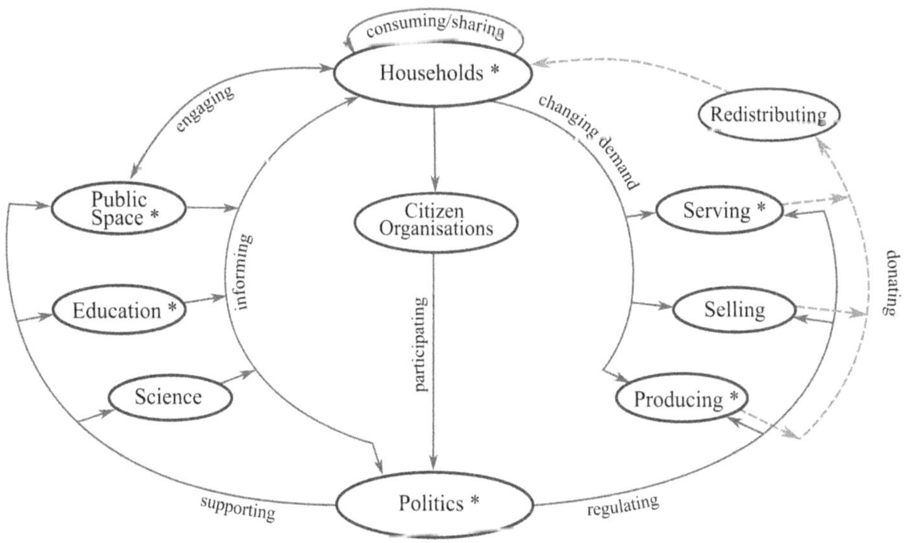

*Figure 1. Nodes in the city food system. The nodes represent groups of actors who need to be involved in a city-wide strategy to reduce food waste. Each node can be a working group for the organizations, businesses, etc. within, to share knowledge and strategies for their particular challenges. The network serves to coordinate those strategies. The arrows indicate some of the major avenues for changing the city food system along strategies #1 and #2. * indicates student research projects related to the node.*

example in for-profit biogas companies). Student projects consequently focused on strategies #1 and #2 to minimize surplus while maximizing use, thereby also reducing inedible food waste and avoiding food in regular garbage. The goal of such a systemic, strategic approach is to connect nodes and their actors in a synergetic effort, with strategies coordinated towards sustainable local-global solutions.

Strategies and visions

In this section, nodes are underlined for easy reference to Figure 1. The students' work included here refers to their final paper, which is cited with the research group's name; all participating students are listed in Appendix A. Short vision narratives show the goal for 2030.

One central node in the city food system is households, which stands for the daily connection of every person to procuring and consuming food. It is also a placeholder for different social groups and their needs. In Germany, households account for about 6.7 million tons of food waste. This node is the main focus of the government's strategy to reduce waste through information aimed at individual behaviour-change (https://www.zugutfuerdietonne.de). Rather than on the moral appeal not to waste, the research group 'Tafelrunde' focused on reviving an enjoyable meal-culture combined with food sharing as a strategy to reduce waste. They connected with the developers of yumwe.de, an internet platform that allows people to connect for cooking sessions, and proposed the idea to students as a sample group. Their survey (122 students) found that 94% would like to improve their cooking skills and 75% liked the idea to make new social connections through cooking (Tafelrunde: Appendix A). The vision: a 'connected city', where people share food through different avenues, including the use of digital tools. This brings different people together who would otherwise not meet, strengthens social bonds in neighbourhoods, and encourages meal-culture as well an ongoing, enjoyable learning process to cook and fully use food.

This connectedness extends beyond private space: research group 'Foodturetown' envisioned public space as an edible city that reconnects citizens with food and with each other. They researched a successful model, the city of Andernach, through an actors-analysis followed by interviews. They found that an edible city indeed increases quality of life, local food security, and social connections. A combination of bottom-up and top-down approach is important: citizens need to feel ownership, and city government needs to provide access public space and resources needed for implementation (Foodturetown: Appendix A). The vision: an 'edible city', where most public spaces are urban gardens/farms that many different groups of people maintain. There is intergenerational and intercultural exchange of knowledges and practices, and explicit opportunity for different foodways (Alkon *et al.*, 2011) to be a visible part of the city. In this manner, Lüneburg supports its new populations (for example refugees) to connect with other residents and contribute new tastes to 'the local', creating a food-polyculture (*ibid.*). Involving city services in maintenance also creates new jobs. The edible city contributes to local food security through 'growing and knowing' (how to grow, including bio- and seed-diversity conservation), as well as by creating a more just local food system through sharing both across diverse groups.

Learning about food is also part of formal education. It is an important node for changing valuation and behaviours around food, yet it is not an integral topic in all curricula. The research group 'Schule' examined a model project at a primary school (interviews with leadership and participating teachers) to see if and how it could be scaled up to become part of all schools in Lüneburg. They found that this successful project covers its cost by including a student company, where pupils not only learn about food, cooking, and avoiding food waste, but also about good business practices. Doing this in other schools is then mostly a matter of engaged staff (Schule: Appendix A). The vision: an 'educated city', where valuation of food in all its aspects starts early, so that all children have the opportunity to engage in growing and cooking food and to learn about sustainable food systems.

Such education could foster a change in demand and therefore in practices in the nodes of producing, selling, and serving food. The research group 'Gastronomie' connected with an initiative that offers a sustainability seal to restaurants that fulfil certain criteria. The group conducted a survey (292 Students as a sample group) and found that 68% would prefer a certified restaurant. For reducing food waste, 76% would like 'doggy bags' to take leftovers home, and 56% would like to be able to order smaller portions. For a tourist-centre like Lüneburg, engaging local hotels and restaurants in a certification that includes food-saving practices would have a significant impact (Gastronomie: Appendix A). The research group 'BackAware' interviewed bakery-owners about their strategies to reduce waste, and then conducted a survey with their customers. A surprising result is that 96% of customers do not expect a full selection at the end of the day – an argument widely used to justify producing surplus. They also found that about 50% would be willing to pre-order, so that bakeries could produce more on-demand and calculate better (BackAware: Appendix A). The vision: a 'smart-use/produce-city'. Customers use virtual technologies (among other methods) to pre-order so that stores (and other sellers/producers) can calculate better. In addition, the city fosters connection of all three nodes to local farmers through policies that support farm-to-table and CSA models. This avenue increases local food security as well as reconnecting people to the concept of sustainable livelihoods and fair working conditions along the food-production chain; a part of the social re-valuation of food and of creating a 'fair-trade-city'.

The research group 'Politik' engaged the importance of the node politics for this vision. They compared political approaches to reducing food waste, and found that in England and France, political regulation quickly produced significant successes. Therefore, the educational and voluntary approach in Germany should be complemented by more active regulation, as well as changes in law where needed, to achieve better results (Politik: Appendix A). The vision: a 'responsible city' that employs its regulatory function to provide the right incentives for all nodes to develop better practices, and its supportive function to aid places of influence like education in their efforts. And a 'knowledgeable city': collaborating with science to produce data on the effects of strategies to reduce food waste, providing feedback to all participating nodes, as well as motivation through reporting successes! And finally, it is a 'democratic city': citizen organizations, including a food policy council to represent different (vulnerable) groups, actively participate in creating a sustainable and just food system.

Conclusions

Feeding cities sustainably is one of the major challenges of this century, and zero-foodwaste-cities are one avenue to meet it. The paper has argued that systemic food waste is directly connected to hunger and unsustainable agriculture, and therefore constitutes a violation of the human right to food. It has also argued that cities have a particular opportunity and responsibility to reduce food waste, through their regulatory, supportive, spatial, and social functions. And it argued that through systematically reducing food waste, cities can contribute much to reaching SDG2, zero hunger and sustainable agriculture, locally as well as globally. Manifesting this requires four key components: First, since food waste is a systemic issue, resolving it requires a systemic approach in a network that includes and coordinates actors from all nodes. Second, it takes a shift from the moral appeal to 'not do' to a positive approach of 'doing': creating different praxes around food in the city that increase the quality of life along with food security for all, thereby encouraging wide participation through positive experience. In order for this to work, it takes, third, a shift from solely ecological reasoning to food justice: recognizing the right to food locally, i.e. providing access to food beyond charity. Here, food justice also relates to food sovereignty: it is 'communities exercising their right to grow, sell, and eat (food that is) fresh, nutritious, affordable, culturally appropriate, and grown locally with care for the well being of the land, workers, and animals' (Alkon *et al.*, 2011: 5). This connects to recognizing the right to food (and food sovereignty) globally: supporting local growing that provides sustainable livelihoods instead of exploiting labour elsewhere, and that uses local resources sustainably instead of depleting them in other countries. 'Local food' is then not a niche-market for the rich but a matter of local-global social responsibility. And fourth, since markets don't respond to moral appeals, it is necessary that cities take a political stand – and that we as citizens do not allow ourselves to be reduced to consumers who 'vote with our wallet', but reclaim citizenship to democratize our food system. Many of the participating students are well on the way: as their projects continue beyond the semester, they become part of creative, collaborative transformation processes that connect reducing food waste with food security, locally and globally.

References

Alkon, A.H. and Agyeman, J. (2011). Cultivating food justice: race, class, and sustainability. The MIT Press, Cambridge, MA, USA, 404 pp.

Bello, W. (2010). Food wars. Verso Books, New York, NY, USA, 176 pp.

Chapagain, A.K. and Hoekstra, A.Y. (2008). The global component of freshwater demand and supply: an assessment of virtual water flows between nations as a result of trade in agricultural and industrial products. Water International 33(1): 19-32.

De Schutter, O. (2011). The transformative potential of agroecology. In: Holt-Gimenez (ed.) Food Movements Unite! Food First Books, Oakland, CA, USA, pp. 223-242.

FAO (2013). Food wastage footprint; impacts on natural resources. Available at: http://tinyurl.com/hao9who.

FAO (2015). Key facts on food loss and waste you should know! Available at: http://tinyurl.com/krrrnge.

Gottlieb, R. and Joshi, A. (2013). Food Justice. MIT Press, Cambridge, MA, USA, 304 pp.

Kreutzberger, S. and Thurn, V. (2011). Die Essensvernichter. KiWi Verlag, Köln, Germany, 368 pp.

Viertel, J. (2011). Beyond voting with your fork: from enlightened eating to movement building. In: Holt-Gimenez (ed.) Food Movements Unite! Food First Books, Oakland, CA, USA, pp. 137-148.

Weis, T. (2010). The accelerating biophysical contradictions of industrial capitalist agriculture. Journal of Agrarian Change 10(3) 315-341.

Weltagrarbericht (2015). Land grabbing. Available at: http://tinyurl.com/ze54ajd.

Appendix A. Student research groups

BackAware:
Luisa Firsching, Karin Havemann, Leonie Jurowski, Patrick Rodenborg, Franziska Welzien, Fentje Zum Felde

Foodturetown:
Kachina Barmbold, Juliane Gallersdörfer, Stascha Kirchner, Violeta Lüdtke, Monisha Moreau, Verena Nüchter

Gastronomie:
David Aksöz, Nick Dreyer, Maya Jaeger, Seher Koç, Antonia Rohweder, Lotte Tetzlaff

Politik:
Lucie Hain, Greta Karcher, Jan Ole Krüger, Anne Paul

Tafelrunde:
Heide-Sophie Benz, Leon Handke, Emily Roberts, Nadine Steinhof

Schule:
Bettina Piontek, Linnea Rüter, Lisa Schulz, Fenia Vesper

17. Policy design for food waste reduction: conflicting political and economic interests

V. Sodano
Department of Agriculture, University of Naples Federico II, Italy; vsodano@unina.it

Abstract

Over the last few years, the food waste issue has received growing attention from governments and international agencies. In the European Union (EU), attention has been focused on waste generated by households and at consumer level. As a result, many awareness raising initiatives targeted at consumers have been carried out, in order to change consumer behaviour and attitudes. Further interventions were aimed at supporting donations of discarded food to charity organizations. The assumption underlining these policy strategies is that food waste is mainly an ethical and organizational/efficiency problem. The paper suggests instead that the most important causes of food waste are mainly political and economic, such as the excessive market power along the supply chain and the rise of neoliberal economic policies. The main conclusion of the paper is that the current forms of interventions are inadequate to thwart food waste since they are based on neoliberal political choices that prevent the state from implementing more effective forms of regulation.

Keywords: food policy, neoliberalism, corporate power

Introduction

Over the last few years, the food waste issue has received growing attention from governments and international agencies, and many reports and blueprints have been published. In all the released documents, food waste reduction is generally portrayed as an objective shared by society as a whole, given the fact that it is a means by which to increase food supply efficiency and to reduce pressures on resources and the climate. Moreover, the analysis of the causes of food waste focuses on the efficiency issue and does not take the political forces that affect the phenomenon as well into due account.

The paper seeks to show that, when all causes other than the lack of technical and organizational efficiency are taken into account, the most common implemented interventions for the reduction of food waste turn out to be inappropriate. The first section of the paper frames the topic of food waste, tackling the issues of how to define and measure food waste and of what its main negative effects are. The second section critically reviews the policy interventions suggested so far, highlighting how they fail to deal with some of the most important causes of food waste. Special attention is paid to the case of the European Union (EU). The main conclusion of the paper is that the current forms of interventions are inadequate to thwart food waste since they are based on neoliberal political choices that prevent the state from implementing more effective forms of regulation. It is worth noticing that the paper, given also the limited length, does not want to provide a thorough analysis of the food waste issue. Instead, it endeavours to show that there are some important political issues that should be seriously taken into account in order to effectively tackle the problem. Therefore, the paper is only a preliminary explorative study of the political underpinnings of the food waste issue.

Framing the issue of food waste

Japan is maybe the first country to recognize the problem of food waste and attempt to deal with it. In Japan in 2001 a law was passed regulating the disposal of food waste produced, for varying reasons,

EurSafe 2016 – DOI 10.3921/978-90-8686-834-6_17, © Wageningen Academic Publishers 2016

along the entire food supply chain. At an international level the issue was launched by the Food and Agriculture Organization of the United Nations (FAO) in 2011, with the publication of a report (FAO, 2011) which estimated food waste at a global level. In Europe the problem was recognized for the first time in 2010, with the release of the 'Preparatory study on food waste across the EU 27' (EU Commission, 2011). More recently, the Organization for Economic Co-operation and Development (OECD) (OECD, 2014) and the US Government (Gunders, 2012) offered new figures and analysis of the phenomenon. Notwithstanding the great interest on the parts of both governments and public opinion, currently the issue of food waste is still neither well known nor addressed. In particular, there is a lack of an internationally uniform definition, reliable estimates, and a general framework of its causes and effects.

The most general definition is offered by FAO, which distinguishes between food waste and food loss as follows:

> Food loss refers to a decrease in mass (dry matter) or nutritional value (quality) of food that was originally intended for human consumption. These losses are mainly caused by inefficiencies in the food supply chains, such as poor infrastructure and logistics, lack of technology, insufficient skills, knowledge and management capacity of supply chain actors, and lack of access to markets. In addition, natural disasters play a role. Food waste refers to food appropriate for human consumption being discarded, whether or not after it is kept beyond its expiry date or left to spoil. Often this is because food has spoiled but it can be for other reasons such as oversupply in relation to the market, or individual consumer shopping/eating habits. Food wastage refers to any food lost by deterioration or waste. Thus, the term 'wastage' encompasses both food loss and food waste. (FAO, 2013)

Although this definition tries to be very detailed, encompassing also possible causes of food waste, it nevertheless still leaves some questions unanswered. For example, it does not unequivocally specify how to express food waste, whether in caloric equivalents, weight, volume or value. It also does not take into account the difference between edible-avoidable and inedible-unavoidable food waste, a difference underlined instead by the UK charity WRAP, which gives the following definitions: edible/avoidable food waste, which refers to food that is thrown away that was, at some point prior to disposal, edible (e.g. slices of bread, apples, meat); inedible/unavoidable food waste which refers to waste arising from food preparation that is not, or has not been, edible under normal circumstances (e.g. bones, egg shells, pineapple skins) (WRAP, 2009). Moreover, when trying to suggest possible causes of food waste, the definition incorporates value judgments since it specifically mentions the consumers as sole responsible individuals. It does not instead directly refer to subjects such as entrepreneurs and public managers, preferring in this case to quote anonymous entities such as the markets.

Estimates made so far of food waste are scarce and their results are not comparable with each other, partly because of the aforementioned lack of a clear definition of the phenomenon. Currently there are three main sources of information coming from the European Commission (EC), the OECD and FAO. An EC report estimated that food waste in EU27 in 2007 amounted to 89 Mt of food waste/year and 179 kg per capita (EC, 2011). The report tried to quantify food waste along the supply chain, but it considered exclusively food waste at household level. In fact, it used Eurostat data on the share of bio-waste in municipal solid waste and only scattered second hand information (taken from different sources) on the manufacturing industry and food services sectors, while excluding the agricultural sector from its scope. In its 2011 report the FAO estimated that roughly one-third of the edible parts of food produced for human consumption gets lost or wasted globally, which is about 1.3 billion tons per year (2007 estimates). On a per-capita basis, FAO estimates showed that the per capita food waste in 2007 by consumers in Europe and North-America was 95-115 kg/year, while the same figure in Sub-Saharan Africa and South/Southeast Asia was only 6-11 kg/year (FAO, 2011). Also in the case of FAO the estimates were indirect; losses and waste were estimated using FAO's Food Balance Sheets from the year 2007 and results from a thorough literature search on the topic of global food waste. The latest estimates of food waste were offered by the OECD (OECD, 2014). Also in this case they are indirect estimates, made through the statistics on municipal waste, data collection from available reports and studies and an ad hoc request for information on food waste sent to the members of the OECD Working Party on Environmental Information (WPEI) (the rate of answer to such a request was anyway very low).

As regards the analysis of the causes and the effects of food waste, while a general framework is still lacking, nevertheless some key points have been already identified (FAO, 2011; Gunders, 2012; OECD, 2014). In developing countries three main causes of food loss and waste have been suggested: poor storage facilities and lack of infrastructure causing postharvest food losses; inadequate market systems and lack of processing facilities also causing food losses in developing countries; poor quality due to premature harvesting. Instead, in developed countries food waste has been deemed to happen when: production exceeds demand; disposal is cheaper than a 'using or re-using attitude'; supermarkets require large quantities on display and a wide range of products/ brands in supply; supermarkets stick to a policy of high 'appearance quality standards' for fresh products; abundance and consumerist attitudes lead consumers to purchase food in amounts higher than required.

As regards the negative impact of food waste, this is generally viewed in terms of economic, environmental and poverty effects. Economic losses are roughly assessed using either the price value of food discarded or the value of the resources used to produce it. FAO has estimated it as about USD 1 trillion each year (FAO, 2014). Environmental effects have been recently monetized by the FAO through a full-cost accounting methodology specifically developed to assess the food wastage footprint. Examples of the cost estimate from FAO include (FAO, 2014): USD 394 billion per year due to greenhouse gas emissions; USD 164 billion per year due to increased water scarcity; USD 35 billion per year due to soil erosion; USD 32 billion per year due to risks to biodiversity; USD 729 billion per year due to soil erosion; USD 153 billion per year because of the adverse health effects due to pesticide exposure. The poverty effect is generally envisioned in the case of developing countries where food losses may exacerbate the lack of food availability.

Policy intervention

All the documents produced so far by the EU and international organizations, such as FAO and OECD, suggest policies able to prevent and mitigate the negative impacts of food waste. In this section, I show how the proposed interventions tend to address only some of the causes of the problem. Moreover, some interventions may have a counter effect due to the fact that they strengthen the same neoliberal political stances that are at the core of the food waste issue.

Table 1 summarizes the main generally recognized types of causes of food waste together with suggested interventions. The first set of causes is ascribed to handling and distributing

Inefficiency due to lack of infrastructures and/or adequate storage and supply management technologies. Corrective interventions include technological innovations (such as inventory management information devices, smart refrigerators, smart packaging and so on) and a stronger vertical integration of the supply chain (Lipinski *et al.*, 2013; WRAP, 2015). The second set of causes refer to those consumer behaviours and attitudes responsible for food waste at household level, such as: lack of knowledge of food preparation and storage; misunderstanding of label information on food expiration date; poor shopping planning; over-preparation; consumerist attitudes; poor ethical concerns. The suggested

Table 1. Causes of food waste and proposed interventions.

	Efficiency/technological causes	Consumer behaviour	Company strategies
Causes	• lack of infrastructure • poor storage facilities • poor procurement coordination and stock management	• poor knowledge • poor shopping planning • overpreparation • consumerist attitudes	• high aesthetic standards • rigid procurement contracts • impulse/bulk promotions • display and assortment policies
Intervention/prevention	• technological innovation • smart packaging • information devices	• consumer education • awareness campaigns	• appeal to CSR
Mitigation	• food banks • supporting charity sector in diverting discarded food to the poor • incentives to collaborating firms		

interventions include consumer education and awareness campaigns, but also technological innovation (also in this case smart refrigerators and smart packaging are mentioned). The third set of causes refers to business practices all along the supply chain. In particular, those at the foodservice level are often quoted: standardized portions, too wide a range of offered products, cost saving stack procurement policies. In the retail sector there is reference both to downstream marketing policies (such as: large food display policies, overlarge package sizes, impulse/bulk promotions, increased sale of rapidly perishable ready-made food) and upstream procurement policies (such as: excessive cosmetic standards responsible for non-harvest and culling of edible produce upstream; overly rigid contract terms causing growers to overplant). With regards to business practices, no specific intervention is suggested; there is instead a generic appeal to corporate social responsibility (CSR) which should make firms put food waste reduction as first priority after profits. Besides the quoted interventions aimed at reducing the amount of food waste, there are proposals for interventions able to mitigate its negative impact, based on re-use and wise waste treatment warnings. The focus is generally on the possibility of redirecting the discarded food to poor people through the involvement of the charity sector, including giving fiscal incentives to firms offering their collaboration to this end.

Consistent with the views expressed in the official documents, the only interventions promoted so far by the EU have been educational communication policies aimed at consumers and the solicitation of businesses to collaborate with charitable institutions seeking to divert the waste food to food banks.

No mention has been made of those regulatory interventions able to correct the negative business practices, such as among others: lowering aesthetic standards; prohibiting inequitable procurement contracts; regulating promotion strategies at retail level; imposing taxes on the food discarded.

Overall, both the currently implemented and the suggested policies to counter food waste seem to be insufficient, since they do not address the causes which stem from company behaviours. They might also be ineffective. In fact, the much vaunted innovations (most of which include possibly risky bio-nanotechnologies) may have negative environmental and health effects which are able to more than outweigh the proposed benefits in terms of waste reduction. Moreover, the actions which target consumers, in addition to being steeped in paternalism, may not have the desired effects because of the three well known obstacles to sustainable food consumption (Gorgitano and Sodano, 2014): the rebound effect, which refers to a behavioural or other systemic response to a measure taken to reduce environmental impacts that offsets the effect of the measure; the knowledge-to-action gap, exhibited when the knowledge of environmental problems may not be sufficient to change consumer behaviour and lifestyles; the behaviour-impact-gap problem, confronted whenever the required behavioural change is achieved, but the observed ecological effect is minor or missing. Finally, the alleged CSR may turn out to be too weak an instrument for managing environmental risks (Sodano and Hingley, 2013).

The inability of EU to affectively tackle the food waste issue depends on the endorsement of a strict neoliberalist view. There are at least four components of EU choices, with respect to the food waste issue, that bear witness to its neoliberal political stances. First, the rejection/renouncement of forms of direct regulation is typical of the neoliberal approach which relies on the fiction of perfectly self-regulating competitive markets. Second, the appeal to consumers as those ultimately responsible for food waste and thus the main target of corrective policies is an example of the contradictory neoliberal version of the consumer sovereignty narrative. The latter, on the one hand, accepts the neoclassical consumer sovereignty credo that consumers command the economy through their rational choices, which are deemed neither to be influenced by company strategies nor based on cultural constructed preferences. On the other hand, by suggesting communication policies to change consumer behaviour, the neoliberal version describes consumers as dependent on the surrounding informative and cultural environment. Third, the appeal to CSR is consistent with the neoliberal choice of substituting the regulatory role of the state with forms of soft regulation based on the concept of governance and on the stakeholder approach, which substitute the power of the state with the power of corporations. Fourth, the reference to technological innovation as an external force, independent of political and economic power, is also part of the neoclassical economics which is the backbone of the neoliberal ideology.

Figure 1 reassumes the neoliberal stances underlying the current EU strategies against food waste and shows how they reinforce instead of correcting the causes of food waste. In the figure it is shown how the further technologization and integration of the food supply chain push towards higher concentration and higher corporate power, which means the strengthening (illustrated by the vertical thick arrow) of those firm strategies responsible for food waste. At the same time, consumer education policies tend to lower consumer choice autonomy and reinforce the idea in society that the current neoliberal food regime (McMichael, 2009) is the only possible way to meet the world's future food needs. Because the neoliberal food regime entails a complete commodification of food, it nourishes the roots of food waste. In a capitalist economy, overproduction and over purchasing, which are the main causes of food waste, are at the core of the processes of capital accumulation, and there are no motivations for production activities other than profit maximization. When food is nothing but a trivial market good, food waste is a false problem. Capitalistic systems eagerly seek new market opportunities; why should we care for still edible food being discarded when we do not care, for instance, about perfectly good new shoes or cell phones discarded only because unfashionable? Either we assume that food is not a commodity (but a human right, for example) or it is nonsense to talk of food waste; we should instead simply talk, as in

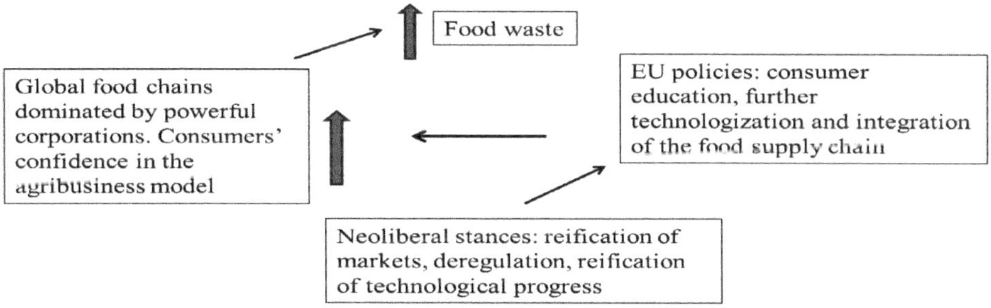

Figure 1. Interventions driven by neoliberal stances exacerbate food waste.

the case of other goods, of negative externalities associated with its production and consumption. In this regard the rhetoric of unethically acceptable food waste used by neoliberal governments is intended to soothe those people who care about the poorest who suffer from hunger and are outraged by good food being thrown into garbage bins. The association often made between reducing food waste and alleviating poverty is deceitful. In the world there are no hungry people because the food gets wasted. If anything, the opposite is true: the food is wasted because there are poor and hungry people. As a consequence, the best way to fight food waste should be to combat poverty, and therefore that high rate of economic inequality which instead is steadily increasing in neoliberal states.

Conclusions

In the EU, strategies to fight food waste have been focused at consumer level with awareness campaigns and consumer education initiatives. Further interventions have been aimed at supporting donations of food fit for human consumption to food banks, linking the objective of food waste reduction to poverty reduction strategies. Instead, causes of food waste stemming from the current agribusiness model (such as: excessive concentration and market power, unequal bargaining power in procurement relationships, and the structural and strategical dynamics of the global value chains) have not been addressed. The paper suggested that the EU policy choices depend on the endorsement of strict neoliberal stances which not only prevent the implementation of more effective regulatory interventions, but also reinforce the causes of food waste. As long as food waste is tackled from a neoliberal perspective it is very unlikely that the fight against food waste may bring benefits to society at large.

References

European Commission (EC) (2011). Preparatory study on food waste across EU 27 Technical Report 2010-054. Brussels, Belgium.

Food and Agriculture Organisation (FAO) (2011). Global food losses and food waste. Extent, causes and prevention. Rome, Italy.

Food and Agriculture Organisation (FAO) (2013). Food wastage footprint: impacts on natural resources. Rome, Italy.

Food and Agriculture Organisation (FAO) (2014). Food wastage footprint: full-cost accounting. climate, energy and tenure division. Rome, Italy.

Gorgitano, M.T. and Sodano, V. (2014). Sustainable food consumption: concept and policies. Quality-Access to Success 15: 207-212.

Gunders, D. (2012). Wasted: how America is losing up to 40 percent of its food from farm to fork to landfill. Natural Resources Defense Council (NRDC) Issue Paper iP: 12-06-B.

Lipinski, B., Hanson, C., Waite, R., Searchinger, T., Lomax, J. and Kitinoja, L. (2013). Reducing food loss and waste. Working paper, Installment 2 of Creating a Sustainable Food Future. World Resources Institute, Washington, DC, USA.

McMichael, P. (2009). A food regime genealogy. The Journal of Peasant Studies 36: 139-169.

Organisation for Economic Co-operation and Development (OECD) (2014). Food waste along the food chain. OECD Food, Agriculture and Fisheries Papers 71.

Sodano, V. and Hingley, M. (2013). The food system, climate change and CSR: from business to government case. British Food Journal 115: 75-91.

Waste and Resources Action Programme (WRAP) (2009). Household food and drink waste in the UK. Banbury, UK.

Waste and Resources Action Programme (WRAP) (2015). Food futures-from business as usual to business unusual. Banbury, UK.

Section 4. Sustainability

18. A sense of planetary urgency: an environmental index for an informed food choice

Y. Goossens[1], J. de Tavernier[2] and A. Geeraerd[1]*
[1]KU Leuven, Department of Biosystems (Division MeBioS) and Ethics@Arenberg, W. de Croylaan 42 – box 2428, 3001 Leuven, Belgium; [2]KU Leuven, Ethics@Arenberg, Sint-Michielsstraat 4 – box 3101, 3000 Leuven, Belgium; annemie.geeraerd@biw.kuleuven.be

Abstract

Sustainability can only be achieved if we live within the ecological carrying capacity of our planet. At this very moment, our consumption patterns are exhausting the environment by depleting more resources and by creating more waste than our planet can sustain. In 2009, a group of renowned scientists presented a planetary boundaries framework to define a safe operating space for humanity with respect to the earth system. The framework was subsequently updated in 2015 and is based on nine key earth system processes with boundary levels for each process (the biophysical limits) which should not be transgressed in order to maintain a low risk of destabilising the earth system. In parallel to these academic analyses, recent years have shown increased consumer concerns on the sustainability of their food. To increase consumer autonomy and enable them to make an informed food choice, consumers should be informed on how to better respect the biophysical limits of our planet through their consumption patterns. This paper therefore proposes communicating life cycle assessment (LCA) results for food products in an understandable way to the general public, by translating them via the use of the planetary boundaries framework. The resulting environmental index enables comparison of alike products, and can be applied to a wide range of food products, based on the 'planetary urgency' of the environmental impacts associated with those products or processes.

Keywords: informed choice, planetary boundaries, food consumption, life cycle assessment

Introduction

Although the concept of sustainable development has been on the agenda for a while now, it is often hard to translate to concrete actions at consumer level. Current consumption patterns are exhausting the environment by depleting more resources and by creating more waste than our planet can sustain. On the other hand, consumers are increasingly concerned about the sustainability of food consumption and, next to food safety and nutritional content of food, some consumers now also value food attributes such as origin of foods, the way it was produced (e.g. organic food), the conditions under which it was produced (e.g. fair trade) or the impact their food consumption has on the environment (Pretty *et al.*, 2005; Salaün and Flores, 2001). This article focusses on responding to consumer concerns related to the ecological burdens associated with food products, with the aim of empowering consumers in making an informed food choice.

Sustainability and respecting the biophysical limits of our planet

According to Moore and Rees (2013), sustainability requires respecting the ecological carrying capacity of our planet. In 2009, a group of renowned scientists presented a planetary boundaries (PB) framework to define a safe operating space for humanity with respect to the earth system, which was subsequently updated in 2015 (Rockström *et al.*, 2009; Steffen *et al.*, 2015). The framework is based on nine key earth system processes, with control variables and boundary levels for each process which should not be transgressed. These nine earth-system processes are: climate change, biosphere integrity, stratospheric

EurSafe 2016 – DOI 10.3921/978-90-8686-834-6_18, © Wageningen Academic Publishers 2016

ozone depletion, ocean acidification, biogeochemical flows (of nitrogen and phosphorous), land-system change, freshwater use, atmospheric aerosol loading and introduction of novel entities. The boundaries set by the authors relate to levels of anthropogenic perturbations below which the risk of destabilization of the earth system is likely to remain low: 'a safe operating space'. For each PB, a zone of uncertainty highlights the area of increasing risk of destabilisation of the earth system. Recent findings show that we have already crossed four boundaries: climate change, change in biosphere integrity, biogeochemical flows, and land-system change. Although the PB framework is a global framework and is not designed to be downscaled to smaller levels, the PB thinking approach may appeal to consumers concerned with living within the biophysical limits of our planet.

Empowering consumers

Informed food choice

Similar to the concept of 'informed consent' in medical ethics, autonomy for consumers refers to 'informed choice', to a choice about food consumption which is not made blindly but based on information (Coff, 2006, 2014). As a growing concern on environmental issues can be observed, also the demand for informed food choice in this particular field rises. Enabling consumers to make an informed food choice and make food purchasing decisions in line with their personal values and preferences, is directly linked to respecting consumer autonomy and to stimulating and building up consumer's capacities for autonomous food choices (De Tavernier, 2012).

Coff (2014) presents a list of major consumer concerns that can be considered as drivers for the demand for informed food choice, with environmental concerns being one of them. However, at this moment, there seems to be insufficient or inadequate environmental information at hand for consumers to base their informed choice on. Without knowledge on the potential environmental loads associated with food products, choices made by consumers cannot be considered as informed food choices. Transparency in the food chain and the provision of reliable information on those food aspects the consumers value, is thus needed to increase autonomy and empower consumers (Coff, 2014; De Tavernier, 2012). This is a process in which food producers and the supply chain have a crucial role to play as they can facilitate individual choices of consumers by labelling food products. It is however important to note that information is likely to be effective only when it addresses specific information needs and when it can be easily understood, processed and used by the consumers (Verbeke, 2008). In addition, too much and too detailed information could lead to an information overload which in turn could result in consumers losing confidence or becoming indifferent (Salaün and Flores, 2001; Verbeke, 2005).

Ecological burdens and human health impacts associated with food products

Environmental and human health impacts are generated throughout the entire food chain, at every stage from production of the raw material or food product, through distribution, and up until its final consumption or wastage. A well-recognised instrument to assess impacts connected with the complete life cycle of a product, is the life cycle assessment (LCA) approach, as developed under the International Organisation for Standardisation (Hauschild and Huijbregts, 2015).

Within an LCA, one of the first steps is compiling a life cycle inventory of all resources used and all emissions produced throughout the entire product chain. Next, the inventory data are multiplied by characterisation factors, resulting in indicators for the so called midpoint impact categories, which could then further be characterised to endpoint impacts (Brentrup *et al.*, 2004). Optionally, the LCA midpoint or endpoint impacts can be normalised to better understand the relative magnitude for each indicator result and to place the impact values in an adequate environmental context. To do so, the impacts are presented relative to reference impacts, such as the impact of one person living in Europe, which leads to a dimensionless value (Benini *et al.*, 2014; Bjørn and Hauschild, 2015; Brentrup *et al.*, 2004). Subsequently, to take into account the potential harm to the environment of each of the impacts, the dimensionless normalized results can be multiplied by weighting factors and aggregated into one single dimensionless environmental index (Brentrup *et al.*, 2004). Weighting factors within an LCA are often based on value choices or on political targets, and are thus not without its critics. Nevertheless, without weighting, the user of the impact assessment would anyhow weigh or choose between the different mid- or endpoint impacts on his own, while using a set of generic weighting factors would result in a more unbiased and objective aggregation of the impacts (Brentrup *et al.*, 2001). Similarly, aggregation of different indicator values into one composite index also has its limitations, as it may lead to simplistic conclusions or disguise serious failings in some dimensions (OECD, 2008). In academic circles, LCA results are usually expressed using midpoint impact category indicators, indicating for example the product's climate change, acidification or human toxicity impact. Information on how a product performs in each of these categories, would in theory enable consumers to make environmentally sound purchase decisions. However, it seems this information, and in particular its level of detail, is perceived as too complex to support sustainable purchasing decisions, and consumers face difficulties in interpreting the results (Leire and Thidell, 2005; Nissinen *et al.*, 2007). As LCA results can be expressed in over ten different impact categories, consumers may not know which impact category to prioritise, while this is particularly relevant when comparing alike products: should we choose the product with the lowest climate change impact or the one with the lowest toxicity impact? Additionally, in the absence of benchmarks or target impacts for each product (group), consumers cannot easily judge which results are worryingly high, or which results can be considered as acceptable, or even better, as sustainable. To facilitate communication to the general public, LCA results are therefore often aggregated into one single environmental index, using normalisation and weighting procedures as explained above. The key is thus to develop a single environmental index to guide consumers in choosing the most sustainable option. And this is where the planetary boundaries concept comes in.

Building a planetary boundaries-based index

We propose building a single environmental index based on LCA midpoint results and using the planetary boundaries framework proposed by Rockström *et al.* (2009) and Steffen *et al.* (2015). The first step in building the index, is a thorough analysis of what is behind each of the nine earth system processes and each of the LCA midpoint impact categories. Based on this, we link the PB framework with the LCA framework through selecting those LCA impact categories which represent the earth-system processes from Steffen *et al.* (2015) and the thresholds and boundaries set within each of them. In some cases, the linkages are rather obvious, for example in the case of the PB and the LCA impact category on climate change; while in other cases, the scope of an LCA impact category may be slightly different than what is covered by the PB, or vice versa. Furthermore, some PB earth-system processes could not be linked to LCA impact categories either because there was simply no match between the two frameworks, or either because in some cases, no control variable or no boundary value was yet identified within the PB framework and thus no weighting factors could be calculated (see further). An example is the novel entities earth system process for which no control variable was defined, and thus no linkages could be made with the LCA impact categories.

After identification of the appropriate LCA impact categories, the LCA results are normalised and weighted. Normalisation is done using the corresponding normalisation factors set by the LCA method. As weighting factors, we propose making use of the PB framework and more specifically, of the boundaries set for each control variable within the nine earth system processes and the current values of these control variables. The calculation of the weighting factor follows the distance-to-target approach whereby a factor above 1 means that the current value of the control variable has surpassed the boundary value. We are then no longer in a safe operating space, and have arrived in a zone of uncertainty with an increasing risk of destabilisation. The weighting factor thus indicates a sense of urgency as high weighting factors refer to high human perturbations beyond the safe boundary. Finally, by summing the weighted LCA impacts, we aggregate the weighted impacts into a single impact score or index, called the 'planetary boundaries-based impact'. The resulting index enables comparison of alike products, and can be applied to a wide range of food products, based on the 'planetary urgency' of the environmental and human health impacts associated with those products or processes.

Figure 1 illustrates the use of such an index for the comparison of 1 kg of broccoli, white cabbage and cauliflower (available in the supermarket). While the LCA impacts show that the identification of the best or worst performing vegetable varies between the impact categories, the aggregated PB-based index clearly shows that the highest impacts, as based on the 'planetary urgency', are observed for broccoli. The present study is not the first in its kind to link LCAs with the biophysical limits of our planet (Bjørn and Hauschild, 2015; Doka, 2015; Tuomisto *et al.*, 2012). However, our study further refines the existing methodologies and contributes to knowledge on communicating complex LCA results to the general public. We further note that the PB framework is not fully developed yet and for example does not consider toxicity or other ecosystem and human health related impacts. We are however hopeful that refinements to the PB concept in the future will introduce control variables that address these issues, for example in the novel entities boundary.

Conclusions

To empower consumers and enable them to make an informed food choice in line with their values, consumers shall be provided with adequate information on the ecological burden associated with the food chain of the food products for sale in our shops. Following the planetary boundaries framework from Steffen *et al.* (2015), which defines a safe operating space for humanity with respect to the earth system, this paper proposes to build an environmental index based on the 'planetary urgency' behind

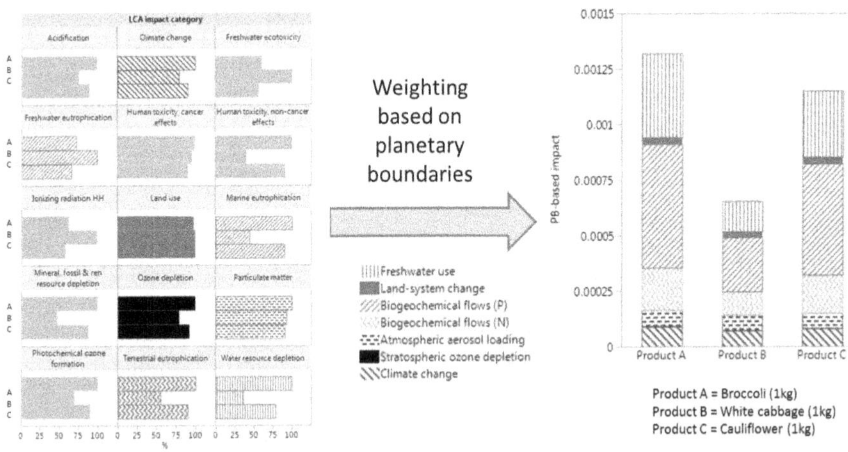

Figure 1. Visual representation of the conversion of relative life cycle assessment (LCA) results (based on ILCD impacts taken from the Ecoinvent database) into a planetary boundaries-based impact index. Reading key: As an example, the LCA impact categories for freshwater and marine eutrophication (as marked with '/////') contributed to the earth system process of biogeochemical phosphorous flows within the planetary boundaries (PB) framework and are thus part of the PB-based impact (also marked with '/////').

the environmental impacts associated with food products. We believe such an index can provide useful information to concerned consumers, in particular for making food purchasing choices for alike products, as illustrated for three vegetables.

References

Benini, L., Mancini, L., Sala, S., Schau, E., Manfredi, S. and Pant, R. (2014). Normalisation method and data for environmental footprints. JRC Technical reports. European Commission, JRC, Ispra, Italy.

Bjørn, A. and Hauschild, M.Z. (2015). Introducing carrying capacity-based normalisation in LCA: framework and development of references at midpoint level. International Journal of Life Cycle Assessment 20(7): 1005-1018.

Brentrup, F., Küsters, J., Kuhlmann, H. and Lammel, J. (2001). Application of the life cycle assessment methodology to agricultural production: an example of sugar beet production with different forms of nitrogen fertilisers. European Journal of Agronomy 14(3): 221-233.

Brentrup, F., Küsters, J., Kuhlmann, H. and Lammel, J. (2004). Environmental impact assessment of agricultural production systems using the life cycle assessment methodology. I. Theoretical concept of a LCA method tailored to crop production. European Journal of Agronomy 20(3): 247-264.

Coff, C. (2006). The taste for ethics. An ethic of food consumption. Springer, Dordrecht, the Netherlands.

Coff, C. (2014). Informed food choice. In: Thompson, P.B. and Kaplan, D.M. (eds.) Encyclopedia of food and agricultural ethics. Springer, Dordrecht, the Netherlands. pp. 1249-1255.

De Tavernier, J. (2012). Food citizenship: is there a duty for responsible consumption? Journal of Agricultural and Environmental Ethics 25(6): 895-907.

Doka, G. (2015). Combining life cycle inventory results with planetary boundaries: the planetary boundary allowance impact assessment method PBA'05. Doka LCA, Zurich, Switzerland.

Hauschild, M.Z. and Huijbregts, M.A.J. (2015). Introducing life cyle impact assessment. In: Hauschild, M.Z. and Huijbregts, M.A.J. (eds.) Life cycle impact assessment. LCA Compendium – The Complete World of Life Cycle Assessment. Springer, Dordrecht, the Netherlands, pp. 1-16.

Leire, C. and Thidell, Å. (2005). Product-related environmental information to guide consumer purchases – a review and analysis of research on perceptions, understanding and use among Nordic consumers. Journal of Cleaner Production 13(10-11): 1061-1070.

Moore, J. and Rees, W.E. (2013). Getting to one-planet living. In: Worldwatch Institute (ed.) State of the world 2013: is sustainability still possible? Island Press/Center for Resource Economics, USA, pp. 39-50.

Nissinen, A., Grönroos, J., Heiskanen, E., Honkanen, A., Katajajuuri, J.-M., Kurppa, S., Mäkinen, T., Mäenpää, I., Seppälä, J., Timonen, P., Usva, K., Virtanen, Y. and Voutilainen, P. (2007). Developing benchmarks for consumer-oriented life cycle assessment-based environmental information on products, services and consumption patterns. Journal of Cleaner Production 15(6): 538-549.

Organisation for Economic Co-operation and Development (OECD) (2008). Handbook on constructing composite indicators. Metodology and user guide. OECD and EU JRC.

Pretty, J.N., Ball, A.S., Lang, T. and Morison, J.I.L. (2005). Farm costs and food miles: an assessment of the full cost of the UK weekly food basket. Food Policy 30(1): 1-19.

Rockström, J., Steffen, W., Noone, K., Persson, Å., Chapin, F.S., Lambin, E., Lenton, T.M., Scheffer, M., Folke, C., Schellnhuber, H.J., Nykvist, B., De Wit, C.A., Hughes, T., Van der Leeuw, S., Rodhe, H., Sörlin, S., Snyder, P.K., Costanza, R., Svedin, U., Falkenmark, M., Karlberg, L., Corell, R.W., Fabry, V.J., Hansen, J., Walker, B., Liverman, D., Richardson, K., Crutzen, P. and Foley, J. (2009). Planetary boundaries: exploring the safe operating space for humanity. Ecology and Society 14(2): 472-475.

Salaün, Y. and Flores, K. (2001). Information quality: meeting the needs of the consumer. International Journal of Information Management 21: 21-37.

Steffen, W., Richardson, K., Rockström, J., Cornell, S.E., Fetzer, I., Bennett, E.M., Biggs, R., Carpenter, S.R., De Vries, W., De Wit, C.A., Folke, C., Gerten, D., Heinke, J., Mace, G.M., Persson, L.M., Ramanathan, V., Reyers, B. and Sörlin, S. (2015). Planetary boundaries: guiding human development on a changing planet. Science 347(6223): 12.

Tuomisto, H.L., Hodge, I.D., Riordan, P. and MacDonald, D.W. (2012). Exploring a safe operating approach to weighting in life cycle impact assessment – a case study of organic, conventional and integrated farming systems. Journal of Cleaner Production 37: 147-153.

Verbeke, W. (2005). Agriculture and the food industry in the information age. European Review of Agricultural Economics 32(3): 347-368.

Verbeke, W. (2008). Impact of communication on consumers' food choices. Proceedings of the Nutrition Society 67(3): 281-288.

19. Sustainable aquaculture governance: challenges to participatory standard setting

*S. Bremer** *and M. Kaiser*
Centre for the Study of the Sciences and the Humanities, University of Bergen; Allegaten 34, 5020 Bergen, Norway; scott.bremer@uib.no

Abstract

Global aquaculture governance is increasingly steered by frameworks of certification and standards, which act to create markets for aquaculture products by building trust through traceability and transparency (Hatanaka *et al.*, 2005). These notions of control over the whole production chain have been accompanied by calls for standards with a more holistic appreciation for the complexity and plurality characterising international aquaculture value-chains, including through more participatory standard setting. This paper reports on research into barriers to this participatory standard-setting, conducted under the auspices of the European Commission-funded SEAT Project: 'Sustaining Ethical Aquaculture Trade'. It presents discussions in a Delphi Process undertaken in 2013, which engaged an anonymous group of 12 European experts in extended peer review of the quality of sustainable aquaculture indicators put forward by stakeholders along the Bangladeshi shrimp value-chain in developing a so-called 'Ethical Aquaculture Food Index'. Rather than report on the appraisal of the indicators, this paper presents the participants' side reflections on six challenges to the participatory compilation of holistic aquaculture standards. First, a value-chain perspective can prove more holistic, but the danger is that broad indicators are not meaningful relative to different nodes in the chain. Second, a holistic view of the value chain, implicating the full diversity of stakeholders is hindered by an asymmetry of knowledge. Third, there are challenges to developing clear indicators that are both meaningful and measureable. Fourth, while respecting the need to be clear and measureable, indicators also need to reflect complexity, and a more nuanced perspective. Fifth, inevitably when considering different perspectives of sustainability, indicators will not always be complementary, such that high performance relative to one indicator may necessarily imply poor performance relative to another. Finally, it must be recognised that in negotiating the fast growing web of standards, this places a significant burden on poor aquaculture farmers in Asia. We ought not further expand this complex terrain of standards, but rather consolidate it.

Keywords: aquaculture, standard setting, participation

Introduction

The rapid development and globalisation of aquaculture over the past 40 years has been supported by a quickly expanding framework of more than 400 voluntary aquaculture standards, and auditing institutions. This emergence of a pervasive 'audit culture' in aquaculture echoes similar influences in other globalised food markets (Konefal and Hatanaka, 2011), and has acted to both shape the sustainable development of aquaculture – mitigating its social and environmental impacts – and create markets for aquaculture products by building trust through traceability and transparency (Hatanaka *et al.*, 2005). However, these standards have also drawn criticism, with some (see e.g. Bush *et al.*, 2013) arguing that standard setting is the closed domain of a small and powerful community of scientists, policy-makers and industry leaders, resulting in a narrow framing of sustainability as compartmentalised 'best practice'. Related critiques have been levelled at standards for imposing an 'objective and universal' framing of sustainability, which disregards the highly political negotiation of what constitutes sustainability, and the specificity of this negotiation to each aquaculture farming context (Konefal and Hatanaka, 2011; Lebel *et al.*, 2008).

EurSafe 2016 – DOI 10.3921/978-90-8686-834-6_19, © Wageningen Academic Publishers 2016

These critiques have motivated a number of initiatives to produce aquaculture standards that are more holistic in their scope, more participatory in their production, and more specific to sustainability challenges facing aquaculture in different countries. We see a growing set of country-specific standards, such as the Thai Quality Shrimp initiative or the Bangladesh Shrimp Seal of Quality, and efforts to include a diversity of stakeholders in their production, such as through the World Wide Fund for Nature's 'aquaculture dialogues' initiative back in 2004, or more localised efforts with Thai Shrimp Growers Associations (Lebel *et al.*, 2008). Indeed these compounding initiatives could suggest a future of co-constructed standards of increasing sophistication; able to reflect the multifaceted trade-offs specific to the particular complexities and challenges of sustainability of each aquaculture product. However, the reality is that there remain some challenges that frustrate attempts at implementing such participatory, holistic and context-sensitive standards or guidelines, as witnessed in the recent SEAT project.

This paper presents some findings from the four-year European Commission-funded Sustaining Ethical Aquaculture Trade research project (SEAT) (2009-2013), which sought to better understand the aquaculture value-chains sending farmed-fish products from Asia to Europe, in order to promote more sustainable and ethical trade. Here 'value chain' refers to the production and transport of aquaculture products from farm to fork, comprising different actors working at different 'nodes' from the hatchery, to the farm, the processing plant, and transport to retailers. In particular, the paper focuses on the results of a Delphi process (see next section), which assembled 12 European experts to appraise the quality of sustainable aquaculture indicators compiled in an 'Ethical Aquaculture Food Index' (EAFI), as put forward by stakeholders along the Bangladeshi shrimp value-chain. Rather than look at the EAFI itself, this paper concentrates on the reflections of the experts on the barriers to instituting an EAFI, as paradigmatic of other participatory and holistic standard setting initiatives.

This paper draws heavily on material written in a SEAT policy report titled: 'Deliverable 8.12: Delphi recommendations' (Bremer *et al.*, 2013). The following section begins by introducing the context and method for the Delphi Process, before the Delphi experts' reflections on six challenges to the participatory compilation of holistic aquaculture standards are presented in the section thereafter. The final section contains the conclusion.

The Ethical Aquaculture Food Index and the Delphi process

Central to the SEAT project was the development of an EAFI as a tool to support the governance of Asia-Europe value-chains, specific to four seafood species – pangasius, prawns, shrimp and tilapia – from the four Asian countries Bangladesh, China, Thailand and Vietnam. The EAFI was designed to have general applicability for different 'end-users' operating in different institutional spheres, acting equally as a resource for government regulators, as for industry actors or non-governmental organisations. As such, it was not itself intended as a certification scheme or industry standard, but aimed to be a reference point for the development of such documents.

The EAFI adopted a holistic and aggregated 'whole-of-value-chain' perspective that recognised the multifaceted nature of sustainability across environmental, social, economic and governance dimensions. It recognised that concepts of sustainable aquaculture mean different things to different stakeholders at different nodes of the value-chain, facing different challenges. The EAFI therefore aimed to be co-constructed, with indicators put forward both top-down by scientific experts in the SEAT consortium (ranging from chemistry, applied aquaculture science to economics and social sciences), and bottom-up by stakeholders all along the different value-chains, including those who are not often given a voice such as (among others) hatchery owners, farmers, processors, and day labourers. Moreover, these indicators were weighted by stakeholders in recognition that some sustainability measures are more important than others when negotiating what constitutes sustainable aquaculture.

The final phase of EAFI development saw a refined list or index of indicators presented for 'extended peer review' by a group of European experts in a Delphi process. The Delphi method was originally developed as a forecasting method according to an assumption that where forecasts are made in the context of significant uncertainty, they are more accurate when reached by a diverse group of experts rather than an individual (Dalkey and Helmer, 1963). The approach is iterative and anonymous, progressing in two or more rounds of questionnaires sent out individually to each of the experts. A Delphi seeks to (1) overcome inter-personal challenges associated with convening expert panels (particularly power asymmetries between individuals and/or disciplines) to enable a critical voice, and (2) either encourage consensus within a diverse group, or try to unpack the diversity of perspectives within that group (Custer *et al.*, 1999). To the former point, experts engaged in a Delphi process remain totally anonymous to each other, and communicate their answers remotely through a facilitator, usually via email. To the latter point, after each round the facilitator provides an anonymous summary of the experts' replies, as well as the reasons they provided for their judgments, and supplies this summary to all experts to support the following round of questions and answers. In this way, experts are encouraged to revise their subsequent answers in light of the replies of other members of their panel. During this process, the group either converges towards a consensus, or the divergences of perspectives are made explicit.

Today, the Delphi process has expanded beyond technological forecasting into public policy-making around health, education and the environment, but also having a pervasive influence in the scientific community (Adler and Ziglio, 1996). This has seen two important innovations in certain fields: first, a shift from forecasting to discussing current issues as a decision-support tool; and second, a shift from focusing on consensus to a focus on nurturing plurality. For instance, we now see 'Argument Delphi' processes used in bioethics, to map the diverse ethical considerations relative to a particular technology. SEAT followed this tradition in participatory practical ethics, by engaging an extended peer community of experts to consider current issues in aquaculture, appraise possible indicators to incorporate in an EAFI, and weight them by importance.

The SEAT Delphi process assembled a panel of 12 experts from across Europe. The experts came from diverse backgrounds, offering diverse perspectives on sustainable aquaculture trade, and ensuring a robust peer review process. Table 1 lists the panel experts according to anonymous descriptors. All but one expert continued their participation throughout the full duration of the Delphi process.

Table 1. Descriptors for the 12 Delphi experts.

1.	Research scientist of sustainable aquaculture production systems at a university
2.	Scientist of aquatic veterinary studies at a university
3.	Managing director of an independent aquaculture management consultant
4.	Manager within an aquaculture farm assurance and food labelling scheme
5.	Seafood operations manager at an international aquaculture certification company
6.	Managing director on an NGO certifying fisheries and aquaculture
7.	Aquaculture director in an NGO promoting corporate responsibility in industry
8.	Program manager in an NGO coordinating collaboration between industry, NGOs and government
9.	Fisheries and aquaculture manager for a major supermarket chain
10.	Purchasing manager for a large multinational corporation in food services
11.	Account manager for an association of sustainable restaurants
12.	A certified sustainable aquaculture producer in Europe

The Delphi process began in August 2013, and progressed in three iterations over the three following weeks, with Delphi experts presented with a subset of indicators each week to peer review. As part of their appraisal of the EAFI indicators, Delphi experts provided a rich side-commentary on six challenges facing any attempt to develop holistic sustainability indices for an aquaculture value-chain in a participatory way. There was no deliberate move to elicit commentary on challenges; rather the discussion naturally emerged as auxiliary to the critique of the indicators, and evolved as experts read the summaries of other panel members points.

Six challenges to holistic and participatory standard setting

Challenges with adopting a holistic value-chain perspective

For Delphi experts, the holistic value-chain perspective adopted by the EAFI offers both opportunities and challenges. On one hand, it provides a greater recognition of the links between the actors and institutions all along the value-chain, and emphasises that a sustainable value-chain cannot be measured according to a fragmented focus on a few nodes. On the other hand, some experts argued that a holistic perspective acts to dilute the effectiveness or meaningfulness of indicators. Indeed, imposing blanket indicators along the whole value-chain ignores the contingency of each node, and the unique role of actors in each node. The danger is that by instituting indicators that are measureable all along the value-chain, we create indicators that are not meaningful at any part of the chain. For example, some experts asserted that measurement of CO_2 emissions should be targeted to the nodes and actors that are more likely to have significant emissions; it would be impractical and uninteresting to measure the emissions of a small farmer in Bangladesh. Finally, some experts argued that certain indicators may in fact be immeasurable at an aggregated scale, or that the 'poor practice' of some farmers will hide the good practice of others. Many current standards focus on practice at the level of each individual farm because when we begin to take an aggregated view of practice in a cluster of farms, or along the value-chain, then we are forced to create an abstract construction of sustainability as the percentage of farmers practicing sustainably. Both good and bad practice can be hidden in these percentages.

Challenges associated with an asymmetry of knowledge and power to act

Some of the Delphi experts considered the EAFI a 'Gods-eye view' of the value-chain, including an array of indicators that no one actor could reasonably measure. They argued that any actor's knowledge is dependent on where on the value-chain they sit, and the resources that they can draw on: 'A retailer with access to considerable resources enables a deep and thorough appreciation of food safety and traceability (and could) consider a greater number of (indicators) than say a small and medium enterprise importer with more limited resources'. Instituting more holistic standards demands thorough knowledge of activities all along and between the nodes of a value chain, but given the complexity and length of these global chains, this is an unrealistic expectation.

In addition, some Delphi experts argued that the EAFI implicates too many different actors in the governance of the value-chain. For example, some experts considered that measuring progress toward sustainability in the value-chain should be limited to those changes that the industry itself can actually affect. From this perspective, measuring an indicator of local food security 'cannot be judged as it depends on external factors'. The aquaculture industry may bring with it a number of secondary impacts on those communities within which it is nested, but these impacts are also affected by a host of other causes independent from aquaculture, which introduce too many uncertainties to be measureable as part of a standard.

Challenges to crafting simple, practical and 'objective' indicators

The Delphi experts levelled criticism at some of the draft indicators as 'utopian ideals' that could not be given practical effect to. Some indicators were seen as too complex or opaque, leading one expert to recommend: 'make it SMART' – referring to the mnemonic that encourages indicators that are 'Specific, Measureable, Attainable, Relevant and Time-bound'. Related to this, others emphasised the importance of having a solid 'baseline' against which progress is measureable. Moreover, some experts argued that implementing these indicators would be prohibitively expensive and cumbersome. Finally, there was some criticism that many of the indicators were subjective or 'very social science', showing a clear preference for quantitative and 'objective' scientific measures.

Challenges with reflecting complexity and context in the value-chain

Some of the Delphi experts emphasised the complexity inherent to the social and ecological systems that comprise the value-chain, and the importance of their context. Many experts wrote about the need to look at the indicators in their particular context, rather than indiscriminately applying them to the whole value-chain. For example, when asking whether workers are free from discrimination, one expert wrote: 'This must be compared with the local norm. We would all probably aspire to equality, but should aquaculture be damned if it happens in an inequitable society?'

Relative to complexity, one expert argued that many of the indicators were in danger of oversimplifying what is in reality a complex situation, and in so doing, potentially confusing causality. For example, they asserted that survival rate was a poor indicator of animal welfare unless grounded in an appreciation for the mortalities that occur naturally, and is particular to seasons, localities and the duration of the production cycle: 'How on earth would you disentangle the impact of aquaculture from all the other dynamic processes affecting biodiversity?'

The challenge of mutually exclusive indicators

One expert raised the potential for internal contradiction within a framework of indicators, such that a high performance relative to one indicator may preclude high performance in another; making a 'perfect score' impossible. For instance, they highlighted the difference between one indicator looking at labour and production efficiency, and another looking at the number of people employed, noting: 'these might be seen as sending a contradictory message'. While inconsistencies may stand in the way of a 'perfect score', we can equally argue that this better reflects the reality that faces any decision-making for sustainability: that trade-offs are inevitable, and that the values and interests around them should be made explicit and discussed.

The dilemma of consolidation for sustainability

One expert concluded the process with a commentary on a dilemma they perceived in the current framework of sustainability certification. This is re-produced here because it presents a conflict to any measurement of sustainability: 'Ethical and food safety certifying agencies are likely to continue to play a major part in (aquaculture governance), with export requirements the primary driver. This mostly means the international certifiers (e.g. ASC, GAA, GlobalGAP, BRI, Friends of the SEA), although there are a few national alternatives (e.g. Thai quality Q). The considerable costs for (these certification schemes) falls upon producers and processors, which can be prohibitive and effectively excludes many smaller, extensive (low input/low impact) producers, who are perversely some of the most sustainable players, environmentally and socially. Reducing costs of such certification is then critical and *inter alia* this means two things: (1) limiting the range of aspects covered by certification to those more easily and cheaply assessed, and (2) promoting and supporting the harmonisation of all the different certification schemes.' Ironically, while standards were developed to inspire trust in an aquaculture product's quality and traceability, the competition between ever-increasing numbers of standards has actually affected distrust between standard setters and stakeholders (see Bush *et al.*, 2013).

Conclusions

This paper analysed the Delphi experts' discussion of the barriers to introducing an EAFI to draw broader lessons for implementing and institutionalising participatory and holistic aquaculture standards. While we may criticise current aquaculture standards for their simplification of sustainability according to a few quantitative indicators, the SEAT project found that there are at least six serious challenges to attempts at a more participatory and holistic appreciation of the whole value-chain. Such standard setting was seen to outstrip our knowledge of these complex chains, or the ability of science (and institutions) to measure the indicators it puts forward. Paradoxically, the EAFIs attempts at more holistic and context-sensitive standards were seen by many experts to detach its indicators from context; replacing an appreciation for each individual aquaculture farm with a more aggregated perspective of practice along the whole value-chain. Moreover, these experts pointed out that any attempts to tailor standards to each value-chain would contribute to their multiplication, which we have seen has shaken the trust of stakeholders in standards.

References

Adler, M. and Ziglio, E. (1996). Gazing into the oracle: the Delphi method and its application to social policy and public health. Kingsley Publishers, London, UK.

Bremer, S., Johansen, J.C., Haugen, A.S. and Kaiser, M. (2013). Deliverable 8.12 – Delphi recommendations: report on the SEAT Delphi process and outcomes. University of Bergen, Bergen, Norway.

Bush, S.R., Belton, B., Hall, D., Vandergeest, P., Murray, F.J. and Ponte, S. (2013). Certify sustainable aquaculture? Science 341: 1067-1068.

Custer, R.L., Scarcella, J.A. and Stewart, B.R. (1999). The modified Delphi technique – a rotational modification. Journal of Vocational and Technical Education 15(2): 50-58.

Dalkey, N. and Helmer, O. (1963). An experimental application of the Delphi method to the use of experts. Management Science 9(3): 458-467.

Hatanaka, M., Bain, C. and Busch, L. (2005). Third-party certification in the global agrifood system. Food Policy 30(3): 354-369.

Konefal, J. and Hatanaka, M. (2011). Enacting third-party certification: a case study of science and politics in organic shrimp certification. Journal of Rural Studies 27(2): 125-133.

Lebel, L., Lebel, P., Garden, P. and Giap, D.H. (2008). Places, chains and plates: governing transitions in the shrimp aquaculture production-consumption system. Globalizations 5(2): 211-226.

20. Unveiling the normative dimension of food sustainability: synergies and clashes of law

A. Bessa
Geneva Academy of International Humanitarian Law and Human Rights, Villa Moynier, Rue de Lausanne 120B, 1202 Geneva, Switzerland; adriana.bessa@geneva-academy.ch

Introduction

Perhaps the major challenge for humanity in the 21st century will be to find ways to feed the whole of the world's population and at the same time cope with the capacity of Earth to produce food for us all. There is wide consensus that this aim can only be achieved by a reorientation of global food policies and laws relating to it so as to take full account of the complex interactions among food production, environmental impacts, and social justice outcomes.

At international level, policy and law making relevant for food systems remain largely fragmented. While the United Nations Food and Agricultural Organization has adopted a number of documents aiming to foster global food security, international human rights mechanisms have concentrated on the obligations of States to respect, protect and promote the fundamental right of every human being to adequate and nutritious food. From the environmental perspective, international environmental agreements have established regulations governing nature resource exploitation as an attempt to strike the balance between nature carrying capacity and human needs.

This study is part of a six-year transdisciplinary research project which aims at contributing to food security-related academic discussions, as well as law and policy-making, through the analysis of the interrelations between coexisting food systems. It accounts for the outcomes and trade-offs in terms of the individual and aggregate contributions of food systems to social, economic, political and ecological effects. In this context, this paper will seek to discuss the manner in which international and policy oriented frameworks concerning the right to food, food security and environmental integrity may contribute or not to the sustainability of food systems, by shedding light on eventual synergies and clashes at stake.

In terms of organization, the paper will start with a discussion of the concept of right to food and food security and the ways its promotion has been pursued in the framework of international human rights agencies and treaty monitoring bodies and the Food and Agricultural Organization. Then, it will analyse how environmental protection and sustainability has been incorporated to this discourse. Finally, some considerations will be drawn.

Right to food and food security: the definition of global aims

Along the last decades there have been a number of attempts to provide a working definition of the human right to food and food security. In 1999 the Committee on Economic Social and Cultural Rights (CESCR) – the monitoring body of the International Covenant on Economic Social and Cultural Rights – adopted the General Comment No. 12 on the right to food and proposed the following definition: '[Right to food is] the right [of] every man, woman and child, alone or in community with others, [to have] physical and economic access at all times to adequate food or means for its procurement'(UN, 1999).

EurSafe 2016 – DOI 10.3921/978-90-8686-834-6_20, © Wageningen Academic Publishers 2016

The CESCR proposition has been elaborated upon by UN special rapporteurs on the right to food, who attempted to provide a comprehensive definition of the right to food that captured the different elements and dimensions (Konstantinov, 2011). The most recent definition was adopted by former special rapporteur Olivier de Schutter in his report to the Human Rights Council:

> The right to food is the right of every individual, alone or in community with others, to have physical and economic access at all times to sufficient, adequate and culturally acceptable food that is produced and consumed sustainably, preserving access to food for future generations. (De Schutter, 2014)

Clearly, the quantitative aspect of nutrition, that is to have access to 'sufficient' food that allows for the full development of the human capacities, is included in this new definition. In addition, this concept has incorporated two further dimensions to previous concepts: cultural and environmental. While acknowledging the importance of the former – human cultural diversity is also reflected in their culinary habits – this study will focus on how the environmental dimension of the right to food has been taken into account. In effect, the concept takes into account general concerns on the impact of food production and consumption patterns to the right to food, and suggests that environmentally unfriendly food production and consumption patterns might have a negative effect on the realization of the right to food. Likewise, while reaffirming the principle of intergenerational equity contained in the notion of sustainable development proposed by the United Nations in 1988, the concept highlights that global food production and consumption must account for the needs of future generations and their right to feed themselves.

At FAO level, the concept of food security has been defined as follows:

> Food security exists when all people, at all times, have physical and economic access to sufficient, safe and nutritious food to meet their dietary needs and food preferences for an active and healthy life. (WFS, 1996)

This notion was proposed in the Rome Declaration on World Food Summit and Plan of Action adopted by FAO Member States in 1996. While the concept itself largely replicated proposed notions to the right to food, the document further elaborates upon factors to be considered on the pursuit of global food security. This study will focus on two of them. The first factor corresponds to poverty alleviation. As the document emphasizes, 'Poverty eradication is essential to improve access to food. The vast majority of those who are undernourished, either cannot produce or cannot afford to buy enough food.' Here, the document proposes a crucial divide in global food security policies. One the one hand, measures to promote food security should aim at enhancing people's capacity to produce their own food. On the other hand, States should focus on increasing family incomes and their ability to buy food.

The second factor – environmental degradation – has been clearly indicated as one of the major contributors to global food insecurity. And in connection to this, the document has called for more sustainable management of natural resources and the elimination of unsustainable patterns of production and consumption. In this context, the declaration acknowledges the contribution to food security of small-scale food producers, in particular farmers, fishers, foresters, indigenous peoples and other rural communities by carrying out sustainable methods of food production. In the document, FAO Member States committed to promote their full participation in development policies and to foster the empowerment of these communities (WFS, 1996).

At this point, the question arises as to how to strike the balance between these two avenues for the promotion of global food security – poverty alleviation and environmental protection – which have been traditionally seen as opposing goals. The remaining of this study will discuss this apparent antagonism.

References

United Nations (UN) (1999). Committee on economic, social and cultural rights. General Comment 12: The right to adequate food (Art. 11). Adopted during the twentieth session. April 26- May 14 1999. Geneva, Switzerland.

Konstantinov, B. (2011). Invoking the right to food in the WTO dispute settlement process: the relevance of the right to food to the law of the WTO. In: De Schutter, O. and Cordes, K.Y. (eds.) Accounting for Hunger: The Right to Food in the Era of Globalisation. Bloomsbury Publishing, London, pp. 211-238.

De Schutter, O. (2014). Report of the special rapporteur on the right to food. Final report: the transformative potential of the right to food. UN Doc A/HRC/25/57, Geneva, Switzerland.

World Food Summit (WFS), 1996. Declaration on world food security and plan of action. World Food Summit. November 13-17, 1996. Rome, Italy.

21. Method in practical ethics: a call for a paradigm change

M. Kaiser
Centre for the Study of the Sciences and Humanities, University of Bergen, Bergen, Norway; matthias.kaiser@uib.no

Abstract

In practical ethics the dominant approach to chart ethical concerns in a specific field of practice is typically a variant of principlism. Going back to the Belmont Report of 1976 and the successful textbook by Beauchamp and Childress (2008) on biomedical ethics, the set of four dominant ethical principles seemed a very useful, if not exhaustive, tool to deal with ethical concerns in concrete decision making and policy. Inspired by this success, Ben Mepham (1996) suggested a slight extension of this method (including several stakeholders) to be used in agricultural policy and decision making: the ethical matrix. Some methodological discussions have ensued, also within EurSafe, whether autonomy, beneficence, non-maleficence, and justice (or variant formulations of these four) would be the most adequate set of practical ethics, or, as Mepham suggested, a reduced set of autonomy, welfare, and fairness (Mepham, 2005; Mepham *et al.*, 2006). This approach is basically founded on a belief in Common Morality Theory. Some slightly alternative variants have been pursued, as e.g. the 'Norwegian approach' (Forsberg, 2007a,b; Kaiser *et al.*, 2007; Kaiser and Forsberg, 2001) based on a radically participatory and bottom-up approach. Other contributors to this discussion include K. Millar, D. Schroeder, M. Cotton, C. Gamborg, V. Beekman, K. Jensen, T. Whiting, C. Croney, R. Anthony, J. Dietrich and others. This paper purports to challenge (1) the usability of various forms of principlism, including the ethical matrix, in culturally sensitive ethical issues, and (2) the foundations of principlism as expressed by Common Morality theory. It argues for a radical separation of a principle-based approach from a value-based approach on the basis of (some) empirical data. It is the latter that the author wants to argue for as a tool for assessing ethical landscapes pertaining to culturally sensitive questions. Value-based approaches as heuristics for ethical principles are, it is claimed, still in their infancy in ethics. As practical ethics is foremost seeking to ease our moral conduct with each other, also inter-culturally, empirical research will be crucial as justification for developing value-based tools in practical ethics.

Keywords: practical ethics, principlism, moral norms, values, moral foundations.

Background

Since the 1970's, ethics developed mainly along two paths: first a more theoretical path of general foundations of normative ethics – meta-ethics – as stimulated by the works of e.g. John Rawls, Robert Nozick, Martha Nussbaum, Hans Jonas, Karl-Otto Apel or Jürgen Habermas, and second the path of 'practical ethics' as exemplified in such fields as medical ethics, prominently promoted by the Hastings Centre, bioethics, as promoted by Peter Singer and others, and environmental ethics as promoted by Garrett Hardin, Arne Næss and others. Later this practical ethics movement diversified even more into a range of sub-fields, including feminist ethics, business ethics, geo ethics, research/science ethics, and now, nano-ethics. Practical ethics is always issue-driven and potentially instrumental. It provides the toolbox for ethicists doing ethical assessments.

Unlike meta-ethics or the more theoretical parts of normative ethics, which could thrive sheltered in the halls of academia, practical ethics was soon discovered as a promising tool to manage our complex social and technological realities. This alone bespeaks the fact that ethical and value considerations, if cast in instrumental mode, are – at least in principle – recognized as decision-important input. Again, it was medical ethics, which was first used as an instrument to address imminent practical needs and concrete

EurSafe 2016 – DOI 10.3921/978-90-8686-834-6_21, © Wageningen Academic Publishers 2016

dilemmas of our social and technological life, i.e. with societal impact. Beauchamp and Childress (2008) set a new and influential standard in 'Principles of Biomedical Ethics', resulting in the 'Four Principles'.

This was the start of an impressive development in the popularity of practical ethics. Better governance of the rapid development of science and technology brought with it the need to conduct ethical assessments in policy contexts. The development of the ethical matrix by Ben Mepham in the mid-1990s transformed the 'Georgetown Mantra' into a more volatile tool applied to a greater variety of cases (Mepham 1996, 2005), mostly within food and agricultural ethics but not exclusively so. The undersigned and co-workers have later developed the ethical matrix further (e.g. Forsberg, 2007a,b; Kaiser, 2005, 2009; Kaiser and Forsberg, 2001; Oughton *et al.*, 2004), and several of these articles are among the most cited articles of the undersigned author. This is not the only example of a framework which aspires to be able to give an ethical assessment of an issue which could work in a policy context or any other practical decision-making. In fisheries, the ecology-based 'Rapfish' tool (Pitcher *et al.*, 2013) has been propagated as an assessment tool to evaluate fisheries along six dimensions, including the ethical status of fisheries. Recently, it has been combined with the ethical matrix to assess these ethics of fisheries (Lam and Pitcher, 2012) and also aquaculture (Lam, 2013). The call for practical assessments of emerging technologies has not diminished, but rather, has increased, as arguably in the case of food security.

While it would be short-sighted to claim that empirical research can establish the ultimate (moral) validity of either the moral principles or the values (it cannot!), it can however clarify how commonalities and differences in ethical principles can be embedded within very different value landscapes. But this is the argument I want to spell out in the remainder.

Definitions

Values are here defined as (definition from the Value-isobars project; www.value-isobars.eu): 'Values are reference points for evaluating something as positive or negative. Values are rationally and emotionally binding and they give long-term orientation and motivation for action.' This conception of values obviously opens a path for empirical research, and places values somewhere between attitudes and preferences on the one side, and norms and principles on the other. (Moral) principles are defined (*ibid.*) as: 'Principles are normative statements that are meant to guide action without prescribing specific actions. All applications of principles need to be contextually embedded and interpreted'. In brief, there is no necessary logical connection between values and principles, while one may tentatively assume that values serve as heuristics to interpret principles for concrete issues and recommended actions. If, for instance, personal security is seen as a value, then this would help interpret the principle of beneficence in the sense that it would support actions which increase a person's sense of security.

The metaphor of a value landscape is meant to convey the insight that (a) values relate to each other in terms of proximity, i.e. there exist pairs of values such that they may differ in their relative distance (for clarity: if Δ signifies a distance operator between two elements), then: Δ (a,b) $\leq$ Δ (b,c); (b) different values may exhibit different intensity (defining peaks in the landscape): $\Phi(a)=\alpha$, i.e. the intensity being a function of the value in question; (c) each value acquires various meanings dependent on where it is perceived from (contextualism); (d) values have an inertia but are malleable over time, in particular in interaction with belief states.

It should be noted that one of the results of the Value Isobars project was that values come to life only in relation to concrete cases, and they may take on different shadings or connotations, depending on these cases. Speaking in terms of logic: values belong to an intensional logic, not an extensional one, which implies that the substitutivity of identicals does not go through for values (=referential opacity). Thus the study of values and principles needs to be linked to specific issues and problems.

Common Morality or Evolutionary Moral Foundations or Moral Relativism?

Practical ethics as developed today operates through ethical principles. Beauchamp and Childress, and also Mepham made much of the 'Middle-Kingdom' of ethical principles, while their justification remained unclear. Beauchamp and Childress tried more or less unsuccessfully to link them to Rawls' reflective equilibrium (2008; Beauchamp 1995). In later editions they clearly embrace the Common Morality view. Certainly, critics of this step abound (cf. e.g. Karlsen and Solbakk, 2011). Mepham clearly favours the common morality approach (Mepham, 2005). Turner (1998, 2003) has described the discussions about the foundations of principlism as competing 'schools of thought': on the one hand the common morality approach, and on the other hand the approach which takes it as a given that society is characterized by a multitude of competing values which are brought to the fore. The former lends itself to a form of universalism, while the latter school stresses cultural and intra-societal diversity which lends credit to the involved principles of practical ethics only to the extent that they can be empirically shown to be endorsed by relevant stakeholders. These conflicting views can also be described as top-down versus bottom-up approaches. The view that we have to face up to a great variety of moral viewpoints and values is often seen to lead to moral relativism. There is – at least in theory – nothing constraining radically different moral viewpoints.

In contrast, (Evolutionary) Moral Foundation Theory argues for a common foundation for human morality founded in evolutionary processes (cf. De Waal, 2009; Graham *et al.*, 2012; Tomasello and Vaish, 2013). Morality is here conceived as cooperation and pro-social attitudes among individuals. The argument is that evolutionary processes have brought about a number of propensities to act in certain ways, which are shared among humans, and also with other primates and higher mammals. De Waal and colleagues have developed for instance impressive research on attitudes of justice and empathy among higher mammals (cf. Brosnan and De Waal, 2003). Common Morality Theory has tried to establish a core set of shared attitudes or values, but this is very much in dispute still. What evolves, though, is the view that certain core propensities may define the evolutionary stable background for moral attitudes, while sociocultural contexts, beliefs and interactions define the surrounding moral landscape. In other words, this view would allow for a core identity of moral principles, and would embed them in culturally dependent framework of other moral principles and values. This is plurality within unity.

What interests me here is not so much the merits or shortcomings of the above theories, but rather the epistemological question of how we could know which of them is better. In other words, how are we to argue for or against any of these theories?

The justification of a moral framework

I do not believe that philosophy can offer any specific form of justification, which could not be said to apply – at least in principle – to the sciences as well (Kaiser, 2000). I furthermore believe that the epistemological classification of justifications are well captured by just three well-known principal forms of argument: the axiomatic, the infinite regress, and the circular/coherentist argument. This is also known as 'Münchhausen's trilemma' after a term introduced by Hans Albert in 1968 (though he had a forerunner in Sextus Empiricus). The question is how we can know that a statement is true? One can (1) provide a proof of the statement from other premises. But then we can continue and ask how we know that this proof is true? We have to stop this process at some point and accept the used proof and premises as axiomatic. We chose a 'dogmatic' justification. We have deduced our statement from higher principles we feel we have to accept. Another alternative is (2) to justify the statement with ever more problematic assertions, i.e. assertions we immediately see need further justifications: 'how do we know these things then?' We can continue forever and end up in an infinite regress. This is a typical problem of empiricism: if we need to refer to our observations, how do we know they are true, and so on? We end up in infinite regress, unless we stop dogmatically and assign certainty to these statements. The third alternative (3) is a circular argument, also referred to as holism or coherentism. We move from one underlying justification to another until we eventually come back to the starting point, which then indirectly justifies itself through the web of other beliefs which it relates to. Willard van Orman Quine favoured this latter approach, as did Pierre Duhem and Otto Neurath. – My basic point here is that we are always caught somewhere in this Münchhausen trilemma, if we are looking for a fundamental justification of same theoretical framework.

Let us approach this issue a bit more pragmatically, and use the moral principles as the starting point for our inquiry. How could I argue that these four principles (beneficence, non-maleficence, autonomy, justice) are indeed core principles for all moral behaviour and attitudes across human cultures? We are caught in this trilemma, we face three principal ways to justify this claim.

First, we could argue that these principles are the very essence of all over-arching valid ethical theories people recognize. We could try to deduce the principles from these theories, or if we could not 'deduce' them, then at least argue that they are convenient short-hands for the content of these theories. The problem is that we need to assume we know which theories are valid/true. We do not, not even in Western philosophical traditions, as one can easily make the point that virtue ethics is nowhere to be found among these principles. If we move east towards other cultures we will face even bigger difficulties. There is no general consensus on which of all the ethical theories is the valid one, or which of the sets of ethical theories cover all valid moral judgments. If we want to follow this line of argument we have to end up with dogmatism – which, of course was the preferred framework of colonialism.

Second, we could try to argue that these principles capture the majority of ethical intuitions, which people bring to the fore in their ethical decision-making. The obvious problem is that this is certainly not empirically true, otherwise we would not need any practical ethics in the first place. We could try to sort the intuitions of people in those that are right and those that are misguided – i.e. wrong. But that is begging the question. How do we know they are wrong? We end up in an infinite regress, unless we stop by dogmatism.

Third, we could try to move up and down between principles and higher theoretical axioms and lower individual intuitions. This roughly is the idea behind Rawls' reflexive equilibrium. Sometimes we need to adjust our concrete intuitions since our theorems can tell us they are remnants of prejudices and bad traditions. Sometimes we need to adjust our theorems, though, particularly if there are very strong intuitions, which we do not want to give up, and then we need to make adjustments in our theoretical superstructure in order to avoid cognitive dissonance. This is obviously a circular procedure, but the claim goes that if the circle is wide enough then the sheer amount of interconnectedness provides a good argument to believe in the truth of the statement. This then amounts to a holistic epistemology for practical ethics.

The scientific approach to ethics

The above was a brief excursion into traditional epistemology. It showed three ways to argue for a certain claim, but it does not in itself explain which of the three one should prefer in a certain situation. Arguments can be made that even scientific scholarship cannot be uniquely classified, as e.g. the foundation of mathematics may be basically holistic (starting from number theory, logically moving up and down) while neo-classic economy has withdrawn into the dogmatic corner. I would still maintain that most empirical science – to the extent that it aspires to adequately represent reality – rests on a bottom-up justification. In other words, empirical data are – in principle that is – epistemologically prior to theoretical statements, laws or theorems. The data are to justify the theory, not the other way around.

Philosophy and ethics cannot claim to be exceptions from this, I believe. A philosophy or ethics which claims to have factual insights (e.g. about the basic moral principles) and which does not stand the test of reality, is not worth its name. I am well aware about the so-called naturalistic fallacy (we cannot deduce ought from is), but before we even start to discuss normative validity we should have a good grip on the moral views people actually hold or do not hold. And certainly in practical ethics we cannot operate with a framework of principles which lacks justificationary underpinnings. Then we slide into the undemocratic practice of letting the elite decide what is the moral right.

My claim is that as long as no ethical theory is established as undoubtedly superior, we need to recognize the epistemological priority of empirical data on moral views of people, inter- and cross-culturally. This then necessitates empirical research when talking about ethics. This may then also provide the basis for a rational discourse about which moral principles we want to endorse as valid in our global interactions.

The ethical landscape

Let me summarize some of the points I have been trying to establish by theoretical reasoning:

1. There is empirical evidence, which indicates that at least some moral principles, attitudes and propensities are founded in evolutionary processes.
2. So far, no evidence exists which shows that the principles deployed by principlism are universally shared. The evidence that does exist shows that they are not, or that they are only shared differentially so (cf. Bremer *et al.*, 2013, 2014; Kaiser *et al.*, 2007; Page, 2012). Autonomy is clearly a problematic example.
3. Thus, a picture emerges with a core of universal moral principles, supplemented by a layer of socio-culturally founded moral principles.
4. The socio-culturally moral principles are typically justified by reference to a landscape of basic values.
5. Empirical studies show that people apparently do hold a variety of values. These values can then be used to underpin moral principles, given one has good data on these values.
6. If we want to make progress in practical ethics moving it closer to global relevance as ethical assessment tools as support for policies and decision-making, then we need to provide the empirical research which can justify the crucial elements (principles) in our ethical toolbox.
7. The empirical research designs should – among other things – be bridging values with principles in specific decision-contexts.

Conclusions

In this brief text I have been arguing that practical ethics needs a clear and radical move towards empirical research, inspired by empirical science in general. I have been criticizing both the 'Georgetown Mantra' and the ethical matrix approach for being in effect dogmatist (and in effect then possibly being culturally 'imperialistic') when applied to culturally sensitive issues or populations. A turn to sober gatherings of data and methodological inquiries in communities is needed to turn practical ethical on its feet. Presently, practical ethics is still discussed with too many nice words and too little empirical data. The important issue should therefore be first to develop one or several methodological approaches how to get good data and insights into the nature of people's values. Only when good empirical insights are available can we start discussing which moral principles we might want to establish globally. Practical ethics is in dire need of empirical respectability with regard to moral values of people. As practical ethics is seeking to ease our moral conduct with each other, empirical research will be crucial as justification for developing value-based tools in practical ethics.

References

Albert, H. (1968). Dogmatismus – unendlicher Regreß – Psychologismus. In: Traktat über kritische Vernunft, Tübingen, Germany.

Beauchamp, T. and Childress, J. (1977). Principles of biomedical ethics. Oxford University Press, Oxford, UK.

Beauchamp, T. (1995). Principlism and its alleged competitors. Kennedy Institute of Ethics Journal 5: 181-198.

Bremer, S., Johansen, J., Øyen, S., Kaiser, M. and Haugen, A.S. (2014). Mapping the ethical terrain of Chinese aquaculture. In: Brautaset, C. and Dent, C.M. (eds.) The great diversity: trajectories of Asian development. Wageningen Academic Publishers, Wageningen, the Netherlands.

Bremer, S., Haugen, A.S. and Kaiser, M. (2013). Whose sustainability counts? Engaging with debates on the sustainability of Bangladeshi shrimp? In: Röcklinsberg, H. and Sandin, P. (eds.) The ethics of consumption: the citizen, the market and the law. Wageningen Academic Publishers, Wageningen, the Netherlands.

Brosnan, S.F. and De Waal, F.B. (2003). Monkeys reject unequal pay. Nature 425(6955): 297-299.

De Waal, F. (2009). Primates and philosophers: how morality evolved. Princeton University Press, Princeton, NJ, USA.

Forsberg, E.M. (2007a). Value pluralism and coherentist justification of ethical advice. Journal of Agricultural and Environmental Ethics 20(1): 81-97.

Forsberg, E.M. (2007b). Pluralism, the ethical matrix, and coming to conclusions. Journal of Agricultural and Environmental Ethics 20(5): 455-468.

Graham, J., Haidt, J., Koleva, S., Motyl, M., Iyer, R., Wojcik, S.P. and Ditto, P.H. (2012). Moral foundations theory: the pragmatic validity of moral pluralism. Available at: http://tinyurl.com/jmup856.

Kahane, G. (2013). The armchair and the trolley: an argument for experimental ethics. Philosophical Studies 162: 421-445.

Kaiser, M. (2009). Ethical aspects of livestock genetic engineering. In: Engelhard, M., Hagen, K. and Boysen, M. (eds.) Genetic engineering in livestock – new applications and interdisciplinary perspectives. Springer-Verlag, Berlin, Germany.

Kaiser, M. (2005). Assessing ethics and animal welfare in animal biotechnology for farm production. Scientific and Technical Review of the World Organisation for Animal Health 24(1): 75-87.

Kaiser, M. (2000). Hva er vitenskap? Universitetsforlaget, Oslo, Norway.

Kaiser, M. and Forsberg, E.-M. (2001) Assessing fisheries – using an ethical matrix in a participatory process. Journal of Agricultural and Environmental Ethics 14: 191-200.

Kaiser, M., Millar, K., Thorstensen, E. and Tomkins, S. (2007). Developing the ethical matrix as a decision support framework: GM fish as a case study. Journal of Agricultural and Environmental Ethics 20: 65-80.

Karlsen, J.R. and Solbakk, J.H. (2011). A waste of time: the problem of common morality in principles of biomedical ethics. Journal of Medical Ethics 37: 588-591.

Lam, M.E. (2013). Comparing the ethics of capture fisheries and aquaculture. The Ethics of Consumption: The Citizen, The Market, and The Law. Proceedings from the 11th EurSafe Congress. Wageningen Academic Publishers, Wageningen, the Netherlands, pp. 305-312.

Lam, M.E. and Pitcher, T.J. (2012). The ethical dimensions of fisheries. Current Opinion in Environmental Sustainability 4: 364-373.

Mepham, T.B. (1996). Ethical analysis of food biotechnologies: an evaluative framework. In: Mepham, T.B. (ed.) Food Ethics. Routledge, London, UK, pp. 101-119.

Mepham, B. (2005). Bioethics: an introduction for the biosciences. Oxford University Press, Oxford, UK.

Oughton, D., Forsberg, E.-M., Bay, I., Kaiser, M. and Howard, B. (2004). An ethical dimension to sustainable restoration and long-term management of contaminated areas. Journal of Environmental Radioactivity 74(1-3): 171-183.

Page, K. (2012). The four principles: can they be measured and can they predict ethical decision-making? BMC Medical Ethics 13: 10.

Tomasello, M. and Vaish, A. (2013). Origns of human cooperation and morality. Annual Reviews in Psychology 64: 231-255.

Turner, L. (1998). An anthropological exploration of contemporary bioethics: the varieties of common sense. Journal of Medical Ethics 24: 127-133.

Turner, L. (2003). Zones of consensus and zones of conflict: questioning the common morality. Presumption in bioethics. Kennedy Institute of Ethics Journal 13: 193-218.

22. Fair water in a changing climate

S. Meisch
International Centre for Ethics in the Sciences and Humanities (University of Tuebingen), Junior Research Group: 'Ethics of Science in the Research for Sustainable Development', Wilhelmstr. 19, 72074 Tuebingen, Germany; simon.meisch@uni-tuebingen.de

Abstract

Climate change will heavily impact on water and food and aggravate existing inequalities. These inequalities result importantly, but not exclusively, from actual physical shortages of water and food, but not alone. Quite often, they are also the result of social conditions. The UN Sustainable Development Goals (SDGs) repeatedly address the issue of water or food such as SDG 2, 6 or 14. This paper deals with normative standards for a fair distribution of water. By doing so, it is critical and constructive contribution to the debate about water invigorated by the SDGs. It aims to identify potential injustices (critical) and argues for just solutions in the face of changing environmental conditions (constructive). The paper starts by outlining that the aim of sustainable development is about safeguarding the right to live in dignity for all present and future generations. Moreover, it obligates that the natural and social preconditions for such a life are to be protected and supported. Yet, the difficulty in protecting a life of dignity lies in defining it by way of universalistic ethical principles without ignoring the diversity of particular ways of living it. This is why this paper, secondly, draws on the deontological approach by the social ethicist Alan Gewirth in order to determine what people need to live a life of dignity. Finally, the paper applies these insights to water governance.

Keywords: water, ethics, Alan Gewirth, sustainable development goals, Paris agreement

Pope Francis, Paris and the sustainable development goals – So, everything settled then?

In 2015, the discourse on sustainable development witnessed a remarkable year. In June, Pope Francis published the encyclical *Laudato si'* criticising non-sustainable development including environmental degradation, anthropogenic climate change, global hunger and consumerism. He argued that 'to protect our common home includes a concern to bring the whole human family together to seek a sustainable and integral development, for we know that things can change' (Francis, 2015). In Radio Vatican, climate scientist Hans Joachim Schellnhuber, who was part of the team presenting the encyclical, stressed that it reflects the state of the art in science, combines religion and science in a consistent way and even offers a new ecology in a poetical language (Radio Vatican, 2015). Besides its inner-church functions, *Laudato si'* was meant to influence two major global political events that took place in late 2015: the UN Sustainable Development Summit (New York, Sep 25-27) and the Paris climate conference (COP21) (Paris, Nov 30-Dec 11). The UN Summit agreed on Sustainable Development Goals (SDG) consisting of 17 goals and 169 targets. Together with financial and governance mechanisms, the outcome document defines a global development agenda which recognises 'that eradicating poverty in all its forms and dimensions, including extreme poverty, is the greatest global challenge and an indispensable requirement for sustainable development' (UN, 2015a). COP21, the second global sustainability event of 2015, was hailed for agreeing on a globally binding climate treaty aimed at 'holding the increase in the global average temperature to well below 2 °C above pre-industrial levels and pursuing efforts to limit the temperature increase to 1.5 °C above pre-industrial levels', at reaching 'global peaking of greenhouse gas emissions as soon as possible' and at undertaking 'rapid reductions thereafter in accordance with best available science' (UN, 2015b). Both UN outcome documents refer to each other. Each believes that its

EurSafe 2016 – DOI 10.3921/978-90-8686-834-6_22, © Wageningen Academic Publishers 2016

respective success depends on the success of the other. A recent report from the scientific community shares this belief (ICSU and ISSC, 2015).

Though processes were (with some reason) celebrated as a breakthrough in global sustainability politics, one might ask: Was that it? Scientists get the facts straight; the Pope gets the values straight; the United Nations get political targets and policies straight. Now, science, technology, and economy go to work and end hunger and poverty in the next 15 years. Such a scenario is simplistic. On the one hand, we might welcome that the global political community agreed on sustainability goals. On the other hand, however, the process raises more issues than it settles. For sure, with regard to social ethics questions remain. The fact that global leaders agree does not make the result morally legitimate (O'Neill, 2009: 221-223). Moreover, the goals as well as the means to achieve them might be questioned.

An ethical approach to do so is the concept of sustainable development – still enjoying great acceptance but becoming ever more contested for its alignment with neoliberalism and techno-science. I will show why it is necessary to address these criticisms if we aim to build our normative arguments on sustainable development and address the questions left open (or opened up) by the remarkable international consensus on global sustainability. In doing so, I will refer to Alan Gewirth's social ethics as a promising approach to spell out rights and duties resulting from the concept of sustainable development. With this, we can use sustainable development with the critical intent to analyse existing policies or with constructive intent to argue for a specific sustainable option. Using the example of SDG 6 ('Ensure access to water and sanitation for all'), I will sketch what the implications of such an approach would be. I do not argue for a specific moral option but call for critical and constructive ethical engagement with a global process that pretends to have settled normative issues.

Sustainable development as neoliberalism and techno-science

Sustainable development enjoys wide acceptance. Yet, some say, it earned its acceptance by becoming conceptually meaningless. I do not agree with this line of critique (cf. e.g. Voget-Kleschin and Meisch, 2015). On the contrary, in the following I address criticism claiming that sustainable development acquired a particular meaning but the wrong one – or as Reid (2013: 354) put it: 'Sustainable development and neoliberalism are not the same, nor is the former simply a proxy of the latter, but they do come into contact powerfully on the terrains of their rationalities of security.'

Leading global environmental scholars keep reminding us that the future is bleak, and we have to act now. Against this background, global scientific and economic elites joined forces and demand solution-oriented science helping to avoid trespassing planetary boundaries and safeguarding 'a safe operating space for humanity' (Rockström *et al.*, 2009; for a critical view, cf. Lövbrand *et al.*, 2015). This approach is criticised for reasons that will not be comprehensively discussed here: Altogether, it is seen as yet another narrative claiming control and mastery over nature (cf. Benessia *et al.*, 2012; Lövbrand *et al.*, 2015; Reid, 2013; Swyngedouw, 2010). Scholars from critical social sciences and humanities argued that political and scientific approaches to complexity (Chandler, 2014; Saltelli and Giampietro, 2016) and resilience (Reid, 2013; Walker and Cooper, 2011) rendered environmental governance into an issue of neoliberal and techno-scientific practices. This paradigm eventually produces post-politicisation of sustainability politics (Beck, 2010).

We are told that in order to prevent the looming apocalypse the biosphere has to be made resilient against imminent shocks from climate change, population growth, etc. A particular focus rests on the world's poorest as most vulnerable. Apparently, this aim is achieved best by building a green economy (e.g. Folke *et al.*, 2002; Rockström *et al.*, 2014): The biosphere is turned into a provider of services on which humanity vitally depends, in particular for agriculture and food production. Yet, biosphere resilience is threatened by human ecological ignorance unable to account for its dependence on ecosystem services. If these services had a real price, so we are told, we would value and protect them adequately.

This paradigm commodifies our natural livelihood and turns its protection into a field of business opportunities (Folke *et al.*, 2002; for a critical view, cf. Reid, 2013). In the face of global environmental threats, biosphere resilience, and poverty reduction are the common sense in sustainability politics that is turned into management in need to find the right technocratic means (Lövbrand *et al.*, 2015; Swyngedouw, 2010). States are seen as inefficient, corrupt and unable to deal with the complexities inherent in global environmental problems. Therefore, they should refrain from intervening in society and leave the production of progress and general welfare to the people and business (Chandler, 2014). The resulting subjectivity of this paradigm is the 'resilient subject' that in the face of the apocalypse constantly feels individually responsible while struggling to adapt to unstable (natural and social) environments. It gives up political ambitions to politically shape the future while preparing to catastrophic external shocks and acquiring services on green markets (Beck, 2010; Reid, 2013). Congruously, in spite of the magnitude of expected global change, political action seems perfectly possible within the political status quo. Evidently existing injustices are ignored or settled by implementing better governance practices and techniques (Lövbrand *et al.*, 2015; Swyngedouw, 2010).

Solution-oriented sustainability research fits into this paradigm. While rhetorically embracing complexity, it rests on the 'modern framing of sustainability, essentially founded on three kinds of belief: techno-scientific control (of the future), power (to cure), and evidence (of data)' (Benessia *et al.*, 2011: 87). As such, it provides standardising one-size-fits-all solutions that cannot account for moral and epistemological plurality (Benessia *et al.*, 2011; Meisch, 2015a). While stressing the significance of social innovations with regard to environmental challenges, mainstream sustainability sciences stay conspicuously silent on meaningful social reform and instead aim to produce marketable solutions within the existing status quo of a more or less green economy (Lövbrand *et al.*, 2015; Saltelli and Giampietro, 2016). Water governance is by no means different (for critiques of hegemonic approaches to water governance, cf. e.g. Leese and Meisch, 2015; Meisch, 2016; Munck, 2015).

Sustainable Development Goals and the neoliberal, techno-scientific sermon

In 2015, Johan Rockström, director of the Stockholm Resilience Centre, and Peter Bakker, president of the World Business Council for Sustainable Development, clearly have tidings of joy (Rockström and Bakker, 2015):

> Did you hear the news? The world has a plan ... This plan is the Sustainable Development Goals (SDGs).' 'Governments and, financial institutions and businesses must all plan to align their strategies and investments with the plan.' And they must do so now because time's running out: 'We are the first generation who can eradicate poverty and ensure abundance for all. We are also the last generation with the chance to turn humanity away from the destructive path that is leading to a planetary crisis. Let ours be the generation to bring harmony to both people and planet. Let's all seize the full opportunity presented by these 17 SDGs, and create our common future through a journey of sustainable prosperity for all. We have a plan.

As this shows, Rockström and Bakker are quite enthusiastic about the SDGs as a package that will deliver prosperity to all. Instead of discussing the SDG as a package, I here focus on SDG 6: 'Ensure access to water and sanitation for all.' This goal is subdivided into seven targets addressing access to freshwater, sanitation and hygiene for all, water safety and security, efficiency and management, protection of water-related ecosystems and social organisation, i.e. international cooperation and local participation.

'Transforming our world,' the UN report introducing the SDGs, mentions water many times stressing its importance for human wellbeing and the ecosphere and refers to legal issues, e.g. the human right to water.

ICSU and ISSC provide a scientific perspective on the SDGs offering 'rigorous analysis of the proposed goals and targets, collectively and individually, assessing whether they are backed up by evidence, whether they address the economic, social and environmental dimensions of sustainable development in an integrated way, and whether they are sufficiently specific to be effectively implemented and monitored.' (ICSU and ISSC, 2015) Two of many possible queries to this approach shall be mentioned here. First, science is confident that it can produce the knowledge needed for the implementation of the SDGs. Apparently, complexities and uncertainties can be managed by the right technologies and sets of data. Second, though the review implies that it is only about technical issues of implementation, it explicitly raises deeply ethical questions, e.g. about the relationship between universal claims ('elimination of poverty in all its forms everywhere', *ibid.*: 8) and local contexts or about the need for an overarching goal (e.g. 'a prosperous, high quality of life that is equitably shared and sustainable') in order to settle conflicts between SDGs (*ibid.*: 9). The assessment of SDG 6 in the ICSU/ISSC report does not seize on these thoughts. It falls back in the narrative of techno-scientific control of water equals promoting sustainability. This confirms Munck's (2015: 15) picture of the water sector 'with its abiding belief in science, technology and engineering as drivers of progress. One need only add the right mix of private and state investment and generate the right attitude among population for success to be achieved.' The scientific perspective does not discuss whether technical means can provide fair distribution of water; it seems to assume that they do so. Characteristically, the authors suggest moving the only two targets concerned with social organisation to other SDGs, thus draining the water SDG completely of its social context. Ensuring access to water and sanitation for all would be reduced to a technocratic exercise, thus completely clouding the sight that water is a moral, cultural and political issue (Meisch, 2015a).

Sustainable development as a theory of justice

The criticism of sustainable development mentioned above is a thorn in the side of all those believing that the concept is a useful guide for action to a more just future. As there is not one uniform sustainability discourse, we might caution against homogenising discourses. While this caution is well-founded, it is only a very first step because we still need to take the critical perspectives mentioned above seriously and elaborate an alternative conception of sustainable development (Sneddon *et al.*, 2006: 260). In a way, this is more than just reviving the initial emancipatory potential of sustainable development; at least conceptually, it might as well be its humanist reinvention (Beck, 2010; Benessia *et al.*, 2011). A different conception of sustainable development does not guarantee change in actions, institutions, and power but it provides an ethically justified basis of a normative argument that can inspire change.

First of all, we have to learn that we do not live in end times (as we are often told), but we need to conceive the future as open and thus complex. We can politically deliberate on what kinds of environments we wish to live in (Lövbrand *et al.*, 2015; Chandler, 2014). Meanwhile, governance needs to be conceptualised in a way that it does not, through the back door, promote neoliberal practices and the resilient subject as their subjectivity. Alternatively, social ethics can inspire debates on a just political order allowing humans the freedom and responsibility to cope with emergent complexities and simultaneously providing help and solidarity. Sustainable development can be this inspiring socio-ethical approach. In this paper, it is conceived as the right of all humans today and in the future to live a good and succeeding life. For that, they ought to have the natural and social preconditions to do so. This approach builds on a wide anthropology, accepting that humans also depend on non-human nature and have more than instrumental relationships to their natural environment (e.g. also aesthetic or all kinds of emotional attachments). Specifying what people need for such a life, requires a universal ethical approach allowing for different notions of a good succeeding life. In order to constitute moral judgments and spell out what we ought to do in specific social contexts, we need a theory of justice (Voget-Kleschin and Meisch, 2015). In this respect, Alan Gewirth's social philosophy is a promising approach.

Alan Gewirth' social ethics

Gewirth provides valuable answers to the challenges of sustainability politics mentioned above. I cannot reconstruct his philosophical approach in detail here (cf. Gewirth, 1978; Steigleder, 1999). But two crucial aspects of his work suffice in our context. First, he argues for a supreme moral principle, the Principle of Generic Consistency (PGC) based on which moral rights and corresponding duties with regard to individual and collective action can be justified. Second, Gewirth employs the PGC as a criterion to settle conflicts between duties.

Firstly, in order to morally justify action, Gewirth introduces the PGC as a supreme moral principle (Gewirth, 1978: 129-199): 'Act in accord with the generic rights of your recipients as well as of yourself' (*ibid.*: 135). The principle is built on the understanding of what it means to be a prospective agent. Gewirth focuses on human agents, what they need for action and their reciprocal rights and duties. Generic rights are rights to the necessary conditions an agent needs for purposeful action (*ibid.*: 64). These rights are freedom and well-being. The right of freedom consists of non-coercion and the ability to act according to one's own choice, while the right of well-being encompasses those general abilities and conditions that are necessary for an agent to reach the purpose of his/her action (*ibid.*). Both rights are to be respected, protected and supported by other agents and by (political) institutions. Thus, the PGC legitimises protective and supportive state action. Gewirth distinguishes direct and indirect applications of the PGC (*ibid.*: 200). The direct application refers to individual action. An agent acts freely without interference by others (negative rights) or legitimately claims those goods required to realize prospective agency (positive rights). It implies that an agent fulfils his/her obligations with regard to other agents' generic rights. The indirect application refers to collective institutions insofar as they protect the negative rights of individual agents and promote the positive rights for developing basic capabilities and basic goods. With this, Gewirth distinguishes a static and dynamic justification of social rules. The first focuses on the protection and restoration of every person's equal possession of generic rights (minimal state). The second acknowledges disproportional inequality in people's ability to exercise their generic rights and thus aims to remove this inequality (welfare state) (*ibid.*: 292).

Secondly, based on the PGC, Gewirth offers three criteria to settle conflicts of duties (*ibid.*: 338-354). These are (a) prevention or removal of inconsistency, (b) degrees of necessity for action, and (c) institutional requirement: (a) demands that actions conform to the PGC, (b) ranks goods with regards to an agent's requirements to act, and (c) deals with the question to what degree indirect applications modify direct ones (cf. also Steigleder, 1999: 179):

> ... it is important to keep in mind that the PGC's standard or central requirement is the equality of generic rights and hence mutual respect for freedom and well-being among prospective agents. Departures from this mutual respect are justified only where they are required either to prevent or rectify antecedent departures, or avoid greater departures, or to comply with social rules that themselves reflect such respect in the ways indicated above. (Gewirth, 1978: 345)

With this, it becomes clear why the PGC can settle conflicts of duties and in what way it can serve as a guide for a critique of the status quo. With all due caution, it can tell us when it is morally legitimate to limit the rights of other prospective agents: when we harm an agent's prospective agency. By stressing the importance of the indirect application of the PGC, we enable humans to act freely but also avoid producing the over-responsible 'resilient subjects' the sustainability discourse is rightly criticised for (Beck, 2010; Meisch, 2015b; Reid, 2013).

Fair water in a changing climate

I started with a critical perspective on sustainable development and sustainability governance and aimed to explain how current political approaches such as the SDGs fit into this paradigm that I characterised as neoliberal and techno-scientific. In doing so, I particularly focussed on SDG 6: 'Ensuring access to water and sanitation for all'. The adoption of the SDGs and the Paris Agreement were celebrated for having settled normative and political issues on global environmental politics. Now, scientists, engineers, and managers can go to work and find socio-technical and techno-scientific solutions for our common future. In this paper, I pointed out that much needs to be done at least on the normative level. We still need ethics and a normative conception of sustainability able to constructively argue for better solutions and to critically engage with existing social orders. I suggested Gewirth's moral philosophy to re-conceptualise sustainable development. His approach 'reflects the complex structure of morality itself. Any attempt to deny this by trying to derive all moral requirements in one simple way from one simple principle, as act-utilitarianism aims to do', runs into difficulties (Gewirth, 1978: 339). Arguing for fair water in a changing climate, we should start with an ethical perspective on the SDGs and SDG 6 in particular. As a result, we would be able to argue constructively and critically for legitimate claims to water and a just political order.

References

Beck, U. (2010). Climate for change, or how to create a green modernity? Theory, Culture and Society 27(2-3): 254-266.

Benessia, A., Funtowicz, S., Bradshaw, G., Ferri, F., Ráez-Luna, E.F. and Medina, C.P. (2012). Hydridizing sustainability: towards a new praxis for the present human predicament. Sustainability Science 7, Suppl. 1: 75-89.

Chandler, D. (2014). Beyond neoliberalism: resilience, the new art of governing complexity. Resilience 2(1): 47-63.

Folke, C., Carpenter, S., Elmqvist, T., Gunderson, L., Holling, C.S. and Walker, B. (2002). Resilience and sustainable development: building adaptive capacity in a world of transformations. Ambio 31(5): 437-440.

Francis, P., (2015). Encyclical letter *Laudato si'* of the Holy Father Francis on care for our common home.

Gewirth, A. (1978). Reason and Morality. Chicago, USA.

ICSU and ISSC (2015). Review of the sustainable development goals: the science perspective. International Council for Science (ICSU), Paris, France.

Leese, M. and Meisch, S. (2015). Securitising sustainability? Questioning the 'water, energy and food-security nexus'. Water Alternatives 8(1): 695-709.

Lövbrand, E., Beck, S., Chilvers, J., Forsyth, T., Hedrén, J., Hulme, M., Lidskog, R. and Vasileiadou, E. (2015). Who speaks for the future of earth? How critical social science can extend the conversation on the Anthropocene. Global Environmental Change 32: 211-218.

Meisch, S. (2016). Resilience and water governance in the Anthropocene. Paper presented at the 3rd European Workshops in International Studies., April 6-8, 2016. Tuebingen, Germany.

Meisch, S. (2015a). Water ethics – reflections on a liquifying topic. In: Meisch, S., Lundershausen, J., Bossert, L. and Rockoff, M. (eds.) Ethics of Science in the Research for Sustainable Development. Baden-Baden, Germany, pp. 353-376.

Meisch, S. (2015b). Knowing one's food – making food a public issue. In: Dumitras, D., Jitea, I.M. and Aerts, S. (eds.) Know your food. Food ethics and innovation. Wageningen Academic Publishers, Wageningen, the Netherlands, pp. 306-311.

Munck, R. (2015). Water, development and good governance. In: Munck, R., Asingwire, N., Fagan, H. and Kabonesa, C. (eds.) Water and Development: Good Governance after Neoliberalism. London, pp. 11-29.

O'Neill, O. (2009). Applied ethics: naturalism, normativity and public policy. Journal of Applied Ethics 26(3): 219-230.

Radio Vatican (2015). Encyclical's launch highlights science, theology, business. Available at: http://tinyurl.com/jkrqbhr.

Reid, J. (2013). Interrogating the neoliberal biopolitics of the sustainable development-resilience nexus. International Political Sociology 7: 353-367.

Rockström, J., Steffen, W., Noone, K., Persson, A., Chapin, S., Lambin, E.F., Lenton, T.M., Scheffer, M., Folke, C., Schellnhuber, H.J., Nykvist, B., De Wit, C.A., Hughes, T., Van der Leeuw, S., Rodhe, H., Sörlin, S., Snyder, P.K., Costanza, R., Svedin, U., Falkenmark, M., Karlberg, L., Corell, R.W., Fabry, V.J., Hansen, J., Walker, B., Liverman, D., Richardson, K., Crutzen, P. and Foley, J.A. (2009). A safe operating space for humanity. Nature 461: 472-475.

Rockström, J., Falkenmark, M., Folke, C., Lannerstad, M., Barron, J., Enfors, E., Gordon, L., Heinke, J., Hoff, H. and Pahl-Wostl, C. (2014). Water resilience for human prosperity. Cambridge, UK.

Rockström, J. and Bakker, P. (2015). We have a plan for our planet. Huffington Post, 19 October.

Saltelli, A. and Giampietro, M. (2016). The fallacy of evidence-based policy. In: Benessia, A., Saltelli, A., Funtowicz, S., Giampietro, M., Pereira, A.G., Ravetz, J., Strand, R., Van der Sluijs, J.P. and Sarewitz, D. (eds.) The Rightful Place of Science: Science on the Verge. Tempe, AZ, USA, pp. 31-70.

Sneddon, C., Howarthb, R.B. and Norgaardc, R.B. (2006). Sustainable development in a post-Brundtland world. Ecological Economics 57: 253-268.

Steigleder, K. (1999). Die grundlegung der normativen ethik. München, Germany.

Swyngedouw, E. (2010). Apocalypse forever? Post-political Populism and the Spectre of Climate Change. Theory, Culture and Society 27(2-3): 213-232.

United Nations (UN) (2015a). Transforming our world: the 2030 agenda for sustainable development (A/RES/70/1).

United Nations (UN) (2015b). Adoption of the Paris agreement (FCCC/CP/2015/L.9/Rev.1).

Voget-Kleschin, L. and Meisch, S. (2015). Concepts and conceptions of sustainable development: a comparative perspective. In: Meisch, S., Lundershausen, J., Bossert, L. and Rockoff, M. (eds.). Ethics of Science in the Research for Sustainable Development. Baden-Baden, pp. 45-72.

Walker, J. and Cooper, M. (2011). Genealogies of resilience: from systems ecology to the political economy of crisis adaptation. Security Dialogue 14(2): 143-160.

23. Conceptions of a sustainable food system

E. Pitkänen and A.-L. Elorinne*
School of Applied Educational Science and Teacher Education, University of Eastern Finland, Savonlinna, Finland; ematilai@student.uef.fi

Abstract

The goal of a sustainable food system is to secure current and future generations' access to nutritious food without destroying the environment. The term 'sustainable food' is widely used in both scientific and everyday language in reference to people basing their food choices on sustainability. However, there is no single generally accepted definition for this term. Therefore, it is important to determine what meanings sustainability has in the food system context because pursuit of a sustainable food system affects both food choices and policies. This multidisciplinary research has two aims: (1) to determine possible different meanings of the concept of 'sustainable food system' and (2) to analyse the suitability of the four classical principles (autonomy, beneficence, non-maleficence, and justice) in the context of the sustainable food system. By elucidating different meanings of the concept and bringing potential characteristics to light, it is possible to promote mutual understanding and reduce disagreements due to different uses of the concept. The aim is not to create a recommendation for a particular use of the concept but rather to clarify the concept of a 'sustainable food system'. The research method used was a combination of the concept analysis developed by Walker and Avant and philosophical concept analysis. The material used in this theoretical study was applied from home economics, philosophy and multidisciplinary databases and from other relevant literature. Sixty-eight definitions and descriptions were selected and analysed via theory guiding content analysis. Preliminary results indicate that the concept of a sustainable food system is ambiguous. All four classical principles were found in at least some of the definitions, and at least some of the principles were found in every definition or description. In conclusion, the four classical principles can operate as a guide for the pursuit of sustainability in the food system context.

Keywords: sustainability, content analysis, principlism, nutrition

Introduction

Food choices are made every day, and taken together those choices may have economical, environmental and health effects. When considering the effects of food choices, the entire food system must be included. A food system is generally defined as a system that includes food production, processing, distribution, preparation, and consumption. Sobal *et al.* (1998) introduced an integrated conceptual model of the food and nutrition system. In this model, the food and nutrition system is linked with other systems, such as health, economic and environmental systems, at both the local and global levels. The integrated model considers subsystems (producer, consumer and nutrition) and nine stages in the food chain: production, processing, distribution, acquisition, preparation, consumption, digestion, transport, and human metabolism.

Several global issues can be connected to the food chain including inequitable distribution of food, malnutrition, obesity and environmental pollution. Additionally, global food markets have had deleterious effects on food production in different countries and subsequently on social justice. Therefore, to solve problems like inequitable food distribution, environmental problems and ensure the public's welfare in the future, the notion of sustainable food systems has become a priority.

EurSafe 2016 – DOI 10.3921/978-90-8686-834-6_23, © Wageningen Academic Publishers 2016

The concept of sustainable food systems is widely used in policy documents, scientific research projects, media, by different organizations, in practice and in everyday conversation. There is no official or generally accepted definition for the concept of a 'sustainable food system' available; we consequently found several definitions. According to one of the many definitions used in the literature, a sustainable food system is described as 'a food system that conserves and renews natural resources, advances social justice and animal welfare, builds community wealth and fulfils the food and nutrition needs of all eaters now and in the future' (Tagtow and Harmon, 2009). This widely used definition is used as a rough starting point for the study. The diversity of definitions and descriptions of a sustainable food system may lead to confusion in or prevention of discussions; lack of conceptual clarity can cause misunderstandings and prevent effective practical application of the concept. Similarly, a request to act according to a sustainable food system's goals may not necessarily bring the desired change in behaviour. The pursuit of sustainability affects people's food choices and policies; conceptual confusion can further complicate political decision-making, particularly in conflicts. The fear is that, due to confusion, this well-intentioned concept often becomes an empty slogan. Based on the results of an international study concerning the obstacles to implementation of sustainable food systems, there is 'limited understanding of what constitutes a sustainable food system' (United Nations Environment Programme, 2015).

Sustainable food systems are positive because they have a common goal, and their ideals regarding sustainability guide at least some human activities. It is also easy to find material that recommends acting according to the principles of sustainable development. Building a sustainable future is an important goal taught in Finland (Board of Education, 2014). Sustainability is also guided by political means (Finnish National Commission on Sustainable Development, 2006; Sipilä, 2015). Therefore, the notion of sustainability/sustainable food systems can be seen as a moral ideal. Moral ideals are values that highlight desirable future states of affairs that are worth striving for (Beauchamp and Childress, 2009: 44-45; Meijboom and Brom, 2012).

However, sustainability is not the sole ultimate moral value that guides ethical decision-making and action; many values guide us. Therefore, the starting point of this study is a pluralistic vision. Principlism, an ethical approach described by Beauchamp and Childress (1977), serves as a background theory in this study. According to their approach, ethical decision-making does not have to be based on one moral theory or sole ultimate moral value. Instead, mid-level principles can guide choices and morally right courses of action. These *prima facie* principles are autonomy, beneficence, non-maleficence and justice (Beauchamp and Childress, 2009: 13-16).

This multidisciplinary research combines home economic science and philosophy. Concept 'sustainable food system' is widely used in the home economic science but it has remained unclear. This study seeks to clarify the concept with philosophical method. This study has two aims: (1) to determine possible different meanings of the concept of a 'sustainable food system' and (2) to analyse the suitability of the four classical principles in the context of the sustainable food system. To do so, we strove to identify ethical values connected to the concept of a 'sustainable food system' and to determine what those values mean, followed by comparing them to the four classical principles: respect for autonomy, non-maleficence, beneficence and justice. By establishing the different meanings of the concept of 'sustainable food system' and sharing them with the public, it is possible to promote mutual understanding and reduce disagreements due to inaccurate uses of the concept. The aim is not to create a recommendation for a particular use of the concept but rather to clarify the concept of a 'sustainable food system'. As a result of this conceptual clarification, the discussion can be conducted accurately and consistently, leading to a more comprehensive understanding of sustainable food systems.

Materials and methods

The research method used was a combination of philosophical concept analysis and the concept analysis developed by Walker and Avant (1988). The general purpose of the combination of concept analysis is to break concepts into simpler elements and to highlight the defining attributes of the concept. Concept analysis is useful for clarifying vague concepts so that a consensus can be reached regarding their meanings (Beaney, 2015; Walker and Avant, 1988: 35-36).

The data analysed in this theoretical study was retrieved from home economics and multidisciplinary databases (eViikki, CAB, AGRICOLA, ARTO, DOAJ, EBSCOhost, ERIC, JOSKU, MELINDA, ScienceDirect, Sustainability Science Abstracts, Wiley Online Library, Social Science Citation Index, Art and Humanities Citation Index, Web of Science and Scopus). The search query used was 'sustainable' OR 'sustainability' AND 'food system'. The search was restricted to keywords. One hundred six articles were retrieved and reviewed. The definitions and descriptions were searched where the goals of the sustainable food system were defined or described. A manual search was also conducted via scientific journals, Google Scholar and Google. A total of 68 definitions/descriptions were selected for the analysis.

Definitions and descriptions were analysed via theory guiding content analysis. References to ethical value terms that indicated sustainability were searched for in the definitions and underlined. Both the Oxford English Dictionary and Collins Dictionary were used to identify the different conceptual meanings of the value terms, which were then compared to the four classical principles (respect for autonomy, non-maleficence, beneficence and justice).

Preliminary results

The preliminary results indicate that the principles of justice, beneficence and non-maleficence occur the most in sustainable food system definitions and descriptions. Respect for autonomy was also found in the definitions and descriptions, yet to a lesser extent.

Principle of beneficence requires to provide benefits to others (Beauchamp and Childress, 2009: 202). In one of the definition, the goal was to 'provide benefits to the community where food was grown' (Koc *et al.*, 2008). This states clearly that the goal of the sustainable food system is to benefit that community where food was grown. The goals like health of the human beings (Hamm, 2008; Pothukuchi and Molnar, 2014; Sonnino, 2010), nutritional health (Harmon *et al.*, 2011), and well-being (Bacon *et al.*, 2014) were mentioned in several definitions. These are all goals of the beneficent action. In some definitions, the aim of the sustainable food system was to benefit global population, while in others, the benefits were restricted to affect people from particular region or community. The well-being of the farm animals was also mentioned in some of the definitions. If principles of beneficence and non-maleficence are seen to concern also animals, this goal can be understood to fall under them.

Preserving or maintaining resources for the future generations came up in several definitions (Hamm, 2008; Harmon *et al.*, 2011; O'Kane and Wijaya, 2015; Shimokawa, 2015; Story *et al.*, 2009). The goal can be understood to fall under principles of beneficence, non-maleficence, and justice. Sustainable food system contributes to welfare of the future generations by preserving resources for them. Action that prevents harm from happening to future generations can be seen as beneficent. Preserving or maintaining resources can also include intentional avoiding of action that cause harm for future generations. In this case the goal falls under the principle of non-maleficence. If we accept the idea of justice between the generations, preserving or maintaining resources for the future generations can be seen as an action that is guided by the principle of justice.

Sustainable food system goals, where the principle of justice appeared, included factors such as social justice (O'Kane and Wijaya, 2015; Pothukuchi and Molnar, 2014) and equity (Pothukuchi and Molnar, 2014). Food security which was mentioned in many definitions (Carlsson and Williams, 2008; Koc *et al.*, 2008; Lang and Berling, 2012; Zanella *et al.*, 2015) falls also under principle of justice. According to the principle of justice, all similar cases must be treated in the same way and unequal must be treated unequally (Beauchamp and Childress, 2009: 250). A sufficient amount of safe and nutritious food is basic human needs. In the light of the principle of necessity people can be seen as similar cases, which all have the right to adequate food. After identifying the conceptual meaning of the term 'food security', we can say that food security falls under all four principles. Respect for autonomy appears on construction conditions in which the individual has the possibility of autonomous choice. By making safe and nutritious food available to all, enables that autonomous decision-making. Access to adequate food can contribute to human well-being and therefore goal falls under principle of beneficence. On the other hand, insufficient food or lack of food can cause harm to people, such as malnutrition and, in extreme cases death. As a result of this conclusion, the principle of non-maleficence can also be seen as a guide for the food security goal.

Providing nutritious or healthy food for all eaters is mentioned in several definitions (Goggins and Rau, 2015; Hamm, 2008; O'Kane *et al.,* 2015; Story *et al.*, 2009). This goal can be seen under the principles of beneficence and justice. Goal include the idea that all the similar cases are treated in the same way and the intention to benefit other persons by providing nutritious or healthy food. In some definitions benefits were targeted to affect all Americans and in the others beneficent action covers all eaters now and in the future.

There were also goals, such as a vibrant economy (Hamm, 2008), local production and distribution infrastructures (Story *et al.*, 2009), and local economies (Pothukuchi and Molnar, 2014) that did not return any of the principles.

According to the preliminary results, the four classical principles can operate as a guide for the pursuit of sustainability in the food system context. Analysis of the study is ongoing and incomplete.

Acknowledgements

I thank Helena Siipi for supervision and comments regarding the earlier version of this paper.

References

Bacon, C.M., Sundstrom, W.A., Gómez, M.E.F., Méndez, V.E., Santos, R., Goltoftas, B. and Dougherty, I. (2014). Explaining the 'hungry farmer paradox': smallholders and fair trade cooperatioves navigate seasonality and change in Nicaragua's corn and coffee markets. Global Environmental Change 25: 133-149.

Beauchamp, T.L. and Childress, J.F. (2009). Principles of biomedical ethics. Seventh edition. Oxford University Press, New York, USA, pp. 3-16, 44-45.

Beaney, M. (2015). Analysis, the Stanford Encyclopaedia of Philosophy. Available at: http://tinyurl.com/z8p52eb.

Board of Education (2014). The basic education curriculum. Juvenes Print–Suomen Yliopistopaino Oy, Tampere. Available at: http://tinyurl.com/nkolpbq.

Carlsson, L. and Williams, P.L. (2008). New approaches to the health promoting school: participation in sustainable food systems. Journal of Hunger and Environmental Nutrition 3(4): 400-417.

Collins Dictionary. Available at: http://www.collinsdictionary.com.

Finnish National Commission on Sustainable Development (2006). Towards sustainable choices: a nationally and globally sustainable Finland. The national strategy for sustainable development. Prime Minister's Office Publications 7/2006. Available at: http://tinyurl.com/jqlqqdg.

Hamm, M.W. (2008). Linking sustainable agriculture and public health: opportunities for realizing multiple goals. Journal of Hunger and Environmental Nutrition 3(2-3): 169-185.

Harmon, A., Lapp, J.L., Blair, D. and Hauck-Lawson, A. (2011). Teaching food system sustainability in dietetic programs: need, conceptualization, and practical approaches. Journal of Hunger and Environmental Nutrition 6: 114-124.

Goggins, G. and Rau, H. (2015). Beyond calorie counting: assessing the sustainability of food provided for public consumption. Journal of Cleaner Production: 1-10.

Koc, M., MacRae, R., Desjardins, E. and Roberts, W. (2008). Getting civil about food: the interactions between civil society and the state to advance sustainable food systems in Canada. Journal of Hunger and Environmental Nutrition 3(2-3): 122-144.

Lang, T. and Barling, D. (2012). Food security and food sustainablity: reformulating the debate. The Geographical Journal 178(4): 313-326.

Meijboom, F.L.B. and Brom, F.W.A. (2012). Ethics and sustainability: guest or guide? On Sustainability as a Moral Ideal. Journal of Agricultural and Environmental Ethics 25: 117-121.

O'Kane, G. and Wijaya, S.Y. (2015). Contribution of farmer's markets to more socially sustainable food systems: a pilot study of a farmer's market in the Australian Capital Territory (ACT), Australia. Agroecology and Sustainable Food System 39: 1124-1153.

Oxford English Dictionary. Available at: http://www.oed.com.

Pothukuchi, K. and Molnar, S.A. (2014). Sustainable food systems at urban public universities: a survey of U-21 universities. Journal of Urban Affairs 37(3): 341-359.

Shimokawa, S. (2015). Sustainable meat consumption in China. Journal of Integrative Agriculture 14(6): 1023-1032.

Sipilä (2015). Finland, a land of solutions. Strategic Programme of Prime Minister Juha Sipilä's Government. Government Publications 12/2015. Available at: http://tinyurl.com/jhd5vge.

Sobal, J., Kettel Khan, L. and Bisogni, C. (1998). A conceptual model of the food and nutrition system. Social Science and Medicine 47(7): 853-863.

Sonnino, R. (2010). Escaping the local trap: insights on re-localization from school food reform. Journal of Environmental Policy and Planning 12(1): 23-40.

Story, M., Hamm, M.W. and Wallinga, D. (2009). Food systems and public health: linkages to achieve healthier diets and healthier communities. Journal of Hunger and Environmental Nutrition 4: 219-224.

Tagtow, A. and Harmon, A. (2009). Healthy land, healthy food and healthy eaters: dietitians cultivating sustainable food systems. Available at: http://tinyurl.com/jadfecx.

United Nations Environment Programme (2015). Sustainable food systems programme. Available at: http://tinyurl.com/zqz32a6.

Walker, L.O. and Avant, K.C. (1988). Strategies for theory construction in nursing. Second edition. Appleton and Lange, CA, USA, 35-36.

Zanella, M.A., Rahmania, M., Perch, L.N., Callenius, C., Rubio, J.L., Vuningoma, F., Rist, S. and Mapfumo, P. (2015). Discussion: food security and sustainable food systems: the role of soil. International Soil and Water Conservation Research 3: 154-159.

24. Out of feta in 2050? Sustainability and substitution in future food

T Kortetmäki
University of Jyvaskyla, P.O. Box 35, 40014 University of Jyvaskyla, Finland; teea.kortetmaki@jyu.fi

Abstract

How should one think about the sustainability of food systems from the viewpoint of food availability? What kind of food must there be available in future to allow stating that today's activities have been sustainable? Some accounts of sustainability allow the substitution of natural capital by human capital within sustainability. Food is however vital and cannot be substituted by human capital. This tenet can be called the principle of non-substitutability. It is an important guide for food policies and food ethics but lacks an examination of its contents. What, exactly, should be preserved for future generations: the opportunity to be adequately nourished or something more? Are there particular foods that future generations should be entitled to enjoy, and how far can this idea be extended? In this paper, I address the non-substitutability principle and future food availability with food ethics. I examine the principle in more details in a European context, where the vulnerability of small-scale food production and new phenomena like artificial food turn it very topical (yet the argumentation is applicable to other contexts as well). I begin by arguing, with reference to food security, that interpreting the non-substitutability principle in merely nutritional terms is unjustifiable and a socio-cultural interpretation should be favoured instead. This however raises further problems and questions that are discussed in the second part of the paper through four cases: religious and ethical food preferences, traditional and local speciality foods, native varieties, and quasi-meat. Finally, I reflect shortly on the implications of this discussion for conceptualising sustainable food systems and practices.

Keywords: food availability, food security, ethics

Introduction

Sustainability of food systems is an endorsed goal in food policies and academic discussions. The availability and substitutability of food is an interesting viewpoint for addressing food and sustainability issues. If we consider the future, what kind of food must there be available to allow stating that today's activities have been sustainable, or that they have not compromised the needs or opportunities of future generations?

It has become common to distinguish two accounts of sustainability: weak and strong. The weak account focuses on the total aggregate of natural and man-made capital (Neumayer, 2003): decline in arable land or food production resources may be sustainable if balanced by growth in man-made capital. Yet food is vital for humans and agriculture is rooted in natural capital: we cannot eat notes, nor grow food from them. Food related sustainability discussion must hence rely on the strong account of sustainability. I call this basic tenet the principle of non-substitutability of food. A parallel argument has been made in a broader context regarding the effects of food systems on the environment (Scoville, 2015), but contents of the non-substitutability principle within food systems have not been discussed to date.

An examination of the non-substitutability principle has practical and theoretical relevance. The threatened availability of particular food items and various artisanal, traditional and local products is a problem that has already been realised in the loss of agricultural diversity and hardships of small-scale food production. Further threats to food availability include various factors from environmental changes to power and market concentration and changes in food policies (see Grando *et al.*, 2016). It is important to ask what kind of 'food extinction' is sustainable and whether there are obligations to

I. Anna S. Olsson, Sofia M. Araújo and M. Fátima Vieira (eds.) ***Food futures: ethics, science and culture***
EurSafe 2016 – DOI 10.3921/978-90-8686-834-6_24, © Wageningen Academic Publishers 2016

preserve certain foods and food practices. The discussion also illustrates the theoretical importance of framing food as a socio-cultural rather than merely nutritional issue and highlights the ways in which the cultural significance of food also invokes conflicts between food preferences and sustainability.

The difference between nutrition and food

Humans have a complex relationship with food, but a primary form of that relationship is biological: food is vital for human functioning. This remark invoked the non-substitutability principle, and it also offers the simplest way to interpret that principle. According to the biological interpretation of non-substitutability, sustainable food availability in the future requires ensuring that the available resources satisfy the nutritional needs of future generations.

Biological interpretation has two benefits: it approaches sustainability with universal nutrition calculations and allows a high degree of substitution. As long as the available stuff satisfies energy and nutrition needs, it is irrelevant whether needs are met by conventional food or, for example, by enriched flour and protein supplements – or even by 3D-printed artificial food. (In the last case, the non-substitutability principle might actually become rejectable: decline in food-related natural capital can be compensated by an increase in man-made capital that enables producing artificial nutrition substances.) This interpretation would not inflict significant demands for the sustainability of food systems, as there is practically no need to worry about the scarcity or depletion of particular food items or food types.

Justifiability of the biological interpretation can be evaluated against the notion of food security, which is a prominent concept in evaluating the prevalence of hunger and food insecurity and closely connected to discussions on food entitlements. I follow the definition offered by the United Nations Food and Agriculture Organization that is institutionalised and applied in global agreements. Food security is a situation where 'all people, at all times, have physical, social and economic access to sufficient, safe and nutritious food which meets their dietary needs and food preferences for an active and healthy life' (FAO, 2015). This defines the minimum food security: full food security might involve for example empowerment. The definition refers to food access, which requires food availability. FAO's definition thus implies that the availability of safe and nutritious food that satisfies dietary needs and food preferences is a necessary condition for minimum food security (for alternative approaches with higher standards for food entitlements, see Gottlieb and Joshi, 2013).

The biological interpretation of non-substitutability cannot guarantee food security due to the neglect of food preferences. The interpretation would consider future food availability sustainable even if the only available foods were nutrient-enriched artificial products that hardly resemble today's food. Food security requires the availability of real food that satisfies at least some food preferences: availability of 'some real food' would not be enough. Would a food system be sustainable if it allowed a world in which a Muslim or a vegetarian had to choose between protein pills and pork to avoid protein deficiencies? With reference to the definition of food security, I maintain the answer is negative. Such a narrow food selection would discriminate against different worldviews and prevent certain groups from accessing real food unless they compromise their central values. Speaking about met food preferences in such circumstances would be hollow.

As the criteria of food security already imply, the biological interpretation of non-substitutability is unjustifiably narrow. Food has rich social and cultural meanings; it is embedded in and constructs our identities, ways of life, and communities (cf. Thompson, 2015: 42-44). Respect for human dignity and cultural equality requires acknowledging the multiple meanings of food and preserving the opportunity to food choices and practices that carry those meanings. This stance can be called the socio-cultural interpretation of non-substitutability. It acknowledges our complex relationship with food and defines sustainable food availability in a way that meets the conditions of minimum food security as well.

The socio-cultural interpretation is a complex idea and raises further questions. How extensively should various socio-cultural meanings be taken into account? For example, what about locally important protected designation of origin (PDO) products: if Greece runs out of feta by 2050, is it a problem of sustainable food availability? Is the problem fixed if Greeks can access substitutes made elsewhere? Or, considering the significant impacts of animal protein production, will there have to be all sorts of animal protein available in future or can it be replaced by quasi-meat? I will next discuss the complexities of socio-cultural interpretation in the light of some real-world examples.

The socio-cultural interpretation of non-substitutability

The socio-cultural interpretation can be used for evaluating the justifiability of various claims for preservation. These claims, whether potential or actually made, are claims for ensuring the availability of particular kind(s) of food for the future generations, on the basis of reasons related to food preferences and social or cultural meanings of food. As the concern is for future generations, the personal tastes of present individuals are irrelevant. What matters are preferences that can be assumed to be commonly held, and socially or culturally essential, for the future human beings.

Justificatory grounds for the preservation claims vary greatly. They include various religious and ethical belief systems that will likely be espoused by future humans as well. Religions with dietary regulations and different forms of vegetarianism are examples of these. Other justificatory grounds involve community level food preferences and practices such as culinary traditions and food practices that are relevant to communal identity. The equal recognition of various cultures and values requires an equal respect and consideration of these grounds for preservation claims: it is unjust to arbitrarily favour some preferences at the cost of disregarding others. It is probably impossible to retain all possible food opportunities for the future, and considered reasons must be given why some preferences are weightier than others. Some claims for preservation may also turn out to have unjustified grounds, in which case they do not invoke obligations of preservation. Next, I will evaluate these justificatory grounds with four cases: dietary preferences based on religious or ethical views, traditional foods and local specialities, native varieties, and quasi-meat. These reflections also help clarify the meaning of the socio-cultural interpretation.

Dietary preferences related to belief systems typically involve preferences to abstain from consuming certain foods and to still have satisfying food opportunities. The abstinence from some common food items, like meat, invokes a need for alternatives: at stake is the availability of satisfactory alternatives, rather than of a particular product. Safeguarding sustainable food availability is relatively easy in these cases: versatility in the selection of food products in future suffices for that – but the availability of alternatives is required. A hot debate took place recently in France as some cities decided to cut out any alternatives for pork-based school lunches (The Guardian, 13.10.2015). Religious and ethical views are intimately connected to one's identity and self-esteem, and respecting them is central to one's dignity. Many belief systems are indeed so important for one's identity that many people are willing to encounter discomfort or suffering rather than violate the core principles of their belief system. The preservation claims related to religious or ethical views have then heavy grounds.

Traditional foods and local specialities (such as feta) comprise foods whose production is regionally restricted and whose qualities owe to regional features or traditional production methods. Technically, 'identical' food products may be made elsewhere and they may taste more or less the same. Traditional and local specialities are a concern for 'gastronationalism', a form of identity politics or collective identity claims that oppose the homogenising tendencies of globalising food markets (DeSoucey, 2010: 433).

Specialities are endangered for various reasons from environmental degradation to changes in food policies and market relations that put small-scale food activities under pressure (Gellynck *et al.*, 2010; Gottlieb and Joshi, 2013). A relevant question for this paper is whether the availability of traditional products should be protected for future generations and whether they can be replaced with similar or like items within the limits of sustainability. If the availability of such items must be ensured, this creates an obligation to protect traditional foods and production methods.

The claims for preservation are here grounded in the importance of those foods in local communities or culinary traditions. Although these products may be loved elsewhere as well, they have particular cultural significance in communities where they are traditional. The more a community values a product as culturally essential, the weightier are the grounds for preservation claims. Although we cannot know the future generations' preferences, it can be argued that the foods essential to a particular community and its cultural identity are something our descendants should have the opportunity to enjoy. If this argument is plausible, it follows that the sustainability of food systems involves safeguarding the cultivation and production of at least some traditional items to avoid their extinction (a relevant question is whether current trade legislation allows measures required for this). It is more important to protect the availability of feta in Greece than to ensure its place in the Swedish groceries. Whether substitutes made elsewhere can be considered as satisfying the availability of these kinds of products, is a more difficult question to which I will return later in this paper.

A moral and practical challenge is related to the protection measures that might be taken to protect the future availability of traditional products. Safeguarding sustainable food availability does not justify any means. This has already become a hot topic in Italy, where some city centres have banned ethnic fast food restaurants in the name of protecting traditional cuisine. Critics have denounced the measures as xenophobic. Cases like this would deserve a separate evaluation.

Native varieties (or otherwise valuable varieties) may be threatened due to agricultural intensification and economic competition, races where varieties with smaller yields tend to lose. An example case is the one of Kyoto heirloom vegetables whose production collapsed after the Second Word War due to lower yields and sensitivity. The production was revived as a co-operation between local chefs, farmers, and city government, and 'Kyoto vegetables' (Kyōyasai) now have the reputation of being more nutritious, 'authentic', and tastier to conventional market vegetables (Rath, 2014).

Socio-cultural claims for preserving native varieties can be grounded on traditions, culinary traits, or cultural significance of these varieties. Diversity-related arguments do not, instead, invoke an obligation of preservation in this framework: diversity deserves overall protection, but that is a different issue from the non-substitutability of particular items. The weight of other grounds for preservation vary in each case: while some (native) varieties may be essential for particular cuisines, there are also varieties whose substitution cannot be told by a layperson. How should we distinguish the varieties that deserve particular preservation?

I propose that this question can be approached with what I call a 'semi-blind review for cultural essentiality'. First, a community determines which foods it considers culturally so essential they should be preserved for future generations. Next, the importance of different varieties can be evaluated with a 'blind review' of those varieties: can people tell a difference between different varieties when they prepare or consume culturally essential foods? If they can, does this difference give rise to a claim for preservation? I exemplify the use of this method with the case of Finnish potatoes.

Potatoes are essential in Finnish cuisine. The cultivars (roughly 40, though not all native) are classified according to their starch content into four categories. Traditional potato foods use potatoes of a particular category: mash is made with high-starch and potato salad with low-starch varieties. The difference in starchiness can be told in cooking and it is hard to make proper food with potatoes of a wrong category. Yet, differences within each category mainly relate to cultivation properties and could not be told in a blind review (with some exceptions like the PDO variety *Lapin puikula*). Hence, the non-substitutability applies between the four main categories but not within them (excluding some exceptional varieties). This 'blind review' can also be applied to evaluating the earlier question of traditional products.

Quasi-meat poses an interesting case. Environmental concern has generated aspirations for a 'less meaty' Western world in future. How can this be evaluated from the non-substitutability viewpoint? How far can quasi-meat go in answering the preferences of future generations? This problem, like the previous case, can be approached with a semi-blind review. First, the socio-cultural meanings of various meat products must be considered. Then, a subsequent question can be asked concerning the culturally essential products: can a layperson tell the difference (in favour of meat) between the meat product and its quasi-meat 'counterpart'? If the difference can be told, are there weighty reasons to claim for protecting the availability of such meat in future? If food industry innovates quasi-prosciutto that cannot be distinguished from prosciutto by an Italian, it would be a suspicious argument that raising and killing a pig is essential for the social and cultural meanings of prosciutto.

There are also other criteria for the justifiability of any preservation claims. Social or cultural significance can hardly justify claims about foods whose production is unsustainable or ethically unviable. Consider foie gras: its production is condemned due to animal welfare issues, and demands for protecting foie gras production conflict with animal welfare claims, creating a hard controversy (DeSoucey, 2010: 442-445). A more problematic question concerns foods with heavy ecological impacts, such as beef: at least demands that would require large-scale beef production would be environmentally unsustainable, and such claims cannot be justified as they conflict with sustainability. It is however impossible to create universal rules for drawing the line here: context-sensitive evaluation will be necessary.

Discussion

This paper makes an argument with demanding implications: the socio-cultural interpretation of non-substitutability raises the threshold of sustainability in food systems much higher. Yet I assert that the reasons given in the favour of socio-cultural interpretation are reasonable. Thinking about food in merely nutritional terms disregards for multiple meanings of food and ways in which humans make and constitute their personal and cultural identities through food practices. If thinking about sustainability involves asking 'whether the kind of environment that made us who or what we are today can and should endure' (Thompson, 2015: 192), these considerations should involve our food culture and food practices more thoroughly and ask which elements of them we want to preserve for future generations.

This question is not just theoretical. The availability of many culturally essential food products has become threatened. This is not only due to the environmental changes (that play a key role in the overall food availability question) but also to food policies as well as market and power disparities that put many traditional products, local specialities, and native varieties to the edge of extinction. Discussion on sustainable food availability illustrates how the environmental, economic and social sustainability are entangled in food systems.

Different interpretations of the non-substitutability principle reveal how different framings of food alter the interpretation of sustainability. In this paper, I hope to have shown that a biological framing of food is untenably narrow. However, the socio-cultural framing has its downsides as well: it creates extensive demands for protection, but the overall challenge of sustainability sets limits on how many things can merit special protection. Reasoning and choosing what should be preserved for future generations requires argumentative tools that help address complex cultural and value-laden questions. In this paper, I have attempted to provide some conceptual ideas for addressing the problematics of (non-)substitutability that concern our common food future.

References

Food and Agriculture Organization of the United Nations (FAO) (2015). Food security statistics. Available at: http://tinyurl.com/8ypauyh.

DeSoucey, M. (2010). Gastronationalism: food traditions and authenticity politics in the European Union. American Sociological Review 75: 432-455.

Gellynck, X., Banterle, A., Kühne, B., Carraresi, L. and Stranieri, S. (2012). Market orientation and marketing management of traditional food producers in the EU. British Food Journal 114: 481-499.

Gottlieb, R. and Joshi, A. (2013). Food justice. MIT Press, MA, USA, 304 pp.

Grando, S., Rahmanian, M. and Colombo, L. (2016). Deliverable D2.5 NxM matrix on the multidimensionality of vulnerabilities and their drivers. TRANSMANGO: EU KBBE.2013.2.5-01. Available at: http://tinyurl.com/zqyewrt.

Neumayer, E. (2003). Weak versus strong sustainability: exploring the limits of two opposing paradigms. Edward Elgar, Cheltenham, UK, 304 pp.

Rath, E.C. (2014). New meanings for old vegetables in Kyoto. Food, Culture and Society 17: 203-223.

Scoville, J.M. (2015). Framing food justice. In: Dieterle, J.M. (ed.) Just food: philosophy, justice and food. Rowman and Littlefield, London, UK, pp. 3-20.

Thompson, P.B. (2015). From field to fork: food ethics for everyone. Oxford University Press, New York, NY, USA, 344 pp.

Section 5. Animal perspectives

25. Animal subjectivity: evolving ethics in animal studies

F. de Giorgio
Learning Animals, Località La Scatta, Frazione San Massimo, Piana Crixia (SV), Italia; Institute for Animal Ethics, Animals Studies and Zooanthropology, Achterstraat 64, Nistelrode, the Netherlands; degiorgio@learning-animals.org

Abstract

In recent years, the Animal Question has come up, topical but complicated, as it concerns all of us, and it raises many questions related to animal ethics, subjectivity, and cognition. Within the field of animal studies, the consideration of animals, their individuality and their being creators of their own experiences, is becoming more and more crucial. There is an emerging need to re-assess the standing of non-human animals in the interaction with humans, to create better living conditions as well as to develop full recognition of their moral status. In animal ethics, to have a moral status means to be worthy of moral consideration. A re-assessment of the status of non-human animals calls for a stop to treating them as instruments and calls for us attributing weight to their needs, their intrinsic motivations, and their subjective pleasures. Not for a particular human benefit, but of importance for the non-human individuals themselves, bearer of their own intrinsic value. The aim of this paper is to emphasize some contemporary questions about animal ethics in animal science and animal studies; questions which are supported by multidisciplinary knowledge, both with theoretical as in practical dimensions.

Keywords: animal ethics, animal science, ethology, animal rights, animality, antispecism, post humanism, zooanthropology

Introduction

Animal ethics is an intricate topic. Ethical issues are part of study fields such as applied ethology, but increasing attention is coming from the growing public awareness regarding the life of nonhuman animals. Attention concerning the Animal Question – referring to the necessary awareness of how animals are treated and morally considered and the necessity to evolve – is more and more of concern of various study fields. There is an emerging need to create better conditions of life for non-humans in interaction with humans, as well as to develop full recognition of their moral status.

The argument regarding how animals are treated is made more complex by the deep routed human habits, practices and beliefs that have been taken for granted for a long time. From a philosophical point of view, this raises essential concerns about the foundation of moral rights. Philosophers question the position animals have in an acceptable moral system. Currently, animals exist on a norm of human moral concepts. Consequently, sometimes people find themselves granting the animals a strong ethical status; while at other times they deny them any kind of ethical status at all. This is partially caused by a normative/functional approach to ethics, rather than using ethics, or learning from ethics, as guidance towards change and growth. Translating ethical theories into practice still has a bridge to overcome.

The aim of this paper is to emphasize some contemporary questions about animal ethics in animal science and animal studies, questions supported from multidisciplinary knowledge, both in theoretical and in practical dimensions. Animal ethics should be a beacon to further evolve and develop our current way of interaction with other animals and integrate this in daily life, as our coexistence with other animals is an important part of who we are as humans.

EurSafe 2016 – DOI 10.3921/978-90-8686-834-6_25, © Wageningen Academic Publishers 2016

Times of change

Animal science is increasingly coming under the pressure of a more sensitive public opinion, with people being interested in a different perspective on our coexistence with other animals. This development is enforced by interdisciplinary debates about animal rights, veganism, antispeciesism, and studies regarding animal agency, critical animal studies, post-humanism (Wolfe, 2009) and other related topics. The animal question represents the most important aspect of contemporary critical thought about non-human animal ethics. To allow the animal question a more concrete contribution to further development, a more structured attention within mainstream fields of animal studies is necessary. For example, although applied ethology is a relative recent discipline, it has moved away from the original meaning of classical ethology. Over the years, it has gained a focus on what people can do with animals, how their welfare can be controlled in non-natural habitats, rather than learn from them and put into question what we ask from them, what we do with them. It is thereby connected more with the exploitation of the animal for human purposes, as opposed to the exploration of animals.

Research cannot evolve as long as the harm done to animals remains hidden within non-transparent ethical statements, coming from a normalized belief that suffering is acceptable for a greater good, that researchers have a free right to research, that confining animals and making them suffer is inevitable for development. What has been improving is a more sophisticated animal usage; abuses that are less visible such as those perpetrating on the mind and moral status of animals. Examples are present in farm-industry, in equestrian or dog-training industry, and even in research to 'test' the cognitive abilities of animals, showing that it is necessary to step out of the anthropocentric paradigm, rather then understanding of how human can coexist with nonhuman animals preserving their quality of life, as individuals, with their own subjectivity. Consideration of these ethical questions starts in our daily interaction with non-human animals.

Essentially, the traditional approaches to animal ethics mainly emerged from one of the two most influential ethical theories: Peter Singer's preference utilitarianism or Tom Regan's deontology/neo-Kantianism (Abbate, 2014). Both theories are committed to repercussions that are undesirable for humans. Abbate argues that the theories should be rejected, as they are unsuccessful in providing sufficient protection for animals. Abbate believes that any theory that seeks to protect animals should meet one of the three conditions he stipulates in his work. First, the theory must condemn industries and practices that cause superfluous animal suffering and death. Second, the theory must prevent individual moral agents from partaking in industries and practices that culminate in unnecessary animal suffering and death. Third, the theory must instruct moral agents to thwart moral disasters by at least diminishing or reducing the suffering, pain, and death of animals if the elimination of suffering, pain, and death is not feasible.

Attempts to formulate ethics for animals such as the move made by Abbate have increased in recent times. These attempts are made mostly based by applying foundational principles drawn from ethical theories that have previously been formulated. Among the theories that contemporary theorists build on to formulate their own theories is Singer's utilitarianism to animals that experience pains and pleasure. Other theories that have been formulated include the neo-Aristotelian approach that advocates for animals to be handled in ways that demonstrate respect for their 'natures' or theories that allow them to grow their 'capabilities' (Nussbaum, 2004). However, translating the understanding of animal otherness, thinking in terms of nonhuman individuals, each with their own unique way of being in the world, into a changing attitude in daily practice, is something that needs dedicated attention. It requires the ability to step outside the human frame of reference, as discussed in the paragraph below, if not, these theories risk to continue to contribute to a more sophisticated usage of nonhuman animal. We are morally obliged, in our choices, to give weight to an individual animal's needs, to intrinsic motivations, and to subjective pleasures, not for a particular human benefit, but as important for the non-human animal, bearer of his own intrinsic value.

From theory to application

Changing our own interaction with other animals requires a deconstruction of a system of values and habits. This is something that lacks in current ethical theories. The practical application and the impact of the human belief system is not sufficiently considered. What is necessary is to put in discussion the ethical implications of manipulating animal minds and the constant interfering with their behaviour, and their own life, also through with operant conditioning.

It requires dedicated attention throughout all levels of society, to cultivate the awareness that each individual we meet has his or her own needs, preferences, emotions, mindsets, and pleasures. It is necessary to first of all deconstruct the current human inability of recognizing otherness and subjectivity. It is necessary to understand how to step out of the anthropocentric paradigm to be able to apply an ethics that starts from the nonhuman animal point of view. This requires a dedicated study and understanding of how to recognize and preserve their subjective experience.

Many developments regarding animal wellbeing are still based on an animal welfare perspective and do not take into account the individual needs, preferences, pleasures, and mental health of animals. They are still far from considering animal autonomy, dignity, integrity or animal privacy. Understanding that nonhuman animals are in dialogue with the world – that they ask their own questions of the world and look for their own answers and information – is the central idea when it comes to preserving their quality of life and considering them as the owners of their own experiences and initiatives. This, of course, requires humans to stop making nonhuman animals objects of predefined behavioural protocols rather than what they were meant to be: protagonists of their own lives.

Interactions between human and nonhuman animals are often based on either a reactive interaction where the nonhuman animal is a stimulus causing a reaction, a projection where the nonhuman animal is read as a symbol (i.e. a horse seen as freedom, a dog seen as loyalty), or an interaction based on usage where the nonhuman animal becomes a tool to exploit or manipulate (examples of animals living close to us: the guardian dog, the service dog, the hunting dog, the jumping horse, the dressage horse, etc.) (Marchesini, 2005). Similar exploitation of others actually happens as well in human-human interaction in our modern economy driven society, resulting in burn-out symptoms based on an incapacity to listen to one's own limits, a form of learned helplessness due to social pressure. However, while we often are aware of the moral problems inherit in such depleting interactions involving humans, we ignore the fact that this is often caused by an artificial situation, lacking intrinsic motivation of the human or non human animal involved, and focuses instead on exploiting potential, with implications on an individual's integrity, dignity and privacy.

The concept of progress can be seen as a means to guide change in the direction of greater democracy, freedom, ecological balance, and respect for nonhuman animal life and the earth as a whole (Best, 2009). Putting in Helmuth Plessner's perspective (Rasini, 2015); every living being is a subject, and every subject 'is' not a body, but he 'has' a body that he owns. Together with this own body he is also the holder of his own mind, used in dialogue with the world in a more or less active way, in a more or less complex way. Animal science should question itself in this perspective. Considering non-human animals as owners of their own intrinsic value, as moral agents (rather than moral patients), and active protagonists on the scene of subjectivity and morality. Subjectivity and morality are biological elements shaped by evolution, just like an arm or a tail or a mind, following the Darwinian paradigm. This asks for a further developed ethical consideration of them, both in animal studies, as well as in practical applications in daily life, and specifically in the direct interaction with human animals, as explained in the next paragraph: Vanguard Ethics.

When the concept of moral agency is connected to the concept that a subject is not a body for the use by someone else but, instead, has a body useful to itself, it provides space for other theories about animality. A change in the day-to-day interaction with a nonhuman animal asks for more than scientific acknowledgements. The awareness of nonhuman animal subjectivity, of the fact that our interaction or mere presence is a cognitive experience in the nonhuman animal's life is something that is not easily taken into account. The example of Derrida's encounter with his cat shows this difficulty, as in the description of this experience, Derrida does not explicitly refer to his interest or understanding of the cat's subjective experience, but rather to his own sense of discomfort by the acknowledgement of being looked at by the cat (Derrida, 2008: 59).

Therefore, seeing another nonhuman animal as an individual other – someone with a unique subjective experience of the world – necessarily requires the understanding of changing values. A paradigm change is required in which the human is no longer central, so we can make room for a new paradigm of our coexistence with animality. Such an understanding must be based on reciprocal relations and on the comprehension of the concept of subjectivity.

Vanguard ethics – animal integrity and animal privacy: to train or not to train

Regarding contemporary human coexistence with non-human animals, although times change, concepts evolve and awareness increases, I argue that a vanguard animal ethics should also question why we still train them, in research as well as in other dimensions of human society. Hence, apart from the anthropocentric reasoning connected with using animals, why do we train nonhuman animals rather than trying to understand them? Very often, animal training is confused with animal learning. However, intelligence is not the same as trainability. What seems like learning during training exercises is essentially an automatic execution of conditioned behaviours. Unlike conditioning, learning means becoming the owner of an experience. It is a proactive ownership of the experience that arises from a cognitive perception of that experience. Learning thus does not come from being conditioned or desensitized.

The concept of operant conditioning, which is generally the base for animal training, lies in a behaviourist matrix. For many years now, the behaviourist paradigm, both in its interpretative models and its various forms of application, is discussed as a Cartesian-like paradigm. However, although the scientific insight into the emotional-cognitive side in animals continues to increase, the trend of applying operant conditioning in day-to-day practice that focuses on stimulus–response protocols has become the predominant language and approach when interacting with animals. The tendency is to create a more 'humane' interaction, with a risk to integrate modern knowledge about animal cognition into a more sophisticated manipulation of the animal's mind.

To create room for the expression and perception of alterity, we should focus on preserving a cognitive-affiliative approach to learning, preserving the subjectivity and the mental and emotional heritage of other animals, instead of training them in order to control their behaviour. We should focus on determining how to preserve spontaneous expressions that would preserve the right social context for individual animals, preserve their curiosity toward the world, and preserve the ability to make choices through one's own authentic internal sources. Understanding nonhuman animal subjectivity, and animal cognition in general, means understanding that they have a world to discover. An environment that does not fixate on fitting them into our ideas of coexistence, but focuses on their alterity, on their attention, on shared experiencing, allowing our society to develop in a way that includes all different ways of being in the world, not by adaptation, but by inclusion (De Giorgio, 2015).

Philosophical challenges

The animal question within animal studies is growing stronger, with different positions between an Anglo-American philosophy and a continental philosophy that stress the importance to break the barriers between humans and the other animals (Derrida, 2008). While the latter seems more open towards animal individuality, it is still also stuck inside an anthropocentric perspective (Calarco, 2008). In order to overcome the anthropocentric perspective both in science as in philosophy, as well as in politics, a substantial shift of paradigm is necessary to gain a more objective exploration of the world. This needs to take into account also the perspective of other species, as the anthropocentric bias is something intrinsically connected with humanity (Marchesini, 1996). Decentralizing the human perspective is necessary to embrace a development in animal ethics, which goes beyond a theoretical level.

Bringing the animal question into animal studies forces us to ask ourselves what kind of ethics we want to take into account: an ethics for human purposes, as is still common (e.g. when however humanely treated animals are still killed in the farm industry (Johnson, 2015)), or an animal ethics that challenges animal suffering and behaviour manipulation. We need to ask ourselves if we want to question data coming from inappropriate treatment of animals, and continue to make interpretations based on aged animal science with its own anthropomorphisms, as in the case of the dominance hierarchy interpretations of social dynamics in horses and in other domestic and wild animals, still present in scientific publications (De Giorgio and De Giorgio, 2015).

The zooanthropologic horizon

One of the coordinates of this vanguard paradigm shift is represented by the discipline of zooantropology that includes several scientific, philosophical and political disciplines. Zooanthropology defines and gives direction within the interpretation of phenomena related to our relationship with other animals, stepping outside the anthropocentric viewpoint. It is not just a matter of study and application, but it gives an ethical direction and inspires a movement towards years to come. It refers to the cognitive ethology dimension, and does not fall into the dimension of classical ethology or in behaviourism.

Zooanthropology recognizes the nonhuman other as a subject and for this clearly disapproves the use of coercive instruments, whether physical, emotional or mental, and certainly calls into question all tools, methods and approaches regarding the interpretation and application of operant conditioning. It questions animal training, and opens up a new model of interpretation and application in the dynamics of animal learning, recognizing animal subjectivity and alterity (De Giorgio and De Giorgio, 2012).

Conclusions

A re-assessment of the standing of nonhuman animals requires that the understanding of nonhuman otherness be the foundation of any interaction with nonhuman animals. This means that in the day-to-day practice, we must consider the nonhuman animal's subjectivity with intrinsic socio-cognitive abilities to create cognitive maps, to search for information, to elaborate and process knowledge, to follow their own inner motivation, to express emotions or intentions, to solve problems, to adapt to changes, and especially to develop relationships based on affiliative expressions. All these elements should form the core of any experience, leaving behind the assumption that they should be trained and conditioned, for any kind of human reasoning.

Hence, there is a world to discover and to explore when it comes to modern ethics, to interspecies interactions and the related socio-cognitive dynamics, as for example in the human animal relationship. We need to create space for these dynamics in order to be able to study them, so that we can learn how to take care of these dynamics, for the quality of life of everyone involved. The research possibilities, thesis topics and study developments are enormous, if we can step outside the current restriction of cultural norms. Create research set-ups that can further study and help to restore and preserve all the nuances of animal otherness and preserve the individual expressions and development each sentient being has.

By studying the dynamics of an interspecies relationship development, based on the preservation of a cognitive-affiliative approach to learning, and the preservation of the mental and emotional heritage of other animals, we can learn again to understand the true value of otherness, helping ourselves as well to preserve our own true socio-cognitive abilities. Understanding nonhuman otherness might be our compass to improve many types of social interactions, between different species, between different cultures and between individuals, allowing a further development of ethics within the human species and concerning other species. Such a development could thus be potentially liberating for all.

From today towards the future, cognitive ethology, the antispeciesist philosophy, vanguard animal ethics and a more and more post-human oriented legislation, can indicate important coordinates of change in the relationship and interaction between human and nonhuman animals, in which will lie the key to the process of decision-making of animal caregivers, professionals, academics, animal rights organizations, political movements and research.

References

Abbate, C. (2014). Virtue ethics and animals: a minimally decent ethic for practical living in a non-ideal world. Journal of Agricultural and Environmental Ethics 27(6): 909-929.

Best, S. (2009). Genetic engineering, animal exploitation, and the challenge for democracy. In: Gigliotti, C. (ed.) Leonardo's Choice: Genetic Technologies and Animals. Springer Press, New York, NY, USA.

Calarco, M. (2008). Zoographies: the question of the animal from Heidegger to Derrida. Columbia University Press, USA, 184 pp.

De Giorgio, F. and De Giorgio, J. (2012). Zooanthropology. Available at: http://tinyurl.com/zm75axl.

De Giorgio, F. and De Giorgio, J. (2015). Comprendere il Cavallo. Edizioni Giunti. Italy, 224 pp.

De Giorgio, F. and Schoorl, J. (2014). The spontaneous horse. Relations Beyond Anthropo-centrism 2(1).

De Giorgio, J. (2015). Protagonists of their own life: understanding non-human Otherness. Proceedings Anthrozoology Student Conference 2015: Moving Beyond the Boundaries. Exeter University, Exeter, UK.

Derrida, J. (2008). The animal that therefore I am. Fordham University Press, Bronx, NY, USA, 192 pp.

Johnson, J. (2015). Humanely killed? Journal of Animal Ethics 5(2): 123-125.

Marchesini, R. (1996). Il concetto di soglia: una critica all'antropocentrismo. Edizioni Theoria, Italy, 237 pp.

Marchesini, R. (2005). Fondamenti di zooantropologia. Alberto Perdisa edizioni, Italy, 727 pp.

Nussbaun, M. (2004). Beyond 'compassion and humanity': justice for nonhuman animals. In: Sunstein, C.R. and Nussbaum, M.C. (eds.) Animal rights: current debates and new directions. Oxford University Press, Oxford, UK.

Rasini, V. (2015). Soggetto vivente. Un confronto con la filosofia dell'animalità. Ethics and Politics 17(3): 195-204.

Wolfe, C. (2009). What is posthumanism. University of Minnesota Press, MN, USA, 400 pp.

26. A philosophical and technical critique of zooanthropology as a moral and practical paradigm

S. Aerts[1], B. Bovenkerk[2] and S. Brando[3]*
[1]*Odisee University College, Ethology and Animal Welfare, Hospitaalstraat 23, 9100 Sint-Niklaas, Belgium;*
[2]*Social Sciences Group, Wageningen University, Hollandseweg 1, 6706 KN Wageningen, the Netherlands;*
[3]*AnimalConcepts, Zoom 1813, 8225 KM Lelystad, the Netherlands; stef.aerts@odisee.be*

Abstract

An addition to the existing schools of thought in the ethological and cognitive science communities is the zooanthropological approach. While grounded in these sciences, this view on the human-animal relation is taken further and strong moral objections to any type of animal training are formulated, but arguments for these objections are still underdeveloped in our view. In this paper we critically assess the zooanthropological paradigm, while acknowledging that it is highly original and innovative work. Applied zooanthropology combines different fields of study exploring the development and dynamics of our relationship with other animals, and animals are considered sentient beings in a cognitive-affiliative paradigm. The opposition to consider animals in a stimulus-response paradigm is understandable. It is in line with contemporary developments of holistic approaches to animals that consider behavioural, physiological, psychological, social, and affective aspects. However, we argue that a clarification of what a modern animal training paradigm entails is desirable. Clarifications on zooanthropological methods and their underlying cognitive, physiological and affective processes are needed, as well as elucidations of practical implications and/or solutions, e.g. human-animal interactions, living with pets, caring for horses or non-releasable wildlife. From a philosophical perspective the rejection of 'outdated' animal training approaches by zooanthropology appears based on the assumption that all animals (including humans) are morally equal, because they possess 'subjectivity'. This concept of subjectivity is not based on Regan's concept of 'subject-of-a-life', nor on that of cognitive ethology, as consciousness is said to be irrelevant for subjectivity. If not based on consciousness, it remains unclear what morally relevant criteria subjectivity is based on. Moreover, it remains unclear why the presence of subjectivity should necessarily lead to a rejection of training. Furthermore, this strong abolitionist stance conflicts with the way the approach is framed in practice.

Keywords: human-animal relation, posthumanism, subjectivity, animal training

Introduction

An addition to the existing schools of thought in the ethological and cognitive science communities is the zooanthropological approach, propagated by for example Roberto Marchesini and Francesco De Giorgio. While clearly grounded in these sciences, this view on the human-animal relation is taken further and strong moral objections to any form of animal training are formulated. In this paper we critically assess the philosophical assumptions of zooanthropology as well as its rejection of animal training approaches that it takes to be out-dated and incompatible with a zooanthropological approach.

Zooanthropology and posthumanism

Zooanthropology is based on a posthumanist vision on the human-animal relationship, moving away from speciesism and anthropocentrism. Posthumanism can be understood 'as a deconstruction of symbolic, discursive, institutional and material arrangements that produce the category 'human' as something unique, distinct, and at the centre of the world' (Pederson, 2011: 68). Zooanthropology shares

EurSafe 2016 – DOI 10.3921/978-90-8686-834-6_26, © Wageningen Academic Publishers 2016

with post- or transhumanism its technoscientific optimism, and both want to undermine the deeply rooted opposition between humans and non-humans (for example other animals) in Western thought. It aims for an 'anthropodecentrism' that does not aim to depreciate humanity, but to favour 'anthropo-poiesis' (Marchesini, 2008a: 5). The anthropocentric heritage is very hard to overcome, because it has such a long history, not only in philosophical thinking – with amongst others the protagonists Descartes, Wittgenstein, Gehlen, and Nagel – but also in biology and classical ethology, with von Uexküll, Skinner, Lorenz, and Pavlov (Marchesini, 2015a). Roberto Marchesini rejects the assumption underlying classical ethology and behaviourism that animals are like machines, automatons that can be conditioned:

> The scientific explanation of animal behaviour, in searching to define 'how the animal machine functions' and restricting itself to defining features of animal mechanisms, has never problematized the Cartesian paradigm of the animal as automata. The only thing that has been problematized is whether that mechanism is an instinct or a conditioning.
> (Marchesini, 2014a)

Marchesini calls this assumption 'ontological anthropocentrism'. Like other posthumanists, he argues that the criterion of consciousness that is often used to determine whether an animal has moral status is just another way of making human beings the standard by which all other beings need to be measured; animals are then only morally relevant when they resemble human beings. Therefore, even in a zoocentric approach an anthropocentric paradigm is still functioning in the background. For example, Marchesini (2008a: 1) refers to the philosophy of Singer and Regan as consolidations of 'the concept of non-human otherness'. In a posthuman perspective one can only understand the human dimension by leaving the dichotomy between humans and non-humans (Marchesini, 2008a: 12).

To Marchesini zooanthropology emerges as a discipline that analyses the human-animal relational interaction and the contributions and results of this interaction. It is explicitly posthumanist in the sense that the animal can play a referential role in that interaction, and not (only) the human (Marchesini, 2008b). He stresses the importance of this type of interaction where the animal becomes a partner that is not reduced to object or instrument or category, although it does not imply symmetry in the relation. Others have described zooanthropology as 'combination of the latest scientific insights and developments, especially regarding cognitive capabilities in animals, culture and philosophy.' (De Giorgio and Schoorl, 2013).

Marchesini as well as De Giorgio and Schoorl have long focussed most of their activities on the practical applications of zooanthropology after the publication of 'Fondamenti di zooantropologia Vol. 2: Zooantropologia applicate' by Marchesini in 2005. Only recently, a theoretical background is emerging (Marchesini 2014a, 2015a,b,c), with 'La zooantropologia teorica' (Marchesini, 2014b) currently available.

Subjectivity

Zooanthropologists take all animals to be subjects. What do they mean by 'subjectivity'? Subjectivity is the state of an animal once we deny that they are simple automatons, 'predetermined by a set of prefixed incentives' (Marchesini, 2015: 80). When they can take ownership of their own 'endowments' (or resources) and exhibit creativity they are subjects. Animals are in 'control of their own endowments' that they can use freely and flexibly, in response to changing external circumstances. Marchesini deduces this from the fact that environmental circumstances change constantly and animals therefore have to respond to novel situations. This is reminiscent of the view that many biologists hold that animals have consciousness or awareness when they exhibit flexible learning. However, making consciousness such an important criterion to attribute subjectivity is problematic: 'Subjectivity precedes consciousness and is the foundation from which consciousness can emerge, not vice versa.' (Marchesini, 2015c). Otherwise 'the human being can always say: 'yes, but we have this characteristic that the non-human animal does not have' and he can base on it the subjectivity, consigning in a tautological way only to the human being' (Marchesini, 2014a). No, 'subjectivity is first of all desire.' (Marchesini, 2015c).

In Marchesini's view, the difficulty to know what it is like to be another animal is overestimated by Nagel, Wittgenstein and others. There is 'biocentric resemblance': universals that are common to all different species (e.g. the capacity to feel pain), homologies, which are characteristics that different species have in common due to their common ancestors, and analogies, which are common characteristics because species have to respond to similar selective environmental pressures. This 'common way of existence' that all animal species share enables mutual recognition and a real partnership between humans and other animals (Marchesini, 2008b). Yet, these common characteristics also make it possible to recognize diversity among species, and to recognise the 'Otherness of the non-human' (*ibid.*). Nevertheless, Marchesini (2008a: 4) stresses the importance of the plurality of the perspectives of the different species, and not the proximity to the human perspective.

It seems, then, that animal subjectivity is not so much something that is based on a specific criterion such as sentience or awareness, or being a subject-of-a-life (Regan, 1983), but something that is simply recognized in others. We can recognize ourselves in other animals by recognizing our common 'being-animal'. There are different dimensions to our commonality: our common vulnerability, common functionality, and taxonomic similarities. Because of these, we simply recognize them as a subject (Marchesini, 2015a).

Marchesini terms his outlook 'functional biocentrism', but it is a different type of biocentrism than used by environmental philosophers such as Taylor (1986) or Attfield (1983). For one, it only seems to apply to animals and not to plants. He takes it to mean 'a recognition of 'similar feelings' in non-human animals, even though they are expressed differently' (Marchesini, 2015a: 80). Yet, while not referring to plants as such, he does seem to take life as the criterion for moral status: 'This means that all life – living as being a body – has its own coordinates of pleasure, suffering, satisfaction, fulfilment, frustration and so forth (*ibid.*)'. Interestingly, ticks, bees or flies, animals that are not commonly regarded as having consciousness, are often-used examples. This view seems based on a rejection of the mind-body dualism and therefore consciousness as such is not that important. Instead, it is posed that animals have an individual 'perspective on the world' and are not completely 'defined by their own needs' as they are often pictured in classical ethology (Marchesini, 2015a: 78).

Marchesini attributes many mental capacities to animals about which there is still a lot of dispute among biologists, philosophers of mind and cognitive ethologists. For example, memory, awareness, even self-awareness, flexible responses to their environment. Moreover, he does so indiscriminately for all animals, also invertebrates. As a matter of fact, he takes all living beings to share certain needs, such as 'feeling pleasure, reproduction, avoiding suffering, and expressing one's own dispositions' (Marchesini, 2015a: 79). He does not seem to feel the need to base this on empirical knowledge, because in his eyes the dispute about animal subjectivity is in the first place a philosophical dispute and anyone who disputes animal subjectivity is put in the camp of anthropocentrists who have a Cartesian worldview.

Zooanthropology and training

Animals, both domesticated and captive wild animals, are cared for and housed in a wide variety of settings such as sanctuaries, zoos, research facilities, as well as in and around our homes. Animals are given different roles and are kept and/or used for different reasons such as: conservation, service animals, companion animals, and entertainment animals. There is little sign of any of these practices diminishing, meaning animal welfare aspects need to be seriously considered and addressed. Contemporary animal care and human-animal interaction approaches, frameworks and paradigms consider behavioural, cognitive, physiological, psychological, social, and affective aspects. The provisions to facilitate optimal animal welfare are (should be) at the forefront of any human-animal interaction, anywhere.

In the context of the zooanthropological rejection of animal training a clarification on what a modern animal training entails is needed. Early animal training was deeply rooted in classical and operant conditioning, offering mechanistic explanations for behaviour and behaviour changes. Contemporary animal training approaches and methods (in both domestic and captive wild animals) are based on a two-way communication and dialogue, choice and control. Humans and non-humans interact by exploring, playing, watching, listening and learning together. They can develop trusting relationships, bonds and friendships. These rewarding and empowering interactions can happen without reinforcers such as food, but through intrinsic motivations such as pleasure and comfort experienced, or information obtained. A bear pushing unwanted hay to a specific place, communicating that new hay would be appreciated is a refinement of caregiving (Poulse, 2009), instead of just cleaning and replacing hay when we think it should be done. This contemporary approach seems similar to what is expected in practical zooanthropology as described above, i.e. coexisting and social relationships, considering the animal's perspective.

Many domestic and captive wild animals are fearful or exhibit aggression especially when they perceive little to no control or choice over their environment and life. Training/learning has shown to be helpful in the reduction of inter-dog aggression, using desensitising and counter-conditioning (Orihel and Fraser, 2008), increasing the adoptability of shelter dogs (Luescher and Medlock, 2009).

Zooanthropology holds that when engaging with animals it is important to evaluate and consider our human expectations and impacts on an animal's behaviour. 'The reduction of the spontaneous behaviour often happens already during the initial training of young horses ... In the zooanthropological approach, especially when working as a facilitator in the horse-human interaction, it is fundamental to give the horse the possibility to express his own world and spontaneous behaviour' (De Giorgio and Schoorl, 2014). However, training does not necessarily reduce animals' spontaneous behaviour. For example, training pace and the horse's control over its response to the stimuli provided by the trainer resulted in a greater degree of explorative behaviour toward human beings when training horses (Baragli *et al.*, 2011). Using alternative methods, such as social learning, can speed up acquisition of complex actions and sequence actions in dogs (Fugazza and Miklosi, 2014). Animal caregivers/trainers of domestic and captive wild animals have to be aware of the effects their approaches and methods might have on the learning, curiosity, exploratory behaviour, and the two-way communication when engaging with animals. In this respect, zooanthropology could well find itself closer to contemporary animal training/learning paradigms than expected.

Zooanthropology as ethics

Zooanthropology makes several strong claims about animals and their mental capacities, often even stronger or broader in its application than those of Tom Regan (cf. supra). The goal in zooanthropology is to reach an ethic of empathy (Marchesini, 2015b). Marchesini distinguishes zooanthropology from Singerian utilitarianism, as the former is an operational ethics, while the latter is said to lead to 'an ethics of abstention'. 'Life is not welfare but well-being. In other words, being able to express all the power that each individual has deeply in her being-alive that is not simply a being exposed to life. Otherwise, we claim to realize a life ethics via an unsupportable mortification of life itself' (*ibid.*).

Presumably, this results in strong claims about how humans should behave and relate to animals (i.e. ethics). Reading the 'Manifesto teorico della zooantropologia' (Marchisini, 2008b) this is certainly the case in the first chapters where the importance of a (possible) 'referential role of the animal' in the relation is emphasised. Zooanthropology does not denounce the use of animals by humans, as long as the animal's Otherness is recognised. The Manifesto mentions pet therapy, but as an activity where one 'does not have to worry only if the person is interacting properly with that animal, but ask first if the configuration of the relation (the set of activated dimensions) develops beneficial or impairing referential content for that particular user.' (Marchesini, 2008b).

Therefore, the Manifesto's final conclusion is surprising. We expected to read that all animal use is wrong unless it meets the zooanthropological standards (as would be expected from an ethical theory), but the conclusion is that one should never mix zooanthropology and animal production; any mixed activity is to be considered 'harmful' for the zooanthropological activity (Marchesini, 2008b: 19). This begs the question whether the many images of production animals in Marchesini (2015b) constitute a change of view and a genuine broadening of the zooanthropological paradigm, or an unfortunate choice of imagery.

Conclusions

Overall, (applied) zooanthropology combines different fields of studies exploring our relationship with other animals in a cognitive-affiliative paradigm. While clearly grounded in the philosophical, ethological and cognitive sciences, this view on the human-animal relation is taken further and strong moral objections to the training of animals are formulated.

The opposition to the stimulus-response paradigm is understandable. It is in line within current holistic training approaches that consider behavioural, physiological, psychological, social, and affective aspects. However, clarifications are needed on why animal training (learning) is incompatible with zooanthropology, as well as on the practical approach, e.g. human-animal interactions, living with pets, caring for horses or non-releasable wildlife.

On a philosophical level, zooanthropology appears based on the assumption that all animals (including humans) are morally equal, because they possess 'subjectivity' The rather elusive concept of subjectivity is something at the base of consciousness, bringing out consciousness, not the other way around. Furthermore, the supposition that all animals have subjectivity and the strong call for an 'ethic of empathy' seem to inevitably lead to a strong abolitionist stance with regard to training and all other (instrumental?) uses of animals. This is certainly idealistic and, as Marchesini (2015b) admits himself, it 'inevitably asks us not small restyling operations ... but a real revolution as paradigmatic as the one carried out in order to overcome theocentrism'. Nevertheless, the way the approach is framed by Marchesini in practice is dramatically less ambitious.

Our conclusion is that zooanthropology certainly contains interesting and promising scientific as well as intuitively appealing ideas, but we question the apparent lack of philosophical groundwork or consequence analysis. Especially the radical limitation of zooanthropological practice to situations that meet the zooanthropological standards is disturbing if zooanthropology is to be accepted as a new kind of animal ethics. It is a positive development that we are starting to see the theory behind zooanthropology being presented and published. This will enable a more thorough discussion and evaluation in the future.

References

Attfield, R. (1983). The ethics of environmental concern. Basil Blackwell, Oxford, UK.
Baragli, P., Mariti, C., Petri, L., De Giorgio, F. and Sighieri, C. (2011). Does attention make the difference? Horses' response to human stimulus after 2 different training strategies. Journal of Veterinary Behavior: Clinical Applications and Research 6: 31-38.
De Giorgio, F. and Schoorl, J. (2013). The cognitive animal. The importance of understanding cognition in animals, in daily life and in the animal-human relationship. International Institute for Zooanthropology, Erp, the Netherlands.
De Giorgio, F. and Schoorl, J. (2014). The spontaneous horse. Understanding how to look at the horse without expectations. Relations. Beyond Anthropocentrism 2: 111-113.
Fugazza, C. and Miklosi, A. (2014). Should old dog trainers learn new tricks? The efficiency of the Do as I do method and shaping/clicker training method to train dogs. Applied Animal Behaviour Science 153: 53-61.
Luescher, A.U. (2009). The effects of training and environmental alterations on adoption success of shelter dogs. Applied Animal Behaviour Science 117: 63-68.
Marchesini, R. (2005). Fondamenti di zooantropologia. Vol. 2. Zooantropologia applicata. Perdisa, Bologna, Italy.
Marchesini, R. (2008a). Alterità non umane. In: Liberazioni. Available at: http://tinyurl.com/gtt9dtm.
Marchesini, R. (2008b). La SIUA – Manifesto teorico zooantropologia. Istituto di Formazione Zooantropologica, Galliera, Italy.
Marchesini, R. (2014a). Critical Anthropomorfism. Looking for an Animal Dasein. ASA conference, Association of Social Anthropologists of the UK and Commonwealth, Edinburgh, UK.
Marchesini, R. (2014b). Fondamenti di zooantropologia. Vol. 1. Zooantropologia teorica. Apeiron, Bologna, Italy.
Marchesini, R. (2015a). Against Anthropocentrism. Non-human Otherness and the Post-human Project. Nanoethics 9: 75-84.
Marchesini, R. (2015b). From an ethic of sympathy to an ethic of empathy. In: Animal publics: emotions, empathy, activism. Australasian Animal Studies Association Conference, 12-15 July, 2015. University of Melbourne, Australia.
Marchesini, R. (2015c). What is philosophical ethology? Symposium at Curtin University, Perth 22 July 2015. Perth, Australia.
Orihel, J.S. and Fraser, D. (2008). A note on the effectiveness of behavioural rehabilitation for reducing inter-dog aggression in shelter dogs. Applied Animal Behaviour Science 112: 400-405.
Pederson, H. (2011). Release the moths: critical animal studies and the posthumanist impulse. Culture, Theory and Critique 52: 65-81.
Poulse, E. (2009). Smiling bears: a zookeeper explores the behaviour and emotional life of bears. Greystone Books, Vancouver, Canada.
Regan, T. (1983). The case for animal rights. University of California Press, Berkeley, USA.
Taylor, P. (1986). Respect for nature. Princeton University Press, Princeton, USA.

27. The kind of perception that humans and animals share

U. Müller
International Centre for Ethics in the Sciences and Humanities (IZEW), Department Ethics and Education, Wilhelmstr. 19, 72074 Tübingen, Germany; uta.mueller@uni-tuebingen.de

Abstract

This article aims to show a special similarity between humans and animals, from which one can derive ethical consequences: We share with animals a kind of perception of objects in the surroundings, more precisely, we share non-conceptual perceptions of outer objects. But animals and humans differ in other regards – also with relevant moral implications: The perceptions of persons capable of speech have in contrast to those of animals (and small children) a conceptual content that enables persons to get knowledge of the outer world and draw logical and epistemic consequences. Combined with the capacity of reasoning is the capacity to act morally – to express wishes, to accept reasons for actions and to doubt some reasons for some actions. Animals do not have these abilities of reasoning and cannot act rational – according to this understanding of rationality at least. It might be objected that claims concerning the similarity between animals and humans concern rather empirical questions that biologists or cognitive psychologists have to answer and do not concern philosophical or ethical questions. One could argue that empirical studies which try to describe animal behaviour have to interpret this behaviour with (theoretical) terms which have different meanings: In animal studies 'perception', 'reasoning', 'language' and 'concepts' are notions which philosophers discuss controversially in various argumentations and traditions, also the attribution of certain characteristics has to be reflected and justified by the means of philosophical analysis. Certain results of empirical research may play a major role in ethical debates. In this field interdisciplinary research in cooperation with the natural sciences and the humanities could provide necessary knowledge and arguments with the objective of prescriptive ethical-epistemic judgements.

Keywords: moral respect, similarity, fellow creatures

Introduction

The central ethical thesis that human beings should not only respect their fellow humans but also non-human life is widely accepted. The fundamental notion is 'respect' and my aim is to argue for a respectful treatment of animals as living beings with certain interests. But I will not draw attention on the notions of 'suffering' or 'interests' and thus I do not discuss utilitarian arguments here (cf. Singer, 1976).

The article will focus on the ability of animals to perceive their surroundings, more exactly to perceive their surroundings and to discern certain objects and living beings. My argumentation is meant to support the pathocentric approach in animal ethics, which is a fundamental ethical approach that argues for the respect for animals (cf. Wolf, 1997). Crucially, I do not attach to my argument the idea that we as humans who are capable of language and reasoning do not differ in other essential respects from beings who are not capable of language or reasoning based on a certain level of abstract thinking. It must be conceded, however, that the interpretation of 'language', 'reasoning', 'concept' and others is controversial; I try to clarify some of these notions in the following paragraphs.

Adult human beings share a kind of perception with animals that one can describe as non-conceptual, that means we and small children and animals see objects as objects, and persons as persons, although animals and small children do not have well-founded knowledge. The perception of adult persons as opposed to animals' (and small children's) perception also has a conceptual and epistemic dimension.

EurSafe 2016 – DOI 10.3921/978-90-8686-834-6_27, © Wageningen Academic Publishers 2016

This gives adults the ability to gain conceptual knowledge of the world, of persons and of theoretical phenomena in the sense that they are capable of engaging in logical reasoning, formulating desires and accepting or rejecting rational reasons for actions. In my interpretation of these abilities, animals apparently do not have these abilities and small children do not have them yet (Bremer 2005: 296). This definition of 'knowledge' is narrow or strict, which might be criticized; my position refers to philosophers like D. Davidson in the tradition of L. Wittgenstein, who claim that knowledge presupposes a whole web of belief, including some theoretical beliefs and justifications of beliefs (Davidson 1984: 155).

With regard to the ethical relevance of this proposition, the direction in which I will argue can be described as follows: If other living beings share perception features with human beings then they share certain capabilities of experiencing the world. In one specific respect, they are like us, they are similar to us, but in other respects they are different.

The role of cognitive episodes

The presented argumentation claims – on the one hand – that perceptual episodes of human adults have a non-conceptual content. This differs from two other prominent theoretical approaches: One is the so called conceptualism, which claims that only living beings who possess the relevant concepts can be said to properly perceive objects, persons, etc. This implies that living beings that do not have such concepts, cannot perceive things or 'anything'. The other approach claims that we have unordered sensual impressions (which animals do have) and which are so ordered and systematised by our conceptual system that we gain knowledge, i.e. that we are capable to perceive things in the world as things. Amongst others, W. Sellars and D. Davidson criticise this understanding for falling victim to the 'myth of the given' (cf. Davidson, 1984: 183; Sellars, 1963). The label 'myth' is supposed to emphasize that the non-conceptually, sensually 'given' cannot serve as a justification for our conceptual perceptions. The 'given' is non-conceptual and it remains unclear how concepts are supposed to be capable to order purely sensual impressions.

The defenders of conceptualism would like to hold on to the notion that perceptions of the external world or general sensual experiences form a substantial part of our perception and our knowledge. However, this can only be the case if they already are conceptually structured. This is why experience or receptivity itself is conceptually structured (in allusion to Kant's conception of spontaneity, Spontanität: McDowell, 1996).

The central argument for conceptualism in the vein of McDowell's can be summarised like this: There are inferential relationships between beliefs, values, doubts, wishes, etc. We can draw inferences from certain claims, judgements or certain convictions that allow us to formulate certain desires. The defenders of conceptualism claim that such relationships also exist between perceptions and beliefs, insofar as certain beliefs are established and justified by certain perceptions: I see that it is raining outside and deduce that I should take an umbrella with me if I do not want to get wet. However, inferential relations can only obtain between conceptual contents: I know that it is raining, and I know that an umbrella protects me from wetness and I do not wish to be wet. If the sequence of beliefs is also supposed to incorporate the truth-content that it rains, then the theoretical conclusion must be that: perceptual contents are conceptual.

The opposing position of epistemological foundationalism – perceptual contents are non-conceptual and are expressed by so-called 'observations sentences' – would, according to the conceptualists, be an instance of the 'myth of the given'. However, if perceptual contents are conceptually structured, then only those beings that also have concepts can have perceptions along with their corresponding contents. I will argue against this notion as follows: Perceptions have an informational content, also when the perceivers

– for example animals – do not have concepts as understood here. One could concede that animals have 'concepts' in the sense of specific perceptions of objects with a certain informational content (e.g. 'bone', 'mistress', 'mouse'), but this interpretation differs from that of concepts according to Davidson and other philosophers, which allows to distinguish between humans and animals.

The notion of non-conceptual perception

The claim that there is a non-conceptual content to perception is not supposed to be made against the notion that there can only be inferential relationships between conceptual mental states (beliefs, desires, claims, justifications, etc.). If perceptions should justify our knowledge and our reasoning, they must have conceptual content (and in fact for us mature adults they mostly have such a content). However, this does not imply that perceptions only have conceptual contents (cf. McDowell, 1996). According to the position I defend it can be granted that when we hear a certain ringing sound, we hear our mobile phone ring, or see a certain piece of paper, we see 'The Times', etc. But, nevertheless the perception has a content which is non-conceptual and has qualities other than epistemic content.

Arguments in favour of the notion of non-conceptual perception

The claim that there is non-conceptual perception is of great importance because it states that animals also have perceptions of something. And this claim is, if one observes animals and small children, rather intuitive: The dog behaves as if he sees his owner and the bone in its hands, and the little child behaves as if it sees its mother and its father. This suggests that they see humans and things, also when they do not know what a human is or what a man or a woman is or what (in the case of the dog) a bone is: 'Animals do not have concepts in the sense in which we attribute them to humans, especially no theoretical concepts. However, they have systematic capacities of discrimination that function like primitive forms of concept-control. To discriminate a cat does not mean to have the concept CAT but a dog is well capable of differentiating a cat from a kennel' (Bremer, 2005: 297). To use concepts presupposes the capacity of having beliefs and other propositional attitudes (cf. Davidson, 2004: 155).

So, this approach claims that there are perceptions of something, which at the same time are not perceptions of something in conceptual relationships as something and are therefore non-conceptual: Living beings with certain biological features, such as human beings and other animals, can also see something if they do not (always) see it as something (in the mentioned strict sense). So perception does not function in such a way that purely qualitatively given perceptions are being systematised and ordered by the concepts of speech-possessing, reasonable beings. It is exactly this idea that there is a purely sensual given without informational content that is only systematized by conceptual thinking that Sellars and Davidson have convincingly criticised with their arguments against the 'myth of the given'. In the following, I will gather some arguments to support the idea of non-conceptual perceptions.

Independence of non-conceptual from conceptual perception

Non-conceptual perception's independence from conceptual perception could also be demonstrated from the conceptual perspective. To do so, all kinds of false perceptions can be appealed to. Let us assume a red light shines on a white wall: I perceive the wall as red, even if I know that the wall is not red but white. The crucial point of non-conceptual seeing lies in the fact that objects can be seen in a certain way without conceptually knowing what one has seen. There is no conceptual seeing without non-conceptual seeing, but – as the examples have shown – there is non-conceptual seeing without conceptual seeing. As stated above, one could argue that (some) animals are capable to use concepts, because they see objects and persons; here, the important point is that this interpretation of 'concept' differs from the interpretation of this term in human communication: We use concepts in epistemic contexts to justify, doubt, know, reason about the facts concepts are referring to; and animals probably do not have these abilities (admittedly, this is an empirical question which appropriate research has to decide).

Perceptions have a non-conceptual content

According to the argument developed, the conceptual content of our perceptual episodes is not their only content: We adults usually perceive the things that we perceive as determined things and know what kind of things they are and which purpose they have. But the embedding of perceptions into the space of concepts does not exclude that perceptions have other qualities. G. Evans explains the connection between non-conceptual and conceptual information as follows:

> The informational states which a subject acquires through perception are non-conceptual, or non-conceptualized. Judgements based upon such states necessarily involve conceptualization: in moving from a perceptual experience to a judgement about the world (usually expressible in some verbal form), one will be exercising basic conceptual skills. (Evans 1982: 124)

Evans emphasizes that the non-conceptual informational states are those that we share with animals; these states allow the explanation of perception of something and memory of something in the case of animals (it may be conceded that they have 'concepts' in a very restricted sense, see remark above). This is relevant because the behaviour of animals, pets for example, suggests that they can remember certain events or persons and – even more plausibly – that they can perceive certain objects and persons as such, for example their mistress or master.

The richness of non-conceptual perception

According to the theory of non-conceptual perception, the content of non-conceptual perceptions is much richer than the concepts that adults have to describe their perceptual impressions (or to provoke them in the first place). Essential elements of the perceptual experience would be neglected if people were to limit themselves to the experiences that they can describe conceptually (cf. Peacocke, 2001: 240). The conceptual processing cannot be adequate to the fineness of the perception: 'Do we really understand the proposal that we have as many colour concepts that we can sensibly discriminate?' (Evans, 1982: 229).

Perception is natural

If adults do have non-conceptual perceptions, one could claim that they have perceptions that are similar to those of beings who have no concepts or rather have primitive concepts or very few concepts: There lies a tomato on the table and the infant sees the tomato and not some other thing. However, it does not have not be capable to know what a tomato is and for example that it comes from Italy.

One might interject here that this may be only the case for children and more complex animals because these are biologically similar to us humans. To answer these questions further conceptual and empirical research is needed; many philosophers work together with natural scientists (especially cognitive psychologists) to get insights into cognition, in particular into learning skills and memory of animals. The research in cognitive ethology shows promising results by analysing the evolutionary development of cognitive capacities of human and animals (cf. Bekoff *et al.*, 2002). Philosophers may support research by clarifying the notions the researchers use (for example which meaning of 'concept' is at stake here). We do not know what it is like to be a bat but we know that bats have a perceptual apparatus that enables them to differentiate some things in their environment from other things – for instance, they are capable of locating fluttering moths to catch and eat them. What matters here is that there are good reasons to assume that many living beings have pre-discursive and pre-intellectual perceptions of discriminate things in the surrounding.

Concluding remarks

The goal of the preceding considerations was that it is plausible to assume that – at least in one respect – we have the same perceptual experiences as animals. As indicated above, we can draw ethical conclusions from this idea insofar as that the basis for moral respect is our similarity with other living beings. A related argumentation is presented by C. Diamond, who introduced the notion of a 'fellow creature' to emphasize the respect for animals (Diamond, 2004: 105).

The argument presented may support the pathocentric approach for treating animals not as non-sentient objects but as living beings who experience partially like humans. It is a possible way to argue with the pathocentric approach against torturing animals and for a respectful treatment of non-human animals. It states that the moral respect for living beings is connected to the corporal-sensual welfare of them (cf. Wolf, 1997).

However, I have emphasized that the perceptual contents for us humans often contain a non-sensual, theoretical content because we are beings capable of language and reasoning. So there is indeed a difference between humans and animals. These capacities, language and reasoning, which define us as human beings, do justify certain obligations: '... the moral expectations of other human beings demand something of me as other than an animal' (Diamond 2004: 106). Rather, these capacities are (in a sense inverted) what turns humans into morally (or immorally) acting beings; for example, the fact that humans are capable of recognising reasons for certain actions (or dismiss such reasons). We can decide how to treat animals, including animals we use for food or as laboratory animals.

It is a promising first step towards finding convincing reasons for respecting other, non-human living beings to begin arguing from the notion of partly similar experiences of animals and humans. And it is a popular and – in my view – convincing opinion that our moral obligations concerning human beings are different from and go further than the obligations we have towards animals. However, this debate is beyond the scope of this essay, in which I have tried to sketch a framework within which to situate my notion of a similarity between humans and animals. Let me conclude with a quote from E. Tugendhat. Regarding the different moral importance of humans and animals, he notes: 'Our moral obligation towards the sheep has a different weight that the one towards the human. Not because the sheep lacks certain qualities (for example those, which it has to have as a subject of a moral community) but because the sheep is a sheep and the human is a human and because we are humans.' (Tugendhat, 1997: 107).

References

Bekoff, M., Allen, C. and Burghardt, G.M. (2002). The cognitive animal. Empirical and Theoretical Perspectives on Animal Cognition. MIT Press, Cambridge, MA, USA.

Bremer, M. (2005). Tierisches bewusstsein als testfall für die kognitionswissenschaften. In: Herrmann, C., Pauen, M., Rieger, J. and Schicktanz, S. (eds.) Bewusstsein. Stuttgart, Fink, Germany, pp. 286-308.

Davidson, D. (1984). On the very idea of a conceptual scheme. In: Davidson, D. (ed.) Inquiries into truth and interpretation. Clarendon Press, Oxford, UK, pp. 183-198.

Davidson, D. (1984). Thought and talk. In: Davidson, D. (ed.) Inquiries into truth and interpretation. Clarendon Press, Oxford, UK, pp. 155-170.

Diamond, C. (2004). Eating meat and eating people. In: Nussbaum, M. and Sunstein, C.R. (eds.) Animal rights. Oxford University Press, Oxford, UK, pp. 93-107.

Evans, G. (1982). The varieties of reference. Clarendon Press, Oxford, UK.

McDowell, J. (1996). Mind and world. Harvard University Press, Cambridge, MA, USA.

Peacocke, C. (2001). Does perception have a nonconceptual content? Journal of Philosophy 98(5): 239-264.

Potthast, T. (2015). Ethics in the sciences beyond Hume, Moore and Weber. In: Meisch, S., Lundershausen, J., Bossert, L. and Rockoff, M. (eds.) Ethics of science in the research for sustainable development. Nomos, Baden-Baden, Germany, pp. 129-151.

Sellars, W. (1963). Science, perception and reality. Routledge, London, UK.

Singer, P. (1976). All animals are equal. In: Regan, T. and Singer, P. (eds.) Animal rights and human obligations. Prentice-Hall, Upper Saddle River, NJ, USA, pp. 73-86.

Tugendhat, E. (1997). Wer sind alle? In: Krebs, A. (ed.) Naturethik. Texte der gegenwärtigen tier- und ökoethischen diskussion. Suhrkamp, Frankfurt/Main, Germany, pp. 100-110.

Wolf, U. (1997). Haben wir moralische verpflichtungen gegen tiere? In: Krebs, A. (ed.) Naturethik. Grundtexte der gegenwärtigen tier- und ökoethischen diskussion. Suhrkamp, Frankfurt/Main, Germany, pp. 47-75.

Section 6. Corporate social responsibility and public procurement

28. Norwegian salmon farming and the chase for social legitimacy

S.G. Carson[1] and K. Rønningen[2]*
[1]NTNU, Faculty of Natural Sciences and Technology, Department of Chemistry, 7491 Trondheim, Norway; [2]Center for Rural Research, 7491 Trondheim, Norway; siri.granum.carson@ntnu.no

Abstract

Norway is a world-leading producer of farmed Atlantic salmon, but despite economic success the industry is controversial, and public opinion and local communities are increasingly sceptical or even hostile. The industry claims to provide a sustainable answer to challenges of food security, and aims for an extensive growth in the years to come, but arguably lacks the social legitimacy needed to achieve this goal. Threats to the social legitimacy of a company or an industry are today typically addressed by help of corporate social responsibility (CSR) initiatives. In this paper, we describe how structural and technological changes affect companies' and their stakeholders' perception of corporate social responsibility. We apply Matten and Moon's (2008) distinction between implicit and explicit CSR; while implicit CSR mainly consists in adapting to regulatory system and institutional demands, explicit CSR typically consists in company-specific initiatives expressively communicated to key stakeholders. In the case of the Norwegian aquaculture industry, the rise of explicit CSR can be seen as a response to economic globalization and the expansion of markets across borders and institutional frameworks. But social legitimacy is not simply a matter of how a company is perceived globally, but importantly also a matter of how it is valued by local communities. Many fish farms started as small, locally owned operations and have developed into multinational corporations. We argue that while the transition from implicit to explicit CSR aims at enhancing social legitimacy, it could paradoxically weaken the ties between companies and local communities if focus is only on sustainability in a global perspective.

Keywords: aquaculture, Atlantic salmon farming, Norway, CSR, social legitimacy

Introduction

Norway has built a technologically advanced and economically viable industry as the world-leading producer of farmed Atlantic salmon. However, despite economic success the industry attracts critical media coverage and protests by a number of interest groups, from coastal fisheries to environmental organizations via sports fishing clubs to second home owner associations. Further, local communities are increasingly reluctant towards setting off the required space for the fish farming industry, as they perceive the benefits from these companies declining along with their moving out and centralizing of HQs, jobs and taxes, and because the industry is not seen as aligned with the identity and profile some of these municipalities are building as tourism and recreational heavens (Frisvoll *et al.*, 2014; Torsethaugen *et al.*, 2014).

The aquaculture controversy in Western societies (cf. Young and Matthews, 2007) is rooted among other things in environmental concerns with regards to diseases and escapes and more recently to the production of feed (soy, palm oil, genetic modified organisms). The competing views regarding the overall sustainability of the industry range from the industry's own claim to provide a healthy and sustainable answer to global challenges of food security, to environmental organizations' claim that the industry can never be sustainable, at least without undergoing dramatic changes (FiVH, 2012). Although Norwegian political authorities support the plans for exponential growth in the years to come, the industry must arguably tackle its social legitimacy issues in order to achieve this goal.

EurSafe 2016 – DOI 10.3921/978-90-8686-834-6_28, © Wageningen Academic Publishers 2016

Threats to the social legitimacy of business organizations are today typically addressed through corporate social responsibility (CSR) initiatives. CSR can be considered as 'a strategic approach to legitimacy' (Suchman, 1995), whereby social and environmental values are increasingly used in brand-building in order to strengthen the reputation of an organization. In this paper we distinguish between implicit and explicit CSR (cf. Matten and Moon, 2008); while implicit CSR mainly consists in adapting to regulatory system and clearly stated institutional demands, explicit CSR typically consists in company-specific initiatives expressively communicated to key stakeholders.

We argue that while the Norwegian salmon farming industry increasingly uses explicit CSR to enhance their social legitimacy, in many cases this aim is not achieved. This is of course partly because the environmental challenges of the industry have yet to be solved properly. But an additional reason might be that the CSR initiatives fail to address the level where the challenge against social legitimacy is most precarious – for example the CSR initiatives might have a global scope, while the primary challenges are on a local level, e.g. a lack of access to production sites. More generally, we argue that the explication of CSR in the Norwegian context is driven by globalization, and that the focus on CSR in terms of global sustainability could potentially weaken the ties between companies and local communities.

Methodologically, this is primarily a conceptual paper, aiming for a clarification and systematization of different aspects of CSR in salmon farming companies and in the general CSR debate. The paper draws upon examples from media and company reports as a means of elaborating on the elements of the conceptual framework, and findings from a previous study on locality access for the aquaculture industry in Norway.[1]

Dealing with complexity: from implicit to explicit CSR

Traditionally, corporate social responsibility has not been a big issue for Norwegian companies, one important reason being that issues such as labour rights, environmental concerns and food safety in Norway are firmly regulated by a combination of laws, corporatist institutions and a strong civil society (Gjølberg 2010), limiting the space for the kind of voluntary corporate initiatives often labelled as CSR. Matten and Moon (2008) has launched a distinction between 'explicit' and 'implicit' CSR, where explicit CSR is what we normally perceive as CSR, i.e. companies explicitly expressing social and environmental responsibility in order to protect or build reputation and good relations with stakeholders. This form of CSR originates and is typical of companies in the US, where weaker institutional embedding of the economy opens up a comparatively larger space for self-regulation and company-specific initiatives. Implicit CSR, on the other hand, simply means the social responsibility of companies as expressed in their adaptation to the prevailing regulatory systems and institutional demands. This form of CSR is typical for Europe in general and not least for the highly institutionalized Scandinavian economies.

It is important to emphasize that a high level of explicit CSR does not necessarily mean that the social and environmental performance will be stronger than in a more implicit CSR regime. Even if Scandinavian companies traditionally have not been very explicit about their responsibilities, there is no evidence suggesting that these companies have performed in a less responsible way than corresponding companies within an explicit CSR tradition. Quite the contrary, Scandinavian companies do comparatively very well on international CSR and sustainability performance measurements (Midttun *et al.*, 2006; Strand and Freeman, 2013).

[1] 'Locality access for aquaculture – competition on area access along the coast'. The research project was financed by the Norwegian Seafood Research Fund – FHF. Project partners were Nofima, University in Tromsø, Centre for Rural Research, and Sintef.

The Norwegian labour situation represents the paradigm of implicit CSR: largely institutionalized through laws and agreements, creating a rather strong congruence between the business community and society (cf. Trygstad and Lismoen, 2008). However, when the state's control function is played down due to economic globalization and political liberalization, many Norwegian companies increasingly adapt explicit CSR. This explication process seems to be driven by a general rise in 'organizational expressiveness' with increasing attention to the image and values of organizations, and a need to 'relegitimise' business organizations in light of their altering role in society (Carson *et al.*, 2015).

With increased focus on environmental and social issues in society in general, power and resource accumulation in the private sector and high profiled corporate scandals, CSR emerges as the solution to the corporate world's new need to legitimate itself. The transfer of authority, together with the growing wealth and power of business, creates a legitimacy gap (Beck, 2007). According to a neo-institutional perspective, explicitating social and environmental values is a strategic move by which an organization can gain, maintain, or repair social legitimacy (Matten and Moon, 2008).The rise of explicit CSR has, in other words, to do with the need to stay legitimate within an economy where business organizations have increased their ability to influence their own conditions and constraints. Companies must answer to a new and more complex set of expectations in order to achieve the social legitimacy necessary for their business. Explicit CSR becomes a method by which the companies attempt to fill the legitimacy gap. Companies spend more and more resources on communicating to society that they are socially and environmentally responsible, through value statements as well as through different kinds of reporting, certification, and other kinds of 'soft governance' (Vogel, 2008)

The example of Marine Harvest

This development clearly affects Atlantic salmon farming in Norway, where a rise of explicit CSR can be seen as a response to economic globalization and the expansion of markets across borders and institutional frameworks. Marine Harvest, which started as a small, locally owned operation, is today the world's largest producer of farmed Atlantic salmon, and one of the major aquaculture companies in the world, employing 11,715 people in 23 countries. Marine Harvest puts a lot of effort into building a public image as a socially and environmentally responsible company, e.g. by cooperating with WWF on stakeholder dialogues. They have also teamed up with WWF against genetically modified salmon, thus becoming a voice in the public debate on this as well as other political and environmental issues. Their CSR policy is carried out locally as well, e.g. by supporting community organizations within their production areas.

However, as a multinational company, Marine Harvest struggles with communicating the right message, and with balancing expectations from different sets of stakeholders. In 2014 there was a controversy in Norwegian media concerning a commercial movie directed at the Japanese market. The movie showed gorgeous pictures of salmon bouncing up the Vosso river. This was actually animated salmon, drawn into a water course where wild salmon is threatened by extinction – as it is in the rest of Norway and the world, partly due to diseases and other problems caused by aquaculture. When parts of the film was showed and criticized (or rather ridiculed) in a Norwegian TV show, the national TV company was instructed to remove the clip on account of breach on copyright, leading to accusations against Marine Harvest that they were trying to gag public debate. Further, in 2015 it became known that one of the girls of a soccer club in Vancouver Island supported by Marine Harvest was a wild salmon proponent protesting against the fish farming activities in the area. In response, the Canadian Marine Harvest management threatened to withdraw sponsor funding to the club (CBC News, 2015). The Norwegian owner claimed they would not limit peoples' right of expression, but did not interfere with the site management (NRK, 2015).

Marine Harvest is generally seen as one of the 'good guys' of this industry. The public outcry in these cases can serve as an illustration of the looming social legitimacy challenges of the industry, and the complexity of the social and environmental issues they must tend to.

The social legitimacy of salmon farming

Onshore farming of Atlantic salmon is highly contested, and social legitimacy is increasingly recognized as a key factor for the continued growth of Norwegian aquaculture (Frisvoll *et al.*, 2014; Gullestad *et al.*, 2011; Isaksen *et al.*, 2012; Whitmarsh and Palmieri, 2009). A key question is whether this kind of aquaculture is – or can be – a sustainable form of food production. Marine Harvest argues that it represents part of the answer to the global challenge of food security through a 'Blue Revolution' (cf. Marine Harvest, 2016).

However, there are a number of challenges against this claim, and in Norwegian local communities, the industry faces resistance from several groups. As locally or regionally owned fish farm companies, the ties to the local community and the direct effects in terms of employment, money, activity, and settlement were obvious. The transition towards large, internationally owned company structures with centralized decision structure changes the situation radically when it comes to stakeholder relations and expectations. Through globalization and concessions, parts of the industry is leaving Norwegian coastal districts and coastal culture. Technological, structural and economic changes entail changes to physical installations as well as the sector's organization. For example, due to the organizing of operations on the feeding platforms, personnel are able to live and pay taxes in other municipalities and head offices, and ownership is removed so far from the local communities that the important issues of these communities become less visible and relevant for the companies.

Often, local resistance against aquaculture has been criticized as cases of 'NIMBY' (Not in my backyard), implying that the protesters understand that Norway needs this industry, only do not wish to see the cages from their own kitchen windows. Arguably, however, there are legitimate concerns with regards to environmental and social justice in the cases where local fisheries and tourism suffer but are not compensated, or where local communities see the environmental damage and social risks without receiving economic benefits as profit mainly goes to the corporation or distant shareholders.

The complexity of social legitimacy

Salmon aquaculture affects a complex interconnectedness of eco-systems, economic and social systems at different levels (cf. Bailey, 2014: 35), and while the industry today flourishes at a national level in Norway, the lack of social legitimacy at both local and global level continues to pose a threat against the industry. The sustainability of the industry is not simply a matter of balancing the overall global issues of economy, environment and society, but the same balance must also be achieved at a local level. Social legitimacy is not only a matter of how a company is perceived globally, but also a matter of how it tends to the needs of the communities of which it is a part. If social responsibility becomes an instrumental scheme, managed by consultation agencies with different or generic interpretations of CSR, this could potentially weaken the sense of obligations and interconnectedness between a company and the local stakeholders, as well as the companies' understanding of what builds legitimacy in what contexts. Rather, it must be integrated as an ongoing process, with self-reflexiveness within the company and an embedded reciprocity between company and community.

In the Norwegian setting, the language of CSR is generally connected with globalization of the economy and the changing expectations companies face due to this. In line with this, Marine Harvest claims to contribute to sustainable production and global food security through a 'Blue Revolution'. But even multinational companies depend on the local communities they operate in, and arguably the biggest challenge to the social legitimacy of the salmon farming industry is the continuing resistance within these communities.

References

Bailey, J. (2014). Looking for sustainable solutions in salmon aquaculture. Etikk i praksis-Nordic Journal of Applied Ethics 8(1): 22-40.

Beck, U. (2007). Organizations in world risk society. In: Beck, U. and Holzer, B. (eds.) International handbook of organizational crisis. Sage, Washington, DC, USA, pp. 3-24.

Carson, S.G., Hagen, Ø. and Sethi, S.P. (2015). From implicit to explicit CSR in a Scandinavian context. Journal of Business Ethics 127(1): 17-31.

CBC News (2015). Available at: http://tinyurl.com/gv7gat6.

FiVH (2012). Available at: http://tinyurl.com/jua9apc.

Frisvoll, S., Rønningen, K., Torsethaugen, K. and Sunde, L.M. (2014) Lokalitetstilgang for havbruket – er det grunnlag for en vinn-vinn-situasjon? Arbeidsrapport 1/2014, SINTEF Fiskeri og havbruk og Norsk senter for bygdeforskning. Trondheim, Norway.

Gjølberg, M. (2010). Varieties of corporate social responsibility (CSR). CSR meets the 'Nordic Model'. Regulation and Governance 4(2): 203-229.

Gullestad, P., Bjørgo, S., Eithun, I., Ervik, A., Gudding, R., Hansen, H. and Røsvik, I.O. (2011). Effective and sustainable use of areas for aquaculture – land of desire. Report from an expert panel appointed by Ministry of Fisheries and Coastal Affairs. Oslo, Norway.

Isaksen, J.R., Andreassen, O. and Robertsen, R. (2012). Kommunenes holdning til økt oppdrettsvirksomhet. The municipalities' attitudes towards increased aquaculture. Nofima Rapport 18.

Matten, D. and Moon, J. (2008). Implicit and explicit CSR. Academy of management Review 33(2): 404-424.

Frisvoll, S. and Rønningen, K. (2009). Kampen om gråsonene [The battle for the grey zones]. Report 8/09. Center for Rural Research, Trondheim, Norway.

Midttun, A., Gautesen, K. and Gjølberg, M. (2006). The political economy of CSR in Western Europe. Corporate Governance 6(4): 369-385.

NRK (2015). Available at: http://tinyurl.com/zzpp6nx.

Strand, R. and Freeman, R.E. (2015). Scandinavian cooperative advantage: the theory and practice of stakeholder engagement in Scandinavia. Journal of Business Ethics 127(1): 65-85.

Suchman, M.C. (1995). Managing legitimacy: strategic and institutional approaches. Academy of Management Review 20(3): 571-610.

Torsethaugen, K., Rønningen, K., Frisvoll, S., Andreassen, O., Robertsen, R., Hersoug, B., Johnsen, J.P., Sunde, L.M. and Karlsen, K.M. (2014) Arealmangel eller fordelingskrise i havbruksnæringa? Å navigere mellom ulike verdier, prioriteringer og behov. Norsk Fiskeoppdrett 4: 46-48.

Trygstad, S. and Lismoen, H. (2008). Fagbevegelsen og CSR [The trade union movement and CSR]. Oslo, Norway.

Vogel, D. (2008). Private global business regulation. Annual Review of Political Science 11: 261-286.

Whitmarsh, D. and Palmieri, M.G. (2009). Social acceptability of marine aquaculture: the use of survey-based methods for eliciting public and stakeholder preferences. Marine Policy 33(3): 452-457.

Young, N. and Matthews, R. (2007). Experts' understanding of the public. Public Understanding of Science 16(2): 123-144.

29. The ethical matrix as a potential tool in public procurement of food

C. Brunius[1], *P. Moula*[2] *and P. Sandin*[3*]
[1]*Department of Food Science, Swedish University of Agricultural Sciences, 75007 Uppsala, Sweden;*
[2]*Department of Philosophy and History, Royal Institute of Technology, 10044 Stockholm, Sweden;*
[3]*Department of Crop Production Ecology, Swedish University of Agricultural Sciences, 75007 Uppsala, Sweden; per.sandin@slu.se*

Abstract

Public procurement represents a noticeable proportion of the total consumption of food in society and ethical concerns are increasingly being raised among citizens and politicians. The present study has evaluated the ethical matrix with regard to its ability to aid decision makers in making ethically informed judgements and taking ethical considerations into account in public procurement of food. We present the results of case studies in which small mixed groups of politicians and officials from Swedish municipalities have applied the ethical matrix to a particular procurement issue (e.g. organic vs conventionally produced meat). Whereas both groups were in general positive to the use of the tool in the seminars, politicians were less positive regarding the usefulness of the matrix as a tool for political decision-making. Both groups emphasized that, on a municipality level, the most likely use of the matrix lie in the early stages of policy work to arrive at a common platform within working groups and also at a stage after decision-making in communication with citizens.

Keywords: public procurement, food, ethics, the ethical matrix

Introduction

Each year, Sweden spends about ten billion Swedish Crowns (just over 1 billion Euros) on public procurement of food, which amounts to about 4% of the total market for food and restaurant services in the country (2013 figures; Konkurrensverket, 2015). These services are procured by national government agencies, but most importantly by regional county councils (landsting) and local municipalities/cities (kommuner). The county councils are primarily responsible for the health care system, e.g. public hospitals, while the municipalities are responsible for kindergarten, schools, and nursing homes. To an increasing degree, considerations often characterized as 'ethical' are demanded by the public and included in qualification criteria for public procurement. Those demands might include commitments by municipalities to buy a certain percentage of certified organic products, fairtrade, locally produced, and so on. This trend is of course not limited to Sweden (Kleine and das Graças Brightwell, 2015; Stein, 2013).

However, the discussion on ethics in procurement appears to lack structure, and is often limited to questions of compliance with various codes of conduct. The object of the present study is to evaluate an ethical tool – the Ethical Matrix – with regard to its ability to aid decision makers in making ethically informed judgements and taking ethical considerations into account in public procurement of food. (By an 'ethical tool', we mean 'a practical method and/or conceptual framework with the main purpose of helping the user(s) improve their ethical deliberations in order to reach an ethically informed judgment or decision', Moula and Sandin, 2015: 264.)

I. Anna S. Olsson, Sofia M. Araújo and M. Fátima Vieira (eds.) **Food futures: ethics, science and culture**
EurSafe 2016 – DOI 10.3921/978-90-8686-834-6_29,

Method

We carried out three exploratory workshops in Swedish municipalities using the ethical matrix. The ethical matrix is a tool which applies the common-sense ethical principles of respect for well-being, autonomy, and justice to a number of stakeholders or affected groups with regard to a specific issue. The Ethical Matrix was developed by Ben Mepham and others in the 1990s with the principal aim 'to assist non-philosophers to appreciate the value of ethical insights in arriving at well-considered ethical judgments' (Mepham, 2010: 18), especially in issues regarding the introduction and application of technology in society. It is one of the most well-known ethical tools and it has been used in several case studies in different areas, not only food and agricultural biotechnology (e.g. Cotton, 2009; Kaiser *et al.*, 2007).

The workshops were carried out in medium to large sized municipalities in central Sweden. They were selected on the basis of interest and there is no claim made that the sample is representative of Swedish municipalities in general. The participants were (a) local politicians and (b) officials somehow involved in procurement of food in their respective municipalities. Both categories were present at all workshops.

Before the workshops, a short description outlining the method was sent out the participants. The workshops took a maximum of three hours with a coffee break included. After a short introduction by the facilitators where the matrix was presented, the participants briefly practiced using the matrix together with the facilitators on a topic unrelated to the one treated in the main discussion.. After this the participants discussed an issue of their choice, in all three cases related to meat. The participants were provided large sheet (A3) blank matrices which they filled in during the exercise. The groups managed the discussion mainly by themselves with the facilitators at hand to provide expert knowledge and occasionally moderate the discussion.

After the workshop, the participants (5-6 per municipality) were asked to fill in a questionnaire with one quantitative part where they rated various aspects of the matrix and one qualitative part with open text answers. The quantitative section consisted of five questions (translated from Swedish):

Q1 'What is your overall evaluation of the method (the Ethical Matrix)?';

Q2 'Do you think that the matrix could be useful in discussions of ethical issues in connection with policy work?';

Q3 'Do you think that the matrix could be useful in practical work to develop a policy?';

Q4 'Do you think that the matrix would be useful for the decision-making on policy work?' and

Q5 'Do you think that the matrix could be useful in communication on ethical questions to local residents?'

Ratings were recorded by pen and paper on a 15 cm visual analogue scale (VAS; Maxwell, 1978) reflecting extreme dissatisfaction of the far left and extreme satisfaction on the far right. The questionnaires were completed individually and the participants were asked to state whether they were politicians or officials. There was also a general, informal discussion with the facilitators. The study was approved by the local ethical board (www.epn.se/uppsala) and all participants signed statements of informed consent.

VAS ratings were analysed statistically using R (v 3.2.2) by repeated measures marginal model using the 'geePack' package (v 1.2-0.1) using 'Place' (i.e. municipality A, B or C), 'Type' (i.e. politician or official) and 'Question' (i.e. Q1-Q5) and their two-way interactions as fixed factors, repeated by individual.

Results and discussion

Forms

The main effect 'Question' and the two-way interactions 'Place × Question' and 'Type × Question' were found to be significant: There was a general reported satisfaction with the workshops (average rating=10.1), with a higher general satisfaction expressed for Q2 (discussions of ethical issues in connection with policy work; Table 1). However, the satisfaction with the workshop and ethical tool was also dependent on the dynamics of the occasion ('Place × Question' interaction) and role categories ('Type × Question' interaction). In fact, politicians were inclined to evaluate the Ethical Matrix with lower ratings, especially on Q4 (decision-making on policy work).

A condensation of the most frequent topics of the free-text answers in relation to perceived importance for politicians and officials is represented in Figure 1. Several participants state that the autonomy (or integrity) column is the most difficult to relate to, in particular with regard to animals and the environment. It was suggested that this be made more concrete. The same held for fairness with respect to animals and the environment. These points are also consistent with the facilitators' observations during the workshops.

Table 1. Visual analogue scale (VAS) ratings of feedback questions relating to Overall evaluation (Q1); Ethical discussions (Q2); Practical policy work (Q3); Political decisions (Q4) and Communication with citizens (Q5).[1]

	Q1	Q2	Q3	Q4	Q5	*P*-value
rating	9.8[a]	11.3[b]	9.7[a]	9.5[a]	9.9[ab]	0.0026

[1] Different letters indicate significant differences at $P<0.05$. VAS scale 0-15 cm.

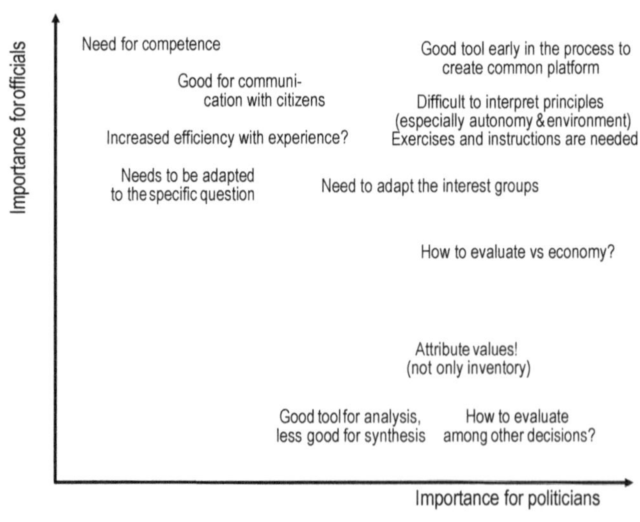

Figure 1. Condensation of free-text feedback, where the position on the axes gives an approximation of the frequency (as a proxy for importance) for politicians and officials, respectively.

Particularly among politicians, there were requests for including an element of weighing or balancing in the matrix, whereas officials more strongly emphasized the need for expertise and efficiency of use as a tool for ethical inventory and communication.

Observations during the workshop

In connection with the workshops and during the discussions, we observed some features that we consider important if the Ethical Matrix is to be further developed for successful use as a tool in the arena of public procurement of food.

First, there is an 'upper-left-hand-corner effect'. This might be due to at least two factors: (1) this is the way texts in our cultural context are read, and the Ethical Matrix is thus approached analogously; (2) as the matrices were presented, the field in the upper left-hand corner contained well-being aspects for production animals. This is an easy-to-grasp category, and thus attractive to start with. The relative contributions of these factors could be investigated by varying the order of principles and stakeholders in the matrix, something which would amount to a form of 'nudging', potentially steering the users in a particular direction.

Second, 'time' is an issue. A major hurdle was in fact the task of finding a time during which the participants were available. This led to notable delays relative to our original plans. It was also remarked by participants that the exercise was very time consuming. This means that in order for the Ethical Matrix to be practically useful, it has to be incorporated in a context where an appropriate amount of time can be allocated. One participant remarked that allocating three full hours for this type of discussion was unrealistic in real political work. However, many participants were of the opinion that increased efficiency would most likely also arise from increased use.

Third, the provision of factual information is a major constraint. The inventory depends on the competence present in the room. Many aspects of food production, procurement and consumption are laden with strong opinions, not always grounded in fact. In order to arrive at a balanced deliberation, participants and facilitators need to be actively engaged in contributing unbiased information. In fact, an increased guidance from the facilitators was sought after by several participants to provide expertise and factual information.

Fourth, discussions were often approaching or crossing into off-topic, requiring moderation. One way of getting the discussion back on track was to emphasize that the matrix is best used as a tool to evaluate alternatives against each other; for example organic vs conventionally produced meat. When the contrast between two alternatives was lost, the discussion ran the risk of losing relevance.

In the feed-back discussions, participants were in general positive to the experience of working with the matrix, which is also reflected in the positive VAS ratings. The overall impression voiced by the participants was that this tool could be useful mostly in early stages of policy work to form a common platform for later, more detailed discussions and also as a tool for communication with citizens in why and how decisions have been made. However, especially politicians hinted that this type of seminar may not be feasible in a concrete political setting. Much discussion focused on future improvement of the tool and suggestions included: (1) development of a web-based tool for pre-meeting inventories as a background for discussions on ethical and political values; (2) A multi-stage process in which officials could prepare an inventory for discussions; (3) inclusion of economy as a principle to gain a more holistic and politically useful inventory.

Conclusions

The high sensitivity to information veracity could perhaps give an indication that the ethical matrix as a tool for public procurement of food may be more suited at a higher expert level, e.g. national agencies for public procurement. Such reviewed and audited inventories could then be used in local agencies, such as municipalities, as background material for discussions, ethical deliberations and more concretely for establishing demands in policy documents and procurements.

We also believe that a tool such as this might be more useful for officials than for politicians. Officials have the task of carrying out political decisions, rather than arriving at those decisions and justifying them, and might therefore be helped by clear guidance.

Finally, the requests for including an element of weighing or balancing in the matrix should be taken seriously. There might be ways of incorporating this information, something that might be facilitated by using other graphic interfaces in addition to the Ethical Matrix.

Acknowledgements

The research was carried out with support from SLU's strategic research platform Future Agriculture. The authors also want to thank Lotta Rydhmer, Elin Röös, Lena Ekelund Axelson and Ingvar Sundh for contributing to the project.

References

Cotton, M. (2009). Evaluating the 'ethical matrix' as a radioactive waste management deliberative decision-support tool. Environmental Values 18: 153-176.

Kaiser, M., Millar, K., Thorstensen, E. and Tomkins, S. (2007). Developing the ethical matrix as a decision support framework: GM fish as a case study. Journal of Agricultural and Environmental Ethics 20: 65-80.

Kleine, D. and Das Graças Brightwell, M. (2015). Repoliticising and scaling-up ethical consumption: lessons from public procurement for school meals in Brazil. Geoforum 67: 135-147.

Konkurrensverket (2015). Offentlig upphandling av mat: en kartläggning av Sveriges offentliga upphandling av livsmedel och måltidstjänster. Konkurrensverkets rapportserie 2015: 1.

Maxwell, C. (1978). Sensitivity and accuracy of the visual analogue scale: a psycho-physical classroom experiment. British Journal of Clinical Pharmacology 6: 15-24.

Mepham, B. (2010). The ethical matrix as a tool in policy interventions: the obesity crisis. In: Gottwald, F.-T., Ingensiep, H.W. and Meinhardt, M. (eds.) Food Ethics. Springer, New York, NY, USA, pp. 17-29.

Moula, P. and Sandin, P. (2015). Evaluating ethical tools. Metaphilosophy 46: 263-79.

Stein, M. (2013). Public sector food procurement in UK local authorities: ethics and sustainability. In: Röcklinsberg, H. and Sandin, P. (eds.) The ethics of consumption: the citizen, the market and the law. Wageningen Academic Publishers, Wageningen, the Netherlands, pp. 395-398.

30. Facilitating decision making in public procurement of food through digital tools

H. Röcklinsberg[1], N. Lindström[2], C. Persson Osowski[3] and E. Röös[4]*
[1]Dept of Animal Environment and Health, Swedish University of Agricultural Sciences, P.O. Box 7068, 75007 Uppsala, Sweden; [2]Dept of Historical, Philosophical and Religious Studies, Umeå University, Biblioteksgränd 3, 901 87 Umeå, Sweden; [3]Dept of Food, Nutrition and Dietetics, Uppsala University, P.O. Box 560, 751 22 Uppsala, Sweden; [4]Dept of Energy and Technology, Swedish University of Agricultural Sciences, P.O. Box 7032, 75007 Uppsala, Sweden; helena.rocklinsberg@slu.se

Abstract

We'll scrutinize value challenges faced by public food procurement managers to ensure multiple values are met within a limited budget and under current regulations. The current highly complex food system makes informed, conscious and morally justified food choices most challenging. As about three million public meals are served daily in Sweden alone, procurement managers' and purchasers' choice of food has considerable impact on worker conditions, animal welfare and the environment. The combination of commodification of farm animals and the environment and the lowest price policy have contributed to downplay ethical values such as animal welfare or sustaining biodiversity in past procurement legislation (2004/18/EG). However, the recent EU directive extends the possibilities for taking such added values into consideration, highlighting the need to include externalised factors in food procurement decisions (Directive 2014/24/EU). Our hypothesis is that even if procurement guides such as the Swedish criteria for public procurement and the ones from the European Commission provide useful factual information about different products and requirements on production it leaves the procurement manager with a difficult task: to balance facts and added values in order to implement an institution's policy or meet multiple values. We see a need of a new set of digital tools to provide guidance, facilitate ethical decision-making and to relieve the moral stress of procurement managers. In the following we will outline nutritional, animal welfare and environmental related values in food systems, describe ethical aspects of guides for public procurement and some ethical decision-making tools, and finally tentatively suggest a set of digital tools to facilitate handling multiple values to ensure the best possible decisions are made to meet citizens' interests and work towards food production systems that stay within the planetary boundaries.

Keywords: animal welfare, Directive 2014/24/EU, environment, Green Public Procurement, nutrition

Background

In Sweden, about three million meals are served daily within the public sector at a cost of approximately 0.8 billion euros for the procured food (NFA, 2013). The purchase of food for providing the meals is governed by the EU regulations of public procurement[1] to ensure fair competition and hence generate business opportunities and drive economic growth (EU, 2016). Since public purchases constitute a large part of the economy (14% of EU GDP; (EU, 2016)), their possibility to influence production in a sustainable direction is regarded as important and powerful. In addition, it is argued that the public sector, representing the public community, has a greater responsibility to show leadership in the field of ethical consumption (Lindström and Röcklinsberg, 2013). In the Swedish Official Government Letter 2006/07:54, 'Environmental Public Procurement', it is highlighted that so-called green or sustainable

[1] Public procurement refers to the process by which public authorities, such as government departments or local authorities, purchase work, goods or services from companies (EU, 2015).

EurSafe 2016 – DOI 10.3921/978-90-8686-834-6_30, © Wageningen Academic Publishers 2016

public procurement is important. At EU-level it is argued that there is potential for cost effective sustainable public procurement (COM (2008) 400 Final). Previous procurement legislation (2004/18/EG) was found to be difficult to interpret as regards the type of green or ethical requirements that could be included, which led to several Swedish municipalities ending up involved in long and demanding lawsuits mainly due to requiring livestock rearing practices that went beyond minimal level in EU animal welfare regulations. Hopefully such problems will be overcome as a recent EU directive extends the possibilities for taking such added values into consideration, highlighting the need to include externalised factors in food procurement decisions (Directive 2014/24/EU). Hence, public procurers now have a clear political requirement to consider externalities of their purchases. However, exactly what that means in terms of actual green and ethical demands to be put on products is still to be defined by the organisation doing the procurement. Cost is often a limiting factor as there is a strong focus on using public money as efficiently as possible. The average cost for a school lunch in Sweden is just above one Euro (Skolmatsverige, 2016). In the following we will first outline nutritional, environmental and animal welfare related values in food systems. Second, by acknowledging procurators' difficult task to meet legislation, their institutions' policy and public expectations within a limited budget in a well-argued decision we look into ethical aspects of guides for public procurement and some decision making tools. Third, we tentatively develop a digital tool to handle multiple values. Hence, the aim of the paper is to suggest a tool to facilitate public procurement: to ensure the best possible decisions are made in order to meet citizens' interests and work towards food production systems that stay within the planetary boundaries.

Nutritional aspects

Food and eating are parts of a complex topic. In addition to added values such as animal welfare and environmental concern, it involves historical, political, socio-psychological, cultural, religious, economic, and sensory aspects, availability and individual choice (Fieldhouse, 1996; Germov and Williams, 2008). As a result, public procurement faces a big challenge in combining these various aspects in a collective meal, at the same time satisfying the nutritional needs of its individual consumers. In Sweden, the National Food Agency (NFA) publishes guidelines regarding public meals, for instance school meals, but these are mainly advisory (see for example NFA, 2013). In compulsory school (year 1-9), however, it is included in the Education Act that school meals are to be nutritious (Swedish Parliament, 2010), i.e. that schools are required to be compliant with the nutrient-based standards in the NFA guidelines (NFA, 2013) or adhere to the nutrient criteria in a web-based tool called 'School Food Sweden' (Patterson *et al.*, 2013, 2014). Compliance is another thing though, and studies based on the web-based tool have shown that only 27% of the participating schools reached all the nutrition criteria. These results apply to the planned portions that the children are intended to eat. According to a national food consumption survey mapping what children actually consume from school lunches, they received about a third of their daily energy and nutrient intake from school meals, but for some essential nutrients the recommendations were not met (Persson Osowski *et al.*, 2015).

Animal welfare aspects

According to e.g. a recent Eurobarometer, citizens are concerned with farm animal welfare (EC, 2007; EC, 2016) and as many as 82% of respondents agree animal welfare could 'probably' or 'certainly' be better protected in their own country (EC, 2016: 9). In line with this, there is an overall increasing demand in 'animal welfare friendly' products in some European countries and many consumers are willing to pay a higher price and look for animal welfare-friendly labels at purchase. In Sweden 52%, in the Netherlands 43% and in Luxembourg 41% of respondents look for such labels 'most of the time', whereas about 51-55% of respondents from Lithuania, Slovenia, Spain, and Slovakia do not look for such labels (EC, 2016).

Turning to the actual situation for animals, recent studies show a discrepancy between farmers and the industry on the one hand, and consumers or citizens on the other regarding what constitutes acceptable welfare for a farm animal (Bergstra *et al.*, 2015; Borkfelt *et al.*, 2015). Irrespectively of whether biological functioning, subjective experience or possibility to perform species specific behaviour is regarded to be the core of an animal welfare definition (Fraser *et al.*, 1997), many industrial housing systems, transportation and slaughter procedures imply welfare impairments. Exemplified here by the life cycle of cows, pigs and poultry one can see a number of welfare challenges they meet all the way from sperm 'donation' from bulls and artificial insemination in cows and pig breeding to slaughter without stunning or use of CO_2 in poultry, well known to cause distress. Chickens are raised without a hen, cow and calf are separated directly at birth, most male chickens in egg industry are killed, day-old chickens are debeaked, piglets are weaned at 3-4 weeks, i.e. before they have an established immune system, pigs are reared on concrete or slatted floors without straw, dairy cows housed indoors with high frequency of mastitis and leg problems, piglets tail docked and teeth cut without analgesics, broilers are living with foot pad dermatitis and/or leg problems. Out of these procedures, only slaughter without stunning (if not religious slaughter) is forbidden in the EU. Hence, any farmer striving for higher welfare levels is bound to choose a private standard according to which the production can be certified to make practices to increase animal welfare pay off. Moreover, animal welfare is related to use of antibiotics, as e.g. stress, early weaning and barren environment increase the risk of health issues, leading to higher use of antibiotics (EFSA, 2007). In the EU antibiotics were allowed for prevention of disease until 2006, but use of antibiotics is still high at many farms, which contributes to increased development of bacterial resistance in society, already leading to lack of treatment in human health care (WHO, 2014). Consumer interest in purchasing products with higher animal welfare standards than EU legislation hence needs to relate both to the actual animals and to human health, as do procurers when trying to mirror consumer interest in their purchase.

Environmental aspects

The current food system is a main driver of many environmental pressures (Foley *et al.*, 2005, 2011). For example, in Sweden 20% of total greenhouse gas emissions come from food consumption (Hjerpe *et al.*, 2013) and agriculture causes about half of the emissions of nitrogen and phosphorous to waterways (SBA, 2015). The rearing of animals for meat, eggs and milk generates some 14.5% of total global greenhouse gas (GHG) emissions (Gerber *et al.*, 2013), occupies 70% of agricultural land and is the main cause of environmental problems such as biodiversity loss and water pollution, due to e.g. monoculture and deforestation (MEA, 2005; Steinfeld *et al.*, 2006). Further, large amounts of non-renewable energy sources are used in post farm gate activities such as processing, packaging, storing and distribution of food. Due to the inefficiency of animals in converting feed to meat, meat is the food product with the greatest environmental impact (Steinfeld *et al.*, 2006; Weidema *et al.*, 2008) and there are major concerns about the depletion of wild fish stocks and the negative effects of overfishing on aquatic ecosystems.

Attention to the environmental impacts of food has increased considerably in Sweden during the last decade. The debate in the media has been intense, environmental NGOs have launched campaigns to promote more sustainable eating and retailers have also taken initiatives to promote less environmentally damaging food products. The increase in the sales of organic foods broke the record in both 2014 and 2015, increasing by 38 and 39% respectively (Ekoweb, 2015). This increased focus on reducing the environmental impacts from food consumption through conscious and informed food purchases has also manifested itself within public procurement in several ways. Many public authorities have taken initiatives to reduce environmental impacts from the food they serve through measures such as, e.g. organic food purchases, reduced waste and serving meals with lower climate impact. These initiatives are driven by either political decisions (some municipalities have political goals to decrease the emissions of GHG that food procurers have to advert to, or that purchases of organic foods should reach a certain percentage of total food purchases), more indirect and diffuse demand from society (e.g. from parents or caretakers or NGOs) or by the engagement of concerned officials.

Formulating strategies to reduce environmental impacts from food is, however, complex, as food production affects the environment in many different ways. For example, the production of meat from ruminant animals such as beef and lamb generate more GHG emissions than do chicken while chicken production is more dependent on soy as protein feed which is associated with deforestation, high use of pesticides and genetic modification. In addition, the nutrient content of different food items needs to be factored in as well as the acceptability of the food. Establishing what food is environmentally 'best' considering all these aspects hence includes many normative choices as regards handling inevitable goal conflicts.

Ethical aspects of criteria for public procurement

Nutrition, animal welfare and environmental sustainability are examples of interrelated value laden issues that need to be handled during public procurement of food. All of these and many more are present in the EU guidelines for Green Public Procurement (GPP)[2] and the more specific criteria supplied by national authorities such as the Swedish National Agency for Public Procurement (NAPP). Taken separately, each of these aspects can be evaluated: calculate food's nutritional value, compile animal health statistics or perform lifecycle assessments. Yet, it is not obvious how these facts regarding the production and consumption of food are related to the values which tend to be associated with them (i.e. health, animal rights and environmental values) and even more open how they in turn are supposed to be weighed against each other in case of conflict. The GPP (EU, 2015) guidelines suggest three main factors when making priorities: environmental impact, budgetary importance and potential to influence the market (c.f. EU GPP, 2008; EU, 2015: 10). Animal welfare and nutritional aspects are hence not suggested important in making priorities. Further, to act within budgetary boundaries and influence the market expresses an economic point of view. Restricting difficult priorities and handling conflicting values to a more or less economical calculus is however problematic for several reasons. It is a very narrow focus and it disclaims dimensions not possible to formulate in economic terms and also values regarded important by citizens.

Within the frame of GPP the Swedish National Agency for Public Procurement has developed a set of more specific criteria for different kinds of food products (e.g. meat, egg, fruit and vegetable). The requirements are divided into three different levels (basic, advanced, and spearhead) and are immediately applicable for municipalities and counties. The requirements basically consist of facts about food production, e.g. use of chemicals or exploitation of land as well as whether certified fair trade or organic (NAPP, 2015). Further, the listed criteria in both GPP and NAPP are presented as parallel rather than hierarchical to each other, leaving it open to each procurator how to choose in relation to a final value. In case of conflicting values or limited resources, typical for public procurement, it is a challenge to weigh different alternatives against each other and know how to relate instrumental values to final values. We conclude that for several reasons there is a need to address questions regarding how to handle multiple values and facts in procurement of public food.

Digital tools to facilitate ethical decision making

In the ethical debate, several suggestions have been presented on how to develop and use methods to balance ethical values (e.g. justice between stakeholders or minimising harm to those affected) with

[2] The other criteria listed in the GPP (pp. 10-11) are: political priorities, market availability, cost considerations, availability of criteria, visibility and practical considerations.

respect to the prevailing circumstances of certain contexts in order to reach well-grounded decisions and develop guidelines for action. Ethicists have contributed to the discussion on how to make priorities in the healthcare sector (Beauchamp and Childress, 2013; Jonsen and Toulmin, 1989), animal husbandry (Lund and Forsberg, 2009), novel foods (Mepham, 2000) and public health (Mepham, 2010). One influential theory is the method of reflective equilibrium considering (a) the considered moral judgments of a certain case; (b) the moral principles we believe apply to the case; and (c) the set of relevant beliefs or theories supporting (b), all duly revised to reach joint coherence (Daniels, 1978; Rawls, 1972). These decision-making models have at least two things in common: they regard ethics to be an eclectic framework containing principles from several normative theories such as utilitarianism, deontology and virtue ethics, and they contain ideas about what gives them content. We suggest that a combination of such a model with criteria frequently used in public procurement, which, as descried above, are basically a collection of facts about different production systems (GPP, 2014; NAPP, 2015), is promising in order to construct an applicable and context-sensitive ethical tool to facilitate public procurement of food. Such a model should incorporate context sensibility with application of ethical principles by developing decision-making tools intended for a certain field. If a practice is used in order to determine ethical principles, which in turn can be used to ethically assess new practical situations in a constantly evolving process, it can be called an interaction model (Lindström, 2012; Röcklinsberg, 2006). As specific cases seem to trigger immediate intuitive reactions, forming a basis for individuals' evaluations, which are then reconstructed as rational reasons in retrospect (*post hoc*) (Haidt, 2012), it is important to combine these intuitions with a procedure where decisions are made with regard to the circumstances of the issue, here production, distribution and consumption of food.

Based on the above, we envisage a digital tool to facilitate ethical decisions in public procurement. Based on an interaction model the digital tool shall guide the procurer in a three step process: from general principles in legislation and requirements and factual information about food product groups to a more detailed level (specific products). The answer on level one (chosen ethical values e.g. maximisation of interests, respect for individuals, or justice) leads to a set of questions on level two on which added values are most central, which in turn lead to a set of questions on level three regarding their most frequently used products. Every choice at each level will be partly reflected in a specific set of questions on the next level. To achieve both context sensitivity and user-friendliness, legislation, other requirements and fundamental ethical values will be visible from the beginning, whereas facts about the institute's context (number of pupils, price per portion, etc.) will be filled for each case. By applying the institute's focus of concern on the value clashes and include the pinpointed facts into the picture the procurator will have a fair chance to decide which product is best suited to meet their own focus of concern.

References

Beauchamp, T.L. and Childress, J.F. (2013). Principles of biomedical ethics. Seventh ed. Oxford University Press, Oxford, UK, 459 pp.

Bergstra, T.J., Gremmen, B. and Stassen, E.N. (2015). Moral values and attitudes toward Dutch sow husbandry. Journal of Agricultural and Environmental Ethics 28(2): 375-401.

Borkfelt, S., Kondrop, S., Röcklinsberg, H., Bjørkdahl, K. and Gjerris, M. (2015). Closer to nature? A critical discussion of the marketing of 'ethical' animal products. Journal of Agricultural and Environmental Ethics 28(6): 1053-1073.

COM (2008) 400 final communication from the Commission to the European Parliament, the Council, the European Economic and Social Committee and the Committee of the Regions Public procurement for a better environment. Brussels, Belgium.

Daniels, N. (1978). On some methods of ethics and linguistics. Philosophical Studies: An International Journal for Philosophy in the Analytic Tradition 37(1): 21-36.

Ekoweb (2015). Ekologisk livsmedelsmarknad ('The market of organic foods'). Ekoweb, Lidköping, Sweden.

European Food Safty Agency (EFSA) (2007). Animal health and welfare aspects of different housing and husbandry systems for adult breeding boars, pregnant, farrowing sows and unweaned piglets. EFSA Journal 572: 1-13.

European Commission (2007). Attitudes of citizens towards animal welfare. Special Eurobarometer, Brussels, Belgium.

European Commission (2016). Attitudes of Europeans towards animal welfare. Summary, Special Eurobaromenter 442. Available at: http://tinyurl.com/huk3low.

EU (2008). Green public procurement product sheet: food and catering. Available at: http://tinyurl.com/jqxmyye.

EU (2014). Directive 2014/24/EU. Available at: http://tinyurl.com/zxerzx6.

EU (2015). Buying green! A handbook on green public procurement, 2nd edition. Available at: http://tinyurl.com/pj7nvev.

Fieldhouse, P. (1996). Food and nutrition. Customs and culture. Stanley Thornes, Cheltenham, UK, 253 pp.

Food and Agriculture Organisation (FAO) (2006). Livestock's long shadow. Food and Agriculture Organisation, Rome, Italy.

Foley, J.A., Ramankutty, N., Brauman, K.A., Cassidy, E.S., Gerber, J.S., Johnston, M. and Zaks, D.P.M. (2011). Solutions for a cultivated planet. Nature 478(7369): 337-342.

Foley, J.A., DeFries, R., Asner, G.P., Barford, C., Bonan, G., Carpenter, S.R. and Snyder, P.K. (2005). Global consequences of land use. Science 309(5734): 570-574.

Fraser, D., Weary, D.M., Pajor, E.A. and Milligan, B.N. (1997). A scientific conception of animal welfare that reflects ethical concerns. Animal Welfare 6: 187-205.

Gerber, P.J., Steinfeld, H., Henderson, B., Mottet, A., Opio, C., Dijkman, J., Falcucci, A. and Tempio, G. (2013). Tackling climate change through livestock – a global assessment of emissions and mitigation opportunities. Food and Agriculture Organization of the United Nations (FAO), Rome, Italy, 115 pp.

Germov, J. and Williams, L. (2008). A sociology of food and nutrition: the social appetite. Oxford University Press, Melbourne, Australia, 425 pp.

Haidt, J. (2012). The righteous mind why good people are divided by politics and Religion. Pantheon Books, New York, NY, USA, 419 pp.

Hjerpe, K., Lundström, A., Markensten, T., Nilsson, K., Pearsson, M., Rundberg, B., Rydberg, I. and Sonesson, U. (2013). Hur liten kan livsmedelskonsumtionens klimatpåverkan vara 2050? ('How small can the carbon footprint of Swedish food consumption get?'). Swedish Board of Agriculture, The National Food Agency and the Swedish Environmental Protection Agency.

Jonsen, A. and Toulmin, S. (1989). The abuse of casuistry: a history of moral reasoning. University of California Press, Oakland, CA, USA, 432 pp.

Lindström, N. (2012). Förhållandet mellan praxis och teori inom etiken. Lund Studies in Ethics and Theology 16.

Lindström, N. and Röcklinsberg, H. (2013). School meals – bridging the gap between citizen expectations, procurement skills and legislation. In: Röcklinsberg, H. and Sandin, P. (eds.) The ethics of consumption. The citizen, the market, and the law. Wageningen Academic Publishers, Wageningen, the Netherlands, pp. 423-428.

Millenium Ecosystem Assessment (MEA) (2005). Ecosystems and human well-being: biodiversity synthesis. Millennium ecosystem assessment. World Resources Institute, Washington, DC, USA.

Mepham, B. (2000). A framework for the ethical analysis of novel foods: the ethical matrix. Journal of Agricultural and Environmental Ethics 12: 165-176.

Mepham, B. (2010). The ethical matrix as a tool in policy interventions: the obesity crisis. In: Gottwald, F.T., Ingensiep, H.W. and Meinhardt, M. (eds.) Food Ethics. Springer, New York, NY, USA, pp. 17-29.

National Agency for Public Procurement (NAPP) (2015). Requirement wizard. Available at: http://tinyurl.com/gmcg7se.

National Food Agency (NFA). Den värdefulla måltiden i vården, skolan och omsorgen. National Food Agency, Uppsala, Sweden. Available at: http://tinyurl.com/jd8yrd2.

National Food Agency (NFA) (2013). Good school meals. Guidelines for primary schools, secondary schools and youth recreation centres. National Food Agency, Uppsala, Sweden. Available at: http://tinyurl.com/j499o4e.

Patterson, E. and Elinder, L.S. (2014). Improvements in school meal quality in Sweden after the introduction of new legislation – a 2-year follow-up. European Journal of Public Health 25: 655-660.

Patterson, E., Quetel, A.-K., Lilja, K., Simma, M., Olsson, L. and Elinder, L.S. (2013). Design, testing and validation of an innovative web-based instrument to evaluate school meal quality. Public Health Nutrition 16: 1028-1036.

Persson Osowski, C., Lindroos, A.K., Enghardt Barbieri, H. and Becker, W. (2015). The contribution of school meals to energy and nutrient intake of Swedish children in relation to dietary guidelines. Food and Nutrition Research 59: 27563.

Rawls, J. (1972). A theory of justice. Oxford University Press, Oxford, UK, 560 pp.

Röcklinsberg, H. (2006). Consent and consensus in policies related to food – five core values. Journal of Agricultural and Environmantal Ethics 19: 285-299.

Swedish Board of Agriculture (SBA) (2015). Jordbruket och övergödning av havet. Swedish Board of Agriculture, Jönköping, Sweden. Available at: http://tinyurl.com/zqkcrxc.

SkolmatSverige (2016). Fakta från andra källor om skolmåltider. Available at: http://tinyurl.com/jtuf8xe.

Swedish Parliament (2010). Available at: http://tinyurl.com/noqh6er.

Weidema, B.P., Wesnæs, M., Hermansen, J., Kristensen, T. and Halberg, N. (2008). Environmental improvement potentials of meat and dairy products. Available at: http://tinyurl.com/z8hl9fr.

World Health Organization (WHO) (2014). Anti microbial resistance. Global Report on Surveillance NLM classification: QV 250.

World Organization for Animal Health (OIE) (2015). Terrestrial animal health code. Chapter 7.11. Animal welfare and dairy cattle production systems. Available at: http://tinyurl.com/heq7fhj.

31. Drivers and challenges of animal ethics in companies; a qualitative study of the Western food industry

M.R.E. Janssens[1] and F. van Wesel[2]*
[1]Erasmus University Rotterdam, the Netherlands, Burgemeester Oudlaan 50, 3062 PA Rotterdam, the Netherlands; [2]Utrecht University, Padualaan 14, 3584 CH Utrecht, the Netherlands; janssens@rsm.nl

Abstract

This study sets out to explain differing company positions on animal ethics, using a qualitative study of drivers (stimulatory factors) and challenges (obstructive factors) of animal ethics in large Western-Europe-based food companies. This is done through in-depth interviews with employees who are responsible for animal welfare policy, in which they talk about how the topic of animal welfare entered the company's responsibility agenda, how it evolved, and is still evolving. We identified ten types of drivers and challenges for taking animals into account (from emotional to economic), which occur in ten different levels (from the interviewee as a manager to the government). This paper presents the preliminary results of this study and our first attempts to infer a model for explaining differences between companies. By strengthening drivers and minimising challenges, all parties (companies, non-governmental organisations, governments, business partners, etc.) will be able to use our findings to accelerate the change toward a more widely felt responsibility towards animals.

Keywords: CSR, business ethics, animal welfare, responsibility

Introduction: why study animal ethics in food companies?

Today, concepts like Corporate Social Responsibility (CSR) and Triple P (People, Planet, Profit (Elkington, 1997)) are well known in all industries and offer companies opportunities to define their own responsibilities toward people, nature, animals, and sustainability of the planet and its ecological and social systems. Business ethics offers extra food for thought, not only for philosophers, but also for decision makers in companies. However, business ethics has ignored the ongoing debate in animal ethics for too long (Janssens and Kaptein, 2016; Van Liedekerke and Dubbink, 2008). Furthermore, Janssens and Kaptein (2016) found that some multinationals express responsibilities towards animals, whereas others do not. Although keeping animals and being in the animal-derived food industry are both factors that correlate positively with expressed commitment to animals, this does not account for all differences.

In the Business Benchmark on Farm Animal Welfare (2014) Amos and Sullivan identify reasons for companies to focus on farm-animal welfare: ethical arguments and business arguments (the latter being legislation and voluntary standards, stakeholder expectations, and market opportunities). On the basis of interviews and e-mail surveys they identify customer and client demands as the most important factors of influence. Though this study offers many interesting insights, even more knowledge is needed about the drivers and challenges companies and their responsibility managers encounter in their struggle to act responsibly.

Based on the above, our research question is: What are the drivers and challenges for large Western-Europe-based food companies to take responsibility for animals? Answering this question will not only explain differences between companies; at the same time, it will offer companies and their stakeholders tools for change. We chose the term 'drivers' for stimulatory factors because it is a common term in managerial literature (e.g. Mauser, 2001), and 'challenges' for obstructive factors, for reasons of positive framing: a challenge is surmountable. To explore the drivers and challenges responsibility managers

I. Anna S. Olsson, Sofia M. Araújo and M. Fátima Vieira (eds.) ***Food futures: ethics, science and culture***
EurSafe 2016 – DOI 10.3921/978-90-8686-834-6_31, © Wageningen Academic Publishers 2016

encounter, we conducted a qualitative study amongst nine Western-Europe-based companies producing, processing or selling meat, poultry, fish, or dairy products. By conducting and analysing interviews, we have identified drivers and challenges, and the way they interact.

This paper is organised according to the qualitative study reporting guidelines by Corbin and Strauss (2015). In this section we outlined the significance of animal ethics as a theme in corporate responsibility and presented our research question. In the next section we explore the theoretical background and explain how this study contributes to animal and business ethics. This is followed by a section where we present our research method. In the final section we offer preliminary findings (the study is unfinished as of publication of this paper), and briefly discuss them.

Theoretical background

In this section previous research is presented and related concepts are explained, as a starting point of the research and as an analytical lens.

Status quo of the capitalist system

People have reared and slaughtered animals for ages. In the second half of the 20th century an industrial system emerged that systematically rears as many animals as possible at the lowest possible cost, and sells them at the best possible price. McMullen (2015) argues that the place of animals in our economy is determined by: low-cost technology, competition-driven efficiency and accountability, and specialisation. It is very hard for individuals to change the system. This is why responsibility managers might encounter challenges related to the system. Rollin (2011) adds that a positive correlation between individual animal welfare and individual animal productivity from traditional agriculture, was probably extrapolated to a correlation between overall animal welfare and system productivity in industrial agriculture.

External pressure

Schultz and Wehmeier (2010) see competition, regulative norms, professional norms, and public pressure as triggers for institutionalisation of CSR. One of the regulative norms could be legislation about reporting (Delbard, 2008), as could non-coercive regulations, such as political pressure or the international CSR guideline ISO 26000 (2010). Castka and Balzarova predicted in 2007 that multinationals would adopt this standard to legitimise their policies. It is conceivable that these kinds of external pressure will also influence the way a company takes responsibility for animals.

Managerial decision making

It can also be expected that individual (managerial) decision making plays a role. While arguing how factors from both within and outside the company influence managerial decision making about corporate responsibility measures, Hoffman and Bazerman (2006) also identified external regulatory pressure and internal reward systems, amongst others, as influential. In addition, they argue that institutional resistance can result from resource constraints, fear of the unknown, threats to political interests, and habitual distrust.

Responsibility awareness

Constantinescu and Kaptein (2015) found that under particular conditions, responsibility awareness of corporations and of individuals within these corporations can be mutually enhancing. The most important condition for that effect to emerge is for corporate structure and culture to be disconnected:

then there is room for doing something better or worse than corporate regulations dictate. However, if people in the company act according to culture more than to structure, this stimulates corporate ethical culture, be it positively or negatively.

Communication

Up till now, CSR communication has merely focused on consumer and (external) stakeholder communication, not so much on organisational communication (Golob at al., 2013). Begley and Boyd (2000) make clear that integrating organisational values starts with writing about them in vision documents. The next step is to create more extended documents on these values and applying them in practice. The third step is making them known among employees. That communication is not only supportive of CSR, but can also trigger responsibility, or even be an act of responsibility in itself, is supported by Schoeneborn and Trittin (2013), who identify a strong formative character in both external and organisational communication. Their constructivist view on communication holds that by talking ethically, a company becomes more ethical. This is the called the 'communication constitutes organisations' perspective (CCO), slightly reformulated by Schoeneborn and Blaschke (2014) as the 'communicative constitution of organisations' (CCO).

More institutional determinants

Basu and Palazzo (2008) made a study of internal institutional determinants of CSR. They distinguish between different kinds of dimensions of sensemaking in CSR: (1) cognitive dimensions (identity orientation and legitimacy); (2) linguistic dimensions (justification and transparency); and (3) conative dimensions (posture, consistency, and commitment). Van Tulder *et al.* (2014) identified more than 70 different managerial tipping points for transition to sustainability in organisations, moving from an inactive position, through a reactive, and an active, to a proactive one. Influencers that take companies from tipping point to tipping point include perception, involvement and satisfaction of employees, purchasing policy, supply chain management, organisational communication, internal coherence between departments, openness to stakeholders, the need to set priorities and appointing a responsible manager with excellent communication skills. Mauser (2001) identified intra-organisational drivers for the environmental performance of businesses from a case study of the Dutch dairy sector. She finds four drivers for environmental performance: (1) policy & strategy; (2) communication; (3) organisation structure; and (4) management commitment. To achieve optimal environmental performance, the right balance between these four has to be found. Visser (2014) holds five principles of Transformative CSR (or: CSR 2.0) responsible for its potential success: creativity (creative solutions), scalability (ability to be offered on a large scale), responsiveness (being proactive, in dialogue with stakeholders), glocality (tailor-made adaptations to global solutions) and circularity (renewability and no waste).

We use the definition of animal ethics from a previous study in the field where we define it as being about the moral status of animals as individuals with interests (Janssens and Kaptein, 2016). CSR in this paper is approached as all ethical responsibility of a company beyond what is legally required. Whereas influencers of CSR have been identified by many studies, this is to our knowledge the first addressing animal ethics in food industry CSR.

Method

The current study is explorative in nature, as no research on this specific topic is known to us. Our focus is on the production, processing, and retail sale of animal-derived food, processes in which large numbers of animals are handled in ways that do not always correspond to their interests. Our methodology is based on grounded theory (Corbin and Strauss, 2015). We studied nine large, internationally operating food companies based in Western Europe. Western Europe was chosen because it is where we ourselves are based (the Netherlands) and because there are signs that Europe is one of the areas where animal ethics in businesses is the most developed (Janssens and Kaptein, 2016). The companies were found by searching on the internet for the most important players in the animal-derived food industry and by asking interviewees for more names (snowball method). Seven companies were based in the Netherlands, one in the United Kingdom and one in Switzerland. They all operated internationally and had more than 1000 employees (up to 172,000). Some were very old (the first shop opening in 1887) and some relatively recent (2007). They are producers, processors, wholesalers and retailers. For reasons of data triangulation, we studied the companies' responsibility reports, their websites and interviews we conducted with the managers responsible for animal welfare policy (CSR managers, quality managers, etc.). We used the 'unstructured interview' method, which provides the richest source of data for theory building (Corbin and Strauss, 2015). Each interview was started with announcing the main topic (drivers and challenges for their company to take responsibility for animals) and an invitation to elaborate on it. At the end we checked our topic list to remind them about topics we missed. Guiding topics on this list, partly derived from the theory presented earlier, were views on animals, views on CSR and corporate responsibility, ethics, leadership, economic considerations, identity & internal communication, and reputation & external communication. The interviews took 40 to 90 minutes each and could all be ended in a natural way. We made audio recordings, which were typed up by one of the researchers, or typed up by an assistant and checked by one of the researchers. Trust was created by offering the research plan beforehand by e-mail, promising to use the data for research purposes only, and conducting the interview in person. One interview was done by telephone because of the geographical distance and the lack of experience with Skype of the interviewee. Interviews were conducted by the first author, who has 25 years of experience in journalism and communications. Data were handled in a cyclical process of interviewing, coding, identity, comparison, and analysis, according to the method described in Boeije (2010). Three of the interviews were coded by both authors. Inter-coder reliability was high (around 90%) and differences in coding were agreed upon. Hereafter, the first author coded the rest of the transcripts according to the adapted coding system. All additional text documents (websites and CSR reports) were also coded (with NVivo software, QSR International, Australia) by the first author. After coding the information from the ninth company, saturation was reached.

Preliminary findings and discussion

Although our analyses were not yet finished at the time of publication of this paper, we can say that we found six types of drivers and challenges for taking corporate responsibility for animals, displayed in the first row of Table 1. We also identified ten levels where these drivers and challenges occurred. Moving from the outer shell of the organisational context to the core of the organisation, those levels vary from 'in the governmental area' to 'in the responsible manager as a person and a professional'. The levels can be found in the left column of Table 1. Each box is illustrated with one example of a driver and a challenge (if found). Other drivers and challenges we found are still under examination.

The first results are emerging. An interesting preliminary observation is that some factors appear as both a driver and a challenge. For example, a given ethical influencer such as the 'Five Freedoms' (for animal welfare; Brambell, 1965) can be a driver for improving the welfare of animals, but compliance can also obstruct openness to new insights or practices. Based on these preliminary results, we propose six types of drivers and challenges, as well as ten levels where they can occur, as shown in Table 1. Both these types and places can offer tools for companies in the food industry to be more aware of their responsibility towards animals and act accordingly. We expect there to be more than 20 influencers in some of the boxes after further analysis. This gives us faith that our final results and our discussion of them will offer companies and their stakeholders the opportunity to identify blind spots, overcome them and make a

Table 1. Examples of drivers (+) and challenges (–) for Animal Welfare (AW) in Western Europe based food companies.

Level	Type					
	Emotional/ attitudinal	Communicative	Ethical	Cultural/historical/ geographical	Practical	Economic/ strategic
Government	+ restraint if branch acts responsible	+ campaign	+ set standards – conflicting demands (environment)	– capitalist system	+ law enforcement	+ investment support – rules on competition
Consumer/ public/media	+ awareness – nostalgia about farming	+ reputation – old-fashioned website	+ transparency – fear of paternalism	+ similar values – less interest in AW	+ AW label	+ responsible shopping – cheap shopping
NGOs	+ openness – negative attitude	+ label publicity – quarrelling	– competing values of different NGOs + materiality analyses	– less NGO influence in supply/market country	+ clear AW criteria	+ Business Benchmark on animal Welfare
Research institutions	+ cooperative attitude	+ show cooperation on website	+ animal ethical research	– history of veterinary medicine	+ evidence based practical advice – lack of AW specialists	+ cooperate in innovative research
Partnerships	+ trust	+ speaking on congresses	+ joint tackling of conflicting interests		+ engaged vets and architects	+ joint programs
External stakeholders/ investors		+ responsibility event	+ strong AW in materiality analyses – weak AW in materiality analyses			
Supply chain	+ flexibility – envy amongst farmers	+ farm visits	+ respect for animals – farmer autonomy	+ learn from other cultures – foreign habits	+ strict purchasing criteria – non-match of farming concept	+ sales guaranties – square validation
Retail chain	+ mutual understanding	+ meeting in person	+ strong AW in materiality analyses – weak AW in materiality analyses	+ strict requirements from export countries – historical efficiency focus	+ joint implementation – animal life cycle influence on production time	+ long-term thinking – balancing productivity and AW
Company	+ leadership and courage of board and directors – obviousness of AW	+ proud organisational communication – absence of organisational communication on AW	+ Five Freedoms – Five Freedoms	+ tradition of caring for animals – fixed idea that AW = animal health	+ written principles – slow change	+ small steps *or* radical change – conflicting interests of departments
Responsible manager	+ empathy with animals – obviousness of AW	+ participate in debate	+ view of animals as individuals with own needs – instrumental view of animals	+ background with animals		+ opportunity thinking

difference. The small number of rather diverse companies in our study is one of its limitations and more research is needed, in particular as regards investors and other external stakeholders.

References

Amos, N. and Sullivan, R. (2014) The business benchmark on farm animal welfare. 2014 Report, BBFAW, UK.

Basu, K. and Palazzo, G. (2008). Corporate social responsibility: a process model of sensemaking. Academy of Management Review 33(1): 122-136.

Begley, T.M. and Boyd, D.P. (2000). Articulating corporate values through human resource policies. Business Horizons 43(4): 8-12.

Boeije, B. (2010). Analyses in qualitative research. SAGE Publications Ltd, London, UK.

Brambell, F.W.R. (1965). Report of the technical committee to enquire into the welfare of animals kept under intensive husbandry systems. Cmnd 2836, HMSO, London, UK.

Castka, P. and Balzarova, M.A. (2007). A critical look on quality through CSR lenses: key challenges stemming from the development of ISO 26000. International Journal of Quality and Reliability Management 24(7): 738-752.

Constantinescu, M. and Kaptein, M. (2015). Mutually enhancing responsibility: a theoretical exploration of the interaction mechanisms between individual and corporate moral responsibility. Journal of Business Ethics 129(2): 325-339.

Corbin, J.M. and Strauss, A.L. (2015). Basics of qualitative research, techniques and procedures for developing grounded theory. Sage Publishing, CA, USA.

Delbard, O. (2008). CSR legislation in France and the European regulatory paradox: an analysis of EU CSR policy and sustainability reporting practice. Corporate Governance: International Journal of Business in Society 8(4): 397-405.

Elkington, J. (1997). Cannibals with forks: the triple bottom line of 21st century business. Capstone Publishing Ltd., Oxford, UK.

Golob, U., Elving, W.J., Nielsen, A.E., Thomsen, C. and Schultz, F. (2013). CSR communication: quo vadis? Corporate Communications 18(2): 176-192.

Hoffman, A. and Bazerman, M. (2006). Changing practice on sustainability: understanding and overcoming the organizational and psychological barriers. In: Husted, B. (ed.) Organizations and the sustainability mosaic: crafting long-term ecological and societal solutions. Cheltenham, UK, pp. 84-105.

International Standard ISO 26000 (2010). Guidance on social responsibility. ISO, Geneva, Switzerland.

Janssens, M.R.E. and Kaptein, M. (in press). The ethical responsibility of companies toward animals: a study of the expressed commitment of the fortune global 200. Journal of Corporate Citizenship. DOI: http://dx.doi.org/10.2139/ssrn.2513354.

Mauser, A.M. (2001). The greening of business. Environmental management and performance evaluation: an empirical study of the Dutch dairy industry. Eburon Publishing, Delft, the Netherlands.

McMullen, S. (2015). Is capitalism to blame? Animal lives in the marketplace. Journal of Animal Ethics 5(2): 126-134.

Rollin, B.E. (2011). Putting the horse before descartes: my life's work on behalf of animals. Temple University Press, Philadelphia, USA.

Schoeneborn, D. and Blaschke, S. (2015). The three schools of cco thinking: interactive dialogue and systematic comparison. Management Communication Quarterly 28(2): 285-316.

Schoeneborn, D. and Trittin, H. (2013). Transcending transmission: towards a constitutive perspective on CSR communication. Corporate Communications: An International Journal 18(2): 193-211.

Schulz, F. and Wehmeier, S. (2010). Institutionalization of corporate social responsibility within corporate communications: combining institutional, sensemaking and communication perspectives. Corporate Communications: An International Journal 15(1): 9-29.

Tulder, R., Van Tilburg, R., Francken, M. and Da Rosa, A. (2013). Managing the transition to a sustainable enterprise; lessons from frontrunner companies. Routledge, New York, USA / Oxon, Canada.

Van Liedekerke, L. and Dubbink, W. (2008). Twenty years of European business ethics: past developments and future concerns. Journal of Business Ethics 82(2): 273-280.

Visser, W. (2014). CSR 2.0: transforming corporate sustainability and responsibility. Springer, New York, NY, USA.

Section 7. Food and literature

32. 'Something a little bit 'tasty" – George Orwell, food, politics, and empathy

S. de Melo Araújo
CETAPS/CITCEM/ILC-ML/IF, FLUP/ESE-IPP; sofiademeloaraujo@hotmail.com

> Because a love for good food is the only thing that remains with man when he grows old. Love? What is love when you are five and fifty and can no longer hide the disgraceful baldness of your pate. Ambition? What is ambition when you have discovered that honours are to the pushing and glory to the vulgar. Finally we must all reach an age when every passion seems vain, every desire not worth the trouble of achieving it; but then there still remain to the man with a good appetite three pleasures each day, his breakfast, his luncheon, and his dinner. (W. Somerset Maugham, The Explorer)

The most basic, biological, and essential aspects of human existence extend beyond time and space, and provide a rare insight into history: childbirth, sex, illness, pain, shelter, death, and food show the path of Man and Mankind alike. The fact that production and consumption are so clearly established when it comes to nourishment, and the daily, transient, repetitive nature of the process make food a particular instrument of power. Eating (and drinking) is indeed a daily ritual which sustains human survival. Thus, the struggle for survival is necessarily first and foremost the struggle for food. Lack of food has defeated armies, maddened individuals, moved people across the globe. The biologically mandatory consumption of food isolates from the community those who suspend its need in the extreme, beyond the case of common (temporary) collective experiences (e.g. fastening), as being either psychologically ill or religious oddities, often both – unlike what happens in the case of sex and chastity, for example. On the other hand, temporary deprivation of food has historically been seen as a cleansing far beyond that of the digestive system – be it in detox routines or tribal suspensions of consumption. Sublimation of hunger is the ultimate recognition of the soul, as that which defeats the body's strongest compulsion. Pleasure in consumption is seen as either soulful – in touch with nature, healthy, wise – or, most of the time, as a weakness, an indulgence. Food as pleasure is possibly the most subversive of pleasures – eating knows no age boundaries, it is experienced daily, it consumes what it enjoys, making it unavailable to others, and it alters the essence of the self.

Most societal elements have been used in history to assure the distinction between social classes, and to reassure the ruling classes of their power. Fabrics, colours, festivities, titles, cultural habits have all (and still are, even if less explicitly) been used to define individuals according to their social class and to structure their mutual exchanges. Food – which primitively would have guaranteed status based on the physical prowess of the hunter – has been a centre stage for social division, not only through 'what', but also 'how' and 'why'[1].

When it comes to 'what', certain foods have been a mark of financial power, whether because they were (or are) expensive in themselves due to their natural rarity or to the difficulty in obtaining them (e.g. caviar and truffles today, spices during the Renaissance), or because they require extra expense beyond common ability (e.g. chests of salt, refrigerators, travelling or having food shipped from abroad). In fact, the fisherman's daily seafood is hardly a sign of elite in his milieu, but it becomes so at a distance.

[1] 'Who' is also a fundamental aspect, even within a social class, as shown by Pierre Bourdieu, reflecting on the working-class: 'It is part of men's status to eat and to eat well (and also to drink well) [...]. On Sundays, while the women are on their feet, busily serving, clearing the table, washing up, the men remain seated, still eating and drinking' (Bourdieu, 1984: 195).

EurSafe 2016 – DOI 10.3921/978-90-8686-834-6_32, © Wageningen Academic Publishers 2016

Also, the 'what' can signal the cultivated spirit of the user who has been sufficiently acquainted with the world as to use a rare ingredient or technique, or to have a – emphasis on the adjective – cultivated taste.

The 'how' is the least expensive and most pervasive guarantee of social distinction/discrimination: be it table manners, places of consumption (including not only location, but also, for e.g. distribution of tables, place in the house, seat at the table), labour division or time invested. Even the moderate biting of high classes – Bourdieu calls it a 'gentle, indirect, invisible censorship' (cf. Bourdieu, 1984: 196) –, conveys the unmistakable evidence of an act of consumption made outside a scenery of starvation, and thus a sign of affluence.

'Why' comes as a sign of possession of information – thus, a sign of class also. The rejection of certain foods, not as a reflection of disgust or satisfaction, but rather for intellectual reasons (health, ethics (animal or human), behaviour (e.g. alcohol), or etiquette), implies a second level of choice – whereby the eater can select, rather than fight for food – and a possession of both information and critical judgment, key to social distinction in contemporary settings. Thus, rejecting a tasty treat for its consequences proves the existence of a thinking human ruling over the animal. Here is where we will find many of the great food ethics debates when it comes to cultural issues. Here lies the minefield that has hindered the reputations of possibly well-intended actors like the Portuguese charity president Isabel Jonet (who publically associated poverty to irresponsible choices) or the far more researched case of British chef-star Jamie Oliver (whose struggle for changing eating habits is as well documented as controversially paternalistic). Apart from the occasional delight in superiority, signalled by authors like Kenneth F. Kiple[2], most of these cases make evident, above all, the inability to comprehend a different circumstance and the humane aspect of pleasurable irrationality in a life of toil:

> Members of socialist/philanthropic groups were also at the forefront of dietary reform among the poorer classes ... For example, after repeatedly explaining to poorer families the nutritional benefits of porridge for breakfast, Fabian women were dismayed to find that it did not find its way onto the family menu. This was because it needed constant attention to stop it burning; attention which it did not receive, what with children milling at the mothers' feet. The result was an unpalatable mess. Moreover, families just did not like it. (Coveney, 2006: 20)

This final aspect is of the utmost importance: it is not merely that poor people are not able to eat in a healthier way; it is that they may not enjoy it. In lives as these in the 19th century or those of the miners of Wigan Pier, an instant of pleasure is no mean treat to go to waste. George Orwell understood that.

Orwell wrote politically, whether it was journalism or fiction. But it is in 'The Road to Wigan Pier' (1937) combined with the fictional 'Animal Farm' (1945) and 'Nineteen Eighty-Four' (1949) that George Orwell makes explicit his understanding of how instrumental food can be as a source of power, both physical and mental. In them, one sees the ability for empathy that makes any criticism all the more humane, but also that much more demanding. 'The Road to Wigan Pier' (1937) is, indeed, the core of Orwell's empathic, if demanding, analysis of the lower classes and their use of food as a soother of fate. The result of many, many days spent beside the miners, this journalistic piece portrays the appalling living conditions of these families. On the one hand, Orwell appeals to action and change, and is worried about the future generations and their upbringing, always as a social concern of the collective:

---

[2] Kiple is adamant: 'Some defenders of the system, however, take a perverse delight in pointing out that obesity is more pervasive among the poor than among the affluent, brushing aside that the poor feed disproportionately on cheap and fatty fast foods' (Kiple, 2007: 255).

> A human being is primarily a bag for putting food into; the other functions and faculties may be more godlike, but in point of time they come afterwards. A man dies and is buried, and all his words and actions are forgotten, but the food he has eaten lives after him in the sound or rotten bones of his children. I think it could be plausibly argued that changes of diet are more important than changes of dynasty or even of religion ... Yet it is curious how seldom the all-importance of food is recognized. You see statues everywhere to politicians, poets, bishops, but none to cooks or bacon-curers or market gardeners. (cf. Orwell, 2011b: 84)

On the other hand, his empathy with this 'bag for putting food into', his existential humanism, assure an understanding that encompasses the actions and thoughts that he does not share or defend. Thus, regardless of his own standing, he is able to sympathize (cf. Orwell, 2001b: 88-89). He understands (and values) pleasure as a legitimate pursue, all the more so in distress. This is not to say that he is indifferent as to choices, or that he is too indulgent or acritical. Loraine Saunders recalls how although Orwell defends miners against Rev. Inge's accusations of gluttony, he also opposes their (ab)use of tinned food, in particular milk unfit for children (cf. Saunders, 2008: 134-135).

One can imagine that Orwell's dystopia has been brought to life in a more Huxleyan perspective, based precisely on consumption, as does John Lukacs (LUKACS, 2005: 222-223). However, both novelists would agree on the relevance of food as an instrument of power. Orwell is particularly attune to the subversive, revolutionary power of Oliver Twist's 'Could I have some more?'

Empathy does not come from being the Other, but is rather the ability to experience the Other's Self. Thus, Orwell's humanism is an intellectual and emotional purposeful work of excellence. At a given moment of 'The Road to Wigan Pier' he reflects on the humanizing power of pleasure in food:

> The food crank is by definition a person willing to cut himself off from human society in the hopes of adding five years onto the life of his carcase; that is, a person out of touch with common humanity. (Orwell, 2001b: 162)

In fact, only through a constructively disillusioned perspective of Humankind can one strive for change, while remaining in touch with common humanity, in an eternal, subversive 'Give me more'.

References

Andrews, G. (2008). The slow food story. Politics and Pleasure. Pluto Press, London, UK.

Berning, N. (2011). Narrative means to journalistic ends. A narratological analysis of selected journalistic reportages. Verlag, Berlin, Germany.

Bourdieu, P. (1984). Distinction: a social critique of the judgment of taste. Harvard University Press, Cambridge, MA, USA.

Coveney, J. (2006). Food, morals and meaning. The pleasure an anxiety of eating. Routledge, London, UK.

Kiple, K.F. (2007). A movable feast. Ten millennia of food globalization. Cambridge University Press, Cambridge, UK.

Hitchens, C. (2002). Why Orwell matters. s/l: basic books.

Mclaughlin, N. (2007). 'Orwell, the academy and the intellectuals'. In: Rodden, J. (ed.) The Cambridge companion to George Orwell. Cambridge University Press, Cambridge, UK, pp. 160-178.

Orwell, G. (2000a). A Clergyman's daughter. Penguin, London, UK.

Orwell, G. (2000b). Animal farm. Penguin, London, UK.

Orwell, G. (2000c). Nineteen eighty-four. Penguin, London, UK.

Orwell, G. (2001a). Down and about Paris and London. Penguin, London, UK.

Orwell, G. (2001b). The road to Wigan Pier. Penguin, London, UK.

Patai, D. (1984). The Orwell mystique. A study in male ideology. University of Massachusetts Press, Amherst, USA.
Piper, N. (2014). Audiencing Jamie Oliver: social engagements with food media. PhD thesis department of geography, University of Sheffield, Sheffield, UK.
Said, E.W. (2000). Reflections on Exile and other essays. Granta, London, UK.
Saunders, L. (2008). The unsung artistry of George Orwell – the novels from Burmese days to nineteen eighty-four. Ashgate, Hampshire, UK.

33. The world is out of joint: extreme weather and food crises in early modern poetry

S. Meisch
International Centre for Ethics in the Sciences and Humanities (University of Tuebingen), Junior Research Group: 'Ethics of Science in the Research for Sustainable Development', Wilhelmstr. 19, 72074 Tuebingen, Germany; simon.meisch@uni-tuebingen.de

Abstract

Extreme weather events are among the most severe effects caused by anthropogenic climate change. They negatively affect agriculture and food production. Sustainability sciences are aware that complex environmental challenges such as extreme weather cannot be dealt with adequately by science only, let alone by a single academic discipline. Consequently, there are efforts to promote inter- and trans-disciplinary science. The humanities are called upon to participate in research for sustainable development. Yet what might be their genuine contribution? This paper looks at early modern poetry. People of that time had to cope with (non-anthropogenic) climate change. The 'Little Ice Age' (1450-1850) manifested itself in extreme weather (e.g. heavy rain, hailstorms, floods, storm tides) affecting negatively food production, agriculture and human welfare. An analysis of 17th century poems addressing extreme weather shows what scientific facts of climatology mean in cultural terms and in people's daily lives. The paper focuses on how the Protestant hymn writers Paul Gerhardt, Simon Dach and Johann Rist verbalise these events and try to attribute meaning to them. In these texts, we not only find manifestations of a climate-induced food crisis, but we also see normative and evaluative strategies to deal with this crisis. The paper analyses strategies of cultural adaption to changing environmental conditions.

Keywords: humanities, climate change, food crises, hybridised knowledge, extreme weather

Telling different stories

Sustainability sciences are aware that complex environmental challenges cannot be dealt with adequately by science only, let alone by a single academic discipline. Consequently, there are efforts to promote inter- and trans-disciplinary science presumably better equipped for the creation of integrated (policy) solutions (cf. Lövbrand *et al.*, 2015). The humanities are regularly called upon to participate in solution-oriented research for sustainable development (cf. Hulme, 2011). For many reasons, they are sceptic about these invitations. They do not know either how they can or why they should contribute to an academic field already neatly set by positivist sciences. The Humanities, just as the interpretative social sciences, do not consider it their task to better communicate scientific facts. In their view, this establishes an unacceptable hierarchy of different forms of knowledge relevant to address climate change and other global environmental challenges (Hulme, 2011; Lövbrand *et al.*, 2015). This brings us back to the question of what is relevant knowledge. Sachs warns that 'environmental action and environmental discourse, when carried on in the name of 'sustainable development', implicitly and explicitly position themselves with respect to the crisis of justice and the crisis of nature. Different actors produce different types of knowledge, they highlight certain issues and underplay others. How attention is focused, what implicit assumptions are cultivated, which hopes are entertained, and what agents are privileged, depends on the way the debate on sustainability is framed' (Sachs, 1999: 29). Rudiak-Gould (2013) reminds us that the relevance of knowledge depends on the way we perceive problems such as climate change. Is it inherently invisible to lay people but not to scientists with their means to see it (invisibilism)? Is climate change visible and can be seen and experienced by those affected by it (visibilism)? Can climate change be made visible by different (cultural, aesthetic, etc.) means (constructive visibilism)? Rudiak-Gould

EurSafe 2016 – DOI 10.3921/978-90-8686-834-6_33, © Wageningen Academic Publishers 2016

does not opt for one solution, but reminds us that 'the visibility of climate change is partly an empirical question of just how dramatically humans have tampered with the atmosphere; partly an instrumental question of how best to communicate anthropogenic global warming to the public; and partly a moral, political, and epistemological interrogation of the value and role of scientific expertise in democratic society' (Rudiak-Gould, 2013: 129). As the three models have different perspectives on the role of science, inter- and trans-disciplinary cooperation would have different outlooks. In this paper, I will thus reflect potential humanistic contributions to sustainability research. First, I will analyse early modern German poetry dealing with extreme weathers (heat wave, cold spell, heavy precipitation). The poets witnessed the effects of the 'Little Ice Age' (1450-1850), which manifested itself in extreme weather events adversely affecting agriculture, food production and human welfare. Analysing 17th century poems addressing extreme weather shows what scientific facts of climatology mean in cultural terms and in people's daily lives. I focus on how the Protestant hymn writers Paul Gerhardt, Simon Dach and Johann Rist verbalise and give (evaluative and normative) meaning to extreme weather and aim to overcome a life-threating crisis. I will then ask how this humanistic engagement can contribute to research for sustainable development. I will not argue for a food ethical position but reflect (on a more meta-ethical level) discursive contexts of food ethics.

Extreme weathers and food crises in early modern poetry

By dealing with three early modern German poems, I address one possible answer to the question what the humanities can contribute to the research for sustainable development:

- 'In kalter Winters-Zeit' (In cold wintertime) (1643) by Simon Dach (1605-1659);
- 'Das Sechste Buhßlied. Sehr nützlich zu singen/ wen etwan grosse langwirige Hitze/und gahr dürre Zeit einfellt' (The sixth song of confession. Very useful to sing in times of great long heat and drought) (before 1651) by Johan Rist (1607-1667);
- 'Buß- und Betgesang bei unzeitiger Nässe und betrübtem Gewitter' (Song of confession and prayer in case of unseasonable wet and distressed thunderstorm) (before 1648) by Paul Gerhardt (1607-1676).

The three poets had an academic background in Protestant (Lutheran) theology. Gerhardt and Rist worked as pastors and are among the most important Protestant writers of hymns and spiritual poems. Dach made a career as a professor of poetry and rector of the University of Konigsberg. He is also famous for his non-religious lyric poetry. While these poets engage with the effects of climate change, we have to bear in mind that 17th century contemporaries knew nothing about the concepts of Little Ice Age and climate change. Yet they were aware that their 'world is out of joint' with a weather not the same as it used to be (Parker, 2013).

Writing during the 'Little Ice Age'

Dach, Rist and Gerhardt were born at the beginning of the 17th century. During their lifetime, they witnessed not only one of the main phases of the Little Ice Age in Central Europe (1621-1650), but also one that, in the late 1630s, turned from exceptionally dry to exceptionally wet years (Glaser, 2008: 152). The authors lived in Northern Central Europe (Dach: Konigsberg region, Rist: Hamburg region, Gerhardt: Berlin region). The more northward people lived, the more likely they experienced the effects of the Little Ice Age. We can thus expect that the three shared these experiences with their fellow contemporaries.

Before looking at the poems, we should turn our attention to the Little Ice Age, a period of relatively cold climate in the Northern hemisphere between 1400 and 1850. Though the term implies a general cooling, 'viewed hemispherically, the 'Little Ice Age' can only be considered as a modest cooling of the Northern Hemisphere during this period of less than 1 °C relative to late 20th century levels' (IPCC,

2001: 135). It was a period of deteriorating climate interrupted by periods of warming. Yet, climate and weather conditions at the time need to be differentiated spatially and temporally. For instance, years of extreme heat, such as 1540 or 1689, resulting in drought and bad harvest were followed by years with mild summers and normal harvests (Glaser, 2008). When focussing on specific years and regions, one finds times of normality repeatedly disrupted by times of devastatingly extreme weather. Meanwhile, the early modern period was also a period of religious tensions, wars and refugees, hunger and starvation, diseases, energy crises and the like (Parker, 2013).

Without overusing climate deterministic arguments, we can read the poems as representations of crises also caused by climate change. The poems address the issue of extreme weather with regard to food and agriculture, and to a lesser degree to energy. Peasants and shepherds are the only professions explicitly mentioned. Since we are dealing with agrarian societies, this should not surprise us. Due to extreme weathers (frost, drought, wetness), peasants cannot cultivate their fields and crop failures are looming. Wild and farm animals do not find or get feed and suffer from heat or cold. Due to frost, fish and birds cannot be caught and sold; due to heat, they perish. Frost causes energy crises because people have to heat more and find dry wood. Heat increases the danger of (forest) fires. Food and energy crises particularly affect the poorest of the poor. People suffer from corrupt and failing state authorities and wars. Darkness, uncertainty and (food and energy) shortages cause existential fear.

As to the representation of crises, Dach's poem is especially interesting as it deals with the exceptionally cold year of 1643. He accurately describes how a mild winter was followed by a very cold spring. Historical climatology tells us that April 1643 saw a trebling of freezing days, and May a cold spell with frost, snow and storm traveling from North to South and causing considerable damage to vegetation (Glaser, 2008). Dach gives us not only an accurate weather report, but also insights into the everyday life of agrarian societies. He describes how the cattle in the barn are roaring because of hunger and cold and the peasant in his despair cuts the straw from the roofs as a last resort to fodder. This existential threat and desperation seems to have been a general experience of agrarian societies as we also find it in other literary works such as 'Die Wassernot im Emmental' written by Jeremias Gotthelf in 1838. Two centuries after Dach and in another geographical region, Gotthelf reports a flood disaster following a long and snowy winter in the Swiss valley of Emmental in 1837.

Lutheran songs of piety after 1600

We can read these texts as sources of single historical events or more general developments. Braungart (1999) reminds us that we cannot stop here. By depicting a crisis, literary texts are also aware of it and part of it. We can thus read them as an epoch's attempt to come to terms with its own critical moments. Yet we do not find a uniform engagement with climate change and extreme weather. The poems have to be understood against a particular historico-cultural background, i.e. the ecclesiastical and religious crises in 17th century Lutheran territories.

Veit (2005) observes that the years between 1580 and 1700 were the heyday of Protestant hymn writing. There was a thematic shift away from the hardships and persecution of the young Protestant church towards the misery of ordinary people. The proportion of hymns dealing with the distresses of ordinary people, i.e. war, plagues, early death, etc., grew disproportionally. Within this text corpus, the number of songs about weather grew particularly. We do find concerns about a changing climate and extreme weather.

When dealing with weather, the poems follow an analogous line of argumentation (Veit, 2005). First, extreme weather is God's just punishment of human sins against God and other humans. The poems repeatedly stress that humans did not act according to the laws and that God is right to punish them.

Particularly, Rist and Dach refer to Old Testament covenant theology and faithfulness (Lev. 26, 18ff., Deut. 28, 23ff.). There, God promises his blessing in case people conform to the laws, and threatens them with his curses in case they do not. When describing extreme weather such as frost or drought, the poems use biblical imagery, such as 'heaven as iron' or 'earth as brass' (Lev. 26, 19), to linguistically paint the crises and classify it within a normative horizon of meaning. All authors consider it extremely severe but fair punishment that the land cannot be worked and that people find no food. Second, God is the master of the weather and therefore extreme weather is a sign of his strength. The authors believe that human unbelief and misbehaviour cause extreme weather. The Creation is God's domain of action. Even after the end of the Creation story, God continues influencing it. Third, God uses his Creation by means of extreme weather (as announced in the Old Testament) to speak to the people, punish them and tell them to change their way by repenting and confessing their faith. Forth, the hymns demonstrate two dimensions of God: the furious and the merciful. In particular, Gerhardt argues that God is legitimately furious due to humanity's sinful misbehaviour, such as unfaith, war, exploitation of the poor, state failure, etc. At the same time, God wants to love his creatures and impatiently waits for them to return. Departing from strict Lutheran orthodoxy, Gerhardt implies that because of his love, God might be susceptible to blackmail: if his creatures reveal their love to God, he cannot withstand, but must love back (Kemper, 1987). Fifth, extreme weather relates to expectations of the Last Days. We find this explicitly in Gerhardt's poem, where the speaker asks for the daily bread as well as 'the sweet bread of eternity'. Dach's lyrics were sung to Bartholomäus Ringwaldt's tune of 'Es ist gewißlich an der Zeit' (The Day is Surely Drawing Near) (1586) referring to the near coming of Christ. This reminds us that for contemporaries the tunes provided another interpretative layer.

The songs are about the collective effort to deal with and overcome an existential fear. After all, church congregations sang them. The hymns sought for crisis management by combining real world experiences of agrarian societies with Lutheran theology and intertextual references to the bible and a known (non-) religious treasury of songs. In spite of extreme weather, 17th century contemporaries shared the certainty of a divine plan for salvation.

We would not do justice to the poems if we treated them as mere colouring pictures of Lutheran orthodoxy. Thanks to their certainty about the divine salvation plan, Lutherans had great difficulties coming to terms with climate-induced crises affecting water, food and energy. There are laments and horror about the extent of God's damages to his Creation. Dach, Gerhardt and Rist are anxious that God might exaggerate his punishment and drive people away from their faith (Kemper, 1987). By referring to the city of Nineveh in the Book of Jonah, Dach's poem begs God to at least save the innocent animals (Jonah 4, 11). With this, the song reminds God (and the congregation) that he once had mercy even with the sinful inhabitants of Nineveh. Permanent weather invariabilities, together with food crises, wars and plagues, caused sorrow in Lutheran authors. For them, this was an existential problem, as Lutheranism theologically saw no means to influence God except through faith. Calvinists at least had the chance to regard extreme events as extreme proof of faith, while Catholics had the opportunity to do good deeds (Kemper, 1987). Therefore, we might understand why specifically Lutheran authors addressed the effects of the Little Ice Age (Parker, 2013). The poems create a particular knowledge with regard to a crisis between God and humans articulated in the context of weather (metaphors) and according to particular genre rules. They are about extreme weather as well as about just human societies (in the Lutheran view).

Seeing climate change differently

Rudiak-Gould (2013: 120) explains that 'the dispute between climate change 'visibilism' and 'invisibilism' is not scientific so much as political, being a proxy war for a larger debate on scientific versus lay knowledge and the role of expertise in democratic society'. Present climate change debate (and more generally sustainability sciences) is still very much dominated by 'invisibilism', relating closely to the modernist frame of science preaching control over nature and promising universal welfare by providing secure knowledge for techno-scientific and marketable innovations (Benessia *et al.*, 2011; Rudiak-Gould, 2013). This narrative is disputed for reasons that I cannot discuss here (cf. e.g. Benessia *et al.*, 2011; Evans and Reid, 2014; Leese and Meisch, 2015; Lövbrand *et al.*, 2015; Sachs, 1999): The humanities and interpretative social sciences question the role science is ascribed to in solving social issues, its focus on problem-solving, the reduction of sustainable development to neoliberal practices and post-politicisation. Humanistic scholars do not see their role in contributing to solutions within integrated sustainability science (Benessia *et al.*, 2011; Hulme, 2011; Lövbrand *et al.*, 2015) based on 'a homogenizing and reductive simplification of the normative complexity of our environmental situation' (Hailwood, 2016: 60).

To bring it back to Rudiak-Gould's distinction, in this paper, I dealt with a combination of 'visibilism' and 'constructive visibilism', i.e. eyewitnesses reporting that they have seen the extreme with their own eyes by reporting and dealing with it within the context of a specific literary genre. By addressing these texts, traditional humanities open up and venture into aesthetically constructed spaces, which allow us to make new experiences and correlate them with the realities of our life. We face different constructions of social nature and diverse value systems that might challenge, widen and enrich our own horizons. Against this background, what might this mean in the case of early modern German religious poetry? What we get is a privileged space of reflection on many different aspects familiar to us from sustainability discourse of which I only want to mention four. First, we learn about 17th century authors dealing with the effects of extreme weather on agriculture and food production. Just like us today, they believed that they caused the changing climate through social misconduct. While today we naturalise the discourse and look to the natural sciences for solutions, 17th century contemporaries saw it as a social issue. With this, I do not deny anthropocentric causes of climate change. Yet engaging with this poetry provokes thought about what it means that we cause global environmental changes such as climate change (cf. Hailwood, 2016), and what respective interpretations of the problem mean for (democratic) societies (Benessia *et al.*, 2011; Rudiak-Gould, 2013). Second, what we get is a space to come to terms with religious lines of argumentation on extreme weather and (its effects) on food. Today's engagement with climate change takes place in secular societies, and yet there are many religious undertones. In addition, we also have to take into account other parts of the world that are still more religious than us and are becoming more religious. Third, what we get is a space to reflect on the social and cultural embeddedness of different climate change discourses. We read about an influential 17th discourse, which was only one contemporary approach of thinking about extreme weather. In this respect, a comparative study of Catholic, Calvinistic or Islamic literature might be interesting. Fourth, what we get is a space of reflection on doomsday narratives underlying sustainability sciences, and the requirements of conceiving the future as open and shapeable by human action in spite of (physical, social and normative) complexities (Beck, 2010; Lövbrand *et al.*, 2015; Taleb, 2012).

As food ethicists, we need to engage in epistemological and ethical debates, or what Benessia *et al.* (2011: 87) calls the 'extension of epistemic and normative rights'. What are the knowledge and the narratives we need for sustainable agriculture and food? The humanities cannot provide techno-scientific solutions, but they can help to identify relevant questions. Food ethics might want to connect to diverse narratives about food and open reflexive spaces themselves to deliberate about the future natures we wish to inhabit (Lövbrand *et al.*, 2015).

Knowledge about weather and climate change and their impact on food and agriculture is inevitably diverse, socially embedded and historically contingent (*ibid.*; Braungart, 1999). After all, there might not be the relevant knowledge for policy solutions but many hybrid ones (Benessia *et al.*, 2011). Literature provides neither a simple nor the only authoritative representation of nature; however, it can generate and represent specific literary knowledge of the relationship between weather and humans. This relationship is embedded in wider moral contexts. Literature offers a space to reflect how we as humans want to live together on this planet and deal with a complex world that confronts us with harmful events. Engaging with literature allows us to liberate ourselves from 'modern, standardizing narratives of techno-scientific power and control' (Benessia *et al.*, 2012: 76). We can learn to creatively deal with a complex world we might never fully understand (Taleb, 2012), and possibly build 'relative resilience, defined as the capacity to embrace change and complexity and creatively adapt to them, as they unfold' (Benessia *et al.*, 2012: 87).

Acknowledgements

I want to thank Elisabeth Jütten, Lieske Voget-Kleschin, Gero Bauer and Silvio Funtowicz for their helpful comments on earlier versions of this paper.

References

Beck, U. (2010). Climate for change, or how to create a green modernity? Theory, Culture and Society 27(2-3): 254-266.

Benessia, A., Funtowicz, S., Bradshaw, G., Ferri, F, Ráez-Luna, E.F., Medina, C.P. (2012). Hydridizing sustainability: towards a new praxis for the present human predicament. Sustainability Science 7, Suppl. 1: 75-89.

Braungart, G. (1999). Poetische selbstbehauptung. Zur ästhetischen Krisenbewältigung in der deutschen Lyrik des 17. Jahrhunderts. In: Jakubowski-Tiessen, M. (ed.) Krisen des 17. jahrhunderts. Interdisziplinäre Perspektiven. Göttingen, pp. 43-57.

Evans, B. and Reid, J. (2014). Resilient life: the art of living dangerously. Cambridge, UK.

Glaser, R. (2008). Klimageschichte mitteleuropas. 1200 Jahre wetter, klima, katastrophen. Darmstadt, Germany.

Hailwood, S. (2016). Anthropocene: delusion, celebration and concern. In: Pattberg, P. and Zelli, F. (eds.) Environmental politics and governance in the anthropocene: institutions and legitimacy in a complex world. London, UK, pp. 47-61.

Hulme, M. (2011). Meet the humanities. Nature Climate Change 1: 177-179.

Intergovernmental Panel on Climate Change (IPCC) (2001). Climate change 2001: the scientific basis. Cambridge, UK.

Kemper, H.-G. (1987) Deutsche lyrik der frühen neuzeit. Vol. 2, Konfessio- nalismus. Tübingen, Germany.

Leese, M. and Meisch, S. (2015). Securitising sustainability? Questioning the 'water, energy and food-security nexus'. Water Alternatives 8(1): 695-709.

Lövbrand, E., Beckb, S., Chilversc, J., Forsythd, T., Hedréna, J., Hulmee, M., Lidskogf, R., Vasileiadoug, E. (2015). Who speaks for the future of earth? How critical social science can extend the conversation on the Anthropocene. Global Environmental Change 32: 211-218.

Parker, G. (2013). Global crisis: war, climate change and catastrophe in the seventeenth century. New Haven, London, UK.

Rudiak-Gould, P. (2013). We have seen it with our own eyes: why we disagree about climate change visibility. Weather, Climate, and Society 5: 120-132.

Sachs, W. (1999). Sustainable development and the crisis of nature: on the political anatomy of an Oxymoron. In: Fischer, F. and Hajer, M. (eds.) Living with nature: environmental politics as cultural discourse. Oxford, UK, pp. 23-41.

Taleb, N. (2012). Antifragile: things that gain from disorder. Random House, New York, NY, USA.

Veit, P. (2005). Gerechter Gott, wo will es hin/mit diesen kalten zeiten? Witterung, Not und Frömmigkeit im evangelischen Kirchenlied. In: Behringer, W., Lehmann, H. and Pfister, C. (eds.) Kulturelle konsequenzen der 'Kleinen Eiszeit'. Göttingen, Germany, pp. 283-310.

34. Soil as sacred religion: the spiritual dimensions of sustainable agriculture

G. van Wieren
Michigan State University, Department of Religious Studies/C714 Wells Hall, East Lansing, MI 48824, USA; vanwie12@msu.edu

Abstract

This paper traces some of the spiritual dimensions embedded in the modern sustainable agricultural movement. It focuses on the early twentieth century writings of three figures considered by many to be central to its development: Rudolf Steiner, Liberty Hyde Bailey, and George Washington Carver. While these figures are well known for their agricultural scientific contributions, their religious ethical outlooks on nature and the human/nature relation have been overlooked. This is too bad because their views (1) provide interesting and novel examples of how ecological, spiritual, and moral values are interpreted in the context of agricultural production; and (2) help illuminate some of the sensibilities embedded in what I perceive as an emergent form of agriculturally-based spirituality in North America, or, what I am calling, soil as sacred religion. Similar to religion scholar, Bron Taylor's notion of dark green religion, soil as sacred religion is oriented by the belief that earth is in some sense sacred and therefore worthy of reverent care. Different from Taylor's conception, however, soil as sacred spirituality is shaped explicitly by the agricultural context and experience.

Keywords: agro-history, religious aspects, human values, sustainability studies

Introduction

In his 'A Brief History of Sustainable Agriculture,' Fred Kirschenmann observes four categories of response to the industrial agriculture that developed in mid-nineteenth century Europe and the U.S. (2016). The first was the humus farming movement influenced by works such as D. Browne's 'The field book of manures or the American mulch book' (1855), Charles Darwin's 'The formation of vegetable mould, through the action of worms, with observations on their habits' (1881), and, later, Albert Howard's 'An agriculture testament' (1943), which Kirschenmann calls the bible for humus farmers. A second wave of response developed in the early twentieth century around 'complex farming systems'. Key writings included F.H. King's 'Farmers of Forty Centuries' (1911) and 'Soil Management' (1914). Rudolph Steiner's 'Agriculture Lectures' (1924) and his vehement critique of reductionist agricultural science fostered a third movement, which resulted in 'biodynamic' agriculture. Lastly, a philosophy of 'organic' agriculture appeared in the 1940s in the work of Lord Northburne ('Look to the Land', 1940), the first to use the term, and also in the work of Liberty Hyde Bailey ('The Holy Earth', 1915), Lady Eve Balfour's 'The Living Soil' (1943), and Louis Bromfield's 'Pleasant Valley' (1946), all of which viewed the farm as a kind of self-contained organism.

Several points could be make regarding the spiritual dimensions embedded in these more holistic accounts of agriculture, including the religious resembling titles of many of the texts. In the interest of space, I focus in this paper on the views of three figures who incorporated religious ethical outlooks in their agricultural philosophies: Rudolf Steiner, Liberty Hyde Bailey, and George Washington Carver, a figure not included in Kirschenmann's list of sustainable agriculture's key progenitors, though one who is routinely drawn upon in African-American sustainable agricultural thought (Jordan *et al.*, 2007).

EurSafe 2016 – DOI 10.3921/978-90-8686-834-6_34, © Wageningen Academic Publishers 2016

I have chosen to focus on these figures both for their significance in the development of the sustainable agriculture movement in North America, as well as for their overt use of religious and ethical language to describe earth, the universe and people's relationship to the land. Furthermore, I have selected them for how I perceive intimations of their views in emergent forms of soil as sacred spirituality in North America, a topic I have written about elsewhere and reference in this paper's conclusion. Here I focus on texts that most clearly evidence Steiner's, Bailey's, and Carver's spiritual and moral views, particularly those that reference beliefs about the sacred and how they inform understandings of good agricultural practice. I begin with Steiner's and Bailey's pantheistic and deistic views respectively, turning finally to Carver for how he blends such views of the divine sacred in relation to nature.

Steiner's spiritual science of agriculture

Rudolf Steiner is perhaps best well known for his pioneering work in the field of biodynamic farming methods, though he also is recognized for his spiritual interpretation of Ernst Haekel's Darwinian based pantheistic philosophy (Taylor, 2010: 156). Steiner's spiritual approach to agriculture is most clearly seen in his 1924 lectures, 'Spiritual Foundations for the Renewal of Agriculture' – also referred to as The Agriculture Lectures – that were given over a ten-day period in Koberwitz, Silesia, then Eastern Europe, as part of an educational course for farmers who had begun to observe a degradation of land health due to chemical fertilizers. Already familiar with the basic ideas of Steiner's spiritual philosophy of Anthroposophy, they were interested in learning techniques for replenishing the vitality and strength of their livestock and crops. The courses were not meant to teach specific farming practices but to supplement them with a spiritual framework for understanding and connecting with the forces that permeated the Earth and its beings and oriented the cosmos (Shouldice, 1993).

Central to Steiner's thought was the idea that the more obvious outer technical world corresponded with a non-obvious inner spiritual world, which could be discovered through a method of scientific investigation that heightened ones perception of the forces of nature (Thomas, 2005). To teach how this could be achieved, Steiner used detailed drawings in his lectures. In one such drawing, for example, he explains the spiritual quality of soil: 'We usually think of the soil – which I'm going to indicate here with this line [drawing 2] – as being something purely mineral, with organic matter coming into it only incidentally to the extent that humus develops or manure is applied. That the soil might contain not only this sort of life, but also an inherent plant – like vitality, and even something of the nature of soul-qualities, this is not even conceived of, much less accepted as fact' (Steiner, 1993: 28). Other drawings depict the 'cosmic constellation' of individual plants, the 'cosmic force' of the sun in a plant's leaves, the 'ego-potential' of animal manure, and the 'etheric principle' of all living beings.

Through meditation, Steiner believed, farmers could become 'ever more receptive to the revelations of nitrogen,' a central element in biodynamic farming methods (*ibid.*: 29). The type of meditation conducted in the West, in Europe, he wrote, was different than that conducted in the East. Only 'indirectly dependent on the breathing process; we live in the rhythm of concentration and mediation' (*ibid.*: 55). Still, Steiner's method of meditation did have a bodily component, although 'very delicate and subtle' (*ibid.*: 55). 'While meditating, we retain somewhat more carbon dioxide than we do in a state of normal waking consciousness ... We don't thrust the full force of the carbon dioxide out there, into the environment that is filled with nitrogen. We hold some back' (*ibid.*: 55). Through meditating like this, 'Everything becomes known, including everything that lives in nitrogen,' according to Steiner. Once the farmer becomes receptive to nitrogen's revelations, 'Suddenly we know all kinds of things, they are simply there. Suddenly we know all about the mysteries at work on the land and around the farm' (*ibid.*: 29-30).

Steiner saw his spiritual science as providing an urgently needed counterpoint to the materialist agriculture of his day. The latter he equated with an attempt to understand the totality of the human body by studying a little finger. A spiritual approach to agriculture, on the other hand, would seek to understand the interrelationships of the cosmos. So too it would seek connections with the elemental spirits and souls of the plant and animal kingdoms. This way, Steiner believed, a farm could become a thriving, self-sustaining entity.

Bailey's holy earth

Writing some twenty years after Steiner and in the U.S. context, Liberty Hyde Bailey wrote similarly to Steiner about a spirituality of nature that developed through people's direct contact with the earth, rather than through religious dogma or belief. 'The good spiritual reaction to nature is not a form of dogmatism or impressionism,' he wrote, rather, 'it results normally from objective experience, when the person is ready for it and has good digestion.' (Bailey, 1943, 2008: 53). Bailey at times drew on what many might consider more conventional readings of the bible to interpret the human relationship to earth, which he also refers to as creation. In the first section of 'The Holy Earth', for example, he quotes the first sentence of the Hebrew scripture: In the beginning God created the heaven and the earth. He continues: 'This is a statement of tremendous reach, introducing the cosmos; for it sets forth in the fewest words the elemental fact that the formation of the created earth lies above and before man, and that therefore it is not man's but God's. Man finds himself upon it, with many other creatures, all parts in some system which, since it is beyond man and superior to him, is divine' (*ibid.*: 4).

For Bailey, the earth was a holy system 'because the Earth, and everything that was initially in it, was the product of divine creation' (Taggart, 2008: viii). Because of this, earth's sacredness 'is intrinsic and inherent' (Bailey, 1943, 2008: 11). The 'essence of righteousness', wrote Bailey, was to 'live in sincere relations with the company of created things and with conscious regard for the support of all men now and yet to come' (*ibid.*: 11). Different from Steiner's pantheistic view that the divine was in some sense embedded and active in the material universe, Bailey held a deist view of the relationship between God and nature, which allowed an adherence to Darwinian evolution in a way that Steiner's theology did not. Nonetheless, Bailey like Steiner saw the need for a farmer to be a spiritual person in order to rightly see the land and how to cooperate with it. 'A man cannot be a good farmer unless he is a religious man,' Bailey wrote (*ibid.*: 24). 'Into this secular and more or less technical education we are now to introduce the element of moral obligation, that the man may understand his peculiar contribution and responsibility to society; but his result cannot be attained until the farmer and every one of us recognize the holiness of the earth' (*ibid.*: 25-26).

Carver's earth as God's workshop

In 1923, a year before Steiner delivered his 'Agriculture Lectures', George Washington Carver gave to a large audience attending the Women's Board of Domestic Missions at Marble Collegiate Church in downtown Manhattan. There he told attendees, 'I never have to grope for methods: the method is revealed at the moment I am inspired to create something new' (Kremer, 1991: 128). Two days later the New York Times ran an editorial entitled 'Men of Science Never Talk that Way', calling into question Carver's scientific approach, his race, and the Tuskegee Institute where he was a researcher and professor. Carver's response is worth quoting for how it represents the blending of his spiritual and scientific outlook:

> My dear Sir:
>
> I have read with much interest your editorial pertaining to myself in the issue of November 20.
> I regret exceedingly that such a gross misunderstanding should arise to what was meant by 'Divine inspiration.' Inspiration is never at variance with information; in fact, the more information one has, the greater will be the inspiration.
> Paul, the great Scholar, says, Second Timothy 2-15, 'Study to show thyself unto God, a workman that needeth not to be ashamed, rightly dividing the word of truth.'
> And again he says in Galations 1-12: 'For I neither received it of man, neither was I taught it, but by the revelation of Jesus Christ'. (Kremer, 1991: 129)

Carver goes on in the letter to list his credentials as a graduate of two degrees in Scientific Agriculture from Iowa State College of Agriculture and Mechanical Arts, and the scholars in chemistry and botany who had influenced him. He closes the letter with another scriptural text: 'John 8-32, 'And ye shall know the truth and the truth shall make you free." He concludes: 'Science is simply the truth about anything' (Kremer, 1991: 129).

Carver, like Steiner and Bailey, viewed science and religion as compatible, writing often about his spiritual relationship with the natural world, which he saw as infused with divine power. He wrote that he loved 'to think of Nature as unlimited broadcasting stations, through which God speaks to us every day, every hour, and every moment of our lives, if we will only tune in and do so' (*ibid.*: 143). Carver was a strong advocate of the nature-study movement that was emerging in the late nineteenth century in the U.S., and wrote a booklet for children about how to incorporate nature study with the development of school-based vegetable gardens. He believed that deeper knowledge about the intricacies and workings of the natural world would lead to better farming practices. 'The highest attainments in agriculture can be reached only when we clearly understand the mutual relationship between the animal, mineral, and vegetable kingdoms, and how utterly impossible it is for one to exist in a highly organized state without the other' (*ibid.*: 143). Conversely, where ecological relationships were ignored in agriculture, the land suffered. 'Conservation is one of our big problems in this section,' Carver was quoted by an Atlanta newspaper in 1940 (Ferrell, 2007: 23). 'It is a travesty to burn our woods and thereby burn up the fertilizer nature has provided for us. We must enrich our soil every year instead of merely depleting it. It is fundamental that nature will drive away those who commit sins against it' (*ibid.*: 23).

Carver's mystical relationship with the natural world appeared to grow with age. Like Steiner, Carver often referenced the importance of childhood experiences of nature, including his own, and how it impacted a lifelong affinity for nature. In a 1930 letter he wrote: 'We get closer to God as we get more intimately and understandingly with the things he has created. I know of nothing more inspiring than that of making discoveries for ones self ... the singing birds, the buzzing bees, the opening flower, and the budding trees, along with other forms of animate and inanimate matter, all have their marvellous creation story to tell each searcher of truth ... We doubt if there is a normal boy or girl in all Christendom endowed with the five senses who have not watched with increased interest and profit, the various forms, movements and the gorgeous paintings of the butterfly, many do not know, but will study with increased enthusiasm the striking analogy its life bears to the human soul' (Kremer, 1991: 142). He concluded that the more people came 'closer and closer in touch with nature' were they 'able to see the divine.' That way, they became 'fitted to interpret correctly the various languages spoken by all forms of nature ...' (*ibid.*: 143).

Carver's account of nature through agricultural science blends Steiner's pantheism and Bailey's deism into a form of soil as sacred religion where the divine is understood transcendentally and immanently in relation to the earth. In her essay, 'York, Harriet, and George: Writing African American Ecological Ancestors,' Kimberly Ruffin interprets Carver's spirituality as a religious syncretism between Christianity and traditional African religion. She quotes Toni Morrison on the intermingling of scientific and religious thinking in African American culture. Morrison writes, 'I could blend the acceptance of the supernatural and a profound rootedness in the real world at the same time with neither taking precedence over the other. It is indicative of the cosmology, the way in which Black people looked at the world. We are very practical people, very down-to-earth, even shrewd people. But within that practicality we also accepted what I suppose could be call superstition and magic, which is another way of knowing things' (Ruffin, 2007: 47).

Conclusions

Taken together Steiner's, Bailey's, and Carvers' conceptions of earth and the human/nature relation in the context of agriculture suggest the following religious ethical points. First, they suggest a view of earth's intrinsic, sacred, or, in Bailey's terms, holy worth by virtue of earth's relation to a divine being or force in the universe. Second, they suggest that the ongoing conservation of earth's sacred system 'trumps' human use in the sense that the mutual relationships among beings should be considered the more significant unit of concern, ecologically, morally, and spiritually. Third, almost paradoxically to this more holistic view, they see sacred value in and the need for attentive concern toward the tiniest units in nature, in Steiner's case, soil's nitrogen. The ethical upshot of these views suggests a fourth point, which is that people, especially farmers, need to learn to live in 'sincere relations' with the company of others (Bailey). This requires, finally, that agriculturalists develop a spiritual practice, if you will, a practice of meditation in Steiner's terms, routine direct contact with the land, in Bailey's and Carver's conceptions.

Intimations of these views can be seen to varying degrees in some of the forms of soil as sacred spirituality emerging in the sustainable agriculture movement in North America. For example, an African-American congregationally supported agriculture project in Chicago, Illinois, Mother Carr's Organic Farm, sees its organic agricultural methods as mimicking natural process and systems as 'God intended', and providing a faithful practice for serving God and the community. Leichtag Ranch, a Jewish farm in Encinitis, California, combines ancient Jewish rituals and biodynamic farming methods as a way to foster spiritual and agricultural practices that connect people to earth's seasonal cycles. Abundant Table in Thousand Oaks, California holds weekly Farm Church gatherings that emphasize the importance of spiritual contemplation in developing a moral relationship with the land and the production and eating of food (Van Wieren, unpublished data).

Even as these examples, along with Steiner's, Bailey's, and Carver's conceptions, tend to involve supernatural beliefs in a divine creator, the ethical and ecological imperative to conserve the land community appears to be the most significant principle. Furthermore, soil as sacred spiritual sensibilities can also be found, historically and contemporarily, in more secular and atheistic forms. What we will want to know moving forward is how, if at all, spiritual and moral interpretations of land and agriculture are being adapted by individuals and groups to make sense of and live sustainably within shifting ecological circumstances.

References

Bailey, L.H. (1943). The holy earth. Michigan State University Press, East Lansing, MI, USA, 123 pp.

Ferrell, J.S. (2007). George Washington Carver: a blazer of trails to a sustainable future. In: Jordan, J.L., Pennick, E., Walter, H. and Zabawa, R. (eds.) Land and power: sustainable agriculture and African Americans. Sustainable Agriculture Publications, Waldorf, MD, USA.

Kirschenmann, F. (2004). A brief history of sustainable agriculture. Science and Environmental Health Network. Available at: http://tinyurl.com/hvc2m9m.

Kremer, G.R. (1991). George Washington Carver: in his own words. University of Missouri, Columbia, MO, USA, 202 pp.

Ruffin, K. (2007). York, Harriet, and George: writing African American ecological ancestors. In: Jordan, J.L., Pennick, E., Walter, H. and Zabawa, R. (eds.) Land and power: sustainable agriculture and African Americans. Sustainable Agriculture Publications, Waldorf, MD, USA, pp. 33-56.

Shouldice, R. (1993). Preface. In: Steiner, R. (ed.) spiritual foundations for the renewal of agriculture. Bio-Dynamic Farming and Gardening Association, Inc., Kimberton, PA, USA.

Steiner, R. (1924, 1993). spiritual foundations for the renewal of agriculture. Bio-Dynamic Farming and Gardening Association, Inc., Kimberton, PA, USA, 310 pp.

Taggart, R.E. (2008). Preface. In: Bailey, L.H. (ed.) The holy earth. Michigan State University Press, East Lansing, MI, USA, pp. 7-9.

Taylor, B. (2010). Dark green religion: nature spirituality and the planetary future. University of California Press, Berkeley, CA, USA, 338 pp.

Thomas, N.C. (2005). Steiner, Rudolf – and anthroposophy. In: Taylor, B.R. (ed.) Encyclopedia of religion and nature. Continuum, New York, NY, USA, pp. 1596-1597.

Section 8. Aquatic species

35. Addressing socio-cultural values in the use and management of Baltic herring

S. Ignatius[1] and P. Haapasaari[2]*
[1]University of Helsinki, Department of Environmental Sciences, Kotka Maritime Research Centre, Keskuskatu 10, 48100 Kotka, Finland; [2]University of Helsinki, Department of Environmental Sciences, P.O. Box 65, 00014 University of Helsinki, Finland; suvi.ignatius@helsinki.fi

Abstract

Small pelagic fish species are often processed into fishmeal and oil, and used for animal feed. This has been criticised as an inefficient use of resources that could directly nourish humans. Baltic herring is an example of a regionally important catch species that has lost its importance as food, and is mainly used for animal feed and fishmeal and oil. Harmful dioxins in Baltic herring have been offered as an explanation for its reduced human consumption. We argue that dioxins are only part of the explanation while socio-cultural values also influence the use and management of the fish resource. By applying the justification theory of Boltanski and Thévenot to expert workshop discussions and a literature review, we investigate (1) what values people in Sweden, Finland, Denmark and Estonia associate with Baltic herring; and (2) which values determine the way the fish resource is used. Our study indicates that management decisions related to Baltic herring are predominantly guided by economic values and industrial efficiency. Profit optimisation in the fishery together with the image problem that Baltic herring suffers from, seem to have contributed to the devaluation of this fish species as food. The image problem relates to several factors, dioxins being only one of them.

Keywords: dioxin, feedfish, foodfish, stakeholder involvement

Introduction

An increasing demand for fish to feed the world's growing population puts pressure on the sustainable and ethical use of fisheries (Thurstan and Roberts, 2014). Especially small pelagic fish species are often used for the production of fishmeal and oil and for animal feed, which has been criticised for being a waste of a resource that could directly nourish humans (Wijkström, 2009). An excessive reduction fishery can also negatively affect marine food webs and, further, whole ecosystems (Skewgar *et al.*, 2007). FAO (2011) recommends states to consider measures to support the use of feedfish species for human consumption, to increase the supply of cheap fish for food. Still, advocates justify the reduction fisheries by referring to a lack of demand for these fish for human food (Allan, 2004).

Baltic herring is an example of a regionally important catch species that has lost its importance as human food. Baltic herring is a rich source of vitamin D and Omega-3, but it contains harmful dioxins, which complicates its use. However, dioxins do not prevent the consumption of herring, as different ways to manage the risk caused by dioxins to humans exist. The EU (EC, 2011) has set maximum levels of dioxins in food, and based on that, prohibits the selling of Baltic herring that contains dioxins above the limit values for human consumption. Within the EU limitation, different Baltic Sea countries have developed their own ways to manage the dioxin risk. In addition, the fishmeal and oil industry has techniques for removing dioxins from their final products.

Dioxins may look like an explanation for the low consumption of Baltic herring. We argue that they are only part of the explanation while socio-cultural values also influence the use and management of the fish resource. This assumption can be made on the basis of the diverse ways that different Baltic Sea countries have for dealing with the dioxin risk. For example Finland and Sweden have been granted a derogation from the EU ban (EC, 2011) to sell Baltic herring for food without restrictions within their territories, because the exposure of consumers to dioxins from other foodstuffs is lower in these countries than in the rest of Europe. The derogation involves a requirement to inform consumers about the risks by eating recommendations. Although the consumption of herring has decreased to a very low level in both Sweden and Finland, consumer surveys indicate that other reasons than dioxins underlie this (Mononen and Urala, 2010), and that not all consumers are aware of the herring eating recommendations (Kosonen and Ristiluoma, 2014). In Denmark, practically all Baltic herring catches go to fishmeal and oil industry, although the Danish fishery focuses on the Western Baltic herring that is almost dioxin-free. In contrast, Estonia uses all its catches for human consumption, and therefore targets small herring that contain less dioxins, and sorts the catch to comply with the EU regulation.

In this paper we explore how socio-cultural values guide the use and management of Baltic herring in four case study countries: Estonia, Finland, Sweden, and Denmark. We investigate (1) what values people in these countries associate with Baltic herring, and (2) which values determine the way the herring resources are managed and used, and which do not materialise in decision-making. Our final objective is to improve the understanding of the influence of values on the management and use of natural resources. The study is part of the research project BONUS GOHERR that investigates the prerequisites and potential of ecosystem-based management in reducing dioxins in Baltic Sea fish. The ecosystem-based approach requires interdisciplinary understanding of the ecosystem, including human behaviour as well as values related to the use and management of fish resources.

Material and methodology

The analysis is based on discussions in an expert workshop that focused on the dioxin problem of Baltic herring, and a review of articles related to Baltic herring use and management. Twelve experts representing the fishing industry, fishmeal and oil industry, herring research and management, dioxin research, public health authorities, and environmental non-governmental organisations (NGOs) from the four case study countries contributed to the workshop discussions. The topics of the discussions were: the implications of dioxins to herring fishing sector, global prospects for herring fishing industry, dioxins as a public health problem, and consumers' fish eating habits. The discussions were recorded and transcribed, and structured using content analysis.

We interpreted the texts by applying the justification theory of Boltanski and Thévenot (2006). The theory defines values as moral principles and distinguishes seven orders of worth that people utilise for justifying why something is valuable or good for society: (1) market worth is based on price or economic value; (2) industrial worth builds on efficiency, technical competence, and planning; (3) civic worth grounds on equality and solidarity; (4) domestic worth underlies traditions, hierarchies, and local ties; (5) inspirational worth evolves from emotion and grace; and (6) worth of fame from public opinion and renown; whereas (7) green worth reflects environmental friendliness.

The seven orders of worth were used as analytical categories, and identified from the texts as justifications for the question why Baltic herring is seen as valuable. We tracked the keywords and discourses reflecting these orders of worth. For example references to traditions were interpreted as manifestations of domestic worth, whereas arguments about public image were seen as expressions of worth of fame. Moreover, we paid attention to the frequency of occurrence of the values, positive or negative overtones attached to them, and the most prevalent conflicts and compromises between the orders of worth (Thévenot *et al.*, 2000.)

Results: socio-cultural values related to Baltic herring

Market worth relating to the economic value of Baltic herring fishery is commonly viewed as the main driver behind the use of herring catches in both workshop discussions and the literature. The fishing sector even relies on the capability of markets, with the support of managers, to ensure the sustainable and profitable use of the fishery. On the contrary, fisheries scientists tend to subordinate market values to industrial values, that is, to effective control, regulation, and enforcement for ensuring the sustainable and profitable use of fish resources in the long-run (Aps *et al.*, 2007; Kulmala *et al.*, 2007). The fishing sector wants to promote the human consumption of herring, because foodfish is assumed to have a higher economic value than feedfish. The relatively cheap consumer price of herring is seen as an opportunity to promote fish eating and the related health benefits in general, although the dioxin risk complicates this. Thus some of the public health authorities express concern that the cheap price of herring may encourage frequent consumption in certain low-income population groups and thereby increase their dioxin intake. Still, market values guide also the management of the dioxin problem. In Estonia, markets and traditions seem to be major drivers in directing herring catches to food use. In Finland and Sweden market values are compromised by values related to public health, which implies that risks and benefits of herring eating to human health are a focus of interest and that herring is increasingly directed to feed.

Industrial worth, that is, the view of herring as a resource, raw material, and an object of scientific knowledge and management, is also common. The fishing sector stresses the importance of utilising the resource as efficiently as possible within sustainable limits, either for food or feed. The environmental NGOs accept the use of herring for fish feed if it is environmentally certified, as a compromise between industrial and green values, but recommend using catches directly for human consumption. The environmental NGOs do not question the legitimacy of herring fishing or using herring as a resource as long as the stocks are sustainable. Scientists highlight the importance of efficient research and management in ensuring the sustainable use of the fish resource (Aps *et al.*, 2007). Overall, the value of herring as a food source is ambivalent since it contains both healthy and harmful substances. Therefore, efficient research and handling of the dioxin risk is regarded important. Swedish public health authorities stress the importance of minimizing the health risks of dioxins (Swedish National Food Agency, 2013), while Finnish authorities emphasise both minimizing the health risks of dioxins and maximizing the net health benefits of herring eating (Tuomisto *et al.*, 2015).

Civic worth, a view of herring as a common resource and its use as a question of social justice, underlies fishers' claims for respect for their livelihood and for participation in research and management (Salmi *et al.*, 2008a). Especially fishers in Sweden and Finland feel that the government and EU authorities disregard their livelihood. Basically, scientists and managers accept stakeholder involvement in policy making, but fear for the politicisation of the science used for policy (Linke and Jentoft, 2013). Civic values are also behind the discussions about risks of herring for public health: practices to manage the dioxin risk are justified by arguments for not only protecting but also enhancing public health. Here, a collision with the individual freedom to make food choices is unavoidable. Sweden and Finland have tried to reconcile this by the dioxin-derogation and eating recommendations. In Sweden the civic values related to public health protection seem to be stronger, as food safety authorities have suggested to give up the derogation for public health reasons (Swedish National Food Agency, 2013).

Domestic worth views Baltic herring as important for traditions, coastal communities, and food culture (Raid and Järvik, 2008; Salmi *et al.*, 2008b). Herring is the national fish of Estonia, which is a common justification for fostering herring related traditions. However, a distinction is made between traditions worthy and unworthy of preserving. Fish festivals are generally seen as good traditions, whereas the fisheries sector perceives for example the trials to preserve traditional fishing by regulating vessel size as a bad example that weakens safety on board. Modern fishing methods are called for, based on industrial and civic values. Herring is also valued as part of the traditional food culture of the Baltic Sea area, which is worth preserving but almost lost because of changed consumer preferences (Bonow, 2014; Sillanpää, 1999). The workshop participants saw that the dioxin-derogation in Sweden and Finland has been justified by traditions of herring eating and related recommendations. Another, less visible justification for the derogation may be the protection of the national herring fishery.

Worth of fame, that is, the public reputation of herring, revolves around the bad image of herring as food. This has raised concern especially in Finland, where consumers' herring eating habits and the image of herring have been studied (Kosonen and Ristiluoma, 2014; Mononen and Urala, 2010). The bad image has been explained by the dioxin discussion, lack of quality and availability of refined and user-friendly herring products, poor presentation of herring in grocery stores, association with animal feed, cooking difficulties, and view of herring dishes as old-fashioned and mundane. Also the environmental problems of the whole Baltic Sea are associated with the image problem of herring, particularly in Sweden and Finland. The fisheries sector emphasises the negative impact of the dioxin discussion to the image of herring more than the other stakeholders that see other reasons more important. In Estonia the public image of herring seems not that bad and also its linkage to the dioxin problem seems looser.

Green worth sees herring as part of biodiversity and is present in arguments about the eco-friendliness of herring as food and feed. The arguments for using herring as food are based on the assumed sustainable state of the stocks and lower ecological footprint of herring fishery as compared to meat production (Mäkinen *et al.*, 2013). Herring is also considered a better food choice than farmed Norwegian salmon because the latter may pose genetic and disease threats to wild salmon stocks (Thorstad *et al.*, 2008). Arguments for herring as ecological feed relate to reducing the environmental effects of Baltic Sea aquaculture by creating nutrient recirculation. On the other hand, the importance of sufficient herring availability for predators such as salmon, cod, marine mammals and birds, is highlighted. Overall, the fishing and using of herring seems to be generally accepted, and the welfare of individual fish does not seem to be an issue.

Inspirational worth of herring is based on its acclaimed unique characteristics or emotional experiences it awakes. Such values seem to be rare, but still Baltic herring is regarded unique both as a fish species and as food with fat content and taste differing from Atlantic herring. These characteristics are often subordinated to product development (Bonow, 2014) although sensory experiences related to herring as food seem to divide opinions (Roininen *et al.*, 2000).

Discussion

The results indicate that Baltic herring is predominantly viewed through its economic and productive values in Finland, Sweden, Estonia and Denmark. In the workshop, all parties emphasised the importance of the profitable use of catches, and efficient utilisation of the resource. Juxtaposition of using herring for food to feed was avoided. In addition, herring is regarded as a contributor to collective welfare and traditions, but these values are more contested. Herring traditions appear to be important in Estonia and Finland, whereas social justice is an issue for fishers especially in Sweden and Finland. The importance of adjusting traditions related to the fishing and eating of herring to modern times was acknowledged by all parties. The fishing sector calls for recognising their livelihood by involving them in herring research and management. The image of herring is considered poor particularly in Sweden and Finland, and almost missing in Denmark. The bad image has been explained by several factors, dioxins being only one of them. Green values are visible in arguments for the use of herring as eco-friendly food and fishfeed, but also in calls for acknowledging the importance of herring as a prey for other species. Inspirational values seem to be rare and subordinated to product development.

These results are based on expert workshop discussions and literature. Thus, possibilities to make conclusions about the importance of these values in different contexts and among general population are limited, and further research is needed. Our next step is to conduct stakeholder interviews and a consumer questionnaire to deepen our analysis.

Conclusions

Our study suggests that multiple socio-cultural values underlie the current undervaluation of Baltic herring as food and the increasing use of this pelagic species as feedfish. Following the recommendation of FAO (2011) to increase the human consumption of herring would require addressing the dioxin problem but also understanding the values and their contribution particularly to the poor image attached to Baltic herring. Based on the discussions of our workshop, we see an increasing cooperation between dioxin experts, herring researches, fisheries managers, herring fishers, environmental NGOs, and other relevant stakeholders, as a way forward in developing the use and management of Baltic herring to a direction supported by all. This study suggests that understanding values helps to explore if the management of a natural resource is in line with the desirable use of the resource, and to consider their reconciliation.

Acknowledgements

This study is part of the project 'Integrated governance of Baltic herring and salmon stocks involving stakeholders' (BONUS GOHERR). The project receives funding from BONUS (Art 185), funded jointly by the EU, the Academy of Finland, and the Swedish Research Council for Environment, Agricultural Sciences and Spatial Planning. We thank all participants of the expert workshop.

References

Allan, G. (2004). Fish for feed vs. fish for food. In: Brown, A.G. (ed.) Fish, aquaculture and food security: sustaining fish as a food supply. Crawford Fund for International Agricultural Research, Parliament House. August 11, 2004. Canberra, Australia. pp. 20-26.

Aps, R., Kell, L.T., Lassen, H. and Liiv, I. (2007). Negotiation framework for Baltic fisheries management: striking the balance of interest. ICES Journal of Marine Science 64: 858-861.

Boltanski, L. and Thévenot, L. (2006). On justification: economies of worth. Trans. Catherine Porter. Princeton University Press, Princeton, USA. 400 pp.

Bonow, M. (2014). Conflict or convergence? Products of origin. An analysis of the Swedish case of Baltic Sea fish. Baltic Worlds 6(3-4): 48-52.

European Commission (EC) (2011). Commission regulation (EU) No 1259/2011 of 2 December 2011 amending Regulation (EC) No 1881/2006 as regards maximum levels for dioxins, dioxin-like PCBs and non-dioxin-like PCBs in foodstuffs. Brussels, Belgium.

Food and Agriculture Organization of the United Nations (FAO) (2011). Aquaculture development. 5. Use of wild fish as feed in aquaculture. FAO Technical Guidelines for Responsible Fisheries No. 5, Suppl. 5. FAO, Rome, Italy, 79 pp.

Kosonen, A. and Ristiluoma, R. (2014). Silakan käyttö. Taloustutkimus and Finnish Food Safety Authority Evira. Helsinki, Finland, 66 pp.

Kulmala, S., Peltomäki, H., Lindroos, M., Söderkultalahti, P. and Kuikka, S. (2007). Individual transferable quotas in the Baltic Sea herring fishery: a socio-bioeconomic analysis. Fisheries Research 84: 368-377.

Linke, S. and Jentoft, S. (2013). Exploring the phronetic dimension of stakeholders' knowledge in EU fisheries governance. Marine Policy 47: 153-161.

Mononen, R. and Urala, N. (2010). Kuluttajien suhtautuminen silakkaan. Report 25.11.2010. Kuulas Research Agency and Pro Kala ry. Helsinki, Finland, 55 pp.

Mäkinen, T., Forsman, L., Grönroos, J., Kankainen, M., Salmi, P., Setälä, J., Silvo, K. and Vielma, J. (2013). COEXIST – Interaction in coastal waters. A roadmap to sustainable integration of aquaculture and fisheries. Baltic Sea Case Study Report, 68 pp.

Raid, T. and Järvik, A. (2008). Current situation and possible ways of sustainable development of small-scale fishery in Estonian coastal sea. 21 ICES Annual Science Conference 2008. Halifax, Canada.

Rannikko, P. and Määttä, T. (2010). Luonnonvarojen hallinnan legitimiteetti. Vastapaino, Tampere, Finland, 295 pp.

Roininen, K., Lähteenmäki, L. and Tuorila, H. (2000). An application of means-end chain approach to consumers' orientation to health and hedonic characteristics of foods. Ecology of Food and Nutrition 39(1): 61-81.

Salmi, P., Mäkinen, T., Salmi, J. and Kettunen, J. (2008a). Institutional sustainability of aquaculture and fishing – a case of the Archipelago Sea region, Finland. American Fisheries Society Symposium 49: 81-94.

Salmi, P., Salmi, J. and Mäkinen, T. (2008b). Livelihood strategies and survival of small-scale fisheries in the Finnish Archipelago Sea. American Fisheries Society Symposium 49: 517-529.

Sillanpää, M. (1999). Happamasta makeaan. Suomalaisen ruoka- ja tapakulttuurin kehitys. Hyvää Suomesta, Vantaa, Finland, 232 pp.

Skewgar, E., Boerma, P.D., Harris, G. and Caille, G. (2007). Anchovy fishery threat to Patagonian ecosystem. Science 315: 45.

Swedish National Food Agency (2013). Risk and benefit assessment of herring and salmonid fish from the Baltic Sea area. Report no 21-2013. Uppsala, Sweden, 124 pp.

Thévenot, L., Moody, M. and Lafaye, C. (2000). Forms of valuing nature: arguments and modes of justification in environmental disputes. In: Lamont, M. and Thévenot, L. (eds.) Rethinking comparative cultural sociology: repertoires of evaluation in France and the United States. Cambridge University Press, Cambridge, MA, USA, pp. 229-272.

Thorstad, E., Fleming, I., McGinnity, P., Soto, D., Wennevik, V. and Whoriskey, F. (2008). Incidence and impacts of escaped farmed Atlantic salmon Salmo salar in nature. Report from the technical working group on escapes of salmon aquaculture dialogue. NINA Special Report 36, World Wildlife Fund, 110 pp.

Thurstan, R.H. and Roberts, C.M. (2014). The past and future of fish consumption: can supplies meet healthy eating recommendations? Marine Pollution Bulletin 89: 5-11.

Tuomisto, J.T., Niittynen, M., Turunen, A., Ung-Lanki, S., Kiviranta, H., Harjunpää, H., Vuorinen, P.J., Rokka, M., Ritvanen, T. and Hallikainen, A. (2015). Itämeren silakka ravintona – Hyöty-haitta-analyysi. Evira Research Reports 1/2015. Finnish Food Safety Authority Evira. Helsinki, Finland, 56 pp.

Wijkström, U.N. (2009). The use of wild fish as aquaculture feed and its effects on income and food for the poor and the undernourished. Hasan, M.R. and Halwart, M. (2009). Fish as feed inputs for aquaculture: practices, sustainability and implications. FAO Fisheries and Aquaculture Technical Paper No. 518. FAO, Rome, Italy, pp. 371-407.

36. Food or feed? The contribution of Baltic herring fisheries to food security and safety

M. Pihlajamäki[1,2], S. Sarkki[3] and T.P. Karjalainen[1]*
[1]University of Oulu, Water Resources and Environmental Engineering Research Group, P.O. Box 4300, 90014 Oulu, Finland; [2]University of Helsinki, Department of Environmental Sciences, P.O. Box 65, 00014 Helsinki, Finland; [3]University of Oulu, Faculty of Humanities, P.O. Box 1000, 90014 Oulu, Finland; mia.pihlajamaki@oulu.fi

Abstract

The global demand for fish as a source of nutritious food is increasing owing to population growth, but also as per capita consumption of fish is increasing in many developed regions such as the Baltic Sea region. Baltic herring is the key commercial species in the Baltic Sea in terms of volume, but its use for human consumption is low and instead, the majority of the catch is used as feed in fur and fish farming. The sale of large Baltic herring for direct and indirect human consumption in the EU is restricted due to the detected levels of dioxins. The objective of this paper is to identify ways to increase the use of Baltic herring for direct human consumption, and to assess the potential of the fisheries to contribute to food security and safety objectives. To collect empirical materials a stakeholder workshop, which brought experts together from four Baltic Sea countries to discuss the dioxin problem of Baltic herring, was arranged. The results of the analysis suggests that the contribution of Baltic herring to food security and safety could be significantly increased, but this entails greater efforts from the management systems to include food security and safety aspects into the governance of Baltic herring.

Keywords: ethical fisheries governance, Baltic Sea, catch utilisation, dioxins, backcasting

Introduction

The majority of the world's marine fishery resources have been assessed as either fully or over-exploited (FAO, 2011). From the perspective of food security, both capture fisheries and aquaculture are needed to meet the growing demand for fish as a source of animal protein to humans (FAO, 2014; Lam, 2013; Thurstan and Roberts, 2014). However, there is an apparent gap in the demand and supply as many developed countries consume more fish than they produce (Thurstan and Roberts, 2014). Despite this, small pelagic fish, such as herring, which form the largest group in capture fisheries, are used primarily for industrial purposes (FAO, 2014), but in the future the competition between the different uses is expected to increase (Tacon and Metian, 2009). The UN Code of Conduct for Responsible Fisheries (1995, 2011) calls for the promotion of fisheries contribution to global food security and safety, while ensuring that the related policies and practices do not cause environmental degradation or undermine the food security of local communities, especially in the developing countries. Recent study also suggests that the inclusion of ethical issues, such as food security and safety, in the fisheries governance improves the sustainability of the fisheries (Lam and Pitcher, 2014). This paper presents an empirical case study on how the contribution of Baltic herring catch to food security and safety could be increased by 2040.

Background

Human consumption of fish has also been increasing in the Baltic Sea countries over the past five decades (FAO, 2015). Still the per capita consumption of Baltic herring (*Clupea harengus*) continues to decrease (see e.g. Glynn *et al.*, 2013; Natural Resources Institute Finland, 2015) and the fish markets in many Baltic Sea countries depend heavily on imported fish products (EUMOFA, 2015). Baltic herring is the

EurSafe 2016 – DOI 10.3921/978-90-8686-834-6_36, © Wageningen Academic Publishers 2016

most important commercial catch species in the Baltic Sea in terms of volume (ICES, 2015) and has therefore great contribution potential to food security. Currently, however, the majority of the catch is fed as whole to fur animals and reduced to fishmeal and oil (ICES, 2015; Lassen, 2011). The utilisation of the catch is driven by market conditions, i.e. price and demand (see Ignatius and Haapasaari in this EurSafe 2016 volume), while the harvesting strategies, namely the Common Fisheries Policy and the related regional multiannual management plan, do not address the use of the catch at all. From the food safety perspective, the potential to use Baltic herring for human consumption is limited. Since 2001, the Baltic Sea countries have not been allowed to place large over 17 cm Baltic herring on the EU market for food or feed due to detected dioxin levels in the fish (Regulation (EU) No 1259/2011). The fish fed for fur animals is not limited by the regulation, but the feed used in aquaculture is. However, the cost-effectiveness of removing dioxins from the fishmeal and oil during the production process has improved recently, which has increased the use of Baltic herring in aquaculture and has thereby contributed indirectly to food security. Finland and Sweden, which land together over 60% of the total Baltic herring catch (ICES, 2015), have an exemption to sell large Baltic herring on their national markets for direct human consumption, but majority of the catch intended for human consumption has been exported outside the EU. Russia and Ukraine have been important buyers of Baltic herring, but the export to Russia halted in mid-2014 resulting from the sanctions against the import of fresh EU products. This share of the catch has since been used in fishmeal and oil production.

An empirical case study on the desirable future use of Baltic herring catch

The empirical research material is based on a participatory backcasting exercise (see e.g. Robinson, 2003) that was carried out in an international stakeholder workshop that focused on the dioxin problem of Baltic herring. The workshop was organised by the BONUS GOHERR project in Copenhagen in 2016. Eleven stakeholders from four Baltic Sea countries representing fishermen, producers of fish products for direct human consumption, fishmeal and oil producers, researchers, policy makers and NGOs took part in the backcasting exercise. The aim of the exercise was to define the most desirable and undesirable future states regarding the use of Baltic herring and the related dioxin problem and then to develop pathways and milestones backwards from the selected goals to the present to determine the feasibility of the goals and the actions needed to reach them (Robinson, 2003). The stakeholders agreed that the key goal for the future is to increase the use of Baltic herring for human consumption significantly by 2040. Three sub-objectives relating to this target were selected for the backcasting exercise: (1) 35% of the total Baltic herring catch is used for human food in Baltic Sea countries by 2040; (2) 40% of the total Baltic herring catch is exported and consumed as human food; and (3) dioxin level in Baltic herring used for human food decreases to a safe level. The stakeholders were divided into three groups to work on the sub-objectives. First, the development paths and milestones to reach the objectives were identified and then the stakeholder specific governance actions were determined. The groups used backcasting tables to structure their work and to take notes of the discussions, in addition to which all the discussions were tape-recorded. Afterwards, the pathways and actions developed by the groups were presented and discussed in a plenary session. After the workshop, the recordings were transcribed and the backcasting tables combined for the analysis. For the purpose of this paper, the analysis concentrated on categorizing the suggested ways to increase direct human consumption of Baltic herring. There were no conflicting views or disagreements on the pathways and actions, however, some problems or challenges relating to them were brought forward. The main outcomes of the analyses, i.e. how to increase the supply of safe-to-eat Baltic herring to the market and the overall contribution of the Baltic herring fisheries to food security and safety are presented in the next section.

Results

Increasing the supply of compliant Baltic herring to the market

The prerequisite for increasing Baltic herring fisheries contribution to food security is that all the fish entering the market is safe to eat or that risk perceptions relating to dioxins change. Based on this, different pathways to provide compliant safe-to-eat Baltic herring products on market were created. First, it was noted that the Baltic herring fisheries could be targeted towards small under 17 cm fish that is unlikely to exceed the maximum allowable dioxin level. This could be done by sorting out the catches and/or size-selective fishing. Currently, sorting is only a recommendation, but in Estonia, for example, where the Baltic herring fishery targets explicitly smaller fish, sorting is the primary strategy to place the products on the EU market. Size-selective fishing could be done by using certain mesh sizes or fishing in certain parts of the sea, but it was pointed out that to employ size-selective fishing at larger scale requires further development of methods and practices.

Second, the stakeholders identified alternative ways to increase the use of large over 17 cm Baltic herring for human consumption. As regards to Finland and Sweden, it was pointed out that the per capita consumption of Baltic herring could increase significantly within the health authorities' recommendations on the total consumption. An access to other EU markets would require either changes in the perceptions on the dioxin risk or decreased dioxin content in the fish. The former refers especially to the prevailing perception in Finland that as long as national health authorities' recommendations, which concerns only the identified risk groups, are followed, the health benefits of eating Baltic herring exceed the health risks. The identified actions to reduce dioxin emissions include implementing existing regulations, raising public awareness, and introducing regulations on domestic burning, but developing realistic and cost-effective methods to remove dioxins from the sediments was regarded equally important. Relating to the dioxin content of Baltic herring, apart from the Gulf of Bothnia where the concentrations are the highest, changes in the size limit of a compliant Baltic herring from 17 cm to 21 cm were anticipated in the foreseeable future.

Third, the stakeholder also noted that fishmeal and oil produced from Baltic herring could be used directly for human consumption. Since it is possible to remove dioxins during the production process, all sizes of Baltic herring and from the whole Baltic Sea can be used for this purpose. In Germany, one production plant is already producing omega-3 and protein supplements out of Baltic fish. Current EU regulations, however, provide strict rules for such production lines and therefore, launching the production of fishmeal and oil for human consumption requires significant investments from the production plant.

Increasing the contribution of Baltic herring fisheries to food security and safety

The analysis of the backcasting exercise also showed that a more proactive management strategy for the utilisation of Baltic herring catch is needed to increase direct human consumption. The identified cornerstones of such strategy included the capacity to provide fish for direct human consumption, branding, and creating and developing markets, both in the EU and outside. First, to increase Baltic herring available for direct human consumption the fisheries, especially in Finland and Sweden, would have to shift from feed to food directed fishing. This means that vessels would have to be re-equipped to fish primarily for direct human consumption by e.g. adding on-board freezing equipment in the vessels. Second, to meet the increasing demand for local and healthy fish, the development and promotion of new and traditional Baltic herring products was considered essential. Third, the importance and usefulness of certifications and other labelling, especially in the short term, as marketing tools and a way to guarantee that the products entering the market are from sustainable fisheries and safe to eat, were widely recognized. Fourth, to gain an access to potentially large export markets, such as Russia and China for example, to develop existing markets in the EU and Ukraine, and to actively search for new market opportunities were also considered an integral part of the strategy to increase direct human consumption. While the restricted access to EU markets is mainly due to dioxin levels in the fish, the difficulties in accessing large export markets, such as Russia and China have related mainly to political and bureaucratic reasons respectively. However, exporting products that do not meet EU's own food safety criteria outside the EU was considered to be an unethical and unsustainable strategy.

Although the main focus of the discussions was on the use of Baltic herring for direct human consumption, aquaculture was also seen as important solution from the perspectives of food security and safety. Especially in the short term, the contribution potential of Baltic herring fisheries to food security and safety is significant as dioxin free feed in aquaculture. In the long term, synergies between the use of Baltic herring for direct human consumption and fishmeal and oil production were highlighted. It was stressed that even if the catch goes primarily for direct human consumption in 2040, the by-products (the waste that remains after e.g. filleting), which can add up to nearly 50% of the total catch, could be utilised in fishmeal and oil production. It was also noted that the profit from the Baltic herring sold for the fishmeal and oil industry has increased in recent years and is quite competitive to that sold for human consumption.

Discussion

Owing to the quantity of the total Baltic herring catch its potential as a source of human food is significant, but owing to the food safety aspect related to dioxins, the possibilities to market a healthy local product are limited. From the ethical perspective, all the fish entering to the regional, EU and global seafood markets should meet the food safety criteria. Drawing from these, the results of this study are encouraging as there appears to be several ways to increase the supply of safe-to-eat Baltic herring to the markets, which suggests that dioxins are not after all a 'killing factor', but merely a conditioning factor that needs to be taken into consideration in the governance of the resource. It seems that such considerations have so far been inadequate. The results also indicate that in the absence of a proactive management strategy for the use of the catch, the Baltic herring management system has submitted to the existing market conditions, and dioxins are used, at least to some extent, as a reason for not attempting to change the current utilization structure of the catch. For instance, the history of fishing Baltic herring primarily for feed is far longer than that of the EU regulation on dioxins in food and feed, which indicates that the use of the catch is determined first and foremost by the tradition to fish for feed.

The findings of the study also raise the ethical question on the acceptability of using fish that can be used for human consumption as feed. This question relates especially to the use of the resource in fur farming, whereas there is no apparent ethical trade-off between its use in aquaculture and direct human consumption in the short term, as aquaculture provides an indirect way for Baltic herring fisheries to contribute to food security and safety. In the long term, the identified synergies between the uses of Baltic herring for direct human consumption and in aquaculture should be promoted. Thus, from the food security and safety perspectives, there is a call for elaborated discussion on the different uses of the Baltic herring catch and their prioritisation.

Conclusions

The conclusion of this study is that there is a call for elaborated discussions on the dioxin problem of Baltic herring in general and its implications to the use of the catch in particular. Furthermore, greater efforts from the fisheries management systems are needed to increase the contribution of Baltic herring fisheries to global food security and safety objectives. This could be done, for example, by including ethical considerations relating to food security and safety in the governance of Baltic herring resource, instead of addressing them separately and letting the market conditions determine the use of the catch. After all, there appears to be common interest to increase the use of the resource for human consumption.

Acknowledgements

This paper is a contribution from the BONUS GOHERR project, which has received funding from BONUS (Art 185), funded jointly by the EU, the Academy of Finland and the Swedish Research Council for Environment, Agricultural Sciences and Spatial Planning. The authors thank all those stakeholders who contributed to the workshop.

References

European Commission (EC) (2014). Final report of an audit carried out in Estonia from 10 to 20 June 2014 in order to evaluate the food safety control systems in place governing the production and placing on the market of fishery products. DG(SANCO 2014-7132 – MR FINAL. Brussels, Belgium.

European Market Observatory for Fisheries and Aquaculture Products (2015). The EU fish market. 2015 edition. Directorate-General for Maritime Affairs and Fisheries of the European Commission. May 2015. Available at: www.eumofa.eu.

Food and Agriculture Organisation of the United Nations (FAO) (2015). FAOSTAT. Statistics division. Available at: http://tinyurl.com/kzr77pz.

Food and Agriculture Organisation of the United Nations (FAO) (2014). The state of world fisheries and aquaculture. Opportunities and challenges. FAO, Rome, Italy.

Food and Agriculture Organisation of the United Nations (FAO) (2011). Review of the state of world marine fishery resources. FAO Fisheries and Aquaculture Technical Paper No. 569. FAO, Rome, Italy, 344 pp.

Glynn, A., Sand, S. and Becker, W. (2013). Risk and benefit assessment of herring and salmonid fish from the Baltic Sea area. Report no 21-2013. National Food Agency, Sweden.

International Council for the Exploration of the Sea (2015). Report of the Baltic Sea Fisheries Assessment Working Group (WGBFAS). April 14-21, 2015. Copenhagen, Denmark.

Lam, M.E. and Pitcher, T.J. (2014). The ethical dimensions of fisheries. Current Opinion in Environmental Sustainability 4: 364-372.

Lam, M.E. (2013). Comparing the ethics of capture fisheries and aquaculture. In: Röcklinsberg, H. and Sandin, P. (eds.) The ethics of consumption. The citizen, the market and the law. Wageningen Academic Publishers, Wageningen, the Netherlands.

Lassen, H. (2011). Industrial fisheries in the Baltic Sea. Note prepared for DG IPOL.

Natural Resources Institute Finland (2015). Statistics database. Fish used for human consumption (kg/person/year). Updated 12/29/2015. Available at: http://stat.luke.fi.

Robinson, J. (2003). Future subjunctive: backcasting as social learning. Futures 35: 839-856.

Tacon, A.G.J. and Metian, M. (2009). Fishing for feed or fishing for food: increasing global competition for small pelagic forage fish. Ambio 38(6).

Tacon, A.G.J. and Metian, M. (2008). Global overview on the use of fish meal and fish oil in industrially compounded aquafeeds: trends and future prospects. Aquaculture 285: 146-158.

Thurstan, R.H. and Roberts, C.M. (2014). The past and future of fish consumption: can supplies meet healthy eating recommendations? Marine Pollution Bulletin 89(1-2): 5-11.

37. Evaluation of the influence of light conditions on crayfish welfare in intensive aquaculture

T. Abeel[1], H. Vervaecke[1], E. Roelant[1], I. Platteaux[1], J. Adriaen[1], G. Durinck[2], W. Meeus[1], L. van de Perre[2] and S. Aerts[1]*
[1]Odisee University College, Ethology and Animal Welfare, Hospitaalstraat 23, 9100 Sint-Niklaas, Belgium; [2]ESAT/Light and Lighting Laboratory, KU Leuven, Gebroeders De Smetstraat 1, 9000 Gent, Belgium; stef.aerts@odisee.be

Abstract

During the last years, evidence has arisen for the ability of crustaceans to suffer from stress and noxious situations. This has led to an increasing public concern for crustacean welfare. Due to the interest in intensification of crayfish aquaculture, the question arises whether artificial culture conditions may impact these crustaceans' welfare. We evaluated the effect of six light conditions on noble crayfish (*Astacus astacus*) behaviour in a recirculating aquaculture system. Crayfish were cultured under cool (5,500 K), neutral (3,800 K) or warm (2,600 K) white light, using bright (761 lux) and weak (38 lux) intensity. The crayfish's dark-light preferences were evaluated by observing their behaviour in a plus maze. Animals kept under 38 lux spent 27.94±8.95% of their time in the light arms of the maze, while those kept under 761 lux only spent 20.06±9.98% in the light arms. Thus, crayfish from bright light showed a slight trend to avoid the light arms of the maze (P=0.098), an indication of anxious behaviour. The light spectrum had no influence. Following earlier preliminary signs of pain perception and neophobia in *A. astacus* these new observations invite to investigate further indications of a more intensive mental life in crustaceans. Human activities such as the catching, rearing, shipping and cooking of crustaceans may have a considerable impact on the welfare of these animals if these results are confirmed. We argue there are good ethical reasons to re-evaluate the omission of non-cephalopod invertebrates from European animal welfare legislation.

Keywords: *Astacus astacus*, dark-light preference, mental ability

Introduction

Although non-cephalopod invertebrates are not protected by European animal welfare legislation, strong evidence has arisen during the last decade for the ability of crustaceans to suffer from stress and noxious situations (Elwood and Adams, 2015; Magee and Elwood, 2013; Vervaecke *et al.*, 2015). In recent years, the public concern on crustacean welfare has been increasing (Yue, 2008).

There is an increasing interest in the intensive production of crayfish in Europe (Abeel *et al.*, 2014; Franke *et al.*, 2013; Seemann *et al.*, 2015). Considering their detritivorous nature and high market potential, these animals are promising candidates for sustainable aquaculture. In intensive rearing systems, these decapods behave highly cannibalistically, hindering the development of commercially viable production (Abeel *et al.*, 2014). The question arises if intensifying the production of these animals, and other decapod crustaceans (e.g. crabs, shrimps and lobsters), has consequences for their welfare.

Light conditions such as illuminance and photoperiod are known to influence growth, metabolism and behaviour in crayfish (Delabbio, 2011; Fanjul-Moles *et al.*, 1998; Franke *et al.*, 2013; Thomas *et al.*, 2016) and are considered potential factors to improve production in crayfish rearing systems (Franke and Hörstgen-Schwark, 2015; Gonzalez *et al.*, 2010). In this study, we evaluated the effect of white

I. Anna S. Olsson, Sofia M. Araújo and M. Fátima Vieira (eds.) ***Food futures: ethics, science and culture***
EurSafe 2016 – DOI 10.3921/978-90-8686-834-6_37, © Wageningen Academic Publishers 2016

light spectra and light intensity or illuminance in a recirculating aquaculture system (RAS) on noble crayfish behaviour.

To evaluate the welfare implications, we quantified negative affective states related to living in particular light conditions using the paradigm of a subaquatic dark-light plus maze, validated by Fossat *et al.* (2015). He assessed stress by placing red swamp crayfish (*Procambarus clarkii*) in a conflicting situation between its innate curiosity for novel environment and its aversion for light. Under natural circumstances crayfish frequently hide in dark places and explore new environments. Experimentally stressed animals however showed less explorative behaviour and preferentially remained in the dark arms, whereas unstressed crayfish also explore the illuminated arms (Fossat *et al.*, 2014).

Material and methods

Light conditions

468 pond raised noble crayfish (*Astacus astacus* L.) summerlings were obtained from a German farm, and transferred to an indoor recirculating aquaculture system. They were housed under six different light conditions for 191 days. Average initial body weight was 0.66±0.13 g and mean occipital carapace length (OCL) was 10.64±0.10 mm (Figure 1). The animals were divided over 18 tanks, resulting in a stocking density of 52 animals/m^2, or 26 individuals per tank. Sex ratio was 50:50. During the experimental period, water temperature was kept at 21.0±0.35 °C, pH at 8.75±0.07 and dissolved oxygen level was 7.25±0.51 mg/l. The animals were housed under six light conditions, with three groups per treatment (Table 1).

Figure 1. Occipital carapace length.

Table 1. The six experimental light conditions, used in the recirculating aquaculture system.

Light condition	Lamp type	Intensity	Spectrum
cool bright light	Philips T8 36W 865	761 lux	CCT≈5,500 K
neutral bright light	Philips T8 36W 840	761 lux	CCT≈3,800 K
warm bright light	Philips T8 36W 827	761 lux	CCT≈2,600 K
cool weak light	Philips T8 36W 865	38 lux	CCT≈5,500 K
neutral weak light	Philips T8 36W 840	38 lux	CCT≈3,800 K
warm weak light	Philips T8 36W 827	38 lux	CCT≈2,600 K

Dark/light preference

A plus maze (Figure 2) was designed and adapted to the crayfish's size, according to the protocol of Fossat *et al.* (2015). The maze had a 29 cm diameter and was nine cm high. Each arm had a surface dimension of 7×11 cm. The sides of the dark arms were covered with a black foil and the top was closed with an opaque lid.

After the crayfish spent six months under the experimental light conditions, two males and two females were randomly caught from each tank. Hence, we observed 72 crayfish in total, or twelve animals from each treatment. Each crayfish was put in the plus maze (Figure 2) in order to determine its preference for dark or light. Illuminance and colour temperature of the lighting in the plus maze were adapted in accordance with the experimental light conditions in the RAS. Animals from the bright light treatments, were exposed to 547 lux in the lit arms and six lux in the dark arms of the maze. For crayfish from the weak light treatments, illuminance in the lit and dark arms were 21 lux and 0 lux respectively. The maze was filled with water from the RAS. This water was changed after each observation.

Each crayfish was caught separately, just before observation took place. The animals were caught from their tanks using a small fishing net. Catching was performed carefully, in order to avoid potentially stressing escape reactions such as tail flipping. After a crayfish was observed in the plus maze, it was placed in a separate container to assure no individuals were tested twice.

Before the observation, each crayfish was put in an opaque cage (inner dimensions: 6×6×10 cm) in the centre of the plus maze during one minute for acclimation. The animal was subsequently released from the cage, allowing it to explore the entire maze for 10 minutes. During this period, the animal's location was scored every five seconds, resulting in 121 observations per individual. From these data, we calculated the percentage of time spent in the light and dark arms of the plus maze, in order to determine the crayfish's dark-light preference. We averaged over the 4 crayfish in each tank to have independent observations per treatment arm.

Statistical analysis

As the number of observations per treatment arm were small (n=3) we used non-parametric tests. A Kruskal-Wallis test was performed to see if there was a difference in dark-light preference between the six treatments. As we were specifically interested in the effect of light intensity this was tested separately with a Mann-Whitney test and light spectrum with a Kruskal-Wallis test.

Figure 2. Dark/light preference test in a plus maze.

Results

There was no significant difference in time spent in the lit versus dark arms between the six treatments (i.e. three levels of light spectrum combined with two levels of light intensity) P=0.209, Kruskal-Wallis test). Crayfish from weak light treatments showed a tendency (P=0.098, Mann-Whitney-U test) to spend more time (27.94±8.95%, mean±standard deviation) in the lit arms of the plus maze than crayfish cultured under bright light (20.06±9.98%). Thus, crayfish kept under bright conditions tended to avoid light, as shown in Figure 3. Light spectrum did not influence the light-dark preference of the crayfish in the maze (P=0.715, Kruskal-Wallis test).

Discussion

Living in an environment with cool, neutral or warm white light did not affect the behaviour of crayfish in a dark-light preference maze. In weak light, the animals showed a tendency towards showing less anxious behaviour ($0.05<P>0.1$)in weak light. These crayfish generally explored the whole maze, including the illuminated arms. This natural exploratory activity under weak light conditions complies with the definition of welfare of Fraser *et al.* (1997). This is consistent with the results of the zootechnical parameters under the same light conditions (unpublished data).

These results are comparable to the findings of Fanjúl-Moles *et al.* (1998), who found that high irradiance caused increased haemolymph lactate in *Procambarus clarkii* and *Procambarus digueti*, indicating higher stress levels in these crayfish. While *P. clarkii* managed to cope with these artificial conditions, high irradiance caused high mortality in *P. digueti*. Prieto-Sagredo *et al.* (2000) found that these species showed a different metabolic response to high light intensity, illustrating the species-specific reaction to this stressor. In addition, Thomas *et al.* (2016) found that invasive signal crayfish (*Pacifastacus leniusculus*) hide in their shelters more often when exposed to simulated light pollution from a street light during the night.

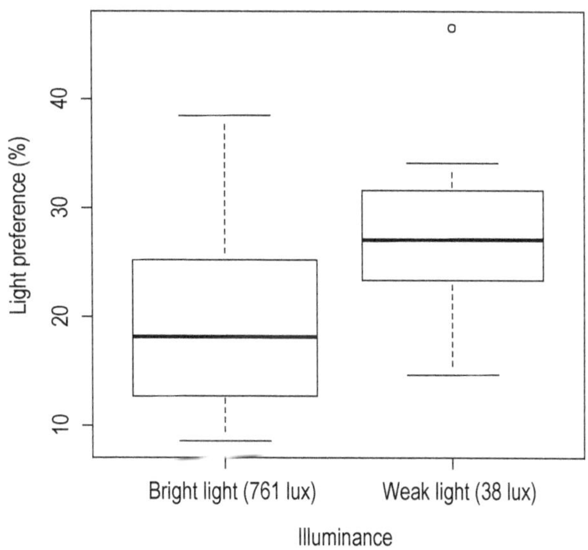

Figure 3. Light preference in the plus maze for crayfish from weak and bright light treatments.

Although the current observed trend does not allow any confident conclusions about the mental capacities of *A. astacus,* following earlier preliminary signs of pain perception and neophobia (Vervaecke *et al.*, 2015), these new observations are a further invitation to investigate the intensity of mental life in crustaceans. Human activities such as catching, rearing, shipping and cooking of crustaceans all are potentially stressing or painful, and may have a considerable impact on the welfare of these animals. This reasoning is based on an argument-by-analogy (Sherwin, 2001) which we believe can be applied to invertebrates as well, although this is certainly made more difficult by the highly different biology between vertebrates and invertebrates. In that sense, the debate about crustaceans may now be at the same point as the fish welfare debate a few decades ago (see e.g. Braithwaite, 2010).

The European welfare legislation on farmed animals (European Union, 1998) is based on the European Convention for the Protection of Animals kept for Farming Purposes (Council of Europe, 1976) that states they should be provided with care that 'is appropriate to their physiological and ethological needs in accordance with established experience and scientific knowledge'. The Lisbon Treaty (European Union, 2007) took this a step further by explicitly asserting 'animals are sentient beings', something that applied to agricultural as well as fisheries policies. From the experimental animals Directive (European Union, 2010), however, it is clear that European legislators are only referring to 'non-human vertebrate animals' and cephalopods.

This legislation clearly mirrors ethical arguments that can be rooted in the utilitarian tradition (suffering, sentiency). From a legislative perspective there is still no 'established experience and scientific knowledge' that proves sentiency in crustaceans, but the results provided here and in Vervaecke *et al.* (2015) provide a moral impetus to embark on a serious endeavour into the mental capacities of crustaceans. A good starting point may be a systematic assessment in crustaceans of the different physiological and behavioural responses listed by Sherwin (2001) when reviewing different studies on invertebrates. The limitations of the current (and previous) study are primarily based on the small number of animals and the isolated nature of the behaviour studied. In order to substantiate claims about the sentience of crustaceans based on the argument-by-analogy a more comprehensive approach is certainly needed. On the other hand, the results of our rather preliminary research can also be said to emphasize the need for more elaborate studies.

A similar and probably even stronger need for clarification arises when approaching this question from a 'subject of a life' approach as suggested by Regan (1983). If these preliminary findings are confirmed, this type of mental activity should be enough reason to include crustaceans (or at least *A. astacus*) in our moral circle of compassion.

Conclusions

As the intensity of white light appears to affect behaviour and welfare of cultured noble crayfish, this factor should be taken into account when setting up a commercial production system. Weak light may lead to a better wellbeing of the cultured animals.

From an ethical as well as legislative view, the current results are a further indication of the need for in-depth research into the physical and mental abilities of crustaceans and possibly other non-cephalopod invertebrates. In relation to EU legislation confirmation would clearly have far-reaching implications (as it might result in a re-evaluation of the omission of non-cephalopod invertebrates from European animal welfare legislation), but it would also bring a sharper focus to the edges of the ethical circle of compassion.

Acknowledgements

This study is part of the research project 'Optimization of *Astacus astacus* culture: influence of production parameters on taste, moulting and welfare', funded by the PWO grant of Odisee University College.

References

Abeel, T., Adriaen, J., Meeus, W., Himpe, W., Roelant, E., Sannen, A., Laevens, H., Van der Elst, J. and Aerts, S. (2014). Eindrapport PWO-project: technische en economische evaluatie van de productie van Europese rivierkreeft *(Astacus astacus)* in recirculatiesystemen. Odisee University College, Sint-Niklaas, Belgium.

Braithwaite, V. (2010). Do fish feel pain? Oxford University Press, New York, NY, USA.

Council of Europe (1976). European Convention for the protection of animals kept for farming purposes. Strasbourg, France: Available at: http://tinyurl.com/h2lmpxs.

Delabbio, J. (2013). Method and system for enhancing growth and survivability of aquatic organisms. U.S. Patent US 2013/0174792 A1.

Elwood, R.W. and Adams, L. (2015). Electric shock causes physiological stress responses in shore crabs, consistent with prediction of pain. Biology Letters 11(11).

European Union (1998). Council Directive 98/58/EC of 20 July 1998 concerning the protection of animals kept for farming purposes. Official Journal of the European Union, L 221: 23-27.

European Union (2007). Treaty of Lisbon amending the Treaty on European Union and the Treaty establishing the European Community. Official Journal of the European Union, C 306: 1-271.

European Union (2010). Directive 2010/63/EU of the European Parliament and of the Council of 22 September 2010 on the protection of animals used for scientific purposes. Official Journal of the European Union, L 276: 33-79.

Fanjul-Moles, M.L., Bosques-Tistler, T., Prieto-Sagredo, J., Castañón-Cervantes, O., Fernández-Rivera-Río, L. (1998). Effect of variation in photoperiod and light intensity on oxygen consumption, lactate concentration and behavior in crayfish *Procambarus clarkii* and *Procambarus digueti*. Comparative Biochemistry and Physiology 119A: 263-269.

Fossat, P., Bacque-Cazenave, J., De Deurwaerdere, P., Delbecque, J.P. and Cattaert, D. (2015). Measuring anxiety-like behavior in crayfish by using a sub aquatic dark-light plus maze. Bio-protocol 5(3): e1396.

Fossat, P., Bacque-Cazenave, J., De Deurwaerdere, P., Delbecque, J.P. and Cattaert, D. (2014). Comparative behavior. Anxiety-like behavior in crayfish is controlled by serotonin. Science 344(6189): 1293-1297.

Franke, R., Wessels, S. and Hörstgen-Schwark (2013). Enhancement of survival and growth in crowded groups: the road towards an intensive production of the noble crayfish *Astacus astacus L.* in indoor recirculation systems. Aquaculture Research 44: 451-461.

Franke, R., Hörstgen-Schwark, G. (2015). Control of activity patterns in crowded groups of male noble crayfish *Astacus astacus (*Crustacea, Astacidea) by light regimes: a way to increase the efficiency of crayfish production? Aquaculture 446: 103-110.

Fraser, D., Weary, D.M., Pajor, E.A. and Milligan, B.N. (1997). A scientific conception of animal welfare that reflects ethical concerns. Animal Welfare 6: 187-205.

Gonzalez, R., Celada, J.D., García, V., Carral, J.M., González, A. and Sáez-Royuela, M. (2010). Shelter and lighting in the intensive rearing of juvenile crayfish (*Pacifastacus leniusculus*, Astacidae) from the onset of exogenous feeding. Aquaculture Research 42: 450-456.

Magee, B. and Elwood, R.W. (2013). Shock avoidance by discrimination learning in the shore crab *(Carcinus maenas)* is consistent with a key criterion for pain. Journal of Experimental Biology 216: 353-358.

Prieto-Sagredo, J., Ricalde-Recchia, I., Durán-Lizarraga, M.E. and Fanjul-Moles, M.L. (2000). Changes in Hemolymph Glutathione status after variation in photoperiod and light-irradiance in crayfish *Procambarus clarkii* and *Procambarus digueti*. Photochemistry and Photobiology 71(4): 487-492.

Regan, T. (1983). The case for animal rights. University of California Press, Berkeley, USA.

Seemann, U.B., Lorkowski, K., Slater, M.J., Buchholz, F. and Buck, B.H. (2015). Growth performance of Noble crayfish *Astacus astacus* in recirculating aquaculture systems. Aquaculture International 23: 997-1012.

Sherwin, C.M. (2001). Can invertebrates suffer? Or, how robust is argument-by-analogy? Animal Welfare 10: S103-118.

Thomas, J.R., James, J., Newman, R.C., Riley, W.D., Griffiths, S.W. and Cable, J. (2016). The impact of streetlights on an aquatic invasive species: artificial light at night alters signal crayfish behaviour. Applied Animal Behaviour Science 176: 143-149.

Vervaecke, H., Roelant, E., Hendrycks, W., Abeel, T., Vermeersch, X. and Aerts, S. (2015). Pain perception in crayfish (*Astacus astacus*): empirical observations and ethical consequences. In: Dumitras, D.E., Jitea, I.M. and Aerts, S. (eds.) Know your food. Food ethics and innovation. Wageningen Academic Publishers, Wageningen, the Netherlands, pp. 157-162.

Yue, S. (2008). An HSUS report: the welfare of crustaceans at slaughter. HSUS Reports: Farm Industry Impacts on Animals. Paper 4. Available at: http://tinyurl.com/zctwd29.

Section 9. *In vitro* meat

38. Meat eating as a practice and the acceptance of radical change

M. Kanerva
Sustainability Research Center (artec), University of Bremen, Enrique Schmidt Str. 7, 28359 Bremen, Germany; m.kanerva@uni-bremen.de

Abstract

It is broadly accepted that current meat production and consumption is in need of changing towards a more sustainable system, at least in the West. This paper views meat eating as a changing social practice, discusses change and some relevant factors regarding meat alternatives, such as insects and cultured meat, including the emotion of disgust that these radically new meats are likely to generate in many eaters. The data analysis, which is to follow, will investigate the response from the public in the United Kingdom to these two meat alternatives to obtain better insights into the acceptance of radically different, but potentially more sustainable, meat eating practices. This project aims to contribute to the understanding of the feasibility of transforming the food system.

Keywords: insects, cultured meat, discourse, social practices, disgust

Introduction

Globally expanding intensive production, and the long-term trend of increased consumption of meat, result in numerous serious problems for the climate, the environment in general, and human and animal welfare. Additionally, there are issues involving ethics and global equity.

From the point of view of the consumer – or the eater – sustainable practices include eating no meat at all; eating no, or radically less meat that has been produced in the intensive meat production system; eating only so called alternative meats; or a combination of these options, with the main focus being a radical reduction of the negative impacts. The alternative meats include meat from traditional (industrial) species produced in organic, or extensive animal agriculture; meat from animals in the wild; meat from (farmed) insects; and meat produced by cell culture methods. The negative impacts from these vary, and are by no means all negligible, but considering the full picture, and including animal welfare, their impacts can generally be considered significantly smaller than from the current intensive animal agriculture. Further, there are the so called 'meat analogues' which include meat-like foods made from plant proteins, such as soy, gluten or pea protein. These analogues are becoming increasingly more sophisticated, and meat-like, also in terms of their texture, and even taste.

As Shove and Spurling (2013) argue, achieving sustainability requires a radical redefinition of what counts as normal social practices, involving not just the consumers, but all other parts of the system as well. Changing social practices forms the foundation for a transformation towards sustainability. Therefore, understanding contemporary social practices, how they have changed and how they might change in the future, is essential. Policymakers, producers, retailers, etc. can create new practice opportunities, i.e. the change may be more about 'push' than 'pull', but people, or consumers will have to accept the radically new ways of doing (e.g. eating meat), bring the new practices from niche to mainstream, therefore forming an important part in the normalization process.

This paper views meat eating as a social practice. Specifically, it will analyse how the eating public in the United Kingdom (UK) sees two unfamiliar options for alternative meats in current Western eating practices: insects and cultured meat. As rather new, radical and controversial topics, they generate

EurSafe 2016 – DOI 10.3921/978-90-8686-834-6_38, © Wageningen Academic Publishers 2016

discussion. Could people in large numbers accept radical change to their everyday eating practices in this respect?

Currently, most people in the West would see both cultured meat and insects as 'non-food', or at most, see insects as animal feed. Therefore, for these to be turned into acceptable human 'food', a significant process of change in attitudes, and perhaps also in values, norms and moral thinking, would have to take place, in addition to people getting over the 'yuck-factor', or disgust, related to such new foods. Further, for these foods to become part of everyday shopping and eating practices, people would have to accept them by actively engaging in such new food cultures, incorporating them in their lives. This process of normalisation might happen in an untidy circular process, with its parts feeding each other, so that, for example, people might gradually accept a non-food as food while actually eating it, and not only so that eating comes after acceptance.

Meat eating as a (changing) practice

A product of history, eating practices have changed enormously over time, and also currently differ significantly between cultures and geographical areas. From a long-term point of view, eating practices – similar to many other social practices – are in a constant process of change.

There is a large number of factors influencing eating practices of individuals (Figure 1). On the one hand, the internal ideology- or value-related side of preferring (or not preferring) to eat, for example, meat; the external cultural side of what is acceptable or preferred; and the habit and routine side of acquiring, preparing and eating food, as well as disposing of spoiled or otherwise inedible food, determine much of our eating practices. Moreover, when an apparent food crisis develops, people may be warned, or they may be wary of eating certain foods for a while, similarly to a longer term concern about the impacts of the intensive meat system, and in opposition to a diet fashion of eating, for example, lots of protein. All these factors can be called soft factors, in that they are somewhat flexible, and their impact determinable on an individual basis. They are mostly, however, not very easily measurable, nor can they be easily controlled at a societal level. Change in these soft factors also takes place more at an individual

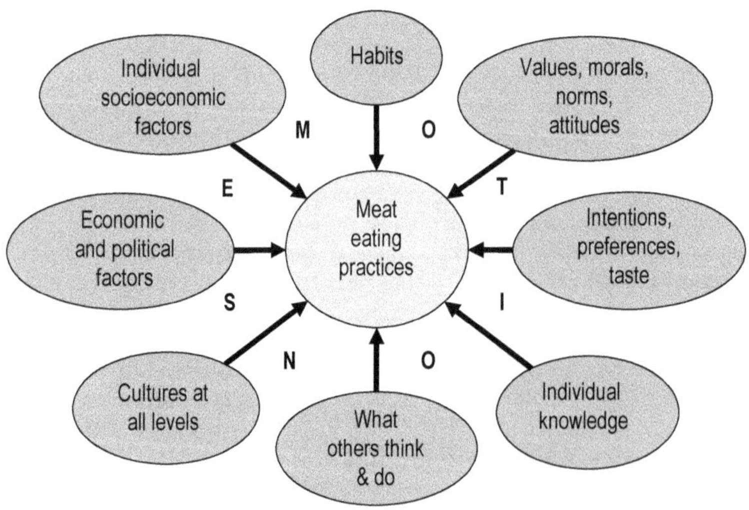

Figure 1. Influences on existing or changing meat eating practices.

level. This study will focus on these soft factors, as opposed to the more concrete and measurable hard factors related to our economic and political environments, or to our individual socioeconomic factors.

Finally, as shown in Figure 1 (with the letters circling the central graphic), emotions – such as fear, curiosity, disgust, guilt, trust, or excitement – can have an impact that reinforces or weakens the impact of other factors, including other emotions. For example, we may buy a more expensive alternative meat product out of curiosity; disgust or neophobia can turn us away from trying insects, although our friends have tried them and liked them; guilt regarding animal welfare can make us reduce our meat consumption, although we have previously enjoyed meat every day; or trust in the authorities keeping what is on offer safe can make us try an insect food product for the first time, although we are suspicious of it.

Changing practices and acceptance of change

Although some suggestions can be made about the impact on practices of the individual factors discussed above, it is difficult, if not impossible to point to any clear overall one-on-one relationships, including clear and direct impacts from the hard factors (see also Kanerva, 2013). In reality, these various factors impact each other, and in a variety of interconnected combinations they together determine our meat eating practices and changes in them. Nor can we probably effect lasting change towards sustainability by just trying to change one of these factors. As practice theories explain, changing values, for example, rarely change practices on their own. Although meat eating practices are often particularly deeply affected by values, political economic factors, or simply the force of habit, have great influence as well. However, every act of valuation of our everyday practices is significant, even if these acts are normally not common, and even if we tend to overlook our own embeddedness (Sayer, 2013).

According to Schatzki (2013), practices form bundles of practice-arrangements. These bundles evolve, but still persist, when relatively small changes take place, whereas when changes are large enough, the bundles dissolve, and new bundles evolve. Changes can also be external or internal to the practice bundle, and they can be abrupt or smooth. An important way to effect change, is to work on changing the ends 'that are acceptable or prescribed in practices, the overall ends that govern what participants do' (Schatzki, 2013: 44). For meat eating practices, such a change in the ends could be, for example, that meat eating would be equated less with power and masculinity, as it traditionally has been, and more with health, both for the individual and the planet, and a good feeling, when animal welfare is taken into account. Sensory qualities of meat (and its alternatives) remain, however, important.

As practice theories also show, the force of habits and routines is a strong barrier for changing a practice. However, it seems that when a certain habit is weak, i.e. it is not very well established, or it includes some variation in the actual behaviour, values may have a stronger impact on practices, i.e. the 'gap' between values and action can be crossed more easily (see e.g. Dahlstrand and Biel, 1997; Matthies *et al.*, 2002). It may then be that the individuals who have weaker meat-eating habits are better able to act according to their values, and can overcome the force of habit more easily. Such can be the case of the increasing number of flexitarians, who incorporate only some meat in their diets. Further, Hekler *et al.* (2010) conclude from their study on the effect of different kinds of food related information on eating behaviour that engaging people with the social, ethical, cultural and environmental aspects of food – in a way, appealing to morals and values, but also providing what can be seen as more neutral information – can be effective in changing eating practices, and significantly more so than mere health education.

What would seem to be crucial to changing practices, would be to question what is normal, and to find opportunities for gradually normalizing the alternatives. For example, the media discourse on eating cultured meat or insects has already made people more aware of their own eating practices. This may be seen as stirring the pot of 'normal', and incorporating new 'normals', such as insects into the pot. Such a process of normalization is nothing new in itself, one only needs to think of sushi, a previously unpalatable 'non-food' to most Westerners. Foods can also turn into non-foods, as has happened in, for example Europe, where people indeed used to eat insects as part of their normal eating practices. This denormalisation process is also increasingly happening in the South, where people are turning away from insects seen as food only suited for the poor. In fact, however, it may be that the 'normal' is defined more by e.g. the media discourse, than by actual practices which may incorporate a much more variety, as a study by Evans and Miele (2012) indicate. They conclude that, at least within Europe, cultural and national differences in what is normal regarding meat eating may be larger than the demographic or socioeconomic differences within cultures or geographical areas. This is also supported by a study by Rettie *et al.* (2012) which concluded that 'normal' is likely to be related to what each person's relevant reference group does, for example, his or her family, neighbours, other young people, or other parents.

Insects, cultured meat vs disgust and fear

When insects in Europe, for example, are currently thought of as food, they often generate a feeling of core disgust, an emotion developed during human evolution and probably meant to warn human omnivores of eating something poisonous, especially in relation to spoiled meat (Pollan, 2006; Rozin *et al.*, 2008). Insects prepared as food also tend to remind us instantly of their animal origin, unlike many other products derived from animals and disguised, intentionally or unintentionally, during food processing (Angyal, 1941, in Rozin *et al.*, 2008).

Further, and linked to core disgust, both insects and cultured meat, as new to the current Western palate, may also generate neophobia. Humans are omnivores, and neophobia is an important part of how omnivores in nature decide what is (acceptable) food, as the choice for omnivores is enormous, unlike for animals that only eat a much more restricted range of plants or animals. With some people, neophobia is only related to certain new foods, but with others, it is a personality trait (Barrena and Sanchez, 2012), and similar to the feeling of disgust. According to Barrena and Sanchez, neophobia can nearly always be overcome, however, with more exposure, or by emphasizing other, positive aspects of a certain food. Fear may also follow intentionally engaging in alternative foods, for example, vegetarians sometimes hide their vegetarianism due to fear of being seen as abnormal (e.g. Taylor, 2012).

Another category of disgust (Rozin *et al.*, 2008) is moral disgust, and rejection of cultured meat tends to fall into this category. Cultured meat is considered by many to be unnatural, and therefore also morally wrong, similar to the cloning of animals, or even humans.

The two alternative meats then both tend to generate disgust, but of a somewhat different nature. Additionally, they also differ at a more fundamental level, as insects are about large-scale use of an existing, but currently rejected food source (in the West), and cultured meat is about fixing problems created by technology (intensive meat production) with even more technology, which may add another layer of suspicion in people's minds regarding cultured meat. Sustainability is usually equated with natural, and cultured meat is a further step away from imagining meat production as natural. This conflict may, however, help people be more aware of how their meat is actually produced.

Another concept described in Rozin *et al.* (1997) is moralization, a process which converts preferences into values, and in which a previously neutral issue becomes morally disgusting. Rozin *et al.* use cigarette smoking in the United States as an example of this process. To turn matters around, if intensive animal agriculture would go through a similar process of moralization, whereby it would eventually be considered immoral, and disgusting, for, in particular, its unethical aspects regarding animal wellbeing, the environment, and global and intergenerational equity, this could in fact help the currently – for the Western majority – disgusting alternative meats become the morally acceptable choices.

Disgust can be seen as a specific form of rejection of a new practice. This is in particular because of the link to moral thinking and values, which can play an important part in the acceptability of new practices. And as Wait (2014) argues, keeping in mind 'gut reactions', of which disgust is one, can be important for policies aimed at more sustainable practices.

Analyzing the meat discourse

The data for this study consists of a number of online newspaper articles and their associated reader comments from the past few years. The United Kingdom is chosen as the origin of the newspapers (the Guardian and the Daily Mail), but the online readership is more global, although most readers still come from the UK. Among European Union countries, the UK is a forerunner, together with the Netherlands and Belgium, in terms of insects being available as food in certain restaurants, and the first cultured meat hamburger was unveiled in London in 2013. In terms of meat consumption trends, the UK has been relatively stable in its total meat per capita consumption over the past half a century. However, in the last 15 years or so, the British have somewhat increased their meat consumption, against the general European trend of stable or decreased consumption.

Analysing online discussion can provide important insight into the social construction of issues such as sustainable lifestyles (Barr, 2011). Further, attempts to change behaviour should 'engage with people's evaluations of behaviours and of the practices in which they are embedded (Sayer, 2013: 172). Observing the discourse around meat can give us insights into what people think (Billig, 2009), with the assumption that discourses shape our understanding of the world; in a Foucaultian sense, language organises practices (e.g. Foucault, 1972).

The study explores how people relate to insects and cultured meat as potential foods by analysing the online discourse related to developments in these areas of food. What are the presented values and attitudes? How do 'eaters' (online commentators) construct the two alternatives, and the possible transition process towards sustainable meat? How do eaters see their own role in the process? What is the role of disgust, or other emotions? Can a process of normalization be seen in the discourse? Finally, does the context matter? The expectation is that a wide variety of views will be expressed, and that the context (e.g. newspaper, topic, framing of the article, point in time) determines the discourse to some extent. It is possible that through the discussion, commentators start to see their own practices in a new light, on the one hand, perhaps confirming what they do as being right for them, or on the other hand, potentially questioning their habits. Responsibility for change will probably be placed largely in the 'system', but also on individuals. It is possible that over time, a process of normalisation of various alternatives to meat is taking place in the discourse, and the emotion of disgust might act as a catalyst in eventually getting people to engage in discussion about what meat is and where it actually comes from, rather than turn away from considering alternative meats.

Conclusions

Understanding contemporary social practices, their history and potential future is essential for a transition towards sustainability (Shove and Spurling, 2013). The main goal of this research is to explore meat eating as a changing social practice, and to analyse real-life public attitudes in connection with the discourse on eating cultured meat and insects, with the goal of obtaining more insights into the acceptance of radically different and 'new' meats. This project aims to contribute to the literature on, and understanding of the feasibility of transforming the food system, and to support policy for sustainable food consumption.

References

Angyal, A. (1941). Disgust and related aversions. Journal of Abnormal and Social Psychology 36: 393-412.

Barr, S. (2011). Climate forums: virtual discourses on climate change and the sustainable lifestyle. Area 43(1): 14-22.

Barrena, R. and Sanchez, M. (2012). Neophobia, personal consumer values and novel food acceptance. Food Quality and Preference 27: 72-84.

Billig, M. (2009). Discursive psychology, rhetoric and the issue of agency. Semen 27: 157-184.

Dahlstrand, U. and Biel, A. (1997). Pro-environmental habits: propensity levels in behavioural change. Journal of Applied Social Psychology 27: 588-601.

Evans, A.B. and Miele, M. (2012). Between food and flesh: how animals are made to matter (and not matter) within food consumption practices. Environment and Planning D: Society and Space 30: 298-314.

Foucault, M. (1972). The archaeology of knowledge. Pantheon Books, New York, NY, USA.

Hekler, E.B., Gardner, C.D. and Robinson, T.N. (2010). Effects of a college course about food and society on students' eating behaviours. American Journal of Preventive Medicine 38(5): 543-547.

Kanerva, M. (2013). Meat consumption in Europe: issues, trends and debates. Artec-paper no. 187. University of Bremen, Germany.

Matthies, E., Kuhn, S. and Klöckner, C.A. (2002). Travel mode choice of women: the result of limitation, ecological norm, or weak habit? Environment and Behaviour 34(2): 163-177.

Pollan, M. (2006). The omnivore's dilemma – the search for a perfect meal in a fast-food world. Bloomsbury Publishing Plc., London, UK.

Rettie, R., Burchell, K. and Riley, D. (2012). Normalising green behaviours: a new approach to sustainability marketing. Journal of Marketing Management 28(3-4): 420-444.

Rozin, P., Haidt, J. and McCauley, C.R. (2008). Disgust. In: Lewis, M., Haviland-Jones, J.-M. and Barrett, L.F. (eds.) Handbook of emotions, 3rd ed. Guildford Press, New York, NY, USA, pp. 757-776.

Rozin, P., Markwith, M. and Stoess, C. (1997). Moralization and becoming a vegetarian: the transformation of preferences into values and the recruitment of disgust. Psychological Science 8(2): 67-73.

Sayer, T. (2013). Power, sustainability and wellbeing – an outsider's view. In: Shove, E. and Spurling, N. (eds.) Sustainable practices – social theory and climate change. Routledge Advances in Sociology, Routledge, Abingdon, UK, pp. 167-180.

Schatzki, T. (2013). The edge of change – on the emergence, persistence, and dissolution of practices. In: Shove, E. and Spurling, N. (eds.) Sustainable practices – social theory and climate change. Routledge Advances in Sociology, Routledge, Abingdon, UK, pp. 31-46.

Shove, E. and Spurling, N. (2013). Sustainable practices – social theory and climate change. In: Shove, E. and Spurling, N. (eds.) Sustainable Practices – social theory and climate change. Routledge Advances in Sociology, Routledge, Abingdon, UK, pp. 1-14.

Taylor, C. (2012). Abnormal appetites: Foucault, Atwood, and the normalization of an animal-based diet. Journal of Critical Animal Studies 10(4): 130-148.

Wait, G. (2014). Embodied geographies of kangaroo meat. Social and Cultural Geography 15(4): 406-426.

39. Art and design visions of future foods: de-extinction and *in vitro* meat

N.S. Vaage[1,2]

[1]*Centre for the Study of the Sciences and the Humanities (SVT), University of Bergen, Allégaten 34, P.O. Box 7805, 5020 Bergen, Norway;* [2]*Centre for Digital Life Norway, Norwegian University of Science and Technology, Norway; nora.vaage@uib.no*

Abstract

Considering that extensive meat consumption will become an increased problem in years to come, alternatives, some more likely than others, have long been under discussion. Multiple scenarios and speculative cultural manifestations for food futures coexist and compete for public and political attention. The proposed paper will discuss speculative design and art approaches to food, focusing on de-extincted meat and *in vitro* meat. De-extinction is treated through the examples of Center for Genomic Gastronomy's 'De-extinction Deli', which ironically suggests that animal species brought back from extinction would again become objects for consumption, and Ai Hasegawa's 'I Wanna Deliver a Dolphin', in which the designer proposes that human women might give birth to endangered Maui dolphins, and, as one of several possibilities, that the animals might then be eaten by their 'mothers'. *In vitro* meat is discussed through the Tissue Culture and Art Project's Victimless Utopia series, which featured the first actual lab-grown meat to be consumed by humans, as well as Next Nature's speculative design prototypes of *in vitro* meat products. I argue that these artists and designers capture what Hal Herzog has called 'the troubled middle', who exhibit most clearly our propensity to put animals to more than one use: we might have a pet bunny, yet eat rabbit in a restaurant. These art and design pieces, therefore, provide the audience with open-ended scenarios, inducing reflection about what our present, and future, meat consumption can be like, as well as what it can entail.

Keywords: food art, speculative design, food futures, the troubled middle, meat consumption

Introduction

Meat consumption will, according to most predictions (FAO, 2011; UNEP, 2010), become an increasing problem in years to come, causing health problems as well as increasing greenhouse gas emissions. Therefore, a number of alternatives are currently up for discussion. From the use of insects as a source of protein to lab-grown meat, the debate revolves around issues of feasibility, sustainability and human emotion (the 'yuck' factor). Speculative art and design approaches offer distinct, materialized visions of potential paths our society might choose. While not mutually exclusive, these visions do not aim to be the most probable or logical. Rather, they present somewhat extreme scenarios that might come to be. This paper will present four examples of such art and design works, all of which explore the rich societal implications of potentially different avenues of future meat production.

Center for Genomic Gastronomy's 'De-extinction Deli' ironically suggests that animal species brought back from extinction might again become objects for consumption. Through cut charts, informative postcards, performative happenings, and a stand offering a menu of 'yesterday's meat tomorrow' (Figure 1), the designers propose that 'de-extincting' extinct animals, whether through cloning, synthetic biology or other means, may give us a chance to use these animals all over again. They point out that the recent hype around the possibility of counteracting de-extinction of species like the Pyrenean ibex, the woolly mammoth, and the passenger pigeon in laboratories may be occluding the reasons why species are going extinct. Through offering up a menu featuring now-extinct animals, the artists ironically suggests that

EurSafe 2016 – DOI 10.3921/978-90-8686-834-6_39, © Wageningen Academic Publishers 2016

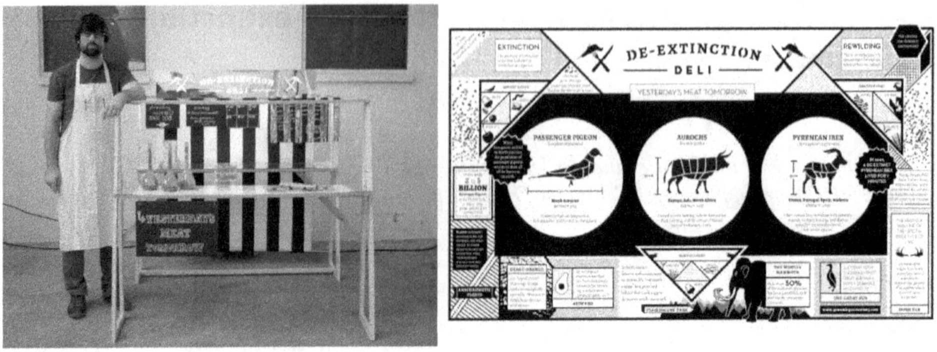

Figure 1. Center for Genomic Gastronomy: 'De-extinction Deli', 2013-2016. Left: Stand featuring 'grow-your-own cell' kits, bottles of 'extinct animal habitats' and polls where audience could give their opinions as to de-extinction (The Center for Genomic Gastronomy 2016). Right: 'Butcher paper' information poster take-aways, featuring more and less recently extinct animals, candidates for different kinds of de-extinction. Reprinted with permission from the artists.

the uses to which the animal brought back from extinction would be put would still reflect the human propensity for consumption.

Ai Hasegawa, in 'I Wanna Deliver a Dolphin', proposed a future scenario where human women might give birth to critically endangered Maui dolphins, through the help of engineered changes to their placentas. This would also have the added 'bonus' of allowing human women a chance at 'motherhood' without contributing to human overpopulation. The Maui dolphin is one of the smallest dolphin breeds, making the size of the foetus close to human foetal size. The rich piece, which included a video of the designer 'giving birth' to a baby dolphin and bottle-feeding it in a swimming pool as well as models and charts with ideas of the 'dolph-human future' (Figure 2), also speculated around how the dolph-human relationship might then develop: would the human woman care for the dolphin as for her own child?

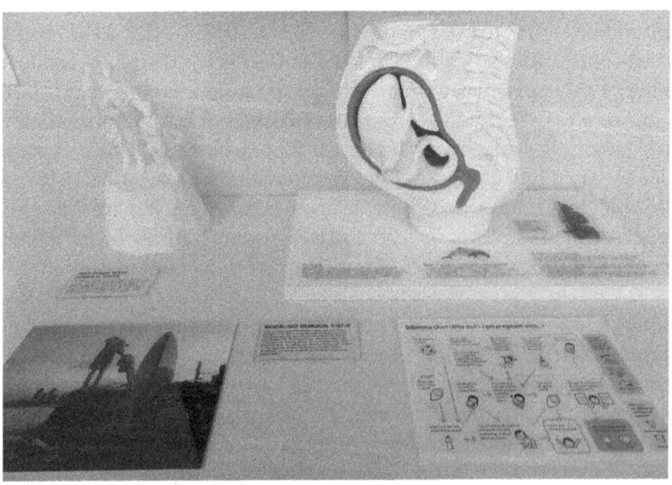

Figure 2. Ai Hasegawa, 'I Wanna Deliver a Dolphin', 2013. Exhibition photo: N.S. Vaage, from the exhibition Grow Your Own... Life After Nature, 2014. Reprinted with permission from Science Gallery Dublin.

Would the in reality transgenic dolphins be released into the environment, as an investment 'in the future food supply' (Hasegawa 2013)? How would their continued welfare be ensured? The designer suggested, as one of several possibilities, that the animals might in the end be eaten by their 'mothers', whose commitment of time and energy from their own body would entitle them to the rare delicacy of dolphin meat.

Artist duo Oron Catts and Ionat Zurr, the core members of the Tissue Culture and Art Project, explored a potentially closer future in their work on *in vitro* meat. Starting in 2000 during a residency at Harvard Medical School, they created the Victimless Utopia series, exploring the idea of creating meat and leather *in vitro*: growing mammalian cells into the shape of a jacket in 'Victimless Leather' (2004), and working towards serving lab-grown skeletal muscle cells as 'meat' in 'Semi-Living Steak' (2000) and 'Disembodied Cuisine' (2003, Figure 3). Due to laboratory regulations, the artists did not get to serve the first 'meat' they cultured in 'Semi-Living Steak', grown from pre-natal sheep muscle cells during a residency at Harvard Medical School in 2000 (Catts and Zurr 2007; Figure 3A). But in 2003, in the course of the 'L'Art Biotech' festival in Nantes, France, they grew, served and ate a small piece of frog skeletal muscle in 'Disembodied Cuisine' (Figure 3B), the first actual lab-grown meat to be consumed by humans. In a makeshift lab environment, the cells multiplied next to the living donor frogs. At the end of the exhibition period, the cultivated frog meat was fried in a sauce of garlic, honey and Calvados served to a small group of diners. The piece played on several aspects of French culinary tradition, from their love of frog legs (a food not widely consumed elsewhere), to scepticism towards engineered food. The two small pieces of meat, each about the size of a coin, and only a few millimetres thick, reportedly tasted quite disgusting, had a gelatinous texture, and the biopolymer on which they had been grown had not entirely dissolved (Gefter, 2011; Hauser, 2005). Several of the eight guests spat out their share.

Through materializing technological utopia *ad absurdum*, sardonically showing what could and could not be achieved with current technology, the TC&A simultaneously performed and critiqued the potential of tissue culturing technology for meat production (Vaage, 2016). Concurrently, as Jens Hauser (2005) has pointed out, through being the first to actually serve *in vitro* meat to the public, the artists might have made it difficult for researchers and corporations to patent and make a profit out of the now emerging methodology for growing *in vitro* meat for consumption. If so, this could have been a real contribution with societal implications. As it happened, researchers in the Netherlands developed a bioreactor vat for growing meat *in vitro* which included a new method for 'exercising' the cells, and

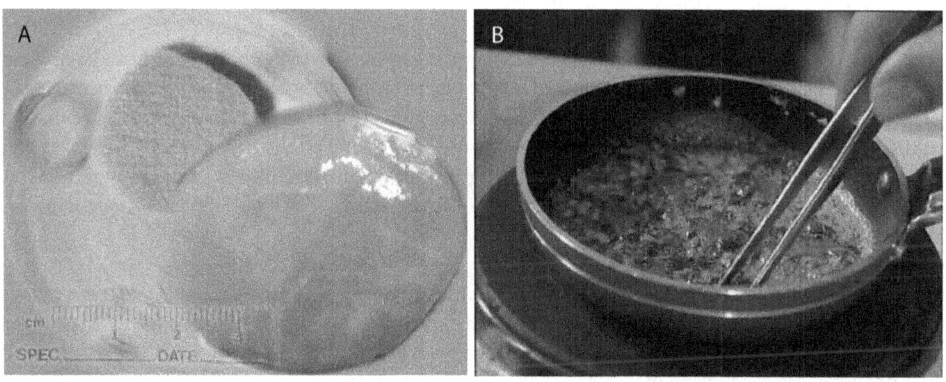

Figure 3. Tissue Culture and Art Project: 'Disembodied Cuisine', Nantes, 2003. (A) In vitro *meat from pre-natal sheep cells. (B) Tissue cultured frog steaks frying. Right: Poster for the event. Reprinted with permission from the artists.*

no actual patenting problems seem to have been encountered. The first *in vitro* burger was unveiled and eaten in London in 2013 (see Maastricht University, 2015). The Maastricht team has also 'rebranded' the product from 'synthetic' or '*in vitro*' meat to 'cultured beef'. In spite of the technical advances and the attempts at creating more positive connotations to the project, *in vitro* meat is, as recent research in the EU-EPINET project stressed, as of now, 'an emergent science in search of funding, in search of users, in search of identity' (Gunnarsdóttir, 2015: 1). Part of the problem is in the technical issues still remaining to be solved, including the addition of fat and haemoglobins to the *in vitro* muscle cells, the use of antibiotics, and the growth conditions of the cells.

In addition, a number of social factors appear to limit the potential of *in vitro* meat. The idea of eating meat grown in a lab tends to invoke the 'yuck factor' (see e.g. Erler, 2012; Goodwin and Shoulders, 2013), a term generally credited to Arthur Caplan. Bioethicist Leon Kass (1997), in discussing human cloning, refers to the idea that such intuitive, negative responses are the result of a judgment that something is intrinsically bad and might cause us harm as the 'wisdom of repugnance'. Kass emphasizes that repulsion is not in and of itself an argument that something is morally reproachable, but claims that in 'crucial cases', such as incest, corpse mutilation, or sex with animals, revulsion serves as 'the emotional expression of deep wisdom', beyond what we can articulate rationally (Kass, 1997: 20).

The artists have themselves repeatedly discussed the ethical problem of using foetal bovine serum (FBS) in tissue culturing eukaryotic cells (see e.g. Catts and Zurr, 2008). FBS is a by-product of the meat industry, produced from the blood of foetal calves taken from the wombs of butchered cows. Meat products created through tissue culturing using FBS will, as such, not be victimless. Although alternatives do exist (see e.g. Hulshof *et al.,* 1995), FBS is still considered the most effective growth supplement mammalian cells *in vitro*. Catts and Zurr (2008: 133) have estimated that 'growing around 10 grams of tissue will require serum from a whole calf (500 ml.), which is killed solely for the purpose of producing the serum'. Yet, FBS is mainly produced in areas where cattle roam free, and an entire herd is taken to slaughter at the same time, including some pregnant individuals. The foetuses, then, are not killed directly for their serum. However, the process in which the serum is extracted may take more than 30 minutes, and may quite possibly be painful to the neonates (Jochems *et al.,* 2002). The artists' use of FBS does pose the question of whether the use of biotechnological animal products in art is moral. However, their explicit approach to creating awareness about the presence of the serum in most tissue cultured products does seem to give their work a valid purpose – if only in a utilitarian framework.

Next Nature's ongoing project 'Nano Supermarket' has featured a number of speculative design prototypes, among others several for *in vitro* meat products. The sub-project *In Vitro Meat* (2014) also included a cookbook and several menus. The prototypes, although considered to have differing likelihoods of being realized, were all selected by the Next Nature designer group as thought-provoking avenues for the potential of nanotechnology. The pieces concerning lab meat focused on what markets and niches *in vitro* meat could aim for: perhaps 'meat pearls' giving all the flavour and juiciness of beef, without the toughness and splinters? Or might 'magic meatballs' make children eat their dinner (Figure 4)? There was also a dish of 'Dinosaur Wings' (mimicking Buffalo Wings), representing the idea of radically de-extincted meat, as well (Myers, 2015).

The glossy and attractive nature of the Next Nature prototypes means that any adverse reaction experienced by the audience as to the thought of consuming *in vitro* meat is not specifically encouraged by the designers, who seem merely to place the options in front of the consumer, making it our choice – but not providing extensive information as to what this meat would actually contain, in terms of additives, nutrients and flavour.

Figure 4. Next Nature, 'Nano Supermarket'. Left: 'Meat pearls'. Right: 'Magic meatballs'. Exhibition photos: N.S. Vaage, from the Article 2015 biennial, Stavanger. Reprinted with permission from the Article biennial.

A hand-painted roadside sign in rural Massachusetts reads:

For Sale:

Rabbits

**Pets
or Meat**

(Gaskell, 2008: 229)

The quote above is a striking example of how animals are used as commodities. The utilizations of rabbits suggested on the sign, as food or pets, appear incompatibly dichotomous. As Ivan Gaskell (*ibid.*) points out, the sign 'starkly evokes the human propensity to put something to more than one use'. The above artists and designers, with their ironic and open-ended approaches, capture the ambivalence of what Hal Herzog (2010) has called 'the troubled middle' (he accredits the term to Strachan Donnelly). Herzog uses 'the troubled middle' to describe people who think animals should be treated well, but still consider it sometimes permissible to use them for research, meat and other purposes that are advantageous to us. He places himself within this group, arguing that 'the troubled middle makes perfect sense because moral quagmires are inevitable in a species with a huge brain and a big heart' (Herzog, 2010: 12). The troubled middle includes those of us who exhibit most clearly the human propensity to put animals to more than one use: we might have a pet bunny, yet eat rabbit in a restaurant.

Whether re-availing ourselves of previously extinct meats or creating new ones through laboratory tissue culturing, these artists and designers explore human impulses relating to meat foods, what we find attractive to eat today, and how this might change in the future. As such, they present a form of materialized technology assessment with an open-ended approach. Even though they of course have their own, differently stated motivations and messages that they might wish to get across, these four art and design works have in common a format that leaves it up to the audience to decide how we relate to these issues personally, and in what direction we would like our society to move regarding future meat consumption.

References

Alexandratos, N. and Bruinsma, J. (2012). World agriculture towards 2030/2050: the 2012 revision. ESA Working paper No. 12-03. FAO, Rome, Italy.

Catts, O. and Zurr, I. (2008). The ethics of experiential engagement with the manipulation of life. In: Da Costa, B. and Philip, K. (eds.) Tactical biopolitics. art, activism, and technoscience. The MIT Press, Cambridge, MA, USA, pp. 125-142.

Erler, A. (2012). *In vitro* meat, new technologies, and the 'yuck factor'. Practical ethics: ethics in the news. University of Oxford. Available at: http://bit.ly/11dCXqP.

Food and Agriculture Organization of the United Nations (FAO) (2011). World livestock 2011 – livestock in food security. FAO, Rome, Italy, Available at: http://tinyurl.com/7f32psx.

Gaskell, I. (2008). Ethical judgments in museums. In: Hagberg, G.L. (ed.) Art and ethical criticism. Blackwell Publishing, Malden, MA, USA, pp. 229-242.

Gefter, A. (2011). Visceral curator: making art out of living tissue. NewScientist, January. Available at: http://bit.ly/1APxF4T.

Goodwin, J.N. and Shoulders, C.W. (2013). The future of meat: a qualitative analysis of cultured meat media coverage. Meat Science 95: 445-450.

Gunnarsdóttir, K. (2015). Case study on *In-Vitro* meat WP5 policy report. Available at: http://tinyurl.com/zpcbpcc.

Hasegawa, A. (2013). I wanna deliver a dolphin ... In: Ginsberg, A.D. (eds.) Grow your own ... life after nature. Exhibition catalogue, Science Gallery, Dublin, Ireland, pp. 27-28.

Hauser, J. (2005). Bio art – taxonomy of an etymological monster. In: Hybrid: Living in paradox. Exhibition catalogue. Ars Electronica, Linz, Austria.

Hertwich, E., Van der Voet, E., Suh, S., Tukker, A., Huijbregts, M., Kazmierczyk, P., Lenzen, M., McNeely, J. and Moriguchi, Y. (2010). Assessing the environmental impacts of consumption and production: priority products and materials. A Report of the Working Group on the Environmental Impacts of Products and Materials to the International Panel for Sustainable Resource Management. United Nations Environment Programme, Nairobi, Kenya.

Herzog, H. (2010). Some We love, some we hate, some we eat: why it's so hard to think straight about animals. Harper, New York, NY, USA, 341 pp.

Hulshof, S.C.J., Figueiredo, J.R., Beckers, J.F., Bevers, M.M., Van der Donk, J.A. and Van den Hurk, R. (1995). Effects of fetal bovine serum, FSH and 17β-estradiol on the culture of bovine preantral follicles. Theriogenology 44(2): 217-226.

Jochems, C.E.A., Van der Valk, J.B.F., Stafleu, F.R. and Baumans, V. (2002). The use of fetal bovine serum: ethical or scientific problem? Alternatives to Laboratory Animals 30(2): 219-227.

Kass, L. (1997). The wisdom of repugnance. The New Republic, pp. 17-26.

Maastricht University (2015). Cultured beef. Available at: http://tinyurl.com/jp4wnue.

Myers, W. (2015). Bio art: altered realities. Thames and Hudson, New York, NY, USA, 288 pp.

The Center for Genomic Gastronomy (2016). De-extinction deli. Available at: http://tinyurl.com/hdj6un7.

Vaage, N.S. (2016). What ethics for bioart? Nano Ethics 10(1): 87-104.

40. Envisioning the futures of animals through *in vitro* meat

A. Ferrari
Institute of Technology Assessment and Systems Analysis (ITAS), Karlsruhe, Germany; arianna.ferrari@kit.edu

Abstract

Although large-scale production of *in vitro* meat is not yet possible, a growing international scientific community has in recent years presented it as a promising solution to ecological, health-related and ethical problems of traditional meat production. These visions of *in vitro* meat elaborate the socio-epistemic practices which simultaneously enable and create social arrangements and knowledge around this innovation. These different visions of *in vitro* meat indicate that ethical reflection on the current situation plays a crucial role in imaging the future of this innovation. In this particular case, the idea of *in vitro* meat has also developed out of a call for changing the current situation of using animals in agriculture. However, what the future will look like for animals in a world with *in vitro* meat remains far from clear. In this paper we will explore how visions of *in vitro* meat propose different relationships of humans with nonhuman animals that are informed by different values. This work has been performed in a project funded by the German Ministry of Education and Research under the title 'Visions of *in vitro* meat. Analysis of the technical and social aspects and visions of *in vitro* meat' (webpage: http://invitrofleisch.info).

Keywords: *in vitro* meat, technological vision, ethics, human-animal relationships

Introduction

Although large-scale production of *in vitro* meat (IVM) is not yet possible, in recent years a growing international scientific community has presented it as a promising solution to ecological, health-related and ethical problems of traditional meat production. The idea is to obtain cells from an animal via a muscle biopsy and to let these cells grow on a petri dish to form a tissue. The production of IVM is an important part of a larger field called 'cellular agriculture' or 'post-animal bioeconomy' which uses technologies such as tissue engineering, stem cell biology, synthetic biology and genetic engineering to produce animal products without animals. IVM is still a proof of concept and an ontologically undefined object (Stephens, 2013), which is trying to create broad support in the scientific and financial communities as well as a potential consumer market. In order to understand its ethical and social dimensions, it is necessary to explore the visions which inform it. As this is innovation in an early stage, the narratives used to explain and present it are essential to its further development. In my paper I concentrate on how visions of *in vitro* meat propose changes in human's relationships with nonhuman animals in the future.

Facing the challenges posed by the use of animals in agriculture

Since the publication of the FAO's report 'Livestock long shadow' ten years ago (FAO, 2006), the detrimental effects of global animal production have been constantly monitored and explained by scientists. Due to the difficulties of modelling and the different global realities connected to meat consumption, there are different estimates on the environmental impacts. Nevertheless solid estimations have pointed out that the so called livestock sector is responsible for approximately 14.5% of all anthropogenic greenhouse gas emissions, approximately equivalent to all the direct emissions from transportation (Herrero *et al.*, 2013; Ripple *et al.*, 2014). Animals used in agriculture in Europe account for 79-88% of the total European emissions of ammonia, nitrates and nitrous oxide (Westhoek *et*

EurSafe 2016 – DOI 10.3921/978-90-8686-834-6_40,

al., 2014), and they occupy 45% of the global agricultural land and nearly one-third of the ice-free land surface of the planet, making it the single largest anthropogenic land use type (Thornton *et al.*, 2011). Meat consumption is also linked with increased health risks: in October of last year the WHO's International Agency for Research on Cancer (IARC, 2015), after evaluating more than 800 different studies on cancer in humans, ranked processed meats as group 1 carcinogens because of a causal link with bowel cancer, and all red meat as group 2A carcinogens (probably carcinogenic). Furthermore, there are other epidemiological studies pointing to a link between meat consumption and higher fasting glucose and insulin concentrations regardless of the glucose and insulin genetic risk scores, as well as ischemic heart disease, stroke and diabetes mellitus (Lippi *et al.*, 2015). The United Nations (UNEP, 2010) has concluded that both human and global environmental health would benefit from a substantial dietary change away from animal products on a global scale. Yet the demand for meat and dairy products continues to rise worldwide (OECD/FAO, 2015).

IVM as ethical and sustainable innovation

IVM is presented by innovators as new solution on the supply chain to the problems posed by animal food production. IVM is not only presented as capable of meeting the traditional standards of higher efficiency, but also as sustainable and ethical because it permits us to save the life of animals, to better protect consumer's health and to dramatically reduce the environmental footprint from the production of animal products.

According to analyses of anticipatory life cycle assessments, IVM presents clear advantages concerning the use of land and water resources; more controversial is its potential to contribute to reducing global warming because of the need for energy (Tuomisto *et al.*, 2014; Mattick *et al.*, 2015). Being grown in the lab, IVM has been also associated with the ideas of safety, cleanness and our ability to control industrialized scientific products. Assembling cells in a petri dish is said to render it possible for us to control the placement of fat within a piece of meat, offering a more customized product. Moreover, it could also open the possibility of optimizing the properties of meat (Pandugaram and Kim, 2015) or even of adding new properties. Last but not least, the ethical value of this innovation is claimed to come from its potential to save animal lives and protect animal welfare. The production of IVM does not necessitate the killing of animals and, once the problem of the growth medium is solved, the innovation will be cruelty-free. IVM scientists are very much engaged into finding alternatives to foetal calf serum, which is currently used as the growth medium in the majority of the cases, not only because it is extracted through a painful method from aborted cow foetuses, but also because animal-derived media may contain complex contaminants which pose dangers for human health (Pandugaram and Kim, 2015).

The claimed ethical advantages of IVM are a very appealing characteristic which has been able to attract different stakeholders, who sometimes hold diametrically opposed positions. IVM should not only succeed because of its appeal to the conscious consumer who is concerned about the environmental footprint of her lifestyle; it can also offer a solution for people who think that the consumption of animal products is ethically wrong because animals suffer and die. Indeed, the research on *in vitro* meat has gained the financial and ideological support of one of the biggest animal welfare organizations worldwide, PETA, which has been supporting the development of IVM since 2008.

Different narratives around IVM

Although IVM has been gaining more and more media and public attention in the last few years, especially after the launch of the first burger made entirely of bovine stem cells in August, 2013, the presentation of cultured 'steak chips' in March, 2014, and the announcement of cultured meat balls in January, 2016, it is still an innovation in an early stage. There is as yet no method which permits its large scale production, mostly due to technical difficulties and to the high price of its fabrication because of the growth factor. Visions of *in vitro* meat develop out of combination of different narratives, which can be defined as key elements of socio-epistemic practices which simultaneously enable and create social arrangements and knowledge. During our project we have identified different narratives in which IVM is presented.

One main narrative on IVM plays with the topic of the novelty versus the continuity of practices in this innovation. Mark Post, one of the leading IVM scientists, is very engaged in underlining the continuity of IVM with traditional meat. IVM is described as exactly the same as meat; the difference lies only in the production process. Meat, instead of being prepared out of entire animals, will be 'directly' produced from cells. However, since novelty/continuity in a narrative around technology work better in pairs, there is always an element of continuity in the novelty of the process and of newness in the sameness of the meat product. The cleanness and controllability of labour conditions make it possible to obtain something different (i.e. better) than the meat traditionally grew in intensive farming. At the same time, it is important that the taste and consistency of IVM will be the same in order to compete with traditional meat (Post, 2016b). As far as the production process is concerned, IVM is described as an evolution in scientific methods and technological processes which have existed for centuries. Datar (2016a) defines the breeding of animals to have selected productive traits as a practice of turning them into living bioreactors; IVM is also produced in bioreactors, but it uses cells instead of whole animals. Datar (2016b) prefers to speak of 'cultured meat' instead of '*in vitro* meat': 'cultured' is not only a more accurate term since this meat will be produced in cell culture in controlled conditions and not necessarily in a glass (as *in vitro* implies), but it is 'a familiar term in food context, since civilization has been culturing foods for centuries through fermentation with a variety of living cultures: bacterial cultures, as used in yogurt, sour cream, cheeses and pickled vegetables yeast cultures, as used in beer, bread and wine; bacterial and yeast co-cultures as used in kefir, kombucha ...' (Datar, 2016b, 19). Cultured meat comes from the development in science and technology that for the first time permits the application of tissue engineering principles to fermentation, and it is a step in the advancement of civilization.

The technological progress attached to the cultural advancement condensed in IVM is the second strong narrative. IVM is seen as an advanced product, the expression of a sophisticated combination of new scientific fields which are able to meet the great challenges of our time. There are different ways in which the ethical value of IVM is expressed: the first is more interested in sustainability, while the second is more in the advancement of our ethical attitude towards animals. IVM is seen as offering the material basis for a profound change in culture, moving away from an inefficient use of resources and of an unjust treatment of animals. In a talk in 2012, Forgacs, co-founder of the start-up 'Modern Meadow' and a main actor in this innovation field, described biofabricating meat and leather as 'a civilized way to move past killing animals for hamburgers and handbags' and concluded with the following statement: 'A natural evolution of manufacturing for mankind: it is environmentally responsible, efficient and human – it allows us to be creative: we can design new materials, new products and new facilities – we need to move past just killing animals as a resource to something more civilized and evolved. Perhaps we are ready for something literally and figuratively more cultured'). Although the innovators describe themselves as non-vegetarian (cf. Hamid, 2014) and praise the high value of meat products (Datar, 2016a), they see IVM as a technology capable of inducing a sort of moral progress in the future humanity. This idea, which is not further articulated by its innovators, has in the field of the ethics of technology been called the possibility of 'techno-moral change'. Precisely because technology and society co-evolve (Rip and Kemp, 1998), a technology is also capable of influencing values while shaping and changing the practices of its use once it is out in society (cf. Swierstra, 2013). IVM not only provokes questions around food, sustainability and human-animal relationships, it affects the answers and may change them. Morality can be described as a force field where conflicting norms and values compete for hegemony (Swierstra, 2013).

Envisioning the futures of animals through *in vitro* meat

Visions, understood as key elements of socio-epistemic practices which create new social arrangements, can be created at different stages in the development of an innovation and at different levels. The different narratives analysed in the previous paragraph are important elements for the construction of different visions surrounding this innovation. The openness of IVM is visible in the way in which innovators present their ideas: which values and messages they use and what they offer as relevant topics and what they do not talk about.

There are different entry points in imaging the future of animals in a society with IVM. One point regards the moment of production of food. IVM pretends to be a substitute for traditional meat since changing the way in which meat is produced will also change the role of the animals in this production process. As the labels 'post-animal bioeconomy' or 'cellular agriculture' suggest, the meat production of the future will take place without animals. Animals will not enter the production chain, and they will simply not be involved *as* animals (as whole living beings) in the process. Forgacs (2014) and Datar (2016) imagine breweries and vertical farms operating in many urban contexts, in which meat can be produced without any contact with animals and in absolute transparency under safe and sterile conditions. Forgacs (2014) openly speaks of a future in which IVM breweries could be toured like beer breweries or ice-cream factories. IVM offers a new method of producing meat without killing, which means without blood and without cruelty; it can then be brought back to the urban centres and even be exposable all the time. These places change the view we have of the people working in meat production: they are no longer exploited slaughterhouse workers, but artisans, craftsmen and microbiologists (Datar, 2016). As far as the production process is concerned, the vision of IVM works by reversing the mechanisms of the distantness and presence of animals: whereas in traditional meat production animals have to be present and the slaughter is mainly kept distant from the consumer, in IVM the production process is present, open and transparent but the animals are absent. The reversal of human-animal relationships in meat production process of IVM constitutes the basis for the ethical added value of this innovation.

The second entry point for the future of animals regards how the animals involved in IVM will live at all. As I previously indicated, the narrative expressing the ethicality of IVM is sometimes linked to the idea of a 'techno-moral change': IVM is seen as a more cultured way of producing food and, therefore, is linked with the idea that a better treatment of animals is a sign of moral progress. If we, however, take a closer look at all stages of the production process, such as how the cells are obtained from the animals and what the lives of the animals providing these cells are like, the picture becomes less clear. Very rarely is the topic of human-animal relationships in a future world with IVM a part of the message of innovators, who prefer to focus mainly on the advantages which directly come from IVM as a product..

For Post, the important issues to consider to further develop this innovation are the 'scalability of the production process, quality control of mammalian cell/tissue cultures, maintaining sterility in the culture, prevention of contamination or disease and *the controlled breeding of stem cell donor animals*' (Post, 2012: 300, italics added). Post is a scientist, interested in the scientific efficiency of the innovation, who does not seem to have intrinsic ethical objections to meat consumption. The future of IVM is perfectly compatible with the selective breeding of animals: animals will no longer be used as direct sources of meat, but they will be optimized in their function as stem cell donors. Post's focus is clearly on sustainability, namely the vision of IVM as offering simply a new production process for a well-known product such as meat, and is attached to what is known as a 'welfarist' position regarding human-animal relationships in the field of human-animal studies. The welfarist position claims that as long as animals are treated well and their welfare is respected, it is ethical to use them (and to kill them) (cf. Francione and Garner, 2010).

Forgacs (2014) spoke of 'Betsy the cow having a perfect happy life' but he never mentioned the necessity of the selective breeding of animals for obtaining cells. The changes in the human-animal relationships which are implicit in the vision of IVM as a sustainable way of producing meat are not substantial changes, but incremental steps for protecting animal welfare. IVM is not primarily aimed at vegetarians and vegans (Forgacs, 2014), and Post excludes vegetarianism as a viable option for the future. IVM is presented as the way to meet different needs of different communities who express different ethical values (concerns for animals and attachment to meat). Negotiating the moral battle field of ethics rests on an anthropocentric focus, in which animal welfare has a place inasmuch as IVM provides other general advantages.

This 'welfarist touch' of IVM, together with the disregard of plant-food diets expressed by IVM innovators who do not consider themselves openly committed to animal rights and/or to antispeciesism, makes IVM ethically suspicious to many critical scholars. These scholars view the utilitarian and pragmatic advantages of this innovation saving millions of lives, as is entailed in the vision of IVM as a sustainable way of producing meat, possibly producing the opposite effect. IVM might possibly reinforce patterns of dominance towards animals because this vision flirts with meat culture and does not call it into question (Miller, 2012). The point being contested lies precisely in the vision of IVM as a further ethical step in human-animal relationships. Whereas the step for some can (and should) only be incremental, because what counts is sustainability and because the very use of animals as such is not a problem (welfarism), for others IVM is a creating a strong illusion that represents only a cure for the symptoms of unjust human-animal relationships, but not for its cause.

It is difficult to say which interpretation is right, not because it is not possible to choose between a welfarist and antispeciesist paradigm (the very opposite is true), but because technological developments exercise an ethical force of its own. This is the core message of the idea of 'techno-moral change': the ways in which these changes will happen are open, but they will enact practices which may result in very different human-animal relationships. The analysis of visions can be extremely helpful in reflecting on the world we want to live in. Since a (technological) vision is to be understood as a socio-epistemic practice which has practical consequences for knowledge arrangements, we can conclude that the vision of IVM as providing sustainable meat is linked with the practical interest to develop further animal breeding for this process. If IVM wants to be attached with a different, non-welfarist and non-anthropocentric vision of a technology committed to liberating animals from oppression, it should find a way to conceptualize a different future of animals and to develop the available alternatives to using animal-based serum as preconditions for continuing this research.

References

Datar, I. (2016a). On animal products without animals. Bitten: a food conversation. Available at: http://tinyurl.com/zzpruvu.

Datar, I. (2016b). *In vitro* meat is cultured. In: Post, M. (ed.) What is *in vitro* meat? Food Phreacking 2: 16-21.

Forgacs, A. (2014). The benefits of *in vitro* meat. Jason Point in MIT Technology Review, EmTech, September 23, 2014. Available at: http://tinyurl.com/j9so3r5.

Francione, G. and Garner R. (2010). The animal rights debate: abolition or regulation? Columbia University Press, Columbia, USA.

Hamid, T. (2014). Animal products created in the lab through cell multiplication. The National Business, February 3. Available at: http://tinyurl.com/zbrrq42.

Herrero, M., Havlík, P., Valin, H., Notenbaert, A., Rufino, M.C., Thornton, P.K, Blümmel, M., Weiss, F., Grace, D. and Obersteiner, M. (2013). Biomass use, production, feed efficiencies, and greenhouse gas emissions from global livestock systems. PNAS 110(52): 20888-20893.

International Agency for Research on Cancer (IARC) (2015). Press release No. 240. Available at: http://tinyurl.com/ot5nffq.

Lippi, G., Mattiuzzi, C. and Sanchis-Gomar, F. (2015). Red meat consumption and ischemic heart disease. A systematic literature review. Meat Science 108: 32-36.

Mattick, C.S., Landis, A.E. and Braden, R.A. (2015). A case for systemic environmental analysis of cultured meat. Journal of Integrative Agriculture 14(2): 249-254.

Miller, J. (2012). *In vitro* meat: power, authenticity and vegetarianism. Journal for Critical Animal Studies 10(4): S41-63.

OECD/FAO (2015). OECD-FAO Agricultural Outlook 2015. OECD Publishing, Paris, France.

Post, M. (2016). *In vitro* meat is an emerging technology. In: Post, M. (ed.) What is *in vitro* meat? In: Food Phreacking 2: 10-15.

Post, M.J. (2014). Cultured beef: medical technology to produce food. Journal of Science, Food and Agriculture 94(6): 1039-1041.

Post, M.J. (2012). Cultured meat from stem cells: challenges and prospects. Meat Science 92(3): 297-301.

Rip, A. and Kemp, R. (1998). Technological change. In: Rayner, S. and Malone, E.L. (eds.) Human choice and climate change. Battelle Press, Columbus, OH, USA, pp. 327-399.

Ripple, W.J., Smith, P., Haberl, H., Montzka, S.A., McAlpine, C. and Boucher, D.H. (2014). Ruminants, climate change and climate policy. Nature Climate Change 4: 2-5.

Stephens, N. (2013). Growing meat in laboratories. The promise, ontology, and ethical boundary-work of using muscle cells to make food. Configurations 21(2): 159-181.

Swierstra, T. (2013). Nanotechnology and techno-moral change. Etica and Politica / Ethics and Politics 15(1): 200-219.

Thornton, P., Herrero, M. and Ericksen, P. (2011). Livestock exchange issue brief 3. ILRI, Nairobi, Kenya. Available at: http://tinyurl.com/q9gwmzj.

Tuomisto, H.L., Ellis, M.J. and Haastrup, P. (2014). Environmental impacts of cultured meat: alternative production scenarios. In: Proceedings of the 9th International Conference on Life Cycle Assessment in the Agri-Food Sector. October 8-10, 2014. San Francisco, CA, USA, pp. 1360-1366.

UNEP (2010). Assessing the environmental impacts of consumption and production: priority products and materials. Available at: http://tinyurl.com/op6x4we.

Section 10. Veterinary ethics

41. Role of moral values in the trade-off between animal welfare and food safety risks in broiler husbandry

M. van Asselt[1], E.D. Ekkel[1], B. Kemp[2] and E.N. Stassen[2]*
[1]CAH Vilentum, University of Applied Sciences, Department of Applied Research, Drieslag 4, 8251 JZ Dronten, the Netherlands; [2]Wageningen University, Adaptation Physiology Group, P.O. Box 338, 6700 AH Wageningen, the Netherlands;m.van.asselt@cahvilentum.nl

Abstract

To cope with societal concerns regarding animals kept in intensive systems, alternative broiler husbandry systems that comply with above-legal animal welfare standards, such as the free-range barn system have been introduced. However, broilers in free-range barn systems show higher *Campylobacter* contaminations than broilers in conventional systems, which might entail a higher risk of *Campylobacter* infections in consumers. This causes a dilemma between animal welfare and food safety. To make a trade-off, moral values concerning humans and broilers must be weighed. Moral values are deeply rooted and shared in society, but in a specific case they depend on the context and are weighed against each other. To get insight into the role of moral values in the dilemma between broiler welfare and food safety, citizens (n=2,259) and poultry farmers (n=100) were surveyed. First, participants were questioned about 13 moral values regarding chickens and consumers. Next, they were presented a case concerning a broiler farmer who would like to change from a conventional husbandry system to a free-range barn system that complies above legal welfare standards, but which entails higher *Campylobacter* contamination of broilers. Respondents were asked whether they agreed with the change from the conventional husbandry systems to the free-range barn system. Additionally, respondents scored the importance of arguments based on moral values and relevance for the case. Citizens and farmers scored equally only the statements regarding chickens being sentient and human health. Both citizens and farmers who agreed with the change of the husbandry system considered the arguments concerning broiler welfare more important than respondents who did not agree with the change of system. This study provides input for deliberation on broiler husbandry especially on the dilemma between broiler welfare and food safety. With the information form this study a broad moral deliberation could be set up between the different stakeholders to define characteristics of future husbandry systems that can count on support from both the poultry sector and citizens.

Keywords: moral convictions, *Campylobacter*, public health risks, broiler production, moral dilemma, survey

Introduction

During last decades, in reaction to the intensification of livestock husbandry, attitudes towards animals have changed and led to increased societal concerns about the welfare of intensively kept animals such as broiler chickens (Rollin, 2007). To cope with these societal concerns, alternative broiler husbandry systems, which comply with above-legal animal welfare standards, have been introduced. Broilers kept in alternative systems, such as the free-range barn system, have more space and opportunities to express natural behaviour, are provided enrichment such as straw, are from a slower growing strain, and slaughtered at an older age. Hence, broiler welfare in free-range barn systems is considered better than broiler welfare in conventional systems (Bessei, 2006). However, broilers kept in free-range barn systems show higher *Campylobacter* contaminations compared to those kept in conventional systems (Bouwknegt *et al.*, 2004; Sommer *et al.*, 2013). Consequently, meat from these broilers might entail a

EurSafe 2016 – DOI 10.3921/978-90-8686-834-6_41, © Wageningen Academic Publishers 2016

higher risk on *Campylobacter* infections in consumers, causing a dilemma between broiler welfare and food safety.

Public debates about broiler husbandry are based on perceptions of animal husbandry and issues, such as broiler welfare, food safety and price of meat. Perceptions are influenced by moral values, norms, knowledge and interests (Te Velde *et al.*, 2002). Moral values are deeply rooted beliefs, which have been formed by knowledge, personal experience, and multiple social, cultural and religious aspects (Cohen *et al.,* 2009). In a society, people share moral values concerning animals and humans. For example, in the Netherlands people consider animals to have value, but the majority of people consider the moral importance of humans superior to animals (Cohen *et al.,* 2012). In a dilemma, such as the dilemma between animal welfare and food safety in broiler practice, moral values become more practical and are shaped to the specific context (Cohen *et al.,* 2016). In such a dilemma the range and the weight of moral values can vary, which may result in people making different trade-offs (Childress *et al.,* 2002; Cohen *et al.,* 2009).

It has been shown that perceptions of citizens and farmers regarding broiler husbandry differ (Van Asselt *et al.,* 2015). Consequently, it is difficult to reach consensus between citizens and farmers about husbandry systems. Because society, and thus citizens and farmers may share moral values, moral values may be a starting point for deliberation. Insight into the role of moral values and the arguments based on moral values in the dilemma between broiler welfare and food safety may provide input to design husbandry systems that address the dilemma between broiler welfare and food safety, and which are supported by both citizens and the poultry sector.

The objective of this study was to get insight into the role of moral values regarding the dilemma between animal welfare and public health risks. Therefore, we surveyed citizens and poultry farmers regarding (1) moral values concerning chickens and consumers and (2) arguments based on moral values concerning the judgement of the dilemma between broiler welfare and food safety.

Methodology

Survey

A survey was done among Dutch citizens and poultry farmers through March and April 2014. The citizens were approached by CentERdata (www.centerdata.nl), a research institute specialized in online survey research by means of the CentERpanel. The CentERpanel is a representative sample of the Dutch population. CentERdata approached 3,344 CentERpanel participants, of whom 2,373 (71.0%) started to fill in the questionnaire and 2,259 respondents (67.6%) completed the questionnaire. None of the CentERpanel citizens was poultry farmer. Poultry farmers were invited to fill in the questionnaire by a digital newsletter of the Dutch organization of poultry farmers (NOP) and an article on the website of a Dutch poultry magazine (Pluimveehouderij). One hundred poultry farmers who kept poultry professionally completed the questionnaire.

Moral values

In judgements regarding animal husbandry various moral values and interests of different entities are considered. Mepham (2000) developed a framework that includes principles of common morality, i.e. wellbeing, autonomy, and justice, for the perspective of various stakeholders, i.e. animals, consumers, producers and environment. Because judgements regarding animals depend also on whether or not people consider animals to have moral value, Cohen *et al.* (2009) added to their version of such framework human-animal hierarchy and animal value. Based on the frameworks of Mepham (2000),

Cohen *et al.* (2009) and Bergstra *et al.* (2015) we formulated 13 statements regarding moral values concerning chickens and consumers for the dilemma between broiler welfare and food safety (Table 1). Participants were asked to score to what extend they agreed with these 13 statements on a five-point Likert scale ranging from strongly disagree (1) to strongly agree (5).

Arguments

A case, describing a dilemma between broiler welfare and food safety was formulated: 'A poultry farmer keeps his broilers in a conventional husbandry system: on litter in a barn, and the broilers are slaughtered at 42 days of age. The farmer would like to change from the conventional system to a free-range barn system. In a free-range barn system the broilers are offered more space, enrichment (straw) and will be slaughtered at 56 days of age. 10% of conventional broiler meat is infected with the intestinal bacterium *Campylobacter*. In the new system, the *Campylobacter* contamination of broiler meat will increase to 30%, due to the longer growing period of the broilers. In humans contamination with *Campylobacter* may cause intestinal infections. Contamination can be prevented through hygienic food handling and thoroughly cooking broiler meat.'

Based on the selected moral values (Table 1) and the relevance for the case, six arguments were formulated (Table 2). Participants were presented the case and were first asked to make a judgement about the case (agree, nor agree/nor disagree or disagree). Second, they were asked to rate the importance of the six

Table 1. Mean scores for statements regarding moral values according to (subgroups of) citizens (n=2,259) and (subgroups of) poultry farmers (n=100)(1=strongly disagree, 5=strongly agree).

	Citizens				Poultry farmers			
	Total	F[1]	N	A	Total	F	N	A
Chickens experience pain	4.2	4.1^{a}	4.1^{a}	4.4^{b}	4.2	4.1	4.2	4.4
Chickens experience pleasure and boredom	3.8	3.6^{a}	3.7^{a}	4.0^{b}	3.6	3.5	3.5	3.9
Humans are superior to chicken	3.9^{*}	4.0^{a}	3.8^{a}	3.8^{b}	4.7^{*}	4.6	4.7	4.6
Function of chicken is meat and eggs for humans	3.3^{*}	3.4^{a}	3.4^{a}	3.0^{b}	4.3^{*}	4.6	4.4	4.1
Chickens have intrinsic value	3.5^{*}	3.3^{a}	3.4^{a}	3.7^{b}	3.3^{*}	2.8^{a}	3.3	3.6^{b}
Chickens should reach their natural lifespan	3.4^{*}	3.2^{a}	3.4^{b}	3.5^{b}	2.2^{*}	2.0	2.2	2.5
Chickens should express natural behaviour	4.0^{*}	3.7^{a}	3.9^{b}	4.2^{c}	3.5^{*}	3.3^{a}	3.3^{a}	3.9^{b}
Chickens have right to a natural life	3.9^{*}	3.6^{a}	3.8^{b}	4.0^{c}	2.9^{*}	2.9	2.7^{a}	3.2^{b}
Consumers themselves should reduce health risks of chicken products	4.0^{*}	3.9^{a}	3.8^{a}	4.1^{b}	4.5^{*}	4.4	4.6	4.3
Consumers are by purchasing behaviour co-responsible for the welfare of chickens	3.9^{*}	3.7^{a}	3.8^{a}	4.2^{b}	4.6^{*}	4.3	4.6	4.8
Chickens or chicken products should not affect human health	4.4	4.6^{a}	4.4^{b}	4.5^{a}	4.4	4.6	4.4	4.3
A farmer should treat a chicken when it is ill	3.9^{*}	3.9	3.9^{a}	4.0^{b}	4.2^{*}	4.3	4.3	4.0
A farmer should respect the intrinsic value of chickens	4.0^{*}	3.8^{a}	3.8^{a}	4.1^{b}	4.2^{*}	3.9	4.3	4.2

[1] F=in Favour of the change; N=Neutral to the change; A=Against the change from a conventional to a free-range barn system.

[*] Means of citizens differ from means of farmers at $P<0.05$.

[a,b,c] Means within a row with different superscripts indicate a significant difference ($P<0.05$) in means between groups of citizens or between groups of farmers (Post-hoc Games Howell multiple comparisons test).

Table 2. Importance of arguments in the judgement of a dilemma in broiler husbandry according to groups of citizens (n=2,259) and groups of poultry farmers (n=100).

	Citizens			Poultry farmers		
	F[1] n=286	N n=1,023	A n=950	F n=23	N n=47	A n=30
In free-range systems broilers live longer	3.0[a]	3.4[b]	3.7[c]	1.7[a]	2.4[b]	3.1[c]
In free-range systems broilers can express their natural behaviour better	3.1[a]	3.7[b]	4.3[c]	2.6[a]	2.7[a]	3.6[b]
Free-range systems are better for broiler welfare	3.1[a]	3.7[b]	4.3[c]	2.3[a]	2.8[a]	3.7[b]
The infection risk is small if during preparing chicken meat one is working hygienic and cooking well the meat	3.7[a]	4.0[b]	4.3[c]	3.3[a]	4.1[b]	4.5[b]
Human health is more important than chicken welfare	4.1[a]	3.7[b]	3.4[c]	4.6[a]	4.1	3.8[b]
Consumers have right to chicken meat with minimal bacterial contamination	4.3[a]	3.9[b]	3.5[c]	4.7[a]	3.9[b]	3.5[b]

[1] F=in Favour of the change; N=Neutral to the change; A=Against the change from a conventional to a free-range barn system.
[a,b,c] Means within a row with different superscripts differ indicate a significant difference ($P<0.05$) in means between groups of citizens or between groups of farmers (Post-hoc Games Howell multiple comparisons test).

arguments between 1 and 5 (1=very unimportant, 5=very important for their judgement) in favour of or against changing the system.

Statistical analyses

Data were analysed using SPSS 19.0. Analysis of variances was used to explore the association between both the mean level of agreement with statements regarding moral values and the mean importance scores of the arguments and (1) stakeholder group, and (2) judgement. One-way analysis of variance (ANOVA) was done if variances were homogeneous according to Leven's test. If variances were not homogeneous, the Welch test was used. If the effect of the judgement was significant, the post-hoc Games-Howell test for multiple comparisons was done to analyse differences between the in Favour of change, Neutral to the change, and Against the change group.

Results

Moral values

Table 1 shows the mean scores for the statements regarding moral values in poultry husbandry according to citizens and farmers. Citizens scored the statement 'chicken or chicken products should not affect human health' highest, while farmers scored the statement 'humans are superior to chicken' highest. The statement 'function of chicken is meat and eggs for human' was scored lowest by citizens whereas farmers scored 'chickens should reach their natural lifespan' lowest. Citizens and farmers did not score the statements 'chicken experience pain', 'chicken experience pleasure and boredom' and 'chickens or chicken products should not affect human health' differently ($P>0.05$). The mean scores for all other statements regarding moral values differed significantly ($P<0.05$) between citizens and farmers.

Of the citizens 286 (42.1%) favoured the change from a conventional husbandry systems to a free-range barn system, while 950 (12.7%) were against the change (Table 1). The remainder 1,023 (45.3%) of the citizens were neutral to the change from a conventional to a free-range barn system. Relatively less farmers (30.0%) than citizens favoured the change and more farmers (23.0%) were against the change, while almost half (47.0%) of them were neutral to the change. The group of citizens in favour of change scored all statements differed significantly (P<0.05) between citizens who disagreed and citizens who agreed with the change to a free-range barn system. Citizens and farmers who agreed with the change agreed significantly stronger with 'chickens have intrinsic value', 'chickens should express natural behaviour', and 'chickens have right to a natural life'. Citizens who disagreed and who agreed with the change did not score the statement 'chickens or chicken products should not affect human health' differently (P>0.05).

Arguments

The importance of arguments in the judgement of the dilemma as presented in the case, are shown in Table 2. Respondents who favoured the change from a conventional system to a free-range system scored the importance arguments concerning broiler welfare higher than respondents who were against the change of system. This 'in favour of change' group scored also the argument regarding diminishing the infection risk by adequate food preparation higher than the disagree group. Respondents who were against the change scored arguments regarding human health and consumers as more important than respondents who favoured the change from a conventional to a free-range barn system.

Discussion

This study is a first step to get insight into moral values regarding broiler husbandry and regarding arguments based on moral values involved in a dilemma between broiler welfare and food safety. The results show that citizens and farmers shared moral values regarding chickens being sentient animals: chickens experience pain, pleasure and boredom. This may be explained by citizens' and farmers' shared social and cultural background and knowledge about mental capacity of chickens (Cohen *et al*,. 2012). However, most moral values were scored differently between citizens and farmers, and within citizens between the group that disagreed and the group that agreed with the change from a conventional to a free-range barn system. These differences in moral values might be caused by the fact that we described the moral values regarding chickens and not in general regarding animals. By adding a specific animal species, i.e. chickens, we added a context. A context shapes and re-values moral values (Cohen *et al.*, 2016) and respondents may have weighed values regarding chickens against personal values. For example, for farmers, chickens are means to achieve their income and therefore they may have valued broilers mainly for their functional value. Respondents who favoured the change from a conventional systems to a free-range barn system scored moral values related to wellbeing, intrinsic value and justice regarding chicken higher than respondents who were against the change. Bergstra *et al.* (2015) showed that moral values of citizens and farmers could not predict their attitudes towards sow husbandry. However, our results show that in citizens moral values were associated with the judgement of the dilemma between broiler welfare and food safety.

In both citizens and famers the judgement of the dilemma was associated with the importance of the arguments in the context of the case. On the one hand, respondents who favoured the change of husbandry system scored the arguments concerning broiler welfare as more important than respondents who were against the change. On the other hand, respondents who were against the change scored the importance of human health and right on minimal bacterial contamination higher than respondents who favoured the change.

Current broiler husbandry systems are developed by those directly involved in the sector, such as farmers, veterinarians and animal scientists. A gap between perceptions of citizens, farmers and veterinarians concerning broiler husbandry exists (Van Asselt *et al.,* 2015). To gain more societal support, concerns of citizens should also be considered (Weary *et al.,* 2016). This study provides input for deliberation on broiler husbandry especially on the dilemma between broiler welfare and food safety. However, moral values and arguments are not shared by all respondents. This study should be further analysed to reveal whether various clusters of citizens exists (Cohen *et al.,* 2012). With the information form this study a broad moral deliberation could be set up between the different stakeholders to define characteristics of future husbandry systems that can count on support of both the poultry sector and citizens. For deliberation, participants should be carefully chosen from the various stakeholder groups to cope with value diversity.

References

Bergstra, T.J., Gremmen, B. and Stassen, E.N. (2015). Moral values and attitudes toward dutch sow husbandry. Journal of Agricultural and Environmental Ethics 28: 375-401.

Bessei, W. (2006). Welfare of broilers: a review. World's Poultry Science Journal 62: 455-466.

Bouwknegt, M., Van de Giessen, A.W., Dam-Deisz, W.D.C., Havelaar, A.H., Nagelkerke, N.J.D. and Henken, A.M. (2004). Risk factors for the presence of Campylobacter spp. in Dutch broiler flocks. Preventive Veterinary Medicine 62: 35-49.

Childress, J.F., Faden, R.R., Gaare, R.D., Gostin, L.O., Kahn, J., Bonnie, R.J., Kass, N.E., Mastroianni, A.C., Moreno, J.D. and Nieburg, P. (2002). Public health ethics: mapping the terrain. Journal of Law, Medicine and Ethics 30: 170-178.

Cohen, N.E., Brom, F.W.A. and Stassen, E.N. (2009). Fundamental moral attitudes to animals and their role in judgment: an empirical model to describe fundamental moral attitudes to animals and their role in judgment on the culling of healthy animals during an animal disease epidemic. Journal of Agricultural and Environmental Ethics 22: 341-359.

Cohen, N.E., Brom, F.W.A. and Stassen, E.N. (2012). Moral convictions and culling animals: a survey in the Netherlands. Anthrozoos: A Multidisciplinary Journal of The Interactions of People and Animals 25: 353-367.

Cohen, N.E. and Stassen, E.N. (2016). Public moral convictions about animals in the Netherlands: culling healthy animals as a moral problem. In: Meijboom, F.L.B. and Stassen, E.N. (eds.) The end of animal life: a start for ethical debate. Wageningen Academic publishers, Wageningen, the Netherlands.

Mepham, B. (2000). A framework for the ethical analysis of novel foods: the ethical matrix. Journal of Agricultural and Environmental Ethics 12: 165-176.

Rollin, B.E. (2007). Cultural variation, animal welfare and telos. Animal Welfare 16: 129-133.

Sommer, H.M., Heuer, O.E., Sørensen, A.I.V. and Madsen, M. (2013). Analysis of factors important for the occurrence of Campylobacter in Danish broiler flocks. Preventive Veterinary Medicine 111: 100-111.

Te Velde, H., Aarts, N. and Van Woerkum, C. (2002). Dealing with ambivalence: farmers' and consumers' perceptions of animal welfare in livestock breeding. Journal of Agricultural and Environmental Ethics 15: 203-219.

Van Asselt, M., Ekkel, E.D., Kemp, B. and Stassen, E.N. (2015). Best broiler husbandry system and perceived importance of production aspects by Dutch citizens, poultry farmers and veterinarians. In: Dumitras, D.E., Jitea, I.M. and Aerts, S. (eds.) Know your food: food ethics and innovation. Wageningen Academic Publishers, Wageningen, the Netherlands, pp. 138-143.

Weary, D.M., Ventura, B.A. and Von Keyserlingk, M.A.G. (2016). Societal views and animal welfare science: understanding why the modified cage may fail and other stories. Animal 10: 309-317.

42. Justifying veterinary interventions: the normativity of health concepts

M. Huth
Messerli Research Institute, Unit Ethics and Human-Animal-Studies, University of Veterinary Medicine Vienna/University of Vienna/Meduni Vienna, Vienna, Austria; martin.huth@vetmeduni.ac.at

Abstract

The Messerli Research Institute in Vienna is establishing courses in veterinary ethics. One central concern is the analysis of the normative implications of core concepts in veterinary medicine. Veterinary medicine is – like human medicine – not to be considered as applied science but as practical discipline with an immanent normative dimension. The paper will clarify the moral significance of concept of health in its diverse conceptualizations. As disease is usually regarded as deviation from a certain normality that is embedded in implicit frames of interpretations (conceptions), health builds a powerful teleological figure. It is the aim and legitimization of any veterinary intervention and shows a multi-dimensionality that raises moral questions. First, it comprises different groups of subjects of care: different animals but also humans qua consumers of animal products, persons concerned with public health, etc. Ethical problems can emerge because of tensions between different needs, interests and values. Secondly, the human perspective frames the problem of health in different fields of human-animal interaction diversely. Different conceptions of health legitimize different kinds of veterinary practices and interventions: (A) The functionalist model relates health to productivity. The normality of e.g. a certain quantity of milk serves as basis for the assessment of dairy cows as ill qua 'unproductive'; (B) The model of damages/diseases focuses particularly on objective physiological functions in their normal state; (C) The sentientistic paradigm focuses mainly on the perspective of the animal patient. Particularly in pets, the indication for veterinary intervention is the alleged experienced well-being and not the objective damage. In this case, well-being trumps over functioning and even life if we think about the indication or alleged obligation for euthanasia. An analysis of the internal differentiation of health reveals different implicit assumptions about veterinary intervention and could support ethical decision-making.

Keywords: ethics, responsibility, internal normative infrastructure

Introduction

The need for ethical reflection in and regarding veterinary practice is shown impressively by the qualitatively surveyed data that over 90% of the veterinarians experience moral dilemmas at least once a week (Magalhaes-Sant'Ana, 2014: 350). In most crucial fields of human-animal interactions veterinarians are dealing with ethically relevant questions – they are at the sharp end when it comes to therapeutic decisions, welfare problems, euthanasia, culling, and other ethically relevant issues.

But as Gray indicates, the implementation of this subject in veterinary education 'has been a creeping progression rather than a huge leap' (Gray, 2014: 590). One can identify at least two reasons for this phenomenon: First, there occurs an alleged gap between 'hard' sciences and 'soft' humanities so that the two spheres of knowledge appear disconnected from each other (cf. Rollin, 1998). The self-understanding of veterinary medicine seems therefore often to rest on the self-misunderstanding of being applied science. But the notion of medicine as more than applied science (due to the central role of clinical encounters and complex responsibilities) and thus as a practical discipline that has an immanent normative dimension is part of the inventory of virtually all current theories of biomedical ethics (cf. Huth, 2011). Second, we can see that veterinary ethics is often in reduced to considerations

EurSafe 2016 – DOI 10.3921/978-90-8686-834-6_42, © Wageningen Academic Publishers 2016

of animal welfare and of health interpreted as prerequisite for welfare (cf. Broom, 2007). One of only a few existing textbooks on the topic contains the following sentence: 'So core aims of veterinary work overlap considerably, if not entirely, with animal concerns' (Yeates, 2013: 1). Notwithstanding that animal concerns must play a crucial role in veterinary medicine one has to acknowledge that these might be in tension with other concerns like human (public) health (e.g. zoonoses), practical limitations in dealing with animal pests, economy and others (cf. Mossop and Cobbs, 2013). Ethical considerations focusing too much or even exclusively on animal welfare are therefore too narrow and do not support ethical decision-making in veterinary practice sufficiently.

Health and disease

A broader and more fruitful approach to deal with veterinary ethics is the analysis of the internal normative infrastructure (*ethos*) of veterinary medicine that is present in the notion of health – and its opposite disease/illness/sickness (for a disambiguation of these concepts see the next paragraph). As Nordenfelt indicates, the theoretical and ethical knowledge of health and disease in animals is neither in veterinary medicine (cf. the quote by Yeates above) nor in related disciplines very much elaborated: 'The animal science analysis has focused on the notion of welfare instead of the notions of health and quality of life.' (Nordenfelt, 2006: ix)

As prominent theorists in biomedical ethics have pointed out – and this applies to veterinary medicine, too –, health can be regarded as a 'teleological' figure because it builds the aim and basis for the obligation to respectively legitimation of any kind of treatment and intervention in the body of the one who is recognized as patient (cf. Engelhardt jr., 1975; Paul, 2006; Schramme, 2012). In human medicine, health and disease are controversial issues: What is a pathology? How can we deal with the gap between the generalization of symptoms to the definition of a certain disease and the individual manifestation? Is the crucial basis for medical interventions the respective disease (the biological dysfunction, cf. Boorse, 1975; for animals cf. Broom, 2007) or the illness (the individual lived experience of feeling ill, cf. Clouser *et al.*, 1981)? Is there a positive definition of health that relates to a particular kind of wellbeing (cf. Gadamer, 1993: 8; Hunter *et al.*, 2013) or is it just the absence of any disease (cf. Boorse, 1975)?

These controversies are also crucial in the field of veterinary medicine but are to be seen in an even more complex setting as 'health' and 'disease' are embedded in very different practical contexts. As Schramme indicates (for human medicine), one can distinguish between the concept as bundle of criteria for the definition of a pathology and conceptions of the pathology as to be treated, to be ignored, to be dealt with in special ways, etc. (cf. Schramme, 2012: 11) In human medicine we see one big controversy between naturalistic and normative conceptions of disease. The naturalistic paradigm interprets disease as value-free term that describes it as deviation from a biological or physiological norm (cf. Boorse, 1975). For the normative approach, disease (or lived illness) as deviation from normality is normatively significant because it presents an existential impairment and should thus be avoided (cf. Engelhardt, 1975; Schramme, 2012; Huth, 2011).

In veterinary medicine we have to deal with the coexistence of different conceptions or implicitly powerful models of health: There are differences regarding the recognized obligations and legitimizations of medical interventions in livestock, wild animals, companion animals that seem to stem from the different human-animal-relations we live in. The normality that is constitutive for our notion of health, the obligation of care and the responsibility for humans and animals is in itself multi-dimensional.

One health – many healths?

As the rise of the concept of One Health in the last decade shows, there is an increasing notion of the connection of human and animal health. Basically, One Health was an initiative to focus on the transferability of medical knowledge from animals to humans (AVMA, 2008). In the last years this has been shifted at least by some authors to mutual transfer and as well to a deeper ethical analysis of health in humans connected to animals (cf. e.g. Nieuwland and Meijboom, 2015; Verweij and Meijboom, 2015). The shared environmental context, our moral obligations towards animals and their well-being but also the inevitable instrumentality of animal health for human health (cf. Nieuwland and Mejboom, 2015: 134) are the background for veterinary medical interventions and any kind of clinical encounter vets are dealing with. There are at least potentially tensions between the different aims of veterinary medicine like food safety, public health, animal welfare or environment protection; obviously we do not deal with conflicting unidimensional interests but with coexisting and conflicting values in our lifeworld (cf. Verweij and Meijboom, 2015: 144-146.). The manifest inherent normative infrastructure of our lifeworld and the often implicit ethical knowledge (Varela, 1999) predetermines the structure of veterinary medicine and the nature of possible ethical questions. Medicine, science and society are accordingly products and producers of the scientific, medical and cultural context in which clinical interventions are embedded (cf. Paul, 2006: 66). The mentioned differences of diverse human-animal relations constitute an allocation of the awareness and recognition of animal vulnerability and health that is pluri-dimensional and has always the potential of shifting from tension to conflict.

Interpersonal can turn into intrapersonal conflicts (cf. Verweij and Meijboom, 2015: 148) and put kind of a moral pressure on veterinarians. The profession constitutes a certain perspective of the ones who are at the sharp end of treating animals extremely differently as meat providers, data providers, family members, therapeutic tools, etc. knowing very well about 'natural needs' of these beings that exceed the lifeworld significance for us. Different values are not always in a clear hierarchy, even not for one and the same person. To support deliberation and decision-making against the background of these conflicts it is indicated to reveal the inherent, often implicit structure of the different conceptualizations of health that are at stake in veterinary medicine. Veterinarians would not understand themselves as mere hench(wo)men of the societal conditions. But to go into deliberation requires an awareness of the respective own viewpoint (cf. Maghalaes-Sant`Ana, 2014). This awareness is only possible if the pluridimensionality of health in veterinary medicine is made explicit. In the last chapter of this paper three powerful models of health will be presented and analyzed regarding their content and their tension between them.

Three concepts of health in veterinary medicine and their ethical implications

It is most crucial to make clear that the interpretations of health and disease presented in the following are not distinct and cannot be seen as clearly separated from each other. They serve as models (understood as conceptions) for a concise description and interpretation of the inherent normative infrastructure of veterinary medicine, a kind of tacit or implicit morality (cf. Varela, 1999) that has (ethically) important influences on different justifications and consequently different treatments of animals. It is neither intended to describe health as it 'is' or as veterinarians would intentionally stick to nor to elaborate the proper definition of health. Basically, the aim of the descriptions is to show that the motifs of veterinary interventions are embedded in different societal contexts constituting different normalities of animal health and predetermine the individual responsibility in the respective situation. The tensions between the three conceptions of health/disease do not lead to the assumption that one of them is (morally) most adequate but should only be seen as given fact ethical consideration of veterinary practice should deal with. Moreover, this is not a finite list of health conceptions but presents only three crucial (structural) models instead of empirical findings.

The functionalist model (A) relates health to productivity, food safety, data validity, etc. In the context of food production the mentioned instrumentality of animal health is crucial and overrides the patient's disease (the biological functionality) and illness (the experience of being ill) in different respects. Notwithstanding that e.g. livestock farmers would also consult a veterinarian when an animal appears as suffering (cf. model C), many interventions are made to support and increase outcomes; the conceptualizations are not mutually exclusive. One example would be the treatment of the bovine ovarian cyst that hinders dairy cows from getting pregnant and thus producing milk. The cyst does obviously not lead to any experienced illness. The legitimization of clinical interventions (surgery) is neither provided by animal welfare nor by an obligation to support and prolong the cow's life in terms of biological functions but by the keeper's economic interest respectively the normality of the average lactation performance.

The support of function/productivity can conflict with questions of well-being and of 'natural functions'. The example of the bovine ovarian cyst shows one kind of conflict between different interests and/or values because here an invasive intervention (the cow's cost) can solve the farmer's problem (the farmer's benefit). Similarly, we find value conflicts when performance-related diseases are treated only symptom-oriented or even only in regard of the support of productivity. Sticking to the example of dairy cows, the increase of the average lactation performance from the 1950's to current is roughly 80% (cf. Martens/ Breves, 2005). This leads to production-related increase of diseases (and accompanying illnesses) like mastitis, fertility disorders, foot disorders, or metabolic disorders. Furthermore, in environments that are very production-oriented (in conflict with an orientation towards animal welfare and natural needs) we find an increased number of technopathies and e.g. stress-related disorders of behaviour.

One can say that the respective animal appears as patient (and as being-to-be-treated) particularly because of her instrumental value. The addressee of the clinical intervention is of course e.g. the dairy cow but the justification for the intervention and the treatment mandate stems first and foremost from the animal owner's need to produce on a normal level. This is the case in any kind of supportive treatment but also if the decision is made for euthanasia.

The objective model (B) (as opposition to lived illness) is – within biomedical ethics – often criticized as reification of the lived body, of the patient and of the nosology (cf. Engelhardt, 1975). Disease is the (value-free) dysfunction of a natural capacity (as described and defended by Boorse, 1977). Basically, one has to acknowledge that modern Western medicine relies on the objectification and 'scientification' of the body that provides a certain methodology and the well-known progresses (Huth, 2011; Schramme, 2012). Regarding animal health, the objective model is widely accepted (cf. Broom, 2007; Nordenfelt, 2006). One can suspect that this is the case because of the scientific common sense Rollin detected in 'The Unheeded Cry' (1998). Knowledge about animal mental states is considered as vague or unscientific and might thus be pushed to the margins when it comes to the legitimation of clinical interventions. In this understanding, the treatment is oriented towards physiological functions.

But this is of high significance if not necessity when we deal with questions that relate to public health (particularly zoonosis), food safety or environmental protection. As the example of culling shows, the disease is not only generalized in terms of statistical definition of normality/pathology (cf. Canguilhem, 2012; Schramme, 2012) but even abstracted from the respective animal body – and *a forterori* from the animals' lived experience of illness. The individual does not appear as patient anymore insofar the aim of the intervention might neither be cure nor euthanasia in terms of a merciful death (although one can respect the animals' suffering in mild killing) but the eradication of the disease per se. The individual might even be healthy but only in danger of an infection to be a candidate for culling.

The sentientistic model (C) differs fundamentally from the ones described above. Especially in companion animals, the treatment of individuals is often justified by the assumed suffering or wellbeing of the individual – the *telos* of interventions is maximizing pleasure and minimizing pain. The centre of the clinical intervention is the patient's perspective. One has to acknowledge that the limitation of this concept is to be found not at least in the economic situation of the animal owner but basically one should care and pay for treatments for her animal as long as it is affordable and as long we would not consider the animal better off dead. And even if the decision is made towards euthanasia as the ultimate veterinary treatment the highest aim is killing without suffering. This is shown impressively by the increasing literature regarding an ethically justifiable practice of euthanizing with strict normative implications like the ones to prevent the animal from noticing the dying process in any way, to use the proper active substances, prepare a comforting surrounding, and also how to behave as animal owner during the procedure (cf. Hofmann, 2013). Whereas in the models A and B the alleged patient will play a minor role due to the fact that the individual is in tendency reified as a resource (A) or the disease is separated from the individual so that it hardly occurs as patient anyway (B), in model C the perspective of the animal is the centre of the justification of therapies or euthanasia. While in livestock production the role of mental states and interests in animals is sometimes estimated very little, in the sentientistic paradigm it is presupposed that we know how to act in the patient's best interest – and be it the decision to kill the patient particularly to be morally good, to fulfil a widely recognized obligation. (cf. Grimm and Huth, 2016). It might be unclear and debatable whether a certain constitution of an individual legitimizes euthanasia but in general it is not questioned that euthanasia is a proper means to end suffering in animals and that there is even an obligation to redeem the terminally ill and suffering patient. This builds a crucial part of the normative infrastructure (normality) of treating companion animals but can of course potentially play a role in livestock or laboratory animals, too.

Conclusions

The three models of health/disease do neither serve as description of the reality of different perspectives veterinarians take nor do they – although it is clear that there might be a strong tension between them – mutually exclude themselves. The aim of the explication of the inherent normative infrastructure of veterinary medicine is to show the presuppositions of taking responsibility as a veterinarian in different contexts. These contexts exist due to the fact that the tasks, obligations and responsibilities of vets are very diverse and dependent on the societal reality of different human-animal relations and the underlying, often implicit values. Thus, a relational account of responsibilities for health in its different notions is required to support deliberation and decision-making. Otherwise the moral claim for stringency in veterinary practice would lack the power to support practically and ethically. The benefit of an analysis of the different conceptions of health (as *telos* of medicine) is the increased awareness of the internal normative infrastructure of veterinary medicine to enable veterinarians to reflect about their moral problems they face on a regular base and to enable them to qualify their social role and responsibility as professionals properly.

References

AVMA (2008). One health: a new professional imperative. Available at: http://tinyurl.com/z2wrn3a.

Boorse, C. (1977). Health as a theoretical concept. Philosophy of Science 44: 542-573.

Broom, D. (2007). Welfare in relation to feelings, stress and health. Revista electrónica de Veterinaria 8(12B).

Canguilhem, G. (2012). Das Normale und das Pathologische. August Verlag, Berlin, Germany, 348 pp.

Engelhardt jr., H.T. (1975). The concepts of health and disease. In: Engelhardt jr., H.T. and Spicker, S.F. (eds.) Evaluation and explanation in the biomedical sciences. Springer, Dordrecht, pp. 125-141.

Gadamer, H.-G. (1993). Über die Verborgenheit der Gesundheit. Aufätze und Vorträge. Suhrkamp, Frankfurt am Main, Germany, 215 pp.

Grimm, H. and Huth, M. (2016). The 'significance of killing' versus the 'death of an animal'. In: Meijboom, F.L.B. and Stassen, E.N. (eds.) The end of animal life: a start for an ethical debate. Ethical and societal considerations on killing animals. Wageningen Academic Publishers, Wageningen, the Netherlands, pp. 79-101.

Hofmann, H. (2013). Tieren beim Sterben helfen – Euthanasie in der Tierarztpraxis. Vetpress, Butzbach, Germany, 96 pp.

Hunter, J., Marshall, J., Corcoran, K., Leeder, S. and Phelps, K. (2013). A positive concept of health – interviews with patients and practitioners in an integrative clinic. Complementary Therapies in Clinical Practice 19(4): 197-203.

Huth, M. (2011). Den Anderen behandeln und betreuen. Phänomenologische Ansätze zu Grundfragen der Medizin. Alber, Freiburg im Breisgau, Germany, 336 pp.

Maghalaes-Sant'Ana, M. (2014). Ethics teaching in European veterinary professionalism. Veterinary Record 175: 592.

Martens, H. and Breves, G. (2005). Physiologische Grenzen der Hochleistungskuh. In: Engelhardt, W. (ed.) Physiologie der Haustiere. MVS Medizinverlage, Stuttgart, Germany, pp. 648-652.

Mossop, L.H. and Cobbs, K. (2013). Teaching and assessing veterinary professionalism. Journal of Veterinary Medical Education 40(3): 223-232.

Nieuwland, J. and Meijboom, F.L.B. (2015). One health as a normative concept: implications for food safety at the wildlife interface. In: Dumitras, D.E., Jitea, I.M. and Aerts, S. (eds.) Know your food. Food ethics and innovation. Wageningen Academic Publichers, Wageningen, Netherlands, pp. 132-137.

Nordenfelt, L. (2006). Animal and human health and welfare. a comparative philosophical analysis. Cabi, Wallingford, UK, 186 pp.

Nordenfelt, L. (2007). The concepts of health and disease revisited. Medicine, Health Care and Philosophy 10: 5-10.

Paul, N. (2006). Gesundheit und Krankheit. In: Schulz, S., Steigleder, K., Fangerau, H. and Paul, N. (eds.) Geschichte, Theorie und Ethik der Medizin. Eine Einführung. Suhrkamp, Frankfurt am Main, Germany, pp. 131-142.

Rollin, B.E. (1998). The unheeded cry. Animal consciousness, animal pain and science. Iowa State University Press, Ames, USA, 330 pp.

Schramme, T. (2012). Einleitung: die Begriffe 'Gesundheit' und 'Krankheit' in der philosophischen Diskussion. In: Schramme, T. (ed.) Krankheitstheorien. Suhrkamp, Frankfurt am Main, Germany, pp. 9-37.

Varela, F. (1999). Ethical know-how: action, wisdom, and cognition. Stanford University Press, Stanford, USA, 85 pp.

Verweij, M. and Meijboom, F.L.B. (2015). One health as collective responsibility. In: Dumitras, E.D., Jitea, I.M. and Aerts, S. (eds.) Know your food. Food ethics and innovation. Wageningen Academic Publichers, Wageningen, the Netherlands, pp. 144-149.

Yeates, J. (2013). Animal welfare in veterinary practice. Wiley-Blackwell, Oxford, UK, 200 pp.

43. Conflicting norms as the rule and not the exception – ethics for veterinary officers

C. Dürnberger[*] *and K. Weich*
Messerli Research Institute, Unit of Ethics and Human Animal Studies. VetMedUni Vienna, University of Vienna, MedUni Vienna, VetMedUni Vienna, Veterinärplatz 1, 1210 Vienna, Austria; christian.duernberger@vetmeduni.ac.at

Abstract

Veterinary officers can be considered as an epitome of an ethically challenging profession. They regularly have to deal with decisions that incorporate moral aspects. Against this background, the Messerli Research Institute carried out a project in order to reflect the ethical framework of veterinary officers. The paper presents and discusses selected results of this project, which brought together veterinary officers and representatives of various scientific disciplines. The paper works out the conflict between 'animal's interests' and 'public interests' as one of the vital moral conflicts in the daily work of veterinary officers. It reflects on the question of how to deal with this conflict by suggesting that it must be accepted and understood as essential part of the job – and not e.g. as a deficit of personal skills. Finally, it reflects on the role of ethics in this field by suggesting that a descriptive approach could be helpful and appropriate especially when ethics have to deal with ethical dilemmas of veterinary officers.

Keywords: veterinary ethics, human animal interaction, professional ethics

Vethics – ethics for veterinary officers

Veterinary officers (in German: *Amtstierärzte*) are veterinarians working for the veterinary administration. Based on specific education for the civil service, their scope of duties includes all areas of the veterinary service, such as animal disease control, animal protection (according to the Austrian animal protection law), official controls on products of animal origin intended for human consumption, etc. In this context, veterinary officers can be considered as an epitome of an ethically challenging profession. They regularly have to deal with decisions that incorporate moral aspects; one need only think of the euthanasia of domestic animals, the monitoring of animal welfare of farm animals from stable to slaughter or the culling of livestock when faced with an animal disease. Moreover, the daily work of veterinarians takes place against the background of conflicting developments and perspectives. Some animals are considered family members but in other cases animals are treated as mere production units, as in the context of farming. Thereby, veterinary officers often find themselves in a field of tension between economic limitations, animal welfare, political interests, legal requirements and last but not least very emotional public debates (cf. Tannenbaum, 1995). In this field of tension, an ethical reflection of the profession and its corresponding responsibilities and values is becoming more and more important. In general, teaching ethics to veterinarians is gaining importance since ethics allows them – or should allow them – to 'understand their different roles and responsibilities as well as appreciating the expectations of different stakeholders in modern society' (Magalhaes-Sant'Ana *et al.*, 2009: 197). However, veterinary ethics (as moral philosophical reflection on the professional field of veterinary medicine) must still be regarded as a rather new discipline. Against this background, the project 'Vethics – Professional Ethics for Veterinary Officers' was launched at the Messerli Research Institute in Vienna in 2012. The Austrian Ministry of Health funded the project. The ambition was not only to identify the wide range of ethical issues in the professional life of veterinary officers and explore the ethical dimensions of this profession, but also to create an 'open space' where veterinary officers have the possibility to reflect on their work

EurSafe 2016 – DOI 10.3921/978-90-8686-834-6_43, © Wageningen Academic Publishers 2016

away from the daily stress in the office and in exchange with other disciplines (such as Philosophy, Psychology or Law).

During the project's planning process, the project was presented at three conferences in different parts of Austria. The audiences were invited to discuss and identify pressing issues. The official veterinarians were invited to hand in further relevant issues per e-mail. Within a timeframe of six months, three main topics have been identified together with Austrian veterinary officers:
1. killing animals (euthanizing, slaughtering and culling);
2. using animals (farm animals);
3. anthropomorphizing animals (psychosocial reasons for animal welfare problems, animal hoarding).

During the project, eight one-day workshops were organized. As a rule, five to ten veterinary officers and two to five representatives of other disciplines (including at least two ethicists who were chairing the meetings and/or giving inputs) were participating. Further, three events open to the public were held, including an international two-days-conference in September 2015. The workshops were audio recorded and the material was analysed in orientation on qualitative content analysis (cf. Mayring, 2003) by identifying relevant topics and corresponding argumentations in the specific sessions. Analytic questions of the interpretation were amongst others: Which arguments occur in the discussions of a specific case study? Which norms and values are inherent in these arguments? Further, which cultural frameworks, narrations, images or metaphors can be identified?

Dealing with conflicting norms

In order to reflect on the role of veterinary officers in the mentioned three fields, case studies were developed in the project's planning process by the project team. These case studies were based on issues, experiences and challenges the official veterinarians presented as pressing (see above). One of the case studies in the context of 'topic 1 – killing animals (subcategory: euthanizing)' told the story of an 'aggressive dog':

> A dog repeatedly exhibits unusual aggressive behaviour. All therapeutic attempts failed. After the dog seriously hurt its owner, the animal is taken in custody by the veterinary officer – and is euthanized.

The main task of the participants was to structure and analyse this decision from an ethical perspective, taking into account the professional responsibility of veterinary officers and posing the question if a decision against euthanasia would have been the better one. For this purpose, the Ethical Matrix (Mepham *et al.*, 2006) was used, especially in order to identify the relevant stakeholders such as the affected dog, the veterinary officer involved in this case, the management and the staff of the animal shelter, which would accommodate and look after the dog if the veterinary officer is doing without euthanasia, the public (e.g. neighbourhood), the owner or the taxpayers.

The summary analysis of the session, done by the Messerli Research team subsequent to the discussions, made clear that veterinary officers are essentially guided by two main values in the exercise of their profession in regard to their societal mandate. They not only have to consider the 'interests' of animals, but also specific 'public interests' such as safety in the case of dangerous dogs, spending tax money on shelter animals, public health and so forth. In other words: Contrary to popular opinion, veterinary officers are not something like a 'police unit to control and support animal protection'. This is only one of their tasks. Ideal-typically, the debate identified two vital norms: Norm 1 is about the individual animal's wellbeing and respect for the animal's life concerning the Austrian legislation, e.g. the Austrian Animal Protection Act asks for a 'justifying reason' to legally kill animals; Norm 2 is about 'public interests'.

Thereby, the term 'public interests' must not be misunderstood. Of course, the public can also have the interest that an animal's wellbeing is respected. This is the main reason why Norm 1 is part of the job description of veterinary officers. Nevertheless, the public also have interests, which can conflict with Norm 1. For example, a big part of the public is probably interested in public safety, which can be at risk by dogs who repeatedly exhibits unusual aggressive behaviour.

A solution without euthanasia, such as bringing the dog to a specific animal shelter where he can live the rest of his life, shielded from the public and supervised by experts, would live up to the individual animal's wellbeing, however, such a solution would pose the question of public financing. To what extent is it justified and fair to use public funds for those purposes?

When it comes to a decision, it is very clear, that a general judgement is not possible. Weighting which norm is more important is only reasonable when taking into account specific characteristics and the facts of the case. The complexity of pending decisions does not allow for general decisions. However, in the context of the outlined case study, veterinary officers must consider public interests as justifying reasons for euthanizing the aggressive dog if they want to live up to their social mandate (In this context, clarifying the different roles of veterinary officers can be crucial: As civil service they must judge on the basis of law, nevertheless, a veterinary officer is not only a representative of the State but also for example a citizen. As a citizen with expertise, he could also contribute to overwork the existent law if he sees potential or necessity for improvement.).

So within the interpretation it became clear that veterinary officers are not only held responsible for animal wellbeing and the life of animals, but also for 'public interests' if they want to carry out their profession in legal terms. Therefore, the place of veterinary officers is neither on the site of 'animal's interests' nor on the site of 'public interests' ... the profession's appropriate place is between the norms. In every case, a veterinary officer has to weight the affected values based on his knowledge and expertise.

On a general note. In public debates about questions of human-animal interaction there are a lot of loudly communicating votes, which pretend to know what the only morally right decision would be. As a result of the evaluating discussions of the workshops among the project managers at the Messerli Research Institute, the following thesis was developed: Veterinary officers seem to be well advised to not join this choir but to be aware that the decisions ahead in their daily job are regularly and by definition morally difficult decisions beyond any 'black-and-white'-thinking. Conflicts of norms imply the basic situation that independently from what measures are taken, an important value will be impaired. Exaggerating a bit, working as a veterinary officer often means searching for a 'best possible' in a 'worst case'.

It is comprehensible that veterinary officers faced with such dilemmas sometimes wish for a clear solution without ambiguity. However, it seems advisable to accept and understand the fundamental difficulty as part of the concept of their profession: Conflicting norms and values are the rule in veterinary officers' daily job – not the exception. They are essential and unavoidable and not the result of a lack of individual competence or personal skills. This thesis was presented to the veterinary officers in subsequent workshops. According to the participants, keeping this in mind can bring a certain sense of relief.

Role of ethics

Projects like 'Vethics', which bring together veterinary officers and ethicists, comprise the question concerning the role of ethics in this context. Since the project was focusing on problems arising in practice, the ethical reflection can be summed up under the term 'applied ethics' (cf. Bayertz, 1991:

23). However, the specific tasks of applied ethics are understood rather differently. Is it to reconstruct something like 'minimal morals' on which a vast majority is able to agree (cf. Ropohl, 1998: 281)? Should applied ethics 'provide answers to problems' (Chadwick and Schroeder, 2002: 1) of practitioners? Should it, validly and with recourse to philosophical concepts and theories, answer the question what to do in a morally relevant situation (cf. Zichy, 2008: 93)? Or is it less about answering and more about describing moral systems, conventions and norms and making clear the relevant arguments in a debate and their tradition (cf. Fischer, 2001; Lesch, 2006)?

It is not possible to submit an elaborated answer to the question of the tasks of applied ethics within the given framework. However, it is possible to reflect on the experiences and discussions about the role of (applied) ethics within 'Vethics'. Firstly, applied ethics – in the form of the project – created a platform for discussions. The participants emphasised not only that there is hardly time for this kind of structured discussion from an ethical perspective in their daily job, but also that the normal office's structures and hierarchies sometimes hinder open debates. In this context, applied ethics makes available an 'open space' and is something like a facilitator, bringing together experts, initiating a dialogue, and structuring the discussions. This outlined role reminds of convictions in medical ethics, which understand the high importance of ethicists not least because of their 'outsider'-status to the health system (cf. Anderlik, 2001: 166).

A few of the above quoted tasks of applied ethics focus on a normative concept, trying to answer what people should do in a specific situation from an ethical point of view. Magalhães Sant'Ana *et al.* (2009) also identified in their attempt to examine how ethics is taught to veterinary students an approach that understands ethics as a normative discipline to evoke the difference between 'right' and 'wrong'. In this context, the ethicist is something like an expert for answering moral questions. However, within the debates of the project and as a result of the analysing process of the sessions, there were signs that also a descriptive approach is able to contribute a genuine added value. For example in the discussion of specific measures facing animal diseases, the 'Animal Disease Intervention Matrix' (ADIM) (Aerts, 2006) was presented to the participants. This tool visualizes the difficulties of weighing conflicting values and norms and illustrates that the achievement of one justified objective puts a burden on achieving another also justified objective. Although the ADIM does not present a solution, the participating veterinary officers provided very positive feedback to the tool since it provides a description of the difficulties facing animal diseases. In short and from the participants' perspective: 'The tool doesn't tell me what's right or wrong, but it provides me with a better understanding of why I have trouble to know what's right or wrong'. Feedback like this can be seen as a very illustrative example for the potential of descriptive approaches in applied ethics.

In a more general form: normative and descriptive approaches don't need to be exclusive or conflicting, however, there is a difference in the emphasis. Describing a moral problem not only in its ethical framework (referring to utilitarian, virtues, contractarian, deontology, etc.) but also in its historical and cultural tradition can be of importance not only for a better understanding but also for dealing with this problem (cf. Fischer, 2001: 165). In this context, the ethicist is less like an expert answering moral questions, but more like a travel guide through the world of moral problems (cf. Reiss, 2005). Following this approach, the focus of the ethicist is not so much on discussing the question, if the 'aggressive dog' in the outlined case study should be euthanized from a moral point of view or not, but to work out the background foils of this scenario: Which moral relevant arguments can be identified in the debate? Which images, narrations and traditions are important in the context of a public debate of euthanizing aggressive dogs? How can the different roles of a veterinary officer in this scenario be clarified (representative of the State, social worker, expert on effectiveness of a therapy, etc.)? In addition, since the participants described the euthanasia in this case as 'stressful' and 'worst case', how would an 'ideal case' of euthanasia look like? Moreover, to what extent must such 'worst' and 'ideal cases' of euthanasia be understood as results of cultural beliefs concerning a 'good life' and a 'good death' in general?

Questions like these make clear that we don't use the term 'descriptive ethics' the way it is used for example by Scarano, who equates an descriptive approach with empirical studies about moral problems (cf. Scarano, 2006: 27) but in a sense like Fischer (2001) and Lesch (2006) are doing it, namely descriptive ethics as a hermeneutical guided project. Describing a moral problem doesn't mean to do a survey, but to evoke a deeper understanding of the problem in different perspectives. If training in ethics should equip veterinarians with 'the necessary knowledge and skills to analyse a difficult situation from different ethical perspectives ... and to take into account different values involved' (Magalhães Sant'Ana *et al.*, 2009: 201) and if ethics in this context can be also understood 'as a means to understand the complexity and ambiguity inherent in human life, rather than internalising codes or values' (Magalhães Sant'Ana *et al.*, 2009: 200), then an explicit descriptive approach and a corresponding self-understanding of the involved ethicists can be helpful. An approach which 'merely' describes and analyses moral relevant situation is often seen as deficient. However, the mentioned questions try to illuminate the whole scenario and are able to lead to a more comprehensive decision, or at least to a better understanding of the values, traditions, arguments and cultural concepts at stake. To put it in words as a thesis on the basis of the interpretation and in the context of the debate, which skills in the context of veterinary education are needed (cf. exemplarily Magalhães Sant'Ana, 2015): especially when conflicting norms are the rule and not the exception and general judgements are not possible, a better understanding of moral dilemmas seems to be a fitting target. That is why one vital role of ethics for veterinary officers could and should be to improve the skills of veterinary officers in understanding and describing moral conflicts in their professional field from different perspectives.

A guide for an ethical reflection

The article only discussed selected results of the project, which ended in early 2016 and is continued by a follow-up until 2019. As a result and a kind of summary which addresses as target audience veterinary officers themselves, a short, German written book with the title 'Ethik in der amtstierärztlichen Praxis. Ein Wegweiser' (Ethics for Veterinary Officers. A guide) was published in 2015 and went out of stock immediately. A second edition follows in 2016. (Weich *et al.*, 2016) The follow-up project focuses on the potential of web-based learning tools to promote ethical decision-making of veterinary officers. All results and project steps are documented on the website of the Messerli Research Institute.

References

Anderlik, M.R. (2001). The ethics of managed care. A pragmatic approach. Indiana University Press, Bloomington, USA.

Aerts, S. (2006). Practice-oriented ethicals models to bridge animal production, ethics and society. Katholieke Universiteit Leuven, Leuven, Belgium, 180 pp.

Bayertz, K. (1991). Praktische Philosophie als angewandte Ethik. In: Bayertz, K. (ed.) Praktische Philosophie. Grundorientierungen angewandter Ethik. Rowohlt Taschenbuch, Reinbek bei Hamburg, Germany, pp 7-47.

Chadwick, R. and Schroeder, D. (2002). General introduction. In: Chadwick, R. and Schroeder, D. (eds.) Applied ethics. Critical concepts in philosophy. Vol. I, Nature and cope. Routledge, New York, USA, pp. 1-20.

Weich, K., Dürnberger, C. and Grimm, H. (2016). Ethik für die amtstierärztliche Praxis. Ein Wegweiser. Harald Fischer Verlag, Frankfurt, Germany, 118 pp.

Fischer, J. (2001). Die Begründungsfalle. Plädoyer für eine hermeneutisch ausgerichtete theologische Ethik. Zeitschrift für Evangelische Ethik 46: 163-167.

Lesch, W. (2006). Hermeneutische Ethik. Narrative Ethik. In: Düwell, M., Hübenthal, C. and Werner, M.H. (eds.) Handbuch Ethik. Zweite, aktualisierte und erweiterte Auflage. Verlag J.B. Metzler, Stuttgart und Weimar, Germany, pp. 231-242.

Magalhães Sant'Ana, M, (2015). A theoretical framework for human and veterinary medical ethics education. Advances in Health Sciences Education.

Magalhães Sant'Ana, M., Baptista, C.S., Olsson, I.A.S., Millar, K. and Sandoe, P. (2009). Teaching animal ethics to veterinary students in Europe: examining aims and methods. In: Millar, K., West, P.H. and Nerlich, B. (eds.) Ethical futures: bioscience and food horizons. Wageningen Academic Publishers, Wageningen, the Netherlands, pp. 197-202.

Mayring, P. (2003). Qualitative Inhaltsanalyse. Grundlagen und Techniken. Beltz Verlag, Weinheim, Germany.

Mepham, B., Kaiser, M., Thorstensen, E., Tomkins, S. and Millar, K. (2006). Ethical matrix. Manual. LEI, The Hague, the Netherlands.

Reiss, M. (2005). Teaching animal bioethics: pedagogic objectives. In: Marie, M., Edwards, S., Gandini, G., Reiss, M. and Von Borell, E. (eds.) Animal bioethics: Principles and teaching methods. Wageningen Academic Publishers, Wageningen, the Netherlands, pp. 189-202.

Ropohl, G. (1998). Technikethik. In: Piper, A. and Turnherr, U. (eds.) Angewandte Ethik. Eine Einführung. Beck, München, Germany, pp. 264-287.

Scarano, N. (2006). Metaethik und deskriptive Ethik. In: Düwell, M., Hübenthal, C. and Werner, M.H. (eds.) Handbuch Ethik. Verlag J.B. Metzler, Stuttgart und Weimar, Germany, pp. 25-35.

Tannenbaum, J. (1995). Veterinary ethics: animal welfare, client relations, competition and collegiality. Mosby, St Louis, USA.

Zichy, M. (2008). Gut und praktisch. Angewandte Ethik zwischen Richtigkeitsanspruch, Anwendbarkeit und Konfliktbewältigung. In: Zichy, M. and Grimm, H. (eds.) Praxis in der Ethik. Zur Methodenreflexion in der anwendungsorientierten Moralphilosophie. De Gruyter, Berlin, Germany, pp. 87-118.

Section 11. Communicating food

44. Misleading advertising on food sector: European and US legal perspectives

A. Arroyo Aparicio
Law Faculty UNED., Obispo Trejo, 2, 28040 Madrid, Spain; aarroyo@der.uned.es

Abstract

In Europe, apart from food sector specific rules, the rules of unfair competition Law must be taken into consideration. Indeed, the Unfair Trade Practices Directive that brought full harmonization of consumer protection laws in its field has a clear relevance to misleading advertising on food sector. In the US, regulation on food advertising is a complex area and recent lawsuits have been filed against companies whose products contain genetically modified organisms and were advertised as 'all natural'. It can be said that in Europe the regulation has a stronger control a priori, while in US the legal control is a posteriori control, through the judicial lawsuits. Some conclusions can be adopted from the analysis of the different legal approaches.

Keywords: unfair trade practices law, misleading advertising legal control, food labelling

Misleading advertising legal control on food sector: European perspective

Food law establishes the rights of consumers to safe food and to accurate and honest information. Apart from food sector specific rules, the general rules on advertising or unfair commercial practices must be considered.

Accordingly, Directive on misleading and comparative advertising must be taken into account. This Directive provided for a minimum level of protection throughout Europe as concerns misleading advertising, and it has harmonized the rules of the EU countries on comparative advertising. However, since the implementation of the Directive on Unfair Commercial Practices (2005/29/EC, UCPD), the scope of application of the Misleading and Comparative Advertising Directive has been restricted to business-to-business (B2B) relations concerning misleading advertising. Its provisions on comparative advertising also apply in the context of advertising directed at consumers.

Therefore, thanks to the UCPD, national enforcers have been able to control a range of unfair business practices, such as providing untruthful information to consumers or using aggressive marketing techniques to influence their choices. It is a horizontal Directive which applies to all business-to-consumer transactions. The system of protection set up by the Directive consists of a four-tier prohibition. First, the Directive establishes a general clause banning all unfair commercial practices which are contrary to professional diligence and that are likely to distort the economic behaviour of the 'average consumer'. Second, the Directive regulates in detail two key categories of unfair practices, namely misleading (including misleading actions and omissions) and aggressive practices which are likely to affect the 'economic behaviour of consumers'. Third, the Directive provides for a 'black list' of 31 particularly harmful practices which are banned up-front, regardless of the circumstances. Finally, the UCPD provides for special protection to 'consumers who are vulnerable' because of their mental or physical infirmity, age or credulity.

The UCPD is based – as said – on full harmonisation, meaning that Member States may not retain or introduce stricter consumer protection rules, except in the areas of financial services and immovable property. All Member States transposed the Directive by the end of 2009.

EurSafe 2016 – DOI 10.3921/978-90-8686-834-6_44, © Wageningen Academic Publishers 2016

In this context, the notion of commercial practice is wider than the notion of advertising. In fact, 'business-to-consumer commercial practices' (referred to as commercial practices) means any act, omission, course of conduct or representation, commercial communication including advertising and marketing, by a trader, directly connected with the promotion, sale or supply of a product to consumers.

It is also important to distinguish between misleading actions and misleading omissions. Misleading actions could be business practices that contain or convey false elements likely to mislead the average consumer or the true elements but presented in such a way that they lead to the same result. In particular with relation to food seems to be important the information about the main characteristics of the product, such as its benefits, risks, execution, composition, method and date of manufacture or provision, geographical or commercial origin or the results to be expected from its use, or the results and material features of tests or checks carried out on the product. The omission of information can, of course, also mislead the consumer.

To give some examples: pastries announced butter in the case they did not contain it (composition). Other examples, in relation to essential qualities: free-range chicken while they are not, homemade bread when it is industrially produced, organic products that are not, etc.

That said, a commercial practice shall be unfair if: (a) it is contrary to the requirements of professional diligence; and (b) it materially distorts or is likely to materially distort the economic behaviour with regards to the product of the average consumer whom it reaches or to whom it is addressed, or of the average member of the group when a commercial practice is directed to a particular group of consumers. In other words, face a practical implementation with regard to a wide audience, the assessment is made in terms of the psychology of the average consumer: it is necessary that the practice affects, or is likely to materially distort the behaviour economic of the average consumer.

It should be emphasized that in the UCPD, although the common definition of consumer is being used in the introductory definitions of the text, the specific legal determinations linked to the protection of the consumer are extremely important. It can be said that the notion of consumer is in this sense 'elastic' (Arroyo, 2003: 99; 2013: 467). Indeed, the notion of consumer is intimately linked to the aim of the specific legal rule ('functional notion'; Arroyo, 2003: 116; 2013: 468). As can be seen, the scope of some legal rules is being determined by the 'economic behaviour of the average consumer' or by the 'consumers who are ... vulnerable'.

'Average consumer' means the consumer 'who is reasonably well informed and reasonably observant and circumspect' according to ECJ case law. Thus, Court of Justice has held that '... in order to determine whether a particular description, trade mark or promotional description or statement is misleading, it is necessary to take into account the presumed expectations of an average consumer who is reasonably well informed and reasonably observant and circumspect'(Case C-210/96 Gut Springenheide and Tusky [1998] ECR I-4657, para 31).

This average consumer is taken as a critical person, conscious and circumspect in his or her market behaviour. For example, in the ECJ case Law, the 'reasonably circumspect consumer' will not believe that the size of a promotional marking on a package corresponds to the promotional increase in the size of that product [Case C-470/93 Verein gegen Unwesen in Handel und Gewerbe Koln e.V. v Mars GmbH [1995] ECR I-01923, para 24]. Or the average consumer will not attribute to goods bearing the marking 'dermatologically tested' any healing effects which such goods do not possess (Case C-99/01 Criminal proceedings against Gottfried Linhart and Hans Biffl [2002] ECR I-09375, para 35).

As underlined in Recital 18 UCPD, the test is based on the principle of proportionality. The Directive adopted this notion to strike the right balance between the need to protect consumers and the promotion of free trade in an openly competitive market.

The trademark case law is used by analogy to delineate factors influencing the level of knowledge of the average consumer (Guidance on the implementation /application of Directive 2005/29/EC on Unfair Commercial Practices, 2009). According to the General Court, in cases of trademark, '... the average consumer normally perceives a mark as a whole and does not proceed to analyse its various details ... In addition, account should be taken of the fact that the average consumer only rarely has the chance to make a direct comparison between the different marks but has to place his trust in the imperfect image of them that he has retained in his mind. It should also be borne in mind that the average consumer's level of attention is likely to vary according to the category of goods and services in question'. And, according to the Court of Justice, 'among the factors to be taken into account in order to assess whether the labelling at issue in the main proceedings may be misleading, the length of time for which a name has been used is an objective factor which might affect the expectations of the reasonable (average) consumer' (Case C-446/07 Alberto Severi v Regione Emilia-Romagna, judgement of the Court of Justice of 10 September 2009, para 62).

Paragraph 29 of the Opinion of Advocate-General in the 'Estée Lauder' case, contains a basic description of how the average consumer test should be structured (Guidance on the implementation /application of Directive 2005/29/EC on Unfair Commercial Practices, 2009):

> ... the test to be applied to any case of restriction on the sale or marketing of a product on the ground of protecting the consumer from misleading labelling or other accompanying information is whether its presence on the market would, in some material respect, be likely to mislead the hypothetical consumer so defined ... The test should enable the national court to assess the facts of each case against this standard on the basis of its own judgment of how such a consumer would be affected. The standard involved, being based on accumulation of four factors, is clearly a high one. Having regard to all the relevant surrounding circumstances of the case, and especially the selling arrangements employed by the vendor, the national court must be satisfied that the average consumer, who is reasonably well informed and observant about the product in question and who exercises reasonable circumspection when using his critical faculties to assess the claims made by or in respect of it, would be confused. The approach is thus not statistical.

As regards the food cases, the Darbo Case can be mentioned (Court of Justice of the European Union, 10 September 2009/ C-366/08 [Wiebe, 2015]). This case concerns the question whether 'extra jam' can be regarded as a 'low sugar jam' within the meaning of Annex III to Directive 95/2. But the most important question was that a consumer association promoted that the Darbo's Jams could not use the words 'naturally pure' on the ground that the pectin gelling agent was an additive used in its composition. ECJ rejected the argument relating to pectin on the basis that it was indicated on the label on the packaging in accordance with the relevant European law. ECJ also pointed out that it is widely understood that lead and cadmium are present in the natural environment as a result of pollution. Lastly, the description 'naturally pure' appearing on the label might indeed be liable to mislead consumers if the foodstuff contained a high level of residues of toxic or pollution substances, but this was not the case (Wiebe, 2015).

There is another important specification in the UCPD related to the different consumer groups. National courts should take into account the fact that, the Directive not only codifies the case law in the sense of the average consumer, but also further refines it by adapting the average consumer test when the interests of specific groups of consumers are at stake (Art 5(2)(b) of the Directive). This means that when the practice is addressed at a specific group of consumers, be they children or rocket scientists, national authorities and courts must assess its impact from the perspective of the average member of the relevant group.

The vulnerability of some consumers is common to other texts and initiatives. For example, the European Parliament Resolution of 8 may 2012 signalled some issues among food sector. The vulnerability of young consumers is evocated in relation to advertisement for food that are particular fatty, sugary and salty (like junk food) in Directive 2010/13 (Friant-Perrot, 2016: 94).

Misleading advertising legal control on food sector: US perspective

The European legal system is totally different from the US system. And the regulation of food advertising in the US is also a complex area (Ernst, 2009). The Federal Trade Commission Act (the FTC Act) and associated regulations are the federal government's primary rules against false, deceptive, or unfair advertising. Section 5 of the FTC Act prohibits 'unfair or deceptive acts or practices in or affecting commerce.' But the FTC's authority over advertising does not cover labelling. The Food and Drug Administration (FDA) regulates claims made on labels of food products while the Department of Agriculture regulates claims made on labels of meat and poultry products. In 2016 FDA is proposing to update the Nutrition Facts label for packaged foods.

It calls attention to the increasing number of claims (lawsuits) involving the food labelling sector and it has predicated violations of state statutes on false advertising, unfair trade practices, consumer protection, fraud, and breach of warranty (Negowetti, 2014). Most of them refers to expressions such as 'all natural,' 'nutritious,' or 'healthful' that are permitted or not regulated by the FDA, but could be misleading. A variety of products such as cookies, ice cream, and cooking spray, containing ingredients like high fructose, corn syrup, alkalized cocoa, ascorbic acid, and genetacally modified organisms (GMOs) are frequently involved.

Taking into account GMO food labelling, the European approach is different from the American model. In a broad sense, whereas the US regulatory focus has been on the end product, the EU model has focused more on process (Bruton, 2013: 1136).

Although the FDA does not recognize any meaningful difference between GMOs and foods developed by traditional plant breeding and therefore does not require labelling of GMOs, these lawsuits allege that GMOs are inherently 'unnatural' US Cases (Negowetti, 2014).

On its website, the FDA has opened a comment period on the use of the term 'natural' on 'food labelling' and has it extended until May 10, 2016 because it received citizen petitions asking the definition of the term 'natural' or the prohibition for the use of 'natural' on food labels. The FDA is taking into account that some Federal courts, as result of litigation, have requested administrative determinations regarding whether food containing ingredients produced using genetic engineering or foods containing high fructose corn syrup may be labelled as 'natural'. And the FDA has provided some explanations:

> Although the FDA has not engaged in rulemaking to establish a formal definition for the term 'natural,' we do have a longstanding policy concerning the use of 'natural' in human food labelling. The FDA has considered the term 'natural' to mean that nothing artificial or synthetic (including all colour additives regardless of source) has been included in, or has been added to, a food that would not normally be expected to be in that food.

Conclusions

Under European approach, misleading advertising or unfair practices on food sector have as central point the average consumer notion and its sub-groups (vulnerable consumers). All competition approach has to take account of the empirical elements and, of course, the circumstances surrounding the case. It is therefore a right moment to question if information instead of prohibition is always satisfactory.

In the opposite corner is the US approach with respect to which it can be argued that in 'the absence of effective regulatory enforcement action against food and beverage manufacturers making misleading claims, consumer protection groups and plaintiffs' attorneys have stepped in to fill a void' (Negowetti, 2014).

One point especially worthy of note is that in both systems the issue is addressed from a competition legal perspective, but in Europe the legal control is *a priori* (through information rights and UPD rules) and the US system is *a posteriori* (ex post control through lawsuits). Both systems are open to improvement: in the American system a stronger legal control would be more efficient as in the European system a more resolute collective redress.

References

Arroyo, A. (2003). Contratos a distancia. Aranzadi, Pamplona, 414 pp.

Arroyo, A. (2013). Who is a consumer on food law ? Some reflections on the notion of consumer and the EU food law. In: Röcklinsberg, H. and Sandin, P. (2013). The ethics of consumption: the citizen, the market and the law. Wageningen Academic Publishers, Wageningen, the Netherlands.

Bruton, S.V. (2013). GMO Food Labeling. Springer, New York, NY, USA.

Ersnst, T.S. (2009). Advertising food products: understanding the regulatory mix. American Bar Association Publications, Chicago, IL, USA.

Friant-Perrot, M. (2016). The vulnerable consumer in the UCPD. In: Van Boom, W. and Garde, A. (eds.) The European unfair commercial practices directive: impact, enforcement strategies and national legal systems (markets and the law). Routledge, London, UK.

European Commission (2009). Guidance on the implementation /application of Directive 2005/29/EC on Unfair Commercial Practices. Brussels, Belgium.

Negowetti, N.E. (2014). Food labeling litigation: exposing gaps in the FDA's resources and regulatory authority. Brookings Institution, Washington, DC, USA.

Wiebe, A. (2015). How much nature for the consumer? Misleading advertising, trademark law and the European average consumer standard in the food sector. Corporate Governance eJournal. Paper 32. Available at: http://tinyurl.com/hbd54hg.

45. Art futuring food

A. Nunes[1,2*]*, J. Borlido-Santos*[1,2] *and M.M. Lopes*[1,2,3]
[1]*i3S – Instituto de Investigação e Inovação em Saúde da Universidade do Porto, Rua Alfredo Allen, 208, 4200-135 Porto, Portugal;* [2]*IBMC – Instituto de Biologia Molecular e Celular, Porto, Portugal;* [3]*ID+ Instituto de Investigação em Design, Media e Cultura, Porto, Portugal; anunes@i3s.up.pt*

Abstract

Through art's history food has been portrayed in different techniques, contexts and differing meanings. In the last few decades the issue of food took a turn in our own lives and that has been reflected upon in many arts projects. In this article we intend to explore some art projects aiming to bring light on how those create utopian and alternative futures for food systems We claim that art is a powerful tool in questioning the present state of food systems, creating utopian solutions and hypothesis and exploring novel social-economic and political scenarios.

Keywords: utopia, participatory art, sustainability, activism

Introduction

What sets apart the roman murals depicting feasts from the Semi-Living Steak of Oron Catts and Ionat Zurr? Throughout history, art has been representing food in its biological, social and cultural dimensions. More than depicting food itself, it is time, fear, sin, pleasure, power and death that are being rendered (Riley, 2015). In ancient times, art represented Humankind's lack of power over nature. The dependence on food to survive, its relation to the life cycle and death, is key to understand the human need to tame forces beyond control. The 'Chauvet-Pont-d'Arc' cave paintings (c. 30,000 BC.), the food offerings to the gods in ancient civilizations and the roman murals of feasts, all reflect a desire for abundance and the human attempt to steer the forces of nature (Bell, 2009; Bendiner, 2004; Riley, 2015). Both the 'Garden of Worldly Delights', painted by Hieronymus Bosch, and Pieter Aertsen's market scenes remind us of the utopian dream of abundance, pleasure and delight. From prehistory to the 20th century, art's portrayal of food depends on social and political context and hints at our inner struggles with mortality, while also representing the eternal battle with the forces of nature (Riley, 2015). Nowadays, famine is no longer the result of a bad harvest but of unequal distribution. The ethical and moral questions are no longer restricted to food as a representation of humanity, thus reminding us of our limitations, our mortal flesh – as in the beef carcass of Rembrandt van Rijn's 'Slaughtered Ox' – but also to the humankind's abuse towards other living beings, including fellow humans (Riley, 2015). Today the question stretches from how to protect humans from nature onto how to protect nature/humans from Humankind. With the increasing unrest regarding human impact in the natural world, several art movements arose. In the 20th century avant-garde food was evoked both physically (banquets with sensory experiments) and metaphorically (excess, indigestion, vomit and excretion), evidencing the semiotic associations of food through real usage of food or language, and of the body by characterizing eating as an interior-exterior manipulative route. By engaging in social actions that were to some degree innovative, revolutionary, and politically charged, new paths of thinking about art and food were developed, which challenged the predominance of a purely aesthetic-based appreciation of art. Eat Art and Fluxus came later, restoring the power of food as art material and actively engaging the audience. The presence of food in the Italian Futurist banquets, Eat Art and among the Fluxus artists, illustrates a desire to redefine art respecting its purpose, materials, production, consumption and appreciation. Claire Bishop (2012) refers that participatory art shows a need to go past the traditional relationships of 'art objects, artists and audience'. The audience's posture in relation to food seems to have morphed, from passive spectators to active participants – and, in some cases, co-authors of the pieces (Bishop, 2012). Art movements linked to

EurSafe 2016 – DOI 10.3921/978-90-8686-834-6_45, © Wageningen Academic Publishers 2016

ecology and sustainability have emerged: first the so-called land art and earthworks, then environmental art and, more recently, ecology and art (Brown, 2014; Thornes, 2008). Furthermore, as the debate around food grows stronger, art's interest in the topic increases.

The present paper seeks to explore some of the ways in which contemporary art is contributing to the discussion surrounding food systems and how it may help to build new food futures. Consequently, the analysis focuses on artworks that tackle the food system: food production, industry and market; economic and socio-political implications; the impact of food systems on the environment; the scientific and technological developments on food production and distribution.

Since this paper intends to grasp how art may create utopian and alternative futures for food systems, artworks were divided into three categories: those which question the current state of food system, document it and allow us to see it with new eyes; those that create utopian solutions and hypotheses for the food system, introducing and discussing ideas deriving from scientific and technological development; and those exploring utopian scenarios and proposing social, economic and political revolutions.

Questioning today's food system

The way art negotiates reality has been deeply influenced by many aspects: the rapid changes in the world within the last few decades, the ripples of the atomic bomb, the collateral damages of wars, of social and political revolutions, the advent of the almighty scientific revolution. Several artists became very socially and politically engaged and art turned into a form of critical comment, a vehicle for questioning society, scientific development, politics and philosophy. 'The artist's role is not, however, to provide definitive answers to the problems. Merely asking questions is often enough' (Brown, 2014).

Therefore, the selected art projects raise questions on how we deal with new scientific methods for food production, from genetic modifications (GMO) to mutation or radiation breeding, as well as on the complexities of food production and distribution.

In the 'Glowing Sushi Cooking Show', the artistic-led group Center for Genomic Gastronomy (CGG) reflected upon GMO use in food industry. Our daily diet is riddled with Genetically Modified Organisms, yet to the general public this is still an almost invisible technology.

By creating glow in the dark sushi using biotechnology, the artists make genetic modifications literally visible. Then serving the sushi to the public and performing cooking shows is a statement that encourages an open discussion and creates a space for dialogue. Another example of this artistic-led group paving the road to discussion is the 'Cobalt 60' project, in which a barbecue sauce is produced from five plants and seeds exposed to mutation or radiation breeding. Once again, the projects goes beyond the physical object that is the 'Cobalt 60' sauce and engages the public in the discussion, using a provocative artefact as a trigger.

The pallet of works from the Center for Genomic Gastronomy is varied, including projects that could easily fit in any of the categories used in this article. The group in itself is definitely a case study about the relationship of contemporary art with food.

The artistic critical view of food is not restricted to scientific and technological advances; it spans every field including food's complex economic systems. Van Brummelen and De Haan developed the project 'Monument of Sugar', which explored sugar trade after realizing European sugar is cheaper outside the European Union. They produced a film essay from the artists' research on sugar trade and an installation with 304 sugar blocks that travelled from Nigeria to Europe (Howells, 2014). This is just an example of artworks dealing with the mechanisms of food trade between powerful economies and third world countries. By questioning the present state of food systems, art forces us to ponder its advantages and disadvantages, frailties and injustices. The critical analysis of the present is the first step in the creation of a utopian vision of the future of food.

Utopian food solutions

If the first step is to look at the reality with a critical gaze, then the second is to imagine alternative options, or utopian scenarios. Art's utopian scenarios of food can be divided into two main groups: utopian technoscientific hypotheses and utopian social, economic and political scenarios.

The overwhelming importance of scientific discoveries over the last few decades inevitably attracted the attention of artists. Hence, many artworks explore technoscientific scenarios around food. Artists walked into laboratories and used techniques and practices from the scientific realm to present alternative technologies to the future of food. However, 'unlike the scientist, who must follow established scientific methods, the artist is free to question and redefine anything or everything at any stage ... Artistic projects are able to withstand a far higher level of risk than scientific experiments, which often come with expectations of tangible results or even profit for their funders' (Brown, 2014). Several artworks debate ethical questions around meat production and alternatives to meat in our diet.

A recurrent topic is *in vitro* meat production; especially after the first *in vitro* beef burger had been created with a team of Dutch scientists, headed by Mark Post, and eaten in 2013 in London. Ten years before, though, Oron Catts and Ionat Zurr created the 'Semi-Living Food: Disembodied Cuisine' project in which they lab-grew a steak made of frog cells cultivated in a bioreactor and later cooked it up and served it as dinner at the National Center for Contemporary Art in Nantes, France. The 'Semi-Living Steak', by Ionat Zurr and Oron Catts, is a utopian technoscientific hypothesis. In this work, as well as in the 'Victimless Leather' piece, where a coat is grown out of living cells, the ethical questions are: what is life? Is it victimless if we do not kill the animal but kill some of its cells? What are the frontiers of the self, of identity? (Catts and Zurr, 2002). Since then several artworks have explored how a future where *in vitro* meat is a reality would look like. That is precisely the case of the '*In Vitro* Cookbook', published in 2014 by Koert van Mensvoort, artist and header of the Next Nature Lab, and Hendrik-Jan Grievink. It suggests 45 recipes of *in vitro* meat and explores how it may look like in a not-so-distant future. Part of the book – the '*In vitro* me', by Chloé Ritzerfeld – is very controversial. '*In vitro* me' is a personal bioreactor-jewel connected to the body that allows the growth of own muscle tissue for meat production and consumption. If the cells are just part of but not the organism, is it ethically acceptable to eat cultures of our own cells? It is important to highlight that most of these works assume a position but without being bias, trying to bring the issue to the surface and engage the public in a discussion. The Waag Society, a platform for artistic research and exploration, which promotes an active involvement of the society by bringing different stakeholders and partners into the process, also explores *in vitro* meat, as well as synthetic biology and biotechnology tools to delve into food futures. The group developed 'The Other Dinner' project, composed of three parts: the first one was a workshop on how to cook a pig nose-to-tail; the second entailed preparing and tasting animals that we do not usually eat, such as mice; and the third was a workshop on *in vitro* meat production from mice cells. 'The Other Dinner' suggests that we can change the meat market by altering some cultural perceptions, thus making a better use of the available meat and introducing in our diet other animals.

Several artworks present alternative food sources that could be more environmentally friendly and facilitate self-sustainability. 'After Agri' (2011), a project by Michael Burton and Michiko Nitta, creates utopian scenarios of 'feast after agriculture', proposing a synthetic relationship between humans and algae, turning humans into semi-photosynthetic and self-fuelling organisms. Also exploring ecological food sources is Erik Sjödin's 'Super Meal', which proposes the aquatic plant Azolla as an ingredient for fast food of the future. Azolla is cheap, easy to produce and grows fast. It is rich in nutrients but can also help reversing greenhouse effects since it has a great capacity to fix nitrogen and carbon.

While these artists explore a food future where humans do not require animals as sources of protein, others propose farming optimization and promote self-sustainability. Anne Marie Maes and the Brussels Urban Bee Lab (artist collective OKNO) have been conducting a project with bees since 2009. Bees are instrumental for many reasons. They not only produce honey, but they are also essential in agriculture due to their role as pollinators. At the Urban Bee Lab, artists, beekeepers and scientists collaborate in order to deal with challenges associated with maintaining and monitoring city honeybees. Bee colonies were placed in urban areas and equipment was also installed to monitor and measure the colonies' activity. The collected data functioned as bioindicators offering a glimpse of the bee ecosystem. Also, the honey they produce varies according to the kind of food they can find in each area. These 'intelligent beehives' exist in several European cities and the data is available online. These projects focused on honey production in urban areas propose different technological approaches to beekeeping, besides the issue of a different form of honey production, in a different setting. These examples of contemporary art projects seam to question food production methods, cultural and personal relations with food, ethical and moral values, scientific developments and even our economic system. In all these projects, however, art does not resort to technology and scientific knowledge to offer viable solutions for the future, but to engage society in the debate, to make us reflect upon what kind of future we want and how we are going to use technology to achieve it.

Changing food changing the world

These artists are no longer raising questions and engaging the public in the debate, they became instruments of change. Such attitude from art and artists can be traced back to Joseph Beuys (1921-1986), a politically active artist that explored the way art could change the world and expand the understanding of the natural and ecological issues while also tackling social, political and economic problems. In order to do so, Beuys defended the concept of Social Sculpture (Adams, 1992; Brown, 2014). In a social sculpture everyone involved is an artist, 'everyone would participate creatively to resculpt the social body' (Adams, 1992), because, advocated Beuys, change had to begin in people's minds. Some very interesting artworks explore these notions of social sculpture to change society and, more specifically, our food system. Nils Norman's 'Geocruiser' (2001) is an open access mobile library and education center in a converted bus. It features books but also endemic plants, vegetables and herbs, a composting unit and a 'wormery' in which worms are used to recycle organic matter. The 'Geocruiser' is simultaneously a piece of art and a device to change people's minds. The resulting social change is the main artwork of the artist.

When Tattfoo Tan, in 'S.O.S. Free Seeds Library', provides the public access to free seeds, to promote self-sustainability and ecological diversity, he is not just debating a timely issue, he is offering society practical tools to change.

Several artistic-led groups use this form of cultural activism to prompt change in the food system and in general they are particularly interested in self-sustainability and environmental responsibility. Such is the case of the Future Farmers, multidisciplinary group formed by artists, researchers, designers, architects, scientists and farmers, founded in 1995, aiming to deconstruct the food system. In 2007, with the project 'Victory Gardens', they endorsed urban agriculture in San Francisco. What began as a utopian proposal at an exhibition morphed into a network of urban gardens and a public participatory project with workshops and demonstration gardens in public spaces. In the 'Soil Kitchen' project, the Future Farmers rehabilitated an abandoned building into a multi-use space where citizens had free soup in exchange for soil samples, and where people gathered to imagine a potential green future, participated in workshops on urban agriculture, soil remediation, composting and cooking lessons. All Future Farmers projects share this public component, this form of activism in which art engages with society and tries to change it. Their revolutionary approach has caught the eye of the art world in such a way that their work has been exhibited in important contemporary art museums, including the Whitney Museum of American Art, the Museum of Modern Art, the Solomon R. Guggenheim, and the Walker Art Center. 'The Future Farmers' are not the only art group exploring this approach. The 'P2P Food Lab' is another example of cultural activism involving citizens in the development of new economic models and incorporates them in every aspect of the food system life cycle. The 'Kultivator', a project in rural Sweden that works with the farming community, promoting several initiatives, including 'Imagine Farm', is yet another example of a project that pushed for a sustainable living by way of discussions, cooperation and sharing experiences.

Although the mentioned projects have their owns specificities and methods we might consider them as small utopian social experiments and, even if they cannot change the world by themselves, they allow the public to immerse in the scenarios, experience them and build knowledge that may instill them to draw a new future.

Final remarks

The feasts on Roman murals painted a utopian dream based on the desire for food abundance. For centuries Mankind craved a world in which food was accessible and plentiful, which was technically accomplished thanks to agricultural and industrial revolutions. However, once the production increase realized, the reality was not quite as expected, as the newfound abundance put social and economic inequalities in display. For the first time, canonical western art began to look into famine as something society must overcome. Artists portrayed covert inequalities so that all society could see it. In today's techno-scientific and globalized world, food is an even more complex issue. Although food is cheaper and more abundant than ever, famine and unequal distribution are still very much a problem, and our production system threatens the environment and its own sustainability. Scientific and technological advances in the food industry raise questions on ethics and safety. Humankind's food future is no longer based on opulence but on sustainability. These complex hidden ecologies in the food's production cycle caught the eye of socially and politically committed artists, which use the power of art as a major driving force to tackle current and future food challenges, although still dependent on context, ideology and culture. The resulting artworks, as discussed in this paper, engage the audiences in the artistic projects directing their gaze to the current problems of the contemporary system and global market. Art, much more than any pragmatic tool, is a conceptual and a viable toolkit for imagination able to able to provoke society into reflecting upon our food systems and where we can envisage new futures both in technological and socio-political terms.

References

Adams, D. (1992). Joseph Beuys pioneer of a radical ecology. Art Journal 52(2): 26-34.

Bell, J. (2009). Espelho do mundo. Uma Nova História da Arte. Orfeu Negro, Antígona, Lisbonne, Portugal.

Bendiner, K. (2004). Food in painting: from the renaissance to the present. Reaktion Books, London, UK.

Bishop, C. (2012). Artificial hells: participatory art and the politics of spectatorship. Verso, London, UK.

Brown, A. (2014). Art and ecology now. Thames and Hudson, London, UK.

Catts, O. and Zurr, I. (2002). Growing semi-living sculptures: the tissue culture and art project. Leonardo 35(2): 365-370.

Howells, T. (2014). Experimental eating. Black Dog Publishing Limited, London, UK.

Riley, G. (2015). Food in art: from prehistory to the renaissance. Reaktion Books, London, UK.

Rutzerveld, C. (2013). The other dinner. Available at: http://tinyurl.com/j8yjglv.

Thornes, J.E. (2008). A rough guide to environmental art. Annual Review of Environment and Resources 33: 391-411.

46. Interrogating ethics in future visions of food at the EXPO 2015 in Milan

Â. Guimarães Pereira[1], A. L'Astorina[2], A. Ghezzi[1] and I. Tomasoni[2]*
[1]European Commission – DG Joint Research Centre, Via E. Fermi 2749, 21027 Ispra (VA), Italy; [2]IREA – Consiglio Nazionale delle Ricerche, Milan, Italy; angela.pereira@jrc.ec.europa.eu

Abstract

In 2015, the universal exhibition in Milan (EXPO 2015) focused on 'Feeding the Planet, Energy for Life' hosting several countries which showcased technologies, innovation, culture and traditions around food. The EXPO seemed to us as a maquette of rather hegemonic narratives about the futures of food and the foods of the future. For a great part, the proposals relied heavily on governmental and techno-scientific initiatives and substitutions to address the challenges posed by the possibility of human population growth forecasted for 2050. We took the opportunity of the EXPO to engage with citizens to reflect on the imaginaries portrayed in food futures through this universal exhibition. Our journey of citizen engagement at the EXPO 2015 was based on material deliberation methodologies. Hence, the activity we will describe in this paper is called 'Food Futuring Tours'; it consisted of a series of walking photographic tours through the Expo pavilions, where participants were invited, both as individuals during the walks and then collectively in workshops, to interrogate the EXPO's futures of food proposals and then re-imagine food futures in the form of visions (*scenari*), considering their possible social, ethical, cultural and environmental impacts. In this paper, we will focus in particular on the ethics dialogues that took place during this citizen engagement exercise, highlighting how ethics has been performed in the ten different visions developed by the participants of the tours: ethics as driving force of future of foods, ethics as a distributed matter and in particular reflecting on the ways imagined by the participants in which human agency can change what is described as an urgent and complex human issue, i.e. quality feeding the planet. The paper will also reflect on how these types of engagement processes disrupt or emphasise current narratives that underlie policy development in the area of food, critically counteracting the prevailing view that describe and perform citizens as just consumers – i.e. choice and behaviour as drivers of food markets. Moreover, stemming from the ten visions developed by the participants, we will argue that current movements of deeper engagement of citizens in reconnecting and transforming the loci of production and consumption reflects desires for greater degrees of agency. In other words, policy making needs to review the relationships people want to have with food and can no longer assume that e.g. food labelling is the boundary object that enables trustful relationships among citizens and loci of food production.

Keywords: futuring, human agency, citizen engagement, imaginaries

Walking the Expo

Between May and October 2015, 90 people of various ages, nationalities and origins took up the invitation of four European researchers to follow them on a unique walk through the Milan EXPO 2015. The participants were asked both to collect suggestions, ideas and visual cues and to develop visions (*scenari*) about food in the future based on the Universal Exposition where, for six months, well over a hundred countries from around the world showcased technologies, innovation, culture, traditions and future proposals centred on food and nutrition.

The purpose of the walks, that we called 'Food Futuring Tours', was to interrogate current narratives about food futures through social research methodologies based on material deliberation practices

EurSafe 2016 – DOI 10.3921/978-90-8686-834-6_46, © Wageningen Academic Publishers 2016

explored elsewhere by Davies *et al.* (2013), where citizen engagement is carried out through 'experiential' approaches.

Discourse on the future of food is often deemed to be the preserve of elites such as scientists, designers, chefs, food companies and other corporations. The neoliberal political and economic perspective describes and performs citizens as just consumers, whose daily decisions are narrowed down to acquisition of goods and services; in this perspective consumption choices and human behaviour are the drivers of the food market. The 'Food Futuring Tours' sought to extend the debate taping into the imaginaries, values, expectations and insights from the people who, every day, choose, prepare, care, eat and feed. Therefore, the ultimate aim of this exercise was to explore other ways in which human agency can contribute to solving what is now described as an urgent and complex human issue: feeding the planet with quality food. In this way, we also hoped to contribute to reach one of the objectives of Expo 2015, that was to 'open up a dialogue between international players, and to exchange views on these major challenges which impact everyone'.

Registration to the tours was free of charge, done through a website set for that purpose and widely publicised (the ticket to the EXPO was provided by the organisers). The majority of the participants were women (75%) from different age groups, nationalities and walks of life; all participants live in the North of Italy. Each tour consisted of a morning and an afternoon session. In the morning participants walked the EXPO in two groups guided by 4 facilitators (2 for each group), following a predefined itinerary – the EXPO was too large a site as well as too crowded to be visited at random; this required from our side several prior visits to identify pavilions that could be interesting for our project. The participants were asked to reflect on the system of food in its private, social, cultural, production-related and economic dimensions, identifying the aspects that interested each of them most. Within the pavilions and spaces visited, they were asked to notice elements of the past, present and emerging futures of food. The material that individual participants collected during the walks, were primarily photos taken from their smartphones or digital cameras to be used in the second part of the 'Food Futuring Tour'. Yet, we encouraged other forms of recording, namely notes about other sensorial aspects, such as taste, odour, sound and texture.

In the afternoon of the same day, the participants would meet again at the European Pavilion, where, using and sharing their pictures and other visual or writing records collected during the walk, they would set on imagining a collective vision of food futures 25 years ahead (2040). In practice, participants gathered into two groups started by each identifying and sharing with others 3 photos representing different points in time. Subsequently, collectively they had to decide on the focus of the scenario, geography, identifying uncertainties, as well as main drivers – that is, social, economic, cultural, ethical and environmental factors – which, in their view, would foster or hinder plausible or implausible food futures.

Ten visions for future foods emerged illustrated with photography, drawings, materials collected at the pavilions together with a narrative. The ten visions speak of well-being, coexistence, quality, awareness, coming together, responsibility, self-sustenance, resilience, technology, inequalities, contamination, consumerism and shared celebration. In other words, they speak of many of the food meanings for humanity and what we can anticipate is that most of the discussions and visions addressed important ethical concerns of our times about food production, distribution and consumption.

The ethical issues in the 'Food Futuring Tours' visions

Moral matters and ethical issues with regards to food consumption are well studied (Coff, 2006; Zwart, 1999). In this section we will focus in particular on the ethics dialogues that took place during this citizen engagement exercise. We will highlight how ethics was incorporated in the ten different visions developed by the participants of the tours: ethics as driving force of future of foods; ethics as a distributed matter; and the ways in which participants imagined autonomy and human agency as drivers of changes to social practice in order to address the challenge of quality feeding the planet. Below we illustrate those ethical issues with selected quotes from the food futures visions and storylines told by the participants of the 'Food Futuring Tours'.

Participants touched often on well-acknowledged ethics aspects, such as food waste, fairness and respect for animals, which is illustrated in the quotes below. The concept of waste is not about just food wasted, but also about lost practices, disconnecting practices and substitutions.

> We imagined putting a stop to waste, and we depicted wastefulness using a tree that is a bit bare and wilted; we said 'let's put a stop to waste and rubbish, useless production, containers and technology gone mad that takes the place of people. *Awareness vision*

> Thinking 'more plants and fewer animal products', a group considered various issues such as 'if everyone in the world ate what we currently eat in Europe, our diets would no longer be sustainable' and 'we should have respect for the suffering of animals'. *Celebration vision*

In several visions ('wellbeing', 'awareness' and 'resilience'), participants addressed issues of fairness, namely with regards to distribution of food, of fertile land, questioning also the beneficiaries of current food production models, lead by corporations. In other words, participants and the visions they developed interrogate the values by which we are eating.

> In the end we thus arrived at several macro factors ...: population density, the increasing world population, overcrowded urban areas, the human factor, i.e. whether production is serving the interests of humanity only or also the interests of profit, the climate factor, the ethical factor, and thus the question of who has the primary and natural resources and how they are distributed. *Wellbeing vision*

> It is important to understand how many hectares per head are needed to live, how much one country consumes as compared to another, and how resources could be redistributed from one country to another. All these things are crucial. We thus agree that a great deal of focus should be placed on the ethical aspect, which corresponds to the roots and trunk of the tree. *Awareness vision*

> Another objective of a resilient society is to have a situation in which resources are fairly distributed. For example, apartment buildings or neighbourhoods should have their own vegetable gardens, and thus the capacity to be autonomous, to produce for themselves. *Resilience vision*

Human agency with regards to food production, distribution and consumption has strongly emerged both in the discussions across all groups but also in final visions' storylines.

> The uncertainties are legion and we imagine that scenarios might coexist. In all this, policy can play a crucial role in tipping the balance towards one or other scenario. People themselves can make a difference, but policy too is important; not necessarily a policy

> dictated from the above, but perhaps something can develop from below such as popular pressure impacting on political choices. And that's what we're hoping for!
> *Co-existence vision*

Regained connectedness between production and consumption is seen presented as enhancing agency.

> We have put this photo of the balcony with vegetable gardens and crops in boxes to show that, even in spaces where the presence of nature cannot be taken for granted, we must make every effort to introduce it, make the most of it and maintain it for future generations. This is because, while it is true that there are general policies at world level, tending one's own little garden in a positive sense triggers this mechanism of awareness.
> *Awareness vision*

> We should regain a sensory connection and physical contact with food and recognise the smell of the fish we used to eat when we went to granny's house. All of this is also linked to technology, as seen in the German Pavilion, which we should put to the best possible use. To reach such a scenario technology is fundamentally important. However, on the other hand technology scared us a bit, such as the Coop fridge that says everything, and it should thus be used with care and awareness. Awareness can be a factor allowing us to use technology in such a way that it is within our own reach, rather than serving the market, while respecting the environment and our lifestyle. *Celebration vision*

In this vision, cooperation and technology are presented as the means to empower communities that traditionally are totally dependent from external food sources.

> In the village in Burundi we have imagined that technology will be an element making it possible, for example, for new forms of packaging to combine something very old such as wicker baskets with a very modern system for packing and thus conserving food. Technology will also make it possible to have new forms of energy, for example maize will be used to produce bioethanol. Therefore, in our African village, where the real driver for development is cooperation and a network of villages, we wanted to draw a high-tech hut with a satellite dish and solar panel. *Convergence vision*

> The 'slow' philosophy, the fact that we are all connected, the recovery of time and space, specifically understood as land. This is what we want, whereas what we would like to overcome is what exists now, which can be summarised as 'too much of a good thing' ... We have imagined Europe, but I could speak about my neighbourhood instead for which I have imagined a future as a 'transition town'. I think that all neighbourhoods should aim for autonomy, for example by rethinking the landscape in 'food forest' terms. But that is just a dream ... *Resilience vision*

> Then we asked whether, in these choices, every individual can really choose freely? We were talking about this a bit earlier. It's not only a question of Volkswagen fiddling with how it calculates the filth emitted by cars, there are also the influences that already affect our lives today, and that in a five-year-old child, as she grows to be 10, 15 or 20 in five, ten or fifteen years, will be even more significant. Thus, what will technology be like? How well off will Experia (a character invented for this vision) be? And the environment she finds herself living in, the resources she has access to, and naturally the society she lives in and will live in, and the information available ... all these things are perhaps filters in

> a way, but do they make us truly free in our choices? Perhaps not, in fact probably not, but in the end choices are made. *Self-sustenance vision*

Human agency is taken further in some visions where ideas of food sovereignty are explored.

> The slogan is: 'a city that cultivates!' A city that cultivates, within the city limits, urban vegetable gardens on rooftops or around buildings, and that also cultivates old traditions, training, knowledge and also sensibility towards the evocative aromas of a certain past ... *Cultivation vision*

> We even imagined that school curricula should and must be revolutionised. We devised brand new subjects and also went as far as imagining that the first aid courses that are available today should go hand in hand, at the very least, with courses for the public on permaculture and related issues. *Resilience vision*

Another key dimension portrayed in many of the visions was 'responsibility' linked to autonomy and human agency. For example, this story suggests education and awareness as means to allow citizens to take matters in their hands.

> One path will involve education from a very young age, which will guide people towards conscious choices and the use of products with a short supply chain. *Responsibility vision*

Care is central for the majority of the visions, care about resources, care about other human beings.

> We imagined the table of the future, where dining companions include: care for the climate, environmentally-friendly policies and correct use of resources; social responsibility; redistribution; and respect for ourselves and nature, which in a way is the support holding up the legs of the table. *Awareness vision*

One of the most discussed dimensions by the participants was the deepening of the disconnect between production, distribution and consumption of foods. Many did not see this gap disappearing in the future; others suggested ways to regain connectedness between citizens as consumers with the production cycle.

> The market (which we depicted with the food market listings in Pavilion Zero) and large-scale retail (which we depicted with Coop's supermarket of the future) ... will still exist in 2040 and we can't foresee how they will develop. We know that they are and will remain a problem, but a problem that cannot be overcome. However, we would like them to change and be offset by small-scale production and renewed contact with food. *Celebration vision*

> We imagined there being loads of people of different origins, different types of food – would there be insects, yes or no, or GMOs or the presence of greenery, and direct contact with food rather than everything coming from large retailers. *Celebration vision*

Discussion

We find it remarkable that across the visions for 2040 developed by the participants of the 'Food Futuring Tours', all visions for food futures resonate well with the concepts of food citizenship (De Tavernier, 2012) and food constitutionalism (Escajedo San-Epifanio, 2015). By this we mean, that the preferred governance models of food systems are based on individual and/or collective agency. This is not so surprising, as for many participants, un-discussed corporate and governmental derived normativities which disregard received social norms about food, are root causes of distrust and anger with current models of food production, processing, distribution and consumption.

We will not debate here the ethics of food intake matters, as these have been deeply addressed by the literature (see for example: Beekman, 2000; 2008; Cheeke, 1999; Coff, 2006; Zwart, 1999, 2005) but we would like to note the evolution of literature on other ethical issues, namely the ever-growing disconnect between production and consumption loci and in particular ways in which these models have been fiddling with individual and collective agency (see for example De Tavernier, 2012 and Escajedo San-Epifanio, 2015). Labelling has been considered the instrument by which citizens can become co-creators of markets and exert their political responsibilities in a free market (Brom, 2000), as once we 'know', we can decide on the moral acceptance of what we eat. But with what choice is one left, when all packed pizza dough contains palm oil? De Tavernier (2012) argues that labelling information no longer meets the full complexity of motivations and intentions of individuals in relation to food consumption. We suggest that food labelling could also be seen as another attempt to describe and prescribe European citizens' moral identity according to values endorsed by the EU (Tallacchini, 2009).

The visions developed during the 'Food Futuring Tours' tell that there are some movements that attempt to disrupt hegemonic normativities about food governance models. Indeed, these imagined futures resonate with existing social practices and what De Tavernier (2012) described as 'food citizenship', i.e. the process in which an individual food consumer is much more than a mere final link in the nexus of production, distribution and consumption of foods, seeking empowerment in governance of food systems. For example, in Italy the quite active 'GAS' (groups of solidary shopping) network emerged when groups of people decided to get together to buy products that respect social justice and solidarity values. Other expressions of this emerging movement that tries to reconnect the production and consumption realms include Do It Yourself activities, which challenge existing agricultural practice and food production in general, by e.g. developing open-source tools for agriculture[1], developing online direct farmers to consumers networks[2], tools for individuals and communities to produce their own foods[3] seeking to integrate mature and urban lifestyles and hacking practices as described in Holt (2013).

Escajedo San-Epifanio (2015) points out that critical analyses of current food systems advocate that transformations of food governance models need to take place in order to empower citizens. The author notes that such transformations demand 'a review of our individual and collective relationship to food in the broadest sense, including fairness, cultural diversity or in the development of personal identity'. The 'Food Futuring Tours' engaged citizens in reflecting on the ethics by which we wish to govern our food systems; the resulting visions point to a need for greater citizen agency as well as the need for an ethics of care.

References

Beekman, V. (2000). You are what you eat: meat, novel protein foods, and consumptive freedom. Journal of Agricultural and Environmental Ethics 12(2): 185-196.

[1] Farm Hack: http://blog.farmhack.org and http://www.latelierpaysan.org/?lang=fr

[2] https://openfoodnetwork.org in Australia but coming also to many countries in Europe.

[3] http://aquaponicslab.org i.e. systems (hardware and software) for agriculture production in Manchester, UK. Also http://tinyurl.com/jcp8yd5 in Denmark and see also Tornaghi (2014).

Beekman, V. (2008). Consumer rights to informed choice on the food market. Ethical Theory and Moral Practice 11(1): 61-72.

Brom, F.W. (2000). Food, consumer concerns, and trust: food ethics for a globalizing market. Journal of Agricultural and Environmental Ethics 12(2): 127-139.

Cheeke, P. (1999). Contemporary issues in animal agriculture. 2nd edition. Interstate Publishers, Danville, IL, USA, pp. 248.

Coff, C. (2006). The taste for ethics: An ethic of food consumption. Springer Science and Business Media, New York, NY, USA.

Davies, S.R., Selin, C., Gano, G. and Guimarães Pereira, Â. (2013). Finding futures: a spatio-visual experiment in participatory engagement. Leonardo 46(1): 76-77.

De Tavernier, J. (2012). Food citizenship: is there a duty for responsible consumption? Journal of Agricultural and Environmental Ethics 25(6): 895-907.

Escajeda San-Epifanio, L. (2015). Challenging food governance models: analyzing the food citizen and the emerging food constitutionalism from an EU perspective. Journal of Agricultural and Environmental Ethics 28(3): 435-454.

Holt, S. (2013). Hacking the fields: crowdsourcing DIY tools for sustainable farming: in the brave new world of farm technology, communities are inventing new tools and cutting out Big Ag. Takepart. Available at: http://tinyurl.com/hnb2xeg.

Tallacchini, M. (2009). Governing by values. EU ethics: soft tool, hard effects. Minerva 47(3): 281-306.

Tornaghi, C. (2014). How to set up your own urban agricultural project with a socio-environmental justice perspective. A guide for citizens, community groups and third sector organisations. Available at: http://tinyurl.com/o72gfmr.

Zwart, H. (2000). A short history of food ethics. Journal of Agricultural and Environmental Ethics 12(2): 113-126.

Zwart, H. (2005). Food consumption in the genomics era: a foucauldian perspective. Tailoring Biotechnologies 1(2): 31-44.

47. Plants in food ethics: a critical approach

A. Kallhoff
University of Vienna, Department of Philosophy, NIG, R. A 0306, Universitätsstraße 7, 1010 Vienna, Austria; angela.kallhoff@univie.ac.at

Abstract

Recently, plants have received new attention in ethics. Differently from earlier approaches to the life of plants in philosophy, authors now claim moral respect for the life of plants. Arguments for this approach are derived from various sources: from new insights in the complexity of the life of plant, from approaches to the good life of plants, from virtue ethics and from a reassessment of biotechnological applications. This contribution presents some core arguments in each area and discusses consequences from plant ethics for food ethics.

Keywords: plants, plant ethics, ethics of plant cultivation

Introduction

In ethics, plants were usually regarded as items that do not deserve a moral status. Exemptions of this general observation are reserved to some biocentric positions in ethics (Attfield, 1991; Taylor, 1986). One key argument for not regarding plants as moral patients is the lack of capacities that support a view according to which plants should not be harmed. Plants lack sentience and they lack conscience. From this, philosophers have concluded that there are no interests of plants in either reduction of suffering or in a good life that need to be considered.

Recently, the assessment of plant life has changed. Even though authors do not revise basic empirical classifications of the life of plant, they argue that plants deserve moral respect (Kallhoff, 2014; Pouteau, 2013). This contribution discusses these recent turns in plant ethics and scrutinizes the arguments. It then discusses the consequences for food ethics.

Arguments from complexity

Plants constitute a biological kingdom whose boundaries are not particularly clear-cut. Within this kingdom, there is a rich diversity of life-forms. In this contribution, I shall focus on 'higher plants' – they form a group with distinct qualities and are particularly important as a source for human food. Higher plants, including trees, have recently been discussed as particularly complex beings. They are regarded as entities whose capacities are comparable to the capacities of animals. In particular, plants are able to communicate (Baluška and Volkmann, 2010). Plants also have senses that – even though exclusively working on a biochemical level of exchange – are highly differentiated (Mancuso and Viola, 2015: 45). Even though plants are completely different from animals regarding their organization and their functional units, the assessment of their complexity contributes to rethinking the value of plants. Mancucso and Viola (2015) go so far as to claim respect for a type of living organization that is particularly intelligent, even though in a specified way.

As for the normative side of the argument for plant respect, the approaches that highlight the complex capacities of plants resonate with former approaches to biocentrism that have been particularly fruitful in plant ethics (Attfield, 1981, 1991; Sterba, 2011; Taylor, 1986; Varner, 1998). In addition to claiming respect for life, authors now claim respect for an underrated life-form. Even though plants differ from humans and from animals, they deserve appreciation and respect not only because of fine-grained

EurSafe 2016 – DOI 10.3921/978-90-8686-834-6_47, © Wageningen Academic Publishers 2016

interactions with the environment. It appears as if plants – even though not self-conscious in any way comparable to people or animals – possess a rich diversity of strategies for developing a broad range of interactions with other living entities.

Arguments from the good life of plants

Stronger arguments for addressing anew the life of plants in ethics result from an interpretation of the life of plants as 'flourishing'. This neo-Aristotelian concept has received some prominence in the debate of the good life of people and of animals (Nussbaum, 2006). It is now extended so that it encompasses the good life of plants too. Actually, it appears to fit much better to the life of plants than to that of other living beings. 'Flourishing' emphasizes that an entity has the capacity to unfold its specific characteristics in environments that do not hinder them from unfolding. In particular, 'flourishing' stresses the capacity of an entity to run through a life-cycle. In particular, this concept does not support a rather anthropomorphic interpretation of plants in terms of 'interests' or even 'consciousness'. Instead, it highlights the capacity of plants to actively support self-evolvement in terms of occurring typical characteristics of a species while reacting to stressful environments and while assuring reproduction at least once. As a consequence, this concept helps to individuate incidents of severe stress and of harmful effects on plants. It does not in and of itself include a reason for why harm on plants should be reduced – yet, it provides a first step in reasoning direct moral respect for plants (Kallhoff, 2002, 2014). Once the 'flourishing' of plants has been addressed as an inherent goal of plant activities, it is also possible to support moral claims that claim respect for the well-being of non-human entities including plants.

The arguments include the following steps: firstly, within a framework that defends the claim not harming the well-being of entities arbitrarily, plants need to be included. Following an interpretation of the good life of plants as an actively pursued goal of the functional unit of a plant, an objective notion of 'harming' that relates to 'flourishing' can be established. secondly, the argument addresses the values of flourishing from various perspectives. In addition to arguments on respect regarding well-being, respect for flourishing gets additional support from an aesthetic, a resource-oriented and an ecological approach to nature. Thirdly, arguments need to be provided in order to explore conflicts between respect for plant life and respect for human interests. They include an approach to limited rights of people once areas of conservation have been declared. They also include approaches to respect in cultivating plants (Kallhoff, 2002, 2014).

Arguments from the ethics of plant cultivation

Differently from re-addressing moral concern in terms of a moral status of plants or in terms of respect for their capacity to flourish, authors who have been interested in plant ethics have focused on values in cultivating plants. Following virtue ethics, good habits in cultivating plants have been addressed as 'virtues'. In particular, the sources of virtue ethics have been diversified in recent approaches. Starting with a re-appraisal of Aristotelian ethics, recent contributions include virtues that resonate with claims in environmental ethics (Walker, 2007). Moreover, respect for the environment now figures as an important part of virtue ethics (Cafaro and Sandler, 2010; Hursthouse, 2007). Different from an approach to the moral status of plants and different from an approach to the ethics of the good life, virtue ethics aims at reasoning attitudes to nature that pay tribute to the needs of natural beings. In particular, recent approaches to virtue ethics depart from a framework that relates virtues to human flourishing. Instead, it argues that virtues are an important ingredient in a new relation to nature. They are necessary in order to overcome a situation of dominion that threatens the life of non-living beings by exploiting practices of civilization (Treanor, 2014).

This line of thought has particular implications for respect of plant life. The ethics of gardening illustrates the application regarding plant life particularly well. The ethics of gardening works on a relational level in that it considers the activities of gardeners in the garden as a paradigmatic case of 'caring for nature' (Brook, 2010; Hall and O'Brien, 2010). In particular, the practice of gardening is a type of cultivated activity that presupposes a virtuous gardener. Being capable of enduring volatility of natural developments and prudence are part of the set of virtues that gardeners should have in order to succeed in gardening.

Plants in applied ethics

Finally, ethics has engaged in debating technical modifications of plants including genetic engineering, the creation and sustenance of so-called 'Biofakte', an assemblage of botanic and technical structures, manipulation of seed and more recently approaches to synthetic biology on plant cells. Authors in applied plant ethics have provided critical approaches in order to justify respect for the integrity of plant life (Heaf and Wirz, 200; Rehmann-Sutter, 2001). This critical investigation of technological applications to plant life cannot be portrayed in full length here (For a thorough interpretation of plants in applied ethics, see Attfield, 1981; Kallhoff, 2002). Yet, some important issues shall be recalled. A reassessment of technical modifications of plant life includes a debate about the meaning of 'naturalness' regarding the life of plants; it also includes a focus on risks both for the environment and for human health and a debate of the meaning of interference with nature in diverse scenarios.

Overall, in this context of research, plants are also not regarded as entities whose moral status needs to be rejected due to a lack of significant capacities – capacities that render an entity closer to human beings. Instead, they are at the centre of concern in environmental ethics. In addition to the arguments that have been recalled there are new emerging contexts of plant ethics, including the debate on eco-centrism and forest ethics more specifically, and a debate on the role of plants and vegetation as the debate on climate justice sinks in.

Consequences for food ethics

What does this reconsideration of plant life in ethics mean for food ethics? One consequence is not far-fetched: against the background of plant ethics, it must be emphasized that plants are more than food. Even though plants are needed for animals as well as for people as a necessary ingredient in the food chain, plant ethics reminds us of the many facets of plant life. In particular, it addresses values of plants that cannot be reduced to the value of food.

Yet, plant ethics does not only argue in favour of respect for plant life. It also helps to spell out some of the normative principles that should guide agriculture and techniques of food supply by means of cultivating plants. Stemming from the insight that plants are capable of a good life, philosophers have explored categories that are important in protecting plants from harm and that are also important in supporting good plant life. Plants differ from other life-forms in that they are immediately tied to a place where they live. Moreover, they interact with other plants and organisms in a multi-faceted and intense way. They build communities; some argue that some plant species are even capable of organizing themselves in a quasi-prosocial way. I shall not discuss this claim here.

Following a proposal by Paul Taylor (1986), the relation between human civilization and vegetative areas of life can be split into various types. This interpretation of nature can be applied to plant ethics (Kallhoff, 2002). Overall, three different types of relationship can be distinguished: firstly, plants belong to 'wild nature' in that they live in spaces that are to a high degree free from being designed to serve human interests. When plants have an interest in their good life that is expressed by ways of flourishing, the claim for sparing some places for a free evolvement of flourishing is a starting-point for rethinking moral claims regarding plants; secondly, plants are cultivated. Yet, when respect for the flourishing of entities that differ from human life is justified, cultivation should also be adjusted to respect for flourishing. Even though it is not a moral duty to respect the conditions for flourishing, it can be argued that – overall – plant cultivation should be exempt from doing arbitrary harm to plants (this has been argued in length in Kallhoff, 2002). Fortunately, conditions of flourishing exemplars and the flourishing itself do not themselves clash with interests in cultivation; thirdly, plants are used for human interests. When it is to some degree justified that plant life is instrumentalised for human interests, people and civilizations are also free to use plants in this way.

Overall, in order to respect plant life, it is necessary to reserve areas and land for plant life and to protect areas from direct impact of plant cultivation. As for areas of wild plant life, non-interference is a leading principle (Taylor, 1986). Another is the restoration of conditions that contribute to areas in which plants have a chance to flourish. This does not say that plants are not highly adaptable. It has been demonstrated that areas such as old military areas or cities are particularly rich in terms of diversity. I simply wish to highlight that if plants deserve respect and if diversity is a matter of concern, it is important to protect areas of 'wild' nature too. This is also a place where food chains among living organisms are not interrupted or interfered with by human beings.

Even though plants cannot be hurt by cultivation in greenhouses, respect for plants includes practices of cultivation that resonate with the integrity and the striving for flourishing of plants. The study of stress behaviour of plants reveals that plants have an inherent impetus for realizing their full set of characteristics, both as a plant and as a species-member. Exploring their behaviour in terms of 'stress-aversive strategies' contributes to interpreting plants' behaviour as an active process of protecting life-processes from harm and of supporting the vital functions and the process of development of the typical characteristics of a life-form (Kallhoff, 2002: 37-69). Even though stress is a permanent phenomenon in plant life, both qualities and quantities of stress can be divided into stimulating factors of plant life and in destructive forces. Respect for plant life includes respect for the limits of adaptation and for 'good' environments according to the parameters that ecological studies claim as essential for the development of plants. This does not mean that cultivation needs to provide 'natural' circumstances for plants. Yet, its practices should be oriented along the lines of stimulating and healthy environments for plants.

As for food ethics, this critical approach to plant life has severe implications. Instead of taking it for granted that plants can be used as 'food' for people, plant life needs to be accepted firstly as a type of life that also deserves respect. As a consequence, the space for agriculture needs to be limited. It has to be respected that plants need space to unfold their capacities in order to flourish. An approach to the flourishing of plants as in itself valuable also forestalls practices of cultivation that do not pay respect to the needs of plants in order to flourish. The argument for paying tribute to good living conditions of plants does not result from claims of human healthiness or an overall better way to integrate civilization into the natural world. In particular, the value of the practices of cultivation is not the core argument against the background of plant ethics. Instead, it is important to take into account that the flourishing of plants can be harmed. Moreover, respect for plant life includes the claim to leave enough space to plants so that they can unfold freely.

References

Attfield, R. (1981). The good of trees. Journal of Value Inquiry15: 34-54.

Attfield, R. (1991). The ethics of environmental concern. University of Georgia Press, Athens and London, 280 pp.

Baluška, F., Mancuso, S. and Volkmann, D. (Eds.) (2010). Communication in plants: neuronal aspects of plant life. Springer, New York, NY, USA, 438 pp.

Brook, I. (2010). The virtues of gardening. In: O'Brien, D. (ed.) Gardening – philosophy for everyone. Cultivating wisdom. Wiley-Blackwell, Oxford, UK, pp. 13-25.

Cafaro, Ph. and Sandler, R. (2011). Virtue ethics and the environment. Journal of Agricultural and Environmental Ethics 23(1-2).

Hall, M. and O'Brien, D. (2010). Escaping Eden. Plant ethics in a gardener's world. In: O'Brien, D. (ed.) Gardening – philosophy for everyone. Cultivating wisdom. Wiley, Oxford, UK, pp. 38-47.

Heaf, D., Wirz, J. and Ifgene (2001). The intrinsic value and integrity of plants in the context of genetic engineering. Proceedings of an Ifgene workshop on 9-11 May 2001. Goetheanum, Dornach, Switzerland, 66 pp.

Hursthouse, R. (2007). Environmental virtue ethics. In: Walker, R.L. and Ivanhoe, Ph.J. (eds.) Working virtue. Virtue ethics and contemporary moral problems. Oxford University Press, Oxford, UK, pp. 155-172.

Kallhoff, A. (2002). Prinzipien der pflanzenethik. Die bewertung pflanzlichen lebens in biologie und philosophie. Campus, Frankfurt and New York, 163 pp.

Kallhoff, A. (2014). Plants in ethics: why flourishing deserves moral respect. Environmental Values 23(6): 685-700.

Mancuso, St. and Viola, A. (2015). Brilliant green. The surprising history and science of plant intelligence. Island Press, Washington, DC, USA, 184 pp.

Nussbaum, M.C. (2006). Frontiers of justice. Disability, nationality, species membership. Harvard University Press, Cambridge, MA, USA, 512 pp.

Pouteau, S. (2013). Beyond second animals: making sense of plant ethics. Journal for Agricultural and Environmental Ethics 27(1): 1-25.

Rehmann-Sutter, C. (2001). Dignity of plants and perception. In: Heaf, D. and Wirz, J. (eds.) Intrinsic value and integrity of plants in the context of genetic engineering. Ifgene, Dornach, Switzerland, pp. 4-8.

Sterba, J.P. (2011). Biocentrism defended. Ethics, Policy and Environment 14(2): 167-169.

Taylor, P. (1986). Respect for nature: a theory of environmental ethics. Princeton University Press, Princeton, USA, 329 pp.

Treanor, B. (2014). Emplotting virtue : a narrative approach to environmental virtue ethics. SUNY Press, New York, NY, USA, 258 pp.

Varner, G. (1998). In nature`s interest? Interests, animal`s rights and environmental ethics. Oxford University Press, Oxford, UK, 168 pp.

Walker, R.L. (2007). The good life for non-human animals: what virtue requires of humans. In: Walker, R. and Ivanhoe, Ph.J. (eds.) Working virtue. Virtue ethics and contemporary moral problems. Oxford University Press, Oxford, UK, pp. 173-189.

48. Media framing and perceptions of risk for food technologies: the case of 'pink slime'

J. Chung[1], K. Runge[2], L.Y.-F. Su[1], D. Brossard[1] and D. Scheufele[1]*
[1]Department of Life Sciences Communication, University of Wisconsin-Madison, 1545 Observatory Drive, Madison, WI 53706, USA; [2]University of Wisconsin Extension, University of Wisconsin-Madison, 610 Langdon Street, Madison, WI 53703, USA; jennifer.chung@wisc.edu

Abstract

On March 7, 2012, ABC World News Report aired the first in a series of reports on 'lean finely textured beef' asserting that 70% of ground beef sold in United States stores contained '(pink slime) beef trimmings ... once used only in dog food.' The resulting response from social media, television stations and newspapers was strong, even though this product had been widely sold since 2001. By the end of 2012, a total of 728 news stories about 'lean finely textured beef' were broadcast on American television stations, and 360 news articles were published in US newspapers. In the aftermath, there have been broad discussions within the meat industry about the effects of message framing on public attitudes and acceptance towards food technologies and production methods. Using this incident as a case study, we embedded an experiment in a survey on consumers' perceptions of food and food-related issues. We created two versions of the survey, one using the media preferred term 'pink slime' and the other using the industry standard term 'lean finely textured beef,' to better understand the effects of media framing of this issue. We found that this variation in terminology was enough to produce significant differences between groups. Those exposed to the 'pink slime' condition reported paying more attention to media stories about the issues as well as perceiving greater risk regarding ground meat. In our analysis we discuss the implications of media framing of food issues and potential consequences for consumers' perceptions of future food innovations and technologies.

Keywords: ground beef, meat production, consumers, experiment

Introduction

'Lean finely textured beef' (LFTB) has been a meat product that has been available and widely used throughout the industry since 1991 (Greene, 2012). By 2011, LFTB was estimated to be in as much as 70% of the ground beef sold in the U.S. (Adams, 2014). It wasn't until shortly after this time that the American public was first presented with a negative portrayal of the product. The issue gained considerable traction in 2011 following an episode of Jamie Oliver's show, 'Food Revolution,' which was viewed by 5.4 million people (Detre and Gunderson, 2012). Oliver, a popular chef and television personality, demonstrated the process of making LFTB by mixing beef trimming and household ammonia cleaner in a washing machine in an attempt to dispel the truth of what Americans were really eating. This created a wave of public outcry as concerned citizens relayed this message on social media and petitioned for its removal from grocery stores and school lunches. Following suit, major grocery retail chains, fast food chains such as McDonald's, Burger King, and Taco Bell, and public school lunch programs across the country stopped selling ground beef with LFTB (Avila, 2012; Detre and Gunderson, 2012). Many meat producers were forced to significantly cut back on production, which resulted in thousands of layoffs and facility shutdowns. Beef Products Inc. (BPI), the main producer of 'LFTB,' as well as other large producers like Cargill, were heavily affected as production was cut by more than half (Kay, 2012). This move affected suppliers downstream resulting in additional shutdowns and essentially collapsing the entire LFTB market (Greene, 2012).

EurSafe 2016 – DOI 10.3921/978-90-8686-834-6_48, © Wageningen Academic Publishers 2016

While much of the aftermath from the controversy has already taken an irreversible toll on the industry and on consumers' perceptions of the product, it is useful to re-examine the impact of media coverage on this issue. The issue may also stem from a simpler problem of wording. LFTB had been a noncontroversial process for many years. It wasn't until the emergence of the term 'pink slime' that the controversy gained momentum. The purpose of this study is to understand the factors that contribute to food controversies such as the one presented here. We want to examine how framing issues and, more specifically, the language that is used to provide context to these issues has an impact on resulting perceptions. The experimental design we have created for this study highlights the importance of something as basic as word selection. We use the terms 'lean finely textured beef' and 'pink slime' to mimic the approach of the media when portraying this issue to the public. Finally, we consider additional factors such as media use, values and knowledge related to food, and trust in organizations to shed more light on what factors may explain consumers' perceptions of risk regarding ground meat.

Literature review

Perceptions of risk in meat and food technologies

Negative judgment on the issue stems from two sources: consumers and the media. Both have the power to create and perpetuate a controversy in a cyclic nature as one leans on the other to keep the issue going. It can be argued that the media's main role is to provide information to the public. However, in the current environment of politicized programming and selective audiences, the media may play an even greater role in influencing public attitudes and dictating the course of discussion (Dudo *et al.*, 2011). This can even be accomplished without including any factually based evidence as media have been known to be amplifiers of misinformation and risk (Bobo and Chakraborty, 2015). Not only are consumers reliant on the media in making informed food decisions, but they are also more interested in receiving negative than positive news about food from the media (e.g. food outbreaks and product recalls) (Colman and Vink, 2005; Yadavalli and Jones, 2014). In conjunction with the presence of social media and the 24-hour news cycle of the internet, both food producers and consumers are in danger of miscommunication on many issues.

These effects have even stronger implications for the meat industry. Meat is a category of food that has a heightened relationship with consumers. People with high proportions of meat in their diet are more concerned about meat products than any other food, which makes them even more sensitive to perceptions of risk (Verbeke *et al.*, 2007). Research on consumer attitudes toward beef processing have found that meat should be minimally processed, and any production should be transparent to the customer or left to be prepared at home; the main goal is to keep it as 'simple and natural' as possible (de Barcellos *et al.*, 2010). In the past, the introduction of new meat production techniques has been less accepted by the public compared to other food technologies (da Costa *et al.*, 2000; Korzen *et al.*, 2011; Van Kleef *et al.*, 2005). Therefore, meat producers should be aware that future innovations are likely to be met with high resistance from the public.

In the case of 'lean finely textured beef,' the level of perceived risk by consumers and the amplification of the risks by the media have created doubt and fear towards a widely accepted industry practice. LFTB processing yields a greater amount of usable meat from each animal, decreases the number of animals that are slaughtered per given volume of beef, lowers the overall fat content of ground beef, and creates a finished product that is deemed to be safe for human consumption by the Food and Drug Administration (FDA) and the US Department of Agriculture (USDA) (Adams, 2014). Despite these benefits, the overwhelmingly negative portrayal of this product in the media has perpetuated the notoriety and public rejection of LFTB.

Product framing: 'Lean finely textured beef' or 'Pink Slime.'

The success of the media in framing 'lean finely textured beef' as 'pink slime' may be explained by our tendency to rely on cognitive schemas and heuristics. New information can be processed by making links between pre-existing and newly formed schemata (Greehy *et al.*, 2013). This has been applied many times before in framing controversial and potentially unfamiliar issues, such as stem cells (Nisbet *et al.*, 2003) and biofuels (Cacciatore *et al.*, 2012), in contexts that are beyond or unrelated to the actual science. One way to achieve this is by focusing on terminology, as wording can have the potential to change the contexts in which people think about these issues (Tversky and Kahneman, 1981). Examples include: 'ethanol' as chemical processing of corn, 'food irradiation' as poisoning by radiation or cancer, 'genetic modification' as injecting foreign substances, and 'frankenfood' as experimentation gone wrong much like Frankenstein's monster (Cacciatore *et al.*, 2012; Greehy *et al.*, 2013; Leydesdorff and Hellsten, 2006).

The primary goal of this research is to explore the factors that people use to evaluate LFTB. First, we assess the role that wording has on consumers' evaluation of the technology. We predict that using the terms 'lean finely textured beef' or 'pink slime' will function as a frame in judging identical products and expect perceptions of risk to be higher for those exposed to the issue as 'pink slime.' In addition, we look at the impact of other variables (i.e. demographics, media use, values, knowledge, and trust) to provide additional support that these factors do not have as much impact on evaluations as the presentation of the messages themselves. In this manner, we plan to better detail the contribution that framing has on shaping perceptions of risk for this issue and related food technologies.

Methods

The data for this study were collected from a state-wide mail survey of Wisconsin residents. The survey was conducted by the University of Wisconsin-Madison Survey Center from April 10, 2015, through July 7, 2015. A random sample of addresses in Wisconsin was selected for participation in this study. Selected homes were sent a copy of the survey for completion by one person in the household, 18 years or older. Each survey also included a $2 cash incentive. Once completed, respondents were instructed to send back their forms in the postage-paid envelope. Participating households also received two additional reminder notices during the first two months of the survey period. Only those surveys that were completed and mailed back to the Survey Center by the required date were included in this analysis. The final sample used for this study included 931 respondents, which represents a 50.3% response rate.

The dependent variable, levels of risk associated with ground meat, was assessed by using the questions, 'How risky are the following foods to you, personally?' and 'How risky are the following food to the average American?'. Respondents were presented with a list of 13 food items, one of which was ground meat. As part of our experimental manipulation, we employed two variations of this item. In one version, participants were asked these questions using the phrase 'ground meat containing lean finely textured beef' while the other version used the phrase 'ground meant containing 'pink slime.'' The variable Survey Version determined whether respondents viewed question wording as 'lean finely textured beef' (51%) or 'pink slime' (49%). The remaining dependent variables are listed in Table 1 in the following section.

Results

Table 1 provides the standardized regression coefficients of all final models. The terminology manipulation was significant in predicting perceptions of risk toward ground beef. In particular, those who received the 'pink slime' wording were significantly more likely to perceive risk associated with ground beef than those who viewed the 'lean finely textured beef' version. This variable alone accounted for 28.6% of the total variance in risk associated with ground beef. Other significant findings include

Table 1. Hierarchical ordinary least squares (OLS) regression predicting risk associated with ground meat (Standardized regression coefficients).[1,2]

	Zero-Order	Model 1	Model 2	Model 3	Model 4	Model 5	Model 6	Model 7
Block 1: Survey Version								
(Pink slime = high)	0.53***	0.53***	0.54***	0.54***	0.47***	0.47***	0.47***	0.47***
Incremental R^2 (%)		28.6***						
Block 2: Demographics								
Age	0.06		0.06	0.06	-0.00	0.02	0.02	0.04
Gender (female = high)	0.07*		0.11***	0.11***	0.09**	0.07*	0.06*	0.06*
SES	-0.07*		-0.01	-0.01	-0.01	-0.01	0.00	-0.01
Children under 18	-0.00		0.03	0.03	0.03	0.03	0.03	0.03
Incremental R^2 (%)			1.5**					
Block 3: Values								
Religiosity	0.03			0.02	-0.00	-0.01	-0.02	-0.01
Ideology (liberal low)	-0.02			-0.04	-0.02	-0.01	-0.01	0.01
Incremental R^2 (%)				0.10				
Block 4: News media use								
Attention to food-related news	0.20***				0.08*	0.02	0.02	0.01
Attention to news stories about pink slime/ LFTB	0.40***				0.23***	0.21***	0.21***	0.21***
Incremental R^2 (%)					7.3***			
Block 5: Food values								
Organic and GMO-free	0.25***					0.12***	0.11**	0.09*
Freshness and locally grown	0.24***					0.10**	0.11**	0.12***
Incremental R^2 (%)						2.9***		
Block 6: Knowledge								
Perceived food production knowledge	0.02						-0.04	-0.03
Perceived food-related knowledge	0.11**						0.01	0.01
Food-related knowledge	-0.06						-0.04	-0.03
Incremental R^2 (%)							0.30	
Block 7: Trust								
Trust in for-profit food institutions	-0.10**							0.16***
Trust in government-related food institutions	-0.06							0.09*
Trust in media for information on science, health, and nutrition	0.06							0.07*
Incremental R^2 (%)								1.4***
Total R^2 (%)								42.1***

[1] Cell entries for all models are final standardized regression coefficients.

[2] *P<0.05; **P<0.01; ***P<0.001.

higher risk towards ground meat for females, those who pay greater attention to news stories about 'pink slime' and 'LFTB,' those who place more value in purchasing foods that are organic, non-genetically modified organisms, fresh, and locally grown, and those with less trust in for-profit food organizations, more trust in government-related food institutions, and more trust in the media. Overall, the regression models predicted 42.1% of the variance.

Discussion

The 'pink slime' debate captures an interesting area of work on the effects of framing messages to the public. An analysis of previous framing research showed the complexity and nuance of human thought processes in helping understand the importance of something as detailed as the selection of words and has the potential to produce radically different responses even when presenting the same set of information (Tversky and Kahneman, 1981). When considering the benefits of this type of work, it would be most useful to discuss future applications. There is a lot of work on the framing of messages but less analysis on the effect of individual words. By changing the language that was used from a longer, more technical name to a short, descriptive moniker displays the impact of wording choices. This approach to framing can be a valuable tool not only for mass communicators but for the industry as well. If using different labels has the impact to sway public opinion in such a negative way, then it could be possible to use them to illicit the opposite response. In the least, it could be useful in countering attacks or crises. Our study provides substantial evidence for the strength of words in framing an issue. However, we also found several variables beyond the frames to have a significant impact on risk perceptions, which suggests the importance of considering factors related to and beyond the message itself. In moving ahead, subsequent work should continue to test better measures for assessing public perceptions of risk and explore additional factors that could be relevant.

References

Adams, R.J. (2014). Consumer deception or unwarranted product disparagement? The case of lean, finely textured beef. Business and Society Review 119(2): 221-246.

Avila, J. (2012). 70 percent of ground beef at supermarkets contains 'pink slime.' ABC News. Available at: http://tinyurl.com/7bez7yo.

Bobo, J. and Chakraborty, S. (2015). Pink slime, raw milk and the tweetification of risk. European Journal of Risk Regulation 6(1).

Cacciatore, M.A., Scheufele, D.A. and Shaw, B.R. (2012). Labeling renewable energies: how the language surrounding biofuels can influence its public acceptance. Energy Policy 51: 673-682.

Colman, D. and Vink, N. (2005). Reshaping agriculture's contributions to society. Proceedings of the 25th International Conference of Agricultural Economists. Durban, South Africa.

Da Costa, M.C., Deliza, R., Rosenthal, A., Hedderley, D. and Frewer, L. (2000). Nonconventional technologies and impact on consumer behaviour. Trends in Food Science and Technology 11(4-5): 188-193.

De Barcellos, M.D., Kügler, J.O., Grunert, K.G., Van Wezemael, L., Pérez-Cueto, F.J.A., Ueland, Ø. and Verbeke, W. (2010). European consumers' acceptance of beef processing technologies: a focus group study. Innovative Food Science and Emerging Technologies 11(4): 721-732.

Detre, J.D. and Gunderson, M.A. (2012). Did the 'Pink Slime' controversy influence publicly traded agribusiness companies. Choices 27(4).

Dudo, A., Choi, D.-H. and Scheufele, D.A. (2011). Food nanotechnology in the news. Coverage patterns and thematic emphases during the last decade. Appetite 56(1): 78-89.

Greehy, G.M., McCarthy, M.B., Henchion, M.M., Dillon, E.J. and McCarthy, S.N. (2013). Complexity and conundrums. Citizens' evaluations of potentially contentious novel food technologies using a deliberative discourse approach. Appetite 70: 37-46.

Greene, J.L. (2012). Lean finely textured beef: the 'pink slime' controversy. Congressional Research Service.

Kay, S. (2012). Industry was too late on LFTB issue. Beef Magazine. Available at: http://tinyurl.com/gqtb6l6.

Korzen, S., Sandøe, P. and Lassen, J. (2011). Pure meat – public perceptions of risk reduction strategies in meat production. Food Policy 36(2): 158-165.

Leydesdorff, L. and Hellsten, I. (2006). Measuring the meaning of words in contexts: an automated analysis of controversies about 'Monarch butterflies,' 'Frankenfoods,' and 'stem cells.' Scientometrics 67(2): 231-258.
Nisbet, M.C., Brossard, D. and Kroepsch, A. (2003). Framing science the stem cell controversy in an age of press/politics. The International Journal of Press/Politics 8(2): 36-70.
Tversky, A. and Kahneman, D. (1981). The framing of decisions and the psychology of choice. Science 211(4481): 453-458.
Van Kleef, E., Van Trijp, H.C.M. and Luning, P. (2005). Consumer research in the early stages of new product development: a critical review of methods and techniques. Food Quality and Preference 16(3): 181-201.
Verbeke, W., Sioen, I., Brunsø, K., De Henauw, S. and Van Camp, J. (2007). Consumer perception versus scientific evidence of farmed and wild fish: exploratory insights from Belgium. Aquaculture International 15(2): 121-136.
Yadavalli, A. and Jones, K. (2014). Does media influence consumer demand? The case of lean finely textured beef in the United States. Food Policy 49: 219-227.

49. Which types of knowledge about organic products are consumers interested in?

T. Christensen[1], M. Thorsøe[2] and K. Klitgaard Povlsen[3]*
[1]Department of Food and Resource Economics, University of Copenhagen, Denmark; [2]Department of Agroecology – Agricultural Systems and Sustainability; Aarhus University, Denmark; [3]School of Communication and Culture – Media Studies; Aarhus University, Denmark; tove@ifro.ku.dk

Abstract

Knowledge and trust are important concepts which help to explain consumers' engagement with organics. It has often been stated that knowledge leads to trust. However, by digging deeper into the relationship between knowledge and trust in relation to organics, a recent Danish study found that trust to a certain extent seemed to replace knowledge and thereby offered consumers a choice between knowledge and trust in their relationship with organics. Indeed, they found that the majority of consumers trusted organics despite knowing little. They also found that a significant part of consumers seemed to have a reflexive and deliberate lack of knowledge in their engagement with organics. As a consequence, an increase in the supply of information is not likely to fundamentally change the configuration of knowledge and trust in the food system. The purpose of the present paper is to investigate which types of knowledge about organics that consumers would be interested in, with particular focus on information concerning differences between organic and non-organic products. A nationwide web-based consumer survey was combined with face-to-face interviews in terms of a focus group and four semi-structured individual interviews recruited among the survey participants. Our preliminary analysis indicated that a lot of people were interested in learning more about the difference between organic and non-organic production. We found that additional information about animal welfare, health, and general differences between rules and characteristics of the different production systems were the most often stated areas – and also information on climatic and environmental impacts and traceability scored high. However, our results also indicated that at least 40% of the respondents were not interested in increasing their knowledge – and that knowledge concerning the values of the producer/seller might be just as important for consumers as specific knowledge concerning the production methods. An important next step will be to identify how to target interested consumers with the type of information they are motivated to take in. Here, social media provide an opportunity but also a challenge.

Keywords: organic consumption, types of knowledge, motivation

Background

Knowledge and trust are important, interconnected concepts which help to explain consumers' engagement with organics.

According to a general description of knowledge, it can be seen as 'a way that actors ascribe meaning to the surrounding world' (Leeuwis, 2007). When talking about the food system, knowledge thus refers to information and experiences relating to the conditions, requirements and effects of food production, as well as knowledge about food system actors. Information cannot be given or received objectively but is always interpreted, and because of this, knowledge also refers to an actor's ability to process information. Trust is a concept that is frequently used and has been defined in a variety of ways. Rousseau *et al.* (1998) developed this widely supported definition of trust as 'a psychological state comprising the intention to accept vulnerability based upon positive expectations of the intention or behaviour of another'. Following a number of sociologists, trust can be divided into proximity or personal trust versus systemic trust (Giddens, 1990; Luhmann, 1979; Putnam, 2000). Based on these refinements of the concept of trust, it was possible to move from trust being a relation between people as suggested by Rousseau *et al.* (1998) to also include a relation between a consumer and an institution safeguarding organic products. Trust is important because it enables consumers to engage with the food system despite uncertainties associated with modern food production. This applies in particular to organics, because consumer trust in the producer's quality claims is vital to their acceptance of price premiums (DuPuis and Gillon, 2009; Kjærnes *et al.,* 2007; Sønderskov and Daugbjerg, 2011).

It has often been stated that knowledge leads to trust (Aertsens *et al.,* 2009; Daugbjerg *et al.,* 2014; Smed *et al.,* 2013). However, by digging deeper into the relationship between trust and knowledge in relation to organics, Thorsøe *et al.* (2016) found that trust to a certain extent seemed to replace knowledge and thereby offered consumers a choice between knowledge and trust in their relationship with organics. Indeed, Thorsøe *et al.* (2016) found that the majority of consumers trusted organics despite knowing little about them, which could indicate some level of blind trust. Secondly, many consumers stating that they knew little or nothing about organics seemed to be content with their levels of knowledge. Thirdly, a desire to increase knowledge about organics seemed to be strongest among consumers who already had a high level of knowledge or a high level of trust. These results introduced a paradox of consumer engagement with organics: a significant part of consumers seemed to have a reflexive and deliberate lack of knowledge. It seems to be necessary that the type of information that is provided appeal to the consumers and is accompanied by efforts to motivate its uptake and conversion into improved knowledge. The present paper presents preliminary results of a study of which types of knowledge about organics that consumers would be interested in when they decide whether or not to buy organic products.

Material and methods

Quantitative as well as qualitative elements were used to explore which types of knowledge about organic products that consumers' were interested in acquiring. More specifically, a nationwide web-based consumer survey was combined with face-to-face interviews in terms of a focus group and four semi-structured individual interviews recruited among the survey participants.

The web-based consumer survey with 3,788 respondents was conducted in September 2013. Respondents from Userneeds' web panel were invited in order to obtain a final sample that was representative of the Danish population according to 3 criteria: gender, age (18-34 years, 35-49 years, 50-70 years) and geographical location (eastern or western parts of Denmark). The respondents were not invited with any specific attention to their organic profile. After two re-invitations, an overall response rate of 16.4% was obtained. Although the response rate is low, this is not uncommon with web-based surveys. We ended up with a sample with an overrepresentation of women (57 against 50% at the national level); younger respondents were underrepresented (17 against 30% at national level), and older respondents were overrepresented (57 against 39% at national level). In addition to these criteria, people with higher levels of education were heavily represented, as were households with incomes above €80,000. Survey respondents, who stated that they would like to know more about the difference between organic and non-organic products, were given the opportunity to state in free text which types of knowledge they found that they needed the most when choosing between an organic and a non-organic product. The free text was systematically analysed according to whether it addressed specific categories of information. Based on the data material, the categories included were: (1) sustainability; (2) proximity; (3) quality/taste; (4) difference in production methods; (5) climate impact; (6) environmental impact; (7) human health; (8) animal welfare; (9) traceability/production chain/country of origin; (10) trust and credibility in the organic production and labelling system; (11) legal requirements of organics; and (12) 'others' which is a rest group including for instance reasons for the large price difference between organic and non-organic products and more information about foreign organic production.

The interviewees for the qualitative interviews were recruited among survey participants, which reduced the flexibility in recruiting considerably. For logistical reasons the qualitative follow-up included conducting only one five-person focus group in Copenhagen in December 2013 which lasted just under two hours, and four individual, in-depth, semi-structured interviews lasting around one hour in the spring of 2014. Despite a limited amount of interview data, they provided very valuable angles on the survey results. Interview guides for the qualitative inquiry were developed from the initial findings of the survey data. In particular, in the qualitative part of the study, we sought to explore consumer motivation for acquiring or ignoring information about organic products. The interviews and the focus group were recorded and subsequently transcribed verbatim. Transcriptions were open coded (Corbin, 1998). Based on the data material, the interview coding produced a total of eleven main categories (with two sub-categories for knowledge): (1) good idea, but ...; (2) health; (3) animal welfare; (4) Danish/local/seasonal; (5) price; (6) taste; (7) climate and environment; (8) media; (9) values; (10) trust and credibility; and (11) knowledge (factual knowledge and trust-generating knowledge).

The analysis focused particularly on combining information from free text answers provided in the survey with findings from the qualitative interviews. The method explored was to report the results of the initial categorization as outlined above and point towards similarities and differences in the two fundamentally different data collection approaches. A deeper analysis of the arguments in the interviews leading to why or why not the interviewees were motivated to acquire additional information will be included in the final article.

Results

Types of knowledge

All respondents in the survey were asked to state to what extent they would like to know more about the difference between organic and non-organic products. The results are shown in Table 1. As many as 60% stated, that they would like to know more about the difference between organic and non-organic products. Unfortunately, 12% of the free-text responses were not sufficiently informative to include in the analysis. Hence, we ended up using information from 48% of all respondents. Among those who wanted to know more about the difference between organics and non-organic products, the most often stated types of knowledge wanted concerned animal welfare, health, and overall knowledge about the difference between organic and non-organic production and processing methods. Also, a relatively

Table 1. Desire for knowledge about organics.

Do you have an interest in gaining more knowledge about the difference between organic and non-organic products?	Respondents (%)[1]
I would like to know much more	19
I would like to know a little more	41
I don't need any further knowledge	32
I don't know whether I want to know more	8
Total	100

[1] n=3,788.

large segment of respondents was interested in obtaining more knowledge about the rules of organic farming, traceability and environmental impact. Minor fractions of the respondents requested more information about climatic impact of organic farming, information that could increase trust in and credibility of organics, information about quality and taste differences, and sustainability information. Furthermore, some respondents had a specific request for more science-based factual information about the exact difference between organic and non-organic products, as much of the current information was perceived as being one-sided and positive. The results are shown in Figure 1.

The personal interviews revealed a different picture than the online survey did. None of the interviewees were interested in additional factual knowledge on differences between organic and non-organic products. In the interviews, desires were expressed for a much broader type of knowledge concerning whether the producers and/or sellers of the product were able and willing to produce food with care and engagement. Here, people were not so interested in factual information, but classified it as 'boring factual talk', rather they indicated an interest in knowing the values, motivations and beliefs of the producers/sellers.

Motivation for acquiring more knowledge

The survey did not confront respondents with questions concerning the degree of effort they would be willing to put into increasing their knowledge about organics. Neither were the interviewees confronted with questions regarding their motivation to acquire additional knowledge about organics in the structured part of the interviews. Nevertheless, when the interviewees stated that their knowledge about organic products was not excessive, they seemed to feel that they should justify why they did not want to acquire additional knowledge. As a common result, the qualitative interviews clearly suggested that the interviewees would only acquire information about organics if it was easily obtainable and could be integrated in their daily routines. One interviewee, when asked whether she was interested in more knowledge about the produce she purchases, replied:

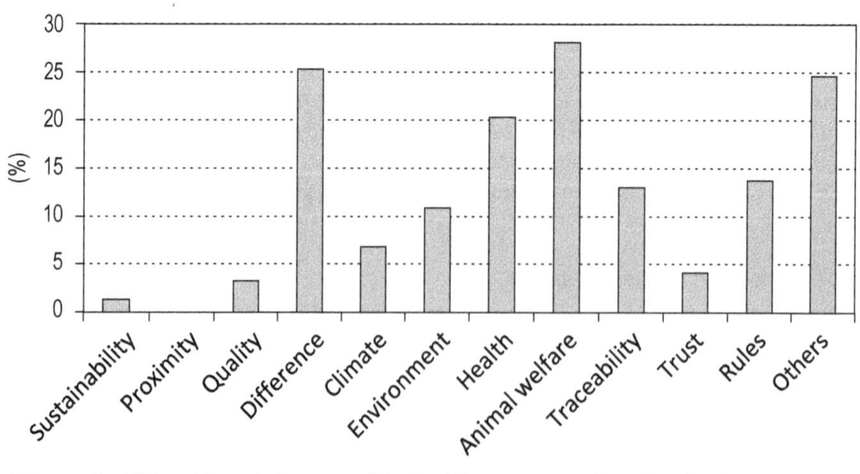

Figure 1. Types of additional knowledge wanted by Danish consumers when choosing between organic and non-organic products. n=1,822 respondents. Online survey among Danish consumers in 2014.

> By god, in my daily life I do not have time to sit in front of the computer to find out who or what it is. I would never get it done. As long as it is certified, then it is fine by me. Then I trust that they (authors: organic producers) have everything in order.

This is a clear example of a consumer who is not motivated to actively search for information about organics. It also shows a person who replaces knowledge with trust in the organic label.

Discussion

Types of knowledge

The types of information that consumers stated to be interested in obtaining in the survey corresponded to earlier findings on which topics consumers link to organic production and products (Aertens *et al.*, 2009). In the individual and focus group interviews were not as interested in factual information as in trust-generating information concerning the producers/sellers. This result supports Meijboom *et al.*, (2006) who argue that it is important to consumers that producers/sellers demonstrate an ability to take consumers concerns into consideration and to make decisions on their behalf. It should be noted though, that the question about acquiring additional knowledge was posed in broader terms. These results suggest that knowledge about the organic producer/seller might be just as important as factual knowledge about production processes and products. They also indicate that differences in the question formulations as well as more general differences between online surveys and face-to-face interviews, might have led the respondents to focus more on product and product-related aspects in the survey than in the face-to-face interviews. These methodological aspects of the conducted study highlight the virtues of using multiple data multiple data sources to increase robustness of the results – and at the same time point towards the dangers of concluding too strongly on too narrow data sets.

Available knowledge

The large number of respondents in the survey who stated their answers in free text reveals a general interest in the topic, but also a general uncertainty about the characteristics and requirements of organic farming. Interestingly much of the requested knowledge regarding the requirements and general knowledge about the difference between organic and non-organic products is readily available on the Internet. It is a fundamental challenge though, that the scientific evidence – the 'facts' – are not unequivocal and therefore difficult to communicate. A recent publication tried to gather all available information regarding differences between organic and non-organic production methods in Denmark (ICROFS, 2015). They found that while organic production seems to have a positive effect on biodiversity in and around agricultural land while the impact on the climate is more complex. Animal welfare in organic production was valued higher in relation to naturalness while higher calf and piglet mortalities in organic production pointed towards lower animal welfare in relation to animal health. No scientific evidence was found to conclude that organic consumption is positively correlated to a healthier life although avoidance of pesticides and food additives were considered as positive contribution of an organic diet. At the same time, there seems to be an agreement that the set of values and rules underlying organic farming are perceived to differ from the non-organic production systems with better values being associated with organic farmers (Thorsøe, 2015). It is a challenge though, how to describe and communicate such differences to potential consumers in a trust-enhancing manner.

Motivation

It is important to bear in mind that only 48% of the respondents in the survey wanted to learn more about the difference between organic and conventional farming. The qualitative interviews added the insight that the interviewees were unwilling to put much effort into increasing their knowledge. This clearly indicates a difficulty in communicating with consumers via pull as opposed to push media as there seems to be a barrier towards actively acquiring more knowledge. Hence, it is still an open question how to design effective ways to communicate the types of information that people perceive that they need and only to those who are motivated to acquire additional knowledge. This task is further complicated by findings dating back to 1944 that knowledge confirming expectations tend to be perceived as trustworthy, while knowledge contradicting expectations is more likely to be seen as untrustworthy (Lazarsfeld *et al.*, 1944).

Limitations

The conclusions that can be drawn from the analysis are tentative as the data have not been analysed in depth yet. A further limitation is that, although the data material from the online survey is extensive, the low response rate and the not-quite representable sample induce a potential bias into the results. The groups that usually are found to be positive towards organics (well-educated and high income) were over-represented in the survey and these groups might also have a different attitude towards their desire to acquire information about organics than other groups. These potential biases have not been dealt with in detail. We ended up with only four personal interviews and a single focus group. Indeed, the information provided from the qualitative part was insightful but too small a sample to provide strong conclusions.

Conclusions

Our preliminary analysis indicated that a lot of people were interested in learning more about the difference between organic and non-organic production. We found that additional information about animal welfare, health and general differences between rules and characteristics of the different production systems were the most often stated areas – and also information on climatic and environmental impacts and traceability scored high. However, we also found that at least 40% of the respondents were not interested in increasing their knowledge – and that knowledge concerning the values of the producer/seller might be just as important as specific knowledge concerning the production methods. An important next step will be to identify how to target interested consumers with the type of information they are motivated to take in. Here, social media provide an opportunity but also a challenge.

References

Aertsens, J., Verbeke, W., Mondelaers, K. and Van Huylenbroeck, G. (2009). Personal determinants of organic food consumption: a review. British Food Journal 111(10): 1140-1167.

Daugbjerg, C., Smed, S., Andersen, L. and Schvartzman, Y. (2014). Improving eco-labelling as an environmental policy instrument: knowledge, trust and organic consumption. Journal of Environmental Policy and Planning 16(4): 559-575.

Corbin, J. (1998). Basics of qualitative research: techniques and procedures for developing grounded theory. SAGE, Thousand Oaks, USA.

DuPuis, E. and Gillon, S. (2009). Alternative modes of governance: organic as civic engagement. Agriculture and Human Values 26(1-2): 43-56.

Giddens, A. (1990). The consequences of modernity. Polity Press, Cambridge, UK.

ICROFS (2015). The organic contribution to societal goods. ICROFS, 405 pp.

Kjærnes, U., Harvey, M. and Warde, A. (2007). Trust in food: a comparative and institutional analysis. Palgrave Macmillan, Hampshire, UK.

Lazarsfeld, P., Berelson, B. and Gaudet, H. (1944). The people's choice: how the voter makes up his mind in a presidential campaign. Columbia University Press, Washington, USA.

Leeuwis, C. (2007). Communication for rural innovation: rethinking agricultural extension. Blackwell Science, Oxford, UK.

Luhmann, N. (1979). Trust and power. Wiley Publishing, Chichester, UK.

Meijboom, F., Visak, T. and Brom, F. (2006). From trust to trustworthiness: why information is not enough in the food sector. Journal of Agricultural and Environmental Ethics 19(5): 427-442.

Putnam, R. (2000). Bowling alone: the collapse and revival of American community. Simon and Schuster, New York, USA.

Rousseau, D., Sitkin, S., Burt, R. and Camerer, C. (1998). Not so different after all: a cross-discipline view of trust. Academy of Management Review 23(3): 393-404.

Smed, S., Andersen, L., Kærgård, N. and Daugbjerg, C. (2013). A matter of trust: how trust influence organic consumption. Journal of Agricultural Science 5(7): 91-106.

Sønderskov, K. and Daugbjerg, C. (2011). The state and consumer confidence in eco-labeling: organic labeling in Denmark, Sweden, the United Kingdom and the United States. Agriculture and Human Values 28(4): 507-517.

Thorsøe, M. (2015). Maintaining trust and credibility in a continuously evolving organic food system. Journal of Agricultural and Environmental Ethics 28: 767-787.

Thorsøe, M., Christensen, T. and Povlsen, K. (2016). Organics' are good, but we don't know exactly what the term means! Trust and knowledge in organic consumption. Food Culture and Society 19(4).

Section 12. Food policy

50. Ethics of dietary guidelines: nutrients, processes and meals

M. Korthals
Free University, Noolseweg 28, 1251 GP Laren, the Netherlands; Wageningen University, Droevendaalsesteeg 4, 6708 PB Wageningen, the Netherlands; michielkorthals@gmail.com

Abstract

Dietary guidelines are mostly issued by agrifood departments or agencies of governments, and are the result of power play between interest groups and values. They have considerable influence in steering food preferences and buyings of consumers. Not only with respect to this power play and their use by consumers are ethical problems at stake. In this paper I will consider three different types of guidelines: a nutrient oriented one (like the Dutch or American ones), a process oriented one and a meal oriented one (like the Brazilian one). In the nutrient oriented guidelines nutrients or sometimes food stuffs are mentioned that excel in containing one or more 'healthy' nutrients. In this variant biological and nutrition scientists have a lot of influence and the producers of the nutrients, like the sugar, dairy and animal industry. The nutrient oriented framing of food is focussed on individual health. Individuals are implicitly addressed to take responsibility for their own long term health, in they have to acquire knowledge skills in interpreting their bodily data, and skills of self-surveillance. In the second process oriented type, foods and ingredients are categorized according to minimally, refined and reconstituted production process. Nutrition scientists have the last say here. In the third one, politicians, social scientists and fresh food producers have the most to say with respect to the formulation of the guidelines. Buying fresh products, preparing and eating together are the main themes. Food is framed in terms of cooking and eating meals. Consumers have to spend time in buying fresh ingredients and cooking. The formulation of dietary guidelines is not a neutral operation, but determined by controversies about framings of food and health and also of integrity and self-confidence. It is striking that the biological and nutrition scientists mostly proclaim that they have the final truth on the healthiness of nutrients, and heap scorn on diet gurus. They do not care that their often incoherent recommendations cause a lot of confusion with consumers and sometimes even lack of trust. I discuss these themes in the paper and will also give some arguments why in the current situation the meal oriented one is ethically more acceptable.

Keywords: dietary guidelines, experts, framing, power, nutrients, meals

Introduction

Dietary guidelines are mostly issued by agrifood departments or agencies of governments, and are the result of power play between interest groups and values. They have considerable influence in steering food preferences and buyings of consumers. Not only with respect to this power play and their use by consumers are ethical problems at stake. Different types of guidelines attribute different responsibilities to stakeholders and define the role of consumers differently; they also presuppose that the evidence their guidelines are based on is reliable and should be trusted. In this paper I will consider three different types of guidelines: a nutrient oriented one (like the Dutch or American ones), a process oriented one (as far as I know not issued by a governmental agency) and a meal oriented one (like the Brazilian one). The formulation of dietary guidelines is not a neutral operation, but determined by controversies about framings of food and health and also about integrity and self-confidence.

Nutrient oriented guidelines

In the nutrient oriented guidelines nutrients or sometimes food stuffs are mentioned that excel in containing one or more 'healthy' nutrients, like animal and vegetable proteins, carbohydrates, saturated

EurSafe 2016 – DOI 10.3921/978-90-8686-834-6_50, © Wageningen Academic Publishers 2016

and unsaturated fats, vitamins and minerals. Sometimes food products are mentioned, like milk, cheese, fruits and meat. These guidelines are mostly formulated by governments or governmental agencies. In this variant biological and nutrition scientists have a lot of influence as well as the producers of the nutrients, like the sugar, dairy and animal industry. It is striking that the biological and nutrition scientists mostly proclaim that they have the final truth on the healthiness of nutrients, and heap scorn on diet gurus. In the Dutch case it is interesting to see that the nutrition scientists recommend to eat fish modestly due to decline of fish species. However, other environmental issues are not taken into account, such as the impact of meat production and eating on climate change. The federal agency issuing dietary guidelines of 2015 was ordered by the US government to pay attention to nutrition exclusively and environmental considerations are not mentioned, although in the first drafts sustainability arguments were incorporated. Despite huge lobbying of the sugar industry it now recommends to drastically cut back on sugar, and it warns not to eat too much chicken, eggs and meat.

The nutrient oriented framing of food is focussed on individual health. Individuals are implicitly addressed to take responsibility for their own long term health, and they have to acquire knowledge skills in selecting, buying and consuming food, in interpreting their bodily data, and to acquire skills of self-surveillance.

Process oriented recommendations

In the second process oriented type, foods and ingredients are categorized in three groups according to minimally, refined and reconstituted production process. Nutrition scientists have the last say here (Scrinis, 2013).

The first group covers a minimally processed of whole foods like fruits and vegetables, by for example cutting, cooking or fermenting. The second group is a refined processing of certain ingredients by selecting them, like flour, olive oil and fruit juice. The third group reconstitute a product by adding ingredients and chemical additives; the product is now reshaped (is now 'ultra-processed food'). The guidelines recommend to eat the first unlimited; the second with care and disapprove eating products of the third type.

Food, nutrition and health scientists, technologists and companies have here the most to say because they have to give reliable information on the type of processing of food products. Consumers have to acquire skills in distinguishing between the different types of processing, to pay attention to labels and other information types to find in what category a food product belongs. Producers of fresh food are encouraged to produce more; producers of the third type are not stimulated.

Meal oriented guidelines

A good example of this orientation are Brazil's new dietary guidelines (Monteiro, 2013). The guidelines starts with three 'golden rules': Make foods and freshly prepared dishes and meals the basis of your diet; Be sure oils, fats, sugar and salt are used in moderation in culinary preparations; limit the intake of ready-to-consume products and avoid those that are ultra-processed.

These rules are specified in ten Brazilian guidelines:
1. Prepare meals from staple and fresh foods.
2. Use oils, fats, sugar and salt in moderation.
3. Limit consumption of ready-to-consume food and drink products
4. Eat regular meals, paying attention, and in appropriate environments.
5. Eat in company whenever possible.

6. Buy food at places that offer varieties of fresh foods. Avoid those that mainly sell products ready for consumption.
7. Develop, practice, share and enjoy your skills in food preparation and cooking.
8. Plan your time to give meals and eating proper time and space.
9. When you eat out, choose restaurants that serve freshly made dishes and meals. Avoid fast food chains.
10. Be critical of the commercial advertisement of food products.

Brazil is suffering from the double burden of what is called the nutrition transition: on the one hand undernutrition because of lack of micronutrients and overweight and obesity due to processed food. The rules presuppose that eating is a social act (5, 6) and encourage to develop critical skills of handling food supplies (7, 10).

In this type, politicians, social scientists and fresh food producers have the most to say with respect to the formulation of the guidelines. Buying fresh products, preparing and eating together are the main themes. Food is framed in terms of cooking and eating meals. Consumers have to spend time in searching for and buying fresh ingredients. Moreover, they have to spend time with preparing and cooking. To have a healthy and tasty meal is the objective of these guidelines, which also means that social processes of choosing, buying, preparing, cooking and eating together are emphasized.

Fresh food producers are encouraged to produce more fresh food and to take care that this food has still sufficient number of vitamins and micronutrients (there is good evidence that the proportion of healthy micronutrients of fresh products such as spinach is declining due to breeding processes). Politicians and other stakeholders are encouraged to provide time and occasion for meals.

Most guidelines are not motivating people to eat well

It is remarkable that although governments and their agencies spend a lot of time in the process of formulating these guidelines, it is quite clear that most consumers do not live up to those rules. In the Netherlands, only 5% (for vegetables and fruit) to 20% (for fish and salt) of the population orient their food habits on these rules (Nationaal kompas volksgezondheid). In Europe we see a decline in the number of people that follow the rules of their food agencies (Figure 1 and 2).

Nevertheless, although these statistics are confronting, the food agencies do not change their approach, and continue by giving recommendations, mostly according to the nutrition type. (Due to its recent formulation, the third is not analysed on its impact). Why do people not live up to the guidelines? It is amazing that the research on this particular topic is scarce. I couldn't find specific literature on this issue. But in general, social scientific research points to lack of trust in official reports; probably this lack of trust is stimulated by the over the years varying guidelines, by the stories of the lobbying practices of big food (less sugar, than lobby, for more sugar; less meat, than lobbying for more meat; pizza is vegetable, etc.), and also by the incomprehensibility of the official language of the guidelines (proteins, saturated fasts, etc.). Although nowadays words like carbohydrates and proteins are not that often used anymore by the agencies, it is from a common sense point of view clear, that most people don't think in terms of these words; they think in terms of spinach, beef, cucumber and milk.

So power play plays probably a negative role with respect to the reliability of the guidelines and the trustworthiness of the institutions. Another aspect that plays a role is the denigrating tone of most food and nutrition scientists towards the public. Consumers are often treated as (and sometimes even called) ignorant, lazy, hot heads, the weakest link in the food chain and technophobe. Often they exaggerate

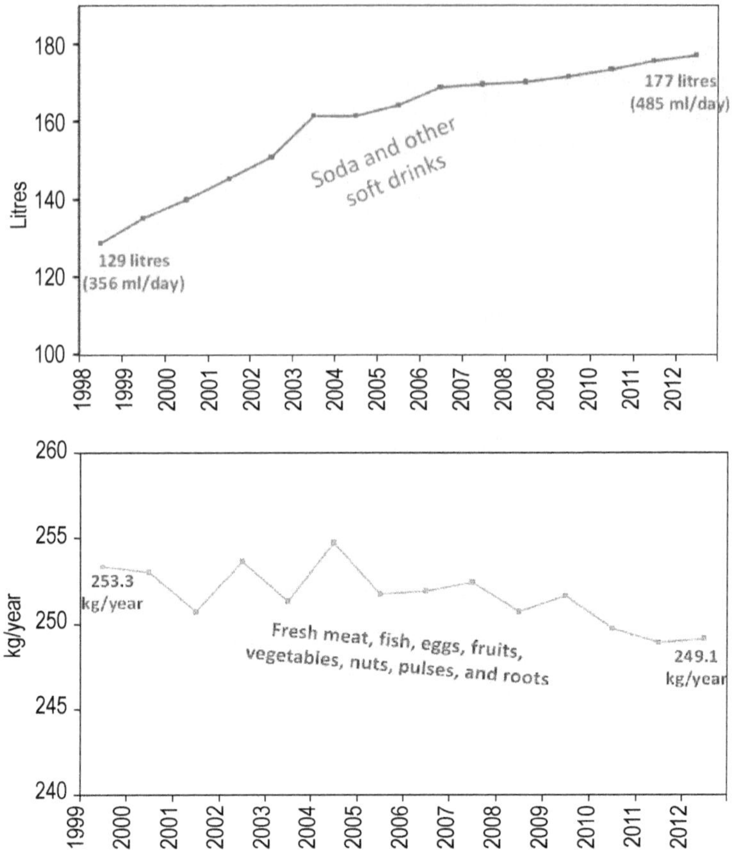

Figure 1. Per capita supermarket selling of ultra processed, ready-to-eat food products and minimally processed food products in Europe, 1998-2012. Euromonitor International.

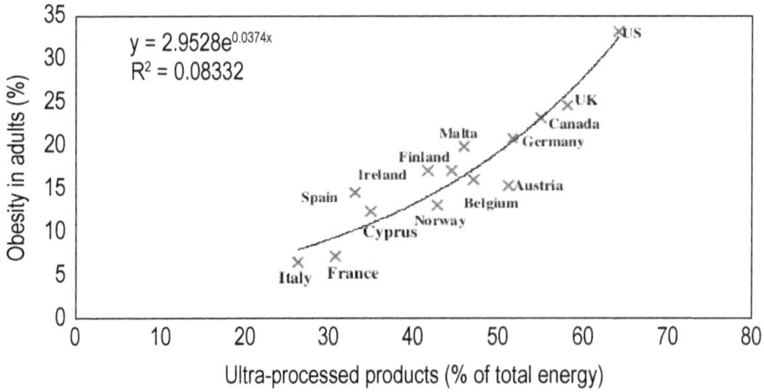

Figure 2. Association between the selling of processed foods in Europa, US, and obesity, 1991-2010 (http://tinyurl.com/hujm9rn).

their health and nutritional claims in a very ostentative way, as the latest truths, and give so-called food gurus a firm slap in the face.

Moreover, the public is influenced from very different spokespersons by environmental reports by FAO and NGOs about depletion of fish resources and the emptying of the seas and oceans, and on the negative impact of meat on global warming. That influence also does not motivate to live up to the guidelines, which disregard this information. These guidelines do not say anything about environmental issues, except the Dutch for fish.

It is striking that a major shift in food preferences the last decade, the increasing demand for meat alternatives (be it fake meat or vegetable alternatives), is not stimulated by these guidelines at all. It comes from non-governmental organisations (NGOs), and their messages. In the end, the guidelines are with respect to consumers having only a very tiny influence (but probably more on school menus and food services). It becomes urgent that these agencies finally reflect upon what they are doing and what their roles are and how to improve their role.

Philosophical considerations

The first strategy considers food as a kind of necessary instrument to stay in life and puts a lot of emphasis to fill the body and to monitor it. However, the last ones starts with a different framing of food, as a social and cultural device, that has personal and social identity achieving meanings. Cooking, sharing and eating together are the main aims. It takes the body serious, and stimulates people to acquire critical skills about processed and novel foods. Indeed, according to many one of the ways to bridge the gap between production and consumption is not by adding more information but by assisting people in getting food skills, i.e. skills to learn to selectively deal with new trends in food production and to acquire coking and sharing skills (abilities in the sense of Sen, 2009). This type doesn't disregard processed foods, but recommends to be careful with their use.

It is interesting to see that some food writers agree with the third strategy; for example, former food writer with the *New York Times*, Bittman (2014), formulates only two rules for eating (and living) well:

1. Stop eating junk and hyperprocessed food. This eliminates probably 80% of the stuff that is being sold as 'food'.
2. Eat more plants than you did yesterday, or last year.
   If you add 'Cook your own food' to this list, it's even more powerful, but these two steps alone allow you to reduce the amount of antibiotics you're consuming; pretty much eliminate genetically modified organisms from your diet, lighten your carbon footprint; reduce your chances of becoming ill as a result of your diet; save money; cut way back on sugar, other junk and unnecessary and potentially harmful non-food additives; and so on.

Even more simple are the rules of Michael Pollan (2008): 'Eat food, not too much, mostly plants.'

Conclusions

The common type of guidelines do not stimulate people sufficiently to have good food attitudes. Their expert language and their often incoherent way of formulation increases the distrust that people have regarding them. The third type of guidelines incorporates a philosophy of food that takes seriously both the nutritional and sociocultural meaning of food. Meal oriented guidelines are a better impulse to encourage people to acquire and maintain good food skills (abilities). Eating well is a substantial part of living well, and the meal is an important factor in the human condition.

References

Bittman, M. (2014). Only two rules for a good diet. The New York Times, October 22.
Monteiro, C.A., Moubarac, J.-C., Cannon, G., Ng, S.W. and Popkin, B. (2013). Ultra-processed products are becoming dominant in the global food system. Obesity Reviews 14, Suppl. 2: 21-28.
Nationaal Kompas Volksgezondheid. Available at: http://tinyurl.com/q65mys2.
Pollan, M. (2008). In defence of food. Colombia University Press, New York, NY, USA.
Scrinis, G. (2013). Nutritionism: the science and politics of dietary advice. Columbia University Press, New York, NY, USA.
Sen, A. (2009). The concept of justice. Lane, London, UK.

51. Nutrition as public policy: still the guarantee or already the restriction of one's rights?

M.L. Neto
Faculdade de Direito da Universidade do Porto, Rua dos Bragas, 223, 4050-123 Porto, Portugal; lneto@direito.up.pt

Abstract

The issue that we intend to focus upon may seem paradoxical, or at least provocative; however, it aims to answer this basic question: what are (not the obligations) but the limits of the State's public policy *quantum* in what concerns diet and/or nutrition? We will follow through the discussion of international legal constraints, but also provide theoretical grounds for discussing the limits (regarding self-liberty and self-determination) of State's imposition, rejecting paternalistic approaches.

Keywords: state, diet, imposition, paternalism, self-determination

Fundamental rights as a form of capability and achievement of equity

The issue that we intend to focus upon may seem paradoxical, or at least provocative; however, it aims to answer this basic question: what are the limits of the State's public policy *quantum* in what concerns diet and/or nutrition?

In order to alleviate any perplexity caused by a jurist speaking about such issues as diets, I would like to propose that we use the Nobel economics laureate Amartya Sen's idea of freedom as one of the elements of justice. For Sen (2010: 346, 386, 396, 397, 398, 401), freedom should be viewed as opportunity and the process of choice, and is related to capability – enhanced by equity. Another author, Martha Nussbaum (2006, 2007), a disciple of this 'capability approach' of Amartya Sen identifies primary goods (among others, and for what concerns us here), such as life (including being able to live to the end of a human life of normal length or not dying prematurely), bodily health, bodily integrity and practical reason (*id est*, the critical reflection upon an autonomous understanding of good and planning one's own life).

In proposing access to real opportunities, Sen (2010: 323, 325) and Nussbaum go further than the list of primary goods that John Rawls had proposed, to identify that 'the prevention and alleviation of disabilities are fairly central to the promotion of justice' (Sen, 2010: 353). In the sense proposed by Sen and developed by Nussbaum, the above mentioned 'capabilities' should become fundamental rights that the State instigates and implements rather than banally and merely guarantees. One of the consequences of this view is the elimination of the distinction between positive and negative freedom or between the rights of the first and second generation, respectively characteristics of the Liberal State and Social State (that in operational terms, we can identify with the Welfare State, or the minimum ethical State). As such, poverty, for example, can be seen both as capability deprivation but also as the deprivation of freedom (freedom from hunger).

However, it is important to clarify that, within a context of Rawlsian political liberalism (which we could group together with the finding of authors, such as Dworkin, Nagel, Scanlon or James Buchanan for public choice theory), one needs to reject the assimilation between encouraging perfectionism and accepting paternalism (Deneulin, 2002: 6, 7). This is where we aim to focus.

EurSafe 2016 – DOI 10.3921/978-90-8686-834-6_51, © Wageningen Academic Publishers 2016

The right to food, in particular

Let us consider (as we did in Neto, 2010) discussions regarding restrictions on advertising aimed at children or the need to find self-regulating mechanisms that consolidate and clarify the nutritional information created for consumers, or the discussion of legislative proposals (considered by many as fundamentalist) on the amount of salt in bread, or in the measures of the Directorate-General for Health regarding food in school canteens.

As such, and simply for the purposes of example, it demonstrates the existence of a main thread that we wish to underline as a unifying object of analysis. However, the modern definition of public policy cannot ignore the context of how public expenditure is divided and the assessment of constitutional norms that promise social goods (Alexandrino, 2010). Indeed, choosing the extent of intervention may be merely welfare rooted in the social determinants of hunger/malnutrition (which we would identify with a food policy *stricto sensu*) or could have a more obvious *desideratum*, from a nutritional perspective, as an element of health policy.

In other words, we can (gradually) talk about merely facilitating access to food, of a *plus* regarding food safety guarantees, of another plus safeguarding healthy dietary alternatives and/or, finally, and finally of a dubious imposition of dietary habits – whether it is justified in terms of protecting individuals, or justified from a solely economic point of view[1].

International constraints

The discussion of the simple advantage and/or desirability of a public policy in this area, however, gives way to the analysis of the constraints and decisions of basic international bodies (multilateral or bilateral, general/universal or sectorial[2]) that oblige States (e.g. the Portuguese State) in accordance with Articles 8 and 16 of our Constitution.

In Articles 55 and 56, the Charter of the United Nations stipulates a joint guarantee of 'higher standards of living and solutions of international economic, social, health problems ...'. More specifically, the provisions regarding food derive from Article 25 of the Universal Declaration of Human Rights (UDHR), adopted in 1948: 'Everyone has the right to a standard of living adequate for the health and well-being of himself and of his family, including food, clothing, housing and medical care and necessary social services ...'.

In 1966, this provision was fulfilled and implemented by Article 11 of the International Covenant on Economic, Social and Cultural Rights (ICESCR), which determined that States should:

(a) Improve methods of production, conservation and distribution of food by making full use of technical and scientific knowledge, by disseminating knowledge of the principles of nutrition and by developing or reforming agrarian systems in such a way as to achieve the most efficient development and utilization of natural resources;

[1] Indeed, if it is considered that a healthy and balanced diet is a contributing factor to improvements in health, then preventing mistakes in terms of diet can be viewed as a way of avoiding public expenditure.

[2] As a simple example of the sectorial aspect that certain issues merit, it is worth looking at the profusion of World Health Assembly Resolutions (WHA) since the 1980s.

(b) Take into account the problems of both food-importing and food-exporting countries, to ensure an equitable distribution of world food supplies in relation to need'.[3]

It should be clarified, however, that these merely declaratory proclamations do not effectively bind States to achieving such goals: this can be considered an altogether *capitis diminutio* of international law. As such, one cannot say that these sources indicate clearly specified guidelines of a fundamental right to food (adequate)[4] as a subjective legal position.

However, since 1966, there has been some attempt to clarify the minimum content of the above mentioned right (Valente *et al.*, 2007). Indeed, the adoption of the 1978 Declaration of Alma-Ata, the 1986 Ottawa Charter for Health Promotion or the conclusions of the 1988 Adelaide Conference[5] indelibly linked public policies to the creation of a favourable environment that allows people to live healthy lives.[6]

The attempt to oblige States to implement the standards of the protection of enforceability of economic, social and cultural rights (ESCR) would become clear in the Kyoto Protocol, signed in 1988 by Latin American states (*et pour cause*, given the respective shortcomings), which warns of the urgent need to consider ESCR as an indivisible part of human rights, and 'part of the basic values of a true democracy.' The Kyoto Protocol, which does not impose standard formulas but rather requires the State to provide the means so the minimum needs of the population are met and incremental improvements in living standards of citizens are defined, determines the specification of obligations to be respected, protected and satisfied,[7] to be guaranteed in procedural terms via non-discrimination – even by affirmative actions –, the suitability of ordinary legislation, production of public information, promotion of legal measures, and the assumption of progressiveness and prohibition of regression or unreasonable delay in the adoption of measures.

In 1999, the comment by the United Nations Committee on Economic, Social and Cultural Rights (CESCR) stated that 'the right to adequate food is indivisibly linked to the inherent dignity of a human being' and also 'inseparable from social justice, requiring the adoption of appropriate economic, environmental and social policies, at both the national and international levels, oriented to the eradication of poverty and the fulfilment of all human rights for all.' However, the Committee, with respect to Article 11 of the ICESCR, elaborated upon what is meant by 'adequate food':

[3] It is worth remembering that the same document expressly recognizes (in Part I, Article 2) the so-called clause of the possible, which we will re-examine *infra*, limiting States' obligations 'progressively, by all appropriate means' and 'to the maximum of its available resources'. However, and as highlighted by the UN Committee for ESCR in its 1999 Comment on this Article 11, in the cases in which the State claims to have insufficient resources as a justification for not being able to guarantee access to food, it will have to demonstrate that 'every effort was made to (...) satisfy, in terms of priority, this tiny part of its obligations [...] and that it attempted, unsuccessfully, to obtain international help to ensure the availability and accessibility to the necessary food.'

[4] There are other expressions: 'right to a correct diet' and/or 'fundamental right to not go hungry'.

[5] In the sense of a commitment to global public health, the Adelaide Conference defined access to nutritious food and drinking water, amongst other things, as pre-requisites for health and social developments.

[6] This globalized perspective of the 'environment' and 'life quality' also derives from the title and wording of Article 66 of the Portuguese Constitution.

[7] This three-way division was later re-employed and made systematic by the UN's Committee on ESCR in the 1999 Comment on Article 11 of the previously-mentioned ICCPR. Originally, in 'Right to adequate food as a human right, Study Series No.1, 1989, United Nations Publication, Sales No.E.89.XIV, these three levels of obligation were proposed in the following way: respect, protect and assist/fulfill.

(a) The availability of food in a quantity and quality sufficient to satisfy the dietary needs of individuals, free from adverse substances, and acceptable within a given culture;
(b) The accessibility of such food (economic, financial and physical, particularly for vulnerable groups).

Moreover, nutrition as an aspect of public health was included at the United Nations Millennium Summit in 2000, and, in 2002, the UN Special Rapporteur on the Right to Food redefined the human right to adequate food as the ability to 'have regular, permanent and unrestricted access, either directly or by means of financial purchases, to quantitatively and qualitatively adequate and sufficient food corresponding to the cultural traditions of the people to which the consumer belongs, and which ensure a physical and mental, individual and collective', as pointed out by Ziegler (2002). As such, it becomes clear that this right does not focus solely on the aspect of combating hunger, but covers a broader set of issues.[8]

At a regional and European Union level, and until 1996, no efforts seem to have been made to implement a Common Food Policy. Indeed, the issue would arise again as a result of harmonization of food quality and safety monitoring and inspection policies[9] due to the advent of the Single European Market in 1993 and food crises that began with BSE in 1996 (Helsing, 1997). If some of the planned measures are directly connected to how the internal market works (such as labelling requirements, authorisation regarding health claims and food controls procedures), others are part of more specific frameworks, such as the CAP (via the common organization of the market in the fruit and vegetables sector), Commission initiatives in the areas of education, regional policy (structural funds) and, last but not least, the policy on audio-visual and media matters. In this sense, what is fundamentally advocated is that the three aspects of consumer rights (information, security and loyalty) are respected, but not without making it clear that the objectives of State intervention should be to demonstrate health risks and to make the healthy option available. However, it should be noted that the positions of the 2007 White Paper highlight (even regarding what we will see below on the possible restrictions of citizens' rights) that making such an option available does not mean enforcing it, which entails complex checks and balances.

The state of the art in Portugal

Although the sources already analysed supposedly entail obligations for the Portuguese State, at the constitutional level we do not find any direct reference to the issue of food and/or nutrition, although one can consider its application to the principle of the dignity of the human being (Article 1)[10] and even the provisions of Articles 64 (right to health) and 66 (environment and quality of life), which is always the State's responsibility, in terms of real equality – as stipulated in paragraph d) of Article 9 of the Portuguese Basic Law. From another sectorial perspective, there is also the reference to consumer rights (Article 60, and alluded to in other legal provisions), which arise in the abovementioned rights to information, loyalty and security. In addition to this, and in the sense that, in relation to the limitations of the State's power to intervene, the opposite outcome may reveal to be true, it is worth remembering that, alongside the 'right to be let alone' that always defines a sphere of self-determination (free from

[8] Currently, it is actually interesting to foresee the future developments that lead to a new and distinct degree of protection in the area of nutrition as intangible heritage, for the purposes of the UNESCO Convention of 2003.

[9] Far from the 1970s worries, the issue has shifted to new debates, since the advent of other fears about food or health crises, such as genetically-modified organisms or banned food additives.

[10] In specific reference to the right to health, Article 12 of the ICESCR (International Covenant on Economic, Social and Cultural Rights) recognises everybody's right to the highest level of physical and mental health, with number 1 of General Recommendation 14 of 2000 of the Committee on ESCR establishing the link between health and dignity: 'Every human being should have the right to enjoy the highest level of health that leads to a dignified life'. Furthermore, the Committee states that the right to health involves aspects such as availability, accessibility, tolerance and quality.

outside interference), the Portuguese Constitution has guaranteed the right to free development of personality since 1997. These rights, both stipulated in Article 26 as a corollary of the right to freedom, may well impose limits on the State determination of diets.[11]

The limits on diet imposition

Whether directly found in a constitutional provision or in a source of legitimacy in international mechanisms, it is clear that the rights underlying food and/or nutrition policies – a right to adequate/proper food or a subjective legal position generically covered by the right to health, the right to quality of life or, even more generically, the principle of human dignity (and specific rights to dignity and equality arising from them) – are always qualified as ESCR (and, therefore, enshrined in section III of Part I of the Portuguese Constitution).[12]

Currently, traditional doctrine dictates a specific regime for these ESCR, highlighting a quantitative perspective that focusses on the State's degree of achievement; therefore, it is necessary to mention the 'condition of the (financial) possibilities' – with the obvious overlap of the 'cost of rights', dealt with superbly by Holmes and Sunstein –, which however finds its other extreme in the principle of the prohibition of regression.

However, and as the Committee on ESCR emphasised in its General Recommendation no. 3[13] regarding the nature of state obligations, there is a minimum core obligation concerning the ESCR to be implemented by the States, to the extent which they must ensure the essential core of these rights particularly for the most disadvantaged and vulnerable segments of the population.

The theoretical debate regarding the minimum content of ESCR, as well as the intangible essence of each right and respective enforcement mechanisms, is thus not divorced from the idea of activism in the definition of a general public policy and/or sectorial policies. However, beyond the basic limit resulting from the 'condition of (financial) possibilities', there are other factors that must be taken into account (axiologically speaking) when we are faced with the objective of moving from a public policy on access to food/food security (formal) to a public policy of nutrition and combating obesity (material).

Indeed, when discussing the boundaries of this public policy, we obviously do not question the levels we have already identified as basic – non-discriminatory access or the regulation of aspects strictly linked to health and/or safety that essentially overlap with consumer rights (information, security and loyalty) protected by Article 60 of the Portuguese Constitution. In fact, this right to information clearly covers the provision of data and information about healthy and alternative dietary options. In relation to these aspects, which are not only legitimate but also enforced within a Democratic State, it is the responsibility for political power to create mechanisms – some of emergency, others permanent – that ensure access to food as a basic inalienable right of every human being.

[11] And although not everyone agrees with a basic right to decide about one's own body – for all of these aspects, see Neto, L., 2004. *O direito fundamental à disposição sobre o próprio corpo.*

[12] For the benefit of clarity and more direct expression, we focus neither on extreme situations of the possible violation of the right to life due to a lack of access to food nor physical well-being stipulated in Article 25 of the Portuguese Constitution, because it is strictly understood that, and regardless of any competitive situation, it must adhere to the provisions of Article 64. In this area, for precisely the same reasons, the preferred term is 'personal injury' or damage to health (or *danno alla salute*). As guardian of the whole person (and not solely their right to health or physical well-being) it is also worth mentioning the concepts 'personal harm', of *danno alla vita di relazione* or existential harm – see Frada (2008).

[13] Committee on ESCR, General Comment n.3, UN doc. E/1991/23, 1990.

However, the State cannot overstep an extremely important threshold in the field of legal social sciences: where the law aims at voluntary and lawful acts that are legally pertinent, the State must allow the exercise of the power of self-determination in human and behavioural decisions.[14] Furthermore, under the terms provided for in the Portuguese Constitution, the State can only restrict rights – which we have tentatively identified as the rights to the free development of personality (and/or the basic right to decide about our own body)[15] – if (abiding by a triple dimension of the principle of proportionality) such a restriction is necessary to safeguard other constitutionally protected rights or interests, if the same restriction is appropriate to the creation of such a case-by-case assessment and if, in addition to this, it is not excessive nor affects the essential core of rights.

Some people will surely say that any imposition by the State of particular diets would be legitimised by practical agreement that would justify the restriction of the individual's freedom to safeguard another right or interest – health, *brevitatis causae*. However, remember John Stuart Mill, the 19th-century philosopher who posited that the value of autonomy neither derives from nor is even compatible with an external and paternalistic view of preferences, implying a conscience decision. There are indeed limits to the imposition of lifestyles, as, in the field of human relations, it is each individual's responsibility to plough a furrow in which they recognize themselves. In fact, physical well-being can be used as a means of pressuring a particular body, such as the State authorities, by creating public opinion from sympathy and solidarity, or from simple interest – typically in the case of hunger strikes.[16]

A different and controversial issue is knowing whether a person's right to decide about their own body means they can demand that others take risks or are accountable for such causes. Here, this would relate to whether, as a result of that choice (consciously taken based on the information that the state is obliged to provide and having consciously refused other dietary alternatives[17]), the individual may be held responsible or not merit the same treatment by the health authorities.

In other words, will there still be responsibility when the individual is the person that causes the damage[18] almost generating informed consent? It would be therefore constitute self-endangerment, where the victim themself consciously contributes to the result through their own conduct, thus distinguishing it from endangering others. Some will say that this self-endangerment is never acceptable because unavailable goods in relation to consent/self-restraint/abstinence are at stake. Others, who are more open to new paradigms of freewill, will say that there can be no negative discrimination regarding access to (albeit scarce) health resources, due to the intervention of other factors (e.g. genetic predisposition) beyond the control of the individual. If the authority of the State must necessarily be legitimised and tested by participative bodies and those of democratic debate, let us return to the beginning: nowadays,

[14] For the sake of simplicity we also do not focus on the issue of eating disorders, such as anorexia and/or bulimia, on the basis that they are illnesses and not (only) the result of people's expression of will.

[15] These are stipulated in the Portuguese Constitution as rights, freedoms and guarantees and enjoy a specific protection regime. Furthermore, there is the aspect of the right to religious freedom and worship and conscientious objection laid out in Article 41 of the Portuguese Constitution, in the cases where a particular dietary choice is dictated by a religious or ethical belief – e.g. in the case of the option of vegetarian diets.

[16] An obvious example, although this is not the place for an in-depth analysis, is the case of hunger strikes and eventual force-feeding by prison authorities.

[17] Within this context, it is worth stressing the overriding importance of the responsibility of education/training for healthy lifestyles in line with the guidelines from the WHO.

[18] For the sake of simplification, let us not focus on self-endangerment as the result of hetero-determination, *id est*, dictated by another, via order, advice, recommendation (e.g. from a nutritionist) or the mere expression of a desire (for example, the husband that puts pressure on his wife to go on various diets). Worthy of independent study would be that of minors, in relation to parents' powers and educational duties.

what should the goal of State public policy be regarding food? If we want to use a broad definition of food safety, four aspects may be involved: the dimension of the quantity and minimum consumption of calories, proteins and vitamins for a healthy and active life; the dimension of quality; the dimension of the regularity and the dimension of dignity.[19] This dimension of dignity involves the idea of justice that Amartya Sen proposes as a true condition of development and one that all of us should be aware of.

References

Alexandrino, J.M. (2010). Controlo jurisdicional das políticas públicas: regra ou exceção? Revista da Faculdade de Direito da Universidade do Porto. Coimbra Editora, Coimbra, Portugal, pp. 147-169.

Deneulin, S. (2002). Perfectionism, liberalism and paternalism in Sen and Nussbaum's capability approach. Review of Political Economy 4(4).

Frada, M.A.C. (2008). Nos 40 anos do Código Civil Português. Tutela da personalidade e dano existencial. Themis, Revista da Faculdade de Direito da UNL (Código Civil Português – Evolução e Perspetivas Atuais), special edition, pp. 47-68.

Helsing, E. (1997). The history of nutrition policy. Nutrition Reviews 55(11): s1-s3.

Neto, L. (2004). O Direito Fundamental à disposição sobre o próprio corpo. Coimbra Editora, Coimbra, Portugal.

Neto, L. (2010). A nutrição como política pública: ainda a garantia ou já a restrição de direitos? Revista da Faculdade de Direito da Universidade do Porto. Coimbra Editora, Coimbra, Portugal, pp. 391-412.

Nussbaum, M.C. (2006). Frontiers of justice: disability, nationality, and species membership. Harvard University Press, Cambridge, MA, USA.

Nussbaum, M.C. (2007). The supreme court 2006 term. foreword: constitutions and capabilities: 'perception' against lofty formalism. Harvard LR 121(4).

Sen, A. (1999). Development as freedom. Oxford University Press, Oxford, UK.

Sen, A. (2010). A ideia de justiça. Coimbra, Portugal.

Valente, F., Franceschini, T. and Burity, V. (2007). A exigibilidade do direito humano à alimentação adequada. Brasília (DF), Brazil.

Ziegler, J. (2002). Report by the special rapporteur on the right to food (Commission on Human Rights, E/CN.4/2002/58/Add.1).

[19] As a result of the Portuguese Constitutional Court Rulings 177/2002 and 509/2002, there is an attempt nowadays to define a minimum core of social rights, linking this with the principle of human dignity as a veto, imposed on the legislator, on the arbitrary reduction of the level of legislative implementation of a fundamental social right.

52. Ethics and law must be inflexible concerning the right to adequate food

I. Portela
Polytechnic Institute of Cávado and Ave, R. do Aldão, 4750-810 Vila Frescainha (São Martinho), Portugal; iportela@ipca.pt

Abstract

This working paper tackles the challenging problematic of food and hunger, which demands mutual responsibility from lawmakers, politicians (as ethical leaders) and the community. Nowadays, food is related with different issues worldwide such as hunger, obesity, eating disorders, abuse of weight reduction drugs, diabetes and metabolic illness. All the citizens of the world, as customers of food and food-related products are in jeopardy to become potential victims. Here it will be discussed the legal and ethical perspectives concerning food and the right to food. This will be the basis to build an epistemological approach to the limits of fundamental constitutional rights. We will extend our discussion to the indestructible limits of the personal human dignity sphere, which in a not-derrogable way enjoys the protection of the Constitution of the Portuguese Republic). From a health perspective, an adequate nutrition is important for our personal and family life, but this discussion is not the aim of this paper. Our first point of analysis is the recognition that 'to eat adequate food' is no more a simple act or material fact. More properly, 'to eat adequate food' is a social right, a fundamental right and even more a human right. The construction of the right 'to adequate food' will be presented in the light of International Law and Law in general. At the same point the epistemological construction of the right 'to adequate food' will be analysed. The misery of the humanity abysm exists without deviations and scientists predict it using the scientific observation and determination. Only twenty years ago we were living decades of 'global hunger', while today the society hyper values the mythological rights, such as the right to drink alcohol or to smoke. Unfortunately, a judge still cannot enforce the right to adequate food.

Keywords: judiciary enforcement, adequate food, soft law, hard law, human right

Introduction

After the Second World War, the international protection of Human Rights has acquired an extraordinary proportion under the aegis of the United Nations. The current protection of Human Rights involves not only national constitutional mechanisms but also a system guided by the United Nations (Rouget, 2000). Among the multitude of Human Rights considered to be fundamental for a full life in the XXI century, this work will focus in particular on 'the right to food' as well as on 'the right to adequate food'. Specifically, herein we aim to define the legal framework of food in two different levels: the hard law and the soft law.

The problematic around food has two different scenarios. On one side, nowadays there are more than 925 million people worldwide suffering from hunger, poverty and diseases caused by starvation. In fact, approximately 20,000 people die each day from undernutrition and related diseases, whereas the majority of them are children. On the other side, there are countries leading with an excess of food, which causes obesity, eating disorders, and chronic diseases such as hypertension, cancer or diabetes mellitus. This paper is divided in three parts. In the first part it will be explained the right to food and in the second part it will be discussed the right to adequate food or nutrition. The adequacy of food is related to the accessibility to food production for the present and future generation. In other words, after the juridical interpretation of the right to food we will expand our view beyond the protection and other nuances of

EurSafe 2016 – DOI 10.3921/978-90-8686-834-6_52, © Wageningen Academic Publishers 2016

this right. In detail, it will also be discussed the physical accessibility to food and also the food production for present and future generations. Finally, it will be illustrated how easy a country can jeopardize the right to adequate food, considering, as an example, the case of Portugal.

The right to food

In the Portuguese constitution the right to food is not an autonomous right. Instead it is a constitutional and legal aspect of the right to 'quality of life' (article 63° of the Portuguese Constitution). The States and International Organizations must have a responsibility to protect and give shape to the human right to food, as well as the rights to which it is connected: how to lead a decent life, develop and retain their physical and intellectual capacities (Nivard, 2012).

At the International level, there are some difficulties in what concerns the identification of a legal basis to take a case to court regarding 'the right to food'. If there is no express constitutional basis for the right to food, and there is no clear statutory basis either, arguing a case on the basis of the text of the International Covenant on Economic, Social, and Cultural Rights (ICESCR) and on soft law documents such as the Comment No. 12 and the Voluntary Guidelines of the Food and Agriculture Organization of the United Nations (FAO, 2011) on the Right to Food, before domestic courts with little or no knowledge on international law, can be a difficult task (Courtis, 2007: 323; Kaltenborn, 2015). According to the prerogatives and faculties of the Human Rights, the freedom and dignity of the person should be protected without discrimination, following the normative and institutional guarantees. Thus, benefited from the human person about the arbitrariness and promote the development of the human person (Mourgeon, 1985: 8). Currently, at the International courts, the right to food (as a human right) represents a challenge. In fact, without access to power, there is no access to proper education or adequate health and there is no chance for a fight against poverty. The truth is that the law should be reviewed and challenged, because there are more than 1/7 of humanity that suffer from hunger. The right to food is underlined in the article 25 of the Universal Declaration of Human Rights' and in the article 11 of the ICESCR (Golay, 2011): but what is the right to food? 'The right to food is a human right. It protects the right of all human to live in dignity, free from hunger, food insecurity and malnutrition. The right to food is not about charity, but about ensuring that all people have the capacity to feel themselves in dignity' (Eide *et al.*, 2010; Ziegler, 2002).

The right to adequate food

What is the right to adequate nutrition? The general Comment 12 of the Committee of Economic, Social and Cultural Rights (CESCR) define that 'the right to adequate food is realized when every man, woman, and child, alone and in community with others, has physical and economic access at all times to adequate food or means for its procurement'. Inspired by the definition of CESCR, the United Nations Special Rapporteur highlighted that the right to food entails: 'the right to have regular, permanent and unrestricted access, either directly or by means of financial purchases, to quantitatively and qualitatively adequate and sufficient food corresponding to the cultural traditions of the people to which the consumer belongs, and which ensures a physical and mental, individual and collective, fulfilling and dignified life free of fear'(Ziegler, 2002).

From a constitutional perspective, the right to food has to be interpreted by the judges, because the fundamental right to food don't exist *per se* as an individual right in all its potentialities. The right to adequate nutrition is a result of a cumulative joint of some elements. In this sense, these gives space to the appearance of indeterminable concepts under the Law, such as 'availability', 'adequacy', and 'accessibility' of food.

The concretization of the right to adequate food is dependent on the long-term 'availability of food', on the quality of enough food production for the present and for the future generations, and by taking in account the protection of the environment. The quality of 'adequacy of food' depends on the type of dietary needs of an individual regarding the quantity and the nutritious quality of the food (the value of nutrients in the food) (Miles and Frewer, 2001). The quality of 'accessibility of food' (economic) implies that the financial costs should satisfy the basic needs of the person (health, education, housing), with the acquisition of food for an adequate diet, without deprivation. Physical 'accessibility to food' implies that everyone, including physically vulnerable individuals, (children, older people, persons with physical disability, terminal illness, and people with persistent medical problems and mental illness), should be ensured with access to adequate food, with no discrimination.

A case study: 'the Portuguese food basket'

The right to food is protected by the ICESCR, which was signed on September, 24, 2009 and ratified on January, 28, 2013. This Treaty is directly applied between the covenant.

The Portuguese food basket indicates the monthly budget required for an adequate food intake by three reference households (consisting of children and people of working age, in good health, without disabilities and living in the capital city). The basket includes a budget for food and for the kitchen equipment required to prepare, serve, consume and preserve this food. Furthermore, it takes into account the necessary budget for physical activity and other functions of food, such as its social function.

The monthly budget required for a healthy diet in Portugal is EUR 173 for a single person and EUR 606 for a family of two adults and two children. If the budget needed for physical activity and the other functions of food (eating out, holidays, etc.) is taken into consideration, the total monthly food budget amounts to EUR 204 for a single person and EUR 766 for a family of two adults and two children. From 2008 to 2012, during the period of the economic crisis, food availability data suggest a hypercaloric food pattern. The Portuguese tend to consume too few fruits and vegetables and to have a too high intake of meat, fish, and eggs, as well as of oils and fats. Furthermore, food insecurity became a matter of concern for public health authorities, which led to the creation of a monitoring system in 2011: Infofamilia Survey. The most recent data show that, in 2013, about 17.3% of the Portuguese population has reported changes in eating patterns or a reduction in food intake due to economic constraints. Data from a food consumption study conducted by the Institute of Public Health at the University of Porto also gave useful information on dietary habits, even though the sample used is not representative of the Portuguese population as a whole.

Conclusions

At these times, the world of food has a new image and conception, as a result of marketing, communication and media. Unfortunately, the food organizations, such as FAO and World Health Organization (WHO) cannot enforce or change any law in what concerns 'the right to food'. The production and the commerce depend on the governmental ratings that are not negotiable with the neighbours. It is impossible to enforce the agreements about the security of the food products and about dietetic treatment without any laboratory testing. In this sense, until now 'the right to adequate food' does not have a peremptory body of principles and norms that would not permit derogation. The legal norms recognized by the international community as a whole would be fundamental to maintain an international legal order or, at least, a minimum of security. The agreements between the states are based on the soft law level and some International Conventions.

Looking for a future perspective, there is the possibility to design a guide to good practices in order to create ethic rules about the use of 'the right to food'. Why we don't try to create a set of ethic rules in this sector that will be applied by our citizens? While fundamental right inherent in human nature and showing connections with rights classified intangible or hardcore of Human Rights. According to international human ethics and morals, 'the right to food' should not be susceptible to any form of limitation.. Thus, the waiver of any provision, which guarantees the right to food, should be prohibited (The Geneva Conventions of August 12, 1949, exclude all and any derogation) or subjected to extremely strict conditions. An example is the case of refugees, given the intrinsically elemental and universal nature of the right (Convention relating to the status of refugees of 28 July 1951 and its Protocol Additional on 18 November 1966).

The right to adequate food should be a priority for governments in order to promote a balanced nutrition for all its citizens. However, it comes the question: which is the correct position of the Portuguese Constitution and how many other European countries acceded to the International Covenant on Economic Social and Cultural Rights? The voluntary policies adopted in November 2004 by the FAO Council state that the right to adequate nutrition security are implemented 'when all human beings have, at every moment, a physical and economic access to sufficient food, as nutrient solution allowing meet all their energy needs and their food preferences to have a healthy life and active' (FAO, 2016). This perspective implies that each person should access the abundance of quality food products for an indeterminate length of time, taking in consideration the accessibility and acceptability of food mentioned above.

References

Constitution of the Portuguese Republic. (2005). Assembleia da República. Available at: http://tinyurl.com/j6pylw3.

Courtis, V. (2007). The right to food as a justiciable right: challenges and strategies, Max Planck Yearbook of United Nations Law Vol. 11. Brill, Amsterdam, the Netherlands.

Eide, A., Hospes, O. and Hadiprayitno, I. (2010). State obligations for human rights: the case of the right to food. Governing food security. Law, politics and the right to food, pp. 105-121.

FAO (2011). Constitutional and Legal Protection of the Right to Food around the World. Available at: http://tinyurl.com/gptvz5n.

FAO (2016). Regional overview of food insecurity Europe and Central Asia. Available at: http://tinyurl.com/jca8xk6.

Golay, C. (2011). Droit à l' alimentation et accès à la justice. Bruylant, Brussel, Belgium, p. 356.

Kaltenborn, M. (2015). In Social Rights and International Development. Springer Berlin, Germany.

Miles, S. and Frewer, L. J. (2001). Investigating specific concerns about different food hazards. Food quality and preference, 12: 47-61.

Nivard, C. (2012). Le droit à l'alimentation. La Revue des Droits de l'Homme. Available at: http://tinyurl.com/gmea9bp.

Rouget, D. (2000). Le guide de la protection internationale des droits de l'Homme. La pensée sauvage, Grenoble, France.

Ziegler, J. (2002). Le droit à l'alimentation doit être rendu justiciable. Available at: http://tinyurl.com/hobwnv9.

53. The human right to food and the role of University research in food security in Kenya

J.L. Omukaga
University of Kabianga, P.O. Box 2030, 20200 Kericho, Kenya; jlikori@yahoo.co.uk

Abstract

The entrenchment of the human right to food into the Kenyan Constitution 2010, did not only enjoin Kenya to the core concerns of international community to eradicate hunger, it also rendered the country's commitment to food security inescapably enforceable. While this double fortune embodied an imminent departure from under-nutrition, Kenya's disastrous performance by the close of the Millennium target brings into even sharper focus the country's entire food security apparatus. Over and above the known loopholes of the legalistic approach to hunger, critics point at the loose connections of the constitution, rampant impunity and corruption as the hurdles in the implementation of the right to food. This paper highlights these hurdles as a basis of caution against overreliance on the human right to food law. It explores the moral-ethical potential of the research and University institutions in Kenya to advance innovation, education and outreach as corporate social responsibility in the renewed initiative against hunger. Consequently, it recommends a tri-partite administrative structure that harnesses and translates this potential into practical food security measures.

Keywords: Kenyan food security, exploring alternatives

The right to food law

The normative content of the Right to Adequate Food was established by General Comment 12 (3). Thus:

> The right to adequate food is realized when every man, woman and child alone or in a community with others, have physical and economic access at all times to adequate food or means for its procurement. The core content of the Right to Adequate Food implies the availability of food in quantity and quality sufficient to satisfy the dietary needs of individuals free from adverse substances, and acceptable within a given culture, (and) the accessibility of such food in ways that are sustainable and that do not interfere with the enjoyment of other human rights. (Eide *et al.*, 2005)

This legislation provides a clear definition of the right to food structured from the existing UN norms, especially the economic, social and cultural rights. Many other pieces of legislation exist though experts consider Article 11 of the International Covenant on Economic, Social, and Cultural Rights to covers the basic content (Alston, 1984). Locally, articles 43 and 53 of Kenya's constitution provide the core content of this right almost as elaborated on in the international law. Thus: 'Every person has the right ... to be free from hunger and to have adequate food of acceptable quality,' and 'Every child has the right ... to basic nutrition ...' Other articles corroborate this basic provision in an attempt to fully domesticate the international provisions. Their content includes empowering individuals to claim their right to food when denied, securing court jurisdictions to prosecute cases of abuse or denial of this right to take recourse to a court of law claiming that a right has been denied, and giving a high court jurisdiction to hear and determine such applications. This basic legislation is further supported by various other policy instruments on national food security initiatives in Kenya, the latest being the Kenya vision 2030. It advocates devolvement of funds to support food security among the poor and agricultural reform policies

I. Anna S. Olsson, Sofia M. Araújo and M. Fátima Vieira (eds.) ***Food futures: ethics, science and culture***
EurSafe 2016 – DOI 10.3921/978-90-8686-834-6_53, © Wageningen Academic Publishers 2016

including the development of irrigation schemes to enhance food and nutrition security (Strategy for Revitalising Agriculture (SRA), 2004-2014).

The current situation of hunger in Kenya and its impact

As the target year of the Millennium Development Goals (MDG) came to close, the world registered a general decline in both the number and the prevalence of the undernourished between 1990-1992 and 2012-2014 (FAO, 2015). In Kenya, progress against hunger is in the reverse direction. With regard to the WFS target of 1996, Kenya's situation is classified as 'no progress or deterioration.' With regard to the MDG 1c, the report considered Kenya's progress as 'insufficient.' Generally, the FAO report considered these targets to be within reach, on condition that considerable efforts are invested, 'particularly in countries where progress has stalled.' The question is whether the country can even just reverse the trend from the current situation. The prevailing situation of under nutrition in Kenya suggests a big gap between the standard food requirement for normal operations and the actual food accessed by each individual reducing general life expectancy at birth to 60 year (United Nations International Children Emergency Fund (UNICEF), 2015). Hunger deprives people of their energy and skill to work productively. The effects of poor nutrition reflect in the low birth weight, wasting or stunting of the children below the age of five, exposing them to higher mortality in Kenya now standing at 80 infants per 1000 live births (FAO, 2014). Those who make it into adulthood risk impaired physical and cognitive growth, adult men risk reduced working capacity while women risk continued births of babies with low birth weight. Due to its quality, the labour force available does not meet the demands of sufficient food production due to lack of skills or physical inability. Dependency is high and exposes the masses to manipulation either by the political class or the unscrupulous business class. This situation triggers endless social inequalities.

Some weaknesses of the human rights approach

The weaknesses associated with the right to food approach include its failure to address the question of responsibility. The historical circumstances surrounding the human rights theory after the Universal Declaration of Human Rights (UDHR) accord the State institution conspicuous centrality. It is the primary locus within which human rights are defined, interpreted, implemented and evaluated. So to speak, 'only in the context of an organized society with public authorities does the notion of "human rights" make sense' (Eide, 1984). In fact it is the State that designs measures to promote observance, on the part of individuals, of their relevant duties with respect to the right to food (Alston, 1984). In this situation experts argue that one's obligation to perform certain duties is at the same time one's right. But given the individual's moral command over a person's own claims, one can suspend any right, and therefore any duty, by choice, one can renege it whenever one finds it convenient. The disproportionate centrality of the State and the explicit failure to be definitive about responsibility renders the idea of ownership of the right to food, and therefore the legitimacy of the claim to the said food consequently rather vague. The normative aspect of this right does not immediately secure the benefit of the individual in the sense of handing free food to everyone, rather it identifies the state as the key duty-bearer obligated to fulfil the demands of this right. In the event of uncertainty with regard to the rightful owner of the right to food, its implementation is compromised. Thus, where the understanding of responsibility is ambiguous, the pursuit of the human right to food will tend to emphasize the pragmatic utilitarian aspect of enjoyment of the food available probably at the expense of basic human social values that last and build mankind's welfare and dignity (Goodin, 2006).

The Structure and the methodology of the human rights approach also pose concern. The existential motivation of the UDHR was notably to prevent recurrence of war and large-scale human suffering. This background establishes the human rights approach as a universal concern implementable through international co-operation. It is therefore argued that the Declaration itself is not a legal instrument, its force is more visible in its moral strength rather than in its legal enforceability. At the implementation level however, this moralistic view is arguably more prone to volunteer burnout. Continuous contributions to relief aid that are not supported with mandatory statutory obligations towards the needy easily generate fatigue among the perennial contributors. On the other hand, as the legal character of the human rights gradually influences the legislative processes in various States, their integration within respective legal frameworks implies all the hurdles involved in the interpretation and implementation of laws. Furthermore, the normative-activist methodology often in use implies both ideological interpretation and formulation of legal technicalities as well as the design of appropriate advocacy. All these concerns imply adaptation and regular reviews of international laws for relevance. The implied complexity furthermore, requires definitive centralized authority that determines the rules and power of both interpretation and enforcement. Such authority is absent on the global level (Nagel, 2008).

Exploring new opportunities against hunger

At the expiry of the MDG era and considering the demands posed by major global trends in Agricultural development and the world's under nutrition, the international community is keen to redraft means and ways of sustaining people's welfare and progress. FAO, for example, has set strategic objectives to forge partnerships in order to mobilize the best available knowledge and capacities to advance a global vision against hunger, food insecurity and malnutrition. It particularly established linkages with a number of universities worldwide with the intention to '... align higher education to further achievement of food security for all and ensure people have regular access to enough high-quality food to lead active, healthy lives.' This motive inspires this paper, thus: 'academia and research institutions nurture critical thinking and generate knowledge and state-of-the-art innovations that are essential for eliminating hunger and food insecurity.' Strengthening FAO-academia partnerships envisions using each partner's comparative advantage to make practical impact against food insecurity. While FAO and other like-endowed organizations can provide financial support, technical and organizational expertise; university institutions can nurture academic research talent, cultivate habits of philanthropy through altruistic extension services to communities as a corporate social responsibility. Ultimately, FAO seeks to bridge the gap between the development and adoption of research outcomes and innovations. Coincidentally, 'the idea of the university' today fits this merger effort promisingly.

The postmodern view of university education triumphs in specialization and vocational training. University systems are more driven by targets, focus is directed more on the measurable impact. Instead of entering higher education to become a 'general scholar' immersed in becoming proficient in the entire curriculum, we now meet a type of scholar whose concern is relevance of university education to the needs of society. However, as the universities expand to admit an ever wider number of students the issue of formation and competition should find a balance that defines a university of our time. Interestingly, the recent development in the structuring of university education attests to the desire for this balance. Comparing 'the *Magna Carta* of the European Universities' in the Bologna declaration, 1988, and the priorities of the Commission of University Education in Kenya, the striking feature is three fold, namely: the interest in quality teaching, the central place of research and innovation and the imperative of extensions or community service. These three aspects today define the potential of a university institution with emphasis on the extension programmes.

Capacity of the university institutions: a case of Kabianga University in Kenya

The University of Kabianga was recently established with a view of creating more opportunities of higher learning and productive research. It is situated in the south-western end of the Rift Valley province well known for its good climatic condition traditionally favourable for commercial tea growing and large scale dairy farming. With the population of 8,000 students, the university's programmes are dominated by courses directly linked with food security and nutrition (Figure 1).

Figure 2 captures the pattern of enrolment in these agriculturally related courses in the last four years. From this comparison, the relevant science courses do not only occupy the central concerns of the university training offered in this period, most importantly, they have attracted a consistently growing enrolment. Students headed to this university (by choice and government placement) seem to be set for a conscious option for agriculture related courses. From Figure 2, it is visible that enrolment into the courses in Agriculture and Agriculture economics remained highest throughout the four-year period. Course in Agro Forestry and Rural Development enrolled the next highest number of students, while programmes in Agriculture Education and Extension, and Biochemistry are picking up rather consistently. The performance in human resource management exhibits similar behaviour. This potential grants the university the position of full command of all the instruments necessary in the war against hunger that is, the capability of creativity, agricultural technology and extension and above all, personnel management and consequent outreach into the rural community environment.

Tri-partite extension model: towards innovation for relevance

This paper proposes a tri-partite administrative structure to anchor and synergize outreach operations within the management. It comprises of the academic, moral and social pillars. The academic pillar

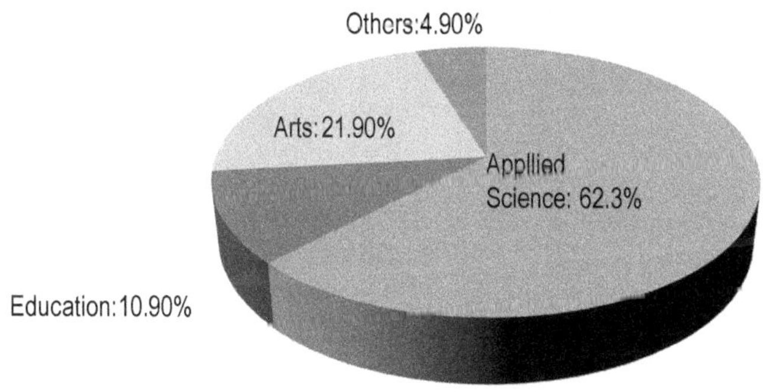

Figure 1. Pie chart showing student enrolment in the four available areas of study.

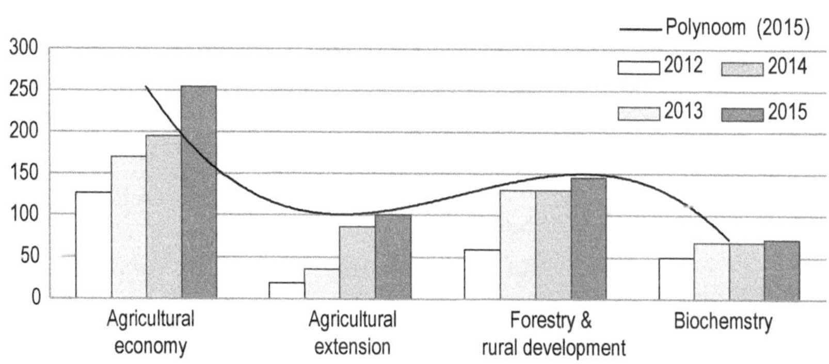

Figure 2. Student distribution in agriculturally-oriented (applies sciences) programmes since 2012.

shall provide the first level of administration concerned with academic and support operations of the university life handled by respective qualified experts. It originates, implements, supervises and evaluates the approved academic curricula and the research undertakings and guides them to a successful and competitive completion. The university administration superintends academic excellence and innovation, and advances research to an ever higher level in competition with other universities. The moral pillar comprises of the university chaplaincy, students' club and society directorates, the university college of deans and counsellors and the support staff directorate. It will marshal the entire potential of the university, moral and academic, deemed as related to food security and channel it towards society through the student groups established from the students' initial free choice of an extracurricular undertaking. The core role of this unit is to draft, discuss and implement outreach programmes aligned to food security based on the students' abilities. The social pillar constitutes the third administrative level in which the students organize their own welfare as a community. Students' administration gathers individual bio data, interest in club and societies, religion, hobbies and sports. These constitute mandatory extracurricular undertaking intended to project the student's attention to community service in outreach. The target in this arrangement is to develop students' sensitivity to the needs of community life. The students will ensure thorough compliance to group membership and strict adherence to group obligations.

The arrangement of these administrative units underscores autonomy of each as much as it fosters creative synergy between these parts. Autonomous in terms of operations and obligations but tightly bound together to one goal in terms of service to society demands, namely: bringing into effect the practical relevance of university education to society. Bearing in mind that university operations are a familiar phenomenon, the three pillars propose a 'personalized' matrix in administrative procedures tailored to specifically pilot an integral university outreach mechanism that engages the entire campus lifestyle.

Conclusions

This paper has observed the Kenyan case where hunger still affects all aspects of human progress. As the international community concludes the MDG era and adopts a new approach to progress through The Sustainable Development Goals the paper likewise proposes partnerships with institutions of higher learning as a new potential to be explored. While the paper recognizes the progress made in Kenya especially in entrenching the human right to food through enabling legal and policy measures, it nevertheless cautions against overreliance on this approach. The paper cites the weaknesses implied in this approach as the possible reason for poor performance and therefore the basis for the pursuit of the institutional approach. It underscores the need to bridge the gap between the development and adoption of research outcomes and innovations and therefore argued in support of the new FAO paradigm that 'academia and research institutions nurture critical thinking and generate knowledge and state-of-the-art innovations that are essential for eliminating hunger and food insecurity.' Citing a case of an up-coming University of Kabianga, this paper has underlined the potential of research and innovation accompanied by appropriate character formation in alleviating hunger. It is only important that universities adjust and integrate their administrative structures to accommodate the implied demands of this undertaking.

References

Alston, P. (1984). International law and the right to food. In: Eide, A. (ed.) Food as a human right. United Nations University, Tokyo, Japan.

Commission for University Education (2014). University standards and guidelines. Available at: http://tinyurl.com/henhrb6.

Eide, A. (1989). Right to adequate food as a human right. Human Rights Study Series. UN Center for Human Rights, Geneva, Switzerland, 115 pp.

Eide, W.B. and Kracht, U. (2005). Food and human rights in development. Vol. 1. Intersentia, Antwerp, Belgium, 217 pp.

Food and Agriculture Organization (FAO) (2015). The state of food insecurity in the world. Available at: http://tinyurl.com/qa5u6kg.

Nagel, T. (2008). The problem of global justice. In: Brooks, T. (ed.) The global justice reader. Blackwell, Oxford, UK, 418 pp.

Robert, E. and Goodin, R.E. (2006). Utility and the good. In: Singer, P. (ed.) A companion to ethics. Blackwell, Oxford, UK, pp. 242-245.

Sen, A. (1990). Food entitlement and economic chains. In: Lucile, F. (ed.) Hunger in history. Basil Blackwell, Cambridge, UK, pp. 374-386.

United Nations (2016). Sustainable development goals. Available at: http://tinyurl.com/nftb9yh.

54. Conscious consumption, respects the environment

B. Oliveira and E. Vasconcelos[*]
Eurest Portugal, lda Avenida da Quinta Grande n° 6, 2610-156 Amadora, Portugal; egidia.vasconcelos@eurest.pt

Abstract

Eurest is a catering company, which operates in Portugal since 1974, in the public and collective catering sector. To the core business of our organization – providing meals – the production of waste, including organic, paper, glass, plastic and metal, is inherent. The waste theme figures in several meetings of organizations worldwide, as a way of rationalizing natural resources consumption, and also, raising the awareness of the population in general. To fulfil its social role in bringing information to our consumers and working towards sustainable development, Eurest developed the CARE campaign 'Consumer Awareness, Respects the Environment'. The aim of the CARE campaign is to encourage the fight against food waste/leftovers, combined with a social action. The campaign aims to raise Eurest Portugal consumers' awareness of how important it is to reduce everyday waste, in order to assure the sustainability of our planet, reduce the use of our natural resources and, at the same time, reduce the quantity of waste produced. The campaign dynamics is based on incentivizing the consumers to place on their tray only the amount of food strictly necessary to meet their nutritional needs and/or appetite. The consumer who at the end of the meal shows an empty the tray with no leftovers, receives a card, equivalent to and representing 10 g of non-perishable foods. At the end of the campaign the cards are accounted for, and the amount is converted into non-perishable food products, to be donated to a charity institution, chosen by the client. After implementing this campaign in 56 catering units, we can confirm a reduction of 30% in the production of organic waste. To this day, more than 4 tons of non-perishable foods have been donated to solidarity institutions. This campaign is exceptionally well received both by the clients who contract our catering service and the individual consumers who eat their meals there. To this day, we have an overall compliance rate by consumers of 59%, whereas in the latest campaign action the compliance rate was 76%. We believe that the success of the CARE campaign is mainly due to its scope: environmental, social and nutritional.

Keywords: food waste, sustainability, social responsibility

Introduction

Food is essential to our survival, yet the Food and Agriculture Organization of the United Nations (FAO) estimates that about 805 million people are undernourished in 2012-2014. About 14.6 million undernourished people lived in developed countries (<5% of their overall population), whereas 790.7 million undernourished people lived in developing countries, corresponding to 13.5% of their overall population (FAO, IFAD and WFP, 2014). On the other hand, food scarcity coexists with excessive consumption. Globally, 35% of adults aged 20 and older were overweight, with half of a billion of them obese (WHO, 2011). At least 2.8 million people die each year as a result of being overweight or obese.

Furthermore, rising population levels combined with shifting dietary patterns in emerging economies will put increasing pressure on global food supply: more food is necessary to feed people. The United Nations predicts that the world population will reach 9.6 billion by 2050 (UN, 2012) and this growth will require at least a 70% increase in food production, excluding crops used for biofuels (FAO, 2009) or a more efficient use of natural resources and food production (European Commission, 2014). In this context, food waste generation is an ethical issue of particular relevance: wasting food means missing opportunities to feed the growing world population (FAO, 2012a), and consuming scarce resources,

EurSafe 2016 – DOI 10.3921/978-90-8686-834-6_54, © Wageningen Academic Publishers 2016

like land, water and energy used in the production, processing, distribution and consumption of food (Bräutigam *et al.*, 2014).

According to FAO, one-third of all edible food produced for human consumption is wasted or otherwise lost from the food chain per year, or about 1.3 billion tonnes (Gustavsson *et al.*, 2011). The food wastage is particularly severe in industrialized countries: following the FAO's food balance sheet for 2007, the food waste in North America and Europe is about 95-115 kg/capita/year, whereas in South/Southeast Asia and Sub-Saharan Africa is 6-11 kg/capita/year (Gustavsson *et al.*, 2011). Considering the whole food supply chain, with the exception of agricultural production, the generation of food waste across the EU-27, based on the EUROSTAT database, in 2006, accounted for 89 million tonnes, corresponding to 181 kg per capita (Monier *et al.*, 2010).

Food waste generated within the EU-27, during 2006, generated greenhouse gas emissions equivalent to 170 million tonnes of CO_2, considering the full life cycle of food, from the agricultural sector to the final consumer. Due to the increasing quantities of food waste, emissions estimates for 2020 rise to about 240 million tonnes of carbon dioxide equivalent gases (Monier *et al.*, 2010).

In addition, food waste causes, at individual and national economic levels, significant monetary losses and other resources invested throughout the initial production of food through the disposal of any uneaten food (Buzby *et al.*, 2011). Buzby and Hyman' results indicate that in the United States, during 2008, food waste at consumer level translated into 124 kg of food lost from human consumption, per capita, at an estimated retail price of 390 $/capita/year. This is roughly 10% of the average expenditure of the food consumed in 2008 and over 1% of the average disposable income (Buzby and Hyman, 2012). It is estimated that resource efficiency improvements, all along the value chains, could reduce material needs inputs by 17-24% by 2030 and a better use of resources could represent an overall savings potential of € 630 billion per year for European industry (European Commission, 2014).

In developing countries, food waste arose mostly during the early and middle stages of the food chain (production, harvest, processing, storage and transportation stages), due to lack infrastructures within the food chain, and a lack of knowledge or investment in technologies (FAO, 2012b; Godfray *et al.*, 2010). By contrast, in medium and high income countries, food is, to a significant extent, wasted at both retail and consumption stages, both at the household and at food service (Monier *et al.*, 2010).

Food service activities generated, in 2006, 12.2 million tonnes of food waste for the whole EU-27, the equivalent to 14% of all the food waste produced by food value-adding chain, excluding agriculture sector, an average of about 25 kg per capita, which is quite substantial (Monier *et al.*, 2010) (Table 1). The food and beverage service sector was identified as the third largest source of food waste based on food input at each stage of the value chain.

In addition, the variation in food waste (in particular the proportion of 'avoidable' and 'unavoidable') and wastage rate can in part be explained by differences in the food service characteristics across the sector. In fact, the edible and inedible food waste is generated before, during or after food preparation phases: (1) planning; (2) storage and preparation; (3) serving; and (4) eating (Figure 1).

A serving (the amount of a single portion in volume and/or weight) must be of sufficient size to ensure that customers get the nutrition level they need and that they are not left hungry at the end of the meal. However, if portions served are too large, more waste will be produced by the customer. Betz *et al.* (2015) reported serving waste to be the greatest part of total food waste in the catering segment (education and business service).

Table 1. Total food waste generated in manufacturing, household, wholesale/retail and food and beverage service/catering sectors in EU-27 (adapted from Monier et al., *2010).*

Sectors	Food waste		
	kt/year	kg/capita	share
Manufacturing	34,756	70	39%
Households	37,702	76	42%
Retail/wholesale	4,433	9	5%
Food and beverage service	12,263	25	14%
Total	89,154	181	100%

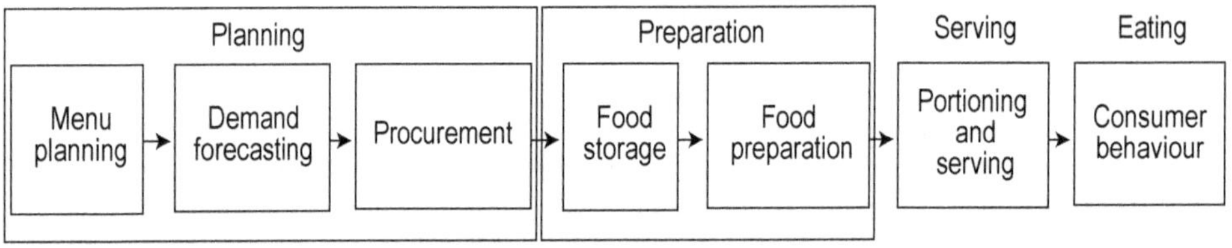

Figure 1. Waste generation in the food service setting(adapted from Canali et al., *2014; Engström and Carlsson-Kanyama, 2004; Parfitt* et al., *2013).*

Eurest is a catering company, operating in Portugal since 1974, in the public and private catering sectors. As a market leader, we believe that the impact we have on our partners, customers and clients must be guided by a high degree of social responsibility and strength in our performance. We believe that our investment in this behaviour is fundament to achieve a balance between economic development and its social and environmental impact.

Based on consolidated information, from 2015, it was possible to determine that we produce around 3.6 tons of organic waste every day, totalling 1,316 tons a year. In order to reduce this type of waste, Eurest coined the idea of the Consumer Awareness, Respects the Environment (CARE) campaign. CARE was born in 2007. This campaign resulted from the need to reduce food waste in our food service units, by simultaneously considering several concepts: environment, social responsibility and healthy eating. We believe that this triad is essential for the success of the campaign.

The aim of the 'Consumer Awareness Respects the Environment' campaign is to encourage the fight against food waste/ leftovers, combined with a social aspect. The campaign aims to raise Eurest Portugal consumers' awareness of how important it is to reduce everyday waste, in order to assure the sustainability of our planet, reduce the use of our natural resources and, at the same time, reduce the quantity of waste produced.

Methodology

The campaign dynamics is based on incentivizing consumers to place on their tray only the amount of food strictly necessary to meet their nutritional needs and/or appetite. If at the end of the meal the tray (soup, main dish, dessert and bread) is shown empty/ no leftovers, the consumer receives a card, equivalent to and representing 10 g of non-perishable foods. At the end of the campaign the cards are accounted for, and the amount is converted into non-perishable food products, to be donated to a charity institution, chosen by the client.

The deployment process is relatively simple (Table 1). The first step is to submit the campaign to our consumers/clients, either by meeting directly with them or through a Customer Inquiry (online).

Once approved, the internal process is triggered which relies on the interaction of different departments within Eurest. On week 1, field work starts in our catering unit, in which waste production is monitored and its source and quantity identified. At the same time, we start an environmental awareness process through training and workshops. In these sessions, teams are made aware of the basic aspects of the environmental business perspective of the campaign. This stage is crucial to the success of the campaign.

In this first stage, all the support and communication campaign materials are distributed (aprons, caps, table displays, result banners and coupons). All materials are reused in subsequent campaigns, unless it is no longer in good condition.

The campaign has its own image; hence when the campaign starts the restaurant is decorated and 'front office' employees are properly uniformed with the CARE emblems. In practical terms, consumers go through the self-service line selecting the food/products they wish to eat. At the end of their meals, before the consumer placing their tray in the rack, a Eurest employee verifies if the tray is empty of food products and, if so, gives the consumer a card to be placed in a box. The card is equivalent to 10 g of non-perishable food products to be donated to a social institution. At the end of each campaign week, results are consolidated and communicated to consumers through a poster in which we announce the amount of food (kg) donated to charities.

In terms of campaign cycle, it appears that the ideal situation is to carry out CARE campaign for 4 weeks and then have a break for another 4 weeks, in order to maintain the interest of consumer and adherence. This cycle allows for considerably higher levels of participation, while maintaining the consumers' interest.

At week 14 a new waste production monitoring is made and subsequent comparison with the initial assessment is made. Finally, on week 15 a campaign report is made, taking into account the participation rate and the evolution of consumer waste production, as well as, any relevant information.

Table 2: Implementing stages.

Stage	Beginning
Campaign presentation to client	day 1
Monitoring of waste production – before deployment	week 1
Team training	week 1
Uniform and support material delivery (coupons, banners, roll ups, ...)	week 1
Campaign deployment – catering unit	week 2
Break	week 6
Campaign redeployment – catering unit	week 10
Monitoring of waste production – after redeployment	week 14
Campaign report	week 15

Results

This campaign is exceptionally well received both by clients and consumers. Currently, we have a compliance rate by consumers of 59%. It is pertinent to note that in the latest campaign action the compliance rate was 76%.

We believe that the success of the CARE campaign is mainly due to its scope. The CARE campaign has a three-dimensional scope, and it comprises 3 main objectives: environmental, social and nutritional education. After implementing this campaign in 56 editions, we can report a reduction of 30% in the production of organic waste. To this day, more than 4 tons of non-perishable foods have been donated to institutions.

According to studies carried out by our Brazilian counterpart, each kg of organic waste produced represents a cost of 1.31 €. This means that each year Eurest spends/ loses, in waste, 1.7 million euros. Through the implementation of the CARE campaign, across Eurest Portugal's catering units, we expect to reduce this number to 1.2 million euros. This reduction could be even more significant when implemented together with other already developed campaigns, such us: full use of food (use of peels, seeds and stems), choose beans (mobilization for the consumption of pulses), choose veg (mobilization for vegetarian food).

The reduction of waste production is crucial to our organization.

Conclusions

This campaign has a great and direct impact on our clients and consumers and it shows our great concern towards society. It is an innovative solution to reduce the production of organic waste, in particular waste originating from meals which have not been fully eaten. With the implementation of this campaign, consumers take only the amount of food they want to consume. This goal is easily achievable since it co-exists with a social purpose. In nutritional terms, people are encouraged to consume the adequate amount of food. The campaign is daily monitored by calculating the adherence (% consumers with no leftovers vs total consumers who eat at canteen).

We have already invested around 28,000 € in the production of communication materials: table displays, banners, uniforms, etc. for the CARE Campaign. We believe that, faced with what we can save in terms of waste cost, the residual investment in the CARE campaign is about 0.005% (campaign cost vs cost of food waste).

Apart from the immediate recognition, Eurest has been publicly acknowledged with several awards. Last year we won the Observatório de Comunicação Interna Awards in the 'Glocal' category. This campaign was also recognized by Green Project Awards 2011 in the communication category. In 2011, we were one of the 3 finalists of the European Week for Waste Reduction, category Business and Industry.

The action is easily reproducible in other European countries. It requires the approval of the client for its implementation and reproduction of promotional materials and communication.

This project amplified our work in the sustainability area. This is a campaign to keep in coming years.

References

Betz, A., Buchli, J., Göbel, C. and Müller, C. (2015). Food waste in the Swiss food service industry – magnitude and potential for reduction. Waste Management 35(8): 218-226.

Bräutigam, K.R., Jörissen, J. and Priefer, C. (2014). The extent of food waste generation across EU-27: different calculation methods and the reliability of their results. Waste Management Research 32(8): 683-694.

Engström, R. and Carlsson-Kanyama, A. (2004). Food losses in food service institutions. Examples from Sweden. Food Policy 29(3): 203-213.

European Commission (2011). Report from the Commission to the European Parliament, the Council, the European Economic and Social Committee and the Committee of the Regions on the thematic strategy on the prevention and recycling of waste: roadmap to a resource efficient Europe. Brussels, Belgium.

Food and Agriculture Organisation (FAO) (2009). How to feed the world in 2050. FAO, Rome, Italy.

Food and Agriculture Organisation (FAO) (2012a). Food wastage footprint an environmental accounting of food loss and waste. Concept Note, Natural Resources Management and Environment Department. FAO, Rome, Italy.

Food and Agriculture Organisation (FAO) (2012b). The role of producer organizations in reducing food loss and waste. FAO, Rome, Italy.

FAO, IFAD and WFP (2014). The state of food insecurity in the world 2014: strengthening the enabling environment for food security and nutrition. FAO, Rome, Italy.

Monier, V., Shailendra, M., Escalon, V., O'Connor, C., Gibon, T., Anderson, G., Hortense, M. and Reisinger, H. (2010). Preparatory study on food waste across EU 27. European Commission (DG ENV) Directorate C-Industry, final report.

UN (2012). World population prospects, the 2012 revision. United Nations Department of Economic and Social Affairs, Population Estimate and Projections Section, Rome, Italy.

WHO (2011). Global status report on noncommunicable diseases 2010. World Health Organization, Geneva, Switzerland.

World Economic Forum (2009). Driving sustainable consumption value chain waste driving sustainable consumption value chain waste. Cologny, Switzerland.

55. The politics of food waste and food poverty in the EU: some ethical reflections

L. Escajedo San-Epifanio
Department of Constitutional Law and History of Political Thought, University of the Basque Country, Bilbao, Spain; leire.escajedo@ehu.eus

Abstract

The coexistence of scandalous rates of waste and a growing number of people without adequate access to food poses old and new dilemmas with regard to the way in which human beings administer their food resources. 800 million people are currently enduring food poverty all over the world while, according to the FAO report 'Global Initiative on Food Loss and Waste Reduction' (2015), just half of what it is lost or wasted could feed the world alone. This paradox is also present in the European Union. According to the data published by the European Parliament, 79 million people were living below the poverty line in 2011, a figure which according to Caritas Europa rose to 123 million in 2015 (1 in 4 Europeans). Meanwhile rates of food waste are estimated at 179 kg per capita per year, without taking into account food loss in the agricultural sector or fish catches thrown back to the sea. This situation has been described as absolutely unacceptable, with ethical and environmental problems, along with the social and economic cost, making it imperative to reduce food waste. This work focuses on the manner in which the EU is addressing the struggle against food waste and the connections established between that objective and improved access to food for people living in a situation of social exclusion. There is a critical analysis of the conceptualisation of food waste and of food poverty, and the premises upon which the strategies adopted are based. Amongst other initiatives, particular attention is paid to the encouragement of what the EU describes as 'free redistribution' of unsold food to 'citizens who lack purchasing power'.

Keywords: food waste, food poverty, food recovery, first world hunger

The approach via which the EU constructs its response to food waste and poverty

Until very recently food waste has not merited special attention in social sciences (Evans *et al.*, 2012). With regard to the study of European political and legislative measures, analysis has been concentrated on the last five years and adopts as an initial reference point two important documents: firstly, the document 'Roadmap to a resource efficient Europe', (EU Commission, 2011), in which the EU Commission (EC) stresses the need to continue evaluating 'how best to limit waste throughout the food supply chain'; and secondly, 'Strategies for a more efficient food chain' (EU Parliament, 2011). In this second document the European Parliament (EP) provides a series of data describing the reality of food waste. Every year in Europe a growing amount of healthy, edible food – some estimates say up to 50% – is lost along the entire food supply chain. Annual food waste generation in the Member States (MS) is estimated at approximately 90 million tonnes, without even considering agricultural food waste or fish catches thrown back to the sea.

The EP unambiguously rejects this situation. Discarding perfectly edible food is unacceptable. And it underlines as a premise of this rejection the fact that food waste gives rise to both environmental and ethical problems, and economic and social costs. What kind of ethical and environmental problems are caused by food waste and what is the EP referring to when it warns of economic and social costs? Although there is no detailed development of either, we do find in the EP's Resolution some references

EurSafe 2016 – DOI 10.3921/978-90-8686-834-6_55, © Wageningen Academic Publishers 2016

to these problems and costs. Thus, the recitals of the Resolution include a joint presentation of data regarding food waste at EU and world level, and references to the economic and environmental costs of this waste. In addition, various paragraphs in the Resolution offer data regarding food poverty. Waste occurs, warns the EP, while '79 million people are still living below the poverty line in the EU – in other words, more than 15% of EU citizens earn less than 60% of average earnings in their country of residence' and '925 million people around the world are currently at risk of undernourishment'. Immediately after the recitals, the first point of the Resolution clearly states that food security is 'a basic human right' and urges the European Institutions, the MS and the players in the food supply chain to tackle a waste issue that it considers unacceptable. The goal of this work is to analyse in detail the way in which the EU has developed since 2011 its waste reduction policies and its commitment to the food security of EU citizens. First it will address the measures directly aimed at reducing rates of food waste throughout the food supply-chain. Secondly, attention will be paid to a food waste reduction strategy which seems to bridge the way towards the objective of the reduction of food poverty in the EU: the promotion of the donation of safe, edible food. Thirdly, and prior to the conclusions, I will offer a series of considerations vis-à-vis the focus from which in the measures addressing food waste the EU address the plight of those in a situation of food insecurity.

The reasoning behind the measures directly aimed at reducing rates of food waste in the EU

Food waste is the result of many different factors and occurs at various points in the long process from production and distribution to final consumption. This means that the detection and targeting of food waste sector by sector requires the involvement of all the players in the food supply chain. Given that there is no one, single definition of food waste at an international level (Gjerris and Gaiani, 2015), the EP identifies the need to define the concept it will employ as a reference. It has opted for a specific concept encompassing 'all the food stuffs discarded from the food supply chain for economic or aesthetic reasons or owing to the nearness of the use by date, but which are *still perfectly edible and fit for human consumption*' (my italics). For the sake of consistency, and though food waste prevention will also feature as an integral part of the EC's 'Circular Economy Package', this definition establishes a clear distinction. Action against food waste is aimed primarily at preventing the latter, offering alternatives whereby rejected products may be employed as foodstuffs with neither food nor quality being affected (González, 2015). And only with regard to unavoidable food waste, which inevitably generates biowaste, are environmentally-friendly management strategies articulated.

One of the main fields of activity of European strategy against food waste is focused on the detection and classification of food waste. This is very complex research, because projects such as EU FUSIONS have identified over 200 drivers involved in the generation of food waste (Aramyan *et al.*, 2015) and it is not easy to determine when that food waste is avoidable – and how – and when it is not (WRAP, 2009). We are seeing, however, strategies aimed at addressing some of the causes identified or proposals to provide an outlet for food removed from the commercial market but still within its shelf-life. It is on these that I shall focus my attention. I believe it is possible to classify into three groups the strategies being implemented in order to avoid food waste. Firstly we find strategies the direct aim of which is to reduce the amount of perfectly edible food being removed from the commercial food supply chain. Many of these strategies are concentrated upon improving food labelling, because it is estimated that it is retailers and consumers who generate the highest rates of avoidable waste in the EU. The primary function of labelling is to inform consumers with regard to the characteristics and safe use of products, but it has been shown that in some cases labelling creates barriers which hinder the responsible use of food (Lissel, 2015). Thus, for example, the 'best before dates' may be interpreted as expiry dates and result in consumers discarding safe, edible food. Without compromising food safety or consumers' rights, both EU institutions and MS have been working on proposals to improve the interpretation of 'best before dates', one option being the introduction of dual-date labelling (EC, 2013). A second group of strategies addresses the case of food that has been removed from the commercial market, 'safe unsold food' in the words of the EC. The goal is for those foods which are still safe and edible to be destined for their primary use; in other words, that they be employed to feed people or, at least, to feed animals. Therefore, one of the strategies consists in facilitating donation of these foodstuffs, although one also observes the alternative possibility of their being used as by-products in feed production. Although this type of use of unsold food is comprehensible within a logic of reduction of food waste, the fact is that at present there exist legal barriers hampering both the donation of food and the use of former foodstuffs in the preparation of by-products for feed production. For those cases in which a flexible interpretation of the laws does not provide an alternative outlet for unsold food, institutions like the European Federation of Food Banks is seeking to promote legal reforms similar to the one approved in Italy some years ago (Arroyo, 2015). I shall discuss food donation in the next section. Finally, and in relation to that food waste regarded as unavoidable, there is a third alternative. The EC's Communication 'Closing the Loop' 2015 describes food waste as one of the priority areas of the circular economy, referring above all to food waste which has become – or is becoming – organic waste with the potential to be employed as bio-based material.

The donation of safe, edible unsold food in the EU

The primary objective of food production and distribution is the commercialisation of food, but there is a certain quantity of edible food that is designated to be wasted due to various reasons such as wrong size or shape, overproduction, low prices, etc. (Schneider, 2013). One of the strategies proposed by the EP with regard to unsold edible food is its redistribution via donations. The EP's Resolution has referred to this practice on several occasions, employing in each case expressions such as 'redistribution' of food or access 'free of charge' to surplus food. This idea of redistribution also appears to lie behind the EP's allusions to the principle of solidarity when it refers to the need to promote a culture opposed to food waste. Moreover, it openly welcomes initiatives which in various MS are aimed 'at recovering, locally, unsold and discarded products through the food supply chain in order to redistribute them to groups of citizens below the minimum income threshold who lack purchasing power'. It stresses, in this sense, the importance of carrying out an exchange of best practices and underlines the contribution being made by both volunteers and professional companies in this respect. The work of food banks is also highlighted in various EC documents related to the prevention of food waste and retailers are called upon to commit to the redistribution programmes and implement measures permitting discounted access to products nearing expiry.

The implementation of this redistribution, nevertheless, is not only a question of willingness being shown by those in possession of edible, safe unsold food, but in many cases requires potential donors to attempt to overcome different legal barriers. Under these circumstances, donors, food banks and non-profit organizations are interested in obtaining a legal framework that will clarify above all two important elements of donation: liability and fiscal issues (especially the fact that in some countries food donations are not VAT-free). The creation of legislation favourable to a donation upon the basis of the 'Good Samaritan Laws', the reinterpretation of the concept of the end-consumer or the introduction of tax incentives are some of the proposals under consideration. These are incentives designed to promote the voluntary action of potential donors, far therefore from imposing obligations upon retailers in the strict sense, as is the case of the obligation to donate imposed upon supermarkets in France (Arroyo, 2015). Note that this voluntariness in redistribution along with allusions I shall make to the principal of solidarity, anticipates demanding of the players in the food supply chain a responsible use of food, in keeping with its enormous importance for human beings.

Is the expression 'citizens below the minimum income threshold who lack purchasing power', a euphemism to avoid speaking of food insecurity in the EU?

Although nowhere is there any literal reference to what is understood by the 'ethical dimensions' of food waste, there are many reasons to believe, reading between the lines, that the EU is referring to the fact that throwing away food while others are starving is something regarded by most citizens as immoral. This consideration seems to be present when in the context of a Resolution 'on avoiding food wastage 'it is admitted, amongst other things, that in 2011 79 million people were still living below the poverty line in the EU (more than 15% of EU citizens). According to the Crisis Monitoring Report by Caritas (2015) this figure has increased. 123 million EU citizens are currently living in poverty, including a third of under 14s in the 28 EU countries. How do these people obtain food? According to figures obtained by the EP in 2011, 16 million received food aid from charitable institutions; or put another way, lacking sufficient economic access to food obtain it by non-economic means. That sufficient economic access, it should be noted, is a fundamental premise in the human right to food described in the Declaration of Rome on World Food Security (FAO, 1996). In fact the second of the commitments included in this Declaration warns that the signatories commit to 'policies aimed at eradicating poverty and inequality and improving physical and economic access by all, at all times, to sufficient, nutritionally adequate and safe food and its effective utilization'. In other words, according to the Declaration of Rome, people who have access to food but in a 'non-economic' manner are not in a situation of food security. However the EP Resolution on avoiding food waste, by expressing that it believes 'food security to be a basic human right', conspicuously omits any reference to economic access. It declares that food security is 'achieved through the availability, accessibility, use and temporal stability of healthy, sufficient, adequate and nutritious food'. Consequently, it does not classify this situation as 'food insecurity', despite its acknowledgement of the existence of 'citizens who lack purchasing power' who may be – and it seems desirable that they should continue to be- the recipients of the redistribution.

The omissions and euphemisms in the construction of this discourse are in my opinion striking and to a certain degree worrying. The notion of 'redistribution' takes two facts for granted. Firstly, it appears automatically to accept the existence of citizens who have no economic access to food without inviting deeper analysis of the factors that generate this 'lack of purchasing power'. Secondly, it identifies these people's access to food upon a basis of solidarity (upon a principle of the willingness of the potential donor), without consideration of approaches based on social justice or recognition of rights. The State (in this case both the EU and the MS) does not appear to find it necessary to assume any responsibility regarding this lack of purchasing power or with regard to attaining access to 'healthy, sufficient, adequate and nutritious food'. There is evidence here of a trend which in other regions of the world such as North America – in particular Canada – and Australia has been criticised by various experts (Riches, 2002; 2014; Vlaholias *et al.*, 2015). The solidarity shown by companies and citizens is proving indispensable in order to address the current social emergency. Governments have entrusted social recovery to economic recovery, basing the latter on austerity and cutbacks. Cutbacks in what? Above all, in social spending. Redistribution has its virtues in exceptional circumstances, but in my opinion we should not run the risk of institutionalising hunger 'as a matter for charity' (Riches, 2014). Europe can learn from the experience of other countries, like Canada, which have demonstrated that in the long term institutionalisation of charitable food redistribution generates structural problems, consolidating two channels of access to food: a primary channel, the market; and a secondary channel, charity, for those who lack sufficient purchasing power. In a society committed to human rights this secondary channel should be exceptional and as far as possible, temporary, allowing all people autonomous access to decent living conditions in accordance with the exercise of their fundamental rights. A second reason for not placing all hope of alternative access to food in redistribution is to be found in EU anti-waste rhetoric. For if the aim is to achieve zero waste, the amount of surplus food will be drastically reduced – provided all the participants in the food supply-chain fulfil the commitments they are assuming. What will we redistribute then?

Conclusions

Food waste and hunger are nothing new; some historical studies trace their origins back to ancient times, although never before in history had such scandalous figures been reached as today. Manifest in the EU's approach to food waste is the condemnation of a situation in which high levels waste coincide with the fact that there are millions of citizens in this region who do not have adequate access to food. Declaring food waste to be something unacceptable and making a firm commitment to the human right to food is something which in my opinion offered the European Union a very interesting opportunity. It offered an opportunity to review the incoherence of considering that foods are fundamentally commodities despite the fact that they are indispensable resources for people's survival and health. The EP Resolution seemed to open the way to responsible use – a sustainable and even caring use – of food, but it strikes me as wrong that in its adoption there has been avoidance of a human rights approach Throwing away food feels morally wrong, especially when there are human beings who do not have adequate access to food.

In that invitation to make environmentally-friendly use of food –avoiding waste- and in the appeal to solidarity, it appears as if food waste has arisen 'spontaneously', just as there seems to be no cause of food poverty in the EU. I regret the absence of a clear, unambiguous rejection of an unsustainable, unfair and exclusive food system. Redistribution of surplus food does not right this injustice, it merely seeks to conceal some of its consequences. How is it possible that, in the current situation of emergency and need, there is no critical review of contemporary food charity? Yes, let us provide it with channels, but let us think too in the mid-and long term. I agree with Professor Graham Riches: food is a social and cultural good, essential to health and societal integration. We need a governance of food systems coherent with what food means for the human being (Escajedo, 2015), we need a commitment tosustainability that will reduce the current distance between people and food sources. Let us work towards a guarantee of access to food, a guarantee that is worthy of human beings, committed to their fundamental freedoms and rights.

Acknowledgements

Project US14/19, Improving access to Food in Urban Societies, funded by the University of the Basque Country (University and Society Research Grant Call for 2014-2016); Research Team Grant of the Directorate on Scientific Policy of the Basque Country IT 743-13 'Multilevel Constitutionalism and Integration of Diversity'; URBAN ELIKA Cross-Disciplinary Studies on Food and Society, University of the Basque Country.

References

Aramyan, L.H., Valeeva, N.I., Vittuari, M., Mahon, P. and Balazs, C. (2015). Potential of market based instruments and economic incentives in food waste prevention and reduction. In: Escajedo San-Epifanio, L. and De Renobales, M. (eds.) Envisioning a future without food waste and food poverty. Wageningen Academic Publishers, Wageningen, the Netherlands, pp. 171.

Arroyo, A.A. (2015). Edible but unmarketable food: some legal problems to be solved on food waste prevention. In: Escajedo San-Epifanio, L. and De Renobales, M. (eds.) Envisioning a future without food waste and food poverty. Wageningen Academic Publishers, Wageningen, the Netherlands, pp. 33.

Caritas Europe (2014). The European crisis and its human costs – a call for fair alternatives and solutions. Available at: http://tinyurl.com/puz3r55.

Caritas Europe (2015). Crisis monitoring report 2015 – poverty and inequalities on the rise; just social systems needed as the solution. Available at: http://tinyurl.com/oo4ll65.

EC, EU Commission (2011). Roadmap to a resource efficient Europe. Brussels, Belgium.

EC, EU Commission (2014). Towards a circular economy: a zero waste programme for Europe. Brussels, Belgium.

EC, EU Commission (2015). Closing the loop. Brussels, Belgium.
EU Parliament (2011). Motion for a European Parliament resolution on how to avoid food wastage: strategies for a more efficient food chain in the EU. Brussels, Belgium.
Escajedo, L. (2015). Challenging food governance models. Journal of Agricultural and Environmental Ethics 28(3): 435-454.
Evans, D., Campbell, H. and Murcott, A. (2012). A brief pre-history of food waste and the social sciences. Sociological Review 60(S2): 1-22.
FAO (2015). Global initiative on food loss and waste reduction. Available at: http://tinyurl.com/hcr47gw.
Gjerris, M. and Gaiani, S. (2015). Values in the trash: ethical aspects of food waste. In: Escajedo San-Epifanio, L. and De Renobales, M. (eds.) Envisioning a future without food waste and food poverty. Wageningen Academic Publishers, Wageningen, the Netherlands, pp. 55.
González Vaqué, L. (2015). Food loss and waste in the European Union: a new challenge for the food law?. European Food and Feed Law Review 1: 20-33.
Lissel, E. (2015). Food laws and labelling as a contributor to food waste. In: Escajedo San-Epifanio, L. and De Renobales, M. (eds.) Envisioning a future without food waste and food poverty. Wageningen Academic Publishers, Wageningen, the Netherlands, pp. 47.
Riches, G. and Silvasti, T. (2014). Hunger in the rich world: food aid and righ to food perspectives. In: Riches, G. and Silvasti, T. (eds.) First world hunger revisited. Food Charity or the Right to Food? Palgrave McMillan, London, UK.
Riches, G. (2002). Food banks and food security: welfare reform, human rights and social policy. lessons from Canada? Social Policy and Administration 36(6): 648-663.
Schneider, F. (2013). The evolution of food donation with respect to waste prevention. Waste Management 33: 755-763.
Vlaholias, E.G., Thomson, K., Every, D. and Dawson, D. (2015). Reducing food waste through charity: exploring the giving and receiving of redistributed food. In: Escajedo San-Epifanio, L. and De Renobales, M. (eds.) Envisioning a future without food waste and food poverty. Wageningen Academic Publishers, Wageningen, the Netherlands, pp. 271.
WRAP (2009). Household food and drink waste in the UK. Available at: http://tinyurl.com/qd3mb25.

Section 13. Animal ethics in practice

56. May we eat our fellow creatures? Virtues and animal ethics

B.K. Myskja[1] and M. Gjerris[2]*
[1]Department of Philosophy and Religious Studies, Norwegian University of Science and Technology, 7491 Trondheim, Norway;[2]Department of Food and Resource Economics, University of Copenhagen, Rolighedsvej 25, 1958 Frederiksberg C, Denmark; bjorn.myskja@ntnu.no

Abstract

A fundamental question in animal ethics is whether it is morally acceptable to eat animals. Cora Diamond rejects the relevance of moral status-approaches, arguing that the crucial fact is that we have moral relations with animals as our fellow creatures. This resonates well with virtue ethical approaches, understood as a matter of having the right attitude towards ourselves and others. But can Diamond's approach provide us with an answer to the question whether it is wrong to eat non-human animals or – if it is acceptable – under which circumstances we may eat them? A virtue ethical argument for veganism holds that industrial meat production is by necessity cruel and partaking in it is not virtuous. Diamond agrees that fellow creatures should not be regarded as stages in the production of a meat product but she admits that this attitude to animals as worthy of respect and compassion does not necessarily lead to veganism. We suggest that one possible case of virtuous non-veganism is eating domestic animals that lead good lives and are humanely slaughtered. Another could be eating wild animals that could be worse off unless some of them are killed through humane hunting practices. But is that really compatible with relating to them as fellow creatures? Two seemingly incompatible implications can be drawn from Diamond's analysis of the special character of human-animal relationships. Either: animals should never be eaten because we recognise them as fellow creatures and potential companions. Or: we should only eat meat under circumstances where we can pay respect to the kind of relationship we have with the particular creatures. We work out these two positions, and leave it open for further discussion whether both of them can be expressive of a virtuous life with animals as fellow creatures.

Keywords: animal consumption, veganism, flexitarianism, moral relations

Introduction

There is wide consensus that cruel treatment of animals is either wrong or only acceptable as the unwanted side-effect of a necessary activity, such as medical research where there are no alternatives to using animals. It is a fact that this consensus is fictitious in the sense that unjustified cruelty to animals is widespread in research and in food production. If, however, we bracket the issue of willed blindness to cruelty and environmental costs (Gjerris, 2015), assuming that using animals for human consumption is possible without subjecting them to suffering or undue restrictions on their ability to flourish, our question here is: Is it then morally defensible to eat them?

The standard way to approach this question is by showing how animals have or lack certain characteristics considered crucial for having moral status giving us duties towards them. The paradigmatic examples are Peter Singer's (1990) utilitarian and Tom Regan's (1984) deontological approaches, but also neo-Kantian approaches such as Christine Korsgaard's (2004) share this basic argumentation pattern. Cora Diamond, however, rejects the relevance of such moral status approaches, because the basis for the moral distinction between humans and animals is not biological and does not translate into interests or rights, typical for moral status thinking. Moral judgements are based on what is of significance in human life, for example that we do not eat dead humans although no interests are affected by this. Animals do not have the same significance. The moral status approaches attack this significance of human life (Diamond, 1979: 471). Diamond suggests that the experience of animals as our 'fellow creatures' (Diamond, 1979:

EurSafe 2016 – DOI 10.3921/978-90-8686-834-6_56, © Wageningen Academic Publishers 2016

474) is a crucial fact indicating a different way of thinking about animals. Her text is not conclusive, and explicitly rejects turning our experience of this reality into what she calls the 'therefore-arguments' on how to treat animals (Diamond, 2003: 4) We will attempt to draw some implications for the question of whether it is ethically acceptable to eat animals from this analysis, which resonates well with some virtue ethical approaches (Hursthouse, 2000: 165-166), understood as a matter of having the right attitude towards others and ourselves.

This paper discusses whether Diamond's approach can be developed into a virtue ethical (understood in a broad sense) answer to the question whether it is wrong to eat non-human animals or – if it is acceptable – under which circumstances we may eat them. We discuss this issue by exploring two possible answers; one flexitarian, suggesting that such use is acceptable provided it shows due respect for animal welfare and the need for leading a full life in accordance with the capacities of the animals for enjoyment and self-expression; the other vegan, suggesting that we cannot use animals as food for humans. Both answers states that the commodification of animals as units in the industrialized production of food is morally unacceptable. This divided conclusion expresses the disagreement of the authors, indicating that the discussion is not completed.

Fellow creatures

Immanuel Kant writes a hypothetical history of how humans, using reason elevated themselves above the immediate and instinct-driven nature shared with the rest of the animal kingdom. In the final step of this process, humans realised that they were 'the true end of nature'. He continues: 'When [man] first said to the sheep: 'the pelt which you wear was given to you by nature not for your own use, but for mine' and took it from the sheep to wear it himself, he became aware of a prerogative which, by his nature, he enjoyed over all the animals; and he no longer regarded them as fellow creatures, but as means and instruments to be used at will for the attainment of whatever ends he pleased' (Kant, 1991, cited from Korsgaard, 2004: 89). Kant imagines that this altered treatment of animals from fellow creatures to expendable goods is communicated to animals in speech, illustrating our ambiguous relationship with them. We do not speak to things. If they are not our fellow creatures, why does Kant feel the need to explain himself to them? This ambiguity is displayed by Kant also in other contexts, as he puts limits to the infliction of pain on animals in work and scientific experiments, and bids us to show gratitude to working animals by treating them 'as if they were members of the household' (Kant, 1996: 193). The moral duty is, however, ultimately towards ourselves; we must not weaken our natural moral feelings by mistreating animals as it can make us morally insensitive towards other humans.

Both Kant and Diamond use the term 'fellow creatures' about non-human animals, although Kant states that this fellowship is not possible (any more) due to the unique position of human beings in nature, as the only rational being. His view of the moral status of animals is based on this fundamental distinction in rational capacity, making humans the only being capable of being an end in itself. But his rejection of animals as fellow creatures is clearly wrong, as is evident by his statement of the duties we have to treat them with compassion and respect, in analogy with how we are obligated to treat our fellow humans. Kant's description of how we should treat animals is held in a language of virtues, even though his justification is clearly deontological and capacity-oriented. His wavering between these two accounts on how to relate to animals is descriptive of our modern debate. His theoretical account of humans and non-human animals does not fit his prescriptions of the right attitudes and interactions with animals. The latter shows that 'fellow creatures' is still a key term in capturing what is at stake in this morally contested group of questions.

Virtues and attitudes

A virtue ethical argument for veganism holds that 'the practices that bring cheap meat to our tables are cruel, so we shouldn't be party to them' (Hursthouse, 2006: 143). Diamond emphasises in a similar vein that our notion of animals as fellow creatures is not compatible with regarding them 'as stages in the production of a meat product' (Diamond, 1979: 475). This is a thick notion of the moral reasons for refraining from partaking in industrial use of animals for food as consumer. It is the wrong attitude to take to animals. However, stating that something is wrong is merely part of the way to finding out what the right thing to do is. This is the challenge we should present to Hursthouse's and Diamond's moral notions. Cruelty is not a virtue but a vice, and can therefore not be considered an adequate description of a virtue ethics for animal relations. Most will readily agree that we should not be cruel to animals, but without an understanding of what the right relationship is, different understandings of what constitutes cruelty, will compete and allow for e.g. a continued factory production of animals – which is clearly not what Hurtshouse or Diamond have in mind. What are the right virtues concerning animals, and how should we regard animals? These are decisive questions we must answer in order to know whether – or under which conditions – we may eat animals.

Michael Slote says that virtues are concerned with the inner traits, motives and dispositions of the individual. In addition, virtue ethics is concerned with excellence; with what is good or admirable ways of acting and living (Baron *et al.*, 1997: 177). Thus, we must discuss what the good attitude or way of relating to animals is: what is the virtuous counterpart to cruelty? Diamond states that taking animals as our fellows in mortality, in life on this earth ... depends upon a conception of human life (Diamond, 1979: 474). A reasonable interpretation of this is that we should derive our attitudes to animals from (or in analogy to) our attitudes to humans. On this view, morality is something that belongs to the human world, to the kind of shared life that we can only have with beings that interact on basis of a shared understanding of life and on how to live it. If we understand Diamond correctly, this is not a theoretical account of morality, giving a set of criteria to decide what acceptable behaviour towards animals is. It is an account of how we should experience animals, which can be expressed in certain attitudes.

Diamond mentions some attitudes that we should take towards animal based on this extension of the way we relate to our fellow humans. The words she employs are usually part of a virtue catalogue. These include charity, justice, friendship, companionship, respect and pity (Diamond, 1979: 474-475). She also talks of some people's experience of the indignity of training animals to do tricks in circus. Although her account is not primarily intended to be about animal ethics, but about our experience of life in a world with animals, and certainly is not meant to answer the kind of questions that Singer and Regan ask, her account of our experience of animals suggests a virtue approach. It is important to keep in mind that the virtues are words usually used to describe how we might or ought to relate to other humans who share our particular form of communication and morality. It is meaningless to claim that these words have meaning in relation with animals without their meaning-giving context of human fellowship. There are moral expectations in our relationships with humans that obligate us and we read that into our relationships with animals (Diamond, 1978: 478). We should therefore highlight the difference between our use of these concepts on humans and on animals, which is one of Diamond's concerns with classical arguments based on moral status or relations. Later she writes of the experience of animals as connected with a sense of astonishment and incomprehension that there should be beings so like us, so unlike us, so astonishing capable of being companions of our and so unfathomable distant (Diamond, 2003: 14).

Moral relationalism

Todd May classifies Diamond as a 'moral relationalist', which means that the moral status of a living being is determined not solely by its particular characteristics but more importantly by the relations it has with human beings (May, 2014: 155). It is reasonable to interpret Diamond's fellow creatures approach as relationalist, but it is wrong to derive a moral status approach from this, as May does. Diamond states explicitly, when discussing how we treat pets, that there is not a class of beings, pets, whose nature, whose capacities, are such that we owe it to them to treat them this way (Diamond, 1979: 469). There is no biological argument at the basis of Diamond's position, but an extension of a non-biological notion of what human life is (Diamond, 1979: 474), as May rightly observes. However, he is wrong when he concludes that pets are more fellow creatures than wild animals (May, 2014: 162). There is a difference in the sense that pets are closer to our lives, more integrated part of it, and therefore not something we normally eat (Diamond, 1979: 469), but they are no more or less fellow creatures due to that. The proper attitudes, virtues, we should have towards our pets are different from those we can and should have towards cows or lions. This is not a matter of different moral status, but of us having different responsibilities towards them based on the impact of our actions on their lives.

May argues that there are problematic tensions within Diamond's account of the human-animal relationship. He thinks it is misguided to point to the difference between our conceptions of eating dead animals and cannibalism based on something deeply aversive about cannibalism that is not present in the eating of non-human animals (May, 2014: 163), and that eating animals should then be equally distasteful. In our interpretation, Diamond's argument works the opposite way. The aversion against cannibalism is expressive of a fundamental distinction between humans and animals. Our moral views are derived from this basic understanding of what it is to be a human being: We learn what a human being is in – among other ways – sitting at a table where WE eat THEM (Diamond, 1979: 470). The problems May finds are a consequence of his attempt to fit Diamond's analysis into the 'moral status'-way of thinking about animals. He suggests that we should start to think differently about animals because of their richness of life, but that is exactly what we cannot do. We can become vegetarians for many reasons, for example due to our experience of the horror in realizing how we treat animals (Diamond, 2003), but that is not due to their characteristics or our particular relationships with them. It is because they are our fellow creatures and we may think that we should not have the attitude to them that is expressed in turning them into food. But is that a necessary part of the experience of reality as sketched by Diamond?

Flexitarianism or veganism

Two seemingly incompatible implications may be drawn from Diamond's analysis of the special character of human-animal relationships. Either: We should only eat meat under circumstances that pay respect to the kind of relationship we have with the particular creatures. Or: animals should never be eaten because we recognise them as fellow creatures and sometimes even companions. For simplicity's sake, we can call them the flexitarian and the vegan position. We present both, and leave it open whether there are decisive arguments for one of them if we share the idea that animals are and should be treated as our fellow creatures.

A flexitarian or semi-vegetarian is someone who only occasionally eats meat. It is not clear whether flexitarianism implies a specific view on eating egg and dairy products, as the main ideal is to drastically reduce meat consumption. Dairy and egg production raise their own moral problems partly distinct from those of meat consumption, so we will leave that aside. Flexitarianism differs from the common eating practice in the Western world where the main meals of the day are built around meat or fish with vegetables as supplements. As the word is fairly recent, there are no definite definitions and people may practice it differently, but the crucial point here is that it is a practice that follows from a view indicated by Diamond as a possible way of understanding animals as fellow creatures. It is clear that the industrial production of meat is unacceptable, due to the attitude it expresses (Diamond, 1978: 475). It follows that the sources for meat will be drastically reduced. We can only eat meat from animals that have been treated as fellow creatures; with respect, compassion, charity and justice. Therefore, flexitarianism is the only alternative to veganism as a virtue ethical response to the experience of animals as fellow creatures.

'It does normally, or very often, go with the idea of a fellow creature, that we do eat them. But it then characteristically goes with the idea that they must be hunted fairly or raised without bad usage' (Diamond, 1978: 475). Accepting this, the argument for a flexitarian position would hold that it is an integrated part of the value in relations between humans and animals that we do hunt or raise them in a fair way, with a minimum of pain. The good hunter is someone who has a deep knowledge and respect for the animals that they hunt. This hunter arguably has a deeper connection with and respect for the hunted animals than the utilitarian vegan who relates to animals as units of pleasure and pain. When there is overpopulation, animals suffer due to hunger and painful diseases. We could increase the number of non-human predators, but they certainly do not kill in a fellow creature manner. The predators are not blameworthy, but we are if we choose this kind of wildlife management rather than humane hunting.

It is also possible to raise and then kill domestic animals with respect and compassion, as Diamond (1978: 471) indicates by a story about a friend who involves the whole family in the project. 'Never eat animals you do not know personally' may not be possible for most of us, but it indicates an ideal for a virtuous life as meat eater and put severe limits to our consumption of meat. Eating meat from an animal you know have had a life of flourishing is morally different from eating meat bought in a supermarket. An animal virtue ethics drawn from the fellow creature experience does not necessarily imply that we should refrain from eating them, but that we should respect their ability to lead a full life in accordance with their nature. That does not necessarily mean dying from old age, predators or diseases rather than from humans. From this perspective, it is part of a good human life to live with domestic animals and interact with wildlife. This includes sometimes killing them for food.

By veganism we understand to refrain from eating any animal products, including egg and milk. Veganism is, from the virtuous fellow creature perspective connected to two arguments. The first is that most consumption of animal products are associated with unnecessary suffering through confinement, breeding practices, production efficiency, etc. The suffering induced is seen as unnecessary as humans can live a full and healthy life without consuming animal products in most circumstances. Things change, if the choice is between malnutrition or hunger and eating animals, but that is not the situation for most Western consumers.

The second is that taking the life of an animal, even to spare it from experiencing the sometimes harsh conditions of a life in the wild, is not compatible with virtues such as respect, compassion, charity, and justice. As we saw, Kant connected the human decision to use animals for our benefit with ceasing to regard them as fellow creatures. Obviously, we can be in a situation where we have the choice of allowing the animal to die a natural death while suffering or putting it out of its misery. But such crisis situations are not determinant of what a respectful attitude towards animals in general is, but rather a situation of the lesser evil brought on by the shared tragic basic conditions of life as a sentient being. Thus, such situations cannot be used to justify e.g. culling of deer to avoid over-population and then eating the meat. Rather such situations should be avoided to the extent possible – and when not possible allowed to play out as a sign of respect for the nature of the life given to the animal and the ecological processes that, even though harsh, have brought forth this animal and its kind – and in the future will bring forth new individuals. Within this line of thought the idea of utilizing the animal after its death, however much it is inflicted to spare the animal for suffering, is not a sign of respect. The animal should rather be left to predators.

Incarnating the virtues of e.g. respect and compassion, is to share their lives and death based on an understanding of the shared tragic character of human and animal lives. Connecting with them in the wild rather than domestication and hunting is the virtuous way to connect with animals from this perspective, as fellow creatures with a moral capability do not eat other fellow creatures, but share the joys and pains of existence.

Conclusions

A virtue ethical approach to animals as fellow creatures is an alternative to the dominant moral status approaches. We have argued that this approach better captures our experience of sharing the world with non-human animals. We have explored how this basic conception can be spelled out in two different ways of living with animals – either a flexitarian life, allowing the killing of animals under strict conditions, or a vegan life, refraining from any killing of these fellow creatures.

References

Diamond, C. (1978). Eating meat and eating people. Philosophy 53(206): 465-479.

Diamond, C. (2003). The difficulty of reality and the difficulty of philosophy. Partial Answers 1(2): 1-26.

Hursthouse, R. (2000). Ethics, humans and other animals. Routledge, London, UK.

Hursthouse, R. (2006). Applying virtue ethics to our treatment of other animals. In: Welchman, J. (ed.) The practice of virtue. Hackett, Indianapolis, USA, pp. 136-155.

Kant, I. (1991). Conjectures on the beginning of human history. In: Reiss, H. (ed.) Kant:political writings. Cambridge University Press, Cambridge, MA, USA.

Kant, I. (1996). The metaphysics of morals. Cambridge University Press, Cambridge, MA, USA.

Korsgaard, C. (2004). Fellow creatures: Kantian ethics and our duties to animals. Tanner Lectures on Human Values 24: 77-110.

May, T. (2014). Moral individualism, moral relationalism, and obligations to non-human animals. Journal of Applied Philosophy 31(2): 155-168.

Regan, T. (1983). The case for animal rights. University of California Press, Berkeley, USA.

Singer, P. (1990). Animal liberation. Avon Books, New York, NY, USA.

57. Ethical acceptability of recreational hunting – does the motive of the hunter matter?

C. Gamborg[1*], *F.S. Jensen*[2] *and P. Sandøe*[1]
[1]*University of Copenhagen, Department of Food and Resource Economics, Rolighedsvej 25, 1958 Frederiksberg C, Denmark;* [2]*University of Copenhagen, Department of Geosciences and Natural Resource Management, Rolighedsvej 23, 1958 Frederiksberg C, Denmark; chg@ifro.ku.dk*

Abstract

Recreational hunting can be a way of taking responsibility for acquiring one's own meat. However, many recreational hunters focus instead on hunting as a hobby or sport. This distinction, between two rather different motives for hunting, is relevant to the activity's moral justifiability. The public appear to be more critical of the killing of wild animals for sport or pleasure than they are about hunting based on necessity or subsistence. A number of philosophical studies of hunting motives have appeared to date, but we have few empirically grounded analyses. A leading aim of this paper is to assess the extent to which the perceived motive for recreational hunting plays a role in its public acceptance. We also compare public perceptions of the importance of motive with those of hunters. We conducted a nationally representative survey (web-based questionnaires) of the general public (n=1,001) and hunters (n=1,130) in Denmark. In this survey just under half of the general public indicated that the hunters' motives affected their attitude to the acceptability of hunting. A significant difference in wildlife value orientations was found between the two groups. Motives relating to nature, the social aspect of hunting and escape from everyday life were stated as the most important by hunters. Nature and social aspects motives were also perceived by the largest proportion of the general public as important motives for hunting. However, the general public tended to ascribe motives such as 'the trophy', 'the sport', 'the excitement', and 'to kill' to recreational hunting to a much greater degree than the hunters themselves; and these motives were associated with lower assessments of the acceptability of recreational hunting among the public. The mismatch between presumed and professed motives among hunters may have arisen because the hunters we surveyed tended to answer tactically. Given the difference in importance attached to the various motives, there seems to be room for improved dialogue between hunters and the general public.

Keywords: public attitudes, Denmark, sport, wildlife values

Introduction

In many European countries recreational hunting is increasingly seen as a way to take responsibility for acquiring food (Ljung *et al.*, 2012). At the same time, however, the ethical acceptability of hunting in general, and the legitimacy of recreational hunting in particular, are being questioned (Cohen, 2014). 'Recreational hunting' here refers to all recreational activities which involve the pursuit and killing of wild animals, fish excluded (Wightman *et al.*, 2002). Many studies have shown that the way hunting is organised and carried out affects attitudes to its acceptability (Woods and Kerr, 2010). For example, when hunting is carried out on farm-reared animals that are released to provide quarry, public support decreases (Gamborg *et al.*, in press). However, research also suggests there are marked differences in attitudes depending on the underlying motives for hunting (Fischer *et al.*, 2013).

Motives are reasons people hold for initiating and performing voluntary behaviour. They point to the meaning of the behaviour for the agent and may say something about his or her values (Reiss, 2004). In recreational hunting the motives of hunters explain why they hunt. Although they are connected, motives are not the same thing as justifications (Wood, 1997). In recreational hunting, the latter seek

EurSafe 2016 – DOI 10.3921/978-90-8686-834-6_57, © Wageningen Academic Publishers 2016

to show that hunting, as opposed to other practices where (wild) animals are killed, such as slaughtering or trapping, is acceptable. Such arguments are not subject-dependent in the way motives are. Roughly speaking, motives are subjective reasons for undertaking a specific activity.

There are several ways of defining and classifying motives. A notable early example, from Aristotle, focuses on ends – i.e. engaging in an activity for no apparent reasons (other than this being an expression of a personal desire) – versus means – where the act in question has purely instrumental value. Another classification focuses on biological, or primary, motives, sometimes called drives, which relate to biological survival needs such as hunger or escape from pain. Two types of social or secondary motive can also be separated. These are in some ways akin to motives as a means to achieve something. Such motives may go beyond one's own achievements, in which case they are classified as altruistic, or they may be personal or egoistic motives, also called intrinsic motives, which pertain to one's goals, interests and values. According to some social psychologists intrinsically motivated people experience interest and enjoyment in their activities and may feel competent and enjoy the power of freedom to form their own actions (Deci and Ryan, 1985). Intrinsic motives are characterised by voluntariness. By contrast, extrinsic motives are formed by external factors or pressure.

From a moral point of view an important distinction may be between self-serving or egoistic motives, which have a strong base in feelings of pleasure, and altruistic motives, which have a final goal of increasing the welfare of other individuals (Batson, 2015). An action such as recreational hunting can be judged ethically acceptable (or not) in the sense that it is seen as morally right (or wrong) or good or (bad). In terms of right or wrong, from a teleological (consequentialist) point of view, an action is made right by its consequences (including indirect consequences where motives may play a role). Deontologists, on the other hand, would claim that an action can be right or wrong for reasons other than its consequences. Thus the moral value could hinge on the person's motives. Emphasis may be placed on 'good' or 'virtuous' motives relating to, for example, benevolence or caring about others (Slote, 2003). From this perspective actions are judged by their motives rather than according to whether they promote happiness or result in 'good' consequences. Spielthenner (2005) provides an illustrative example of a deontological view: 'If a person spends money in order to help innocent civilians in a war-torn country, her motivation tends to make her action morally good. But if she spends the money only because she regards it as a lucrative investment, her actions may be prudent but it would not be seen as morally praiseworthy.'

Two interrelated issues are raised by recreational hunting and motives. One reflects a normative ethical standpoint and requires us: (a) to consider arguments that motives should or should not matter for the judgement of an action, here recreational hunting; and (b) to discuss what kinds of motive ought to be considered conducive to a favourable moral judgement. The other issue arises in the field of empirical ethics. This requires us (c) to examine if, and to what degree, non-hunters think motives matter in assessment of the acceptability of recreational hunting (cf. point (a) above); and (d) to explore what motives hunters have, or at least state that they have, and what non-hunters believe hunters' motives actually are (cf. point b) above). In connection with (c) and (d) we can also ask what the implications are for the ethical acceptability of recreational hunting. In this paper we focus on (c) and (d).

There are several philosophical studies of hunting motives (Cartmill, 1995; Causey, 1995; Kheel, 1996; Wood, 1997). There are also a number of descriptive, empirically grounded analyses of hunters' motives, as expressed by hunters themselves, taking an internal perspective, and examining a mainly Western hunting culture. Some of the pioneering, and often cited, studies are those by Kellert (1978) and Decker and Connelly (1989). Many subsequent studies focus on recreational hunting, including trophy hunting (Radder, 2005), and other types of hunting in Africa (Radder and Bech-Larsen, 2008), as well as on the motives of recreational hunters of different types in North America (e.g. Bhandari *et al.*, 2006), New

Zealand (Woods and Kerr, 2010), Australia (Finch *et al.*, 2014), and Europe (Grandy *et al.*, 2003); and a number of these studies show that, to hunters, hunting is a 'cultural activity with significant social contexts' (McCorquodale, 1997: 568).

Specific research in Europe includes qualitative studies such as Fischer *et al.* (2013) and Delibes-Mateos *et al.* (2015) and, to a much lesser extent, quantitative surveys such as the one conducted by Lundhede *et al.* (2015). Broadly speaking, from these studies it appears that acceptance of recreational hunting is negatively affected when hunting is viewed as a recreational pastime in which wild animals are killed for sport or pleasure, as opposed to when it is undertaken out of 'necessity', for subsistence purposes, for example (Ljung *et al.*, 2012). In a similar vein, when hunting is conducted to cull individuals or populations, say, to prevent disease spreading or for ecological population control, it is viewed as more acceptable by the general public. However, we possess no research which measures the general public's beliefs about hunters' motives and hunters' self-professed motives, and which examines the link between these beliefs and people's wildlife value orientations, where the latter tell something about the place of wildlife in one's ideal world, including relations between humans and wildlife (Gamborg and Jensen, 2016).

The main question this paper addresses is: How and to what extent do the motives of the hunter play a role in the general public's acceptance of recreational hunting? This gives rise to four sub-questions: (1) Does the motive of the hunter affect public attitudes to recreational hunting?; (2) What motives do members of the general public ascribe to hunters?; (3) To what degree do the motives ascribed by members of the general public match the motives the hunters ascribe to themselves?; (4) What are the implications of these findings for the ethical acceptability of recreational hunting?

Methods

Our work was part of a research project gauging experiences with, and attitudes to, wildlife, hunting and wildlife management among the general public, hunters and landowners. A national survey of the general public (n=1,001) and hunters (n=1,130) was carried out. The data were collected in late 2012 and early 2013 in two separate web-based questionnaires prepared in SurveyXact (a data-gathering system designed for web-based questionnaire surveys, http://www.surveyxact.com). The results were weighted for gender and age. The part of the survey which is presented here consisted of a set of questions about hunters' motives for recreational hunting. Respondents among the general public were asked if hunters' motives for hunting affected their attitude to recreational hunting. They were invited to answer on a gradual scale from 'no, not at all' to 'yes, very much so'. Thirteen motives were presented to the respondents, who were asked how important they thought each motive was on a 4-point scale, ranging from 1 'not important' to 4 'very important'. There was also an option to state motives 'other' than the ones listed. Respondents also had the option of answering 'don't know'.

Results

Just under half of the general public indicated that hunters' motives had, to some degree, an impact on their overall attitude to the acceptability of hunting (Table 1).

Persons categorized as having a 'mutualist' value orientation to wildlife – who consider the coexistence of humans and animals in a community of sorts to be fundamental – were especially likely to assert that motive matters very much for the acceptability of hunting. The opposite was the case for persons with a 'distanced' orientation – i.e. for those who were not particularly interested in wildlife and related topics. Here 41% answered that hunters' motives played no role at all in their attitude to hunting.

Table 1. Importance of hunters' motives for the general public's overall attitude to hunting by wildlife value orientations (%).*[1]

	Wildlife value orientation				n	χ^2	*P*-value	Cramer's V
	Utilitarian (22.9%)	Mutualist (31.7%)	Pluralist (13.0%)	Distanced (32.4%)				
						58.1	<0.0001	0.139
No, not at all (33.6%)	24.2	21.8	12.8	41.2	336			
No, just a little (20.7%)	24.5	31.0	9.6	34.9	207			
Yes, to some degree (30.5%)	22.7	33.0	14.5	29.8	306			
Yes, very much (15.2%)	18.3	51.7	15.1	14.9	152			

[1] In the questionnaire hunting was confined to the context of outdoor recreation, and the question was: 'Does the hunters' motive for hunting play a role in your attitude to hunting?'

Most hunters (54%) stated that 'experiencing nature' was their most important motive for going hunting, followed by 'the social aspects', which is also what the largest proportion of the general public believed to be the main motive for hunting (32%). However, motives such as hunting for 'the trophy', 'the sport', 'the excitement', and 'to kill' were all believed to be more important motives for hunting by members of the general public than the hunters themselves stated.

Discussion

For many hunters a deeper meaning seems to be connected with (recreational) hunting. Some non-hunters, however, find it difficult to understand what the point of going around the countryside and shooting at the other living beings is, and it might be hard to guess what motivates hunters and to judge the importance of various reasons for hunting. The majority of Danes accept recreational hunting, and earlier studies have shown that people might accept recreational hunting to avoid accusations of double standards because they themselves eat meat. The degree of acceptance has also been shown to be related to the form of hunting where, notably, (bigger) hunting parties – often involved in rented day-hunting or -shooting of farm reared and released birds – seem to be accepted least. The hunter is also expected to behave 'responsibly' towards the game and the rest of nature – whatever that may mean (Jensen, 2009). However, acceptance also seems to be connected, as the present study shows, with what is believed to be the underlying motives of the hunter.

Perhaps this aspect was recognized when, in the Danish Act on Hunting and Game Management, which came into force on 1 April 1994, a standard for good hunting practices was put forward. The standard is not legally binding but nevertheless seems to have played an important role in legitimizing and justifying recreational hunting in a broader societal context. The principles are often referred to in discussions with other outdoor stakeholders, including animal welfare and rights organisations and members of the general public. The standard for good hunting practices not only specifies 'good and right' hunting behaviour but also discusses the 'right' motives of hunters. It states that all hunting must have a purpose which must mean more than simply killing wildlife (for the fun or excitement of it), and that all killed animals are to be collected and used for food, fur and other recognised purposes. Although a purpose is not the same as a motive, it is clear that the purposes here will correspond to motives for the hunter, and that the latter will include the wish to make good use of the products derivable from hunted animals.

When recreational hunting is justified with reference to utility, let alone necessity, a discussion of what it means for hunting to be of use, or necessary, opens up. These days, it is hard to argue that eating (a lot of) meat is good for one's health or necessary for survival. Behind the justification based on utility there is clearly a cultural aspect. Producing livestock and hunting wildlife may be accepted because they are part of a culture in which these activities have traditionally played a major role. But references to culture and traditions cannot serve as a substitute for the hunter's ability to explain and be answerable for his or her actions, and such explanations often entail clarifying one's motives, not just the consequences of the action.

The results indicate that motives play an important role in public discussion of the ethical acceptability of recreational hunting. Given the difference between the motives attributed by the public to hunters and the motives hunters declare, there seems to be room for improved dialogue between hunters and the public. On the one hand, hunters need to do better in communicating their motives, and more importantly, perhaps, in making it clear what motives they do not have or do not see as the most important ones. (Here we assume that hunters will profess their real motives, at least as far as they are aware of these, and not simply play a game of tactics). It may be the case, however, that when hunters are confronted with the motives ascribed to them by the public, especially those related to killing for fun and excitement, they will be unable to convince non-hunters that they really have their other motives, and that they are sincere about the importance of these. This might be because some practices, such as the shooting of farm reared and released birds, may not be considered representative or in line with some of the ascribed motives that correlate positively with hunting's acceptability, such as 'peace and quiet' or 'experiencing nature'. In the course of time, a mismatch between stated motives and practices, or misrepresentation of the importance of motives and practice, may undermine the legitimacy of recreational hunting.

References

Batson, C.D. (2015). What's wrong with morality: a social-psychological perspective. Oxford University Press, Oxford, UK, 272 pp.

Bhandari, P., Stedman, R.C., Luloff, A.E., Finley, J.C. and Diefenbach, D.R. (2006). Effort versus motivation: factors affecting antlered and antlerless deer harvest success in Pennsylvania. Human Dimensions of Wildlife 11: 423-436.

Cartmill, M. (1995). Hunting and humanity in Western thought. Social Research 62: 773-786.

Causey, A. (1995). Is hunting morally acceptable? High Country News 49.

Cohen, E. (2014). Recreational hunting; ethics, experiences and commoditization. Tourism Recreation Research 39: 3-17.

Deci, E.L. and Ryan, R.M. (1985). The general causality orientations scale: self-determination in personality. Journal of Research in Personality 19: 109-134.

Decker, D.J. and Connelly, N. (1989). Motivations for deer hunting: implications for antlerless deer harvest as a management tool. Wildlife Society Bulletin 17: 455-463.

Delibes-Mateos, Viñuela, J. and Arroyo, B. (2015). Game managers' views on the release of farm-reared red-legged partridges in hunting estates within central Spain. Journal of Nature Conservation 26: 1-8.

Finch, N., Murray, P., Hoy, J. and Baxter, G. (2014). Expenditure and motivation of Australian recreational hunters. Wildlife Research 41: 76-83.

Fischer, A., Sandström, C., Delibes-Mateos, M., Arroyo, B., Tadie, D., Randall, D. and Reljić, S. (2013). On the multifunctionality of hunting – an institutional analysis of eight cases from Europe and Africa. Journal of Environmental Planning and Management 56: 531-552.

Gamborg, C. and Jensen, F.S. (2016). Wildlife value orientations: a quantitative study of the general public in Denmark. Human Dimensions of Wildlife 21: 34-46.

Gamborg, C., Jensen, F.S. and Sandøe, P. (in press). A dividing issue: attitudes to game bird shooting among landowners, hunters and the general public in Denmark. Land Use Policy.

Grandy, J.W., Stallman, E. and Macdonald, D.W. (2003). The science and sociology of hunting: shifting practices and perceptions in the United States and Great Britain. In: Salem, D.J. and Rowan, A.N. (eds.) The state of the animals II. Humane Society Press, Washington, DC, USA, pp. 107-130.

Jensen, F.S. (2009). Befolkningen og jagt. In: Kanstrup, N., Asferg, Flinterup, M.T., Thorsen, B.J. and Jensen, T.S. (eds.) Vildt & Landskab. Skov- og Naturstyrelsen, pp. 46-49.

Kellert, S.R. (1978). Attitudes and characteristics of hunters and antihunters. In: Transactions of the Forty-third North American Wildlife and Natural Resources Conference, March 18-22. Wildlife Management Institute, Phoenix, AZ, USA, pp. 412-423.

Kheel, M. (1996). The killing game: an ecofeminist critique of hunting. Journal of the Philosophy of Sport 23: 30-44.

Ljung, P.E., Riley, S., Heberlein, T.A. and Ericsson, G. (2012). Eat prey and love: game meat consumption and attitudes towards hunting. Wildlife Society Bulletin 36: 669-675.

Lundhede, T.H., Jacobsen, J.B. and Thorsen, B.J. (2015). A hedonic analysis of the complex hunting experience. Journal of Forest Economics 21: 51-66.

McCorquodale, S.M. (1997). Cultural contexts of recreational hunting and native subsistence and ceremonial hunting: their significance for wildlife management. Wildlife Society Bulletin 25: 568-573.

Radder, L. (2005). Motives of international trophy hunters. Annals of Tourism Research 4: 1141-1144.

Radder, L. and Bech-Larsen, T. (2008). Hunters' motivations and values: a South African perspective. Human Dimensions of Wildlife 13: 252-262.

Reiss, S. (2004). Multifaceted nature of intrinsic motivation: the theory of 16 basic desires. Review of General Psychology 8: 179-193.

Slote, M. (2003). Morals from motives. Oxford University Press. Oxford, UK, 236 pp.

Spielthenner, G. (2005). Consequentialism or deontology? Philosophia 33: 217-235.

Wightman, A., Higgins, P., Jarvie, G. and Nicol, R. (2002). The cultural politics of hunting: sporting estates and recreational land use in the Highlands and Islands of Scotland. Sport in Society 5: 53-70.

Wood, F. (1997). The delights and dilemmas of hunting: the hunting versus anti-hunting debate. University Press of America, New York, NY, USA, 237 pp.

Woods, A. and Kerr, G.N. (2010). Recreational game hunting: motivations, satisfactions and participation. Land Environment and People research report, no. 18. Lincoln University, Canterbury, New Zealand, 55 pp.

58. Eating to save wild-life: is a truly conservation-minded zoo/ aquarium a vegan zoo/aquarium?

M. Gjerris[1], M. Birkved[2], C. Gamborg[1] and S. Brando[3]*
[1]Department of Food and Resource Economics, University of Copenhagen, Copenhagen, Denmark; [2]Department of Management Engineering, Division for Quantitative Sustainability Assessment, Technical University of Denmark, Lyngby, Denmark; [3]AnimalConcepts, Zoom 1813, 8225 KM Lelystad, the Netherlands; mickeygjerris@gmail.com

Abstract

According to the European Association of Zoos and Aquaria their mission is 'to facilitate cooperation ... towards the goals of education, research and conservation'. Livestock production is one of the leading causes of often-irreversible land use changes, greenhouse gas emissions, loss of biodiversity and different types of environmental degradation – all affecting wildlife negatively, and hence undermining conservation policies that aim to protect individuals, populations and species. But what is the link between livestock production and zoos and aquariums? One link, putting it a bit boldly, could be: Does it make sense to work for conservation by preserving animal species in captivity while selling food to visitors that may be undermining this effort? Complicating the issue is that zoos and aquariums are dependent on generating a profit from 'non-core' services such as cafeterias and the like to generate funds for running the zoo, and conceivably, in turn for conservation purposes – funds that might diminish if zoos and aquariums do not sell a variety of food products, including animal-based ones to their visitors. The main question addressed by this paper is: If zoos and aquariums are to work for sustainability and species conservation – should food served in zoos be part of considerations – and to what extent? To answer this question the paper presents the goals of EAZA along with environmental impact profiles, relying on previously published life cycle assessments of the entirety (i.e. from cradle to gate) and across a multitude of impact categories (i.e. including and beyond climate change), of typical food items sold in zoos and aquariums. It describes the impacts on wildlife and nature that these products may have. Further we link this analysis to different ideas of sustainability, addressing the issue of how to balance positive and negative impacts of zoos and aquariums. Finally we discuss the educational opportunities that arise if food served in zoos and aquariums is seen as part of a conservation strategy – and the possible challenges such an approach faces.

Keywords: climate change, ethics, food, habitat loss, life cycle assessment

Introduction

Climate change has, and is most likely to have, widespread effects on wildlife due to especially changing weather patterns, environmental degradation and habitat loss (Bellard *et al.*, 2012). Human population growth and logging, fossil fuel extractions, as well as mining are known to threaten wildlife and their habitats. Climate change, however, has added a new urgency into the efforts to protect and conserve the biodiversity of the planet. One organization that for many years has worked with species conservation and raising public awareness around these issues is the European Association of Zoos and Aquaria (EAZA) through the collaboration of 377 zoos and aquariums in Europe and the Middle East, with an estimated 140 million people visiting EAZA member zoos and aquariums annually. The goal and strategy of EAZA regarding species conversation are very visible on their website:

> Our approach to species conservation, called the One Plan approach, recognises that zoos and in situ conservationists need not only to work together to protect animals, but also

EurSafe 2016 – DOI 10.3921/978-90-8686-834-6_58, © Wageningen Academic Publishers 2016

> to engage the public of their communities to take the lead in demanding action from authorities, governments, corporations and themselves so that together we can reduce the stress on endangered species and their habitats. (EAZA, 2016)

EAZA's ideals of collaboration and reducing the stress on species and their habitats through the work of zoos and aquariums within conservation policy and efforts may contribute to help to protect animals and their environment. Our question is: Does the food served in zoos and aquariums have anything to do with this? This paper examines what a sustainable zoo or aquarium might entail and gives a short overview of food typically served in zoo and aquarium restaurants and snack stands. With Life Cycle Analysis of these commonly served food items we intend to shed light on the effects of production and consumption of animal protein (including fish) on wildlife and their habitats. We conclude with discussing how to meet the aforementioned ideals of protecting animals and their environment could be linked to food choices presented to visitors at zoos and aquariums.

What does being a sustainable zoo/aquarium include?

The notion of sustainability has been widely accepted as a broad and inclusive political (value-based) as well as managerial (practical) framework (Gamborg and Sandøe, 2005a). One of the reasons for this is that it seemingly represents a more unified view rather than polarized discussions with either a one-sided focus on effectiveness or profit on the one hand, or uncompromising demands for conservation, animal welfare – or what other concern may be felt relevant. In many ways, sustainability has become another word for 'the good' or the ethical 'right' thing (although, naturally what is good and right depend on one's values and ethical positions).

Zoos and aquariums could be seen as a kind of animal husbandry, albeit with a focus on edutainment and breeding animals for conservation, instead of production. To further animal stewardship, including inspiring to sustain species, they seek to create a reverence for wildlife populations and more broadly to care for nature (Conway, 1995). There are many ethical parameters to ponder in the running of a sustainable *sensu lato* zoo or aquarium ranging from profitability, animal acquisition, care, breeding, culling, and individual and population level welfare to wildlife, nature conservation and environmental impact and social responsibility. Clearly, a lot of these issues have been, and are addressed in attempts to build sustainable zoos and aquariums.

However, problems abound when trying to decide which elements to include, and possibly 'tweak', when to trying to become more sustainable. Although sustainability is presented as the amalgamation of three main spheres: economic, environmental, and social, there are different interpretations and different ways to balance these areas (Söderbaum, 2014). Some have argued that the concept of sustainability is prone to being shaped according to the interests at play (Maxey, 2007). In any event, it seems fair to say that what is considered sustainable depends on the underlying 'weak' or 'strong' interpretation, which in turn rest on a number of underlying values, including who (or what) to include as to be of concern (Gamborg and Sandøe, 2005b). One of the many things to discuss would be the wider environmental impact of the kinds food served in zoos and aquariums.

Many zoos and aquariums have chosen to enter certification and accreditation schemes and to some extent (at least indirectly) the balancing has been decided upon by the certifying organization. Often, it entails promoting local, organic, seasonal, fair-trade products, and certified chains of custody, such as MSC seafood and FSC wood. The question is, however, whether reliance or at least inclusion of animal protein, 'sustainably' sourced or not, is, in the broader perspective sustainable – enough – given the documented environmental impacts?

Catering in zoos/aquariums

A statement sometimes uttered by zoo directors or finance executives responsible for the economic running of a zoo suggests that zoos can be seen as a restaurant with animals. Money is made by sales of food, drinks and snacks while animals and staff only cost money. Zoos and aquariums have many costs, such as hiring staff, feeding and caring for the animals, advertising, research, education and conservation programs. The sales of food, drinks and snacks as well as interactive programmes, photo and feeding opportunities, and souvenirs are important sources of income. Today, many zoos and aquariums offer a wide variety of food choices, from small snacks (e.g. hot dogs and popcorn), fast food such as hamburgers and fries to seated dinners and fancy buffet catering for wedding and dinner parties. Menus are often adapted to reflect a theme within the zoo, like 'Asian', 'South American' or 'African', including meat, fish, vegetables, eggs, dairy products and even tofu with flavours spanning from curry to chilli. The food served varies from country to country: from baked potatoes with beans, to pizza and pickled herring. It is important to note that many zoos and aquariums outsource the catering; from the restaurants to the hotdog stands. Therefore, many zoos and aquariums have only little or no control over what types of food are offered to the visitors. The question is, however, whether the wide variety of food sold in zoos and aquariums could be used more actively to promote the outspoken goal of EAZA of contributing to species conservation?

Life cycle analysis of catering in zoos/aquariums

Evaluating the environmental performance of food possesses a considerable challenge in terms of assessment approach. All steps in the value chain of food from farming, processing, transport, use and disposal to different extent and in different ways impact on the environment (i.e. some steps may mainly contribute to global warming while others may primarily contribute to toxicity related impact categories). To account for the impacts related to food it is hence necessary to account for as many as possible of the activities in the value chains of food items – and also to account for the diversity of impacts related with food consumption. The methodology of life cycle analysis seeks to account for all this and includes a multitude of environmental impact categories (Goedkoop *et al.*, 2008), of which most are relevant for food consumption. Despite the fact that the life cycle of food contributes to all impact categories typically accounted for in LCAs, most often only the contribution to global warming is reported; which is the reason for focussing on this impact category in this paper.

Food is an important contributor to the overall impact of consumption. As presented by Kalbar *et al.* (2016) the contribution from food consumption to the overall impact pattern from consumption among Danish citizens ranges across the 18 covered impacts categories between 0.3 and 94.6% of the total consumption related impact profiles on person level. The results presented by Kalbar *et al.* (2016) indicate that food consumption may be a significant segment to consider when trying to reduce the overall environmental burden resulting from consumption. For global warming, food consumption accounts on average for 16.4% of the burden resulting from consumption (covering impacts related to consumption of food, various transport forms, accommodation and various energy carriers), and food is hence in the top 3 of the consumption segments with the largest global warming contributions. Different food items are associated with different environmental burdens. Foster *et al.* (2006) reveal that the burden difference between e.g. potato and beef compared on a mass basis amounts to a factor >60 in terms of climate burden, meaning that the contribution to global warming of 1 kg beef is 60 times larger than the contribution from 1 kg of potatoes in terms of global warming accounted for over a 100 year period.

Meat and in particular beef are some of the most burdensome meal ingredients from a climate perspective. Consumers may therefore ask themselves if switching meat types may improve the environmental performance of an average meal or if they should fully or partially (e.g. as known from week day vegetarians) abandon meat consumption in order to reduce the environmental impacts related with food intake. The results presented by Foster *et al.* (2006) indicate that comparing the global warming performance of sheep, beef, pork and poultry on a weight basis sheep and beef are the most burdensome while poultry is the least burdensome meat type. As the results presented by Foster *et al.* (2006) also indicate, the global warming contribution of poultry is still 6-22 times higher than that of plant protein based food items such as potatoes and bread when compared on weight basis. This indicates that selecting between different meat types may reduce the impacts from meals to a considerable extent, however not as much as leaving out the meat items completely.

Since different food items obviously have quite different impact profiles the question remains how these differences are manifested in our choice of food items or rather through our diets. As presented by Goldstein *et al.* (2016) there are considerable (approximately 50% reduction) global warming savings to be achieved by changing from an omnivorous diet to a vegetarian diet. Moreover, there appears to be a saving by changing from a vegetarian to vegan diet but this saving is less significant.

Overall, meat consumption is associated with considerably higher global warming contributions than consumption of food items based on plant protein (Foster *et al.*, 2006; Goldstein *et al.*, 2016). The issues relating to the environmental burden of food presented here all indicate that the relation between human nutrition and wildlife conservation is complex, to say the least. It involves a number of trade-offs between life cycle stages of food. Saving food waste reduces the demand for agricultural production – hence a trade-off between the disposal and manufacturing life cycle stages and impact categories. Reduced meat consumption compensated for by increased vegetable consumption will lead to a lower climate burden from human nutrition, but as argued by Goldstein *et al.* (2015) will most likely be counterbalanced by an increased/altered pesticide use potentially leading to an increase in toxicity related impacts related to conventional agricultural production. As zoos and aquariums generally seek to attract people with an interest in wildlife and conservation, it may be interesting for zoos and aquariums to not only address the general conservation interest of their audience, but also appeal to the role of visitors as part of reducing the environmental burden related with food consumption.

Effects of production and consumption of animal protein on wildlife

Meat production substantially influences the environment (Gamborg and Gjerris, 2012) in terms of direct and immediate effects (e.g. clearing of woodland for pastures) or indirect and long-term effects (such as subsequent land use changes or emission of greenhouse gases) at local and global scales (Ilea, 2009). Studies have shown that agriculture is responsible for 70% of global freshwater consumption, which thus puts scarce water resources under pressure. Livestock production takes up more than 65% of all arable land and hereby contributes to deforestation, changes in savannahs, drainage of wetlands and desertification (Norris *et al.*, 2010). Evidently some of these areas affected host a number of species and thus biodiversity conservation is related to climate change both directly and indirectly. The estimated contribution of the livestock sector to the humanly caused GHG emissions is 18% according to FAO (Steinfeld *et al.*, 2006), while others estimate a contribution as high as 50% (Goodland and Anhang, 2009). Educated guesses are that almost a doubling in consumption of animal products can be expected by 2050.

Including catering in the sustainability concept: opportunities and challenges

Assuming that zoos and aquariums, besides entertaining, wish to promote conservation of wildlife, they often do this by seeking to create understanding and awe in the visitors – and, in turn, may raise money for conservation efforts. Zoos and aquariums work together on breeding programmes for endangered species, conduct research and participate in local attempts to preserve wildlife habitats, etc. However, as one of the threats to current wildlife populations is the loss of habitats, land use changes and changing climate patterns, and it is hard to deny that human diet plays a role in this, it may seem counter-productive to try to engage visitors in efforts to save wildlife at the same time as offering hamburgers and other animal protein based food products.

First, as previously shown, a higher environmental impact arises from eating meat and dairy products than from products based on plants. Still, considering the amount of meals served in zoos and aquariums each year compared to the total amount of human emissions, the human climate impact would not change significantly, even if all zoos and aquariums only served vegetarian or vegan meals. In the current situation with rapidly rising temperatures, however, one could argue that any contribution, however small, is important. Here, however, we will draw attention to the educational potential of including the food served to visitors in zoos and aquariums as part of a conservation strategy. Through what is served, it is possible to highlight that a diet based more on plant protein is necessary if we are to lower emissions and not run the risk of seeing the animals that we have just admired in the zoo and aquarium go extinct because of climate change and habitat loss. The educational value and effect of such experiences at the zoo or aquarium might result in visitors consuming less meat and fish and/or eat more sustainably at home too.

There are obvious challenges in adopting such a strategy. Zoos and aquariums are, as mentioned above, dependent on the income from food sales to visitors. If the visitors respond negatively to initiatives that link food consumption to wildlife conservation and simply bring their own food or choose to visit other places, obviously nothing is gained. We realize this is a risk both because of expectations and dietary habits and because people might find it uncomfortable to be confronted with the negative effects of their own lifestyles on a recreational visit. Thus such initiatives need to be well-planned and communicated clearly. Further, as mentioned previously, many zoos and aquariums have outsourced food services to private companies and might find it difficult to or cannot influence what is on the menu. Thus in those situations it will take a re-thinking of the business model of catering to gain influence on the food selection available to visitors.

Nevertheless, it could be an opportunity to let visitors know about the connection between wildlife conservation and livestock production – and, at the same time, provide them with a tool to act in a situation where many express hopelessness and confusion on how to make a difference (Kim *et al.*, 2015). Zoos and aquariums could play a role here. And although they should probably not 'go vegan' from day 1, they could consider taking the opportunity to inspire people to a change in diet by presenting them with information and alternative food choices.

References

Bellard, C., Bertelsmeier, C., Leadley, P., Thuiller, W. and Courchamp, F. (2012). Impacts of climate change on the future of biodiversity. Ecology Letters 15: 365-77.

Conway, W. (1995). Zoo conservation and ethical paradoxes. In: Norton, B.G., Hutchins, M., Stevens, E.F. and Maple, T.L. (eds.) Ethics on the ark. Zoo, animal welfare and wildlife conservation. Smithsonian Institution Press, Washington and London, pp. 1-12.

Doherty, T.J. and Clayton, S. (2011). The psychological impacts of global climate change. American Psychologist 66(4): 265-276.

European Association of Zoos and Aquaria (2016). Conservation. Available at: http://www.eaza.net/conservation.

Foster, C., Green, K., Bleda, M., Dewick, P., Evans, B., Flynn, A. and Mylan, J. (2006). Environmental impacts of food production and consumption: a report to the

department for environment food and rural affairs. Defra. Manchester Business School, Manchester, UK.

Gamborg, C. and Sandøe, P. (2005a). Applying the notion of sustainability – dilemmas and the need for dialogue. In: Holm, S. and Gunning, J. (eds.) Ethics, law and society. Hants, Ashgate, UK, pp. 123-130.

Gamborg, C. and Sandøe, P. (2005b). Sustainability in farm animal breeding: a review. Livestock Production Science 92: 221-231.

Goedkoop, M.J., Heijungs, R., Huijbregts, M., De Schryver, A., Struijs, J. and Van Zelm, R. (2009). ReCiPe 2008, a life cycle impact assessment method which comprises harmonised category indicators at the midpoint and the endpoint level. Report I: Characterisation. Available at http://www.lcia-recipe.net.

Goldstein, B., Hansen, S., Gjerris, M., Laurent, A. and Birkved, M. (2016). Ethical aspects of life cycle assessments of diets. Food Policy 59: 139-151.

Goodland, R. and Anhang, J. (2009). Livestock and climate change. What if the key actors in climate change are cows, pigs, and chickens. World Watch Institute, Washington, DC, USA.

Ilea, R.C. (2009). Intensive livestock farming: global trends, increased environmental concerns, and ethical solutions. Journal of Agricultural and Environment Ethics 22: 153-167.

Kalbar, P.P., Birkved, M., Kabins and Nygaard, S.E. (in press). Personal Metabolism (PM) coupled with Life Cycle Assessment (LCA) model: Danish case study. Environment International.

Kim, B., Neff, R., Santo, R. and Vigorito, J. (2015) The importance of reducing animal product consumption and wasted food in mitigating catastrophic climate change. John Hopkins Center for a Livable Future, Baltimore, MD, USA.

Maxey, L. (2007). From 'alternative' to 'sustainable' food. In: Maye, D., Holloway, and Kneafsey, M. (eds.) Alternative food geographies: concepts and debates. Elsevier, Oxford, UK, pp. 55-75.

NASA Earth Observatory (2016). Record warmth in February. Image of the day. Available at: http://tinyurl.com/z2cpklr.

Norris, K., Potts, S.G. and Mortimer, S.R. (2010). Ecosystem services and food production. Issues in Environmental Science and Technology 30: 52-69.

Söderbaum, P. (2014). The role of economics and democracy in institutional change for sustainability. Sustainability 6: 2755-2765.

Steinfeld, H., Gerber, P., Wassenaar, T., Castel, V., Rosales, M. and De Haan, C. (2006). Livestock's long shadow – environmental issues and options. FAO, Rome, Italy.

59. Euthanasia in small animal practice – an Austrian survey

S. Springer[1], S. Hartnack[2] and H. Grimm[1]*
[1]Messerli Research Institute, Unit of Ethics and Human-Animal Studies, Vetrinaerplatz 1, 1210 Vienna, Austria; [2]Vetsuisse Faculty, University of Zurich, Section of Epidemiology, Winterthurerstrasse 190, 8057 Zurich, Switzerland; svenja.springer@vetmeduni.ac.at

Abstract

Euthanasia plays an important role in small animal practice. When it comes to killing animals, animals, their owners, public interests, as well as the legal framework, have to be considered by veterinarians. In cases where the interests of veterinarians and owners diverge, ethical conflicts can occur, and different concerns have to be taken into account. Data about the attitudes of Austrian veterinarians with regards to euthanasia have been gathered by means of a 53-item questionnaire. In total, 514 questionnaires were included in the analysis. Among other things, factors that influence the veterinarian's satisfaction with euthanasia, reasons given by the owners for wanting to have their healthy animal euthanized and questions concerning preparation and support regarding euthanasia were addressed. The results of this study emphasize that euthanasia constitutes an important topic for Austrian veterinarians and that acting in the assumed interest of the animal contributes to the veterinarians' satisfaction. Further, the ethical principles concerning the protection of the animal's life and avoidance of suffering play a central role for decision-making.

Keywords: killing animals, questionnaire survey, attitudes, veterinary medical ethics

Introduction

In view of the status of pet animals as family members euthanasia is considered a crucial and sensitive issue within the field of small animal practice. Already in 1989 Edney described euthanasia as the most sensitive situation for the relationship between owners and veterinarians. He mentioned that regularly euthanizing small animals regularly causes problems and ethical dilemmas within the profession (Edney, 1989). The consideration of the animal patient, the owner, public interests as well as the existing legal framework and moral ideas constitute demanding questions relating to euthanasia. Topics like convenience euthanasia, over-treatment of suffering animals and financial expenditure seem to be morally relevant within this context (Fernandez-Mehler *et al.*, 2013). Particularly in cases where the interests of veterinarians and owners diverge, ethical conflicts are likely to emerge (Batchelor and McKeegan, 2012). In these situations various and often conflicting interests have to be considered and the veterinarian is responsible for avoiding the unnecessary suffering of the animal and ensuring the well-being of their patient. On the other hand, veterinarians take responsibility for protecting the animal's life (Fahrion *et al.*, 2011; McMillan, 2003; Yeates, 2010; Yeates and Main, 2011). The principle to protect animal's life is underpinned by the Austrian legal framework, where a justifying reason is legally obligatory to kill an animal. Paragraph 6 of the Austrian animal protection law makes this point explicit and the Austrian Criminal Code (§ 222) sanctioned persons who violate this law. Despite the legal background the phrase of 'justifying reasons' seems to be very open and it raises the question what reasons are good enough to euthanize an animal.

In this survey, attitudes of Austrian veterinarians were investigated. Questions like, 'How do veterinarians experience ethical decision-making when it comes to killing pet animals?' and 'How do veterinarians describe their specifically ethical challenge?', 'What support would be useful for dealing better with ethical dilemmas?', 'What are the most frequent reasons given by the owner to justify an euthanasia of a healthy animal?' provided the framework of the survey. The ultimate aim of this study was – firstly – to

EurSafe 2016 – DOI 10.3921/978-90-8686-834-6_59, © Wageningen Academic Publishers 2016

remedy the lack of data regarding euthanasia in small animal practice in Austria. Secondly, the obtained empirical data should contribute to the body of knowledge on veterinary ethics and provide information about current aspects regarding euthanasia. Thirdly, it is intended that the survey may help to interpret, structure and analyse current moral problems within the veterinary profession.

A quantitative analysis of data has been already published. This analysis focused on the relation of demographic and general variables and attitudes of Austrian veterinarians (Hartnack *et al.*, 2016). Moreover, a further publication has been submitted, which primarily concentrates on open-ended questions and a comparison between Austrian veterinarians and veterinary students regarding five selected case examples of this survey. In the light of the contents of these publications and with the aim of presenting novel results this extended abstract focuses on a number of case examples and 26 statements of the questionnaire. The case examples should provide an insight into how Austrian veterinarians would react in various situations, which exert different pressure on them. Furthermore it should be identified if differences between female and male respondents exist and if the working time spent with small animals have an influence in answering these questions. The 26 statements should clarify attitudes of Austrian veterinarians regarding euthanasia.

Materials and methods

An online survey was conducted using the program 'limesurvey'. The questionnaire was sent to registered veterinarians (n=2,478) via e-mailing list from the Austrian Chamber of the Veterinary Surgeons. The link to the survey was open from 1st to 30th November 2012.

The survey comprised 53 questions and was divided into several sections. In general, one part of the questionnaire included questions about demographic variables (e.g. age and gender) and parameters like duration of professional activity, numbers of euthanasia, etc. In eight case examples the opinion of veterinarians for or against euthanasia and whether they would inform an official veterinarian in particular cases was examined. Furthermore, respondents were confronted with 26 statements to determine personal attitudes about euthanasia. Case examples were developed following the paper of Dürr *et al.* (2011), who used similar scenarios for their questionnaire study. Also the development of the 26 statements can be ascribed to Fahrions' *et al.* (2011), who presented workshop results on the issue of killing animals. Four open-ended questions addressed topics such as reasons for convenience euthanasia given by the owner and whether support in the form of a catalogue of criteria for decision-making is considered helpful. Respondents had the opportunity to answer the open-ended questions in a free text-field. To facilitate the interpretation, given reasons were listed in descending order of frequency. Consequently, it was not differentiated if a reason was mentioned at the first or third place.

The data were recorded and managed with Microsoft Access 2007. Analysis of descriptive questions was performed in Microsoft Office Excel 2007. Commercial statistical software (SPSS, v.20) was used for the results of quantitative analysis. The case examples and variables were divided or transliterated into two groups by using a non-parametric t-test (Mann-Whitney-U-Test). For univariable analysis, distinctions were considered statistically significant if $P<0.05$.

Results and discussion

A quantitative analysis has been already published, which focuses on the graphical modelling of association between demographic variables, statements and case scenarios (Hartnack *et al.*, 2016). A publication referring to the open-ended questions and qualitative analysis is in progress. For these reasons the present extended abstract concentrates on preliminary findings, which will be further explored.

Demographic data

Out of 2,478 contacted veterinarians, 764 returned the questionnaire, 514 were fully or sufficiently completed to enable analysis and corresponds to a response rate of 21% (514 out of 2,478). 167 (40%) men and 251 (60%) women indicated their gender. Equally, 418 persons included their age; the average age of the sample was 45 years. The average time of professional experience in years was 16.2±9.4 (average ± standard deviation). Almost half of the respondents (n=421, 48%) had over 15 years of professional experience whereby 158 respondents have been working in the profession between six and 15 years. Veterinarians indicated that they euthanize three times per month and are requested two times annually to euthanize a (mainly) healthy animal (median values).

Case examples

Veterinarians were faced with seven case examples, which presented external pressure for or against euthanasia. Respondents had the opportunity to rate on a Likert-type scale from one (refuse euthanasia) to nine (affirm euthanasia).

Five case examples presented 'external pressure for euthanasia'. For illustration, two examples shall be given:

- A dog has twice bitten persons. It has attended training courses and animal psychologists have tried to educate it. However, two days ago it severely injured a child that is now in hospital.
- An animal owner comes to your office with a young dog. This dog is severely ill, but therapy is possible. This therapy would be time-consuming, but there is a chance of success. The owner rejects the therapy because he does neither have enough time nor money. He wants you to euthanize the dog.

The first case focused on the public interest and expectation to euthanize the dog. In this case, respondents agreed to euthanize the animal to a great extent. The second case demonstrates pressures that originate from the owner. In this case veterinarians tended to refuse euthanasia.

Three case examples highlighted 'external pressure against euthanasia'. Again, two examples shall be given for illustration:

- A dog sitter comes to your office with a 17-year-old dog that suffers from breathing problems. The owners have left for a trekking tour three days ago and cannot be reached. You removed a malign tumour in this dog six months ago and you are afraid that it has developed lung metastases. The dog sitter refuses to make a decision regarding euthanasia and cannot tell you what the owners might want.
- An animal owner comes to your office with a severely ill Persian cat. You know that he has a very close relationship with his cat and does not want to part with it. In your opinion, euthanasia would be reasonable, but the owner does not agree. You reject any further treatment apart from analgesia.

In the first case, the veterinarian cannot reach the owner to ask for consent, which is a reason not to euthanize the dog under these circumstances. In the second case, the owner has a strong emotional bond with the animal and clearly refuses euthanasia. Even though external pressure against killing the animal exists veterinarians tended to affirm euthanasia in both cases.

Further there was a clear difference between female and male veterinarians in six of eight case examples. All questions with an 'external pressure for euthanasia' resulted in a significant difference ($P<0.001$). On average, female respondents tended to deny euthanasia even if an external pressure existed. Furthermore two groups were defined to relate the case examples with working time spent with small animals. Group 1 included all veterinarians, who spent 60% or less of their working time with small animals. Group 2 covered all veterinarians who spent over 60% of their working time with small animals. In total, four of eight case examples resulted in a significant difference (P<0.001) in both groups, while three cases have an 'external pressure for euthanasia'. Group 2, which spends more time with small animals, had a greater tendency to deny euthanasia compared to Group 1.

Owner-based reasons for euthanasia

A further crucial point was to determine how veterinarians decide in situations when external pressure arises especially from the owners' side. Therefore particularly four of seven case examples illustrated situations in which the request for euthanasia was based on arguments given by the owner: a rabbit breeder asks for euthanasia because some of the young animals' coat colour does not meet breeding standard or a rabbit owner believes the costs for treatment are higher than buying a new rabbit. Another case example describes an owner who wants to have his 15 years old dog euthanized because he wants to travel with his family. In all three cases respondents strongly refused euthanasia. The case example in which an owner requests euthanasia of a young and severely ill dog which could possibly be cured, but the owner rejects the therapy because of lack of time and money, reflects financial issues. Respondents are not entirely sure about refusing euthanasia and show a tendency not to euthanize.

Attitudes of Austrian veterinarians regarding euthanasia

Twenty-six statements were presented to clarify attitudes of Austrian veterinarians regarding euthanasia. Respondents were given the opportunity to rate on a Likert-type scale between one (completely disagree) and nine (completely agree). Statements were divided into three categories

The first category included twelve general statements like 'I am still not used to euthanizing animals.' or 'Retrospectively, it becomes more easier for me to deal with euthanasia.' Veterinarians were hesitant to agree or disagree to these two statements. When asked whether euthanasia is an unavoidable and inherent part of veterinary practice, respondents were more likely to agree. The technically flawless procedure of euthanizing, respectful treatment of the dead animal body, careful planning and the right moment make it easier for veterinarians to deal with euthanasia. Veterinarians disagreed that euthanasia is easier for them if the owner is present or does not have a close relationship to the animal. The statement 'It mostly causes more trouble if the owner is present.' was not generally supported. However, veterinarians showed strong agreement with the statement that understanding the pet owner is an important part of euthanasia.

The second category of eleven statements mirrored the veterinarians' attitudes towards euthanasia against their belief. Respondents indicated that a short expected future life span of the animal, a rich life until death and an advanced age of the animal contribute to their acceptance of euthanasia. Generally, respondents completely agreed that it is a burden to euthanize an animal. However, they are not afraid that owners would see another veterinarian or kill the animal by themselves if the veterinarian did not euthanize according to the wish of the owner. The limited power to convince owners is generally perceived as problematic by veterinarians. The practice of Austrian veterinarians is facilitated if all medical, social and economic options are considered and heard by the owners.

The last category of statements investigated the level of satisfaction in cases where veterinarians find themselves in the situation, where they would prefer perform euthanasia but cannot, because e.g. the owner does not want to part with the animal. Veterinarians can better deal with an animal's prolonged suffering if they know that they have done the best for the animal's well-being, treated the animal with effective analgesia and know that they have properly informed the owner.

In general Austrian veterinarians accept that euthanasia is an unavoidable and inherent part of veterinary practice. They agree, that it is important to understand the owner's perspective. However, by disagreeing that euthanasia is easier for veterinarians if an owner is present or does not have a close relationship to the animal Austrian respondents indicate that such aspects is not perceived as a problem during euthanasia. This comes as a surprise since Fernandez-Mehler *et al.* (2013) consider the owner's presence during the euthanasia as one important factor for clients. At this point veterinarians' and owners' perspectives differ in this respect.

Austrian veterinarians strongly disagree that they are afraid that owners would see another veterinarian or kill the animal alone if they did not carry out the owner's wish to euthanize. In this regard, the findings of this survey are similar with those of Yeates and Main (2011) who show that only a few respondents answered that they have the fear that owners might destroy their animal inhumanely.

Fernandez-Mehler *et al.* (2013) argued that information and preparation are important aspects for a client's satisfaction. Our study confirms this finding. Practitioners can better deal with animals' suffering if owners refuse euthanasia, when they know they have done what is best for the patients' well-being, used effective analgesia and provided information to the owner.

However it can clearly be shown that owners' motives like time problems, financial issues or aesthetic aspects are not justifying reasons for veterinarians to kill an animal. By means of several case examples this study identified veterinarians' attitudes towards convenience euthanasia. Austrian veterinarians strongly disagree with euthanizing an animal based on such reasons or arguments. It can also be shown by the results from Hartnack *et al.* (2016). Furthermore they indicated that female veterinarians and even younger veterinarians are more likely to refuse euthanasia in such situations (Hartnack *et al.*, 2016). Coming to `external pressure for euthanasia', Yeates and Main (2011) draw attention to the fact that several veterinarians made choices that rest upon external pressure from others that is associated with more stress for them. If we compare women and men, women tended to deny euthanasia even if external pressure exists (cf. Hartnack *et al.*, 2016). One possible explanation is that women denied euthanasia in these situations because, although they considered the external factors, they were also more critical. It seems likely that female veterinarians did not accept the arguments of the pet owner immediately. A possible explanation is that women feel less sympathetic to the owner's situation when thinking about the sensitive topic of euthanasia and were not so easily convinced compared to men.

Further, it was indicated that veterinarians who spend more time on small animals have a greater tendency to deny euthanasia compared to veterinarians who spend less time. A possible reason for this is that they have to be more sensitive in the field of small animal practice compared to veterinarians who work more with farm animals. This is particularly noticeable since practitioners in small animal practice are confronted with the individual desires and interests of the owner. Veterinarians who are working with farm animals are usually involved with livestock owners, who do not have a comparable emotional bond between their animals and themselves compared to a pet owner and his dog. Obviously, economic factors play a different and important role in agriculture when it comes to euthanasia.

Conclusions

All in all, our study contributes to the empirical body of knowledge on veterinary ethics within the field of euthanasia in small animal practice. It was shown that euthanasia is a delicate part of Austrians veterinarians' professional life. Acting according to animal based factors cause great satisfaction for the veterinarians although they know about the crucial position of the owners and their interests. Several case examples give an insight into which factors play a vital role in the ethical decision-making process regarding euthanasia. It has been clearly demonstrated that the principle of protecting the animal's life has great significance for Austrian veterinarians and is used against convenience euthanasia.

References

Batchelor, C.E.M., Creed, A. and McKeegan, D.E.F. (2015). A preliminary investigation into the moral reasoning abilities of UK veterinarians. Veterinary Record 177: 124-130.

Dürr, S., Fahrion, A., Doherr, M.G., Grimm, H. and Hartnack, S. (2011). Akzeptanz des tötens von tieren: umfrage bei tierärzten und anderen berufsgruppen. Schweizer Archiv für Tierheilkunde 153: 215-222.

Edney, A.T.B. (1989). Killing with kindness. Veterinary Record 124: 320-322.

Fahrion, A., Dürr, S., Doherr, M.G., Hartnack, S. and Kunzmann, P. (2011) Das töten und die würde von tieren: ein problem für tierärzte? Schweizer Archiv für Tierheilkunde 153: 209-214.

Fernandez-Mehler, P., Gloor, P., Sager, E., Lewis, F.I. and Glaus, T.M. (2013). Veterinarians' role for pet owners facing pet loss. Veterinary Record 172: 1136-1154.

Hartnack, S., Springer, S. Pittavino, M. and Grimm, H. (2016). Attitudes of Austrian veterinarians towards euthanasia in small animal practice: impacts of age and gender on views on euthanasia. BMC Veterinary Research 12: 26.

McMillan, F.D. (2003). Maximizing quality of life in ill animals. Journal of the American Animal Hospital Association 39: 227-234.

Yeates, J. (2010). Ethical aspects of euthanasia of owned animals. Practice 32: 70-73.

Yeates, J.W. and Main, D.C.J. (2011). Veterinary opinions on refusing euthanasia: justification and philosophical framework. Veterinary Record 168: 263-267.

Section 14. Food in communities

60. 'Bread riots' in Spain during the 19th and 20th centuries

J.A. Redondo Cardeñoso
CIDEHUS – University of Évora, Palácio do Vimioso, Largo do Marquês de Marialva 8, Apdo. 94, 7000-809 Évora, Portugal; jredondocardenoso@gmail.com

Abstract

The riot, especially in the form of the subsistence riot (also known popularly as the bread riot), has been one of the most common expressions of popular protest for centuries. Nonetheless, these protests have long been ignored by political authorities and intellectuals alike. In both cases, riots are considered to be an expression of the irrationality of the popular classes, aggravated to extremes by hunger. Fortunately, however, this pejorative image of the popular riot has been reassessed in recent decades by a variety of new sociological, historical and anthropological studies. The riot has come to be understood as an instrument of political action and participation used by social sectors excluded from politics. In the specific case of 'bread riots', for example, the popular classes demanded the enforcement of fair prices from the political authorities and an end to speculative trade in order to achieve a fairer redistribution of staple foodstuffs. This paper analyses the phenomenon of subsistence riots through a study of revolts that took place in contemporary Spain, especially during the second half of the nineteenth century and the first third of the twentieth century, the period most studied by national historiography. This study illustrates the rationale that existed behind these riots and how the people who participated in them articulated their actions around specific political demands that, essentially, proposed an alternative system of food distribution to that established by the capitalist liberal system.

Keywords: social unrest, subsistence conflicts, modern age

Introduction

The riot was the main collective expression of popular protest during the contemporary age in the final years of the 19th century and the first decades of the 20th Century, coinciding with the emergence and growth of the labour movement and its classic protest method: the strike. Originally, riots formed part of the 'traditional' repertoire of collective protest which had an essentially reactive character, but they later became included within the repertoire of 'modern' protest action and were essentially proactive, because it was a means by which people could demand rights that they had not enjoyed up until that moment (Tilly, 1976; Traugott, 1995).

The causes of riots varied, determining its typology. For example, there were tax riots (against the increase or the collection of taxes), antimilitary riots (against the recruitment or the abusive presence of troops in a territory) and anticlerical riots (against the religious ministers or institutions). However, the riots most widespread and feared by the authorities were undoubtedly the subsistence riots, commonly known as 'bread riots', in which people protested against sudden price increases or shortages in the supply of basic foodstuffs, mainly of grain, flour and bread (Lucea Ayala, 2001, 2002).

This was the kind of riot that prevailed in European societies from at least the 18th century, such as the cases studied in England and France by Rudé (1964). A similar phenomenon occurred in Spain. The most notable example of the Spanish eighteenth-century riots were the 'Esquilache riots' of 1766. Although these disturbances were mainly centred on Madrid, they had repercussions in many towns and villages throughout the country. In these cases, rioters demanded reduction in the price of foodstuffs and abolition of the 'Junta de Abastos' (Supplies Board), in the hope that the authorities would restore the ancient system in which supplies depended on local communal institutions. The disturbances were so

EurSafe 2016 – DOI 10.3921/978-90-8686-834-6_60, © Wageningen Academic Publishers 2016

important that Charles III was forced to change his government. Consequently, he replaced the ministers of the 'Italian Party' headed by the Marquis of Esquilache, with ministers of the 'Spanish Party' leaded by the Count of Aranda (López García, 2006; Rodríguez Díaz, 1976).

Similar incidents were repeated in Spain during the 19th century and the first third of the 20th century, either through new waves of riots, as happened in 1856 or 1898 – as we will see later – or as isolated events that they took place in individual towns or villages, as occurred in Madrid in January 1907 and June 1914, when popular protests forced the local government to decree bread price reductions (Sánchez Pérez, 2005).

The main cycles of subsistence conflicts in Spain

As mentioned above, several waves of subsistence riots took place in Spain during the 19th century and the first years of the 20th century. One riot took place in 1847, the repercussions of which were mainly in small and medium-sized towns and agricultural areas such as the grain-producing shire called Tierra de Campos. These conflicts were preceded by sudden increases in the price of grains, particularly wheat, which saw a price increase of 60% between 1846 and 1847 (Díaz Marín, 2003).

In 1856 a new wave of subsistence riots occurred, again preceded by sudden increases in the price of grains mainly as a result of shortages in the international markets caused by the Crimean War between 1853 and 1856. The price of bread in Spain increased from 1854 and it rose especially high in the first half of 1856 as a result of a bad harvest the previous year. In this context, around June 20th riots broke out in towns and villages of Old Castile and Leon in the northwest of the country. An important flour industry existed in these adjacent communities. The protests, although less virulent, spread to the main cities such as Seville, Valencia or Madrid at the month; in July disturbances even occurred in Barcelona. The protests were so significant that, as in 1766, the governability of the country was affected: General Espartero's progressive government was overthrown by General O'Donnell, who established a conservative administration (Moreno Lázaro, 2003, 2009). These riots had an impact in Portugal, where several successive conflicts took place in August. Of particular importance were the events in Lisbon, which obliged the authorities to resort to severe military repression (Palacios Cerezales, 2011).

In 1868 a new subsistence crisis took place, giving rise to new riots such as those that happened in February in a number of Andalusian cities including Seville or Granada (Fuentes, 2007). However, in this case, there are not yet enough studies that assess the true impact of these conflicts or the extent of their link with the political crisis in September 1868 that led to the establishment of the first democratic regime in Spain.

More studies have been done on the subsistence crisis of 1898, where studies at national level have analysed more than sixty conflicts, mainly in small and medium towns (Serrano, 1981). However, as several regional studies have shown, the repercussion of the riots was far greater (Bascuñán Añover, 2008; Baumeister, 1996). Here again, the subsistence crisis was preceded by increases in the price of wheat, which had occurred since 1895. As in 1847, the disturbances were concentrated in medium-sized towns and agricultural areas, while the larger cities (Madrid, Barcelona, Valencia) remained on the side-lines. The conflicts tended to occur in the areas most affected by the agricultural crisis of the end of 19th century that affected large areas of Europe (Plans Maresma, 1992).

In this case, an internal crisis, together with external problems caused by the loss of the old colonies Cuba, Puerto Rico and the Philippines, gave rise to a deep crisis of national identity and of the political system. This led to new political opportunities for the lower classes in national politics. This process has been called 'the people's turn' (Serrano, 2000), in allusion to political alternatives proposed by the lower classes, as opposed to elite politics embodied 'the pacific turn', a political system that consisted of the rotation in government of the two main parties (Conservative and Liberal).

A new wave of subsistence riots broke out barely half a decade after a new increase in the price of wheat, between 1903 and 1904. However, this time a new interlocutor had emerged in the form of socialism. This led to the appearance of new kinds of popular protest. In addition to the 'bread riots', people used strikes to demand higher wages, especially in agricultural regions such as Andalusia and Old Castile (Maurice, 1989; Redondo Cardeñoso, 2013). However, the socialist organizations themselves, and specifically the UGT (Workers General Union), took the lead in demanding a decrease of the subsistence prices during the second half of 1904, and organized a strong nationwide campaign that included meetings and demonstrations (Redondo Cardeñoso, 2013). From this moment on, the labour movement assumed increasing importance in the organization of popular protest.

In spite of this, a new wave of subsistence riots was caused by the shortage of basic foodstuffs at the end of the First World War. In effect, between 1918 and 1920 there were 'bread riots' in several locations throughout Spain, although they were not as numerous as in the previous century. These disturbances were more common in small towns and villages, where the 'traditional' repertoire of collective protest had not been superseded by the new labour movement ideas (Bascuñán Añover, 2008; Baumeister, 1996; Gil Andrés, 2000). In these years, the most important subsistence riots broke out in Portugal, both in large cities like Lisbon or Porto, and medium-sized towns like Coimbra or Beja, in a historical event known as 'A Revolução da Batata' (the Potato Revolution) (Dias Pereira, 2014; Palacios Cerezales, 2011).

From the 1920s on, new forms of labour movement protest became dominant and the large waves of riots largely disappeared, returning only sporadically as part of the repertoire of collective protest in the thirties (Gil Andrés, 1995).

Popular demands in the subsistence riots: campaigning for a new politics of food distribution

The most traditional historiographical interpretations have viewed the 'bread riots' as a natural reaction to famine. Such a perspective interprets the action of those taking part in the riots as a function of their natural survival instinct. This instinct drove people to practice indiscriminate looting, their only objective being to steal the greatest possible quantity and variety of goods[1].

However, this image of irrational rioters wildly looting was long ago refuted. Hobsbawm (1959) already showed not only that famine was not conducive to the formation of social movements, but also that it actually inhibited them, because in times of shortage people are more concerned with their daily sustenance than with protest. Instead of arising from hunger itself, people used protest action when there was a threat of famine (Rudé, 1969). However, there were not conflicts each time there was a threat of famine. As Hobsbawm and Rudé (1969) have noted, if this had been the main reason why the lower classes engaged in conflict, they would have engaged in a quasi-permanent conflict, because their economic situation regularly bordered on absolute poverty. Tarrow (1994), for his part, suggested that social movements began when new political opportunities presented themselves, that is, when the lower classes saw that protest action could result in immediate improvement of their socioeconomic conditions. Therefore, protests based on subsistence were frequently linked to national political crises, because the rioters could easily obtain more concessions from weakened authorities. The rioters aimed to oblige the authorities to take important political decisions that would improve their lives, such as reductions in taxes, limits to military recruitment, or, as occurred in the subsistence riots, the issuing of

[1] For the historiographical debate, see Bascuñán Añover (2009).

decrees or acts that set fair prices for foodstuffs that guaranteed decent survival conditions. This is what Thompson (1971) termed the 'moral economy'.

By focusing more closely on testimonies that described the 'bread riots', we can see that the rioters did not indiscriminately loot stores to steal goods for their own benefit. Nor did they ask the authorities to give them bread charitably. They demanded reasonable prices and a fair redistribution of the existing foodstuffs (Bascuñán Añover, 2008). For example, the main riots that took place in 1856 in the towns of Valladolid, Medina de Rioseco and Palencia broke out after the authorities refused to enact the production of cheap bread (Moreno Lázaro, 2009). Years later, the first riots of 1898 took place in Gijón, a maritime town where the crowd protested against high fish prices (Radcliff, 2004), and in Talavera de la Reina, a small inland town where the rioters demanded 'bread and jobs' (Díaz Díaz, 1994). In 1904, in Segovia, women led a riot following a failure to reduce the price of bread and, later, they forced grocers to sell potatoes at a specific price. Nearby, days later, again in Valladolid, when a town councillor offered free bread to the rioting women, they replied: 'We don't want alms, we demand jobs for our husbands and a reduction in the price of food' (Redondo Cardeñoso, 2013).

On the other hand, the anger of the rioters was not limited to making demands on the authorities; they were also directed against grain traders and flour makers, who were considered the main culprits of the shortages. Thus, it was not uncommon in the riots for people to take their protests to inland or maritime harbours, or to railway stations, in order to block grain convoys that were ready to leave town (Bascuñán Añover, 2008). In 1856, one of the main targets of the rioters in Old Castile and Lion were the flour factories. These were consequently set on fire and destroyed (Moreno Lázaro, 2009). In 1898, rioters in several towns and villages of Extremadura focused their efforts on preventing the departure of grain from their localities (Sánchez Marroyo, 1998), while in 1904 the railway station was one of the first targets of the women who led the riot in Segovia. They set up a rioters' committee there to ensure that no grain wagon left the town (Redondo Cardeñoso, 2013).

In conclusion, if we accept that the riots were a way of 'doing politics' used by the lower classes for a variety of demands (Gil Andrés, 2000), it can be appreciated that the subsistence riots were not a 'primitive', 'instinctive' or 'irrational' expression of protest caused by famine, as traditional historiographical interpretations would have us believe. On the contrary, a thorough examination of the 'bread riots' shows that the rioters demanded specific political measures and held an ethical view of economics that basic foods should be guaranteed to the working classes. In other words, they defended an alternative model of food distribution to that posed by liberal capitalism which did not prioritize individual profit but rather the common interest.

Acknowledgements

Research work carried out within the scope of UID/HIS/00057/2013 (POCI-01-0145-FEDER-007702), FCT/Portugal, COMPETE, FEDER, Portugal2020.

References

Bascuñán Añover, O. (2008). Protesta y supervivencia. movilización y desorden en una sociedad rural: castilla-la mancha. Fundación Instituto de Historia Social, Valencia, Spain.

Bascuñán Añover, O. (2009). Campesinos rebeldes. Las luchas del campesinado entre la modernización y la globalización. Catarata, Madrid, Spain.

Baumeister, M. (1996). Campesinos sin tierra. Supervivencia y resistencia en Extremadura, 1880-1923. Diputación de Badajoz, Badajoz, Spain.

Dias Pereira, J. (2014). O ciclo de agitação social global de 1917-1920. Ler História 66: 44-55.

Díaz Díaz, B. (1994). La protesta popular en Tavalera: el motín del pan de 1898. Cuaderna 1: 76-90.

Díaz Marín, P. (2003). Crisis de subsistencias y protesta popular: los motines de 1847. Historia Agraria. Revista de Agricultura e Historia Rural 30: 31-62.

Fuentes, J.F. (2007). El fin del Antiguo Régimen (1808-1868). Política y Sociedad. Síntesis, Madrid, Spain.

Gil Andrés, C. (1995). Protesta popular y movimientos sociales en la Restauración: los frutos de la ruptura. Historia Social 23: 121-135.

Gil Andrés, C. (2000). Echarse a la calle. Amotinados, huelguistas y revolucionarios (La Rioja, 1890-1936). Universidad de Zaragoza, Zaragoza, Spain.

Hobsbawm, E.J. (1959). Primitive rebels. Studies in Archaic Forms os Social Movement in the 19th and 20th Centuries. Manchester University Press, Manchester, UK.

Hobsbawm, E.J. and Rudé, G. (1969). Captain swing. A social history of the great English agricultural uprising of 1830. Lawrence and Whishart, London, UK.

López García, J.M. (2006). El motín contra Esquilache. Alianza, Madrid, Spain.

Lucea Ayala, V. (2001-2002). Protesta colectiva en Zaragoza en el tránsito del XIX al XX: una tipología. Studia Histórica-Historia Contemporánea 19-20: 129-159.

Maurice, J. (1989). El anarquismo andaluz. Campesinos y sindicalistas, 1868-1936. Crítica, Barcelona, Spain.

Moreno Lázaro, J. (2003). Fiscalidad y revueltas populares en Castilla y León durante el Bienio Progresista, 1854-1856. Historia Agraria. Revista de Agricultura e Historia Rural 31: 111-140.

Moreno Lázaro, J. (2009). Los hermanos de Rebeca. Motines y amotinados a mediados del siglo XIX en Castilla la Vieja y León, Región Editorial, Palencia, Spain.

Palacios Cerezales, D. (2011). Portugal á *coronhada*. Protesto popular e ordem pública nos séculos XIX e XX. Tinta da China, Lisboa, Portugal.

Planas Maresma, J. (1992). La crisis agraria de fines del siglo XIX: nuevas contribuciones y nuevos enfoques. Noticiario de Historia Agraria. Boletín Informativo del Seminario de Historia Agraria 3: 173-180.

Radcliff, P. (2004). De la movilización a la Guerra Civil. Historia política y social en Gijón (1900-1937). Debate, Barcelona, Spain.

Redondo Cardeñoso, J.A. (2013). 1904: Rebelión en Castilla y León. Universidad de Valladolid, Valladolid, Spain.

Rodríguez Díaz, L. (1973). Los motines de 1766 en provincias. Revista de Occidente 122, 183-207.

Rudé, G. (1964). The crowd in history. A study of popular disturbances in France and England, 1730-1848. Wiley & Sons, New York, USA.

Sánchez Marroyo, F. (1998). 1898. Guerra en las colonias y crisis social en España. Anales de Historia Contemporánea 14: 179-193.

Sánchez Pérez, F. (2005). La protesta de un pueblo. Acción colectiva y organización obrera, 1901-1923. Ed. Cinca/Fundación Largo Caballero, Madrid, Spain.

Serrano, C. (1981). Guerra y crisis social: los motines de mayo del 98. In: Castillo, S., Forcadell, C., García-Nieto, M.C. and Pérez Garzón, J.S. (eds.). Estudios sobre historia de España. Homenaje a Tuñón de Lara. I. Universidad Internacional Menéndez Pelayo, Madrid, Spain.

Serrano, C. (2000). El turno del pueblo. Crisis nacional, movimientos populares y populismo en España (1890-1910). Península, Barcelona, Spain.

Tarrow, S. (1994). Power in movement. Collective action, social movements and politics. Cambridge University Press, Cambridge/New York, UK/USA.

Thompson, E.P. (1971). The moral economy of the English crowd in the 18th century. Past and Present 50: 76-136.

Tilly, C. (1978). From mobilization to revolution. Random House, New York, USA.

Traugott, M. (1995). Repertoires and cycles of collective action. Duke University Press, Durham/London, UK.

61. Politics of ecological public health: a Poggean perspective

J. Nieuwland
Institute for Philosophy, Leiden University, Reuvensplaats 3-4, 2311 BV, Leiden, the Netherlands; j.nieuwland@hum.leidenuniv.nl

Abstract

Public health is increasingly understood from an ecological perspective. Emphasis is put on determinants of human health, ranging from socio-economic conditions to the natural environment. I argue that ecological approaches to public health face two issues concerning political justification. First, ecological approaches to public health help to understand the ways in which drivers of disease and determinants of health disregard national boundaries, while at the same time there remains significant disagreement about the extent of transnational duties. I argue that Thomas Pogge has provided a plausible justification for transnational duties with regard to health policy, and will explore implications of his theory by looking at the outbreak of Ebola in West Africa. Second, given the role of human rights and political justification in health policy, what does this entail for the health of animals? This is particularly interesting in the light of the recent political turn in animal ethics. Should non-human animals, for example, be part of health policy as a matter of justice? Based on the idea of sentient rights, the Poggean framework can be broadened to include animals as well.

Keywords: Pogge, ebola, justice, sentient rights

Ecological public health and transnational duties

The idea of ecological public health aims to reify human beings and their health as part of ecosystems. As described by Tim Lang and Geof Rayner, 'ecological public health focuses on interactions, with one strand focusing on the biological world ... another strand centres on material issues such as links between industrial pollution, energy use and toxicity, and the impact on human species and nature' (2012: 3). Ecological processes support human health in complex and various ways, and though often peripheral to the individual, these are vital preconditions.

Ecological approaches to public health help to understand the ways in which drivers of disease and determinants of health disregard national boundaries. The extent and existence of transnational duties, however, are a matter of on-going theoretical disagreement. It is not my aim in this paper to settle such disagreement. Nonetheless, in my view, Thomas Pogge has provided a plausible and modest framework to look at responsibilities across the borders of nation states. I will explore his theory within the context of ecological public health and expand on it by looking at the prospects of developing it further as a distinct approach to animal justice.

Pogge (2008) argues that global institutions are unjust to the extent that they permit significant violations of human rights. The interconnections between institutions bring about responsibility, as they significantly determine the conditions of the world we live in. In that sense, Pogge's approach does not have to rely on claims of positive duties to human beings in developing countries. Rather, the (tacit) endorsement of current policy and the institutional framework makes everyone complicit to some extent. Allowing the present state of institutional order to persist amounts to harming those disadvantaged by it. This harm should be prevented, and otherwise remedied by duties of assistance and support comparative to the extent of harm done. We all have a duty to configure our institutions in a way that safeguards human rights.

EurSafe 2016 – DOI 10.3921/978-90-8686-834-6_61, © Wageningen Academic Publishers 2016

Ebola: drivers and responsibility

The outbreak of Ebola in 2014 in West Africa underlines the need not only for an ecological approach to health but also to look at how the global institutional order affects the health of people living in developing countries and the way this translates in terms of duties.

Many have criticized the late response by the World Health Organization (WHO) to the outbreak and the plight of people living in affected areas. While the outbreak can be traced back to a case in December 2013, Médicins Sans Frontières (MSF) voiced urgent concerns at the end of March of 2014, and pushed forward on its own for a long time (Médecins Sans Frontières, 2015). The WHO waited until August 6th to declare the outbreak of Ebola as a public health emergency of international concern (World Health Organization, 2015).

The lack of vaccines also attracted ample criticism. Although Ebola has emerged several times since its first occurrence in 1976 (Centers for Disease Control and Prevention), Pogge and Abraar Karan (2015) have pointed out that development of vaccines has proven difficult and slow. They therefore propose structural changes to the way medicines are developed and protected by patents in order to safeguard the health of people living in developing countries.

But there is more to the threat of Ebola than the availability of safe and efficient vaccines. Though vital, the discussion about vaccination does not address the underlying factors that drive the emergence of a disease such as Ebola. To do so, one has to look at how economic forces interact with ecological ones. Indeed, as I argue, this is also falls within the scope of Pogge's account.

Pogge (2008) understands the 'uncompensated exclusion from the use of natural resources' as an instance of injustice. Not only do wealthier people have a significantly larger ecological footprint, their consumption also uses up much of the natural resources in countries of poor people. From an ecological perspective, uncompensated resource depletion by wealthy foreigners amounts not only to a material loss. Resource extraction goes hand in hand with habitat destruction of wildlife as well as human encroachment. Effects of these activities include loss of biodiversity and an increase in the change of emergence of infectious disease, both of which disproportionately impact already disadvantaged human communities (Bausch and Schwarz, 2014). In their attempt to unravel the possible drivers of the Ebola outbreak, Daniel Bausch and Lara Swarz bring out the complex relation between economy and ecology. Poverty, as they analyse, 'drives people to expand their range of activities to stay alive, plunging deeper into the forest to expand the geographic as well as species range of hunted game and to find wood to make charcoal and deeper into mines to extract minerals, enhancing their risk of exposure to Ebola virus and other zoonotic pathogens in these remote corners' (Bausch and Swarz, 2014: 4). The lack of functioning public health infrastructures and adequate governance makes things even worse, they add, increasing the changes of a full-blown outbreak of zoonotic disease.

To what extent can the international community be hold responsible? International companies pursue activities such as mining and logging in these countries, exporting the resources to developed countries. Pogge argues that 'citizens and governments of the affluent states are therefore violating a negative duty of justice when they, in collaboration with the ruling elites of the poor countries, coercively exclude the poor from a proportional resource share' (2008: 209). While one might object that international interest in natural resources supports rather than undermines livelihoods by providing jobs, something more appears needed for true compensation. Moreover, Pogge's claim does not incorporate the side effects of uncompensated resource-extraction. The threats disadvantaged communities have to face in terms of loss of biodiversity and increased risk of infectious disease should, as I argue, also be considered a matter of injustice.

These considerations support strengthening public health capacity, and to protect livelihoods not only from uncompensated resource extraction but to develop ways to protect against the side-effects of such activity, or, however difficult this may be, avoid the disruption of animal habitats and ecosystem in the first place. Indeed, following from the framework that Pogge defends, this puts a considerable burden on the shoulders of the international community, in addition to the task of the WHO.

Just animal health?

Shifting perspective to an ecological public health emphasizes the interdependence between human and non-human nature but does not in itself entail an inclusion of health interests beyond human beings. Furthermore, the health of non-human animals or ecosystems (if the latter can be understood in terms of health) may be only of instrumental value to human health. In this sense, the ecological challenge to public health remains empirical, with human rights as an important normative reference point.

However, if moral status is acknowledged beyond human beings, then one could make a case to take the health interests of non-humans into considerations and weigh interests across species. Indeed, animal welfare is gaining ground as an important consideration in public health policy but with animals mostly as bystanders of such policy, perhaps deserving more than what presently falls their way.

This brings us to the realm of politics. The anthropocentrism of public health has been challenged with a call for 'more-than-human' considerations (Rock and Degeling, 2014). In my view such criticism should also address the distinction between ethical debate and political justification. In what way does the ethical challenge to public health relate to political theory? Given the role of human rights and political justification in health policy, what does this entail for the health interests of animals? This is particularly interesting in the light of the recent 'political turn' in animal ethics (Milligan, 2015). Should human animals, for example, be part of health policy as a matter of justice?

I argue that this view holds potential. To substantiate this claim, I will discuss some of the recent developments in political theory. Attempts of, among others, Alastair Cochrane (2012), Robert Garner (2013), Sue Donaldson and Will Kymlicka (2011) all start from interests to develop their own animal rights based perspective. In agreement with Peter Singer about sentience being a necessary and sufficient condition to have interests, they part ways by taking a non-utilitarian route in hashing out political implications. This also explains why attempts to include non-sentient non-human nature are less convincing; interests that qualify for rights lack in non-sentient human nature.

So, interests do not deliver you straight into the hands of the utilitarian. Depending on the plausibility of the route between interests and rights, they support rights-based approaches as well, an insight that can be traced back to the work of Joel Feinberg and Joseph Raz. If, however, one judges a utilitarian route inadequate, a choice-based theory of rights might still stop the interests-based theories in their tracks. An individual should be able to claim their rights, or choose not waive them, as the argument goes (Cochrane, 2013). It is, however, not hard to agree with Donaldson and Kymlicka that 'this was once an influential theory of rights, but it is now widely rejected, since it would not only preclude any idea of animal rights, but also any idea that children, the temporarily incapacitated, or future generations might have rights' (2011: 264). If one believes that human individuals who are unable to waive their rights should bear rights nonetheless, then species-membership should be regarded too ad-hoc in hindering the attribution of rights to sentient animals as well (Feinberg, 1974). Moreover, interests appear to do a better job in capturing the quintessence of certain rights. The protection that human rights provide can for a large part be explained in terms of interests, without the need to reference the cognitive abilities of individuals. In this sense, we might re-conceptualize human rights in terms of sentient rights (Cochrane, 2013).

Though the aforementioned political theorists agree on the basis (interests) of a rights-based approach, they disagree about the specific commitments that follow. This ranges from citizenship-theory to 'animal rights without liberation'. The relation between interests and right thus plays a definitive role. While Donaldson and Kymlicka defend a full-fledged account of inviolable animal rights supplemented by citizenship theory, Cochrane and Garner allow for differential treatment regarding animal life. Indeed, while the interest-theory of rights gains momentum, it does not provide us with an uncontroversial set of animal rights. The route from interests to rights needs further specification and justification, which will inevitably involve disagreement. Which rights do animals possess and what moral weight do these rights impose?

Do these questions throw a wrench in the works of the political turn in animal ethics? They need not do so, especially as these questions might also arise for human beings (Cochrane, 2013). Of course, the challenge at the beginning of any interest-theory of rights is to delineate the interests that are worthy to be protected by rights and impose duties upon others. Both Cochrane and Garner drive home the point that animals should at least have the right not to be made to suffer. Whether animals have a right to life is controversial, a right to liberty even more so. As Cochrane (2013: 659) writes:

> Of course, when it comes to assigning other types of rights to sentient creatures – to liberty, to life, and so on – there will be reasonable disagreements. But disagreements over the precise content of the rights of sentient creatures should not stand in the way of recognizing that some such rights exist. We can reasonably agree that all sentient creatures possess at least some interests which impose duties on us. As such, we can also reasonably agree that all sentient creatures possess at least some basic rights.

If one accepts an interests-based theory of rights for human beings, then it would be coherent to take a similar route for animals, especially if the rights protect interests that are shared by humans with other animals. If so, an ecological public health has to do justice to animals, in addition to human beings.

Poggean account of interspecies justice

What does it mean to do justice to animals living relatively independent from human affairs? To what extent does context matter? This brings to the table another area of disagreement amongst political theorists. While some question the extent to which such contingent considerations bear much normative weight, and argue for a full cosmopolitan approach, others view justice to demand different things along the continuum between domesticated and wild animals. Donaldson and Kymlicka, as an example of the latter approach, propose the idea of wild animal sovereignty to regulate our interactions with them. Several different concerns, however, make attributing sovereignty to wild animal communities rather controversial, as cosmopolitan theorists have pointed out (e.g. Horta, 2013). There might nonetheless be truth to the idea of morally relevant differences between animals based on their environmental context.

Current debates on global justice provide other means to think about doing justice to animals living in the wild, and to develop an alternative account to cosmopolitanism and citizenship models. Indeed, Pogge's institutional account provides a relatively unexplored possibility to do so. Clare Palmer (2010) is a notable exception. She has has applied Pogge's account to human-animal interactions as part of developing her own individual duty-based account, without putting it in terms of justice (Palmer, 2015). To the extent that animals are made vulnerable and dependent, she argues, we hold positive duties towards them. This entails that we should not unnecessarily harm animals that live relatively untouched by human activity but we do not owe them positive duties of assistance (though these may be permitted under certain conditions).

Elizabeth Cripps (2012) takes this thought further and hints at the possibility of a Poggean account of interspecies justice in her discussion of Palmer's work. We should not only frame our duties to animals according to Pogge's model, she argues, but develop institutions that reflect the rights of animals. As Cripps writes, 'these cannot neatly be categorised as positive duties of members of existing institutions, or as negative individual duties, but are rather duties to establish institutions to ensure that negative duties are not collectively violated' (*ibid.*: 204). In addition to an interest-based theory of rights, it provides a framework to structure our institutions with regard to animals in the wild and abroad. One might worry, as Garner (2013: 170) writes, 'that it would involve us having radically different obligations to wild animals since they, as opposed to domesticated animals, are more likely to suffer from the dictates of nature'. But holding positive duties to animals in the wild equal in strength to negative duties should at least be considered just as controversial. Perhaps it is more important to develop institutions that protect animals in the wild from human-caused harm, and demand further attention to their plight when humans are involved. Pogge's approach exactly tries to avoid controversial assumptions in order to have political relevance. Moreover, assuming the priority of negative over positive rights, this theory already bears far-reaching implications. It brings out what justice requires at minimum, without rejecting the possibility of additional relevant considerations.

The loss of habitat, as well as the threats associated with habitat encroachment should then be considered a matter of interspecies justice. This might entail, for example, developing vaccination strategies to protect great apes against Ebola (Capps and Lederman, 2015) due to the ways humans have made them vulnerable. However, justice according to the Poggean account does not require intervention when the harm and suffering is not caused by human action. Ethical considerations might still allow for intervention in certain cases, but whether this should be cashed out in terms of justice as Donaldson, Kymlicka and Cochrane argue, remains to be settled by means of further normative debate.

Concluding remarks

In this paper, I have discussed two political challenges that arise from an ecological perspective on public health. What do we owe, in terms of health policy, to human beings in across the boundaries of nation states and to animals? Regarding the first question, I have briefly explored the implications of a Poggean approach in the context of the outbreak of Ebola in West Africa. Though Pogge's own contribution makes a strong case for structural reform of pharmaceutical drug development, his theory has further implications that follow from the side effects of (uncompensated) resource extraction.

With regard to animals, I have discussed the possibility of broadening the Poggean approach, drawing on the recent political turn in animal ethics. It provides a plausible alternative to cosmopolitan and citizenship approaches to animal rights.

Acknowledgments

I would like to thank Raymond Corbey, Franck Meijboom, Tim Meijers, and Bruno Verbeek for inspiring discussions in preparation of this paper. Arcus Foundation is gratefully acknowledged for funding.

References

Bausch, D. and Schwarz, L. (2014). Outbreak of ebola virus disease in Guinea: where ecology meets economy. PLoS Neglected Tropical Diseases 8(7): e3056.

Capps, B. and Lederman, Z. (2015). One health, vaccines and ebola: the opportunities for shared benefits. Journal of Agricultural and Environmental Ethics 28(6): 1011-1032.

Centers for Disease Control and Prevention (2016). Outbreaks chronology: ebola virus disease. Available at: http://tinyurl.com/nrzolre.

Cochrane, A. (2012). Animal rights without liberation: applied ethics and human obligations. Columbia University Press, New York, NY, USA.

Cochrane, A. (2013). From human rights to sentient rights. Critical Review of International Social and Political Philosophy 6(5): 655-675.

Cripps, E. (2012). Book reviews – Clare Palmer's 'animal ethics in context.' Environmental Values 21(2): 233-250.

Donaldson, S. and Kymlicka, W. (2011). Zoopolis: a political theory of animal rights. Oxford University Press, New York, NY, USA.

Feinberg, J. (1974). The rights of animals and unborn generations. In: Blackstone, W.T. (ed.) Philosophy and Environmental Crisis. University of Georgia Press, Athens, GA, USA, pp. 43-68.

Garner, R. (2013). A theory of justice for animals: animal rights in a nonideal world. Oxford University Press, Oxford, UK.

Horta, O. (2013). Zoopolis, intervention and the state of nature. Law, Ethics and Philosophy 1: 113-125.

Karan, A. and Pogge, T. (2015). Ebola and the need for restructuring pharmaceutical incentives. Journal of Global Health 5(1): 010303.

Lang, T. and Rayner, G. (2012). Ecological public health: the 21st century's big idea? An essay by Tim Lang and Geof Rayner. BMJ 345: e5466.

Médecins Sans Frontières (2015). Ebola: pushed to the limit and beyond. A year into the largest ever Ebola outbreak. Available at: http://tinyurl.com/ncuya64.

Milligan, T. (2015). The political turn in animal rights. Politics and Animals 1(1): 6-15.

Palmer, C. (2010). Animal ethics in context. Columbia University Press, New York, NY, USA.

Palmer, C. (2015). Response to 'vulnerability, dependence, and special obligations to domesticated animals' by Elijah Weber. Journal of Agricultural and Environmental Ethics 28(4): 695-703.

Pogge, T. (2008). World poverty and human rights: cosmopolitan responsibilities and reforms. Polity Press, Cambridge, UK.

Rock, M. and Degeling, C. (2014). Public health ethics and more-than-human solidarity. Social Science and Medicine 129: 61-67.

World Health Organization (2015). Statement on the 1st meeting of the IHR Emergency Committee on the 2014 Ebola outbreak in West Africa. Available at: http://tinyurl.com/mu8cful.

Section 15. Consumers and animal products

62. The milky way upon the heaven? Questioning the normative good of milk

T. Wandegren
Dept. of Theology, Uppsala University, Box 511, 751 20 Uppsala, Sweden; towe.wandegren@teol.uu.se

Abstract

Different opinions on food values are nothing unusual, but interestingly enough, Svensk mjölk (representative of the largest Swedish dairy association LRF Mjölk), has chosen to sue a small oat drink producer, Oatly, for misleading advertisement regarding milk and plant based alternatives to milk. The summon against Oatly in August 2014 was ruled in favour of Svensk Mjölk in November 2015. In this paper, I will perform an argumentation analysis searching for core values in the judicial documents and scrutinize the relevance, validity and coherence of each suitor's arguments. The hypothesis of this paper is that the parties show different perceptions of the question at stake, 'reality', and also use concepts like animal welfare, human health and sustainability so differently that a real dialogue or mutual understanding has not taken place. The aim is to evaluate whether the legal process, as performed, had the potential to contribute to a better understanding of the normative good of milk. In order to do that, I will use three different approaches; history, health and nutrition and animal welfare, to methodically categorize the arguments and values each part gives greatest weight.

Keywords: milk, plant-based drinks, goodness, health, animal welfare

Introduction

For many decades, both dairy and farmers' associations in Sweden have highlighted the positive effects of milk, and the good of milk. Their claim has become the normative hegemony of milk consumption, and parents, schools and health centres take these associations' claims as incontestable truths. More recently, critical voices have questioned the inherent goodness of milk, pointing at climate issues related to cows' methane production and other greenhouse gas emissions in food production. Further, they explicate the fact that cows need to give birth to calves in order to start lactating, but nowadays calves are often taken from their mothers directly after birth, in order to sell the milk to human consumption. Moreover, critics claim, advertising has been blurred with dietary guidelines.

In September 2014, Swedish Milk, legal representative and until mid-2014 trade association for the Swedish dairy industry, sued the small oat drink producer Oatly for misleading marketing. The lawsuit came to be a symbolically important case for the entire food industry regarding basic values and the question of what constitutes 'goodness'. What was the lawsuit all about? Which claims were stipulated and which areas did these claims touch upon? What is it that makes this lawsuit so interesting in terms of values and 'goodness' of food producers, that we believe it is worthy of a special ethical analysis?

Every statement Oatly has made in their campaign has a broad approach, which contains a whole spectrum of ethical assertions. In this paper, however, I will highlight three approaches that appear clearer than others, each of which represents a distinctive variety of good: History, health and nutrition, and animal welfare. These three approaches also reflect what both parties associate with the concept of 'goodness'.

EurSafe 2016 – DOI 10.3921/978-90-8686-834-6_62,

The historical approach

Part of the creation of Swedish welfare vs inventor of tomorrow drinks

In the early 1990s, scientists at Lund University developed a method for extracting milk-like drink from oats. The idea was to create a new type of food that was an environmentally and health-wise alternative to milk and other milk-like beverages. The researchers patented the product and established the company Oatly in 1993-1994. Starting out as an experimental research project, the quality of the oat product has made such progress that Oatly currently accounts for a significant share of the Swedish market for alternatives to dairy products. With products such as Oat Drink Original, 'Havregurt' as an alternative to yogurt, as well as oat-based cooking cream and ice cream, the company has reached out internationally, and their products are currently available in about thirty countries (Marknadsdomstolen, 2014a).

In the first twenty years, Oatly's marketing was relatively modest and low-key, just like the majority of other producers of plant based food. The products were merely presented as valid alternatives to dairy products. In early 2014, however, the marketing strategy changed radically with an extensive profile change of the brand. Oatly was now presented as the underdog, ready to take the fight with the big milk producers, by questioning the dairy industry's hegemonic interpretation of the value of milk and milk production. The packages got a defiant layout with handwritten childlike exclamations and a new font, reminiscent of telegram style. Still, it was the content of the texts and slogans on the new packages that had changed radically, compared with the former ones. A parallel provocative ad campaign was launched at the same time. 'Wow no cow!' became the new slogan along with phrases like 'It's like milk, but made for humans', 'Milk? No, it's not milk. Milk comes from a cow. It was designed for baby cows' and 'Full of goodness'. The marketing idea behind the campaign is indeed quite unique and very smart. With its provocative appeal, the company wanted to get three different reactions: get backed by those already committed to the strict vegetarian cause, arouse curiosity of the general public and, not least, provoke the dairy industry to defend themselves and their values. The latter would be beneficial for Oatly, since it would confirm that they are a company to be reckoned with and give them a lot of free advertising. Sales actually increased after the lawsuit became known (Helsingborgs Dagblad, 2014).

Swedish Milk was founded in 1932. Until 2014, it was the the trade association for dairy companies which together accounted for 98% of Sweden's total milk production. The organization also included the milk lobbying organization Mjölkfrämjandet, formerly known Mjölkpropagandan, whose aim was, and is, to promote the consumption of Swedish milk at the expense of other beverages. They have done this by informing about the connection between milk, diet and health and by stressing the role of milk and milk products in the Swedish food culture. Thus, Swedish Milk became a part of and symbol for the creation of the Swedish welfare state (Jönsson, 2005).

Swedish Milk stresses the important role of milk in Swedish history. It has historically been an important part of the diet; used to make butter, cheese and other dairy products. That milk was not seen as a beverage until the late 1800s is not mentioned. During the last hundred years, through its lobbying and marketing, which was often done in consultation with The Swedish National Food Agency (NFA), Swedish Milk has managed to create a normative hegemony around the milk as something good. Milk is good, wholesome, and even necessary to give children strong bones and good health. Milk is the most Swedish of Swedish, the epitome of Swedish open agrarian landscape, where humans and cattle live in harmony (Jönsson, 2005).

In the early 2000s, the major threat to the Swedish milk industry was considered to be the increase in imports of foreign milk, an effect of the rule adjustment that took place when Sweden joined the European Union (Simonsson, 2016). Similarly, the municipal savings proposal to withdraw milk in school canteens tore up strong emotions among parents. The arguments for why customers should opt out of buying foreign milk in favour of Swedish milk have been that milk is a national symbol, a food like no other that built both a healthy human body and the Swedish welfare system. Milk has symbolized rationality and progress. However, today the threat is not only foreign milk, but also that whole milk may be perceived as boring and out-dated. Profitability has decreased, and the dairy industry has tried to acquire a larger market share with new products adapted to contemporary lifestyle ideals (Jönsson, 2005).

Oatly, on the other hand, emphasize that they are a young company with innovative ideas and products, based on pure Swedish oat. This is in stark contrast to cow milk, which they call out-dated and 'Old school milk'. Instead, Oatly argues, they are part of a future that is already here, with completely different, and much better solutions to human needs for healthy, nutrient, ecological and animal friendly dairy-like products. Milk was something people drank and relied on 'back in the days', but now they are ready for the next-generation products. Oatly's expressions are self-confident, and most likely the result of two intertwined reasons. Firstly, a real increase in the confidence in their products and business idea, and secondly, the result of an advertising agency's smart and intrepid marketing. Worth noting is that Oatly uses the same arguments as Mjölkfrämjandet did, when milk challenged other drinks as the most rational and healthy beverage option in a progress-oriented society. What we can see in this section is that the historical dimension as such is emphasized, and used by the two parties quite similarly. However, by emphasizing and interpreting certain historical events in almost diametrically different ways, they both try to influence the writing of history in such a sense that they come out as having a rightful dominant market position in the future. By virtue of its historically important social role, Swedish Milk contests Oatly's claims and designates them as false. Oatly, on the other hand, uses precisely the argument Swedish Milk itself historically made use of: Innovation, progress and rationality in the choice of drink. 'No Milk, No Soy, No Nonsense', no old school milk like back in the days (Marknadsdomstolen, 2014a).

The health and nutrition approach: goodness and badness questioned

To which extent consumption of cow milk is healthy is a moot point in the lawsuit. Swedish Milk has highlighted the positive effects of milk for decades. Milk is healthy for humans, since it contains a lot of nutrients that are good for the human body (Marknadsdomstolen 2014b). Milk and products made thereof are also natural ingredients. Therefore, milk is 'naturligt gott' ('naturally good' and 'naturally tasty') (Arla Foods, 2016). As already mentioned, Swedish Milk consider itself as a part of the creation of Swedish welfare. They remind the court on several occasions that NFA's recommendations for a long while have been to consume a half a litre of milk a day due to its nutritional value (Marknadsdomstolen, 2014e). Through this marketing, or food information, Swedish Milk has succeeded in establishing the good of milk as an incontestable truth among parents, health centres and schools, etc. This is why they argue that milk and other dairy products are particularly good as such when it comes to improve human health. They mean that Oatly's advertisement deliberately blur this fact by comparing two completely different products – milk and oat drink – by arguing that oat drink is more healthy than milk from cows (Marknadsdomstolen, 2014e). Swedish Milk, however, omits to mention that that NFA in early 2015 launched new recommendations to consume only 2-5 decilitres of dairy products a day, and a maximum of half a litre a day (Livsmedelsverket, 2016d; Sveriges Radio, 2016). Swedish Milk also omits NFA's dietary guidelines for vegans: That it is possible to assimilate the same amount of nutrition, as long as the plant based drink is enriched with essential nutrients (Livsmedelvereket, 2016c).

The oat drink company does not deny that milk from cows contains a lot of nutrients, but argues that oat drink, the healthy and modern alternative to 'temporary inventions' like milk, is even more nutrient rich, and that it contains 'no badness', 'no nonsense' and 'nothing that isn't necessary'. Hence, Oatly's products are good for the human body (Marknadsdomstolen, 2014a).

In the campaign, Swedish Milk was addressing in the lawsuit, Oatly stressed that 'what's amazing' with their products is that they 'can taste this good and still be good for you', that they are 'lighter and cleaner and healthier' and 'not a substitute for cream, It's an upgrade'. They also stress that 'what might be less amazing' is that they 'added some sugar (but) it is just a fraction of what you find in yoghurts, which is good' (Marknadsdomstolen, 2014b). Swedish milk claims that these allegations are false and generalizing. The data Swedish milk presents indicate that Oatly greatly exaggerate how much healthier their products are compared to cow's milk. Swedish milk also means that Oatly's marketing give the impression that the appearance of milk products, unlike plant based foods, are bad and/or harmful to the human body, are not intended for human, and/or contain unnecessary ingredients (Marknadsdomstolen, 2014e). Oatly denies this allegation and argues that it is entirely a matter of product information. The claims on their packaging refer to their own product content, and informs consumers on the absence of allergenic substances such as milk and soy. Likewise, the expression 'but made for humans' should to be understood as Oatly's oat drink is specially designed for humans. In a diet where the whole is important, it is an advantage that a staple food – like oat drink – has a balanced composition in relation to human needs. The terms should therefore not be seen as an evaluation as to whether the milk is suitable for human consumption or not, or whether milk is bad or dangerous for humans and the human body (Marknadsdomstolen, 2014e). However, what Oatly emphasizes, if albeit briefly, is that what they mean by 'goodness' is not primarily about health, but rather about animal ethical issues concerning animal welfare (Marknadsdomstolen, 2015a).

The animal welfare approach: the happy cow vs the ignored baby cow

Both parties' respective opinions of how the concept 'goodness' should be understood are most apparent in questions concerning animal welfare as moral 'good'.

In their advertisement, consumer information and debate articles, the Swedish dairy industry stresses the question of animal welfare as centrally important in their work. Even though cows live most of their lives inside a barn, dairy producers often depict them in idyllic pasture landscapes, connoting animal welfare with the possibility to perform species-specific behaviours (Borkfelt *et al.*, 2015). Swedish Milk presents milk produced in Sweden as good, and far better than milk produced in other countries, thanks to stricter animal welfare regulations. From this perspective, there are only ethical benefits with the consumption of Swedish milk; a statement of moral goodness in the framework of comparison with international dairy production. However, in their claim against Oatly, the producer – the cow, whose milk the whole lawsuit revolves around – is absent. By obscuring the cow in their petitions, the cow's actual living conditions are made invisible. Swedish Milk consistently choose to interpret the expression 'No badness' as purely related to health claims about humans and the human body, and that it is not likely at all that Oatly's intention was to convey ethical values about what's morally good concerning the cow's and their calves standard of living (Marknadsdomstolen, 2014e).

In response to this, Oatly emphasizes it is exactly the dairy cow's living conditions that they are questioning, as well as the right Swedish Milk and large parts of the population claim to have to use the cow's body as a means of production. This unwillingness to admit that there are different definitions of animal welfare among producers (focus on health) and consumers (focus on natural life/behaviour) has a parallel in how the (natural) science discipline is bound to focus on animal welfare whereas the philosophical discipline animal ethics can include wider aspects as ethically relevant (cf. Borkfelt *et al.*, 2015).

In the law suit, Oatly claims, in a much more modest way than in their marketing, that the issue is not about whether milk is healthy or not for humans; Oatly does not value the suitability or non-suitability of milk as food. Hence, the basis for Swedish Milk's claims that Oatly is misleading the general customer with non-specific and false statement about health, is invalid (Marknadsdomstolen, 2014c). Oatly claims that they, in their marketing, contrasts 'goodness' with its opposite 'badness', and that this should be understood as an issue of conscience or 'good karma'. They claim that the typical consumer will understand the term 'goodness' as ethical/moral/biblical 'goodness', and that the concept is part of Oatly's aim to be a 'good' company (Marknadsdomstolen, 2015b). Further, Oatly's products are vegetal alternatives, while cow milk is an animal product. For consumers who choose a strictly vegan diet for ethical or life style related reasons, and especially reasons related to animal ethics, this is of great importance. These reasons, Oatly claims, are also well known for consumers who are not vegans, i.e. 'goodness' can hardly be understood as related to the health of humans. (Marknadsdomstolen, 2015a).

Thus Oatly, means that 'goodness' is part of being 'good' in relation to animals and nature, whereas Swedish Milk understands 'goodness' as a health concept. To make a wise health decision is to make a 'good' choice. That something is 'full of goodness' is equivalent to that it is healthy for humans and the human body. Swedish Milk's target group is the health conscious consumer, while Oatly's target group are the consumers who are engaged in issues of environment and animal ethics. It is very clear that respecting or considering animal welfare means different things to them.

Discussion: misleading marketing or just pointing at facts?

The starting point of this paper was the legal process which followed with Swedish Milk's summon against Oatly for misleading marketing. As Jönsson (2005) and Borkfelt *et al.* (2015) have shown, the dairy industry often depict cows as living there lives in pasture rural landscapes, although most cows are not housed like that, or only during a few months, and Oatly's new marketing challenged this interpretative prerogative. Therefore, the hypothesis was that Swedish Milk and Oatly in the law process, as performed, would show such different perceptions of the question at stake and the 'reality', and also understand the concepts history, animal welfare and human health so differently that a real dialogue or mutual understanding would not take place. By categorizing the arguments and values each part give greatest weight under three different approaches, corresponding to the three concepts mentioned above, it has been showed that each of the parties has chosen to emphasize the facts that strengthen their own case. This is, of course, self-evident. It has also been shown that each party has chosen to exclude the part of the arguments attached to the facts that have not supported their respective views. The weighing of the arguments used in the judicial documents has highlighted that the two parties have filled the different approaches with so different meanings, that they hardly can reach a real dialogue. The court documentation confirm that they have not wished to enter a dialogue or reach mutual understanding. Therefore, no such thing has been achieved. Each party sticks to its own approaches and does not appear to try to understand the other party's arguments, by opening up for the possibility that the case is not about possessing the right worldview. Instead, Oatly has argued that their texts constitute pure product information for those who are likely to choose vegetal products for ethical or lifestyle related reasons. On the other hand, Swedish Milk has disregarded every claim Oatly has expressed, if they have questioned the age-long conception of milk as 'good': good as in healthy, rational and forward-looking. As I have shown, the mutual misconceptions are particularly clear in the animal welfare approach and in the interpretation the parties make about the meaning of 'good'. Swedish Milk claims that, in this context, 'goodness' means nutritional benefits for human beings, while Oatly insists that 'goodness' above all is related to animal ethics.

Conclusions

In conclusion, the law process, as performed, did contribute to a better philosophical understanding of the 'good', but not as a clearly defined concept. Rather, the scrutiny has shown that the very concept 'good' was interpreted in so many ways, and filled with such different meaning, that the kind of understanding we can reach, is that the law process confirmed that the concept 'good' cannot be understood without ambiguity. The parties' lack of a will to understand each other, however, is unlikely to have led to a better understanding among the public about what the term 'good' means and what Swedish Milk and Oatly actually litigated about.

Acknowledgements

The author would like to thank Assoc. prof. Helena Röcklinsberg and Assoc. prof. Per Sundman for valuable comments.

References

Arla Foods (2016). Frågor och svar om Bregott. Available at: http://tinyurl.com/zm7bmle.

Borkfelt, S., Kondrop, S., Röcklinsberg, H., Bjørkdahl, K. and Gjerris, M. (2015). Closer to nature? A critical discussion of the marketing of 'ethical' animal products. Journal of Agricultural and Environmental Ethics 28(6): 1053-1073.

Helsingborgs Dagblad (2014). Lyckad stämning för Oatly! Available at: http://tinyurl.com/jrm5cxu.

Jönsson, H. (2005). Mjölk – en kulturanalys av mejeridiskens nya ekonomi. Brutus Östlings Bokförlag Symposium, Stockholm/Stehag, 231 pp.

Livsmedelsverket (2016a). Barn 2-17 år. Available at: http://tinyurl.com/grrmhbr.

Livsmedelsverket (2016b). Mjölk och mejeriprodukter. Available at: http://tinyurl.com/jnhwuz8.

Livsmedelsverket (2016c). Vegetarisk mat till barn. Available at: http://tinyurl.com/gn32xkj.

Livsmedelsverket (2016d). Mejeriprodukter råd. Available at: http://tinyurl.com/zx7jsvc.

Marknadsdomstolen (2014a). Mål nr C 23/14 2014-10-15.

Marknadsdomstolen (2014b). Mål nr C 23/14 Domstolsbilaga AktID 7, 2014-09-25.

Marknadsdomstolen (2014c). Mål nr C 23/14 Aktbilaga 113, 2014-11-10.

Marknadsdomstolen (2014d). Mål nr C 23/14 Aktbilaga 113, 2014-11-11.

Marknadsdomstolen (2014e). Mål nr C 23/14 2014-12-22.

Marknadsdomstolen (2015a). Mål nr C 23/14 2015-02-02.

Marknadsdomstolen (2015b). Mål nr C 23/14 2015-03-20.

Simonsson, H. (2016). Den moderna mjölkproduktionens etiska problem. En kvalitativ studie baserad på djupintervjuer med 8 nyckelpersoner. SLU, Uppsala, Sweden.

Stolt Kommunikation (2016). Bregottfabriken och de eviga värdenas varumärkeslöfte. Del 7/Bregott. Available at: http://tinyurl.com/zs5qnfw.

Sveriges Radio (2016). Nya mjölkråd får kritik. Available at: http://tinyurl.com/zv3owgn.

63. Vegans and omnivores: differences in attitudes and preferences concerning food

P. Kerschke-Risch
University of Hamburg, Department of Social Sciences, Allende-Platz 1, 20146 Hamburg, Germany; pamela.kerschke-risch@uni-hamburg.de

Abstract

The aim is to show the results from two online surveys in which German consumers were asked among other topics about their attitudes and preferences regarding choices of food. One survey was conducted with a sample of preselected vegans (n=852), which is the first quantitative study in Germany based on sociological research questions while the other group was made up of a sample without any specific conditions (n=1,666). It will be shown how and why the two groups differ in their choice of food depending on their particular internalized norms, trust in organic food, and experiences related to food scandals. Moreover, environmental awareness, as well as animal welfare and protection, play an important role in the decision to consume meat or to refrain from the consumption of flesh as well as any animal products. Against this background, it is very important to consider these different attitudes in order to understand why some people take into account ethical issues when making their food-choices while on the other hand for a large number these issues are less relevant or even negligible. These results can help us to understand how ethical questions can become more relevant for more consumers in the future. In understanding decision processes, we can try to change consumer behaviour towards more consciousness regarding health, environmental questions and of course animal welfare and protection.

Keywords: omnivores, vegans, choices of food, trust in organic food, environmental awareness

Introduction

Current developments in food and nutrition are pointing in different directions. On the one hand, there is an increasing demand for meat and animal products worldwide (WHO, 2016) whilst on the other side in western societies the number of vegetarians is growing (Kerschke-Risch, 2015). Moreover, throughout the last decade the percentage of vegans, who are not only excluding meat from their diet but all animal products is increasing (Kerschke-Risch, 2015). In general vegetarians are people who do not eat any meat, poultry, game, fish, shellfish or by-products of slaughter (Vegetarian Society, 2016) whereas vegans 'object to the use of nonhuman animal products for food, cosmetics, clothing, and vivisection – virtually all invasive activities involving nonhuman animals' (McDonald, 2000: 1).

The main sociological research questions were what factors affected food choices and how vegans differ from omnivores in this regard? Because norms, trust in conventional agriculture respectively organic farming, environmental awareness and questions related to animal welfare issues play a key role regarding food choices (Aertsens *et al.*, 2009; MacNair, 2001; McDonald, 2000; Povey *et al.*, 2001; Zanoli, 2002), it is hypothesized that omnivores and vegans differ in their attitudes concerning these relevant criteria.

Before people decide for total abstinence from consuming any animal products they normally would have taken into consideration environmental issues as well as animal welfare and health before becoming a vegan (Kerschke-Risch, 2015; McDonald, 2000). A logical consequence might be to mistrust conventional agriculture and to adopt ecological ideas.

EurSafe 2016 – DOI 10.3921/978-90-8686-834-6_63, © Wageningen Academic Publishers 2016

Regarding these two groups, one general sample and a vegan sample are compared in the following. The vegan survey is the first quantitative study in Germany which is based on sociological research questions. Possible influences on meat consumption are modeled for the general sample, influences on consumption of organic fruit and vegetables are analysed for omnivores as well as vegans.

Method

In 2013 two quantitative online surveys were conducted in Germany using 'unipark', an academic survey tool from 'QuestBack'. For the general sample (n=1,666) an invitation email with a link to the personalized survey was distributed through a scientific online platform (PsyWeb: https://psyweb.uni-muenster.de/pages/home; University of Muenster) while the vegan one (n=852) was an anonymous so-called 'snowball-sampling' which was spread through social networks and forums. This is an appropriate method for reaching relatively small and difficult to access sub-populations (Baltar and Brunet, 2012; Brickman, 2012). Only people who affirmed the first question that they followed a vegan diet could participate in the vegan survey. Both questionnaires consisted of similar standardized questions, only specific items concerning animal products were adapted (see for further details: Kerschke-Risch, 2015) and analysed using the statistics software SPSS. The scores were identified by using different Likert scales which the respondents could tick.

Results

Table 1 illustrates the proportion of male and female respondents. Women were significantly over-represented in both samples whereby the proportion in the vegan sample with nearly 80% was even more pronounced than in the general one which is similar to other vegetarian studies (Schoenhoefer-Rempt, 1988). Due to the absence of official statistics on vegans, the distribution can only be compared with other studies which show rather similar results. The 'typical' vegetarian is rather young, urban female and well educated (Max Rubner-Institut, 2008).

In the following it will be shown which items differed in both groups. The mean scores of the scales the respondents answered demonstrate specific beliefs and attitudes of omnivores and vegans.

Contrary to the assumptions the comparison of norms regarding illegal spraying or fertilizing as well as intentional mislabelling of conventional and organic food showed no differences between vegans and omnivores. The same also applies in the probability of risk, both samples estimated similarly the risks of the above mentioned illegal acts.

Table 1. Omnivores and vegans: socio-demographic variables (percentages).

	General (n=1,666)		Vegan (n=852)	
	male	female	male	female
	35.3	64.7	20.3	79.8
Age in years (SD)	47.6	43.0	32.4 (10.3)	30.8 (10.1)
Age span	18-84	17-80	15-77	15-72

Furthermore, it was hypothesised that vegans have more confidence in organic farming than the general sample and less in German agriculture as well as in food corporations. Therefore, the mean scores of each group were compared, shown in Table 2.

Contrary to the expectation both samples showed no significant difference in their trust in organic agriculture, the mean score was about 4.4. On the other hand, respondents of the general sample had more trust in conventional agriculture than vegans, whereby the mean score in both samples was lower than the item concerning organic farming. Trust in large food corporations with a score of 2.2 (general sample) resp. 1.52 (vegans) showed the lowest level of confidence in each group with a significantly lower level for the vegans. This result might explain the vegan's preferences for organic products respectively omnivores greater tendency towards conventional agricultural products.

In Table 3 shown below, people were asked about their priorities when choosing food. The purpose of the statement was to use it as a measure for a sign of hedonism. This means environmental sustainability and animal welfare aspects are not taken into account.

The percentage of omnivores who affirm that they eat what is tasty and don't care about anything else is three times as high as the vegans, which means that nearly one fifth of the general sample doesn't care about environmental or animal welfare questions whereas the vegan's proportion is only 6.5%.

As the above results in Tabe 4 demonstrate, vegans believe significantly more than the omnivores that conventional agriculture harms the environment and destroys nature. Lowering meat consumption especially is regarded in both samples as beneficial for the climate, whereas almost all vegans are of this opinion. Moreover, the table suggests that vegans believe that the media does not tend to exaggerate on environmental problems caused by conventional farming. On the whole, vegans show a far greater environmental knowledge and awareness than omnivores of the general sample.

Table 2: Omnivores and vegans: trust in German agriculture, organic agriculture and food corporations: comparison of means (t-test; SD scores in brackets).[1]

Trust in:	General (n=1,666)	Vegan (n=852)	Significance
Range 1-7 (1=no trust at all; 7=very high trust)			
Organic agriculture	4.45 (1.39)	4.43 (1.36)	n.s.
German agriculture	3.79 (1.22)	2.61 (1.21)	***
Food corporations	2.20 (1.11)	1.52 (0.84)	***

[1] n.s. = not significant; *** P<0.001.

Table 3. Omnivores and vegans: hedonistic way of food choice (percentages, Chi^2-test[1]).

	General (n=1,661)		Vegan (n=848)		Significance
	yes	no	yes	no	
I eat food that is tasty and don't care about anything else	18.4%	81.6%	6.5%	93.5%	***

[1] *** P<0.001.

Table 4. Omnivores and vegans: environmental awareness and animal welfare: comparison of means (t-test[1]; SD scores in brackets).

Environmental awareness/animal welfare	General (n=1.666)	Vegan (n=852)	Significance
Range 1-5 (1=strongly disagree; 5=strongly agree)			
Animal protection is exaggerated	1.64 (0.87)	1.09 (0.38)	***
Climate: lowering meat consumption is beneficial	3.84 (1.24)	4.80 (0.69)	***
Climate: conventional agriculture is harmful	3.26 (1.12)	4.44 (0.90)	***
Today's agriculture destroys nature	3.27 (1.07)	4.47 (0.84)	***
Media exaggerates environmental problems caused by agriculture	2.24 (1.12)	1.33 (0.66)	***

[1] *** $P<0.001$.

As previous research has shown, women eat less meat than men and represent the majority of vegetarians (Max Rubner-Institut, 2008). Asked, if they had changed their eating habits as a result of certain food scandals, more than one-third of male (36%) and almost half of the female respondents answered that they had changed their meat consumption.

Based on theoretical considerations as well as on descriptive analyses (Tables 2 to 4) regression models were constructed for both samples to explain the dependent variable 'consumption of organic fruit and vegetables.' Another model was built to show the influence of the independent variables on meat consumption in the general sample.

Results from multiple regression analyses for consumption of organic fruit and vegetables are shown in Table 5. Looking at Table 5 variables which measure environmental awareness and trust in organic farming explain 25% variance for the omnivore sample whilst the explanatory capacity for vegans is low. In both samples trust in organic farming has the strongest effect on choosing organic fruit and vegetables. The influences of hedonism, environmental awareness, and animal protection are only significant in

Table 5. Omnivores and vegans: model 1: explaining consumption of organic fruit and vegetables.[1,2]

Independent variables	Omnivores	Vegans
Organic fruit and vegetables		
Sex	0.02 n.s.	0.08 n.s.
Age	0.06 **	0.15 ***
Hedonism: only taste important	-0.12 ***	-0.04 n.s.
Animal protection is exaggerated	-0.10 ***	-0.04 n.s.
Climate: lowering meat consumption is beneficial	0.07 ***	-0.00 n.s.
Climate: conventional agriculture is harmful	0.14 ***	0.05 n.s.
Trust in organic agriculture	0.31 ***	0.24 ***
Adj. R^2	0.25	0.09

[1] Results from multiple regression analyses for consuming organic food and vegetables (beta values; all beta values in the table are those at entry).

[2] n.s. = not significant; ** $P<0.01$; *** $P<0.001$.

the general sample. This could possibly be explained with the fact that vegans, in general, are aware of environmental problems, an often mentioned reason for becoming vegan (Kerschke-Risch, 2015).

Table 6 shows that hedonistic characteristics, trust in German agriculture, environmental awareness and animal protection have significant effects on the frequency of consuming meat. As hypothesized women consume less meat than men.

Discussion

Consumer choices are often very complex decision processes which depend on numerous variables (Aertsens *et al.*, 2009). As the above results indicate, consumer's decisions depend on, among other things, on environmental awareness as well as trust in organic farming. Norms and estimated risks are less relevant when choosing organic products. People in the general sample are less aware of possible harms to health and the environment caused by conventional farming and high meat consumption as the significant influence of hedonism indicates. This could be an element to explain why people consume products without taking environmental and ethical issues into account. In this study, respondents were asked what they actually bought so the impact of the independent variables could be measured, and so possible problems related to attitude-behaviour-gaps (Padel and Foster, 2005) are reduced.

Regarding the introductory filter question, a possible limitation of the vegan sample has to be kept in mind. It might be possible that respondents incorrectly assessed their diet as vegan even though they ate animal products. A similar problem is known in vegetarian studies, were in some cases participants classified themselves as vegetarians although they consumed meat (MacNair, 2001; Schoenhoefer-Rempt, 1988).

If there could be changes made in a more ecological direction consumers have to be better informed about environmental problems caused by conventional farming methods as well as the high consumption of animal products, especially meat. The analyses of omnivores and vegans show how differences in attitudes and beliefs cause some people to take ethical issues into account when making their food-choices while on the other hand for a large number these issues are less relevant or even negligible. With this knowledge, one implication could be to find ways to make it easier to realise utopian ideas promoting environmental consciousness.

Table 6. Omnivores and vegans: model 2: explaining consumption of conventional meat.[1,2]

Independent variables	Omnivores (n=1,656)
Sex	-0.15 ***
Age	-0.01 n.s.
Hedonism: only taste important	0.11 ***
Animal protection is exaggerated	-0.09 ***
Climate: lowering meat consumption is beneficial	-0.16 ***
Climate: conventional agriculture is harmful	-0.17 ***
Trust in German agriculture	0.10 ***
Adj. R^2	0.21

[1] Results from multiple regression analyses for consuming conventional meat (beta values; all beta values in the table are those at entry).

[2] n.s. = not significant; *** $P<0.001$.

Acknowledgements

The surveys are part of the project 'Geschlechtsspezifische Aspekte bei der Lebensmittelwahl'(Gender-Specific Aspects in Food Choice), which is supported by the Equality Fund from the University of Hamburg and by the Fund for Women from the Faculty of Economics and Social Sciences.

References

Aertsens, J., Verbeke, W., Mondelaers, K. and Van Huylenbroeck, G. (2009). Personal determants of organic food consumption: a review. British Food Journal 111(10): 1140-1167.

Baltar, F. and Brunet, I. (2012). Social research 2.0: virtual snowball sampling method using Facebook. Internet Research 22(1): 57-74.

Brickman Bhutta, C. (2012). Not by the book: Facebook as a sampling frame. Sociological Methods and Research 20(10): 1-32.

Kerschke-Risch, P. (2015). Vegan diet: motives, approach and duration. Initial results of a quantitative sociological study. Ernaehrungs Umschau 62(6): 98-103.

MacNair, R.M. (2001). McDonald's 'empirical look at becoming vegan'. Commentary. Society and Animals 9(1): 63-69.

Max Rubner-Institut/Bundesforschungsinstitut für Ernährung und Lebensmittel (2008). Nationale verzehrs studie II. Ergebnisbericht, Teil 1. Karlsruhe, Germany.

McDonald, B. (2000). Once you know something, you can't not know it. An empirical look at becoming vegan. Society and Animals 8(1): 1-23.

Padel, S. and Foster, C. (2005). Exploring the gap between attitudes and behaviour. Understanding why consumers buy or do not buy organic food. British Food Journal 107(8): 606-625.

Povey, R., Wellens, B. and Conner, M. (2001). Attitudes towards following meat, vegetarian and vegan diets: an examination of the role of ambivalence. Appetite 37: 15-26.

Schönhöfer-Rempt, R. (1988). Gießener vegetarierstudie. Ernährungsgewohnheiten, gesundheitsverhalten sowie einstellung und wissen zu ernährungsbezogenen themen. Wissenschaftlicher Fachverlag, Gießen, Germany.

Vegetarian Society (2016). Available at: http://tinyurl.com/m6oe338.

World Health Organisation (WHO) (2016). Global and regional food consumption patterns and trends. Available at: http://tinyurl.com/d7mnqa2.

Zanoli, R. and Naspetti, S., (2002). Consumer motivations in the Purchase of organic food. A means-end approach. British Food Journal 104(8): 643-653.

64. Veganism as a choice: Experiences and food strategies in transitioning to a vegan diet

A.-L. Elorinne[1*], *M. Kantola*[1], *S. Voutilainen*[2] *and J. Laakso*[3]
[1] *School of Applied Educational Science and Teacher Education, University of Eastern Finland, Savonlinna, Finland;* [2] *Institute of Public Health and Clinical Nutrition, University of Eastern Finland, Kuopio, Finland;* [3] *Finnish Safety and Chemicals Agency, Helsinki, Finland; anna-liisa.elorinne@uef.fi*

Abstract

Adherents to a vegan diet experience their diet as more of an ideology than just a daily dietary practice. The aim of this study was to learn Finnish vegans' underlying reasons for their dietary choices, and to learn about their experiences during the transition period. Compliance with a vegan diet and the meanings given to dietary choice by participants were also studied. The research data consist of nine interviews with voluntary vegans. The vegan participants (six women and three men, 24-50 years) were recruited through an advertisement published in the monthly newspaper of the Finnish Vegan Association and via an online discussion forum. The vegan participants had adhered to their diet for 3-12 years. Interviews were taped, transcribed and analysed using a content analysis approach. All participants prepared food themselves and used recommended dietary supplements. Most of them also consumed fair trade and organic products. The reasons for transitioning to a vegan diet were a concern for animal rights and ecology, as well as concerns for personal health. The most significant external factor in adhering to a vegan diet was the existence of peer support. The vegan diet was experienced as a part of their life and self-identity. Formulating a vegan diet and finding vegan foods to purchase were considered easy, but social gatherings often caused embarrassment or uncomfortable situations.

Keywords: animal rights, social situation, food-related values

Introduction

Being vegan is more than just adhering to a meat-free diet. It is an ideology, a way of life and a value perspective for all decision-making (Fox and Ward, 2008; Ruby, 2012; Whorton, 1994). The transition to a vegan diet takes place gradually from semi-vegetarian or ovo-lacto vegetarian diet to a vegan diet over time (Fox and Ward, 2008). Studies have also identified a variety of motivations for adopting a diet free of all animal products. The main three categories of reasons are health issues, ethical issues and environmental concerns. (Dyet *et al.*, 2013; Fox and Ward, 2008; Jabs *et al.*, 1998; Larsson *et al.*, 2003; Radnitz *et al.*, 2015).

Research has shown that ethical and health reasons are most often cited for choosing to follow a vegan diet (Dyet *et al.*, 2013; Fox and Ward, 2008; Jabs *et al.*, 1998; Ruby, 2012). Ethical vegans are characteristically concerned for the rights of animals more than their own nutrition, and their transition to a vegan diet can take place suddenly. Health vegans are concerned about their own or loved ones' health and they tend to gradually change their diet and make 'trial adoptions' of food choices (Fox and Ward 2008; Jabs *et al.*, 1998). In a study by Fox and Ward (2008), health vegetarians emphasized personal reasons for their diet more than concern for animals, whereas ethical vegetarians considered their practices as fundamentally altruistic and involving personal sacrifice to prevent cruelty to animals. Finnish researchers Lindeman and Sirelius (2001) showed that vegetarians' perspectives may have different ideological bases. Ethical vegetarianism was broadly associated with humanistic commitments, and health vegetarianism with conservative and normative values.

EurSafe 2016 – DOI 10.3921/978-90-8686-834-6_64, © Wageningen Academic Publishers 2016

Experiences on a vegan diet and reasons for becoming a vegan were studied on the basis of two different food choice models (Furst *et al.*, 1996; Leitzmann, 2003). According to Furst *et al.* (1996), life course experiences are major influences on food choice. The Furst *et al.* model has a holistic approach where experience, expectations, resources (tangible and intangible), value-oriented meanings (concerning health, taste, price, facility, ethics, familiarity, etc.) and personal factors (age, gender, family, and state of health) have an impact on the food selection strategy people finally adapt. According to Leitzmann (2003) a vegetarian lifestyle represents the ecological model of food choice. In the nutritional ecology model the entire nutrition system, with special consideration of the effects of nutrition on health, the environment, society, and the economy, are taken into account. To fulfil these demands, the principles of a wholesome ecological diet have been described by the researchers (Leitzmann, 2003)

The aim of this study is to describe vegans' food choices, experiences during the transition period and reasons for becoming a vegan. We were especially interested in long-term vegans' experiences. Our hypothesis was that modern long-term vegans consume a varied diet and they take good care of their nutritional health. We also expected long-term vegans possessing several value-oriented meanings related to a vegan regimen.

Subjects and method

The data were obtained by interviewing nine voluntary vegans who had earlier participated our nutrition-toxicological research project (Elorinne *et al.*, 2016).). Originally vegans were recruited through an advertisement published in the monthly newspaper of the Finnish Vegan Association and via online discussion forum. The participants consisted of five individuals and two couples (Table 1).

Table 1. Reasons for becoming a vegan and experiences of the transition by identified type of vegan life style.

Participant[1]	Reasons for becoming a vegan		Experience of the transition	
	Intrinsic	Extrinsic	Speed	Feelings
An organised vegan (socializes with other vegans, mainly in groups)				
Anni (F, 29 y)	• Ethical reasons • Empathy for suffering animals	• Read a book on animal rights • Joined an animal rights organisation • Peer support	• Got the idea when very young • Quick but step-by-step transition • Smooth and natural progression	• First two weeks were difficult; I just felt hungry all the time • Then I learned how to compose the diet
Ulla (F, 35 y)	• Ethical reasons • Animals and humans have similar value • It is unnatural for humans to drink an animal's milk	• At school, a vegetarian friend	• Slow transition • Went on a 10 day fast before making the decision	• At first it was difficult to find replacement foods of vegan type • Replacing milk was easy; replacing eggs was difficult
Liisa (F, 38 y)	• Ethical reasons	• At school, wrote an essay on animal rights • Info from the Vegan Society • Peer support • Student restaurant menu	• Very quick decision and transition • Afterwards, many readjustments were made to the vegan diet	• It felt socially very difficult at the beginning • It was difficult to learn how to cook vegan foods

Table 1. Continued.

Participant[1]	Reasons for becoming a vegan		Experience of the transition	
	Intrinsic	Extrinsic	Speed	Feelings
Arto (M, 38 y)	• Ethical and environmental reasons • We should not kill animals	• Online information and chatroom discussions • Peer support	• Slow transition	• Easy
Lauri (M, 40 y)	• Ethical reasons • Concern for animal rights and the poor treatment of farmed animals	• Peer support	• Slow transition, although she got the idea when very young	• More difficulties in social life than in preparing or cooking vegan foods • In the 80's and 90's, it was rare to be vegan in Finland
An individualistic vegan (has no need to unify or associate with other vegans)				
Katja (F, 25 y)	• Health effects • Experimenting • Rebellion against parents	• Availability of vegetarian foods at school and in shops • Vegetarian friends	• Slow and step by step	• No special difficulties
Sanna (F, 30 y)	• Ethical reasons • Humans and animals have similar value • Killing and intensive farming of animals is wrong	• Support from boyfriend and best friend	• Quick and natural due to a vegan boyfriend	• No problems at the practical or personal level
Taina (F, 50 y)	• Love for animals • Shocked by the names of meat-based foods that carry the animal's name • Dislike for the taste of meat • It is unnatural to eat meat	• Some vegetarian relatives • Good-tasting vegetarian foods • Vegetarian cooking • Peer support • Joined the Vegetarian Union	• Quick decision • Slow and step by step transition • Availability of tasty foods has helped a lot	• Actual change was easy; socially it was difficult • People think that a vegan diet cannot be adequate and healthy
Teemu (M, 24 y)	• Ethical and ecological • Poor treatment of farmed animals • Environmental burden of animal farming	• At school, vegetarian friend • Vegetarian Union's website • Articles and videos on poor treatment of farmed animals	• The idea developed slowly • Quick transition once the decision was made	• No difficulties • Mc Donald's food strike was much more difficult

[1] All participants had a higher education degree.

The emphasis of qualitative research is on process and meanings (Sale and Brazil, 2004). This study has a constructivist approach, which emphasizes that individuals construct the truth of their lives through their experiences depending on resources and context (Carr, 2003). In this respect it was important to interview participants in order to understand more fully the complexity of the food choice in vegan's everyday practices (Furst *et al.*, 1996).

The interviewees were asked about the factors that had contributed to their becoming vegan, how committed they were to their diet and what kind of experiences they had faced as vegans. In addition, they were asked to identify the value-oriented meanings or important issues they associate with their diet.

Before the interview, the researchers explained the contents of the interview, how the data would be used and for what purpose it would be used. Written informed consent had been given earlier. The study received the approval of the Ethical Committee of Kuopio University Hospital (69//2011). The interviews were recorded; one interview lasted 35 minutes and the remainder lasted from one to one and a half hours. The interviews were transcribed. To increase reliability, the interview questions were pretested, and each respondent checked the transcripts of his or her interview. Data were analysed using theory content analysis (Bazeley, 2013).

Results

Transition to vegan diet

The most significant intrinsic factor contributing to transition was the rights of animals; the most important extrinsic factor was the influence of peers. The subjects had received information mostly from friends, and through online discussion forums, vegan websites and brochures. The transition was generally seen as a natural progression. Most vegans said the idea of becoming a vegan had developed gradually (Table 1).

The participants indicated that preparing and cooking vegan foods were easy to learn. But in contrast, they faced challenges and problems in social situations; in communication with relatives, friends, and previous partners and during participation in a variety of social events both in private life and at the workplace. Several vegans were troubled that their diet caused problems and difficulties for other people.

The experience of difference and abnormality was felt in food-related discussions. Due to their diet, vegans also faced social situations where they were placed in the centre of attention; they were asked to justify their food choices, and explain whether they are getting enough nutrients from their food.

Value-oriented meanings and food choice strategies

In conversations with vegans, veganism was more than just a food choice. For example, one vegan mentioned that she would not go to a circus or a zoo, and many refuse to use woollen clothes. One participant also commented on the political discourse of a sustainable diet: 'That is an ultimate goal in animal rights activities, that all become vegetarians just because that is the best option for animals, humans and nature' (Anni, personal communication).

Value-oriented meanings related to vegan food included justice, ethics, responsible behaviour, health and nutritional value, and mental and physical well-being. Vegans had developed several strategies in their food management. Many vegans mentioned carrying their foods in different situations where there would be no suitable vegan food available, such as in family visits and in kinder garden. Characteristically, vegans would study the package labels, favour certain foods such as fair trade products, and in turn, boycott some foods, such as genetically modified foods (GMO-foods). Some vegans picked berries and/or practiced home farming to gather large stocks of inexpensive healthy food. Different roles for preparing food among the family members had taken shape over the time. In some families, both parents prepared their own meals. In a family with children, the parents' food selection strategy was affected by concerns about their children's health. The mode of interview (individual versus couple interview) didn't have any effect on results. Participants talked freely in both interviews.

Discussion

Vegans participating in this study came from households with different habits in everyday food management. However, all vegans were highly educated, which is known to influence food consumption and dietary habits (Konttinen *et al.*, 2012). All participants prepared food themselves and used recommended dietary supplements. Despite of these actions we found vegans whose dietary practices didn't fill the recommendations and the body concentrations of some key nutrients such as vitamin D and iodine were below the reference range (Elorinne *et al.*, 2016).

Since the vegans were of different ages, they had experiences in consuming a vegan diet at different periods of time. Subjects felt that social climate has changed to a more open and approving environment for vegetarians. Twenty years ago, a vegetarian was a 'strange bird' and a 'UFO'. Changes in peoples' awareness and attitudes, as well as a larger availability of vegetarian foods, have made the dietary change much easier today than it was years ago. Consumption of a vegan diet, however, resulted in experiences of difference and abnormality in vegans. Food is much more than just a source of energy and nutrients for vegans, it is also a unifying factor on one hand and a distinguishing factor on the other. To our opinion observed several value-oriented meanings related to vegan diet cement vegan's adherence to their dietary regimen. There is scanty research on this phenomenon. According to Ruby (2012) vegans hold stronger beliefs about meat eating, animal welfare, and the environment compared to other vegetarians. For a small percentage of vegans, these convictions run so deep that they will not sleep with non-vegans. In the future we need more studies on vegan's experienced social separation.

Our data confirms the conceptual model of food choice by Furst *et al.* (1996). Our study revealed participants' life course affecting their influences such as personal factors, social framework and food context. We also observed how these influences shaped participants' personal systems, including conscious value negotiations and unconsciously operationalized strategies that may occur in a food-related choice situation

The vegans in this study were highly educated. Hence, transferability of the results of this study to other vegans is only partially possible. The reliability of the study was confirmed by describing the steps of the research process in detail and using direct quotes. The researchers also had no personal relationship with the subjects or the studied phenomena. The credibility of the study was high because the vegans were volunteers and thus motivated to respond to the interviewers' questions.

There are Finnish vegan families, like three in this study, where children grow and develop exclusively on a vegan diet. This poses challenges for food services, such as kinder gardens, schools and health clinics, for maternal and child nutrition counselling. The focus in nutrition counselling should be the promotion of health, and the conditions for this are created in a confidential and equal interaction between the client and the employee. The results of this study will help readers to understand the value-oriented background of vegans' food choices and contribute to the interaction between vegans and food service providers and health professionals.

References

Bazeley, P., (2013). Qualitative data analysis practical strategies. SAGE Publications Ltd, Warriewood, NSW, Australia.

Carr, D. (2001). Education, knowledge, and truth: beyond the postmodern impasse. Taylor and Francis Group, New York, NY, USA.

Dyet, P.A., Sabate, J., Haddad, E., Rajaram, S. and Shavlik, D. (2013). Vegan lifestyle behaviors. An exploration of congruence with health-related beliefs and assessed health indices. Appetite 67: 119-124.

Elorinne, A.-L., Alfthan, G., Erlund, I., Kivimäki, H., Paju, A., Salminen, I., Turpeinen, U., Voutilainen, S. and Laakso, J. (2016). Food and nutrient intake and nutritional status of Finnish vegans and non-vegetarians. PLoS One 11(3): e0151296.

Fox, N. and Ward, K. (2008). Health, ethics and environment: a qualitative study of vegetarian motivations. Appetite 50: 422-429.

Furst, T., Connors, M., Sobal, J. and Falk, L.W. (1996). Food choice: a conceptual model of the process. Appetite 26: 247-266.

Jabs, J., Devine, C.M. and Sobal, J., (1998). Model of the process of adopting vegetarian diets: health vegetarians and ethical vegans. Journal of Nutrition Education 30: 196 -202.

Konttinen, H., Sarlio-Lähteenkorva, S., Silventoinen, K., Männistö, S. and Haukkala, A. (2012). Socio-economic disparities in the consumption of vegetables, fruit and energy-dense foods: the role of motive priorities. Public Health Nutrition 16: 873-882.

Larsson, C.L., Rönnlund, U., Johansson, G. and Dahlgren, L. (2003). Veganism as status passage. The process of becoming a vegan among youths in Sweden. Appetite 41: 61-67.

Leitzmann, C. (2003). Nutrition ecology: the contributions of vegetarian diets. American Journal of Clinical Nutrition 78: 657S- 659S.

Lindeman, M. and Sirelius, M. (2001). Food choice ideologies: the modern manifestations of normative and humanist views of the world. Appetite 37: 175-184.

Radnitz, C., Beezhold, B.L. and DiMatteo, J. (2015). Investigation of lifestyle choices of individuals following a vegan diet for health and ethical reasons. Appetite 90: 131-136.

Ruby, M. (2012). Vegetarianism. A blossoming field of study. Appetite 58: 141-150.

Sale, J.E.M. and Brazil, K. (2004). A strategy to identify critical appraisal criteria for primary mixed-method studies. Qual Quant 38: 351-365.

Whorton, J.C. (1994). Historical development of vegetarianism. American Journal of Clinical Nutrition 59: 1103S-1109S.

Section 16. Food in a local context

65. Can we reach food sustainability through local food? Evidence from Romania

D.E. Dumitras, C.B. Pocol and I.M. Jitea*
University of Agricultural Sciences and Veterinary Medicine Cluj-Napoca, Department of Economic Sciences, 3-5 Manastur St., 400372 Cluj Napoca, Romania; cristina.pocol@usamvcluj.ro

Abstract

Food is a major defining element for the cultural identity of a region. Rediscovering and revaluing food as patrimony, is nowadays a major concern for scientists. Preserving the Romanian culinary patrimony requires awareness of local food products connected to tradition, even if they are not yet defined within the national legislation. What is right, just and fair in the case of Romanian local food products? This study critically analyses the issues of sustainability of the local actors of fresh cow milk supply chain in Cluj County, Romania. The performance of the local supply chain was assessed by investigating 18 attributes within five dimensions of sustainability (economic, social, environmental, health and ethical) following the methodology proposed by Kirwan. The use of several research methods allowed an in-depth analysis of the supply chain performance. Based on the findings, a SWOT analysis for the local fresh cow milk supply chain was performed with the aim to emphasize the critical issues that disturb the sustainability of the chain.

Keywords: local food, sustainability, multidimensional assessment, fresh cow milk

Introduction

Local foods are preferred due to environmental concerns, health reasons, perception of high quality, willingness to support local farmers, economies and communities (Kirwan, 2004). The purchase of local products depends on the physical or financial access (Kneafsey *et al.*, 2013), major barriers being lack of availability and challenge in identifying them (Feldman and Hamm, 2015). The most common aspect used to define 'local food' is the geographical area varying from 20 to 100 km (EC, 2013). A local food system exists when production, processing, trading and consumption occur in a relatively small geographical area (EC, 2013). In Romania, traditional and natural are the most frequently attributes people assign to local food products, preferred for the 'specific combination of natural factors like the climate, soil, traditions and methods of fabrication' (Chiciudean *et al.*, 2013). Romanian respondents agree that local markets should be encouraged (91%), that consumers benefit from consuming local foods (79%) and that a label would be useful to identify them (EC, 2011).

Rediscovering and revaluing food as patrimony, is nowadays a major concern for scientists (Grasseni, 2012). The culinary Romanian patrimony is represented by a large variety of traditional food products, each with its own particularities originated not only from the recipe but also from the habits and traditions kept along generations with pride, fame and hospitality by villagers. 'Local food' products are not defined within a national legislative framework, the certification process referring only to traditional products. Since 2013, 124 dairy products were certified according to the new more restrictive legislative framework (MADR, 2013).

What is right, just and fair in the case of Romanian local food products? This challenging question is discussed within a study case: local fresh cow milk supply chain in Cluj County, Romania. The sustainability of the local fresh cow milk chain is analysed by investigating five dimensions: social,

EurSafe 2016 – DOI 10.3921/978-90-8686-834-6_65, © Wageningen Academic Publishers 2016

economic, environmental, health and ethical. Few studies discuss the ethical dimension due to the difficulty in defining ethical issues (Galli *et al.*, 2015).

Material and methods

The study case refers to fresh cow milk produced in Cluj County (North-West Region, Romania), focusing on small farms which sell locally. Quantitative and qualitative data were gathered by reviewing the literature, consulting official reports and field work (visits to farms, markets, fairs). Considering the particularities of the Romanian local fresh cow milk supply chain, the multidimensional sustainability assessment is performed by analysing 18 attributes selected from the multi-criteria performance matrix (Kirwan *et al.*, 2014) (Table 1). The methodology supposes the analysis of the attributes within five sustainability dimensions. Even if in general the economic, environmental and social dimensions refer also to ethical aspects, the method proposed by Kirwan *et al.* (2014) separately discusses health and ethical issues as separate dimensions. Based on the analysis of the attributes, a SWOT analysis (Strengths, Weaknesses, Opportunities, Threats) was performed to identify the critical issues of the local supply chain.

Results

The local fresh cow milk supply chain starts with farmers who need inputs to obtain the milk production (Figure 1). In the local short chain (LSC), farmers sell directly to local consumers (farm gate, door to door, local markets, local fairs). In the local long chain (LLC), the fresh cow milk reaches consumers after it is handled by more intermediaries.

The economic dimension

In 2015, farmers engaged in LLC sold fresh cow milk in large quantities to collectors/processors at a price of 1.40-1.45 lei/litre (0.32-0.33 Euro/litre), the milk being further sold to consumers for 3.97-

Table 1. Selected attributes for sustainability assessment after Kirwan et al. *(2014).*

Sustainability dimension	Attributes
Economic	affordability; connection; contribution to economic development; creation and distribution of added value; efficiency; governance; profitability/competitiveness; resilience
Social	connection; consumer behaviour; food security; information and communication; territoriality
Environmental	biodiversity; resource use
Health	food safety; traceability
Ethical	animal welfare; food security; governance; information and communication; responsibility; territoriality

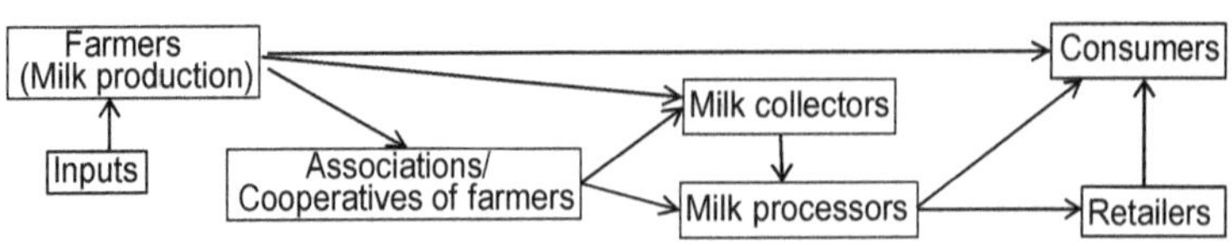

Figure 1. Cluj County local fresh cow milk supply chain.

4.58 Lei/litre (0.89-1.03 Euro/litre). Farmers choose the LSC, get a higher selling price, 3 lei/litre (0.67 Euro/litre) in farmer markets. Consumers also buy from milk vending machines at a price of 3 Lei/litre (0.67 Euro/litre), with nine vending machines being positioned in Cluj County. 'Connection' exists among the chain's actors, with an important role played by the seven associations registered at the Romanian Ministry of Agriculture and Rural Development (MADR) acting in Cluj County, helping farmers to cooperate and build relationships towards more 'economic and environmental efficiency'. Even if at the beginning farmers were reluctant to cooperate under 'cooperative' forms of organization, because of unfortunate situations during the communist period, today they are searching for official forms of association which assures them not only an efficient distribution channel but also selling price negotiation, cash flow, opportunity to buy inputs at a discount and additional points in the evaluation of project proposals under the second pillar of the Common Agricultural Policy (CAP). Moreover, some associations plan to invest in a processing plant, to diversify the production and adapt to consumer needs. Another key actor is AgroTransilvania Cluster, which promotes local products offering the opportunity to producers to sell at their stands. The 'creation and distribution of added value' depends on the size of farm, the distribution channel and the quantity offered. Even if farmers who sell directly to consumer receive more per litre, many prefer to have an immediate cash flow that allows them more time to be involved in production activities. Another factor that may affect local producers in the long run is the competition from international suppliers, as a result of eliminating the milk quota starting April 1st 2015. For instance, in January 2016, the price was 24.76 Euro/100 kg in Hungary and 27.27 Euro/100 kg in Romania, a difference that may attract large processors to import milk (European MMO, 2016). It seems that the ability of the local supply chain to be 'profitable and competitive' faces many challenges. The financial support provided through the National Rural Development Programme Romania (PNDR) (2007-2013, 2014-2020) to develop or start a business in agriculture, encouraging young people, has also a significant 'contribution to the economic development' of the rural areas. The small-size farms that use high nature value (HNV) landscape are heavily financially depended on different types of Common Agricultural Payments. They are not economically self-sufficient under the present market conditions (Jitea and Arion, 2015). Dairy products and other products from cattle breeding are essential for their economic sustainability (Page *et al.*, 2011). Unfortunately, there is lack of accessibility of local dairy products on the market. These types of products can be obtained directly from producers and to a less extent from local stores. Fresh cow milk is available to consumers; however, the origin remains questionable since it is not written on the package. Farm 'resilience' is low especially in environmental sensitive areas. Small size farms are heavily dependent to CAP subsidies (Jitea and Arion, 2015).

The social dimension

Consumers from Cluj County prefer to consume milk from nearby, the main reasons being freshness and taste; they associate large distances between producer and consumer with low quality and lack of freshness (Chiciudean *et al.,* 2013). The average packed cow milk consumption per day per person is very low (12.6 kg/capita/year) as compared with other countries (Grodea, 2015). Governmental measures were taken at national level to stimulate the increase of milk consumption, e.g. the 'Bagel and milk' in schools (since 2002). This program and other measures (e.g. the 'First-buyer') contributed to a better flow of 'information and communication' within the chain. Consumers are also informed through national campaigns, e.g. 'Read the label' (2015). Romanians 'do not associate local products to labels of origin because of their novelty on the food market' (Chiciudean *et al.*, 2013), not surprising since there is no label for local food nor a requirement to state the origin of the product in case of fresh milk. Even if it is a regular food product, milk has a strong link to 'territory' and traditions. Dairy cattle husbandry is one of the main farming activities and it is part of the country patrimony built on traditions kept by farmers, e.g. feeding cows through pasture grazing from March to late-October and the remaining months with hay and silage made during the summer. Tradition is also kept by consumption habits, e.g. hot polenta with cold milk. The high level of self-sufficiency of milk and dairy products at national level, 95% in 2013 (Grodea, 2015), directs that 'food security' is assured.

The environmental dimension

HNV farming represents an important asset for Romania. Accordingly to different bibliographical references it represents between 3.32 and 5.22 million hectares (Keenleyside *et al.*, 2014). It covers from 25 and up to 41% of the total of 13.306 million hectares of the Romanian Utilized Agricultural Area (Eurostat, 2012). Cluj County also has important HNV areas materialised in permanent meadows and pastures and in mosaic landscapes that brings important environmental benefits (Akeroyd *et al.*, 2011). These landscapes are likely to change under current EU regulations and global market pressure, with potentially negative consequences for 'biodiversity' (Loos *et al.*, 2015).

The health dimension

The local chain has flaws as concerns the 'food safety', even if the national and EU legislative framework clearly imposes hygiene norms and quality standards. 'Traceability' is mandatory for all food supply chains in the EU (Regulation (EC) No 178/2002). Unfortunately, the inadequate behaviours of either farmers or intermediaries lead to some extreme negative consequences for consumers' health. Several food alerts came into the attention of governmental bodies. In some cases the real causes were not exposed or it was proved to be false alarms, as it was the case of the presence of 'aflatoxins' in 2013 in Cluj County, as reported by the Direction for Sanitary Veterinary and Food Safety Authority Cluj in mass-media. Some situations created anxiety among consumers and worries among farmers and intermediaries, accentuated by the continuous flow of information and debates in mass-media. Public information campaigns were done mainly when legislation changed. For instance, in 2013 milk producers, collectors and processors were informed that starting January 2014, fresh cow milk production must meet the EU hygiene and quality standards (Regulation (EC) No. 853/2004). In March 2016, the Committee for Agriculture, Forestry, Food Industry and Specific Services of the Chamber of Deputies has adopted the report for the draft law regarding the establishment of additional rules on how to label fresh milk for consumption and dairy products. Changes refer to adding the statement 'Romanian product' on the label only when milk is 100% from Romanian farms.

The ethical dimension

'Animal-welfare' represents a major concern both for consumers and farmers in the EU. In the last years, the EU Commission narrowed the regulations such as to respond to high consumer pressure (EU, 2016). In Romania, it is known that especially the small-size farms do not have any awareness regarding this issue, even though farmers consider that the farming conditions should be improved (Cziszter *et al.*, 2011). Moreover, on-farm assessments of commercial Transylvanian dairy-cow farms show that animal welfare needs improvement (Popescu *et al.*, 2008). At the same time, small-scale HNV farming is important not only for its societal benefits (biodiversity conservation; maintaining cultural landscapes; climate change mitigations; social welfare and household food security) but also for animal-welfare due to traditional management practices closely linked with nature conservation (Redman, 2010). For the Romanian consumers, animal welfare is an important attribute in the buying-in decision (EU, 2007). Based on these previous findings and the consultation of the local expert group, it seems clear that in Cluj County the local producers do not have the power to signal this type of attributes to consumers. Milk is usually collected by large local processors and milk from different farms with different standards is mixed. Practically, consumers do not know the level of animal-welfare standards for the product they buy. 'Governance' attribute was assessed by analysing the relationships among the actors and the regulations. Starting April 2015, new economic entities which buy fresh cow milk from producers for selling and/or processing are obliged to register in the First-Buyer Registry Book at the Paying and Intervention Agency for Agriculture. By this measure, records are kept more accurately about the delivered volumes of milk, assuring an increased transparency and better information within the chain. The European Milk Market Observatory also contributes to the increase in transparency within the supply chain, aiming to assist the actors with information about the milk market (MMO, 2016). 'Responsibility' is a challenging attribute to be assessed at the level of all actors within the local chain. Without a quantitative study, it is difficult to assess to what extent are consumers aware of the rational use of food. However, the ongoing national campaigns function to increase awareness on how to consume responsible. Responsibility at food chain actors is mostly due to the regulations imposed, whereas as regard to the policy and state responsibility, some efforts can be seen through the regulations imposed in the last years. 'Food security', 'information and communication' and 'territoriality' attributes were discussed under the social dimension.

Conclusions

Based on the findings, a SWOT analysis was further developed to emphasize the critical issues that disturb the sustainability of the local fresh cow milk supply chain. 'Strengths' were identified as: preservation of traditional practices of production (e.g. feeding cows on grazed grass) and consumption (e.g. traditional dishes); proximity of farms to markets; favourable conditions for production (HNV farming; specific geographical resources); existence of different associations; improved legislative framework for hygiene normal and quality standards; local traditional fairs. The main 'Weaknesses' of the chain are: fresh cow milk at the milk vending machines does not have the same quality and taste all year around; farmers lack of marketing skills affecting the credibility among consumers; the minimum number of cows required to be accepted as member in associations; food scandals caused by actors that do not follow the hygiene norms and quality standards; lack of added value in LSC as compared to LLC; biodiversity and HNV areas not valorised in the actual market chain; the current chain situation is not ethical both for farmers and consumers (farmers cannot sell their products; consumers do not know the true origin of the product); lack of 'local food' label/certification framework, large quantity of local products for self-consumption; 'Opportunities' were identified as: the financial support through CAP funding 2014-2020; adding-value in the LSC through diversification of production; well educated consumers for localised products specific in short marketing chains; specific EU market for geographical indication products; concern for environmental benefits and multifunctional agriculture in the EU society; work opportunities and increase in revenues on local level; certification of some local products. The main 'Threats' to which the local chain is exposed are: the potential increase in imports due to the international competition causing decreases in revenues at farm level as a consequence of eliminating milk quota starting 2015, since in the majority of Romanian regions milk is produced at higher costs as compared with other EU regions (e.g. Hungary and Poland claim to have subsidies for collecting milk); the migration of young people to urban areas; competition with global retailers in the case of fresh cow milk as final product; EU legislation concerning food safety, animal welfare and environmental protection without harmonizing to the traditions and socio-economic characteristics of Romania and a successful information system at the level of the farmer; farm cross-compliance rules.

References

Akeroyd, J.R. and Page, J.N. (2011). Conservation of High Nature Value (HNV) grassland in a farmed landscape in Transylvania, Romania. Contributii Botanice 46: 57-71.

Chiciudean, D., Funar, S., Arion, F. and Chiciudean, G. (2013). Exploratory study regarding the consumer motivations for purchasing local products and their generally accepted definition. Bulletin UASVM Horticulture 70(2): 282-286.

Cziszter, L.T., Acatincăi, S., Sossidou, E., Szűcs, E., Gavojdian, D., Tripon, I., Erina, S. and Konrád, Sz. (2011). General knowledge of the Romanian farmers about the farm animal welfare. Lucrări Ştiinţifice 55: 88-93.

Department for Environment, Food and Rural Affairs (DEFRA) (2003). Local food – a snapshot of the sector. Report of the working group on local food. DEFRA. London, UK.

European Commission (2007). Attitudes of EU citizens towards animal welfare. Special Eurobarometer 270, Brussels. Belgium.

European Commission (2011). Common agricultural policy. Special Eurobarometer 368, Brussels. Belgium.

European Commission (2016). Communication from the Commission to the European Parliament and the Council on a community action plan on the protection and welfare of animals 2006-2010. Brussels, Belgium.

European Milk Market Observatory (MMO) (2016). EU prices of raw milk. Available at: http://tinyurl.com/h3mzlbn.

Eurostat (2012). Agricultural census in Romania. Available at: http://tinyurl.com/jt82ccy.

Feldmann, C. and Hamm, U. (2015). Consumers' perceptions and preferences for local food: a review. Food Quality and Preference 40(2015): 152-164.

Galli, F., Bartolini, F., Brunori, G., Colombo, L., Gava, O., Grando, S. and Marescotti, A. (2015). Sustainability assessment of food supply chains: an application to local and global bread in Italy. Agricultural and Food Economics 3(21): 1-17.

Grasseni, C. (2012). Re-inventing food: the ethics of developing local food. In: Carrier, J.G. and Luetchford, P.G. (ed.) Ethical consumption: social value and economic practice. Berghahn Books, Oxford, New York, NY, USA, pp. 198-216.

Grodea, M. (2015). Romania's self-sufficiency with regard to milk consumption in the context of domestic supply restructuring – trends and perspectives. Scientific Papers Series Management Economic Engineering in Agriculture and Rural Development 15(3): 153-158.

Jitea, I.M. and Arion, F.H. (2015). The role of agri-environment schemes in farm economic sustainability from high natural value Transylvanian areas. Environmental Engineering and Management Journal 14(4): 943-953.

Keenleyside, C., Beaufoy, G., Tucker, G. and Jones, G. (2014). High nature value farming throughout EU – 27 and its financial support under the CAP. Institute for European Environmental Policy, London, UK.

Kirwan, J., Maye, D., Bundhoo, D., Keech, D. and Brunori, G. (2014). GLAMUR WP2 – scoping / framing general comparative report on food chain performance (deliverable 2.3). Countryside and Community Research Institute, University of Gloucestershire, UK.

Kneafsey, M., Venn, L., Schmutz, U., Balázs, B., Trenchard, L., Eyden-Wood, T., Bos, E., Sutton, G. and Blackett, M. (2013). Short food supply chains and local food systems in the EU. A state of play of their socio-economic characteristics. Joint Research Centre.

Loos, J., Turtureanu, D.P., Von Wehrden, H., Hanspach, J., Dorresteijn, I., Frink, J.P. and Fischer, J. (2015). Plant diversity in a changing agricultural landscape mosaic in southern Transylvania (Romania). Agriculture, Ecosystems and Environment 199(1): 350-357.

Page, N., Popa, R., Gherghiceanu, C. and Balint, L., (2011). Linking high nature value grasslands to small-scale farmer incomes. Târnava Mare, Romania.

Paying and Intervention Agency for Agriculture (APIA). Ghidul prim – cumpărătorului de lapte de vacă. Available at: http://tinyurl.com/gp28ux2.

Popescu, S., Borda, C., Lazăr, E.A. and Hegedüs, C.I. (2008). Assessment of dairy cow welfare in farms from Transylvania. 44th Croatian & 4th International Symposium on Agriculture. Opatija, Croatia.

Redman, M. (2010). Securing public benefits from subsistence agriculture in Romania. Assessing the impact of Rural Development Policies. Deliverable D 8.1: Case Study Report. Centre for Ecological Engineering. Tartu, Estonia.

Romanian Ministry of Agriculture and Rural Development (MADR) (2013). National register of traditional products.

66. Organic food production in Norway: what is it good for?

H. Nilsen
Centre for the Study of the Sciences and the Humanities (SVT), The University of Bergen, Allégaten 34, 5020 Bergen, Norway; helene.nilsen@uib.no

Abstract

It is an explicit goal of the Norwegian government that 15% of food production and consumption should be organic by 2020. The aim of this paper is to question what this goal is good for in terms of what it is meant to accomplish. The 15% goal has two objectives; one is to ensure market diversity, the other is the development of organic farming techniques that will function as a 'spearhead' for an overall more sustainable food production. This paper discusses the preconditions for these purposes and points to some structural challenges connected to the generic marketing of organic food, as well as questioning the conditions for a knowledge exchange between the two modes of farming. It concludes that an extended public debate is needed in order to illuminate the conflicts of interests and values informing the political management of the food system.

Keywords: organic food, agricultural policy, Norwegian agriculture

Background

It is an explicit aim of the Norwegian government that 15% of the production and consumption of food should be organic by 2020. 'Organic' here refers to food produced in accordance with regulations based on those of the EU, certified and controlled by the Norwegian organization Debio. In February 2016, The Office of the Auditor General of Norway (OAG) finalized Document 3:7: 'Investigation of the Authorities' Efforts to Reach the Goals Concerning Organic Agriculture'. The report concluded that while organic food consumption (1.45%) had increased marginally, several indicators suggested that organic farmland (4.7%) and livestock (2.9%) are in decline. There is no statistical overview of the market share of imported organic food (Document 3:7 2015-2016). The main criticisms in the report were listed as follows: the regulations for organic farming are too complex and fragmented, and therefore unclear; the measures of stimulation for organic agriculture are not sufficient to increase the production; the generic marketing of organic food is not comprehensive enough; the authorities have not adequately facilitated for the use of organic food within its institutions (*ibid.*)

OAG provides a thorough account of what has been (or not been) done with regard to organic agriculture. It does not, however, touch upon the whys, i.e. the purpose of this goal. According to the white paper St. Meld. 9 (2010-2011), the 15% goal is meant to achieve two objectives: first, it is a postulate that organic agriculture will function as a spearhead for conventional farming in the sense that the methods of organic farming will feed back into the conventional food production system, creating an overall more sustainable agriculture. Second, the white paper emphasizes organic food as representing market diversity, offering consumers a wider range of choice (St. Meld. 9, 2010-2011). This paper seeks to contribute to the Norwegian discussion of organic food production by touching upon the twofold objective of increased organic production and consumption. The first part of the paper points to some structural challenges in the generic marketing of organic food and notes a conflation in the marketing of organic vs conventional produce. The second part questions the assumption that an increase in organic production and consumption will more or less automatically lead to a transformation of conventional agriculture and takes a critical look at the conditions for a knowledge exchange between organic and conventional farming.

EurSafe 2016 – DOI 10.3921/978-90-8686-834-6_66, © Wageningen Academic Publishers 2016

Creating a market for organic food

According to Storstad and Bjørkhaug (2003), the Norwegian food system holds great trust in the population. This trust, they observe, influence the interest in organic produce, and they attribute it to there having been no major food scandals in Norwegian agriculture, that Norwegian agriculture is based in several smaller scale farms, and that GMO is strictly forbidden (Storstad and Bjørkhaug, 2003: 152).[1] Norwegian agriculture is furthermore often depicted as 'close to organic', particularly with regard to sheep, due to the use of roughage, outdoor access and space (see e.g. Norwegian Scientific Committee for Food Safety, 2014b).

Similarly to Storstad and Bjørkhaug, Vittersø and Tangeland (2014), comparing Norwegians' attitudes to organic food in the period 2000-2013, see the low interest in organic food as a result of the high trust in Norwegian food system, but take on a different stance. They see this development as a result of political management: the political strategy of creating an organic labelling scheme and creating action plans where stakeholders are part of the information and marketing efforts reflect a change of the structure of political management from government to governance (Vittersø and Tangeland, 2014: 97). This means that an undue responsibility is put on the consumer to effect political change. The political strategy they depict is arguably reflected by minister of Agriculture and Food in his answer to the OAG in 2016, where he very clearly states that organic food production should be driven by consumer demand (Document 3:7 2015-2016: 22). Considering that only 30% of the total organic produce is actually sold as organic (Solemdal and Friis Pedersen, 2014: 22), this seems like a valid argument. How to reach the political goal of increasing this consumption, however, is a different matter.

OAG's report (2015-2016) finds that the generic marketing of organic food is insufficient, and sees a need for more knowledge based information in the public. Vittersø and Tangeland, on the other hand, found that consumers are generally satisfied with the availability of and information about organic food. However, the statement 'I see no benefits of organic food' was found to be the most influential perception negatively influencing purchase frequency of organic food, and this effect was significantly stronger in 2013 than in 2000 (Vittersø and Tangeland, 2014,:96). They call for an in-depth discourse analysis and a detection of the conflicts of interest that are at play in the public discourse of organic food. While a comprehensive discourse analysis would be outside of the scope of this paper, it is still worth taking a look at the structure of agricultural promotion in Norway, and how this structure can be said to play out in two marketing campaigns from 2013. In Norway, the mandate and obligation of generic marketing of organic food lie with MatMerk, an independent organization founded by the Ministry of Food and Agriculture in 2007. The board of MatMerk consists of actors from the largest supermarket chains to producers. MatMerk furthermore manages the KSL Quality System for agriculture and the label 'Nyt Norge' ('Enjoy Norway') and is thus responsible for the promotion of Norwegian agricultural food produce (http://tinyurl.com/z26qpas).

In 2011-2012, MatMerk conducted a 10 million NOK marketing campaign for organic food under the title FRI (literally FREE), co-sponsored by the government. Interviews performed by OAG note that conflicts of interest were very visible in the proceedings, and that there were difficulties agreeing upon the slogan 'Chose no artificial additives'. The campaign, intended to last for three years, was concluded after only one year (Document 3:7 2015-2016: 75). Framing the campaign videos was the text 'Indulge in a good feeling today' and: 'You feel a bit fresher. You feel more conscious. You feel free!' This text was posted on MatMerk's site for the promotion for organic food. It is remarkably free of any value

[1] Import of GMO is, on the other hand, not technically prohibited, but any product must be approved by The Norwegian Food Safety Authority, who has yet to allow any such products for import, according to the Government's homepage: http://tinyurl.com/h3mxdza (last updated: 21.08.2015).

judgments or reference to regulations, and situates organic food firmly in a luxury niche directed towards a consumer seeking to purchase a good feeling – in a sense a commodification of conscientiousness.

Around the same time, Nyt Norge, also owned by MatMerk, launched an promotional film featuring a little girl and her father with the text: 'When Daddy and I are at the store, we look for the Nyt Norge label, because then I know that the food is healthy and fresh, and that the farmer is nice to the cows, and also that it tastes good' (http://tinyurl.com/h5x84b6). The Nyt Norge brand also consistently refers to the KSL Quality System as 'strict, Norwegian rules', guaranteeing that the food is made from Norwegian raw materials and produced by companies localized in Norway (http://tinyurl.com/z5os7hg). This label has as such been very successful in promoting the idea of Norwegian agriculture as being of higher quality than in other countries and as such functions as an effective protectionist brand (Myskja *et al.*, 2015). Furthermore, it can be said to directly and indirectly incorporate the organic values of clean, safe, pure, local food with a high regard for animal welfare in the promotion of conventional food produce. The result can be seen as a conflation of the representation of organic and conventional agriculture, rendering organic food marketed as an alternative, unnecessary luxury product for a small consumer niche. Hence, the organization of the generic marketing of organic food may actually function counter-productively in the attempt of increasing the organic consumption rates from 1.45% to 15%. Furthermore, one may ask what it is that this increase is thought to achieve. I will now go on to discuss the role of organic food and farming in the public sphere, and its potential to function as a corrective to conventional farming.

Organic agriculture as a corrective to conventional farming

It is a much repeated political notion that organic agriculture should function as a corrective to conventional practice, a 'spearhead' of innovative and sustainable methods that will be incorporated in the general food system. On the other end of the scale, critics point to a certain degree of 'conversionalisation' where organic food production becomes commodified, in consequence decreasing the difference between organic and conventional food production (Flaten *et al.*, 2005, Guthman 2004). Flaten *et al.* in their 2005 paper, found that those farmers who transitioned from conventional to organic farming were, to a greater extent than their predecessors – the pioneers of organic farming in Norway – motivated by profitability and organic farming subsidies(Flaten *et al.*, 2005: 179). The authors also stress that the organic farmers, regardless of when they converted, are not a homogenous group (*ibid.*: 180). In this context, it is also worth mentioning that this also applies to conventional farmers, particularly in Norway where there are still a number of smaller-scale farmers who do not regard themselves as 'conventional', particularly not in comparison to many European countries (Storstad and Bjørkhaug, 2003: 154). While Norway has one of the world's most protected and subsidized agricultural sectors, due to topographical challenges (Myskja *et al.*, 2015) as well as a political goal of a decentralized population, the two last decades have seen a shift towards higher demands of cost-efficiency and fewer, larger farms both in conventional and organic agriculture (Document 3:7 2015-2016: 22; Storstad and Bjørkhaug, 2005: 152).

While the differences between organic and conventional production methods are not necessarily clear-cut, the debate in the public sphere has been quite polarized. For instance, in 2014 The Norwegian Scientific Committee for Food Safety published a report at the request of the Norwegian Food Safety Authority. The study – a review of regulations and studies on organic food and farming – was divided into three parts, dealing with respectively plant health, animal health and welfare and human health. The Committee concluded that there were small differences between conventional and organic farming and food products. With regard to plant health, it was concluded that the rate of crop loss was higher in organic than in conventional farming, and that there were 'small differences' in nutrient content, etc. Regarding animal health and welfare, it was concluded that the differences in terms of regulations were smaller in Norway compared to 'most other countries', and that there were only 'small and insignificant' differences in terms of antimicrobial resistant bacteria. Furthermore, diseases in conventional livestock were compared to the increased exposure to parasites and predators in organic farming. Finally, the Committee found that there was no conclusive evidence of clear positive or negative effects of organic food on human health compared to conventionally produced food (The Norwegian Scientific Committee for Food Safety, 2014a).

The report resulted in headlines such as 'Organic food is not healthier!' (http://tinyurl.com/jyqxyco). Oikos – the national movement of organic producers and consumers in Norway (http://www.oikos.no/english), strongly contested the report and claimed that it did not conclude in accordance with its own findings. In a letter to The Norwegian Food Safety Authority, they compared the conclusions from the Committee's summary with the results described in the report body, demonstrating how findings in favour of organic food and farming were under-communicated or ignored. In terms of plant health, they showed that over-all, the references in the report exhibited a higher level of nutrients and a lower level of toxins in organic produce (with the exception of patulin in apples). For animal welfare, Oikos argued that the statement on parasites and predators as a risk factor for outdoor space in organic farming was not at all documented. With regards to human health, Oikos found that the Committee's finding that 'Available data suggest that the combined exposure to multiple pesticide residues is not likely to result in increased human health risk' was not supported by evidence, and thus corresponded poorly to the conclusion that there was not enough research to determine whether organic food was healthier than conventional produce (Oikos, 2014).

This fundamental difference in interpretation illuminates the inherent value-ladenness of scientific research and demonstrates that 'facts' cannot be distinguished from 'values'. In the ensuing, polarized debate, however, 'organic' seemed to be conflated with 'idealistic' and 'conventional' with 'evidence-based'. This dichotomy was reinforced in the *pro et contra* correspondence facilitated by the weekly paper Morgenbladet in the wake of VKM's report. Morgenbladet chose organic 'enthusiast' food writer Andreas Viestad with no formal agricultural competence to oppose proclaimed 'sceptic', senior researcher Arne Grønlund, resulting in a conversation where the defence of conventional farming was represented with research, facts and figures, and the case for organic production was represented by observations based in taste, feelings and anecdotes (Morgenbladet 04.07.2014 – 24.07.2014). Grønlund is a researcher at Bioforsk: Norwegian Institute for Agricultural and Environmental Research, who conducts targeted research on 'multifunctional agriculture and rural development, plant sciences, environmental protection and natural resource management' with the objective of providing 'industries, governments and consumers with new knowledge, services and solutions within these scientific fields' (http://tinyurl.com/j5wmoqj). Notably, OAG's report comments that Bioforsk mentions organic agriculture in only 3 out of 8 meeting records, and then in mere administrative terms with regard to Bioforsk Økologisk (organic), a subdivision of the institute (Document 3:7 2015-2016: 92). This finding indicates that organic and conventional farming are not on speaking terms even within their designated research institutions. One might thus ask where and how a knowledge exchange should take place that could facilitate for organic farming techniques to feed back into conventional farming – even if the 15% goal is reached.

Concluding remarks

Agriculture is heterogeneous and complex, and the issue of organic food production is, as I have shown, highly value-laden and filled with dilemmas and conflicts of interests. I have in this paper tried to illuminate two challengers related to the two political objectives of the 15% goal. First, I have attempted to show that the organization of the generic promotion of organic food is problematic due to the double role of MatMerk. Comparing and contrasting two promotion campaigns, I have demonstrated how this double role may lead to a conflation of conventional Norwegian produce with organic produce, reinforcing the public perception of conventional farming as 'close to organic', thus rendering organics in a marginalized position. Second, I have questioned the assumption that an increase in organic production will automatically lead to a more sustainable conventional production. Pointing to the complexities of agriculture and the intrinsic value-ladenness of scientific representation, I have argued that the apparent *pro et contra* argumentation in the public sphere, both in terms of media representation and academic institutions signals poor conditions for the knowledge exchange required for organic farming to have an impact on the overall food system.

Considering the increase in demands of cost-efficiency in the agricultural sector, it is therefore questionable that conventional agriculture will shift towards a more sustainable production mode solely by means of an increased market share of organic produce. Without an active and ideologically based politics directed towards the food system, reaching the 15% goal seems more likely to contribute to the development of an alternative market running parallel to an intensified conventional agriculture. Organic farming would in that case be catering for a 'conscious consumer' rather than influencing the over-all sustainability of agriculture. This paper therefore concludes that an extended debate about what kind of role organic food should play in the Norwegian food system is called for.

References

Document 3:7 (2015-2016). The office of the auditor general's investigation of the authorities' efforts to reach the goals concerning organic agriculture.

Flaten, O., Lien, G., Ebbesvik, M., Koesling, S. and Valle, P.S. (2005). Do the new organic producers differ from the 'old guard'? Empirical results from Norwegian dairy farming. Renewable Agriculture and Food Systems 21(3): 174-182.

Genmodifisert mat (GMO). Available at: http://tinyurl.com/h3mxdza.

Guthman, J. (2004). The trouble with 'organic lite' in California: a rejoinder to the 'conventionalisation' debate. Sociologia Ruralis 44: 301-316.

MatMerk. Available at: http://tinyurl.com/z26qpas.

Myskja, B.K., Carson, S.G. and Ursin, L.Ø. (2015) Fair local and environmentally sound? Corporate social responsibility in Norwegian food production. In: Dumitras, D.E., Jitea, I.M. and Aerts, S. (eds.) Know your food: food ethics and innovation. Wageningen Academic Publishers, Wageningen, the Netherlands.

Norwegian Scientific Committee for Food Safety (2014a). Comparison of organic and conventional food and food production. Overall summary; impact on plant health, animal health and welfare, and human health.

Norwegian Scientific Committee for Food Safety (2014b). Comparison of organic and conventional food and food production, Part II. Animal health and welfare in Norway.

Oikos (2014). Mattilsynets behandling av rapport fra vitenskapskomitéen for mattrygghet om forskjeller på økologisk og konvensjonell mat og produksjon. Available at: http://tinyurl.com/zm4hmfh.

Solemdal, L. and Friis Pedersen, S. (2014). Økologisk mat i de nordiske landene – tilgang på råvarer og faktorer som påvirker omsetning av økologisk mat. Bioforsk Report 9(139).

St. Meld. 9 (2010-2011). Landbruks- og matpolitikken – velkommen til bords (the food – and agricultural policy – welcome to the table). Ministry of Food and Agriculture.

Storstad, O. and Bjørkhaug, H. (2003). Foundations of production and consumption of organic food in Norway. Common attitudes among farmers and consumers? Agriculture and Human Values 20: 151-163.

Vittersø, G. and Tangeland, T. (2014). The role of consumers in transitions towards a sustainable food consumption. The case of organic food in Norway. Journal of Cleaner Production 92: 91-99.

67. An outline of the foodscape of Lisbon: knowledge from the past for a better future

M. Sanchez Salvador
CICS.NOVA, Av. de Berna, Edifício FCSH-ID, room 3.14, 1069 Lisbon, Portugal; marianasanchezsalvador@gmail.com

Abstract

As world population increases, and is becoming progressively more urbanised, cities are turning into the main human habitat. Dense megacities sprawl through the land and contrast with massive scopes of industrialised food production, both contributing to landscape transformation, destruction of ecosystems, climate change, and consumption of natural resources. While urban and rural become more detached, both spatially and functionally, social inequity and hunger aggravate worldwide. Bearing this in mind, how can we conceive more sustainable cities for our future? Food, surprisingly, may have an important role. For millennia, food production, distribution, retailing, cookery and consumption have shaped landscapes, cities, and houses. Food conveys regulations and beliefs, economic and cultural premises, technological and social resources, and imprints them into the territory's physical space by transforming natural resources, organising our built environment. It is this complex and holistic character that makes food a powerful tool for shaping future cities, one that ripples in several different areas of society. For this, however, it is imperative to understand how we got to our present situation, which transformations our urban food systems underwent and how these were translated into the foodscape of cities. The research outlined in this paper focuses on the case of Lisbon and the evolution of its foodscape; specifically, how the activities of food production, distribution and retailing have shaped its landscape and urban form, during the last century – from late 19th century until today. By describing the key concepts and methodological approach adopted – one that privileges historical cartography as the primary source of information, complemented by bibliography, photography and statistical data – this presentation will provide a preliminary contribution to this knowledge. With this ongoing research, we shall understand how Lisbon, like other cities worldwide, has gone from a local food system model – with a significant part of its food supply coming from a close productive hinterland – to one defined by a global network of distribution, that weakens cities' ability for food sovereignty, and how this transformation was imprinted into its urban form. From this analysis, we may draw some potentialities from the past to improve urban food security, by appending food systems as a tool for urban planning and design – in preserving certain productive land uses, or rethinking food retailing networks for a greater proximity between consumers and producers – contributing to more sustainable and resilient cities.

Keywords: urban food systems, foodscape, sustainable city, Lisbon, Portugal

Contextualization

For thousands of years, humans lived primarily in rural environments. Even with the Industrial Revolution and the rise of big cities – quintessential political, economic and creative centres – urban environment still did not represent the predominant human way of living. That reality only changed in 2009 (UNPD, 2010) when not only cities became the most common human habitat worldwide, but also 75% of the staggering 9 billion people thought to exist by 2050 were expected to be urban – in just 40 years.

This new and extreme reality is not innocuous: towns are turning into cities, and cities into 10-million-people-megacities. Aided by railways and highways, urban fabric is unrolling over the territory – often the fertile lands that enabled the city to exist in the first place – clearing woods, transforming landscapes,

EurSafe 2016 – DOI 10.3921/978-90-8686-834-6_67, © Wageningen Academic Publishers 2016

destroying ecosystems, and altering climate. Potentiated by urban living and population increase, our lifestyles and food habits are changing into highly consumptive and costly ones for the planet, regarding the use of energy, water and other resources. At this rate, by 2050 we will need two and half Earths to fulfil our demands (Sande, 2012). Simultaneously, cities are thought to be the most sustainable model for housing our future world population (Kunzig, 2012): there is simply not enough space for all humans to live in a different environment, and cities allow the rationalisation of infrastructures and some resources. So, one of the great questions is: How can we conceive more sustainable and resilient cities to live in? Cities that address these environmental challenges and combine them with social, economic, cultural and health ones?

When thinking about the future of cities, one often forgets the role of food in them. This may not be surprising. François Ascher began his 'The urban new principles (2010) by defining cities as clusters of people who do not produce their means of subsistence themselves, and emphasised a technical, social and spatial division between cities and food production from the outset. When it comes to urban planning, food often is not even mentioned, it's not part of urban green spaces, we don't see it arrive or go away, and most days there are no markets at sight. Despite being confronted with it three to five times a day – and despite all the trendy restaurants, celebrity chefs and TV-shows, hip grow-yourself-movements and so on – most of us living in the city have no idea where our food comes from, how it is produced, or by whom, how it gets to us or how it is prepared.

Of course, there has always been a dualism between city and countryside, even associated with the one opposing culture and nature. What is new, however, is the increasing physical and mental distance between city and its productive spaces, the distance between farm and fork. And this phenomenon, as a general practice in the history of mankind, is quite recent. Once chained to their local hinterlands and their limitations, with the fast connections to remote productive locations and ports made possible by the railways – combined with refrigeration and latter potentiated by airplanes – urban food systems changed forever: provided the money, one can now eat anything from anywhere in the world, at any time (Steel, 2013: 32). Urban 'foodsheds' – the catchment's area of food production required to feed an urban centre (Hedden, 1929) – became global and physical distance irrelevant. Notions like proximity or distance are more readily translated by time and cost than by metric units (Steel, 2013: 39). Food is produced more intensively, extensively and further away from cities, with billions of tons of food being transported worldwide, at any given moment, travelling an average of 2,500 to 4,000 kilometres before reaching its final destination (Vivas, 2011: 188). As urban and rural become more detached – both spatially and functionally – cities rely on this global distribution network to supply their inhabitants and guarantee food security, an extremely fragile link, since not only it relies on cheap fuel and other externalities to operate, but also it is progressively owned by fewer private companies, dominating world food distribution and access (Vivas, 2011: 186).

Although currently 'excluded' as a planning tool, food may play a game-changing role in the design of sustainable cities. It is often said that cities are physical materialisations of immaterial forces, such as politics, economy, knowledge, or culture. Food – or urban food systems, to be more accurate – also operates within all these realms, and roots them into the ground, shaping landscapes, cities and buildings, into what we may designate as foodscapes. Food can be a powerful tool to transform our future cities, if it is understood holistically as a system. But in order to act upon this reality, we must first understand its history, how it evolved, how the ancient interdependence between cities and food production was disrupted, and how we got to this precarious situation. That is the purpose of this on-going research, focusing specifically on the case of Lisbon, simultaneously unique and representative of this widespread phenomenon of detachment between cities and food supply systems.

Studying Foodscapes: concepts and methodology

The approach of studying the history of a place – specifically a city – through its physical materialisation, the form of its matter, has been widely defended by the different Urban Morphology Schools, throughout the 20th century (Moudon, 1997). Cities are by these researchers beheld as palimpsests, resulting of the juxtaposition of different layers over time – as also described by UNESCO (2011) in its 'Historic Urban Landscape' recommendation – that at each moment carry all the previous sets of conditions and history, materialised into their physical form. Therefore, the study of this resulting configuration is defended as a legitimate approach to studying its evolution (Whitehand, 2007). What is proposed here is the interweaving of this methodological approach of urban morphology with the composing elements of food systems, to characterise what may be called a 'Historic Urban Foodscape'.

'Urban food systems' comprise all the different stages and activities required to supply a certain urban area, from food production, distribution and retailing, to cookery, consumption and waste, to which are added, more recently and at times, the stages of processing and packaging. They include not only the activities, but also the biological processes and physical infrastructures inherent, and the human resources that provide labour, research and education. They operate within, are influenced by and interact with social, political, economic and environmental contexts (FAO, 2011: 15). We can translate each of these sets of factors into a specific space – landscape, roads, cities, or buildings – ranging from the regional (or global) scale to the domestic one. A foodscape is, therefore, the materialisation of infrastructures and immaterial factors of the food system into space. This research will focus on food production, distribution and retailing at the urban scale, especially of fresh products – fruits and vegetables, meat and milk – exempted from the intermediate stages of processing and packaging.

To describe the evolution of Lisbon's Foodscape, the layers corresponding to the natural physical support of the city and its urban form will be characterised and mapped, as well as the plots, roads and buildings related to food production, distribution routes and retail, like markets. This methodology will be applied to three different moments of Lisbon's history – late 19th century, mid-20th century and early 21st century. With the juxtaposition of these layers, we will be able to read the city's foodscape at each moment, apprehend its global evolution, and compare transversally the separate evolution of each stage, allowing different combinations, analysis and a wider range of possible conclusions.

The primary source of information for this research is historical cartography. If the various factors and forces present in the city are embodied in its physical form, and this is in turn registered – objectively and rigorously – in cartography, then, cartography becomes a true synthesis of urban history at any given time. The role of cartography as primary source of information was defended by the Anglo-Saxon or Conzenian urban morphology school (Whitehand, 2007). These cartographic databases will be complemented and infused by other data – including bibliography, photography and statistics – attributing to these resulting maps a dynamic character, a possibility enhanced by technological resources like GIS (Moudon, 1997), while making spatially visible data that traditionally didn't have this ability.

This methodology allows, therefore, not only to collect and systematise information, but also to make it readable and operational. Also, it makes it possible to characterise the evolution of the Foodscape of Lisbon and gradually give it more detail, depending on the data collected, being the layers constantly updated. So, departing from the methodology described above, a first approach to the evolution of Lisbon's foodscape will now be drawn.

The evolution of Lisbon's Foodscape: a preview

This primary approach to the evolution of Lisbon's Foodscape will mainly confront data from the early 20th century with contemporary data, highlighting the transformations occurred during the intermediate period.

The early-twentieth-century Lisbon, synthesised in the cartographical survey coordinated by Silva Pinto (1904-1911), shows us a city with some preindustrial features still, characterised by the proximity and interdependence between urban form and urban food supply, operating mostly at a local level. The city's urban area was significantly smaller, being concentrated inside the perimeter defined by the railway and a set of roads. However, this area aggregated the majority of population living in Lisbon and its surroundings, about two thirds (435,359 of about 690,893 people in total, 63%) of the population residing in an area correspondent to the current Metropolitan Area (MEMO, 2014). Due to its natural vocation associated with navigations, up until this moment the city had been spreading along the riverfront. But the introduction of the railway and electric trams offered the opportunity of expansion into the inland. Thus, in this cartography are already clear, starting from the Pombaline centre, the lines of expansion of Avenida da Liberdade and Avenida Almirante Reis, following the natural valley lines towards North.

Regarding the agricultural production spaces, there was a large set of farms that defined a consistent productive ring around this built urban centre, sprawling all the way to the Saloia region, up North. We can observe a predominance of orchards, vineyards and olive groves on the outskirts to the East, while at the West prevailed hewn lands (Marat-Mendes *et al.*, 2014: 73). Built areas were interweaved with food production spaces, with vegetable gardens, cereal crops, olive groves, vineyards and pastures within the city's boundaries, reaching these agricultural areas 'almost to the heart of the city, and still remembered in place names' (Brito, 1976: 62-64). A significant part of Lisbon's food distribution in the pre-industrial period was made by sea or river, which is still evoked in the names of places – particularly at their arrival points – like Cais do Trigo ('Wheat Pier') or Campo das Cebolas (Onions' Field).

The food retailing spaces were practically limited to markets and a few fairs. The Silva Pinto cartography already shows a set of these public buildings, inaugurated during the previous decades, such as the Alcântara Market (1905), located next to the train station 'Alcântara-Terra' (in what is now Avenida de Ceuta), the July 24 Market or 'Nova Ribeira' (1882) at the 'Cais do Sodré', that complemented and later replaced the old 'Ribeira' Market, located next to 'Casa dos Bicos', the Market that existed in the current 'Praça da Figueira' (1885), the 'São Bento' Market (1881) and a few others that complemented this network. For a long time, also, fish, fruit and vegetable were informally traded, by the river.

As the urban pressure intensified, farmlands were reduced, often persisting only along the lines corresponding to the richest soils, such as alluvial valleys (Brito, 1976: 64). The city could not guarantee its food sovereignty, and the produce of its surrounding lands became increasingly reinforced by Estremadura and Algarve productions, competitive because of their similar prices, a situation potentiated by railways and, later, by road transport. Progressively, access would no longer depend on distance. Therefore, transport development played a key role in the transformation of local foodsheds into global ones, by obliterating geographical constraints. Hedden had already referred this when he first coined the term in 1929. When the author adapted the concept of watershed to foodshed, he highlighted an important distinction: while the first one operates according to topographical features, in the second economic conditions prevail. So, regarding food supply, transportation allowed people to emancipate from the territory and its limitations. The price criteria became hegemonic in defining food flows, a situation aggravated after the Second World War, together with technological developments such as the ones in food conservation and packaging. In Lisbon, the impact of these changes became more visible in the city from the 1960s and 70s, a period that coincided with major migratory flows from the countryside to the city (Brito, 1976: 65).

Today, those that had previously been food production spaces became urban and semi-urban areas, being the municipality densely occupied by construction. Within the non-built areas, only a small part corresponds to a food productive use – urban gardens – often with an informal or leisure character, but not systematic. Food products come from everywhere, by all possible means of transport, into a complex network of retailing spaces, in which supermarkets and hypermarkets stand out, being sprinkled throughout the city. The network of Municipal Markets, expanded during the 1940s and 50s to follow the rise of new neighbourhoods, lost its importance as local food retail spaces. They are, however, either being currently recovered by specific publics, although still a minority, or assuming different 'more hip' uses.

Conclusions

Every answer depends on how you pose the question. When collecting the knowledge that will allow us to describe and characterise the evolution of Lisbon's Foodscape, there are several questions whose answers might benefit from this research approach and contribute to the conception of more sustainable urban models: How has the number and area of production spaces in Lisbon evolved in the last century? The ones that still exist, can they still play a role in supplying the city? Can we maximise that role? From where did food historically come to the city? Can we design a more sustainable food distribution network, which relies less on fuel, maybe taking advantage from public transports and other existing infrastructures? Can we work with shorter routes, from production to consumption? Can we do something to consume more local and sustainable products, empowering the local communities? Markets, supermarkets, hypermarkets: how did they evolve, their size and location? What impact do they have in us as consumers, and food systems transformative agents? Can we physically change the world we live in by making different food choices?

The future of Lisbon's foodscape is uncertain, and it may take on different shapes, depending on the decisions made for both its urban development and food system. Making those decisions requires knowledge about the current situation and its history, its evolution: knowledge from the past to shape a more sustainable and resilient future for the city. It is believed that the current detachment between urban dwellers and food related issues strongly contributes to the existing problematic relationship between city and food supply, a model with severe cultural, social and environmental consequences. Therefore, knowledge and awareness about the transformations operated in food systems in the last century – perhaps the deepest in the entire existence of humanity – and the ones occurred in Lisbon's Foodscape, specifically, may become the first step in changing the culture and mentality about these issues, a potentially decisive change in the way we view the city, and how we will design its future.

References

Ascher, F. (2010). Novos princípios do urbanismo; novos compromissos urbanos: um léxico. Livros Horizonte, Lisbon, Portugal.

Brito, R.S. (1976). Lisboa: esboço geográfico. Separata do Boletim Cultural da Junta Distrital de Lisboa, III series, n. 82. Lisbon, Portugal.

Food and Agriculture Organization of the United Nations (FAO) (2011). Food, agriculture and cities: challenges of food and nutrition security, agriculture and ecosystem management in an urbanizing world. FAO, Rome, Italy.

Hedden, W.P. (1929). How great cities are fed. D.C. Health & Company, New York, USA.

Kunzig, R. (2012). A cidade como solução. National Geographic, January 2012, pp. 58-79.

Marat-Mendes, T., Mourão, J., D'Almeida, P., Niza, S. and Ferreira, D. (2013). Água dá, água leva. What the water gives, the water takes. Cidades, Comunidades e Território 28: 56-87.

MEMO (2014). MEMO – evolution of the Lisbon metropolitan area metabolism. Lessons towards a sustainable urban future: results (selected). Available at: http://tinyurl.com/gtkrr3p.

Moudon, A.V. (1997). Urban morphology as an emerging interdisciplinary field. Urban Morphology 1: 3-10.

Sande, B. (2012). Food for the city: a future for the metropolis. NAi Publishers/Stroom, Den Haag, the Netherlands.

Steel, C. (2013). Hungry city: how food shapes our lives. Vintage Books, London, UK.

UNESCO (2011). Recommendation on the historic urban landscape. Available at: http://tinyurl.com/j2sw7z9.

UNPD (2010). World population prospects: 2009 revision. Available at: http://tinyurl.com/haqsmnl.

Vivas, E. (2011). A distribuição moderna: a invasão dos supermercados. In: Vivas, E. and Montagut, X. (eds.) Do campo para a mesa: os circuitos de produção e distribuição de alimentos. Sururu Produções Culturais, Carcavelos, Portugal, pp. 183-199.

Whitehand, J.W.R. (2007). Conzenian urban morphology and urban landscapes. Proceedings of the 6th International Space Syntax Symposium. Istanbul, pp. 1-9.

68. Animal slaughtering in Romanian rural households: between tradition and legal framework

C.B. Pocol[1], D.E. Dumitraş[1] and C. Moldovan Teselios[2]*
[1]University of Agricultural Sciences and Veterinary Medicine Cluj-Napoca, Department of Economic Sciences, 3-5 Manastur St., 400372 Cluj Napoca, Romania; [2]Metro Media Transilvania, 174 Constantin Brancusi Street, 400462 Cluj-Napoca, Romania; ddumitras@usamvcluj.ro

Abstract

Food production and consumption in rural areas of ex-communist European countries were affected by the opposition between understanding modern and traditional practices (Kiliánová and Kanovsky, 2005). The modern practice can be unfriendly to the tradition, but this also depends on what we understand by modern: technical progress? new legislation? ethical actions? The paper focuses on the aspects of animal slaughter (pig and lamb) for Christmas and Easter' celebrations, in Romanian rural areas. The European legislation adopted by Romania supports the preservation of this tradition, but under certain conditions. In the case of pigs and lambs, the owner is allowed to slaughter animals in the household for personal consumption, by respecting the common minimum standards for the protection of animals: to be stunned in advance (Council Directive 93/119/EC). The research questions were: is there legislative pressure on tradition? Can the perpetuation of slaughtering practices be explained by socio-demographic characteristics of rural households? 1,190 households in rural Transylvania were surveyed in winter-spring 2015. Results show high percentages of households partaking in traditional practices: 60% in the case of lambs and 88% in the case of pigs. There is a statistical correlation between slaughtering practices and household revenues (bigger incidence in the case of high revenues). However, the level of education and possession of durable goods do not influence the decision regarding slaughtering. Even if traditional practices related to pig and lamb slaughtering are frequent and part of the local identity, education should not be neglected as a predictor of these actions. More initiatives should be taken to inform the rural population about the meaning of legislation through public policies.

Keywords: traditional practices, legislation, animal slaughtering, rural space

Rurality in Romania

Romania is a country with a high level of rurality. 87.1% of the territory is represented by rural areas (Pocol, 2013). According to data provided by the National Statistics Institute, Romania's rural population registered in 2015 a total of 9,171,608 inhabitants, representing a share of 46.15% of the total population (NSI, 2016). Compared to the 60s, when the percentage of rural population was of 67.9%, by the '90s, it reached 45.7%, due to rural exodus. This percentage has remained relatively constant to this day. According to the National Institute of Statistics (2016), agriculture, forestry and fishing are the main occupations of a significant share of the rural population of Romania (41.23%). Compared with the European average, this percentage is about 10 times higher. Also, at the European level, Romania is the country with the highest number of farms (3.6297 million), with an average area of 3.6 ha/farm (Eurostat, 2016). Three quarters of these farms have less than 2 hectares, which demonstrates their character of subsistence and semi-subsistence. Rural farmhouses typically have a vegetable garden, a small orchard, shelters for cattle, sheep, pigs, chickens and barns for hay and grain (Page and Popa, 2013). Their role is to produce food for the consumption of the household and, sometimes, for the extended family or for commercialization at local markets, generally through unofficial channels. Holding on to small subsistence farms is extremely important for the vitality of rural areas, but it remains a difficult goal to achieve. Szabó (2013) talks also about the marginalization of these farms due to neglectful

EurSafe 2016 – DOI 10.3921/978-90-8686-834-6_68, © Wageningen Academic Publishers 2016

agricultural policies – national and European Directives, predominantly created by outsiders, which try to replicate the rural conditions. The rural population must now respond to new challenges regarding traditional practices, which are subject to various pressures from the external environment: economic, social, legislative and political pressures.

Traditions related to the growth and slaughter of animals for personal consumption

The peasant traditions and significant life events are largely associated with agriculture, which has an ontological significance for them (Galani-Moutafi, 2013). Livestock in Romanian rural farms is a traditional activity, which covers the food consumption needs of the local population. In terms of pork production, Stanciu (2014) shows that, presently, it is based mainly on extensive growth achieved in individual subsistence households, for their own consumption. According to the 2010 Agricultural Census, approximately 80% of farms raise between 1-2 pigs on average (Luca *et al.*, 2012). In terms of internal lamb production, Petroman *et al.* (2014) talk about an annual supply output of about 3 million heads in Romania. Pig or lamb slaughtering in the household on religious holidays – Christmas and Easter – is a Christian tradition, which involves performing several rituals. Regarding pig slaughter, this usually occurs around Christmas, on the day named 'Ignat', on the 20th of December, on Christmas Eve, the 24th of December, or around 1 January, on St. Basil (Schileru, 2015). According to specialists, pig slaughter in Romanian rural areas has a magical function, based on compliance with certain canons, places, consecrated formulas, the aim being to please the gods and the spirits so that they bestow health and luck (Schileru, 2015). Pig slaughter tradition does not only belong to the Romanian people. Kiliánová and Kanovsky (2005) talk about pig slaughtering in the household when organizing a wedding in the 70s and 80s in Poland, presenting it as a traditional activity specific to the Polish people. Just like in Poland, in Romania, traditional pig slaughtering persisted during the communist period (Stan, 2000). Regarding lamb slaughter in Romanian rural areas, it occurs for seasonal consumption, in the months of March, April or May before Easter, when the animal reaches an average weight of 10-12 kg (Petroman *et al.*, 2014). The slaughter of lambs is a ritual with archaic meanings and connotations that go beyond the simple consumption of the meat. The conflict between traditions, religion, national/European legislation is not new. The polemics about slaughtering without stunning in Romania are due to the tradition and rituals of religious holidays. Maintaining and respecting rural traditions is often conditioned by a compromise between dominant interests and forces (Hoggart *et al*, 2014). Political changes can alter the perception of traditional and modern practices, creating a real tension between them (Kiliánová and Kanovsky, 2005). Modern means something that belongs to our times, something new and recent, something that corresponds to the current state of progress and that is different from tradition. The 'modern' can be at odds with tradition, but it depends on how 'modern' is defined: technical progress? new legislation? ethical actions? The present research refers to 'tradition' and 'modernity' as key concepts of rural studies. One of the main functions of rural space is the cultural one, which includes values such as: traditions, preferences, motives, thoughts and beliefs of individual actors (Madsen and Adriansen, 2004). The traditions represent a set of practices governed by accepted rules and rituals, which implies repetition and continuity in rural communities, being real and not invented (Hobsbawm and Ranger, 2012). This is an 'idyllic' representation of rural that is based on the preservation of traditional values. The opposite vision is the 'productivist' representation, that emphases the need of modernization for the development of rural areas. The modernization is related to agricultural and food production practices (Da Silva *et al.*, 2016).

A short history of the Romanian legislation on the household slaughter of animals

A report on 'religious slaughter in the EU member, candidate and associated countries' (Ferrari and Bottoni, 2010) talks about the Second World War period, when in Romania, as in other European countries, animal slaughter was prohibited for religious purposes, except for birds. During the communist period, Animal Health Act no. 60/1974 did not specify anything regarding the subject. After the fall of the communism, the first bioethical concerns in Romania emerged after 2000, thanks to the efforts made by NGOs, and because of the need to adapt to European standards. The academic society showed great interest in the subject but the public interest was low (Barbulescu and Andreescu, 2010). The first legislative act adopted by Romania in this period was Order no. 425 of 19 September 2002 concerning provisions of veterinary health for the protection of animals during slaughter and killing. By implementing this order, Romania aimed to synchronize national legislation with the European one (Council Directive 93/119/EC), the order being poorly known at the time by small breeders. Two other legislative acts were approved in the next period: Law no. 205/2004 on animal protection and Order no. 180 of 11 August 2006 of the National Authority for Veterinary Health and Food Safety concerning the regulation for veterinary health on animal protection during slaughter and killing. Public reactions to these acts came later (2007). Critical voices claimed that this legislation destroyed traditions and that pig slaughter on Christmas and lamb slaughter on Easter being a part of national identity (Ferrari and Bottoni, 2010). Another counter-argument with respect to this legislation argued that animal slaughter during the holidays is a social act which involves the whole family, relatives, neighbours, representing a moment of joy and celebration (Barbulescu and Andreescu, 2010). During negotiations for EU accession, Romania's representatives tried to obtain exemptions from EU law on animal protection at the time regarding animal slaughter or killing, requesting that animals could be slaughtered without previous stunning, in order to preserve and respect the traditions of Easter and Christmas. On the other hand, there were also positive reactions coming from Romanian civil society, the academic society and from specialists in bioethics, who believed that the tradition of slaughter could be continued, but in compliance with more human methods (Barbulescu and Andreescu, 2010). Currently, in Romania, the enactment in force (Order no. 180/2006 on animal protection during slaughter and killing) provides that NSVFSA (National Sanitary Veterinary and Food Safety Authority) may grant exemption for slaughtering outside slaughterhouses for pigs and lambs by their owners only for their own consumption, on the condition that all animals will be stunned before the slaughter.

Who respects the tradition of slaughtering pig on Christmas holidays and lamb on Easter in Transylvania?

A total of 1,190 households from historic regions of Transylvania have been investigated during February-March 2015 on agricultural production practices and food consumption behavior. The research was conducted exclusively in rural areas. The sample had a balanced structure in terms of gender (49.9% men and 50.1% women). In terms of age groups, there were four categories: between 18-30 years, between 31-50 years, between 51-70 years and above 70 years (72.9% of respondents were between 31 and 70 years old). Regarding education, 79.5% of respondents had at least secondary school, secondary or post-secondary, while only 17.9% higher education. In terms of ethnicity, 79% of respondents were Romanian, Hungarian 15.3%, the rest belonging to other ethnicities. A crucial question related to production and consumption practices in the household was the following: 'In your household, do you usually sacrifice lambs on Easter and pigs around the Christmas holidays?' 60% of respondents answered affirmative for lamb and 88% for pork. The percentages indicate a high incidence of these practices in rural areas of Transylvania. The analysis continued with the identification of socio-demographic factors influencing most these practices. Four socio-demographic characteristics were initially tested in relation with pigs and lambs' slaughtering: income, religion, ethnicity and education. The main socio-demographic factors identified were: level of education and income level. For Romanian rural space, education and income are the most important elements which describe the quality of life, also considered as main factors in the construction and validation of 'Index of development of the municipality' by Sandu *et al.* (2009). The results show that the relationship between slaughtering practice and education is not statistically significant. In terms of income, there is a positively correlated linearity with incidence of slaughter practices: it is higher where income is higher. Household endowment with durable goods (e.g. computer, mobile phone, refrigerator, microwave, washing machine, car) was analysed as is an indicator of modernity (Stanciu, 2004), which is an unobservable phenomenon. Associations were noticed between the incidence of slaughtering practices and the possession of such items. From this point of view, analyses show that modernity has not changed very much traditional practices. Households that are equipped with modern goods can be found among those households where slaughtering practices continue. This clearly shows that, in terms of technical progress (household endowment with durable goods), the modern tradition can be friendly with tradition and does not prevent its perpetuation. Household endowment with a refrigerator does not prevent continued use of traditional methods of preserving meat by smoking, frying in lard or drying. It is only those households without any durable goods mentioned above, that the incidence of these traditional practices is reduced. Basically, in this case, the primary determinant is poverty, poverty which does not provide the opportunity to have animals for sacrifice.

Is there a legislative pressure on continuing tradition? Opinions of specialists in agriculture and food sectors

To better respond to the research purpose, it was considered necessary to conduct also a qualitative research based on a debate during a workshop where participants were specialists in the field of food processing, from abroad. They were represented by graduate students with background in food identity, both from European and non-European countries (France, Spain, Costa Rica, Honduras, Ukraine, Tajikistan, Russia, the Philippines, Brazil, Ghana, South Africa, Mexico, Colombia, Brazil, Bangladesh, Taiwan, USA, Bangladesh, Syria). The subjective participation of the specialists provides the research with a deeper understanding of the concept being discussed, generating ideas and hypotheses for future research. The method implied discussions on the following question: Is there a legislative pressure on continuing the tradition of lamb slaughter on Easter and pig slaughter on Christmas in Romania? Participants expressed their views by pointing out the main arguments in favour of tradition/legislation.

The main arguments in favor of continuing the tradition of slaughter without prior stunning were the following: 'Natural, traceable source of meat for rural population, product of local biodiversity', 'Better overall respect for animal as compared to those raised in intensive agriculture', 'Raised throughout year, attentive and individual care for animal up until slaughter', 'Conservation and use of all of the meat of animal. 3-4 months of food: no waste', 'Gesture of respect for rural traditions. Encourages solidarity among farmers and rural inhabitants, encouraging their sense of well being', 'Important to strengthen rural agriculture and to counteract rural exodus trends', 'Potential consequences of banning this traditional ritual', 'Discontent and lack of cooperation among rural Romanians; forced loss of identity and culture', 'Encroaching upon religious freedoms, the tradition being based on biblical practices', 'Disrupting the infrastructure/current means by which rural Romanians obtain their pork for 3-4 months out of the year: these sacrificed animals can feed families for months. If banned, the rural Romanians would need to source their pork from elsewhere, probably creating a dependence on the agro-food industry, the cheapest meat from which is imported. This could also cause even lower rates of self-sufficiency within the Romanian pork sector'.

Arguments for applying legislation were the following: 'Modernization in pork slaughtering practices in Romania would imply the application of technologies that require standards in terms of time, quantity, technique and hygiene', 'Animal welfare- animal ethics. It is not only about man? Practice or tradition, but about the considerations made over another living being. Under this awareness, animals can be respected by what they bring back to man: nutrition as well as identity', 'The science of meat for consumption; after sacrifice and under stress conditions, an animal develops different chemical conditions affecting the quality of its meat', 'Sensory criteria of people consuming this type of meat remains stunted; there are scientifically proven differences in the sensory characteristics of meat from animals undergoing traditional slaughtering. Compared to fast slaughtering and immediate bleeding, the animal is quickly rendered unconscious, therefore allowing blood to keep its flow, facilitating bleeding, and traducing in the tenderness of the final product', 'A law of such nature will allow Romania to consolidate as a country willing to integrate to the EU framework on slaughtering practices', 'In reality, a law of this nature will also allow slaughtering traditions to be placed in a more valuable position. People should prove that they understand which are ancient habits that provide identity to Romania. On the other side, Romanians will show an opportunity to provide an understanding of animal respect, by adapting the law in their system'. Participants' opinions lead to the conclusion that tradition may harmonize with legislation when it is correctly understood and it is not perceived as a negative factor but a progressive one for animal welfare and consumers' health as well.

Conclusions

Traditional practices relating to lamb slaughtering on Easter and pig slaughter on Christmas in Romania are part of the national identity. Rural households should not abandon these practices, growth and slaughter of animals for their own consumption being beneficial both for food security and for maintaining the vitality of these areas. The tradition can be continued also in compliance with current legislation, allowing animal slaughtering in the household, for personal use, provided prior stunning. In terms of this legislation, there is no clear data, no official data, the investigation of these issues being a 'delicate' issue, and the truth being difficult to obtain. Educating the rural population in order to understand the law correctly is still needed. Otherwise, there is a risk that small farms will consider the alignment to European norms and legislation as a threat to the preservation of these traditions. On the contrary, traditions must be perpetuated, their contribution being extremely important for the development of rural tourism in Romania.

References

Barbulescu, I. and Andreescu, G. (2010). Animal stunning, the EU and the Romanian lobby. Revista Română de Bioetică 8(1): 190-199.

European Commission (1993). Council Directive 93/119/EC of 22 December 1993 on the protection of animals at the time of slaughter or killing. Brussels, Belgium.

Da Silva, D.S., Figueiredo, E., Eusébio, C. and Carneiro, M.J. (2016). The countryside is worth a thousand words – Portuguese representations on rural areas. Journal of Rural Studies 44: 77-88.

Sandu, D., Voineagu, V. and Panduru, F. (2009). Dezvoltarea comunelor din România, Available at: http://tinyurl.com/jcjsfca.

Eurostat (2016). Agriculture, forestry and fishery statistics. Available at: http://tinyurl.com/gm8hrwp.

Ferrari, S. and Bottoni, R. (2010). Report on legislation regarding religious slaughter in the EU member. Candidate and Associated Countries, pp. 148-152.

Galani-Moutafi, V. (2013). Rural space (re) produced – practices, performances and visions: a case study from an Aegean island. Journal of Rural Studies 32: 103-113.

Hobsbawm, E. and Ranger, T. (2012). The invention of tradition. Cambridge University Press, Cambridge, MA, USA.

Hoggart, K., Black, R. and Buller, H. (2014). Rural Europe. Routledge, London, UK.

Kiliánová, G. and Kanovsky, M. (2005). Communities in transformation: central and eastern Europe. Vol. 12. LIT Verlag Münster, Germany.

Law no. 205/2004 (2004). Law on animal protection. Published in Monitorul Oficial nr. 531.

Luca, L., Cionga, C. and Giurcă, D. (2012). Consolidarea exploatațiilor agricole. Editura Economică, București, Romania.

Madsen, L.M. and Adriansen, H.K. (2004). Understanding the use of rural space: the need for multi-methods. Journal of Rural Studies 20(4): 485-497.

National Institute of Statistics (2016). TEMPO Online – time series. Population and its demographic structure; labour force. Available at: http://tinyurl.com/gt4pl2z.

National Authority for Veterinary Health and Food Safety (2006). Order no. 180/2006, concerning the regulation for veterinary health on animal protection during slaughter and killing.

Ministry of Agriculture, Food and Forestry (2002). Order no. 425/2002, concerning provisions of veterinary health for the protection of animals during slaughter and killing.

Page, N. and Popa, R. (2013). Family farming in Romania. Fundația ADEPT Transilvania. Available at: http://tinyurl.com/j363xnt.

Petroman, C., Lozici, A., Petroman, I., Benk, I., Balan, I., Marin, D., Chirila C. and Momir, B. (2014). Dynamics of livestock and production of sheep and goats exploited in farms from Romania. Review on Agriculture and Rural Development 3(1): 151-155.

Pocol, C.B. (2013). Economie rurală, identitate și actualitate. Editura AcademicPres, Cluj Napoca, România, pp. 10-43.

Schileru, I. (2015). Crăciunul – tradiție și adaptare. Available at: http://tinyurl.com/zugl87f.

Stan, S. (2000). What's in a Pig? State, market and process in private pig production and consumption in Romania. Dialectical Anthropology 25(2): 151-160.

Stanciu, M. (2004). Caracteristici definitorii ale modelului de consum mediu românesc actual. Calitatea vieții 15, 1-2, București, pp. 25-43.

Stanciu, S. (2014). The Romanian swine market in the EU context. Economics and Applied Informatics 3: 87-96.

Szabó, Á.T. (2013). Does the countryside still feed the country? Producing and reproducing the rural in Transylvania. In: Shaping rural areas in Europe. Springer, Dordrecht, the Netherlands, pp. 165-180.

69. Agricultural and food ethics in Turkey: past, present, and future

C. Talug[1], *N.Y. Yalim*[2*], *P. Ataman*[3] *and G. Celtek Sungur*[4]
[1]*Ankara University, Faculty of Agriculture, Kultur Mah. Cemre S. 4/9, 06420 Cankaya, Ankara, Turkey;* [2]*Ankara University, Faculty of Medicine, Turkish Bioethics Assoc., Kultur Mah. Cemre S. 4/9, 06420 Cankaya, Ankara, Turkey;* [3]*Former Chair of UCTEA, Chamber of Food Engineers, Kultur Mah. Cemre S. 4/9, 06420 Cankaya, Ankara, Turkey;* [4]*FAO Regional Office for Europe and Central Asia, Gayret Mah. İvedik C. No. 55, 06170 Yenimahalle, Ankara, Turkey; yalim001@yahoo.com*

Abstract

Turkey is an agricultural country. The economic size of the Turkish agricultural industry ranks 7th or 8th in the world and 1st in Europe. However, agricultural and food ethics (AFE) have not been on the agenda until very recently. Attempts to remove the veil of ignorance regarding ethical problems in agriculture and food were first introduced by the professional chambers. The chambers identified the ethical principles and rules of conduct that should be followed and encouraged academic stakeholders to add ethics courses to the curricula. Subsequently, public institutions began to establish ethics committees; however to date, there is neither a single academic unit nor specific scientific journal in Turkey in the field of AFE. As an important step in filling this gap, the Turkish Bioethics Association launched a project entitled 'Promoting Agricultural and Food Ethics in Turkey and Improving the Ethical Decision-making Capacity of the Stakeholders in Agriculture' (TARGET). The European Associates of this EU-funded project are EurSafe, Wageningen University Philosophy Group, and the Food Ethics Council, and they will greatly contribute to the transfer of the EU experience to Turkey. The project utilizes an interdisciplinary team consisting of scientists and professionals with high-level academic backgrounds in ethics, agriculture, and food. At the first stage of the project, a survey and a focus group study will be conducted to identify the current value perceptions of the stakeholders in agricultural and food sectors. This will be followed by training sessions in AFE, with the design of the curricula and training materials to be based on the results of the preliminary survey. Additionally, Turkey's first documentation centre will be established and its first scientific congress will be held on agricultural and food ethics. This article will provide a summary of past experiences in the field of agricultural and food ethics in Turkey and will discuss ongoing activities as well as the potential future impacts of efforts in this field.

Keywords: EU projects, ethical analysis, ethical decision-making

Introduction

The world's agriculture and food systems (AFS) have long been in a fragile state, economically and ecologically as well as ethically. AFS involve all vertically related layers, from farm to fork, such as production factors and inputs, production, processing, distribution, marketing, consumption, and even waste management. Although the economic, social, and ecological dimensions of AFS have received a significant amount of attention and have been thoroughly studied; the ethical dimension has long been neglected.

The present authors perceive agricultural and food ethics (AFE) as a sub-discipline of applied ethics, which from an ethical standpoint, is the philosophical examination of particular issues in everyday life that are matters for ethical judgment. While recommending a particular course of action is not a must in ethics, in applied ethics, proposing a course of action that is better than other alternatives (an action that sacrifices the least amount of value and protects the most value) is the primary objective of the discourse

I. Anna S. Olsson, Sofia M. Araújo and M. Fátima Vieira (eds.) ***Food futures: ethics, science and culture***
EurSafe 2016 – DOI 10.3921/978-90-8686-834-6_69, © Wageningen Academic Publishers 2016

(Kucuradi, 1988). Thus, the present authors do not think that AFE only has dimensions related to ethics in general, as its main components (e.g. soil, water, plants, and animals) are constantly analysed whether or not they have an intrinsic value. It also necessitates a systematic ethical discourse because the values on which it is based frequently conflict with each other. As the present authors understand it, AFE studies must suggest a correct course of action when there are conflicts between values.

The objective of this article is to review studies on AFE in Turkey and provide a strong foundation for this new area of discourse in order to promote future studies by harmonizing it with the EU experience.

Agricultural and food ethics in Turkey: the past

AFE is a relatively new concept for Turkey. It is addressed in the fields of agricultural and food engineering as well as veterinary medicine as a professional code of practice, and professional organizations have some experience in this area limited to professional ethics; however, it has not yet been addressed in Turkey from a broad sense as noted above. There is neither a single academic unit in this area nor a specific scientific journal, although a specialized non-governmental organisation (NGO) was recently established and will be discussed below. Because all members of society have strong interests in food and agriculture not only as consumers but also as citizens, improving the ethical decision-making capacity in AFE and preventing corruption in this field will be a benefit to all.

Turkey has been directly affected by global facts of agriculture and food production, as well as its own pitfalls regarding farm structures. The vast majority of farms in Turkey are small. According to data received from the Ministry of Food, Agriculture, and Livestock (MoFAL), the average Turkish farmer has only 6.5 hectares of land and 94% of the farms are less than 2 hectares in size. It is extremely difficult for small farmers to get a good price for their produce. The incomes and living conditions of family farmers are inadequate throughout the country. Although the employment share of agriculture is 25% in Turkey, its contribution to gross domestic income is only 8%. The country's gross domestic income per capita is 10,504 USD; the same income level for the agricultural sector is only 3,622 USD. This data clearly indicates that farming in Turkey brings 1/3 less income than most of the other economic activities (MoFAL, 2016).Producer cooperatives and other forms of farming organizations are insufficient in number and hardly effective. Natural resources are being threatened by increasing demands, and market pressure leads to wild competition.

There are approximately 71,600 businesses producing food of non-animal origin or food contact materials; 311,000 retailers; and 247,000 mass consumption venues registered with the MoFAL. There are more than 10,374 food establishments processing products of animal origin that must be approved by the Ministry. Roughly 80% of these establishments are small and medium enterprises (SMEs) (MoFAL, 2016).

As consumers, Turkish society's first expectation from producers is safe food at reasonable prices. We also expect farmers to follow environmentally friendly farming practices. Therefore, the output of the agricultural production process should not be our only concern. Without providing farmers with higher standards of living and visible social value, it is not fair to expect them to stay in business and meet our expectations. Small family farming is the most sensitive mode of agricultural production in terms of environmental protection. Unfortunately, these farmers are fragile when faced with large industrial agribusiness. To realize, discuss, and overcome this value issue in the agricultural sector, capacity building in AFE can be a great asset. It is necessary to comprehend, clarify, re-structure, and transmit knowledge regarding value changes in the food and agriculture sectors and to raise awareness of the need for an ethically sensitive agriculture and food policy based on these newly defined values.

Despite being a country highly developed in agriculture with a long agricultural tradition, there is little focus on agricultural and food ethics in Turkey. Because the AFS throughout the world are changing dramatically, this ignorance and reluctance to review the values of Turkish farming and food production leads to a moral anomie. This moral anomie leads to corruption, which is one of the primary concerns of the government and society today. The agriculture and food industry is the largest economic and trade industry in Turkey. Historically, it was the first industry to become acquainted with corruption, and public perception about corruption in the food and agriculture industry is intense. Unfortunately, fraudulent and fake food production and misinformation are common. Although there is extensive state control over agriculture and food, food production and associated sanitation regulations are clearly open to corruption (MoFAL, 2016).

Furthermore, as a candidate for EU membership, Turkey must harmonize its policies and legislation with EU principles prior to accession. Food safety is a policy area where the EU has zero tolerance for any practice that could lower the community standards in place to protect consumer health and interests in relation to food while guaranteeing the smooth operation of the market. In order to achieve this objective, the EU ensures that there are well-established control standards and detailed rules to appropriately inform consumers. Turkey is expected to converge to AFE approach to that of the EU.

It is necessary to critically think over and revise the ethical background of food and agriculture policies in Turkey. It is crucial for both service providers and service users to understand the ethical claims behind the legislation and procedures. One of the major ethical issues surrounding agriculture in Turkey is responsibility for the common good such as collective attitudes toward child agricultural workers, or protection of pastures. Because AFE is a new field of study in Turkey, the responsibility factor regarding these issues is rarely discussed. The collective responsibility and the ethical issues related to responsibility such as protecting biodiversity, maintaining pastures, acting in the best interest of future generations, and attitudes toward the agricultural workforce must be addressed and raising awareness is essential (Persson and Savulescu, 2012).

Agricultural and food ethics in Turkey: the present

Over the past two years, the Turkish Bioethics Association (TBA) has realized that AFE, one of the most important fields intersecting healthcare ethics and environmental ethics, is not being addressed academically, politically, or professionally in Turkey. Preliminary research in this area has shown the following: (a) The history and deontology of agriculture as a compulsory subject in undergraduate and graduate curricula (in agriculture and food engineering programs as well as related programs such as forestry, aquaculture, etc.) is typically taught from a historical perspective by senior faculty members, and ethics is primarily taught within the limits of the code of professional conduct. (b) The public authorities that determine the country's agriculture and food policies neither take the ethical dimension of these policies into consideration nor open them up for discussion. They are almost entirely concerned with the economic impacts of the policies or, less frequently and more limitedly, they are interested in ecological and social dimensions. (c) Both chambers of agriculture and food engineers, although extremely effective with active professionals as members, and the rapidly growing private sector are only interested in the professional code of conduct dimension of the ethical issues in the field. The latter is also interested, to some extent, in the context of the rules for fair competition. This negligence leads to inconveniences not only for ethical decision-making capacities, but for monitoring the corruption that is becoming the norm rather than the exception, for making it visible, and for taking necessary precautions and providing community support for the implementation of these measures.

Therefore, the TBA decided to apply the Prevention of Corruption and Promotion of Ethics Grant Scheme to a project in this area. Funded jointly by the EU and the Republic of Turkey, the grant scheme has a total budget of 2.5 million Euros. In order to increase awareness and policy-making capacities regarding the prevention of corruption and to promote ethical behaviour in society, 500,000 Euros of the grant project's funds are made available to NGOs in Turkey. The TBA answered the call with a project entitled 'Promoting Agricultural and Food Ethics in Turkey and Improving the Ethical Decision-making Capacity of the Stakeholders in Agriculture' (TARGET). The primary purpose of the proposal was to bring together agriculture and food professionals and philosophers/ethicists in a multidisciplinary project. The project was officially launched on January 15, 2016 and will continue for 15 months.

In this project, in addition to assessing the current ethics universe and transferring the current core AFE knowledge to key opinion leaders, the main contributors for the future of the field are the establishment of the documentation centre, the foundation of a specific, AFE-related NGO, and the organization of the first scientific meeting to inaugurate the operation of the networks. Altogether, this should trigger and encourage the future scientific studies and new academic programs.

TARGET consists of one national and three international associates. The national associate is the Turkish Society of Philosophy. The international associates are the European Society for Agricultural and Food Ethics (EurSafe), the Philosophy Working Group of the Wageningen University, and the Food Ethics Council.

The general objectives of TARGET are to promote AFE in Turkey, to increase awareness and expose corruption in Turkey's AFS (both existing and potential), and to support preventative corruption measures in this field. The specific objectives are to improve and generalize the value perceptions of environmentally friendly farming systems and practices in order to protect natural resources, preserve biodiversity, combat climate change, underscore the problems of disadvantaged actors in agricultural production such as agricultural labourers and women farmers, and increase awareness on the preservation of local cultures in the agriculture and food sector.

Project activities are grouped into five work packages. The research package consists of two primary actions: a preliminary assessment meeting and a survey. Because AFE in Turkey is still an emerging concept, in order to create a common language and to share information, it is essential to involve key actors. Research methodologies and questions for the survey were developed in accordance with the results of the preliminary assessment meeting. Approximately 800 participants will be surveyed. The sample will be drawn from members of the local units of the Chambers of Food Engineers and Agricultural Engineers of the Union of the Chambers of Turkish Engineers and Architects (TMMOB) from seven geographic regions where educational activities will be implemented. The questionnaires will be sent by ground mail. The results of the preliminary assessment meeting and the survey will be shared with associates of TARGET and other stakeholders at the 1st Evaluation Meeting in June, 2016.

The education package is designed to improve ethical evaluations and the decision-making capacity of agriculture and food experts. A training course will be organized in the seven pre-selected cities. Trainees will participate in two day-long courses with approximately one month between each course and successful participants will be certified. Gender equality and balanced representation of stakeholders will be taken into consideration in the selection of the participants.

The 1st National Agricultural and Food Ethics Congress with international participation will be organized before the end of the TARGET, in March 2017. In addition to sharing the results of the research and education goals of the project, recent global AFE experiences and their influences in Turkey will be discussed, case analyses will be performed, and professional ethics in the agriculture and food sectors and associated liabilities will be discussed.

The fourth work package for the project is the establishment of the Agricultural and Food Ethics Documentation Centre. This centre will be a first of its kind in Turkey and will serve agricultural and food experts and ethicists as well as any individual who wants to learn about AFE. Turkey's first agricultural and food ethics civil society organization will be established through the TARGET project as well.

Agricultural and food ethics in Turkey: the future

The global challenges for agriculture forced the EU to review its oldest and most comprehensive policy, the Common Agricultural Policy (CAP), and it was radically revised in 2013 (EU, 2013). It is important to note the crucial role of ethical considerations in this new agricultural policy. There is a significant history of ethics in food and farming in Europe. This knowledge is the result of serious academic work, quality publications, and strong civil society organizations. It has constantly been nurtured and shared with the public. This experience will be transferred to Turkey starting with the TARGET project and the effects will be closely monitored by the Association of Agricultural and Food Ethics. Initial contact with stakeholders revealed great interest and potential for collaboration.

The future of AFE poses a number of problems to be addressed such as the aging rural population, difficulties in carrying out supportive policies for primary producers because of the lack of education and the plethora of bureaucratic requirements, in addition to existing issues such as climate change, reduced biodiversity, etc. All of these issues must be addressed from a strong theoretical perspective supported by strong organizations. AFS organizations in Turkey are strong to a certain extent but their ethical perspectives are not fully developed. The process discussed above will form the initial stages of future development in this area.

References

European Union (2013).Overview of CAP reform 2014-2020. Fifth report. Agricultural policy perspectives brief. Available at: http://tinyurl.com/pjjh9h8.

Kucuradi, I. (1988). Etik. Meteksan Yayınlari. Ankara, Turkey, 169 pp.

Ministry of Food Agriculture and Livestock (2016). Food and control general directorate.

Persson, I. and Savulescu, J. (2012). Unfit for the future. Oxford University Press, Oxford, UK, pp. 66-72.

70. Good home-grown food? An ethical approach to kitchen gardening

M. Schörgenhumer
Department of Philosophy, University of Vienna, Universitätsstraße 7 (NIG), C0210, 1010 Wien, Austria; maria.schoergenhumer@univie.ac.at

Abstract

Gardens have been serving as an important source of food through the ages and cultures. Yet, in agriculture-centred accounts of food, gardening – restricted to insignificantly small patches of land and personal preferences – seems not to matter. This paper argues that the idea of the garden should not be disregarded when it comes to 'food futures' exactly because the small-scale scope proves valuable. The first section of the paper highlights the specific features of the garden/gardening in contrast to agriculture. Not only does the small-scale setting of a protected space of nature under intensive personal care allow for an interesting diversity of plants and cultivation styles; the practice of kitchen gardening also shapes, or even reconfigures, people's relations to food and to nature more generally. In the second section, the ethical implications of this are analysed in terms of: (1) environmental virtues; (2) autonomy and responsibility; and (3) valuing food. Finally, applying a realistic, non-idealising perspective, the significance of 'growing your own food' in the garden is clarified by contesting the idea of self-sufficiency.

Keywords: philosophy of gardening, environmental virtues, value of food, autonomy, care

Introduction

On a global scale of 'feeding the world' in times of industrial agriculture, biotechnology and a globalised food market, and in theory thinking on such global scale (e.g. Freyer and Bingen, 2015; Marsden and Morley, 2014), hands-on gardening, restricted to small patches of land, seems rather 'primitive' (Pence, 2002: 118) and insignificant. Even if local food production is re-emphasised, it is mainly farming, not gardening, that is taken into consideration (e.g. Norberg-Hodge *et al.*, 2002). Publications on 'urban agriculture', as for instance 'Urban Farms' (Rich, 2012) or 'Food and the City: Urban Agriculture and the New Food Revolution' (Cockrall-King, 2012), contain many examples of what could also or better be called 'urban gardening', yet, the idea of gardening appears to be subsumed and replaced by the somehow bigger, stronger concept of farming, as if 'the garden' seemed too small, too personal, too unprofessional, and not useful enough to matter. However, gardens have been serving as an important source of food through the ages, and in different cultures; they should not be ignored when making sense of 'food futures'. Gardening is a common environmental practice that is, and can be made, accessible to many people in their everyday lives, with growing importance in urban contexts (Hauser, 2011; Müller, 2011). It is a radically local, and political (Lemke, 2012), practice that shapes both landscapes and individual 'foodscapes' – independent of the globalised food market, long supply chains and whatever is going on beyond consumers' control; dependent, however, on nature and skilful care. The following sections examine kitchen gardening and its food-ethical implications from a philosophical point of view.

Specifying small-scale gardening (in contrast to agriculture)

It is crucial to clarify the terms and work out the specifics of the garden/gardening, which differs from agriculture in important aspects, in order to explore its particular ethical implications. Small, personal, unprofessional, not useful enough – these characteristics, as mentioned above, suit the garden just fine. They can be easily framed as reservations against gardening when thinking about food production in

EurSafe 2016 – DOI 10.3921/978-90-8686-834-6_70, © Wageningen Academic Publishers 2016

terms of farming; on different grounds, however, gardening is worth considering and pursuing exactly because it is commonly practiced: (1) on a small scale; (2) on a personal level; (3) unprofessionally; and (4) with regard to more than just the mere use/utility of plants. The following especially refers to the kitchen garden respectively kitchen gardening, which is intrinsically linked to food; most of what is proposed is as well true for other types of gardens, though (and often the kitchen garden forms part of a larger garden with additional features, e.g. areas dedicated to idle recreation or aesthetic pleasure without further utility). The underlying idea is to look at the practice of gardening instead of assuming a ready-made garden: 'Inherent in the idea of a garden is some kind of care and attention beyond the initial design. The actions by a person to nurture plants, to shape and develop, or just to encourage what grows, we call gardening' (Brook, 2010: 14).

1. Small scale: While agriculture covers vast areas of land, organises the landscape on a large scale and cultivates crops in open fields, most effectively in monocultures, a garden, marked by an enclosure, is a protected 'inner space' that comprises a landscape in itself: a quite limited space in which different plants can be assembled and fostered in intensive personal care. Such care is possible due to the manageable scale and supports a high diversity and abundance of plants, which is an essential desideratum not least for the kitchen garden.
2. Personal: The small scale of the garden facilitates direct contact with the plants that grow within its enclosure, and it is also necessary to get involved with them: intensive care is required to create and sustain the humanised landscape that a garden is supposed to be, including artificially arranged combinations of (mostly) cultivated plants that are in need of human support. In repeatedly and constantly caring for the garden over a long time, the gardener personally relates to her plants (this also means getting in touch with where the food comes from and, in a very down-to-earth way, with the 'cycle of life'). The plants become 'my' plants, not so much in terms of possession but in the sense of being concerned with them, feeling responsible for them, pertaining to them, even holding them dear. This goes far beyond instrumental purposes. In agriculture, economically speaking, developing such relationship to one's plants (which is unlikely in large-scale cultivation anyway) seems to be of no use, it may even hinder their effective exploitation.
3. Unprofessional practice: Whereas agriculture is an economic undertaking that is usually, and most efficiently, pursued professionally (within prevalent standards and conventions of production, chiefly based on established professional knowledge), the garden is open for amateur gardening. It provides a situation for experimenting and learning by experience in interaction with nature, for extending the range of possibilities in trying out new ways of cultivation, as well as growing, preserving, breeding new/rare/old cultivars. All of this, of course, can be professionalised, but the focus on amateur gardening highlights two important aspects: firstly, increasing personal experience, skill and knowledge, or even virtue (see section 3) in dealing with plants/nature, and secondly, the sense of 'amateur' that stems from Latin '*amare*', to love – to love what you do, to engage in a practice because you like it, which, again, encourages a positive relation to the objects involved in that practice: the plants, the fruit/the food they yield.
4. Beyond useful: Agriculture is economic. It is gainful work meant to produce something useful – mainly plants and animals for instrumental purposes, especially for food, to sustain human life – and to be effective and economically viable, whilst the garden, even if it is a kitchen garden, is not restricted to that purpose, but importantly also involves aesthetic benefits and diverse garden practices (Cooper, 2006) beyond 'food production', related to embodied experience and well-being, social interaction, playful and recreational endeavour, and the aesthetic appreciation of what can be found and experienced in the garden. A garden is not necessarily supposed to be effective, or economically viable, or at least this is not the main concern of an amateur gardener who does not garden for a living.

What is good about home-grown food? Ethical implications of kitchen gardening

Good practice by environmental virtues

In the kitchen garden, food can be grown in an environmentally responsive and responsible way. Good practice in this respect is characterised by the virtues of gardening (e.g. Brook, 2010; Cooper, 2006). These virtues are necessary for succeeding in creating a flourishing, sustainable garden and, in the end, being able to reap the harvest. They are 'induced' by the practice of gardening (Cooper, 2006: 91) provided that someone seriously engages in it, which involves not ignoring nature's part. Specific attitudes learned in that process can be argued as environmental virtues, that is, as 'right attitudes' regarding the environment, configuring a good human-nature relationship in the specific context of inhabiting and cultivating nature (Kallhoff and Schörgenhumer, unpublished). Important environmental virtues of gardening are: patience, persistence, a realistic grasp of nature, humility in acknowledging the limits that nature puts to any human endeavour, prudence in collaborating with nature rather than imposing unsuitable human ideas on it: 'to attune our desires with nature's ways' (Pollan, 2003: 195), and care for nature, which includes attentiveness and dedication to the flourishing of the other. With regard to food, these virtues matter because they can be obtained and practiced by growing one's own food, and they support an environmentally good way of 'producing' food (taking into account, as the virtuous gardener does, that plants and the food they yield cannot be produced, made like things, but need to grow and be let grow, and their flourishing can be supported). Furthermore, virtues like patience, attentive care, and humility are likely to prompt more careful and appreciative ways of dealing with food.

Do it yourself? Autonomy, responsibility

Growing one's own food means being in charge: in charge of what is cultivated and how it is cultivated. Being able to choose which crops should grow in the garden is an important aspect of designing personal 'foodscapes' (Heistinger, 2011). On a commercial level, food production and consumption are to a large extent limited to a slim selection of economic viable crops (*ibid.*: 313; Cockrall-King, 2012: 27-29). The diversity of food plants, and in consequence of fruit and vegetables, can be increased by cultivating unusual plant varieties – rare, old fashioned crops, local/private breeds, passed on over time and shared between gardeners. This brings about an enrichment of diet, which can be seen as a contribution to a person's good life, as well as higher biodiversity on cultivated areas, a counter-monocultural environmental gain.

Being able to choose how plants are cultivated in the garden is, for one, important in terms of personal health and well-being; organic gardening methods (without use of pesticides) therefore seem preferable. Anthropocentrically speaking, it is in the very own interest of the producer-consumer that the food she grows and eats is good for her. But the choice of the right methods is also a question of environmental ethics, for instance with regard to the good of plants as living beings capable of flourishing (Kallhoff, 2014; Odparlik, 2010), or to the value(s) of biodiversity, or to a well-functioning human-nature relationship (see the virtues mentioned above, and Pollan, 2003).

It is important not to disregard that a person's autonomy in her garden is limited by nature (e.g. weather, climate, natural time cycles and interrelations between plants, animals and other organisms in the garden). Nevertheless, it means increased independence of the industrial food market and its long, mostly obscure, wasteful supply chains as well as its standardisations of crops and products (Cockrall-King, 2012: 23-57). Being in charge as a kitchen gardener goes hand in hand with responsibility: for the quality of the food one grows and eats, for the flourishing of one's plants, for the garden that be sustained, and for how one's choices affect other people, other living beings, and the environment.

Valuing food

Home-grown food is valuable, and the practice of growing your own food changes how you value food more generally. Home-grown food is good because (or when) it is fresh, delicious and beautiful; fresh food is both nutritious and rich in taste, and a personal kitchen garden enables to prepare, process, or directly consume freshly harvested fruit and vegetables. Food from one's own garden is further valuable because one knows where it comes from, and how it comes into being. This involves with acknowledging the fact that it takes a lot of time, skill and personal effort, and nature's 'grace' (Cooper, 2006: 96), like the right weather, until something edible has grown. Experiencing crop failures further gives way to valuing food as something that cannot be taken for granted, something that is not simply available at any time and any cost, but requires nature to play along (as well as we are required to play along with nature). Successfully grown food, in turn, may very well be valued as a 'gift' (Pollan, 1996: 145); the virtuous gardener is grateful for the harvest.

All this, experienced hands-on in the kitchen garden, makes food appear precious, too valuable to waste – an insight that is not restricted to the area and output of 'my' garden, but likely to shape a person's relation to food in general. Appreciating plants and food as too valuable to waste leads to increased diligence regarding their 'production' and consumption, and also in dealing with what is too much for one's personal use. Gardeners commonly share their surplus of home-grown plants and food with other people, giving it away for free, which is owed to gardening not being subject to economic purposes. In this respect, the garden prompts an 'economy of giving' (Bennholdt-Thomsen, 2001), a different and socially fruitful approach to food and what to do with it.

It is further important to point out that in the relation of care the gardener engages in, plants are not instrumentally reduced to 'food objects' ready to be exploited, but valued and treated as living beings in their own right that can grow, flourish, and die (Werner, 2011: 60-61). Furthermore, the attentive gardener also appreciates them for their aesthetic qualities that are intrinsically bound to their flourishing. This again enhances the understanding of the value of food, a rich value that cannot be expressed in numbers. Kitchen gardening is in this sense an antidote to detaching the value of food from the cycles of life it is – and we are – embedded in.

Significance and limitations of 'growing your own food'

In conclusion, it's worth taking gardening into account in respect of food. Kitchen gardens are a source of fresh food with a clean, green history (if the gardener wants so) beyond long supply chains, wasteful transportation and packaging, beyond the standards of the industrial food market that dictate and limit the choices the consumer (as well as the professional producer) can make within that logic, including a narrow range of industrialised standard crops, and beyond any uncertainty about production conditions the consumer can neither control nor keep track of.

That said, it is important not to overestimate (or romanticise) the difference that gardens/gardening can make in the question of food. Authors tend to be quite enthusiastic about autonomous do-it-yourself-gardening and its potential to overcome the restrictions 'the system' imposes on people, and change life for the better (especially in the matter of urban gardening, e.g. Müller, 2011) – which might be just another echo of the idea of paradise that remains intimately tied to gardens and garden thinking. Yet, growing your own food is hard work, and it gets harder the more you depend on the outcome. Food self-sufficiency sounds good as long as it is just the idea, but when put into practice, it requires a lot of space, time, effort, back-wrecking work on a scale that is not so small any more, and it comes along with crucial deprivations due to the strict seasonality of locally grown food as well as crop failures (e.g. for weather reasons), with shortages and monotonous diet in lean times (like winter), and problems of managing the surplus in times of plenty, which again means loads of work (high workload of harvesting, processing, distributing the fruit of the garden; problems of storage and of preserving food).

The particular benefits of growing one's own food in the garden come into bloom on a small scale, without the claim to self-sufficiency; the foodscape-changing role of gardening is limited, but nonetheless valuable. For one, kitchen gardening reconfigures how persons relate to, and how they value food. It further provides the possibility of cultivating plants and producing food in an environmental attentive and responsible way, fostering biodiversity and acknowledging human limits. Last but not least, life is enhanced by home-grown food, for it positively contributes to people's diet in terms of variety, freshness and healthiness of the food, and aesthetic pleasure (the looks and feels, the smell and taste of a perfectly ripened, freshly picked, sun-warm apple), all of which makes home-grown food considerably good.

References

Bennholdt-Thomsen, V. (2011). Ökonomie des gebens: wohlstand durch subsistenz. In: Müller, C. (ed.) Urban gardening: über die rückkehr der gärten in die stadt. Oekom, München, Germany, pp. 252-265.

Brook, I. (2010). The virtues of gardening. In: O'Brien, D. (ed.) Gardening – philosophy for everyone: cultivating wisdom. Wiley Blackwell, Oxford, UK, pp. 13-25.

Cockrall-King, J. (2012). Food and the city: urban agriculture and the new food revolution. Prometheus Books, Amherst, NY, USA.

Cooper, D.E. (2006). A philosophy of gardens. Oxford University Press, Oxford, UK.

Freyer, B. and Bingen, J. (2015). Re-thinking organic food and farming in a changing world. Springer, Dordrecht, the Netherlands.

Hall, M. (2010). Escaping Eden. Plant ethics in a gardener's world. In: O'Brien, D. (ed.) Gardening – philosophy for everyone: cultivating wisdom. Wiley-Blackwell, Oxford, UK, pp. 38-47.

Hauser, S. (2011). City, art, and nourishing gardens. In: Witzgall, S., Matzner, F. and Meder, I. (eds.) (Re)designing nature. Hatje Cantz, Ostfildern, Germany, pp. 38-51.

Helstinger, A. (2011). Leben von gärten: warum urbane gärten wichtig sind für ernährungssouveränität, eigenmacht und sortenvielfalt. In: Müller, C. (ed.) Urban gardening: über die rückkehr der gärten in die stadt. Oekom, München, Germany, pp. 305-332.

Kallhoff, A. (2014). Plants in ethics: why flourishing deserves moral respect. Environmental Values 23: 685-700.

Lemke, H. (2012). Politik des essens: wovon die welt von morgen lebt. Transcript, Bielefeld, Germany.

Marsden, T. and Morley, A. (2014). Sustainable food systems: building a new paradigm. Routledge, New York, NY, USA.

Müller, C. (2011). Urban gardening: über die rückkehr der gärten in die stadt. Oekom, München, Germany.

Norberg-Hodge, H., Goering, P. and Page, J. (2002). From global to local: sowing the seeds of community. In: Pence, G. (ed.) The ethics of food: a reader for the twenty-first century. Rowman and Littlefield, Lanham, MD, USA, pp. 191-214.

Odparlik, S. (2010). Die würde der pflanze: ein sinnvolles ethisches prinzip im kontext der grünen gentechnik? Alber, Freiburg, Germany.

Pence, G. (2002). Organic or genetically modified food: which is better? In: Pence, G. (ed.) The ethics of food: a reader for the twenty-first century. Rowman and Littlefield, Lanham, MD, USA, pp. 116-122.

Pollan, M. (2003). Second nature. a gardener's education. Grove Press, New York, NY, USA.

Rich, S.C. (2012). Urban farms. Photography by Matthew Benson. Abrams, New York, NY, USA.

Werner, K. (2011). Eigensinnige beheimatungen: gemeinschaftsgärten als orte des widerstandes gegen die neoliberale ordnung. In: Müller, C. (ed.) Urban gardening: über die rückkehr der gärten in die stadt. Oekom, München, Germany, pp. 54-75.

71. Diet for a small planet – rich or poor?

C.N. van der Weele
Department of Social Sciences, Philosophy Group, Wageningen University, Hollandseweg 1, 6707 KN Wageningen, the Netherlands; cor.vanderweele@wur.nl

Abstract

In a world that on average grows richer, the consumption of meat has increasingly replaced that of pulses, the 'meat of the poor'. But as the problems of meat pile up, people in Western countries are becoming increasingly ambivalent about meat eating, and while ambivalence is not a great driver of immediate change, it may be a sign that change is going on below the surface and that we are becoming more open to attractive alternatives. What are attractive alternatives? Innovative searches are mostly directed at technologically novel options, often designed to be as meatlike as possible. They generally require elaborate processing, which decreases their sustainability. Pulses beat all the other alternatives in this respect, but they do not resemble meat and they are not novel or surprising. This paper argues that cultural innovation deserves more attention and that pulses need to be the focus of such innovation.

Keywords: pulses, meat alternatives, ambivalence, cultural innovation

Introduction

Now that meat is increasingly associated with a variety of problems, beans, lentils and other pulses are among the possible alternatives. They are attractive from various perspectives: they are rich sources of protein, minerals and fibre and thanks to their alliance with nitrogen fixing bacteria they are a great boost to soil fertility. When it comes to sustainability, no other meat alternative beats pulses. The United Nations have declared 2016 the International Year of Pulses, in order to draw the world's attention to the benefits of these 'nutritious seeds for a sustainable future'.[1]

Pulses can use the extra stimulus; from World War II onward, they have steadily been losing out to meat, in rich as well as poor countries. They are widely regarded as meat for the poor, to be replaced by the real thing as soon as people can afford it. This paper wonders about the present prospects. Starting from existing trends I will reflect on increasing ambivalences about meat and on the search for attractive alternatives. Given the great attractions of meat, that search is most conspicuously directed at technological products that resemble meat as closely as possible. Pulses do not resemble meat. They also do not need technological innovation, at least not primarily; in order to contribute to a sustainable world they need only to be grown and eaten. Will the Year of the Pulses make us do that?

Trends contested

After World War II, production and consumption of animal protein (meat and dairy) started to increase in Western countries. In the Netherlands, for example, the consumption of animal protein first exceeded that of plant protein around 1950 and from then on the gap widened; in 1975, the Dutch ate twice as much animal protein as plant protein (Vijver, 2005). Meat in particular stood for increased prosperity as well as health. Meat consumption grew, actively stimulated by the government, from about 35 kg per person in 1950 to more than 80 kg in 2010. That growing prosperity comes with growing meat

[1] Pulses are seeds of the *Leguminosae* family that are harvested for dry grain. Legume species that are used as vegetables (e.g. green peas, green beans) or primarily for oil (soybean, peanuts) are not classified as pulses by the Food and Agricultural Organisation (FAO), the UN-organisation that helps to implement the International Year of Pulses.

 I. Anna S. Olsson, Sofia M. Araújo and M. Fátima Vieira (eds.) ***Food futures: ethics, science and culture***
EurSafe 2016 – DOI 10.3921/978-90-8686-834-6_71,

consumption is a worldwide phenomenon. Figures from China illustrate that the change can be fast: meat consumption in China rose from 14 kg per person in 1970 up to levels that now approach those of Western countries (FAO, 2012). As the world on average continues to become richer, global meat consumption thus continues to rise; in 2050 it will have increased another 70%, according to widespread expectations.

Pulse consumption, on the other hand, has been in steady and worldwide decline. In developing countries, it decreased from over 9 kg per person per year in 1970 to less than 7 kg in 2005; during that same period, developed countries diminished their average consumption from 3.6 kg to less than 3 kg (FAO, 2012: 44). The Dutch eat even less pulses, about 1 kg per person per year, while some estimates for China are almost as low (Lagos and Zhan Lei, 2013). The decline is naturally associated with a decline in production as well as in research and development: farmers turn to other crops, breeding programmes having been stopped. In the Netherlands, researchers in plant breeding would certainly be interested in pulses, as they see room for improvements in yield, resistance to insects and possibly protein quality to decrease flatulence. However, they currently find it impossible to find money for pulse breeding; grant conditions require the participation of companies, but companies are not interested because of the lack of economic prospects (Wageningen UR plant breeders' personal communications).

Already in the 1960s, the conditions and implications of meat consumption began to be contested. Ruth Harrison's 'Animal Machines' (Harrison, 1964) was the first forceful protest against 'the new factory farming industry', in which animal lives were fully in the service of profit. More objections followed. In 1971, Frances Moore Lappé's 'Diet for a small planet' focused on the losses involved in turning plant protein into animal protein. As she wrote in the introduction of the 20th anniversary edition of this best-selling book: 'In one sense, what motivated me to write 'Diet for a small planet' was simple outrage. We feed almost half the world's grain to livestock, returning only a fraction in meat – while millions starve. It confounds all logic.' (Moore Lappé, 1991: xvii). In search of a more responsible diet, she made the alternative very concrete by offering recipes in which protein was provided by a combination of pulses and grain. In the foreword of the 1991 edition, the author is awed by the rapidity of the change in our consciousness. What was heresy and 'fringe' when she wrote the book, she notes, 'just twenty years later is now common knowledge' (*ibid.*: xv).

When indifference is ambivalence

Knowledge about the problems of meat certainly increased – animal ethicists and animal activists had also been very active meanwhile – but it was not accompanied by a reversal of trends. Neither did such a reversal take place after the dramatic images of the mass killings during outbreaks of animal diseases or the publication of 'Livestock's long shadow' (FAO, 2006), which put the unsustainability of global livestock on the agenda. Global meat consumption continued to rise, pulse consumption continued to fall. Although in Western countries, growth of meat consumption has been stagnating in the last few years and some countries even see a modest fall (e.g. Verhoog *et al.*, 2015), so far there are no signs of a real protein transition. Clearly, common knowledge does not automatically lead to significant change in consumption.

What did change was the moral reputation of meat consumption in at least some Western countries; the best food of the 1950s is now often called 'the new smoking'. Already in 2000, an interview study in Copenhagen showed that most meat-eating people had very mixed feelings about meat and in fact hardly differed from vegetarians in the negative things they had to say about meat. The idea of intensive husbandry in particular caused 'a diffuse sense of disgust' towards meat (Holm and Møhl, 2000: 281). Later studies revealed similar uneasiness and found great ambivalences about meat, again first of all because of intensive husbandry (e.g. Te Velde *et al.*, 2002; Van der Weele, 2013). A common finding was that negative information and feelings did not typically cause people to change their diets. Instead, the studies pointed to various mechanisms with which people distanced themselves from compromising knowledge and uneasy feelings. Holm and Møhl saw that people tended to choose types of meat, such as minced meat, that resemble animal less and 'hides the origin of meat'. The other studies just mentioned found evidence for 'functional ignorance'(Te Velde *et al.*, 2002) or 'strategic ignorance' (Van der Weele, 2013). A similar finding was reported by Bastian *et al.* (2012), who experimentally found that meat eaters tend to adapt their thoughts about mental capacities of animals in order to reduce cognitive dissonance: they ascribe less cognitive capacities to animals they eat, such as pigs and cows, than to animals they don't eat, such as cats and dogs.

Apparently, the abundance of negative thoughts and feelings about meat does not have massive behavioural implications, at least not immediately. Observations of this kind have often led to a certain cynicism about consumer behaviour, articulated in terms of a gap between what people say and what they really do, or as a gap between our citizen- and consumer identities. According to the latter gap, we have altruistic values and feel responsible in our role as citizens, but when push comes to shove in the supermarket, self- interest is decisive. Explicitly or implicitly, the message is that deep down we really don't care. Jonathan Safran Foer summarises this succinctly: 'We can't plead ignorance, only indifference' (Foer, 2009).

That we are really indifferent is increasingly implausible; if we were, we would not need to shy away from burdensome knowledge through complicated and paradoxical psychological mechanisms to reduce tensions. Rather, we seem to be increasingly ambivalent; our values and motivations clash. What cannot be denied, however, is that these clashing values have not resulted in a protein transition so far; established patterns of behaviour mostly win out. The increased knowledge that Moore Lappé observed already in the early 1990s no doubt exists, but results in widespread strategic ignorance and other coping mechanisms rather than dramatic decreases in meat consumption, or increases in pulse consumption, for that matter. A considerable reduction in meat consumption is clearly not easy.

Yet there are signs that many people would welcome change. We found that focus groups about cultured meat were also occasions for meat eaters to express concern about meat. When people compared meat and cultured meat, they regularly wondered whether 'normal' meat is not just as little normal and natural as cultured meat. These negative views did not lead most people to eat (much) less meat, but they did lead to widespread sympathy towards cultured meat. With some caution, this may be interpreted as a sign that quite some people, while they do not see themselves as individual pioneers, do hope for collective solutions and changes (Van der Weele and Driessen, 2013, 2014). Ambivalence may not lead to dramatic change in itself, but it may make us more open to collective game changers. Sufficiently attractive alternatives would evidently be of great help.

Innovation: technological and cultural

Given that meat is so beloved, a plausible requirement for attractive alternatives is that they should resemble meat as closely as possible, which is indeed what studies about the (non)attractiveness of alternatives suggest (Hoek, 2010). In line with this, much development of plant based meat alternatives aims at the perfect imitation of meat, not only in taste, but also in fibrous structure and other characteristics. The Dutch 'Vegetarian Butcher' has been surprisingly successful in making soy-based 'chicken' that can hardly be distinguished from real chicken, while the American company 'Impossible Foods' is working on a hamburger that even contains fake blood. The strength of cultured meat is that it takes the love of meat completely seriously: tissue culture made of animal cells does not just resemble meat, it IS meat, in the sense at least that it is made of animal cells.

Using technology for making meat alternatives catches the imagination of innovators as well as artists. Inevitably, such explorative efforts also venture beyond the idea of imitating meat. Thus, the '*In vitro* meat cookbook' (Van Mensvoort and Grievink, 2014) presents cultured meat as a 'what if game' to explore weird future products and scenarios. Meat may be the golden standard at this moment, but why not develop other delicious protein products? For both innovators and artists/designers, technology is a source of inspiration to explore novel and exciting realms.

Yet these technological ventures also come at a price. Plant-based alternatives may be more animal friendly than meat, but because of the elaborate need for processing, they are not necessarily more sustainable (Apaiah *et al.*, 2006). For cultured meat, life cycle analyses are still tentative, but estimates about energy requirements are developing in more pessimistic directions (Smetana *et al.*, 2015). Processing inevitably takes energy.

All the while, far removed from all technological excitement, pulses are another option. The only form of processing they need is cooking, and from whatever additional angle one looks, they always look like the most sustainable meat alternative. If we take sustainability seriously we should embrace them. But if Moore Lappé's bestseller could not stop their decline forty years ago, and if increased knowledge about meat problems has not helped either, and if innovative energy is mostly directed at technological options, what reason might we today have for more hope?

That the FAO is now promoting pulses is a big leap from the 'fringe' activities that hoped to make us eat beans in the 1970s. Yet that is not a sign of triumph, it may perhaps better be seen as an illustration that pulses need assistance more than ever. There are hopeful signs, in the form of initiatives such as the 'brown bean gang' (http://www.bruinebonenbende.nl), beautiful new cookbooks, bean-loving chefs, increasing varieties of humus in supermarkets, new initiatives of companies that sell canned beans. But when 2016 is over, will it be enough?

This paper does not aim to predict the impacts of the year of the pulses. The conclusion is more restricted. In so far as we link innovation and imagination with technology, we would do well to weaken that link. If pulses are to be a really serious meat alternative, they can do with a lot of innovative help: the reputation of pulses as the poor man's meat calls for very deep social and cultural reinvention.

References

Apaiah, R., Linnemann, A. and Van der Kooi, H. (2006). Exergy analysis: a tool to study the sustainability of food supply chains. Food Research International 39: 1-11.

Bastian, B., Loughnan, S., Haslam, N. and Radke, H. (2012). Don't mind meat? The denial of mind to animals used for human consumption. Personality and Social Psychology Bulletin 38(2): 247-256.

Foer, J.S. (2009). Eating animals. Little, Brown and Co, New York, NY, USA.

Food and Agricultural Organisation (FAO) (2012). World agriculture towards 2030/2050; the 2012 revision. ESA working paper no. 12-03.

Harrison, R. (1964). Animal machines. Vincent Stuart Publishers, Wallingford, UK.

Hoek, A. (2010). Will novel protein foods beat meat? Consumer acceptance of meat substitutes, a multidisciplinary research approach. PhD thesis, Wageningen University, Wageningen, the Netherlands.

Holm, L. and Møhl, M. (2000). The role of meat eating in everyday food culture: analysis of an interview study in Copenhagen. Appetite 34: 277-283.

Lagos, J. and Zhang, L. (2013). China pulse annual 2012. GAIN report no CH 13007. USDA Foreigh Agricultural Service, USA.

Moore, L.F. (1971/1991). Diet for a small planet. Ballantine Books, New York, NY, USA.

Onwezen, M., Snoek, H., Reinders, M. and Voordouw, J. (2013). De agrofoodmonitor: maatschappelijke waardering van de Agro and Food Sector. LEI, The Hague, the Netherlands.
Specht, M. (2009). Anbau von körnerleguminosen in Deutschland – situation, limitierende faktorenund chancen. Journal für Kulturpflanzen 61(9): 302-305.
Smetana, S., Mathys, A., Knoch, A. and Heinz, V. (2015). Meat alternatives: life cycle assessment of most known meat substitutes. International Journal of Life Cycle Assessment 20: 1254-1267.
Steinfeld, H., Gerber, P., Wassenaar, T., Castel, V., Rosales, M. and De Haan, C. (2006). Livestock's long shadow. Environmental issues and options. FAO, Rome, Italy.
Te Velde, H., Aarts, N. and Van Woerkum, C. (2002). Dealing with ambivalence: farmers and consumers' perceptions of animal welfare in livestock breeding. Journal of Agricultural and Environmental Ethics 15: 203-219.
Van Mensvoort, K. and Grievink, H.J. (2014). The *in vitro* meat cookbook. BIS Publishers, Amsterdam, the Netherlands.
Van der Weele, C. (2013). Meat and the benefits of ambivalence. In: Röcklinsberg, H. and Sandin, P. (eds.) The ethics of consumption. Wageningen Academic Publishers, Wageningen, the Netherlands.
Van der Weele, C. and Driessen, C. (2013). Emerging profiles for cultured meat; ethics through and as design. Animals 3(3): 647-662.
Van der Weele, C. and Driessen, C. (2014). *In vitro* meat: animal liberation? In: Van Mensvoort, K. and Grievink, H.J. (eds.) The *in vitro* meat cookbook. BISpublishers, Amsterdam, the Netherlands.
Verdonk, D.-J. (2009). Het dierloze gericht; een vegetarische geschiedenis van Nederland. Boom, Amsterdam, the Netherlands.
Verhoog, D., Wijsman, H. and Terluin, I. (2015). Vleesconsumptie per hoofd van de bevolking in Nederland, 2005-2014. LEI nota 2015-120. LEI Wageningen UR, Wageningen, the Netherlands.
Vijver, M. (2005). Protein politics. PhD thesis, Twente University, Enschede, the Netherlands.

Section 17. Individuals and food choices

72. A sociocultural analysis of Taiwanese married women's perception and management of food risks

Y.C. Chiu[1] and S.H. Yu[2]*
[1]National Taiwan University, Taipei, Taiwan; [2]London School of Economics and Political Science, London, United Kingdom; ychiu@ntu.edu.tw

Abstract

Since food safety issues, in recent years, occur one after another in Taiwan, this research aims to discover risk reduction strategies used by Taiwanese married women who are responsible for household diet. Adopting the perspective of risk as social construction and a qualitative approach, we interviewed 39 married women, ranging from 31 to 67 years old and coming from different regions of Taiwan, for understanding their food risk perception and management by examining how they purchase, consume and handle foods in their daily life. It is found that food practices of Taiwanese married women are influenced by demographic and sociocultural factors, such as age/generation and position in life course, as well as cultures, such as the culture of worshiping and the custom of shopping at traditional markets. In addition, several 'risk philosophy' are observed. To begin with, 'do one's best' is the most adopted philosophy as the majority of our interviewees use as many risk reduction strategies as they can to ensure food safety. Next, unlike Beck's argument that the responsibility for risks, in western societies, lies in humans rather than God, many of our interviewees draw upon the perspective of fate, employing the risk philosophy of 'what will be, will be.' Thirdly, as it is contended in the past literature on 'optimistic bias' denoting a discrepancy existing between food safety knowledge and food risk handling behaviours, some married women in our study also adopt this reasoning by comparing eating out and dining in. Last but not least, 'detoxification' which has its roots in Chinese medicine is a common risk philosophy as well. Results have also shown that pesticide residues in agricultural products and chemical additives in food products are two food issues that married women feel most concerned. Moreover, married women in our study share a distrust in organic certification and government-certified food in general. Furthermore, two discourses appear repeatedly throughout our interviews, including 'black-hearted businessmen' and 'eat foods not food products,' which are explained by the frequent occurrence of food safety issues and the influence of news media as well as TV programs for health.

Keywords: Taiwan, food risk, food safety, household diet

Introduction

A large number of research has contributed to studying either how journalists report on food risks or how governments communicate food safety issues to the public (Chen and Wu, 2007; Hsieh and Hsu, 2011). As this stream of research details risk epistemology of experts in news articles and technocrats in governments, examining how people perceive and manage food risks in their everyday life is needed for reducing the gap between people's understanding of risk and that of governments and experts. Indeed, many Western studies have found that governments/experts tend to adopt a scientific view of risks which is characterised by calculations of probability and assessment of consequences whereas people usually define and evaluate risks in a subjective manner which is influenced by demographic and sociocultural factors, such as age and traditional values (Douglas and Wildavsky, 1982; Draper and Green, 2002; Lash, 2000; Tulloch and Lupton, 2002, 2003). In Taiwan, women, compared to men, have been taking major responsibility for household diet (Lu and Yi, 2005); this paper, therefore, examines how women who are or were married interpret and handle food risks in their daily life.

EurSafe 2016 – DOI 10.3921/978-90-8686-834-6_72, © Wageningen Academic Publishers 2016

Literature review and research questions

Past research has shown that people's understanding and management of risk changes with their age. For example, people in Lupton and Tulloch's (2002) study tend to take more risks when they were younger but avoid risks when becoming parents as they prioritize their family responsibilities. That is to say, age, along with position in life course and family responsibility, influence people's risk-taking-or-avoiding decisions. Although research participants in our study are all married, age, position in life course and family responsibility are still worth of our attention since their age ranges from the early 30s to late 60s, and some have younger children whereas others are grandmothers.

Next, the geographical region also plays a role in influencing people's risk perception and food purchasing behaviours. Caplan (2000) examined how Welsh villagers and Londoners responded to bovine spongiform encephalopathy (BSE) and found that the former tended to continue consuming beef because they knew where the beef came from, who the farmers were, and how beef was produced whereas the latter might stop eating beef or consume only organic beef because they lacked such knowledge. In other words, even during a period of food scare, people living in rural areas can be confident in local foods because of their knowledge about food production and be certain about the sources of foods due to closer social relations. Draper and Green's (2002) investigation into people's perception of BSE and vCJD risk tells the same story. It is found that residents in the south-east England, compared to their London counterparts, had better knowledge of foods' provenance and thus held greater trust in local food produce. Recognising that geographical region brings differences to people's perception of food risks and their risk-handling behaviours, this study recruited research participants from four regions, including North, Central, South and East Taiwan. It also pays particular attention to the interplay between people's knowledge about production or origin, social relations, confidence and trust.

Thirdly, several studies have shown that trust has a negative impact on people's risk perception (Saba and Messina, 2003; Siegrist, 2000). Take organic food consumption for example. Saba and Messina (2003) found that the more trust people have in the institutions responsible for regulating pesticides, the less risks they perceive in pesticides. Nevertheless, what constitutes trust remains unexplored. Finally, as Tulloch and Lupton (2003: 1) advocates for 'a sociocultural approach to exploring the meanings and significance of risk for 'non-experts' or 'lay people'', contending that 'understandings about risk, and therefore the ways in which risk is dealt with and experienced in everyday life, are inevitably developed via membership of cultures and subcultures', a snapshot of Taiwanese food culture is given in the following paragraph.

Two specific features of Taiwanese food culture are under examination in this article. The first one is the custom of shopping at traditional markets. Traditional markets in Taiwan consist of a variety of vendors selling both cooked and raw foods. In the past and nowadays in some rural areas, poultry is slaughtered on the back of stalls. No matter they are produced on the spot or transported from slaughterhouses, meat is placed on the counters of stalls and free for touching. As one may find, traditional markets have little resemblance to supermarkets which are divided into different sections according to types of food and where raw foods are normally already packed and stored in fridges. Even though many more supermarkets have been created since the first supermarket was built in 1989, people continue doing their grocery shopping at traditional markets (Hsu and Chang, 2002; Liu, 2015). Hsu and Chang (2002) found that older housewives prefer purchasing meat at traditional markets. Interestingly, they consider meat at traditional markets better-quality but less safe (*ibid.*). The second feature of Taiwanese food culture is the concept of using foods to enhance health or treat illness. In other words, foods are not merely for relieving hunger. Instead, they have nutritional and medical values (Anderson, 1997; Koo, 1984; Simoons, 1991; Wu, 1995). Food is categorised as 'hot' or 'cold' in Taiwanese food culture. For example, it is a prevalent practice to eat lamb hot pot in winter time since lamb is considered hot. In short, this paper adopts a sociocultural approach toward risk and asks: How do married or once-married women perceive and manage food risks in their everyday life?

Methodology

In-depth interview method is chosen to help researchers understand participants' opinion, attitudes and motivations in their own words (Berger, 2011; Lindlof and Taylor, 2011). Information on recruitment of research participants was posted on PTT, the largest local social media platform, and snowball sampling was used. 39 Taiwanese women (37 married women, 2 divorced women) were interviewed in the summer of 2014. The majority of them (22 women) are living in the northern region which is characterised by business services and considered highly urbanised. Both central and southern regions have big manufacturing sectors while the eastern region features non-toxic farming and tourism. Then, in terms of age, most of our participants are in their 30s or 50s, constituting nearly 67% of the sample. In addition, over 60% of our participants are working mothers (Table 1).

Participants were interviewed individually in their homes or in some coffee shops in their neighbourhoods. Although an interview guideline covering open questions on a series of food handling stages was used, two authors asked follow-up questions according to participants' responses and encouraged participants to make further elaboration. After transcribing all the interviews, two authors collaborated on developing key themes, narratives and so on. In this article, the focus is of the influences which demographic and sociocultural factors have on our participants' perception and management of risk.

Results

Aligned with the past research, this study found that age has an influence on how women perceive and handle risks. The elder women, compared to the younger ones, are more likely to shop at traditional markets. In their eyes, non-frozen meat is fresher and thus safer than frozen meat products in supermarkets. As one 60-year-old housewife living in North Taiwan said:

> One time I bought some pork ribs at A. mart (one of the biggest supermarket chains in Taiwan) and the smell was very strong. They were raw and frozen. But somehow the taste and smell ... um ... they were stinky.

Table 1. Demographics of participants.

Demographics		Number (n=39)	Percentage (%)
Age	30-39	13	33.3
	40-49	10	25.7
	50-59	13	33.3
	Above 60	3	7.70
Region	North	22	56.4
	Central	5	12.8
	South	5	12.8
	East	7	18.0
Personal situation	In paid work	24	61.5
	Doing housework	15	38.5

The elder women in our study happen to have one or two negative experiences concerning those foods they bought at the supermarket. Therefore, supermarkets, to them, are not the first choice for foods, especially meat and sea foods. However, the younger women often hold a contrary viewpoint as they argued that meat in the supermarket is stored in fridges and thus do not easily go rotten. As we can see, the risk perception and food practices of the elder women are experience-based whereas their younger counterparts seem to adopt a relatively scientific view, drawing on storage condition and temperature.

Generational differences can also be discerned in cooking and eating leftovers. The elder tend to cook large amount of foods at a time. Although such a practice might lead to leftovers, they, on the one hand, prefer making foods more than needed to having not enough foods to eat and on the other hand, they do not consider eating leftovers a problem. As long as vegetable leftovers are, according to them, eaten up in next meal whereas meat leftovers are cleared up in two meals, there is no harm done in eating leftovers. Nevertheless, the younger women hold a stronger dislike to leftovers. They contended that eating leftovers not only is bad for health but also has a poor taste. In other words, the elder women do not find eating leftovers as risky as the younger women do and they actually have developed a way of managing the risk of eating leftovers – according to the food type of leftovers, they decide how many meals they can eat with the leftovers.

Nevertheless, all women share some common ground: they grow more concerned and cautious about managing food risks when becoming mothers. As a 35-year-old housewife from the eastern region said:

> When I was single, I used to eat much canned food because of convenience. It's convenient but in a long run, really bad to one's health. So I made a big change in diet after having a baby. In addition to expiration dates, I also read ingredient lists because they have a great impact on children. For instance, some strange food additives will hinder the development of children's intelligence or result in sexual precocity.

A 52-year-old working mother in the central region gave a similar story: 'When I was on my own, I used to eat outside, like eating noodles at street vendors. But after getting married, I go to markets every morning for ingredients, and when my children were little, I cooked rice porridge'. Adopting a healthier diet and being more careful with foods is for children's health and for setting a good example to children. 'Because of him (her eight-year-old son), I eat less junk food and greasy stuff. I used to eat potato chips, but you know, children cannot resist such kind of food, so I tell myself not to eat it', said a 38-year-old working mother in South Taiwan. It is clear to see how participants' food practice is changed by their positions in life course and corresponding family responsibility.

The sense of responsibility often comes as early as pregnancy: 'During pregnancy, I bought a blender and soybeans to make soy milk for my son. Since then, I have been subscribing to Common Health Magazine (one of the biggest health magazines in Taiwan). Raising a child is difficult. Because their mothers' diet is not normal, many babies have congenital disorder' (35-year-old housewife in the southern region). To several women, mothers have an innate ability to deal with food risks. As a 51-year-old working mother in the northern region argued: 'It's mother's potential. When you become a mother, you'll know what to eat and how to handle foods. As an old Chinese saying goes, 'becoming mothers is being strong''. This study, therefore, contends that in addition to responsibility, it is also a matter of identity; to be more specific, of motherhood.

This study observed that participants in East Taiwan tend to have a better understanding of food production and provenance. As a 54-year-old housewife living in this region said:

> Of course I buy pork here. There is one slaughterhouse nearby, so it's very convenient for me to get pork. There are also many fish ponds in our neighbourhood, so we eat a lot of tilapia. Sometimes I don't even have to buy [fish] because people give me some [fish]. Living in the countryside is really nice. People raise fish on their own and they live just nearby, so it's convenient [to buy fish from them]. And their fish is very fresh, alive and kicking.

The eastern region, leaning against the Central Mountain Range and facing the Pacific Ocean, is rich in natural resources, including foods, and it has one of the largest cultivated areas of organic farming in Taiwan. Participants in this area thus can consume local foods more easily. In addition to its advantageous location and the development of farming industry, closer social relations also contribute to confidence and a sense of security in local foods. Participants often purchase foods from their relatives or friends and receive free foods from their neighbours. What's more, many of them engage in team buying on a regular basis. By doing so, they get more expensive and higher-quality foods at a lower price.

Contrary to Saba and Messina's research (2003) on people's perception of organic fruits and vegetables, the majority of our participants do not consider organic foods safe, but hold a suspicious attitude toward organic farming in general. A 58-year-old housewife in the northern region related her own experience for illustrating why organic farming is a 'stunt': 'I don't believe organic foods. Nowadays the industrial pollution is so severe. How can you grow organic foods in such an environment? I think it's just a stunt'.

Moreover, many of our participants share a distrust in organic certification. In fact, they tend to be suspicious of food certificates and labels in general: 'I am suspicious of GAP (good agriculture practice). I suspect that this certificate can be bought because of the complicity between corporations and governments' (37-year-old working mother living in the southern region).

Trust is of great complexity. Although many studies have demonstrated that between trust and risk perception exists a negative relationship, little research discusses the components of trust. In our study, it is found that trust is associated with a number of personal characteristics, such as honesty and friendliness. As a 41-year-old working woman in the northern region said, 'It depends on if the owner is honest. After talking to the owner for several times, you'll know. Like if he's been helpful, if he's willing to talk to you'.

Culture also plays a complicated role in people's perception and handling of risks. In terms of the custom of shopping at traditional markets, the result is consistent with the precious findings that the elder participants, especially those in their 50s, are more likely to purchase meat at traditional markets. To justify their purchasing behaviours, they would draw upon some unpleasant experiences – for example, meat from supermarkets smelled stinky. Then in terms of using foods to enhance health and treat illness, two common practices are discovered: one is cooking with ginger and sesame oil because these two ingredients are classified as warm and hot respectively in Taiwanese food culture. The other one is drinking herbal tea for detoxication. As a 47-year-old housewife in the eastern region noted:

> Because the food safety issues are serious, you sometimes have to detoxicate by drinking Qing Cao Cha (one type of herbal teas). My second son who is now in puberty and has acnes also knows that Qing Cao Cha helps detoxicate, so he asks me to make Qing Cao Cha for him.

Discussion

This article first demonstrates how age influences married or once-married women's perception of food risks. It is discovered that elder women perceive fewer risks in non-frozen meat and leftovers than younger ones. Additionally, some other generational differences are identified. For example, it is found that elder women are also more likely to shop at traditional markets. Then, our research results show that all women, regardless of their age, become more cautious about food choice when they enter motherhood. That is, in addition to a position in life course and family responsibility, identifying themselves as mothers also plays a decisive role in our participants' management of food risks.

Moreover, it is found that participants living in East Taiwan usually know where foods are from and by whom they are produced. On the contrary, participants in other regions appear to have little idea about food origin and production. Accordingly, we suggest that local governments in these more urbanised areas arrange workshops in communities for educating the public to understand how foods are grown and produced as well we how production-related food risks can be reduced, if not avoided. Furthermore, we argue that the distance between food origin and consumption also influences food practice. To participants in the eastern region, their relations to food producers are relatively close, so they hold greater confidence in and buy more local foods. However, the majority of our participants, regardless where they live, show strong suspicion of food certificates and labels. As a result, we suggest that the government should re-evaluate the current system of food certification and improve the public communication. This study also explores the constituents of trust and finds that trust is often based on some subjective judgments. For instance, several participants claim that friendly vendors are trustworthy and consider buying foods from these vendors safe. Last but not least, some cultural practices of managing food risks are discussed, such as drinking Qing Cao Cha for detoxication.

Even though this study is not designed for exploring food ethics, it still discerns that four participants talked about how religion, the relationship between human and nature as well as concern for animal welfare might influence their choice of foods and their perception of food risks. As a result, the ethical dimension of daily food practices needs further examination. Nevertheless, this study is aware of its limitations. Since all participants are married/or divorced women, research results might not be generalized to other populations. Future research can study the perception and management of food risks among different groups of people, such as singles.

References

Anderson, E.N. (1997). Traditional medical values of food. In: Counihan, C. and Van Esterik, P. (eds.) Food and culture: a reader. Routledge, London, UK, pp. 80-91.

Berger, A.A. (2011). Interviews. In: Berger, A.A. (eds.) Media and communication research methods: an introduction to qualitative and quantitative approaches. Sage Publications, Los Angeles, USA, pp. 135-152.

Caplan, P. (2000). Eating British beef with confidence: a consideration of consumers responses to BSE in Britain. In: Caplan, P. (ed.) Risk revisited. Pluto Press, London, pp. 184-203.

Chen, S.Y. and Wu, Y.C. (2007). Exploring the content and quality of food pollution news reports: a case study involving the 2005 reports on 'Dioxin Duck's Eggs' and 'Malachite Green Groupers'. Taiwan Journal of Public Health 26(1): 49-57.

Douglas, M. and Wildavsky, A. (1982). Risk and culture: an essay on the selection of environmental and technological dangers. University of California Press, Berkeley, USA.

Draper, A. and Green, J. (2002). Food safety and consumers: constructions of choice and risk. Social Policy and Administration 36(6): 610-625.

Hsieh, J.W. and Hsu, M.L. (2011). Framing of scientific development and risks: a news analysis of genetically modified foods in Taiwan. Chinese Journal of Communication Research 20: 143-179.

Hsu, J.L. and Chang, W.S. (2002). Market segmentation of fresh meat shoppers in Taiwan. The International Review of Retail, Distribution and Consumer Research 12(4): 423-436.

Koo, L.C. (1984). The use of food to treat and prevent disease in Chinese culture. Social Science and Medicine 18(9): 757-766.

Lash, S. (2000). Risk culture. In: Adam, B., Beck, U. and Van Loon, J. (eds.) The risk society and beyond. Sage Publications, London, pp. 47-62.

Lindlof, R.T. and Taylor, C.B. (2011). Producing data II: qualitative interviewing. In: Lindlof, T.R. and Taylor, B.C. (eds.) Qualitative communication research methods. Sage Publications, Los Angeles, USA, pp.170-216.

Liu, J.J. (2015). The development history of supermarkets in Taiwan. Storm Media Group. Available at: http://tinyurl.com/z2etn5o.

Lu, Y.H. and Yi, C.C. (2005). Conjugal resources and the household division of labour under Taiwanese social change: a comparison between the 1970s and 1990s social-cultural contexts. Taiwanese Sociology 10: 41-94.

Lupton, D. and Tulloch, J. (2002). Risk is part of your life: risk epistemologies among a group of Australians. Sociology 36(2): 317-334.

Saba, A. and Messina, F. (2003). Attitudes towards organic foods and risk/benefit perception associated with pesticides. Food Quality and Preference 14: 637-645.

Siegrist, M. (2000). The influence of trust and perception of risks and benefits on the acceptance of gene technology. Risk Analysis 20(2): 195-203.

Simoons, F.J. (1991). Food in China: a cultural and historical inquiry. CRC Press, Boca Raton, FL, USA.

Tulloch, J. and Lupton, D. (2003). Introduction: researching risk and everyday life. In: Tulloch, J. and Lupton, D. (eds.) Risk and everyday life. Sage Publications, London, pp. 1-15.

Wu, Y.M. (1995). Food and health: the impact of the Chinese traditional philosophy of food on the young generation in the modern world. Nutrition and Food Science 95(1): 23-27.

Section 18. Economy

73. Economic analysis of activities to prevent foot and mouth disease in Denmark

S. Denver[1], L. Alban[2], A. Boklund[3], T. Halasa[3], H. Houe[4], S. Mortensen[5], E. Rattenborg[6], T.V. Tamstorf[2], H. Zobbe[1] and T. Christensen[1]*
[1]Department of Food and Resource Economics, University of Copenhagen, Rolighedsvej 25, 1958 Frederiksberg C, Denmark; [2]Danish Agriculture and Food Council, Axeltorv 3, 1609 Copenhagen V, Denmark; [3]National Veterinary Institute, Technical University of Denmark, Bülowsvej 27, 1790 Copenhagen V, Denmark; [4]Department of Large Animal Sciences, University of Copenhagen, Groennegaardsvej 8, 1870 Frederiksberg C, Denmark; [5]Animal Health Unit, Danish Veterinary and Food Administration, Stationsparken 31-33, 2600 Glostrup, Denmark; [6]SEGES, Knowledge Centre for Cattle Production, Agro Food Park 15, 8200 Aarhus N, Denmark; sd@ifro.ku.dk

Abstract

The latest foot and mouth disease (FMD) epidemic in Denmark dates back to 1982-1983. Hence, Denmark has not experienced an FMD outbreak in more than 30 years. Still this disease poses a serious threat either as a risk of introduction and spread in Denmark or as a risk of a ban on Danish export of pigs, pork, cattle, beef and milk products due to an outbreak in another country within the EU. It is estimated that a middle sized outbreak of FMD would cost around € 1 billion. It is evident that even though the probability of introducing FMD is very low the consequences are devastating for the agricultural sector and society because the expected costs are so enormous. Therefore, the industry and the public authority have implemented a number of mitigating and preventive activities. The costs of FMD and swine fever related activities in Denmark in 2013 were estimated to be approximately € 32 million. The purpose of the present study is to estimate how changes in resources allocated to the FMD related activities may affect the costs of an FMD outbreak. Nine alternative scenarios describing changes in the contingency plan were formulated by a group of experts from the livestock industry, universities and public authorities. A modified version of Davis Animal Disease Simulation model (DADS version 0.05) was used to estimate costs of FMD outbreaks in each of these alternative scenarios. The modified and updated version by the technical university of Denmark (DTU) is called DTU-DADS. The model simulations indicate that some changes in risk-reducing activities may significantly affect expected costs of an outbreak while other changes have no effect. Our results suggest that increased efforts in terms of efficiently restricting low-risk contacts between farms, such as non-professional visitors and trucks, might reduce the size and costs of an FMD outbreak. In addition, simulations indicate that current resources allocated to depopulation and surveillance could – but only to some extent – be reduced without affecting the size and costs of an outbreak.

Keywords: foot and mouth disease, contingency plan, economic analysis, simulation

Introduction

Access to export markets for animals and animal products depends crucially on Denmark being able to stay free from infectious livestock diseases such as food and mouth disease (FMD) and swine fever (SF). Denmark has been free from FMD since 1982-1983 and classical SF since 1933. In recent years, the threat of SF – and in particular African SF – to the Danish pig industry has increased, as a number of outbreaks have occurred along the eastern European borders. Furthermore, a ban from Russia on all imports of live pigs and pork from Europe has been in place in since 2014 due to outbreaks of African SF in Estonia, Latvia, Lithuania and Poland.

Therefore, these diseases pose serious threats to the Danish export of pigs, pork, cattle, beef and milk products; a threat caused by the risk of introduction and spread in Denmark and by the risk of an outbreak in another country within the EU, which can also lead to a ban on Danish export. It is estimated that direct costs and export loses relating to a middle-sized outbreak of FMD in Denmark would amount to € 1 billion (Halasa *et al.*, 2015). Hence, even though the probability of introducing FMD is very low, the consequences are devastating for the agricultural sector and the society, because the expected costs are so enormous.

The costs of FMD/SF preventive activities in peace time, i.e. a time where no outbreaks are detected, were estimated by Christensen and Denver (2015). The total costs of the FMD/SF preventive activities and contingency plans in 2013 were estimated to approximately € 32 million. Of those, € 3.6 million were paid by the public authorities while € 20 million and € 8.5 million were paid by the pig and cattle industries respectively. While the costs of the preventive activities burden the industries' and the public budgets recurrently, the benefits are less transparent. There is, in general, a high level of attention around the use of public funds and thus about the dimensioning of emergency response team. Furthermore, there is a general ongoing focus on cost savings in the field.

The purpose of the present study is to estimate how changes in resources allocated to the contingency plan may affect the costs of an outbreak. Such an assessment of the cost efficiency of the contingency plan is of great practical value to the public authority and the industry.

Material and method

The simulation model

The Davis Animal Disease Simulation model (DADS version 0.05) is a stochastic simulation model developed at UC Davis in California to simulate spread of FMD (Bates *et al.*, 2003). This model has been modified and updated at the technical university of Denmark (Boklund *et al.*, 2013; Halasa *et al.*, 2015), and is now called DTU-DADS. This model was used to simulate how changes in preventive activities affect the costs related to FMD outbreaks. The core of the model is data on Danish husbandry production and trade practices. From the Central Husbandry Register, information on herd sizes and location, as well as movements between herds were extracted for all Danish cattle, swine, sheep and goat herds and used in the model. Furthermore, the model was parameterized by information from the literature, existing data and expert opinion. A detailed description of input-parameters can be found in Boklund *et al.* (2013) and Halasa *et al.* (2015) (supplementary material). The model covers the time from when infection is introduced (the first infected case) until Denmark again is declared FMD-free by the Standing Committee on the Food Chain and Animal Health (SCoFCAH) and the World Organization for Animal Health (OIE). The probability of introduction of virus to the country is not included in the model. Furthermore, analysis of risks and costs associated with reopening of export markets are not included in the model results. Spread of virus between herds is modelled via:

- High-risk contacts (movements of animals).
- Medium-risk contacts (persons visiting a farm and expected to visit another farm after e.g. veterinarians and artificial inseminators).
- Low-risk contacts (feedstuff trucks, rendering trucks, technicians, visitors).
- Local spread (contains all infection that cannot be explained by other means, i.e. rodents, birds and airborne transmission over short distances up to 3 km, unrecorded animal movements and unregistered visits to the herd).

Basic scenario

The duration of the period from infection is introduced into a country until it is detected is called the high risk period (HRP). In the basic scenario, the HRP was simulated to vary between 18 and 23 days. When an epidemic is detected, 3 km protection zones and 10 km surveillance zones are established around all infected premises. Herds in the protection zone were assumed to be inspected twice by a veterinarian within 3-7 days after creation of the zone and again 21 days later, while herds in the surveillance zones were assumed to be inspected once, 3-7 days after creation of the zone. High-risk contacts to and from herds in the protection and surveillance zones were assumed to be prohibited as soon as an outbreak was detected. In the basic scenario, the most likely compliance rate with this rule was assumed to be 98%. In addition, 80% of medium-risk contacts and 30% of low-risk contact were assumed to be restricted during an outbreak. Furthermore, a 3-day national stand-still was implemented from the day of first detection, including all animal movements in Denmark. The risk of local spread of virus was assumed to be a decreasing function of the distance from the herd and was assumed to continue unchanged during outbreak. It was assumed that there are human resources during an outbreak capable of surveillance of 450 herds per day. Also, it was assumed that there are human resources sufficient to cull.

According to Boklund *et al.* (2013), the three most influential parameters for economic losses of an outbreak are: the duration of the HRP, the probability of local spread, and the probability of transmission from low-risk contacts.

Alternative scenarios

Nine alternative scenarios were simulated to assess the effect of changing different elements of the contingency plan. It was assumed that all changes in contingency plans were intended to be decided upon in peacetime. Some changes affected preventive activities that were carried out on a daily basis by the farmers before as well as after detection of an outbreak, while other changes were only enforced after detection of an outbreak.

Scenario 1 to 3 assumed increased level of biosecurity in cattle herds leading to a 25% reduction of transmission of virus by low-risk contacts. The focus is on cattle herds as Danish swine herds generally are characterized by a high level of biosecurity, while Danish sheep and goat herds mainly are hobby-type herds with low level of biosecurity. Furthermore, previous studies have shown that the simulated epidemics are primarily running in cattle herds. Consequently, we focused on investigating the effect of increased levels of biosecurity in cattle. The increased level of biosecurity was simulated to be in place both during the HRP and after first detection.

Scenario 4 assumed increased restrictions on low-risk contacts during an outbreak in all herd types. In particular, the restrictions on low-risk contacts were simulated to correspond to the restrictions placed on medium-risk contact. These restrictions were only enforced after detection of an outbreak.

Scenario 5 to 7 assumed reduced resources allocated for surveillance or culling. These scenarios illustrate that it may take time to gather qualified and sufficient human resources, when an outbreak is detected. These changes were related to the time after detection of an outbreak.

Scenario 8 assumed a reduced probability of local spread in all herd types. This change was effective both in the HRP and after first detection. Scenario 9 analysed the effect of reducing the HRP to 14 days. Table 1 provides an overview of the alternative scenarios.

Table 1. Overview of alternative scenarios.

No.	Type of scenario	Description of scenarios
1	Low-risk reduction	Low-risk contacts are assumed to follow a Pert distribution. 25% reduction in most likely probability of transmission through low-risk contacts, dairy cattle herds.
2		25% reduction in most likely and maximum probability of low-risk contacts and milk tanker, dairy cattle herds.
3		25% reduction in most likely and maximum probability of low-risk contacts and milk tanker, all cattle herds.
4	Human resource reduction	Resources to surveillance during an FMD outbreak are reduced from 450 herds per day from week one to: week 1 (50 herds), week 2 (100 herds) week 3 and onwards (450 herds).
5		Resources to surveillance during an FMD outbreak are reduced from 450 herds per day from week one to: week 1 (25 herds), week 2 (50 herds) week 3 and onwards (100 herds).
6		50% reduction in resources to culling during an FMD outbreak.
7	Local risk reduction	25% reduction in risk of local spread above 100 m., all herds.
8	Low-risk restrictions	About 80% of low-risk contacts restricted during an outbreak (from 30% in the basic scenario).
9	HRP reduction	Reduction of HRP to 14 days.

For each scenario, the consequences of changes in parameters were simulated, focusing on identifying the effects of the scenarios on the duration of the epidemic, number of infected herds and total economic costs (direct costs as well as costs due to lost exports).

Results and discussion

The simulation results of Scenario 1 to 3 indicate that increased levels of biosecurity in cattle herds corresponding to a 25% reduction in the probability of disease transmission due to low-risk contact do not significantly affect the size and costs of an outbreak of FMD. However, Scenario 8, focusing on intensifying the restrictions on low-risk contacts in the protection and surveillance zones after an outbreak was detected, reduced the median size and costs of an epidemic considerably.

Also, Scenario 7, focusing on reductions in the probability of local spread, had a significant effect on all investigated aspects of the outbreak except the number of infected herds during the HRP.

The results of Scenario 4 to 6 indicate that resources for depopulation and surveillance can be reduced compared to the current level without significantly affecting the size and costs of an FMD epidemic but there is a lower limit to the resources below which the size and costs of an outbreak will increase considerably. Finally, Scenario 9 of reducing the HRP through faster detection of an outbreak will significantly reduce the size and costs of an outbreak.

These results indicate that targeting forces towards reducing low-risk contacts during an outbreak might be a more cost-effective use of resources related to an FMD outbreak than improving biosecurity at a general level in peace time. Such restrictions on low-risk contacts can be designed by imposing more intensive bans on movements of lorries and visitors in the zones. As long as the amount of money spent on increasing restrictions on low-risk contacts with the simulated effect are less than the expected reductions in the costs of an outbreak, it will be beneficial from an economic point of view to impose the restrictions.

The results indicate that resources allocated to depopulation and surveillance during an outbreak – but only to some extent – could beneficially be re-allocated to increase the focus on restricting low-risk contacts and local spread.

It should be noted that the analysis was designed to include the time from when an infection is introduced until Denmark is declared FMD free by SCoFCAH and OIE. Hence, it is a partial analysis that points to ways to re-allocate resources to improve cost-effectiveness of the Danish contingency plans for FMD. The analysis does not claim to identify the optimal design of an FMD policy. As a straight forward example, the analysis does not enable us to prioritize between targeting resources towards further reductions in the probability of introduction of FMD, further reductions in the duration of the HRP, further reductions the number of infected farms – or further reductions in the time before re-opening of export markets. In particular, export losses account for the vast majority of the costs of an FMD outbreak due to e.g. takeover of markets from other countries, loss of confidence in the Danish system, strategic policy, etc. Hence, studies that could shed light on faster re-opening of export markets would have great economic value.

As with any model, the results depend heavily on the model assumptions. For example, high-risk contacts were assumed to be stopped very efficiently as soon as an outbreak is detected with a 98% compliance rate being the most likely. That is the reason why scenarios investigating further restrictions or increased compliance were not included. We have shown that increasing the focus on reducing low-risk contacts has a significant and positive effect on the size of an outbreak.

Conclusions

The present study shows that the epidemiological and economic consequences of an outbreak of FMD in Denmark can be affected by changes in some preventive activities while others are without effect.

Our results indicate that increased efforts in terms of efficiently restricting contacts between farms during an outbreak by low-risk contacts, such as non-professional visitors and trucks, might reduce the size and costs of an FMD outbreak. Hence, if low-risk contacts can be as efficiently restricted as medium-risk contacts costs may be reduced.

Our findings also suggest that current resources allocated to depopulation and surveillance could be reduced – but only to some extent – without affecting the size and costs of an outbreak. These resources could be re-allocated to monitoring and enforcing restrictions on low-risk contacts during an epidemic.

References

Bates, T.W., Thurmond, M.C. and Carpenter, T.E. (2003). Description of an epidemic simulation model for use in evaluating strategies to control an outbreak of foot-and-mouth disease. American Journal of Veterinary Research 64: 195-204.

Boklund, A., Halasa, T., Christiansen, L.E. and Enøe, C. (2013). Comparing control strategies against foot-and-mouth disease: will vaccination be cost-effective in Denmark? Preventive Veterinary Medicine: 206-219.

Christensen, T. and Denver, S. (2015). Costs of Foot-and-Mouth Disease related activities in Denmark. University of Copenhagen, IFRO report 238. Copenhagen, Denmark.

Halasa, T., Toft, T.N. and Boklund, A. (2015). Improving the effect and efficiency of FMD-control by enlarging protection or surveillance zones. Frontiers in Veterinary Science 2(70).

74. EU-sustainability reporting requirements – an incentive for more sustainable food retailers?

S. Stirn[*]*, M. Martens and V. Beusmann*
University of Hamburg, Dept. of Biology, Biocentre Klein-Flottbek, Ohnhorststr. 18, 22609 Hamburg, Germany; susanne.stirn@uni-hamburg.de

Abstract

It is a near-consensus that the way we produce and consume goods and services in modern economies is not sustainable. For the necessary shift towards more sustainable production and consumption patterns, all actors are asked to play an active role. Retailers are in a key position being a mediator between producers and consumers. In order to make their business processes more transparent the amended EU regulation on 'disclosure of non-financial and diversity information' (EU-Directive 2014/95/EU) will introduce the compulsory integration of sustainability information into the reporting duty of retailers in 2017. Here, we will study the publicly available information on sustainability of the five largest German retailers today in order to assess the impact the amended directive might have on their reports. We will compare these reports with best practice examples chosen from sustainability rankings and will reflect about the question of the interplay between state, private companies, and consumers and their respective responsibility for sustainable food.

Keywords: German retailers, best practice examples, shared responsibilities, sustainability

Introduction

The necessary change of unsustainable patterns of consumption and production was already one of the key topics at the UN World Summit on Sustainable Development in Johannesburg in 2002. The Johannesburg Plan of Implementation called for governments, relevant international organizations, and the private sector to play an active role in changing unsustainable consumption and production patterns. Recently, the important role the supermarkets play was again acknowledged in the new UN Sustainable Development Agenda. In goal 12, the business sector is explicitly addressed: 'Encourage companies to adopt sustainable practices and to integrate sustainability information into their reporting cycle' (United Nations, 2015).

The latter point was tackled by the EU: The EU directive on 'disclosure of non-financial and diversity information' (EU-Directive 2014/95/EU) was amended and will come into force in 2017. In the reasoning behind the Directive it is stated that large companies should disclose in their management report relevant information on their policies on, inter alia, environmental matters, social and employee aspects, and the respect for human rights in order to meet the needs of investors as well as the need to provide consumers with easy access to information on the impact of businesses on society. In 2016, the EU Commission will publish guidelines to assist companies in the reporting process, providing them with a methodology that will facilitate the disclosure of relevant, useful and comparable non-financial information. In Germany, the German Sustainability Code (German Council for Sustainable Development, 2015) can be used by the companies in fulfilling their reporting duties. Based on 20 criteria, companies submit a summary declaration of the measures relating to the environmental, social and economic dimensions of sustainability. These criteria are comparable to the key performance indicators from the Global Reporting Initiative (GRI) and the European Federation of Financial Analysts Societies (EFFAS).

EurSafe 2016 – DOI 10.3921/978-90-8686-834-6_74, © Wageningen Academic Publishers 2016

We used these indicators to analyse the publicly available reports of the top five retailers in Germany in order to answer the question (a) whether the new EU directive will change the quantity and/ or quality of the data the companies will have to provide and (b) whether this will promote more sustainable production and consumption patterns. As comparators we used the information already provided by Kesko and Ahold, best practice examples from the Global 100, an index of the Global 100 most sustainable corporations in the world and the Dutch Responsible Supply Chain Management Benchmark, respectively. For former analyses in the UK see Jones *et al.*, 2013 with references to preceding reports in 2011 and 2005.

Results

The comparison of sustainability achievements is a complicated task on its own. Therefore, several ranking lists exist which should facilitate such a comparison for investors and consumers, e.g. the platform 'RankingTheBrands.com' offered by SyncForce (http://www.rankingthebrands.com). When referring to these rankings, it became obvious, that only few retailers are listed at all: In the most recent rankings, Walmart, US (ranked 4th in the 'Change the World' global ranking of Fortune), Kesko, FI (ranked 15th in the Global 100 Most Sustainable Corporations by Corporate Knights), and Ahold, NL (ranked 12th in the Dutch Responsible Supply Chain Management Benchmark by the Dutch Association of Investors for Sustainable Development). Since Walmart is based in the US, we chose Kesko and Ahold as best practice examples for European retailers. In Germany, the Institute for Ecological Economy Research (IÖW) and the business initiative future e.V. – verantwortung unternehmen are publishing the biannual IÖW/future-ranking (IÖW/future e.V., 2012). Here, the top German retailers are listed within the middle and lower group of businesses.

For our analysis we chose the top five German retailers Edeka, Schwarz-Group, Rewe, Aldi, and Metro Group, which in 2014 held a market share of 73,33% in Germany, equivalent to 181.132 Mill. Euros (Lebensmittel Zeitung, 2015). Three of them provide some information on sustainability already in their reports of 2014 or 2013/2014, respectively. The respective reports are: Edeka Cooperation Company Report 2014 (http://tinyurl.com/o9qty8g), Rewe Group Sustainability Report 2013/2014 (http://tinyurl.com/hegs4gk) and Metro Group Corporate Responsibility Report 2013/2014 (http://tinyurl.com/hb7hqod). For the Schwarz Group and Aldi no information concerning sustainability reporting is publicly available.

We focussed on the environmental and strategically/social dimensions of sustainability since it is known that for consumers sustainable consumption options are those that reduce the ecological and social problems along the supply chain while still being economically viable (Hanss and Böhm, 2012). When comparing the three reports of Rewe, Metro Group, and Edeka, it became obvious, that the information provided is quite different regarding the aspects which are disclosed as well as regarding the quantity and quality of information.

The Rewe Group was ranked 33rd according to the IÖW/future ranking and held a market share of 15.4% in 2011. Information concerning sustainability can be found in four sustainability reports, in a brochure and online via the menu item 'Sustainability'. One of the key principles of the Rewe Group is 'to bring sustainability out of its niche status' (own translation). The 'Guideline on Sustainable Business' has four major pillars: 'green' products', 'energy, climate and environment', 'employees' and 'social commitment', which are reflected in the reporting details:

> Organic products hold a share of 3.51% of the total sales value. More than 67,000 sustainable products are listed in the German product range. Established labels for

sustainable products within the Rewe product range are: ASC, Cotton made in Africa, Fairtrade, FSC, MSC, PEFC, Rainforest Alliance or UTZ certified.

In 2014, 36% reduction in CO_2 emissions ... The emissions due to cooling systems could be reduced by 23% (in comparison to 2012) targeted to a reduction of 35% up to 2020 ... The power consumption per sales area could be reduced by 4% (in comparison to 2012). The share of CO_2 emissions due to logistics is 10%. We will reduce the driven kilometres, optimize the efficiency of the lorries, rely on modern technology, and adjust the driving behaviour.

In view of the complex supply chain the Rewe Group established a quality assurance and control system. Suppliers of own brands are obliged to prove a certification according to IFS FOOD or the British Retail Consortium (BRC)

Rewe Group Sustainability Report, 2013/2014

The Metro Group was ranked 48th according to the IÖW/future ranking in 2011 and held a market share of 12.03%. Information concerning their sustainability approach is available online and in the Corporate Responsibility Report 2013/2014. The sustainability vision of the Metro Group 'We offer quality for life' is seen as the 'need to ensure our business activities generate additional value and simultaneously reduce any negative impact they may have.' Accordingly, the following spheres of action are of central importance for METRO GROUP:

1. Dealing with employees and shaping their working conditions.
2. Working with suppliers and helping them to comply with quality, social and environmental standards.
3. Procuring resources, using them efficiently and avoiding waste.
4. Providing customers with high-quality products with the soundest possible social and environmental credentials and offering sustainable alternatives.
5. Social commitment with regard to our core business and in areas where help is urgently needed.

But when it comes to actual implementation, targets and figures are given only for climate protection:

METRO GROUP is reducing its greenhouse gas emissions by 20%, from 330 kg/m^2 in 2011 to 264 kg/m^2 in 2020.

For organic products sales figures are given:

We aim to provide our customers ... with safe, high-quality products and goods that have been produced by socially and environmentally sound means with a minimum of resources. Various labels are used to inform customers about the individual sustainability aspects of our products and to help them to make their decisions ... In Germany, products certified in accordance with the EU Regulation on organic farming accounted for around € 96 million in sales at METRO Cash & Carry, Real and Galeria Kaufhof during the reporting period.

Other statements are often more intentional:

> METRO GROUP is initiating and promoting the development of an international, multi-industry technical traceability solution that can be used for all products.

The Edeka Verbund was ranked 90th according to the German IÖW/future ranking and had the highest market share of 21% in 2011. Information concerning sustainability can be found in the Edeka Cooperation Company Report 2014 as well as in the brochure 'Acting responsibly' (own translation), available online.

The main focus of the reporting is on the responsibility concerning products and supply chain, the environment and social commitment. Topics like sustainable fisheries, organic products, Fair Trade, quality assurance as well as environmental and climate protection are explicitly stated. Special emphasis is put on the strategic partnership with the WWF:

> There are more than 300 own brands with the additional label of the WWF. They comply with certified ecological standards like the EU regulation on Organic Farming or the label of the Marine Stewardship Council (MSC). [In addition, a] nearly 100% conversion to certified sustainable palm oil in own brands [was achieved].

> From 2015 on Edeka, together with the WWF, is looking for more ecological friendly packaging in the case of i.e. plastic, glass, and metal.

Concerning climate relevant emissions, the statement is relatively vague and driven by economic considerations:

> Emissions of the company as well as of selected own brands over their respective life cycle are recorded in order to reveal savings potential ...

The Schwarz Group was ranked 113th according to the IÖW/future ranking and had a market share of 18.8% in 2011. The Schwarz Group still belongs to the group of non-reporting businesses. There exist no common principle for the group and only marginal information is available on the websites of the members Lidl and Kaufland.

Aldi (Aldi Süd, Aldi Nord) was ranked 118th according to the IÖW/future ranking and had a market share of 11.14% in 2011. Comparable to the Schwarz group no reports are available for the group. On its website Aldi Süd presents its general positions and actions concerning their range of sustainable products, sourcing of fish and wood as well as regarding the energy management and logistics. Aldi Nord presents some information concerning the buying principles for coffee, tea, and fish.

When comparing the German reports to our best practice example Kesko, Finland, the differences were even more pronounced. Kesko uses the Global Reporting Initiative G4 guidelines in its Corporate Responsibility Report for 2014. However, the General Standard Disclosures according to the GRI G4 guidelines are presented in a much wider scope than required by the Core option. Additionally, the targets as well as the current achievements of the respective targets are given.

> Kesko's long-term corporate responsibility work is based on our strategy and responsibility programme. All our operations are guided by our value: the customer and quality – in everything we do.

> We will improve our annual energy efficiency by 65 GWh by 2016: During 2014, Kesko improved its energy consumption by 59 GWh and achieved 90% of its objective.
>
> We recover the waste generated in our operations to achieve zero landfill waste: In 2014, the waste recovery rate of Anttila's logistics centre was 99%, and that of Keslog's central warehouses and terminals 98%. The recovery rate of waste generated in stores was 96% and the recycling rate was 67%. VV-Auto Group's waste recovery rate was nearly 100%.
>
> We offer a wide selection of Pirkka responsible products; 500 Pirkka products meeting a responsibility criterion by 2015 and 200 Pirkka organic products by 2015: In 2014, the Pirkka range included 44 Fairtrade products, 116 Organic products, 29 MSC certified fishes, 16 UTZ certified products and 13 products containing certified sustainable palm oil (CSPO).

An outstanding specialty of Kesko's sustainability strategy is the AITO bag which was developed in cooperation by Motiva, the Heart Symbol and the K-Group. The AITO bag is a product that combines several different benefits for consumers: faster shopping trips, support for good choices, easier cooking, breaking cooking routines and avoiding unnecessary purchases. Each meal bag contains all the ingredients and the recipe for an easy everyday meal with a focus on health and environmental aspects, in addition to good taste and the ease of cooking (Kesko, 2015).

The differences in the quality of the reports holds also true for the Responsible Retailing Report 2014 of Ahold: Here not only the strategy Ahold is pursuing is disclosed but also a 'Responsible retailing scorecard' showing the progress achieved in the five priority areas against the Group-wide targets and commitments. There one can find e.g. the target/commitment to:

> Source 100% of the six critical commodities for own-brand products in accordance with industry certification standards by 2015' and the corresponding achievement of 'Coffee: 83%, Tea: 82%, Cocoa: 40%, Palm oil: 100%, Soy: 0%, Seafood: 44%.

with the additional explanation:

> that since they cannot yet directly link the amount of segregated palm oil to products with 100% accuracy, 100% of our estimated palm oil use (in 2014, 4,000 tons) has been offset using GreenPalm certificates.

Discussion

When coming back to the first research question: 'Will the new EU-sustainability reporting requirements change the quantity and/or quality of the data the companies will have to disclose?', the answer is certainly yes, because from 2017 on there will be the formal requirement to fulfil reporting duties on a given set of sustainability indicators. Therefore, the new EU directive will certainly change the reporting praxis of the German retailers analysed: two of them, the Schwarz Group and Aldi will have to report on sustainability issues for the first time, but also the three other companies will have to close some gaps to fulfil the requirements.

The second question on whether the amended EU-sustainability reporting requirements will promote more sustainable production and consumption patterns is not that easy to answer because it is directly related to the distribution of responsibility among the different actors, consumers, private businesses and government.

The role of retailers in this transition is twofold: On the one hand they should act responsibly by offering sustainable products at prices fair to producers, their employees and customers. On the other hand, they have to communicate in a transparent and comparable way on their activities in order to improve competition between firms' different approaches towards sustainability.

The role of consumers in the transition is twofold as well: First they have to inform themselves in order to make the right choice at the shelves. Secondly, they must be willing to pay fair, that means higher prices for sustainably produced and distributed products. The common interest of retailers and consumers lies in the fact that retail firms are economically interested in product-differentiation in order to capture part of consumers' willingness to pay for products with higher quality attributes, e.g. sustainability standards. However, this might not be enough from a societal perspective, since retailers have to be economically viable and not all of the consumers are willing to pay more for sustainability attributes.

Therefore, also government action is required to pursue sustainable development. Within the EU governments differ in their self-concept of the role of the political system between liberal positions and more interventionist concepts. A common denominator is improving education and information and investments in public research. This is where the new EU directive comes in. If the guidelines still to be agreed upon increase transparency and thereby promote competition of different approaches towards sustainability then the effect of the EU directive will be positive. However, not every good intention leads to a good result, as the experience with the EU proposal for the new Organic Regulation shows: Stakeholders as IFOAM (2014) criticised: 'The EU Commission wants to strengthen organic in Europe, yet it creates new hurdles.' For imports from low developed countries higher product standards are requested. One lesson one could draw would be that increasing standards too high might discourage producers. In contrast, setting standards too low would not yield sufficient effects for sustainability. Another critical task for governments is the control of the concentration process within the retail sector. In 2012, an internal Task Force dedicated to the food sector was set up by DG Competition within the European Commission (Food and Drink Industry, 2011). In Germany, the Bundeskartellamt (Federal Cartel Office) is responsible. In 2015, it prohibited the acquisition of around 450 Kaiser's Tengelmann outlets by EDEKA (Bundeskartellamt, 2015). However, the Federal Minister of Economic Affairs and Energy, Gabriel, intervened with the argument to safeguard employment. This decision induced the competitors to go to court. This shows the difficult and contested task to balance the different dimensions of sustainability, in this case between higher employment and lower food prices as well as between the positions of competitors. In other European countries governments have intervened in markets by e.g. imposing taxes on saturated fat in Denmark. After one year the Danish government revoked the law because of the intended effects not being reached i.a. because consumer purchases of fats has moved from supermarkets to discount stores (Jensen and Smet, 2012).

A pivotal question remains: can the main burden of the necessary change towards sustainable consumption be placed on the consumers (Sustainable Consumption Roundtable, 2006; Grunwald, 2010 and the responses in GAIA 20/1, 2011)? Responsible consumers can contribute to sustainability by their own buying actions and by showing that it is possible to realise more sustainable lifestyles. This might encourage others to follow this route. However, the shortcomings are obvious: Consumers not convinced of the necessity to change can act as 'free riders' going on to exploit other human beings and nature. In addition, if the group of responsible consumers remains small, their effect on improving sustainability will not be significant. Hence, political action is required complementary to personal action. As was shown above searching for routes towards sustainability is a learning process and requires the constant check of successes and failures in self organisation as well as in governance, properly designed reporting can be a helpful tool.

References

Bundeskartellamt (2015). Bundeskartellamt prohibits takeover of Kaiser's Tengelmann by EDEKA. Available at: http://tinyurl.com/jt33ad6.

European Union (2014). Directive 2014/95/EU of the European Parliament and of the Council of 22 October 2014 amending Directive 2013/34/EU as regards disclosure of non-financial and diversity information by certain large undertakings and groups. Official Journal of the European Union L 330(1).

Food and Drink Industry (2011). Data and trends of the European food and drink industry. Available at: http://tinyurl.com/hu3zdke.

German Council for Sustainable Development (2015). The sustainability code – benchmarking sustainable economy. Available at: http://tinyurl.com/jy73e25.

Grunwald, A. (2010). Wider die privatisierung der nachhaltigkeit. Warum ökologisch korrekter konsum die umwelt nicht retten kann. Gaia 19(3): 178-182.

Hanss, D. and Böhm, G. (2012). Sustainability seen from the perspective of consumers. International Journal of Consumer Studies 36: 678-687.

International Foundation for Organic Agriculture (IFOAM) (2014). Controversial overhaul of EU organic regulation raises red flags. Available at: http://tinyurl.com/zfsecfd.

Institut für ökologische Wirtschaftsforschung/future (Hrsg.) (2011). IÖW/future-ranking der nachhaltigkeitsberichte deutscher unternehmen 2011. Kurzfassung der Ergebnisse, Berlin, Münster, Germany.

Jensen, J.D. and Smed, S. (2013). The Danish tax on saturated fat – short run effects on consumption, substitution patterns and consumer prices of fats. Food Policy 42: 18-31.

Jones, P., Hillier, D. and Comfort, D. (2013). In the public eye: sustainability and the UK's leading retailers. Journal of Public Affairs 13(1): 33-40.

Kesko (2015). Meal bags make customers daily lives easier. Available at: http://tinyurl.com/hbrg4nh.

Lebensmittel Zeitung (2015). Top30 lebensmitteleinzelhandel 2015. Available at: http://tinyurl.com/jrr3bk8.

Sustainable Consumption Round Table (2006). I will if you will. Towards sustainable consumption. Available at: http://tinyurl.com/qjdg39m.

United Nations (2015). Integrated and coordinated implementation of and follow-up to the outcomes of the major United Nations conferences and summits in the economic, social and related fields. Follow-up to the outcome of the Millennium Summit. United Nations General Assembly, A/69/L.85.

Section 19. Animals in research

75. Food, health and animal research: on the need for ethical reflections on uncertainty in animal research

A. Molavi and F.L.B. Meijboom*
Utrecht University, Ethics Institute, Janskerkhof 13, 3512 BL Utrecht, the Netherlands; a.molavi@uu.nl

Abstract

Both food and biomedical research on animals are meant to further our knowledge about improving health. However, due to a generally low translational value, the use of animals for research has become a contentious issue, hence calls for more efficient replacements. However, data obtained through the current alternatives to animal testing, such as *in vitro*, are also deemed insufficient without an *in vivo* component. Because, whatever their translational potential might be, most research models' physiological and genetic traits closely mimic those of humans. This raises the question to what extent factors other than inter-species differences contribute to inefficiency of animal studies and whether improving them can make up for these differences. Moreover, the need to further health-related research on animals should be weighed against the moral mandate to avoid unnecessary harm. However, the problem of uncertainty in determining the translational value of animal studies for clinical trials complicates answers to such questions. In this paper, we argue that a more integrated approach by all the relevant stakeholders to actively improving translatability of animal models within the research chain would help inform responsible research strategies, thereby improving the ethical acceptability of animal research.

Keywords: translational strategies, integration, responsible research

Problems of translation as an ethical problem

Scientific evidence supporting potential health benefits of probiotics have contributed to their dietary and clinical use becoming commonplace all around the world. However considerable uncertainty still remains as to the exact mechanism behind the health effects of these bacteria. Low methodological quality, poor experiment design, and regulatory hurdles have all been blamed for this uncertainty. Moreover, problems associated with applying data from animal studies to clinical trials add to the complexity. This is because proceeding to clinical trials without ensuring reliability of animal data could result in unwanted consequences. For example, in 2008, a Dutch trial of probiotics for acute pancreatitis led to 24 unexpected deaths despite the researchers having followed through preceding studies on both humans and animals (Sheldon, 2008). A two-year investigation by the Dutch authorities found that the flaws with the study's design, approval, and conduct were all contributing factors to its failure in clinical trials. The fact that the product developed for this trial had only been tested on animal models raised the question whether the ethics committees that approved this study had weighed the usefulness of animal data critically enough (Sheldon, 2010). It is important to note that similar probiotic experiments on both humans (Weale *et al.*, 2003) and animal models (Elzaquirre *et al.*, 2002) had yielded promising results prior to this study, but nevertheless fell short of ensuring a reliable basis for an extensive clinical trial.

The shortcoming associated with animal-to-human translation is not limited to the probiotic studies. Recent systematic reviews of experiments on animals have consistently shown that there are different arrays of obstacles to translating data from animal research into the ones that would tangibly contribute to human health. Some argue that such obstacles render animal studies ethically unjustifiable, casting doubt over the use of animals in general. But an assumption underlying the argument for continued use of animals is that the evidence for a minute proportion of animal studies finding their way into clinical trials should suffice to confront the sceptics alone. Such assumption usually focuses on the physiological

EurSafe 2016 – DOI 10.3921/978-90-8686-834-6_75, © Wageningen Academic Publishers 2016

and genetic similarities the human organism shares with those of animals such as mice and rats. However, taking ethics seriously in the face of these obstacles to translation remains an important issue for current innovative projects that aim to replace the inefficient ways with solutions that would improve the translational value of animal models. The idea is that attempting a drastic change in methodologies to determine the translatability of a particular line of research could offer genuine chances for practical debate on ethics. Such a debate would focus on the use of animals in research and the use of animal data in clinical trials.

The background of current problems

There seem to be a clear ethical imperative to strive for optimised methods and models in animal research. Nonetheless, two sets of problems appear to be associated with the process of choosing the right animal models for a better overall efficiency:

The first set arises from the inevitability of uncertainty when it comes to bench-to-bedside translation. This uncertainty arises from biological differences between humans and animals. If we establish that the negative contribution of these differences to the inefficiency of animal research cannot be considerably diminished by changes in methodology, the ethical acceptability of the current laboratory practices would be difficult to defend solely on account of the current level of translational success. On the other hand, it is not certain how dismantling the existing system, which requires a vast quantity of animals, could be justified from prevalent ethical perspectives across the research chain. Thus, the conflict worth exploring in this paper will be between the moral imperative to continue health-related research and the imperative to justify the exploitation of animals' lives in exchange for small chances of clinically relevant input.

The second set of problems is based on a hypothesis that clearly defined measures to regulate the use of animals in laboratories can potentially pose hurdles to scientific innovations that would ultimately improve the situation for animals. That is, being bound by rules on numbers and the choice of animal models, researchers could be dissuaded by what they would see as a flaw in the distribution of responsibility within the research chain. Thus, in light of the first set of problems mentioned above, a second conflict emerges: Is it morally permissible to hinder innovations by controlling quantitative and qualitative aspects of selecting animal models?

As far as the current literature on the efficiency of food and biomedical research on animals is concerned, these problems are shared universally, only in different measures. Nonetheless, practical engagement with a holistic approach to the research chain with a clear aim of increasing translatability and ethical acceptability seems to be scarce. Thus, besides an overview of the existing literature, the multidisciplinary project attached to this paper will also rely on data found by an ongoing systematic review of animal studies for Cystic Fibrosis (Leenars, 2016) and Rheumatoid Arthritis (Leenars, 2015). The protocols of these reviews are designed to connect the dots between shortcomings in animal studies and those in clinical trials. The goal is to identify the break in the research chain and find translational strategies for its amendment. The ideal strategies would entail a fair distribution of responsibility across the research chain and should be applicable to all health-related research. In what follows, we discuss the normative challenges of putting forward a strategy as a morally responsible one.

The problem of uncertainty

Despite the fact that the clinical relevance of animal research has rarely been questioned by the stakeholders within the research chain, evidence suggests that researchers do regularly act upon the assumption that it rarely informs clinical trials with sufficient certainty. For example, systematic reviews of a series of studies on treatment of stroke, wound healing and fluid resuscitation for bleeding trauma patients have shown that the clinical trials were conducted without taking into account the findings from preliminary animal experiments partly because they were judged as being unreliable (Pound *et al.*, 2004). However, as Bakhle (2004) argues, such a finding does not automatically pave the way for questioning animal use in general as it only accounts for a very small proportion of research. This means that selective investigations cannot give a fair idea of the extent to which animal research actually contributes to improving human health. Thus, given the complexity and scope involved in conducting a general overview of this contribution, a coherent discussion on animal use is bound to take place on a case by case basis. However, this also implies that a low animal-to-human translatability within a particular study cannot be weighed against the overall benefits of animal research.

Focusing on the translational success of particular studies can potentially assist researchers in determining the right animal model for their similar projects. For example, BALB/c mouse, one of the most widely used strains in experimentation, was used in a study to measure the influence of a probiotic on its gastrointestinal tract (GIT). The clinical trial that followed this experiment showed a similar functionality of this probiotic on human GIT with comparable success, contributing to this strain becoming a staple probiotic model (Dunne *et al.*, 1999). However, a purely statistical analysis, more often than not, fails to consider the normative aspect of dealing with the problem of uncertainty. This problem stipulates responses to two interrelated questions:

1. Can we morally justify at least some animal experiments when the translational outcome is always uncertain?
2. If yes, which measurable criteria should be taken into account to define a 'responsible' amount of uncertainty?

To answer these questions, we can start by the uncontroversial premise that uncertainty is the very aspect of scientific research that has been driving the search for new knowledge through applying hypotheses to practice. If we accept that the route to medical progress will always have uncertainty as an inevitable component, then at first glance, not much can be said about how we should deal with it from a normative standpoint. Depending on the position we take on the moral spectrum regarding the treatment of non-human animals, we would either have to oppose all kinds of animal experimentation on absolutist moral grounds and jeopardise medical progress, or define an acceptable level of uncertainty for ethical research. If, as Festing and Wilkinson (2007) argue, we grant that stopping all animal testing now would be detrimental to human health, the abolitionist position will be less tenable than those that would allow animal testing conditionally.

One way to tackle unsatisfactory or non-pragmatic aspects of responses from classic ethical positions is to shift the normative spotlight on the intention of the researcher and the goals set out by different players within the research chain. The practical realisation of such move could be as follows. First, we could use empirical data from particular studies to show that robust translational strategies could contribute to efficiency and therefore ethical acceptability of a line of research. A prerequisite for applying such strategies to the function of the research chain is to improve communication between the stakeholders. Because an effective communication between the stakeholders creates room for ethical dialogue on uncertainty across the chain, the chances for a more responsible evaluation of translatability increases. Moreover, a multidisciplinary approach which would integrate clinical and preclinical stages means that the stakeholders will deal with a clearer representation of uncertainty at stake. The idea is that, making decisions on the use of animals based on the level of uncertainty requires a coherent and consistent identification of this problem by both animal researchers and the users of translatable data.

Thus, in addition to the regulative principles such as the 3R, an integrated response to uncertainty based on translatability will help make research more responsible. However, attending to the problem of uncertainty also highlights another moral dilemma we would face in our search for strong translational strategies. Institutional hurdles that could frustrate scientific innovations and effective changes within the research chain could be explored from a normative point of view.

The need of a more integrated approach

Ensuring an acceptable rate of translational success as a component of efficiency in health-related research necessitates innovation and change in the status quo. Since genuine innovation requires sufficient space for researchers to think out of the box, we might have to rethink some of the limitations we impose on the research chain. Most regulatory restrictions on animal use do fulfil their objective of ensuring a better animal welfare in laboratories. However, plain rules to control and assess numbers, models and replacement of animals may not do justice to the translational complexity that is inherent in animal studies. A 2011 survey of Canadian researchers' views on the 3R as the staple welfare component of research regulations demonstrates that the replacement tenant is largely considered unrealistic by the chain insiders (Fenwick *et al.*, 2011). Moreover, it reflects a sceptic view of reduction and refinement. This is because, reducing numbers, reusing models and some refining measures applied to an experiment could potentially create obstacles in achieving clinically relevant results. The first obvious response to this survey is to propose institutional and regulative changes that would capture the complexity of particular research chains and allow for some flexibility for scientific innovation (Tamayo, 2008). Then, the changes which would affect the quantity and the manner of animal use have to be ethically justified. However, these justifications should be offered while taking into account a dichotomy of duties that are usually conferred upon the researchers: To further knowledge that is demonstrably useful for human health, and to avoid unnecessary harm to animals. The following proposal for a revision of the current system could help reconcile the two duties.

As a first step, we could assess whether research regulations focused on directly influencing animal use (e.g. the need for the approval of an ethical committee) can improve animal welfare better than those that are focused on overall efficiency. The advantage of welfare-based rules such as the 3R is that they help incorporate a binding normative component to the decision-making process for research applications. This automatically encourages extra attention to efficiency. Because cost-effectiveness analyses imbedded in welfare regulations seek to justify experiment on animals based on its predicted use and necessity for humans. However, as discussed above, basing justifiability on the translational value of animals can be quite objectionable. This is because normative evaluations are usually expected to rely on at least some invariable factors. One consequence of relying on translatability is that having to make predictions without verifiable certainty becomes inevitable for most health-related research.

A patient-cantered evaluation of efficiency in research chain, however, seems to fill the normative gap left by uncertainty. If we direct the energy and the resources put into ensuring a particular level of welfare for animals into securing consistent clinical outcomes in a research chain, a more holistic approach can be possible. This approach will juxtapose the needs of patients with the interests of animals in a more effective way than the welfare approach does. Because choosing efficiency as the overriding principle would open up the flow of ethically relevant data between the two ends of the chain. Patients and clinicians will be able to achieve a better understanding of the pre-clinical complexities in translation, thereby making a more informed use of the data available. Likewise, animal researchers are more likely to act on the necessity to contribute directly to finding health solutions.

The normative questions applicable to this approach would be asked on a case by case basis, taking into account our ethical duties to both humans and animals. In practice, the need to replace, reduce, or refine will remain as a guiding principle as they can be conducive to efficiency regardless of its ethical relevance. However, the responsibility to fulfil this need should be shared by all the relevant stakeholders. That is, the researchers who directly engage with animals should not have to carry the moral burden to justify their experiment plans isolated from other stakeholders. This view can be defended as follows:

All welfare-based positions on animal experimentation are bound to define in quantitative and qualitative terms what constitutes responsible research. These positions might differ on the use of certain species or models within one specie. However, when an experiment plan fulfils the necessity criterion for human health, what is remained of the ethical evaluation will take the form of an empirical investigation that can confine the researcher to limited choices on methodology and numbers. What is problematic in this system is that applying the necessity check and welfare evaluation will not always mean that a research proposal will have a higher chance of producing clinically relevant data. Whereas, an active push for decision-making based on translational value across the chain will mean all stakeholders will have a shared understanding of what could improve the result. For example, clinicians and patients involved in treating a disease could weigh in the necessity discussion as well as that of the suitable methodology. Similarly, animal researchers could conduct systematic reviews of their line of research to ensure the clinical goals match those found in basic research.

Thus, for all the abovementioned reasons, ethical justification of continuing research on animals in the face of uncertainty have to be backed by concrete strategies to maintain an efficient level of translation between animal studies and clinical trials. Moreover, where an empirical investigation forms the basis of solutions to ethical challenges, the integrated approach discussed above can help increase animal welfare by ensuring a more holistic approach to conflicting moral views across the chain. Because extending a patient-cantered focus to the whole of the research chain will encourage a more robust and consistent understanding of problems of translation and uncertainty in animal studies.

References

Bakhle, Y.S. (2004). Missing evidence that animal research benefits humans: evidence is all around us. British Medical Journal 328(7446): 1017.

Dunne, C., Murphy, L., Flynn, S., O'Mahony, L., O'Halloran, S., Feeney, M., Morrissey, D., Thornton, G., Fitzgerald, G., Daly, C., Kiely, B., Quigley, E., O'Sullivan, G., Shanahan, F. and Collins, J. (1999). Probiotics: from myth to reality. Demonstration of functionality in animal models of disease and in human clinical trials. Antonie Van Leeuwenhoek 76: 279-292.

Eizaguirre, I., Urkia, N.G., Asensio, A.B., Zubillaga, I., Zubillaga, P., Vidales, C., Garcia-Arenzana, J.M., Aldazabal, P. (2002). Probiotic supplementation reduces the risk of bacterial translocation in experimental short bowel syndrome. Journal of Paediatric Surgury 37(5): 699-702.

Fenwick, N. and Griffin, G. (2011). Scientists' views on the three Rs: comparison of Canadian and UK scientists. Canadian Council on Animal Care 1(12): 319-321.

Festing, S. and Wilkinson, R. (2007). The ethics of animal research. Talking point on the use of animals in scientific research. EMBO Reports 8(6): 526-530.

Leenars, C. (2015). Experimental design in methotrexate efficacy studies for rheumatoid arthritis. SYRCLE. Available at: http://tinyurl.com/hcgc9wh.

Leenars, C. (2016). Experimental design in methotrexate efficacy studies for rheumatoid arthritis. SYRCLE. Available at: http://tinyurl.com/zwuej3v.

Pound, P., Ebrahim, S., Sandercock, P., Bracken, M.B. and Roberts, I. (2004). Reviewing animal trials systematically (RATS) group. Where is the evidence that animal research benefits humans? British Medical Journal 328(7438): 514-517.

Sheldon, T. (2010). Dutch probiotics study is criticised for its 'design, approval, and conduct'. British Medical Journal 340: c77.

Sheldon, T. (2008). Dutch trial of probiotics in acute pancreatitis is to be investigated after deaths. British Medical Journal 336(7639): 296-297.

Weale, R. and Edwards, A. (2003). Letter 1: randomized clinical trial of specific lactobacillus and fibre supplement to early enteral nutrition in patients with acute pancreatitis. British Journal of Surgury 90: 122-123.

76. Killing of animals in science – is it always inevitable?

N.H. Franco[1,2]
[1]*IBMC – Instituto de Biologia Molecular e Celular, Universidade do Porto, Rua Alfredo Allen, 208, 4200-135 Porto, Portugal;* [2]*I3S – Instituto de Investigação e Inovação em Saúde, Universidade do Porto, Rua Alfredo Allen, 208, 4200-135 Porto, Portugal; nfranco@ibmc.up.pt*

Abstract

Within the ethical discussion of animal experimentation, the questions of why, how many, and under what circumstances animals are (or should be) used takes precedence over the fact that virtually all lab animals are killed after their scientific utility. When death is indeed an issue, the discussion often concerns the circumstances of death, from a welfare point-of view. This is a likely consequence of two factors: firstly, killing being seen as an inevitable consequence of animal use and, second, a predominantly 'welfarist-utilitarian' influence in the ethical and legal framework on the acceptability of animal research. While the former leads to the killing of laboratory animals being implicitly accepted along with the acceptance of animal research itself, the latter makes death a lesser issue (provided it is carried out humanely), as 'being dead' is not in itself seen as a welfare problem, and the early euthanasia of animal models of disease can moreover prevent avoidable suffering (i.e. by humane end-points). In this landscape, animal experimentation without the burden of killing animals seems unfeasible, if not undesirable. However, while acknowledging that most studies do require killing animals out of scientific (e.g. from the need to extract large-enough samples from small animals) or ethical (when animals would otherwise suffer needlessly) necessity, it remains to be ascertained whether: (1) this is true for all cases; or (2) that curtailing the life of laboratory animals is of little ethical importance. Accepting that – at least some – animal research is relevant, ethically acceptable and presently not replaceable, it should nevertheless be reflected upon whether there can be a scientific, ethical and legal framework within which a 'no-kill' approach may be equated, for some cases. A few examples are herein presented to discuss current possibilities and constraints, and help identify under which circumstances can a new set of '3Rs' – Re use, Rehabilitation and Rehoming – be applied as an alternative to the killing of animals when their scientific usefulness ends.

Keywords: animal research, experimentation, euthanasia, ethics

Killing laboratory animals as a moral issue

This topic is further discussed in Franco and Olsson (2016). Animals used in scientific procedures are typically euthanized, when the experiment ends. In the EU 11.5 million vertebrates (European Commission, 2013) are killed annually for research. This corresponds however to just 2.3 vertebrates – mostly rodents and fish – per 100 EU citizens per year (for a population of 502.5 million EU citizens, Eurostat, 2012) for all basic and applied biomedical science in academia and industry, a figure that gets dwarfed in comparison with the number of animals – mostly cows, pigs, sheep, poultry and fish – killed for food in the EU daily (for statistics see Eurostat, 2008; Eurostat, 2009, 2011). Considering that only between 1-9% (0.3% in Portugal, Centro Vegetariano, 2008) of people in western countries are vegetarians, we can assume most people are not in principle against the killing of animals for food, despite more replacement alternatives might be available for this purpose than for biomedical research. Nonetheless, not only is an *ad populum* argument hardly acceptable, it should also be pointed out that while meat production without killing is for the time being inconceivable – albeit eventually achievable through *in-vitro* meat culture (Weele, 2014) – biomedical experiments may not necessarily require curtailing animal lives. Furthermore, regardless of number, humans have moral obligations towards all animals under our care and responsibility.

EurSafe 2016 – DOI 10.3921/978-90-8686-834-6_76, © Wageningen Academic Publishers 2016

Following Jeremy Bentham – who laid the foundations of utilitarian moral philosophy and for whom animal research was acceptable, provided it had 'a determinate object, beneficial to mankind' (Boralevi, 1984), – acceptability of animal research is currently based on a predominantly utilitarian evaluation between the harm endured by animals and the potential scientific and medical benefits of the experiments. For this reason, greater attention is given to preventing or minimizing animal suffering than to avoid killing, insomuch as to the killing of laboratory animals not being accounted as a procedure if no other interventions are carried out, in current EU legislation (European Commission, 2010). Indeed, animals killed by routine culling are not even accounted for statistical purposes. This landscape reflects a predominantly welfarist view of what constitutes good practice, by which painless killing of laboratory animals poses no ethical problems, as non-existence implies absence of negative experiences.

In the tenet of the 3Rs (Russell and Burch, 1959) the value of animal life is patent in the principle of 'replacement', but probably even more so in the principle of 'reduction', as using – and ultimately killing – as few animals as possible is valued as the most ethical approach, when replacement is unfeasible. In this sense, 'reduction' alludes to the 'badness of killing' argument (Hansen *et al.*, 1999), by which one would as far as possible avoid taking the lives of animals. However, whenever 'reduction' might conflict with 'refinement' – e.g. anytime reducing the number of animals used would result in a higher burden for each individual animal (Franco and Olsson, 2014; Olsson *et al.*, 2012) – the latter is widely viewed as the priority, in line with a 'fairness to the individual' position (Tannenbaum, 1999), as well as animal research regulations in most countries. In a hypothetical, but realistic, 'reduction/refinement' dilemma presented to researchers undergoing mandatory training to work with animals, in a survey carried out in eight European cities (Franco *et al.*, unpublished data) (Figure 1), 70% of respondents would favour pair housing, thus using and killing more animals, to avoid the distress associated with single housing.

However, for a more extreme hypothetical case-study (Franco and Olsson, 2014), where a painful procedure that could either be carried out twenty times in one mouse or just once in twenty mice, scientists were much more divided, with only 53% favouring using more animals. Also, if subjects were companion species or primates the preferred option would be using a single animal. This suggests that for scientists, the perceived value of animal life will depend on the context and the species in question.

Killing might also pose an animal welfare issue that goes beyond the circumstances of death, as it is in the interest of the animal to not only avoid negative feelings but also experience positive ones. Hence,

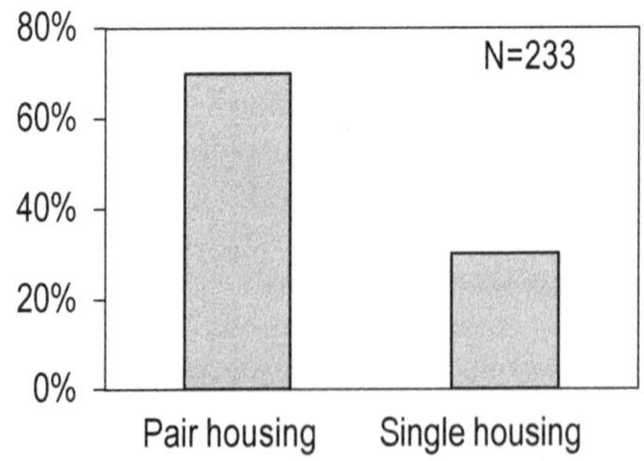

Figure 1. A reduction/refinement dilemma presented to laboratory animal science course participants. Dilemma: Mice are social animals, to which individual housing is stressful. However, in a given experiment, each cage must be considered as a single experimental unit due to 'cage-effects', regardless of the number of animals in each cage. Having no financial or logistic constraints, which of the following approaches would you take, if 20 experimental units are needed?

death deprives an animal of the positive feelings it would otherwise experience in its lifetime, provided it would be a life worth living (Yeates, 2010) This is relevant given the considerable proportion of procedures with little to no welfare impact, in which animals that could make a full recovery and live healthy lives afterwards are nonetheless euthanized at a young age.

For those who follow a *stricto sensu* animal rights perspective – i.e. following Tom Regan (1989) – no deliberate killing of animals for the sake of human interests is acceptable. In fact, even if animals are allowed the 'right to live, an animal rights advocate will not accept animal research on the grounds that I makes animals 'means to an end'. As for a *preference utilitarian*, hardly any animal research is acceptable (Singer, 2006), as direct benefits from any single experiment that could offset the cost endured by animals in that experiment are seldom observed, while killing laboratory animals goes against their interest in keeping on living. However, the case could be made that a preference utilitarian should accept an animal experiment – even if medical benefits are not directly reaped – whenever the benefits for animals (such as 'freedom' from hunger, thirst, cold, predators, or disease) can outweigh the costs (such as captivity or transient, mild harm), and animals are allowed to live a long, healthy life. Indeed, if most people would not accept reduction and replacement as legitimate goals for the keeping of companion animals, or even farm animals (when these are expected to have good quality of life, Olsson *et al.*, 2012), why should laboratory animals not be allowed to live a life worth living when this is possible both during and after their scientific use?

Re-use, Rehabilitate, Rehome: an alternative to killing laboratory animals

Animals used in research are typically killed for three reasons: (1) as a scientific requirement; (2) to prevent avoidable suffering; or (3) for financial, logistic, technical or even cultural reasons, secondary to the scientific process.

Euthanizing animals to spare them from suffering (i.e. humane endpoints, not only considered best practice but also a legal requirement (Morton, 1999)) provides a morally compelling justification for killing. However, killing out of scientific necessity might be avoided by changing methodological approaches. For example, *post-mortem* evaluation can sometimes be replaced by imaging technologies that allow monitoring disease progress in the living animal e.g. tuberculosis infections (e.g. Andreu *et al.*, 2012), hence avoiding killing animals for this purpose. Such methodological adaptations to prevent killing animals are only possible for some procedures, but there is a wider margin for preventing killing as a 'side-effect' of research. For example, the predominance of single-sex studies, i.e. using either females – e.g. to avoid aggression-related problems with group-housed males (Van Loo *et al.*, 2003) – or (more often) males (Wald and Wu, 2010), leads to the culling of animals of the other sex. There are usually no reasons for not using both male and female animals, which moreover allows detecting possible sex differences (insomuch as now being a requirement for all NIH-funded research, Clayton and Collins, 2014). Routine culling may also result from scientists not planning experiments in advance, which makes animal facilities and commercial breeders breed in excess to be able to supply them on very short notice (Hawkins *et al.*, 2004).

One possible alternative for killing laboratory animals is their continuous use (such as in pharmacokinetic studies) or re-use in other experiments, provided it does not have a detrimental effect in their wellbeing. For example, when large volume samples are needed, animals of smaller species – like rodents or fish – usually have to be euthanized for sample collection, with the number of animals needed (and killed) being multiplied when samples must be collected at different time-points. When using larger animal species, however, repeated sampling can be carried out without having to kill different animals at several time points, with the advantage of eliminating inter-individual variability, since each animal can be its own control (e.g. Dell *et al.*, 2002). And while repeated sampling has less impact for larger species, rodents can for instance be re-used in single-dose pharmacokinetic studies of non-toxic substances, with no significant impact on their welfare (Balani *et al.,* 2008).

Laboratory rodents and fish have relatively short life-spans, but larger animal species, such as dogs, may live much longer lives, which with adequate care can be healthy and worth living lives, even in laboratories (Franco *et al.,* 2014). But what to do when their scientific utility has for any reason reached its end? One option is their re-homing as companion animals or to sanctuaries (Wolfensohn, 2010), which must be preceded by their rehabilitation, i.e. adapting them to a life a completely different environment, with different people and different routines. Indeed, in most cases re-use, rehabilitate and rehoming – here proposed as a set of '3Rs' for preventing the killing of animals in research – must necessarily go together, and in this order, as it might be financially and logistically unfeasible to rehabilitate and rehome animals that do not provide an extended scientific usefulness. Rehoming is still far from being common practice, except for the retirement of chimpanzees to sanctuaries in the USA, where a fund has been established for this purpose, since euthanasia of surplus chimpanzees under the care of federal public agencies is forbidden. It might be more difficult to envisage purpose-bred laboratory animals outside their scientific role – and hence to consider them appropriate for a life outside the laboratory (Kerwin, 2006) – but there are several examples of successful programs for re-homing of laboratory cats and dogs into family homes and non-human primates into sanctuaries (e.g. Carbone, 1997; DiGangi *et al.*, 2006; LASA, 2004; Wolfensohn, 2010), showing it can be a feasible option.

In order to make rehabilitation and re-homing of lab animals a feasible option, it must be contemplated in laws and regulations. One example is the German 'Animal Welfare Act' – *Tierschutzgeset* – by which the killing of any vertebrate without a 'sound reason' is forbidden. Although animal research – and many other uses – is deemed a 'sound reason' by the *Tierschutzgeset*, it has served as a legal framework in Germany for rehoming laboratory animals after their use in scientific research (Doehring and Erhard, 2005). The rehabilitation and re-homing of non-human primates, cats, and dogs used in research is also encouraged (although not required) in present EU legislation (European Commission, 2010). A more explicit legal framework to allow rehabilitation of laboratory animals is patent in the 2006 amendment of the Indian law regulating animal experiments, which makes researchers using large animals responsible for what happens to them afterwards (de Boo and Hendriksen, 2005; Pereira and Tettamanti, 2005). Thus, researchers using dogs, cats, sheep, goats, cattle, horses, non-human primates or other large animal species must include in the research budget expenses related to their aftercare and rehabilitation (unless euthanasia is deemed necessary). Rehabilitation of small laboratory mammals is however 'not necessary', but is open to consideration by institutional animal ethics committees.

Even for smaller species, such as rodents, rehabilitation and rehoming may in some circumstances be feasible, and sometimes even easier than for larger animals, as rodents are easy and affordable to maintain. Their short life-span also makes them less demanding in terms of the commitment from potential adopters. Also, it might be easier for rodents kept enclosed with their social group to adapt to a new environment than for a dog that has to adapt from living primarily with other dogs to be the single canine member of a human family. One interesting option for rodents is their use as classroom pets in schools (Baumans *et al.*, 2007). Rodents can be housed in classrooms in highly enriched habitats, and used as an educational resource, including for teaching respect for animals or to deal with loss (Fonseca *et al.*, 2011; Huddart and Naherniak, 2005). In any case, regardless of animals being adopted by families or schools, it is central that adopters are able and capable of caring for them for as long as they live. The rehabilitation of laboratory animals might raise considerable difficulties. Firstly, it might be labour demanding, costly and time consuming. Also, it is difficult to ensure re-homed animals will be housed and taken care of under at least the same standards found in animal facilities; as well as to define what to do in case animals are found to be mistreated outside. Animals might also manifest physical and behavioural abnormalities from prolonged captivity or inadaptation to re-homing conditions (Hubrecht, 2002; Tettamanti *et al.*).

If animal research is portrayed to the public as only being carried out when no alternatives are available, it follows that the same should be applied for killing. However, for the above reasons, avoiding killing might be unfeasible, and humane euthanasia might often be the only available or ethically referable option. However, the goals of re-use, rehabilitation and re-homing deserves more serious thought, whenever these can allow laboratory animals to live healthy, longer, and worth living lives, either within laboratories, or outside them.

References

Andreu, N., Elkington, P.T. and Wiles, S. (2012). Molecular imaging in TB. From the bench to the clinic. In: Cardona, P.-J. (ed.) Understanding tuberculosis– global experiences and innovative approaches to the diagnosis. InTech.

Balani, S.K., Lu, C., Cardoza, K., Berg, C., Zhang, J. and Lee, F.W. (2008). Proposed new addition to 3Rs for ethical and humane use of rats in pharmacokinetic studies-'recycle'. Drug Metabolism Letters 2(3): 193-197.

Baumans, V., Coke, C., Green, J., Moreau, E., Morton, D., Patterson-Kane, E., Reinhardt, A., Reinhardt, V. and Van Loo, P. (2007). Making life easier for animals in research labs. Animal Welfare Institute, Washington, DC, USA.

Boralevi, L.C. (1984). Bentham and the oppressed. Walter de Gruyter, Berlin, Germany, 248 pp.

Broadhead, C.L., G. Betton, Combes, R., Damment, S., Everett, D., Garner, C., Godsafe, Z., Healing, G., Heywood, R., Jennings, M., Lumley, C., Oliver, G., Smith, D., Straughan, D., Topham, J., Wallis, R., Wilson, S. and Buckley, P. (2000). Prospects for reducing and refining the use of dogs in the regulatory toxicity testing of pharmaceuticals. Human and Experimental Toxicology 19(8): 440-447.

Carbone, L. (1997). Adoption of research animals. Animal Welfare Information Center Newsletter 7.

Centro Vegetariano (2008). Portugal: 30 000 Vegetarianos. Available at: http://tinyurl.com/jusmc5v.

Clayton, J.A. and Collins, F.S. (2014). NIH to balance sex in cell and animal studies. Nature 509(7500): 282-283.

De Boo, J. and Hendriksen, C.F.M. (2005). Reduction strategies in animal research: a review of scientific approaches at the intra-experimental, supra-experimental and extra-experimental levels. Alternatives to Laboratory Animals 33(4): 369-377.

Dell, R.B., Holleran, S. and Ramakrishnan, R. (2002). Sample size determination. ILAR Journal 43(4): 207-213.

DiGangi, B.A., Crawford, P.C and Levy, J.K. (2006). Outcome of cats adopted from a biomedical research program. Journal of Applied Animal Welfare Science 9(2): 143-163.

Doehring, D. and Erhard, M.H. (2005) Whereabouts of surplus and surviving laboratory animals. ALTEX 22(1): 7-11.

European Commission (2010). Directive 2010/63/EU of the European Parliament and of the Council of 22 September 2010 on the protection of animals used for scientific purposes. E. Commission. Official Journal of the European Union L276: 33-79.

European Commission (2013). Seventh report on the statistics on the number of animals used for experimental and other scientific purposes in the member states of the European Union. Brussels, Belgium.

Eurostat (2008). Poultry statistics in the European Union: flock numbers, hatcheries, trade and slaughterings. Data In Focus 31/2008.

Eurostat (2009). EU cattle, pigs, sheep and goats: monthly slaughter statistics in 2008. Data In Focus 15/2009.

Eurostat (2011). Fishery statistics. agriculture and fishery statistics – main results 2010-11. Publications Office of the European Union, Luxembourg.

Eurostat (2012). Europe in figures – Eurostat yearbook 2012. Publications Office of the European Union, Luxembourg, 119 pp.

Fonseca, M.J., Franco, N.H., Brosseron, F., Tavares, F., Olsson, I.A.S. and Borlido-Santos, J. (2011). Children's attitudes towards animals: evidence from the RODENTIA project. Journal of Biological Education 45(3): 121-128.

Franco, N.H, Magalhães-Sant'Ana, M. and Olsson, I.A.S. (2014). Welfare and quantity of life. Dilemmas in Animal Welfare: 46-66.

Franco, N.H. and Olsson, I.A.S. (2014). Scientists and the 3Rs: attitudes to animal use in biomedical research and the effect of mandatory training in laboratory animal science. Laboratory Animals 48(1): 50-60.

Franco, N.H. and Olsson, I.A.S. (2016). Killing animals as a necessary evil? The case of animal research. In: Meijboom, F.L.B. and Stassen, E.N. (eds.) The end of animal life: a start for ethical debate, Wagenignen Academic Publishers, Wageningen, the Netherlands.

Hansen, A.K., Sandøe, P., Svendsen, O., Forsman, B. and Thomson, P. (1999). The need to refine the notion of reduction. Humane endpoints in animal experiments for biomedical research. RSM Press, London, UK, pp. 139-144.

Hawkins, P., Morton, D.B., Bevan, R., Heath, K., Kirkwood, J., Pearce, P., Scott, L., Whelan, G. and Webb, A. (2004). Husbandry refinements for rats, mice, dogs and non-human primates used in telemetry procedures. Seventh report of the BVAAWF/FRAME/RSPCA/UFAW Joint Working Group on Refinement, part B. Laboratory Animals 38(1): 1-10.

Hubrecht, R. (2002). Comfortable quarters for laboratory dogs in research institutions. Comfortable quarters for laboratory animals. Animal Welfare Institute, Washington, DC, USA, pp. 56-64.

Huddart, S. and Naherniak, C. (2005). Animals in the classroom. Teaching green. The elementary years: hands-on learning in grades K-5. T. New Society Publishers, Toronto, Canada, pp. 101-107.

Kerwin, A.M. (2006). Overcoming the barriers to the retirement of old and new world monkeys from research facilities. Journal of Applied Animal Welfare Science 9(4): 337-347.

Laboratory Animal Science Association (LASA) (2004). LASA guidance on the rehoming of laboratory dogs. A report based on a LASA working party and LASA meeting on rehoming laboratory animals. LASA, Hull, UK.

Morton, D.B. (1999). Humane endpoints in animal experimentation for biomedical research. Royal Society of Medicine Press, London, UK, pp. 5-12.

Olsson, I.A.S., Franco, N.H., Weary, D.M. and Sandøe, P. (2012). The 3Rs principle – mind the ethical gap! ALTEX 2011: 333-336.

Pereira, S. and Tettamanti, M.(2005). Ahimsa and alternatives – The concept of the 4th R. The CPCSEA in India. ALTEX 22(1): 3-6.

Regan, T. (1989). The case for animal rights. Animal rights and human obligations. Prentice Hall, Upper Saddle River, NJ, USA.

Russell, W.M.S. and Burch R.L. (1959). The principles of humane experimental technique. Methuen & Co. Ltd., London, UK.

Singer, P. (2006). Setting limits on animal testing. The Sunday Times, December 3.

Stephens, M.L., Conlee, K., Alvino, G. and Rowan, A.N. (2002). Possibilities for refinement and reduction: future improvements within regulatory testing. ILAR Journal 43, Suppl.: S74-79.

Tannenbaum, J. (1999). Ethics and pain research in animals. ILAR Journal 40(3): 97-110.

Tettamanti, M., Veeraraghavan, P. and Pereira S. The impact of facts from the rehabilitation of laboratory rabbits on reliability and evaluation of experimental data. Available at: http://tinyurl.com/jndsrkh.

Tettamanti, M., Veeraraghavan P. and Pereira S. The success of the concept of the 4th R: a new era in laboratory animal care. Available at: http://tinyurl.com/z47fovg.

Van Loo, P.L.P., Van Zutphen, L.F.M. and Baumans, V. (2003). Male management: coping with aggression problems in male laboratory mice. Laboratory Animals 37(4): 300-313.

Wald, C. and Wu, C. (2010). Of mice and women: the bias in animal models. Science 327(5973): 1571-1572.

Weekley, L.B., Guittin, P. and Chamberland, G. (2002). The international symposium on regulatory testing and animal welfare: recommendations on best scientific practices for safety evaluation using nonrodent species. ILAR Journal 43 Suppl.: S118-122.

Weele, C. (2014). *In vitro* meat. Encyclopedia of food and agricultural ethics. Springer, Dordrecht, the Netherlands, pp. 1219-1225.

Wolfensohn, S. (2010). Euthanasia and other fates for laboratory animals. The UFAW handbook on the care and management of laboratory and other research animals. Wiley-Blackwell, Hoboken, NJ, USA, pp. 219-226.

Yeates, J. (2010). Death is a welfare issue. Journal of Agricultural and Environmental Ethics 23(3): 229-241.

77. Mouse models: some reflections from the lab

A. Blanchard
Centre for the Study of the Sciences and the Humanities and Centre for Cancer Biomarkers, Allegaten 34, 5020 Bergen, Norway; anne.blanchard@uib.no

Abstract

In biomedical research, reducing complexity is arguably a necessary first step in understanding the complex human biology. One important way of reducing complexity, specific to oncology, is the use of mouse models for understanding how cancer tumours develop and spread in the human body, and for testing the efficacy of treatment regimens. Mouse models are used for cancer research at the pre-clinical level, before drugs are tested in the clinic on humans. Mice are genetically modified so that the variations in their genome are low, allowing for a scientific hypothesis to be tested where most variables can be controlled. For instance, most mice are modified to be 'immuno-deficient', so as to remove one layer of complexity associated with the interaction between the immune system and the development of cancer tumours. However, there are important limits to these models, which have led to a methodological literature on the fine-tuning of mouse models. This paper makes a contribution to these debates from a Science and Technology Studies perspective, elaborating on two key themes of this literature that also emerged as important in discussions between the author and oncology researchers at the Centre for Cancer Biomarkers (CCBIO) in Bergen, Norway. Going beyond technical recommendations, this paper argues for a reflexive move on: (1) the predictive power of mouse models; and (2) the reproducibility of such models. Acknowledging uncertainties, the first theme relates to the predictive power of pre-clinical mouse models for clinical purposes. While animal models are informative at the pre-clinical level, when it comes to validating these animal models in clinical trials, Mak *et al.* (2014) assert that 85% of early clinical trials for new drugs fail; with the largest proportion of these for cancer drugs. The second theme, related to the first, is the problem of reproducibility. There have been some criticisms on the lack of reproducibility of results from pre-clinical models (e.g. Begley and Ellis, 2012; Hayden, 2014; 'Must try harder' Editorial in Nature, 2012), stemming from the mice' sensitivity to the environment and from the sometimes poor quality of pre-clinical data. Despite their weaknesses, mouse models are 'better than nothing', and allow us to make steps into the unknown. But we assert that their use should be accompanied by some deep reflection among users of these models, recognising that no matter how advanced they become, they remain 'models' that are by design fundamentally different from the reality of the human condition. Indeed, these reflections in this paper have been the subject of on-going reflexive discussions with researchers at CCBIO.

Keywords: cancer research, mouse models, uncertainties

Mouse models in cancer research: ethical, scientific and practical reasons

Biomedical research, including oncology research as the focus of this paper, is confronted by the immense complexity of human biology. The various and intricate biological causes of cancer, the mechanisms behind cancer cell growth and proliferation, or the resistance mechanisms to cancer therapeutics are all highly complex processes that are surrounded by significant uncertainties. Achieving robust knowledge on these mechanisms is further complicated by the fact that cancer tumours are heterogeneous entities, both within each patient, and also from one patient to another. Understanding cancer biology thus necessitates reducing its complexity and looking at it through a simplified lens. *In vitro* cell cultures and *in vivo* mouse models are in this respect important tools, with Price and Stevens (1982) noting: 'It is not difficult to appreciate that if we hope to gain a detailed understanding of the behaviour in a complex system ... we must first try to understand its properties in as simple a system as possible.'

EurSafe 2016 – DOI 10.3921/978-90-8686-834-6_77, © Wageningen Academic Publishers 2016

There are hundreds of different mouse models to choose from (see for instance the lists by de Jong and Maina, 2010; or Wartha *et al.*, 2014), ranging from genetically engineered mouse models where a gene is removed or replaced in the mouse DNA to mimic genetic alterations found in human cancers, to transplantation mouse models where human cancer cells or solid tumours are transplanted into mice, and to carcinogen-induced and spontaneous mouse models where mice that are prone to develop cancer are exposed to a cancer-giving agent. All of these models have strengths and weaknesses; some of which are discussed in the next section.

Animal models have been introduced into biomedical research for ethical reasons, and constitute prerequisites to human clinical trials for the assessment of new drugs: 'The experiment should be so designed and based on the results of animal experimentation and a knowledge of the natural history of the disease or other problem under study that the anticipated results will justify the performance of the experiment' (extract of the Nuremberg Code as cited by Mitscherlich and Mielke, 1949).

In addition to ethical concerns, important scientific reasons can be found in the literature to justify the use of mouse models in cancer research; these mice have helped us to learn much about cancer. With the completion of the mouse and human genome sequence, it is known that humans and mice share at least 95% of their genomes (De Jong and Maina, 2010). In addition, mice have a relative physiologic similarity to humans in terms of the organ system, and develop similar diseases (Gould *et al.*, 2015). Mouse models therefore have been an important tool in eliciting some key biochemical and physiologic processes of cancer onset and growth, as well as to predict the activity of an anticancer drug in the clinic (Wartha *et al.*, 2014). In particular, transplantation mouse models have provided key insights into the mechanisms of leukaemia development (Ablain *et al.*, 2013), and genetically engineered mouse models have allowed molecular and cellular manipulations impossible to perform in patients, and thus have given access to an improved understanding of liver cancer (Bakiri and Wagner, 2013).

Finally, there are also practical reasons for the use of mouse models in cancer research. With their small size, their fast breeding rate that permits studies in a short timeframe, their relative affordability and their capacity to be easily manipulated genetically, mice are seen as convenient artefacts in biomedical research (De Jong and Maina, 2010; Gould *et al.*, 2015; Radiloff *et al.*, 2007).

Mouse models are thus central tools to explore the biology of cancer. However, even if mice are very similar to humans at the genomic level, their phenotypes are very different, and the various mouse models all have their weaknesses. Even though researchers of the field know about and reflect upon the limits of their models, these reflections seem, in the literature, to be limited to methodological and design questions.

A focus on methodology, a need for deeper reflection

The literature is extremely rich in articles and books addressing methodological questions around mouse models. There is a general agreement that differences between mice and humans in terms of size and physiology are a limit when trying to translate results from mouse models to the clinic (see for instance De Jong and Maina, 2010). However, the focus of this critical literature is mainly on the practical issues of designing, choosing and improving mouse models (Strand *et al.*, 1996), as well as assessing their advantages and disadvantages (see for instance Gould *et al.*, 2015, and the report by the Mouse Models of Human Cancers Committee (Ittmann *et al.*, 2013), which reviewed more than 40 models for prostate cancer).

Two methodological issues that are often discussed relate to the use of immuno-deficient mice and sub-cutaneous models. Immuno-deficient mice are genetically modified to have no immune system. While this allows the quick development of cancer tumours in mice, and therefore quick observations, these mouse models do not allow for the study of the effects of the immune system with regard to a particular cancer therapy, and to study drugs that depend on functions of the immune system (Aparicio *et al.*, 2015; Dranoff, 2012; Wartha *et al.*, 2014). This is considered an important weakness, as the human immune system is known to play an important role in the rate of cancer growth. The second issue relates to sub-cutaneous models, where tumour cells or bulks are injected under the skin of mice. Such models are frequently used because of their ease to implement (compared to mouse models where tumours are implanted into the organs of mice), and the ease to monitor tumour growth and anti-tumour activity (Dranoff, 2012; Peterson and Houghton, 2004). However, these models do not permit the study of the interactions between the tumour and its microenvironment, despite numerous local processes (such as angiogenesis – the vascularisation of the tumour) having important impacts on the growth and spread of the tumour (Aparicio *et al.*, 2015; Wartha *et al.*, 2014).

As Strand (1999: 277) notes, although very important, these discussions on methodological and design aspects found in the literature 'are often written in the form of recipes, supplied with observed advantages of some details of the recipe as compared to previously preferred choices'. Indeed, the conclusions of such papers and reports point to the fact that despite the impossibility for one single model to capture all different mechanisms of cancer, clinical predictions from mouse models can be ensured through 'a rational consideration of the major/minor strengths and weaknesses of the model' (Peterson and Houghton, 2004: 838), through 'preclinical studies utilising multiple models ... rather than a single model approach' (Merlino *et al.*, 2013: 13), or through the enhancement of current mouse models, such as 'genetic modifications to humanise the host strain further' (Schultz *et al.*, 2007: 127).

Against this background, this paper reports on an embedded study of the ethical, legal and social aspects of research at the Centre for Cancer Biomarkers (CCBIO), elucidated through a dialectic move between the literature and informal one-hour discussions with 15 researchers at CCBIO. Some of these discussions have been about mouse models, and among them, two themes were recurrent, and said to be important for their work: (1) the predictive power of mouse models; and (2) the reproducibility of such models. This paper now presents some early reflections on these two themes, which have provided a starting point for ongoing reflections with researchers at CCBIO about the role and practices around mouse models in their work.

Uncertainties clouding the predictive power of mouse models

One of the key themes discussed by CCBIO researchers is the non-linearity and uncertainties characterising the relation between pre-clinical mouse models and clinical trials on humans. As seen above, the study of complex cancer mechanisms requires, for ethical, scientific and practical reasons, the use of mouse models. However, even though mice and humans are very similar at the genomic level, there are significant differences between these two species; some of the most obvious ones being their size and their heartbeat rate. The heartbeat rate of a mouse is around 600 beats/minute, which is much higher than the average human heartbeat rate of 80 beats/minute. The faster heartbeat rate results in accelerated physiological processes (de Jong and Maina, 2010), and these differences come to challenge the predictive power of mouse models for clinical purposes. Indeed, as Mak *et al.* (2014) note, less than 8% of animal models are translated into successful cancer drugs after clinical trials.

Accepting its exceptional nature, there is a striking example (though not cancer related) discussed in the literature, that illustrate where predictive power has failed, with serious consequences. It is the 1993 clinical trial of Fialuridine, a drug for hepatitis B, that was stopped when five patients died from liver failure, and two others were saved from death only by liver transplants (Manning and Swartz, 1995), even if the drug had been tested on different animals including mice, rats, dogs, monkeys, and woodchucks, and proven safe at doses hundreds of times higher than those administered to the patients.

As de Jong and Maina (2010: 503) explain: 'the reason that small animals can generally tolerate larger doses (in mg/kg of body weight) of a pharmaceutical is that they are able to clear most chemicals from their bodies much more quickly than humans ... A given dose will lead to higher concentrations of the chemical in the tissues of a human than in the tissues of a small animal.' To compound the uncertainties of mouse models, mice used in pre-clinical experiments are often young, and have not been exposed to previous treatments or medical interventions like humans often receive. They are, however, exposed to various anaesthetic agents, which alter their respiration, heart rate, blood pressure and temperature (de Jong and Maina, 2010). This needs to be taken into account when designing clinical trials for humans.

Convincing results from mouse models does not necessarily mean that these results will be successfully translated into clinical studies on humans. In discussions with CCBIO researchers, it emerged that it is important to acknowledge that when working with mouse models, they do interpretations that are subject to uncertainties when extrapolated to clinical conclusions; these uncertainties weaken the predictive power of these models.

The problem of reproducibility as a challenge to scientific quality

The other key theme discussed by CCBIO researchers is the problem of reproducing results from mouse models, even with careful control over the experiment. Mice are genetically modified to have minimal variations in their genome so that a research hypothesis can be tested under conditions where most variables can be controlled. In addition to having the same genetic background, the mice of a model have the same gender, age and weight, and they are bred and maintained under the same conditions (food, cage, temperature). However, the mice's sensitivity to their environment means that different experimental conditions, such as different labs, may lead to different results. Therefore, despite careful precautions, reproducing results from a mouse model in order to validate them, and try them out in the clinic, is very difficult. For instance, Begley and Ellis (2012) and other Amgen scientists tried to reproduce the results from 53 landmark oncology pre-clinical (animal model) studies. In the end, the results of only six of those landmark studies could be reproduced, and the authors claimed: 'even knowing the limitations of preclinical research, this was a shocking result' (*ibid.*: 532). The authors concluded that an important cause of this poor success rate was not only the mice's sensitivity to their environment, but the overall poor quality of published pre-clinical data. This is backed up by other authors who observe 'missing references; incorrect controls; undeclared cosmetic adjustments to figures; ... inaccurate and incomplete methods; and improper use of statistics' in pre-clinical research (Editorial in Nature (2012)), and others who point at the fact that negative results often go unpublished, which leads to animal models overestimating by about 30% the likelihood that a drug will work (Sena *et al.*, 2010).

If we consider reproducibility as a foundation of the scientific method, then these challenges to reproducibility seem to call into question scientific quality. Of course, strict reproducibility is not a precondition to scientific enquiry, with a number of disciplines drawing on rigorous methods that do not attempt to be reproducible. However, through discussions with CCBIO researchers, it was seen as important to be transparent about the barriers to reproducibility met in the lab, and to accept that these barriers imply redefining the character of this research; rather more interpretative than descriptive.

Conclusions

In this paper, we have discussed how mouse models have been useful for investigating mechanisms of cancer, and that there is a substantial literature fine-tuning these models. The paper took up and reflected upon two key themes of this literature, which also emerged as key concerns among researchers at CCBIO: the predictive power and reproducibility of mouse models. These two weaknesses are part of the reason why so few new cancer drugs are translated from animal models to clinical practice. It is known that cancer drug discovery and development is a very slow and expensive process, with 89% of new drugs entering clinical trials failing to get the American Food and Drug Administration's approval after an average investment of US$ 400 million (Sharpless and DePinho, 2006). This high rate of failure represents a significant economic and social cost, as these resources might be used more efficiently for other diseases or other sectors of society. And this is to say nothing for the false hope of patients who are offered last-chance enrolments in clinical trials for drugs that have been poorly studied in pre-clinical animal models: 'I hear too many stories about patients who have used their one shot at getting into a trial on a drug that didn't have enough legs to begin with, and that's a tragedy' (Steve Perrin cited by Hayden, 2014).

In spite of their weaknesses, even those of a more fundamental character, mouse models remain a useful tool that has improved our understanding of cancer, and they ought not be abandoned. However, the current discussion in the literature on how these models can be perfected according to an improved 'recipe' needs to be accompanied by more essential reflections in the lab on how we frame or characterise research guided by mouse models. Given their increasing sophistication it is tempting to think of these models as perfect substitutes for humans. But 'mice are not men', and will never be. Working with mouse models is an interpretative step towards the clinic, with useful though imperfect knowledge about the biology of human cancers.

These early reflections have formed part of ongoing, both formal and informal, reflexive discussions between the author and researchers at CCBIO on mouse models. In more formal settings, this has manifested in debates structured around internal seminars and PhD courses. In a more informal sense, this continues through conversations with cancer researchers, who see these reflexive exercises as an important, but often neglected, part of their research.

References

Ablain, J., Nasr, R., Zhu, J., Bazarbachi, A., Lallemand-Breittenbach, V. and De Thé, H. (2013). How animal models of leukaemias have already benefited patients. Molecular Oncology 7(2): 224-231.

Aparicio, S., Hidalgo, M. and Kung, A.L. (2015). Examining the utility of patient-derived xenograph mouse models. Nature Reviews Cancer 15(5): 311-316.

Bakiri, L. and Wagner, E.F. (2013). Mouse models for liver cancer. Molecular Oncology 7(2): 206-223.

Begley, C.G. and Ellis, L.M. (2012). Drug development: raise standards for preclinical cancer research. Nature 483(7391): 531-533.

De Jong, M. and Maina, T. (2010). Of mice and humans: are they the same? Implications in cancer translational research. Journal of Nuclear Medicine 51(4): 501-504.

Dranoff, G. (2012). Experimental mouse tumour models: what can be learnt about human cancer immunology? Nature Reviews Immunology 12(1): 61-66.

Gould, S.E., Junttila, M.R. and De Sauvage, F.J. (2015). Translational value of mouse models in oncology drug development. Nature Medicine 21(5): 431-439.

Hayden, E.C. (2014). Misleading mouse studies waste medical resources. Nature News.

Ittmann, M., Huang, J., Radaelli, E., Martin, P., Signoretti, S. and Sullivan, R. (2013). Animal models of human prostate cancer: the consensus report of the New York meeting of the mouse models of human cancers consortium prostate pathology committee. Cancer Research 73(9): 2718-2736.

Mak, I.W.Y., Evaniew, N. and Ghert, M. (2014). Lost in translation: animal models and clinical trials in cancer treatment. American Journal of Translational Research 6(2): 114-118.

Manning, F.J. and Swartz, M. (1995). Review of the Fialuridine (FIAU) clinical trials. National Academies.

Merlino, G., Flaherty, K., Acquavella, N., Day, C.-P., Aplin, A. and Holmen, S. (2013). Meeting report: the future of preclinical mouse models in melanoma treatment is now. Pigment Cell and Melanoma Research 26(4): E8-E14.

Mitscherlich, A. and Mielke, F. (1949). The nuremberg code in doctors of infamy: the story of Nazi medical crimes. Henry Schuman, New York, NY, USA.

Nature (2012). Must try harder. Editorial. Nature 483(7391): 509.

Peterson, J.K. and Houghton, P.J. (2004). Integrating pharmacology and *in vivo* cancer models in preclinical and clinical drug development. European Journal of Cancer 40(6): 837-844.

Price, N.C. and Stevens, L. (1982). Fundamentals of enzymology. Oxford University Press, Oxford, UK.

Radiloff, D.R., Rinella, E.S. and Threadgill, D.W. (2007). Modeling cancer patient populations in mice: complex genetic and environmental factors. Drug Discovery Today: Disease Models 4(2): 83-88.

Schultz, L.D., Ishikawa, F. and Greiner, D.L. (2007). Humanised mice in translational biomedical research. Nature Reviews Immunology 7(2): 118-130.

Sena, E.S., Van der Worp, H.B., Bath, P.M.W., Howells, D.W. and Macleod, M.R. (2010). Publication bias in reports of animal stroke studies leads to major overstatement of efficacy. PLoS Biology 8(3): e1000344.

Sharpless, N.E. and DePinho, R.A. (2006). The mighty mouse: genetically engineered mouse models in cancer drug development. Nature Reviews Drug Discovery 5(6): 741-754.

Strand, R. (1999). Towards a useful philosophy of biochemistry: sketches and examples. Foundations of Chemistry 1(3): 269-292.

Strand, R., Fjelland, R. and Flatmark, T. (1996). *In vivo* interpretation of *in vitro* effect studies with a detailed analysis of the method of *in vitro* transcription in isolated cell nuclei. Acta Biotheoretica 44(1): 1-21.

Wartha, K., Herting, F. and Hasmann, M. (2014). Fit-for purpose use of mouse models to improve predictivity of cancer therapeutics evaluation. Pharmacology and Therapeutics 142(3): 351-361.

Section 20. Genetic modifications

78. Gene editing animals – part of a utopian future?

A. Bruce
The University of Edinburgh, Old Surgeons' Hall, High School Yards, Edinburgh EH1 1LZ, United Kingdom; ann.bruce@ed.ac.uk

Abstract

The last few years have seen a resurgence of interest in genetically modified (GM) animals, based around developments in gene editing technologies. Gene editing offers the prospect of transfer of alleles from one breed to another in highly specific and discriminating ways. Additionally the approval for commercial sale of GM salmon by the US Food and Drugs Administration has encouraged developers of GM animals by demonstrating that it is possible to advance through regulatory hurdles to commercial reality. In the rapidly moving animal science world, gene editing technologies have been used to produce cows without horns and pigs resistant to a fatal pig disease, African Swine fever. Perhaps more significantly, a global commercial pig breeding company has announced it has used gene editing to produce pigs resistant to a serious respiratory and reproductive disease. This latter development provides a credible pathway for gene edited pigs to be integrated into global production. Questions remain as to whether these developments will contribute to a utopian future? And if so, whose future? The future of agriculture is already contested, with advocates for sustainable intensification of global production systems, and advocates for whole system change to better reflect social and environmental aspirations in opposition to global production systems. Can gene edited animals contribute to either or both of these visions? I suggest that gene editing is conceptually different from both genetic modification and marker-assisted selection, but has similarities with both. Its acceptability in either future food production scenario will depend on the specific applications being advocated and the social context within which gene edited livestock are being produced. The question is not whether gene editing is natural or not but rather do we want it or not? Do we want the products it offers to deliver? And does it deliver them in the relational contexts in which we want them offered?

Keywords: GM animals, gene editing, community supported agriculture, sustainable intensification

Introduction

Whereas genetic modification has been widely used in food and feed crops, modified livestock have not so far figured to any significant degree. Most of the traits of interest to producers could be achieved more easily by advanced selective breeding. The advent of genome editing techniques, especially CRISPR/Cas, is likely to change that picture, opening up a wide potential for effective and marketable modifications in food animal production traits, hitherto unachievable. In addition the first commercial genetically modified salmon has been authorised for sale by the US Food and Drug Administration.

This paper will explore how these novel developments may impact upon the much contested space of future visions of agriculture, in which some argue for sustainable intensification of global production systems and others advocate whole system change to better reflect social and environmental aspirations, in opposition to global production systems. Can gene edited animals contribute to either or both of these utopian visions?

The first GM animals were produced in 1980s, to considerable fanfare. There followed speculation as to all kinds of possible modifications that would enable advanced genetic manipulation of animals by humans (Bulfield, 1990). Early efforts at genetic modification were inefficient, expensive and there were few single genes identified that would be attractive to transfer. Of these early speculations, only

EurSafe 2016 – DOI 10.3921/978-90-8686-834-6_78, © Wageningen Academic Publishers 2016

one strand, the production of pharmaceutical proteins in the milk of animals, has become commercially available. GM techniques became largely restricted to the production of mouse models of human diseases. Civil society groups acting as advocates for animal welfare were strongly against the use of genetic modification, and surveys of publics found them largely antipathetic to such developments.

The technologists' vision of enabling novel genetic manipulations in livestock continued in fish, however. GM salmon (AquAdvantage™) modified for faster growth entered the regulatory process in the USA, and after over 20 years was finally approved by the FDA in December 2015, acting as a flag bearer, and the lead product in the world for commercialisation (Waltz, 2016). Although awaiting clarity on labelling, the approval has generated optimism among producers of GM animals.

Now, gene editing techniques, in particular the use of CRISPR (Tan *et al.*, 2016) offer a highly specific method of modifying genes with a plethora of possibilities. In the rapidly moving animal science world, these technologies have been used to produce cows without horns (Recombinetics, 2016) and pigs resistant to a fatal pig disease African Swine fever (Lillico *et al.*, 2013). There are reports of gene edited muscly pigs (Cyranoski, 2015), cattle and sheep (Proudfoot *et al.* 2015) and apparently heat tolerant cattle (Aggenetics, 2016). Perhaps more significantly, a global commercial pig breeding company has announced it has used gene editing to produce pigs resistant to a serious respiratory and reproductive disease (Genus, 2015; Whitworth, 2016).

Visions for future agriculture

Numerous different visions are presented for the future of agriculture. For the purposes of this short paper they are broadly conceived of as two alternative systems. One is based on the global need to feed nine billion or more humans, without increasing environmental burdens (Godfrey *et al.*, 2010). The other seeks to promote a transformation in the system linking producers and consumers, such as promoting community supported agriculture (Gebhard *et al.*, 2015). Can gene edited (GM) animals fit into these visions?

Considering first conventional, global production systems. The future vision depends on innovation for smarter agriculture, using technologies associated with precision agriculture (information, satellite and diagnostic technologies) and increasing the efficiency of resource use. Progress is measured in metrics, such as carbon footprints, efficiency of use of animal feed, fertility and growth rate, often summarised by the term 'sustainable intensification'.

Gene edited animals fit this innovation trajectory, because they provide opportunities such as improved efficiency of production by better disease resistance. The announcement by the global pig breeding company Genus that they have developed gene edited pigs with a view to selling within five years (Genus, 2015) indicates a credible pathway for gene editing to enter the conventional livestock market. Since this food production system is subject to global competition, being first in developing desirable gene edited animals could become a driver. But developments are also subject to public acceptance, as transmitted by the behaviour of organisations with agency, notably supermarket chains. Some of the gene edited traits may prove to more acceptable than others such as addressing a fatal disease like African Swine Fever, where no treatments exists, compared with producing more muscly pigs, achievable by conventional genetics and of ambivalent benefit.

The alternative system may be considered as a group of concepts which aspire to transform food production systems. There are many different alternatives, such as community supported agriculture, urban agriculture, food hubs, farmers' markets and intercultural gardens. Together these are expressions of resistance to the prevailing global production system that separates producers from consumers.

Rather than emphasise metrics of sustainability, these alternatives emphasise relationships and advocate education, in the expectation that educating urban citizens about their food will help bring about a transformation in the food system. Thus the emphasis is on developing new social systems for food production reflecting equitable relationships between consumer and producer, among members of communities, and between the earth, livestock and human beings.

An approach presented at the 10th EURSAFE conference, 'Agriculture and Society in 2050: A Students' Vision' (Gebard *et al.,* 2015: 498-499) exemplifies some salient features of this second approach:

- Traditional forms of cultivation adapted to local conditions.
- The cultivation of old as well as modern breeds of crops ensuring variety and security of supply.
- Recognising livestock animals as sentient beings and treated accordingly.

If 'crops' are translated to 'animals' then this vision could value both old and modern breeds of animal, but the methods of cultivation needs to be adapted to local conditions and livestock treated well. Gene edited animals might offer improved adaptation to local conditions but may equally be excluded on the grounds of excessive commodification, reflecting an inappropriate relationship with livestock. The emphasis on resistance to global production systems may also mitigate against developments perceived to be associated with them.

Locating gene editing in food production practices

Previous GM animals were not considered acceptable in most systems of production because of objections to transferring genes across species, and the unknown risks associated with transferring complex gene constructs in unpredictable ways, as well as animal welfare issues around the production of the GM animals (requiring surgery of the parent animals). Gene editing ostensibly offers a solution to the first two objections, although there may still be animal welfare issues associated with producing founder animals. Gene editing offers the prospects of modifying genes within species, and with precision (although reference is sometimes made to off-target events, so additional changes clearly do occur at times).

A key question is whether gene editing constitutes genetic modification or not? In scientific writings, the terms 'GM' and 'transgenic' are often treated as synonyms, such that an organism that does not contain a transgene (a gene from a different species) is considered non-GM. This argument may convince some but not others (see Bruce, 2016).

A key ontological question is what is this 'gene editing'? As described, it consists of making a break in DNA at a specific location and harnessing the genome's natural ability to repair damage and to incorporate a specific change in DNA at that point. This change can consist of a single base pair change. Proponents argue that such change could have occurred naturally by mutation, and indeed, cannot be detected as 'foreign' DNA. In this framing, the change could be understood as induced mutation (rather as radiation has been used to induce mutations in plant genomes), albeit directed and using human artifice to bring it about.

Single nucleotide changes that occur naturally have been utilised in farm animal breeding in the past. For example, visual selection for muscular pigs inadvertently selected for the presence of the halothane gene, with the unfortunate consequence of the sudden death of pigs when stressed. The causal mutation was traced back to a single nucleotide change. A blood test was developed, which allowed this (recessive) gene to be selected out, thus changing the genetic make-up of a pig population based on an existing single nucleotide change, enabled by laboratory analysis. It is clear that other genetic changes (some with deleterious effects) also occur through nature. This could be used as an argument against gene editing (on the basis of failure to understand all the effects of a gene) or for gene editing (on the basis that 'induced naturalness' is little different to selecting on the basis of a natural occurrence identified through laboratory practices).

Advocacy groups for alternative forms of agriculture appear to be comfortable with marker-assisted selection (MAS), that is, using genetic markers. Greenpeace assert that MAS is superior to genetic engineering and perfectly represents the new paradigm in biotechnology (Greenpeace International, 2009), because it respects species barriers, makes traditional breeding more efficient but does not replace it, and does not transfer isolated gene sequences but treats genomes as a coherent entity. Greenpeace believes this raises fewer safety concerns and offers high potential to meet the challenges of future foods, as one part of activities that includes the promotion of biodiverse, low-input farming.

Like MAS, gene editing respects species barriers and treats the genome as a coherent entity, thus differing from former genetic modification. It is different from MAS in that it requires a deliberate intervention rather than relying on increasing the efficiency of traditional practices, but if the mutations it induces could already exist in the animal's genome, whether known or unknown, should this be seen as problematic?

Conclusions

Developments in technology combined with a regulatory 'break-through' has created new interest in developing gene edited livestock. Despite the excitement generated in the scientific world, questions remain as to whether these developments will contribute to a utopian future for food production? And if so, whose utopia? It seems likely that food will continue to be produced at global and local levels, with society (producers and consumers) unlikely to reach a consensus as to what constitutes 'good' agriculture and whether gene editing can be viewed as part of that desirable future or not.

Frivolous or controversial applications have the potential to strangle all gene edited products at birth. Competition to be the first on the market needs to be tempered by consideration of the wider future of the technology. Public engagement is sometimes advocated as a route to securing public support for a technology, however, it is equally possible to engender public disquiet. But, where relationship building is a key focus of food production, the impact of gene editing on relationships also needs to be taken into account.

Perhaps the question is not whether gene editing is natural or not but rather do we want it or not? Do we want the products it offers to deliver? And does it deliver them in the relational contexts in which we want them offered?

References

Aggenetics (2016). Available at: http://tinyurl.com/jp87ted.

Bruce, D. (2016). Genome editing: moving the goalposts on the GM playing field? This volume.

Bulfield, G. (1990). Genetic manipulation of farm and laboratory animals. In: Wheale, P. and McNally R. (eds.) The bio-revolution cornucopia or Pandora's box. Pluto Press, London, UK, pp. 18-23.

Cyranoski, D. (2015). Super-muscly pigs created by small genetic tweak. Nature News, June 30.

Gebhard, E., Hagemann, N., Hensler, L., Scweizer, S. and Wember, C. (2015). Agriculture and food 2050: visions to promote transformation driven by science and society. Journal of Agricultural and Environmental Ethics 28: 497-516.

Genus (2015). Announcement information digest. Genus tackles major pig disease with breakthrough technology. Available at: http://tinyurl.com/zldmz9e.

Greenpeace International (2009). Smart breeding: marker-assisted selection. Available at: http://tinyurl.com/jvcubby.

Godfrey, H.C.J., Beddington, J.R., Crute, I.R, Haddad, L., Lawrence, D., Muir, J.F., Pretty, J., Robinson, S., Thomas, S.M. and Toulmin, C. (2010). Food security: the challenge of feeding 9 billion people. Science 327: 812-818.

Lillico, S.G., Proudfoot, C., Carlson, D.F., Stverakova, D., Neil, C., King, T.J., Ritchie, W.A., Tan, W., Mileham, A.J., McLaren, D.G., Fahrenkrug, S.C. and Whitelaw, C.B.A. (2013). Live pigs produced from genome edited zygotes. Scientific Reports 3, Article number 2847.

Proudfoot, C., Carlson, D.F., Huddart, R., Long, C.R., Pryor, J.H., King, T.J., Lillico, S.G., Mileham, A.J., McLaren, D.G., Whitelaw, C.B.A. and Fahrenkrug, S.C. (2015). Genome edited sheep and cattle. Transgenic Research 24: 147-153.

Recombinetics (2016). Available at: http://tinyurl.com/h4xwl9x.

Tan, W., Proudfoot, C., Lillico, S.G., Whitelaw, C.B.A.W. (2016). Gene targeting, genome editing: from Dolly to editors. Transgenic Research 25(3): 273-287.

Waltz, E. (2016). GM salmon declared fit for dinner plates. Nature Biotechnology 34(1): 7-9.

Whitworth, K.M., Rowland, R.R.R., Ewen, C.I., Trible, B.R., Kerrigan, M.A., Cino-Ozuna, A.G., Samuel, M.S., Lightner, J.E., McLaren, D.G., Mileham, A.J., Wells, K.D. and Prather, R.S. (2016). Gene-edited pigs are protected from porcine reproductive and respiratory syndrome virus. Nature Biotechnology 34(1): 20-22.

79. Genome editing: moving the goalposts on the GM playing field?

D.M. Bruce
Edinethics Ltd., 11/6 Dundonald Street, Edinburgh, EH3 6RZ, Scotland, United Kingdom; info@edinethics.co.uk

Abstract

The rhetoric of claims for genetic modification of unprecedented precision and unlimited scope have proved exaggerated. Public rejection in Europe was in part because of no tangible benefits to consumers, perceived risks, and imposition without choice. Genome editing techniques have emerged as powerful tools capable of achieving the precision and range of application that 'conventional' genetic modification (GM) has not. It also makes possible modifications in food animals and animal disease, not hitherto regarded as worthwhile by the major livestock breeders. This paper explores to what extent genome editing changes the ethical landscape of GM, whether it might be more acceptable to publics, and the implications for policy and regulation. It revisits the insights from our Engineering Genesis study of 1998 on the main value considerations of GM food, crops and animals, and assesses how genome editing might affect fundamental ethical objections. It examines the complex question of what is or is not genetic modification, and assesses what ethical concerns genome editing may or may not satisfy. Avoiding transgenesis might be convincing for those whose basic concern is about mixing genes across species or violating evolved or God-given 'barriers'. The notion that the edited sequence is capable of occurring naturally would be attractive if one's objection was to creating an unnatural gene construct. On the other hand, it would not impress those for whom any genetic alteration beyond selective breeding is unacceptable, or simply fear the uncertainties of scientists 'tampering with genes'. In the light of this, the acceptability is considered of various genome editing applications in animals and crops for food. Recent insights are included from using our Democs card game to include genome editing issues. A more sensitive regulatory system would be essential but should not be expected to solve the GM problem. The largest factor may be the perceived usefulness of the application to humans, against which the perception of risk and uncertainty might in some cases tip the balance of public acceptability.

Keywords: genome editing, GM food, ethics, public acceptability

Introduction

Recently genome editing has emerged as a powerful scientific tool potentially capable of achieving a range of genetic modifications in plants, animals and even human embryos. This paper explores whether in addition to opening up the scientific potential of genetic modification (GM), genome editing might also open its general ethical and social acceptability. It considers the main value considerations of GM food and assesses how genome editing might affect fundamental ethical objections. It examines the complex question of what is and is not genetic modification, and considers what factors will be important for perspectives to be changed by genome editing.

When GM crops began commercial use, the supporting rhetoric was its unprecedented precision and the unlimited scope, compared with the uncertainties and restricted range of selective breeding. When examined more closely, however, methods did not seem precise, and the more appealing applications proved difficult, e.g. nitrogen fixation and stress tolerance. Such second and third generation crops are still few. The vast majority of GM crop use remains the same two 20-year old traits, herbicide tolerance and insect resistance.

 I. Anna S. Olsson, Sofia M. Araújo and M. Fátima Vieira (eds.) ***Food futures: ethics, science and culture***
EurSafe 2016 – DOI 10.3921/978-90-8686-834-6_79,

Moreover, public acceptability suffered greatly, especially in Europe, because applications brought no tangible benefits to consumers and were perceived to carry environmental and health risks. The GM crisis erupted, and GM foods remain largely unavailable in Europe.

Genome editing techniques have emerged as powerful tools with transformative potential. By altering individual nucleotides at precisely specified locations in the DNA sequence, genome editing offers the potential to realise the claims for precision. The new techniques, however, have their own uncertainties, in this case 'off-target insertions' which will need to be overcome

The ease, cheapness and versatility of genome editing techniques also suggest they have the potential to enable the breadth of applications, which were confidently claimed for GM in the late 1990s, but never delivered, and also enable modifications in food animals not hitherto considered worthwhile. These are early days. It remains to be seen whether the commercial aims of the principal actors in this emerging field will once again reduce the field to a few agronomic traits, or whether this time the more imaginative applications will also attend to factors of more value to consumers, and particularly to address the needs of the global poor and malnourished.

Value criteria for considering GM crops and food

From our expert working group study completed in 1998 (Bruce and Bruce, 1998), ten value criteria were identified representing the key arguments presented. They illustrate the 'lightning rod' effect of GM bringing to attention underlying questions beyond the method itself.

1. Scientific rationality – the next step in technological progress.
2. Commercial – economic growth, jobs, competitive advantage.
3. Theological/philosophical – is switching genes across species right or wrong?
4. Ideological – industrial versus organic agriculture.
5. Risk – uncertain outcomes: when do we know enough? 'Will it harm my child?'
6. Resource/food security – do we need GM technology 'to feed the world'?
7. Development – how does GM relate to global justice for the poor?
8. Benefits – goals and interests of companies vs benefits to consumers (if any).
9. Trust – government, regulators, scientists, NGO's; is there any 'honest broker'?
10. Control and participation – who decides if GM goes ahead, against what criteria?

People's response to the GM food question is likely to depend upon which ones amongst these criteria are the most important. That will vary from person to person, and with different social contexts, countries, and organisations.

These criteria can be applied to genome editing of crops and animals. The contextual questions as #6 – #10 will all be relevant, but at this early stage of application, it would be premature to speculate on the impact of genome editing on them. The agricultural system, scientific and commercial implications (criteria #4, #1 and #2) are explored in a parallel paper at this conference (Bruce, 2016). Here the focus is on how fundamental ethical questions (#3) may change people's views and perceptions, and some risk aspects (#5).

Is genome editing 'genetic modification'?

In so far as it does not normally involve transgenesis, the introduction of genes from other species, it is suggested that genome editing should satisfy one of the major ethical and social objections to GM food. But is this the case? Genome editing re-opens the question of exactly what is, and is not, genetic modification? There are three types of intervention, but not all are considered as necessarily 'GM'.

The clearest case is transgenesis, adding a gene from a different species which could not occur naturally. Some make a distinction of a greater degree of 'unnaturalness' if the gene is from a different kingdom, e.g. from fish or bacteria, compared with a gene still within the same kingdom, e.g. a daffodil gene into rice. A second case is adding a gene to an organism from the same species, sometimes called cisgenesis. A third case is creating new combinations of DNA sequences within the species, intragenesis. An older form of the latter is random mutagenesis using irradiation or chemicals. Genome editing now performs targeted mutations within the organism itself, using enzymes or other non-genetic means to cut and re-insert base pairs in its own DNA.

Another type of intervention is to switching off a gene within the organism. This was done using transgenesis methods but can now be done instead by genome editing.

One way of considering these interventions is naturalness. Although questions of natural and unnatural are notoriously difficult in ethics, there are important insights to be had (Bruce and Bruce, 1998; Nuffield Council on Bioethics, 2015). Transgenesis poses ethical objections if it is deemed that it is wrong to mix genetic material 'unnaturally', in the sense of a process of genetic mixing that cannot happen in nature. But can it be thought wrong to create 'unnatural' genetic mixes in the sense of a mixture not known to exist in nature, but by a process within the same organism? Is the ethical objection to the unnatural act of transgenesis, or to a genetically unnatural product?

Genome editing avoids the first problem but not necessarily the second. The mutation achieved by genome editing may exist in the same or a related species. Hornless cattle have been bred by editing the gene known to be responsible for hornlessness in cattle (Recombinetics, 2016). It is possible to achieve the same result by crossing breeds and selecting the hornless animals but would take many years. This might be termed 'selected naturalness'. Genome editing achieves the same result very much faster, in what might be termed 'induced naturalness'. If the process is not seen as ethically objectionable, and if the result is possible to be achieved by selective breeding, it could be argued that there should be little reason to object on principle to such examples of genome editing.

A slightly different case is creating a mutation which confers resistance to the disease African Swine Fever into the genome of a typical UK pig. This mutation is found in some African warthogs, but not in domestic pigs (Lillico *et al.*, 2013). The same could not be achieved by normal breeding, because the two species, while close, do not interbreed. Given that both are subject to the disease, it would seem churlish to object on the grounds of being 'a genetically unnatural product'. Perhaps our human classification into kingdom, genus, species, etc. is not always be helpful as a reflection of our interaction with nature.

But in this case it is possible that the mutation introduced by a genome editing process already exists in the domestic species, but has never been observed (B. Whitelaw, personal communication). We only know the genomes of a very few individuals compared with the whole population, and spontaneous mutations occur all the time. Somewhere 'out there' the mutation might exist in one or more individuals, or may have done in the past, or might do at some time in future. This situation might be termed 'unknown naturalness'.

The ethical objection would be to the deliberate creation of a mutation which could exists naturally but is not known to. One might object on grounds of risk that the researcher does not know that it might be a deleterious mutation, but here we are considering the principle of doing so. It would seem strange to object to this ethically, if one accepts the deliberate direction of selective breeding by marker assisted selection, or random mutagenesis.

The most problematic case of genome editing is the introduction of multiple changes, in which the resulting plant or animal would definitely not exist otherwise nor be unachievable by any natural process. An extreme example is the attempt to switch off a large number of pig genes to avoid a violent rejection of a pig's heart if transplanted into the human body (Reardon, 2015). Here the focus is not on the method but the product. Is it ethical to create such a pig, whether by genome editing or any other method?

Would genome editing find acceptance with publics?

In the light of this complex situation, it is argued that genome edited plants should not simply be treated as GM crops, because there are important ethical differences. An anomaly in EU regulation is that 'Golden Promise' barley, created by radiation-induced mutageneis, was not considered to be genetically modified, because the mutations are within its own genome. This is despite the fact that the technical method used is in several respects far more uncertain and disruptive than conventional genetic modification. In so far as genome editing is in effect a targeted version of mutagenesis, it would seem illogical to regulate it as if it were 'GM'.

But this illustrates a point that whether publics, regulators or Parliaments find genome editing acceptable or not will depend on many more factors than a detailed exercise in principled moral reasoning. One may conclude that avoiding transgenesis might address the concerns of those whose basic concern was about mixing genes across species, or violating evolved or God-given 'barriers'. Similarly, the notion that the edited sequence is capable of occurring naturally would be attractive if one's objection was to creating an unnatural gene construct.

But if one's objection was to any genetic alteration beyond what is achievable by selective breeding, genome editing might be acceptable only under a very limited set of circumstances. It would not be accepted at all by someone whose primary concern is that of scientists 'tampering with our food'.

The latter has elements of both risk, lack of trust and perhaps revulsion, rather than based on a philosophical position or agricultural system. Perceived risk is likely to be a significant barrier, unless a very good case can be made that off-target events pose no significant long-term risk. While genome editing procedures may be more reliable than GM methods, subtle risk distinctions may not easily be conveyed to publics.

Thus genome editing in food crops and animals is not expected to be a panacea which solves 'the GM problem', but rather that it is likely to receive a range of responses, from acceptance to continued opposition, depending on what factors matter most to people. The largest factor may be in the perceived usefulness of the application to humans, against which the perception of risk and uncertainty might tip the balance. In both animals and crops, if the lead applications of genome editing are agronomic traits with no tangible benefit to European consumers, the perception is likely to be that this is just another version of GM. If producers wish to present genome editing as something different which should make a sea change in public acceptance, it will need to look very different in terms of the applications that are offered.

Another important factor is the extent to which those with agency to enable genome edited foods and other products to be marketed, notably the principal retail chains, are swayed more by those who object than those who might be willing to try. One might think a genome edited pig resistant to a deadly disease was a good thing, but would one then buy the ham, if this became a normal item on the supermarket shelf?

At the moment these questions are largely untried with general publics. At previous EURSAFE conference we reported the results of playing a Democs card game on GM crops with MSc students over a period of ten years, which revealed a cautious view but open to the possibility of GM crops (Bruce, 2015). As a preliminary exercise, a set of games played in March 2016 included some additional questions on genome editing. Opinions among the students varied as to whether this was less ethically problematic than GM, and whether examples like the resistant pig were acceptable. This initial result will need further games to explore.

Overall, although genome editing has the potential to be a 'game changer' on biotech interventions in food, from a technical point of view, it will need very careful managing for the goalposts to move very far with publics.

References

Bruce, A. (2016). Gene editing animals – part of a utopian future? This volume.

Bruce, D. and Bruce, A. (1998). Engineering Genesis. The ethics of genetic engineering in non-human species. Earthscan, London, UK.

Bruce, D. (2015). Changing attitudes to GM crops? In: Dumitras, D., Jitea, I.M. and Aerts, S. (eds) Know your food: food ethics and innovation. Wageningen Academic Publishers, Wageningen, the Netherlands, pp. 283-288.

Lillico, S.G., Proudfoot, C., Carlson, D.F., Stverakova, D., Neil, C., King, T.J., Ritchie, W.A., Tan, W., Mileham, A.J., McLaren, D.G., Fahrenkrug, S.C. and Whitelaw, C.B.A. (2013). Live pigs produced from genome edited zygotes. Scientific Reports 3, Article number 2847.

Nuffield Council on Bioethics (2015). Ideas about naturalness in public and political debates about science, technology and medicine. Analysis Paper, Nuffield Council on Bioethics, London, UK.

Reardon, S. (2015). Gene-editing record smashed in pigs. Nature News 6.

Recombinetics (2016). Available at: http://tinyurl.com/hgy3fcj.

80. Naturalness and benefits in the debate on genetic modification

B.K. Myskja
Department of Philosophy and Religious Studies, Norwegian University of Science and Technology, 7491 Trondheim, Norway; bjorn.myskja@ntnu.no

Abstract

In the debate on the acceptability of genetically modified organisms, there is a schism between the dominant scientific view where the relevant issue concerns risks and benefits, and parts of the societal and philosophical discourses where ambiguous concepts such as 'naturalness' create an additional dimension. It is reasonable to regard 'naturalness' in this debate as a gradient rather than as an all-or-nothing characteristic. The public acceptability of GMOs may depend on both the kind and the degree of perceived unnaturalness of the application. I will argue that given that there are reasons for taking the public opinion seriously regarding the significance of naturalness in genetic modification, we should make this issue a part of the risk-benefit discussion. I will connect the degree of naturalness to the distinction between cisgenic and transgenic plants. I will use four different cases of genetic modification to illustrate how we can give meaning to the notion of naturalness in this debate. These include transgenic glyphosate-resistant soybeans and beta-carotene producing Golden Rice, and cisgenic fungus resistant strawberries. In addition, I will discuss whether the perceived naturalness of the technique would make a difference for assessment of the production of enzymes in GM tobacco. These enzymes are used for breaking down cellulose in the production of biofuel. I will discuss to what extent the assumed benefit of the application is relevant for assessing what significance the perceived naturalness of the genetically modified products should have in discussions on ethical acceptability.

Keywords: transgenic, cisgenic, public opinion, biotechnology, risk assessment

Introduction

In the debate on the acceptability of genetically modified organisms, there is a schism between the dominant scientific view where the relevant issue concerns risks and benefits, and parts of the societal and philosophical discourses where ambiguous concepts such as 'naturalness' create an additional dimension. There have been many attempts at moving beyond that schism, but with little success. There have also been both theoretical and practical attempts at overcoming it. Some have claimed to demonstrate the incoherence of using the idea of 'naturalness' in evaluating agriculture (e.g. Borlaug, 2000). Others have tried to show the ideology behind the claims about the significance of lay people's knowledge deficit in biology and technology (Hansen *et al.*, 2003). None of these attempts have proven successful. One can try to educate the public, or engage the public in interactions with scientists to create arenas for 'reciprocal learning', but the effects are relatively insignificant.

In this article, I do not choose between these apparently contradictory views. I will not attempt to correct any of them or to judge their relative strengths. Through discussing four different types of genetic modification in agriculture, the aim is to suggest how these two approaches can be combined in an assessment of the different applications. I will first discuss how 'naturalness' should be understood in this context, and why it is reasonable to pay heed to the concept in assessing GM plants. Following that, I will explore what difference this concept of naturalness combined with a stronger focus than usual on benefits will give to an assessment of four different cases of genetic modification. This will illustrate not only how to integrate the two perspectives, but also indicate how we can give meaning to the notion of naturalness in the context of modern agriculture. The cases include the transgenic glyphosate-resistant soybeans and beta-carotene producing Golden Rice, and a cisgenic fungus-resistant strawberry. In

EurSafe 2016 – DOI 10.3921/978-90-8686-834-6_80, © Wageningen Academic Publishers 2016

addition, I will discuss whether the perceived naturalness of the technique would make a difference for assessment of the production of enzymes in GM tobacco. These enzymes are used for breaking down cellulose in the production of biofuel. I argue that using a gradualist notion of naturalness in combination with a scientific commitment to develop GM varieties that are generally perceived to be beneficial may help to create a more nuanced ethical debate.

Degrees of naturalness

The term 'naturalness' is common in everyday language, and is conceived as meaningful by most of us. It is generally accepted that the word has many meanings and can be both descriptive and normative. The normative aspect of some uses of the term gives rise to a group of argumentative fallacies, called the 'appeal to nature' fallacies. The typical 'appeal to nature' fallacy claims that something is ethically acceptable because it is natural or unacceptable because it is unnatural. However, one can argue that the reference to naturalness is not a fallacy in the cases where one makes it clear why it is ethically better if the condition or entity is natural or 'more natural'. In social intercourse, it is often an ideal that people avoid pretences or trying to appear in certain ways, and praising someone for being natural in this context is at the same time an expression of this ideal and a statement that the person praised did live up to this ideal. Then 'natural' is shorthand for a certain behaviour that is taken to be good.

Some hold that people should avoid using products like silicone implants or Botox in order to change their physical appearance, unless medically required. Calling these products unnatural, and the ensuing change in appearance likewise, is not a fallacy but an expression of an aesthetic or moral ideal. Thus, we can say that this kind of normative use of 'natural' is meaningful and can be understood by the context in which it appears. The fact that some put on make-up or perm their hair is not necessarily considered unnatural in the same manner although these are also 'artificial' changes to physical appearance. Quite a few people would accept this latter kind of 'unnatural' manipulation without thinking that they would be bound to accept Botox and similar interventions in the body. One could defend a claim saying that the interventions in the body makes the former more unnatural in the normative sense, or one could say that this kind of unnaturalness is less acceptable than the latter because it is a kind of intervention that is more radical. From a scientific or philosophical point of view, this distinction may appear irrelevant, since all the changes above are either time-limited or reversible, and the implants or Botox interventions are doing the same thing as the more superficial ones, namely changing the physical appearance. The main difference is the kind and the novelty of the technologies employed (Moen, 2011). But these facts do not undermine the argument that there is a distinction that can be regarded as relevant and make a difference in assessing these different ways of altering the physical appearances. The normative idea underlying this distinction can meaningfully be described as one being more unnatural than the other.

There is a similar disagreement regarding how to understand 'naturalness' in the debate on genetic modification. I have previously argued that that there are reasons for taking the public opinion seriously regarding the significance of naturalness in genetic modification (Myskja, 2006). A large number of lay people reject GM products because they are perceived to be unnatural or more unnatural than non-GM products in the same category. Scientists and philosophers point out that all of modern agriculture is unnatural. All the food plants that we use are developed through systematic breeding and produced by intensive use of modern agricultural technology. This is the case even with food produced through organic or biodynamic methods. Replies of this kind have little impact on the GM opponents. One can assume there are two reasons for this; they do not believe the claims about all food production being unnatural or they consider the reply to be irrelevant for their rejection of GM products. Even if lay people generally have little knowledge of GM technology and modern food production and the similarities and distinctions between them, I will assume that ignorance is not the main reason for them classifying only GM technology as unnatural. My working hypothesis is that people reject GM technology because they perceive the kind of unnaturalness represented by these products to be morally unacceptable. Their rejection is not of unnatural products as such, since they have to use unnatural products in their everyday lives, however they define 'unnatural'. That is part of modern life, and there is no reason to assume that these people are more inconsistent than most of us.

Siipi (2008) has distinguished three senses of unnaturalness: history-based, property-based and relation-based. GM plants are primarily perceived as unnatural due to the human intervention in their development, meaning that they are considered history-based unnatural. In addition, many of them are regarded as property-based unnatural due to the expression of traits that could not appear without the human intervention, such as insecticide produced by Bt-modified cotton and corn, or the beta-carotene enhancement of Golden Rice. In addition, some may be considered relation-based unnatural, due to their unfamiliarity. A typical example is the fluorescent Glo-fish that would be unnatural in all three senses. However, the common denominator for the unnaturalness of all GM organisms is history-based. For most biotechnologists and regulators this is expressive of an irrelevant concern with process rather than product. The acceptability of a product must concern its characteristics, not its origins, they hold. However, this objection seems to miss the point of being concerned with the unnaturalness of the plant. Those who make the naturalness objection are not primarily concerned with the traits even if experts think that they ought to be. They take a wider perspective on the issue, considering how we ought to relate to nature.

One way to make sense of the rejection of some products as unnatural in an unacceptable way, while accepting others that are no less the product of human activities, is to regard 'naturalness' in this debate as a gradient rather than as an all-or-nothing characteristic (Siipi, 2008: 77). Then the problem with GM plants is that they are more unnatural than their conventional counterpart, due to their production history. In addition, some may be problematic due to the kind of unnaturalness they are perceived to represent, affecting their acceptability (Mielby *et al.*, 2013). If we assume this to be a correct analysis of the naturalness objection to GM plants, then people will not reject each and every application of GM technology, but will be willing to find some more acceptable due to being less unnatural or of a kind of unnaturalness that is ethically acceptable. Under an interpretation of this kind, the naturalness objection is not an indication of public ignorance based on irrational misunderstandings of modern food production, but an ethical challenge to the biotechnology proponents. The fact that some products are taken to be unnatural is not a reason for rejecting them, but a way to express that they require extra reasons for being acceptable compared to those that are not taken to represent the same degree or kind of unnaturalness.

Transgenic food plants – naturalness and benefits

Several studies show that people are more willing to accept cisgenic modification than transgenic because the former is taken to be less unnatural. The normal science-based procedure would be to say that perceived naturalness is irrelevant; we should accept any plant product that is tested and found to be within the acceptable range of health risks. But this is the normal argument regarding technologies that are taken to be beneficial, or at least not harmful in themselves. Rightly or not, GM technology is by many not perceived to be a product of this kind. A significant number of consumers regard it with suspicion and reject it as potentially harmful due to its inherent unnaturalness (Kronberger *et al.*, 2014). As we have seen, people cannot consistently be opposed to unnatural products or events as such, making it reasonable to assume that they do accept some degree of unnaturalness if the benefits are obvious and irreplaceable. This is evident from the large general acceptance of biotechnology in medicine (Gaskell *et al.*, 2010). Clearly, it is no less unnatural to modify the genes in your body than to modify the genes in your foods, and it is highly unlikely that people would think so. The main difference is that the former is without question aiming for clear benefits for the ones who use the product.

A typical case of a successful GM product is the Roundup Ready soybean. In 2014, 84% of all soybean in the world was GM (ISAAA, 2014), and the glyphosate tolerant Roundup Ready varieties were dominant. There is little doubt that this GM crop is beneficial for large-scale industrial farmers, especially in terms of lowered risk of loss and easier management. It is also obvious that these benefits fail to reach the consumers beyond being marginally cheaper than conventional soybeans. This makes the assessment fairly easy for most consumers. There is no reason to buy GM soy because there are no benefits for them. It does not matter whether they value naturalness more or less. As long as there are no definite benefits to themselves or to society and environment, the perceived unnaturalness of the GM crop makes it unacceptable. The issue is similar to regulatory authorities; why should they accept a product that has no benefits for their citizens? The matter is different if they have farmers who want these benefits, but then it is a matter of weighing the different groups against each other.

The posterchild case of beneficial use of GM technology is the beta-carotene enhanced Golden Rice (see http://tinyurl.com/jdon5al). This GM rice is promised to help combat vitamin A deficiency (VAD), which is the cause of up to 500,000 cases of blindness and almost 2 million deaths every year. It is arguably even more unnatural than GM soybeans according to the classifications of Siipi (2008), as it can be described under all three senses of unnaturalness. Despite its benefits, Golden Rice has little significance in reducing the 'naturalness'-based scepticism to GM technology. The scepticism is high in the affluent parts of the world, especially in Europe, and there is no market for Golden Rice there; people in this part of the world do not suffer from VAD. The benefits will only affect the significance of the naturalness objection for those who have a concern about VAD and see Golden Rice as a solution. However, many argue that Golden Rice is a replacement for the real solution, namely to provide a sufficiently varied and nutritious diet for people who today suffer from VAD. This can also be related to the naturalness objection, as a varied diet is more in keeping with ideals of a natural human life. Even if one accepts that point, Golden Rice should be considered part of the solution, and as an indication forward for GM technology. There are clear benefits, even when taking the alternatives into account. Those who may benefit should be the ones deciding whether this benefit outweighs any concern about unnaturalness. But the fact that Golden Rice is acceptable, should not impact on other cases of GM that have comparable societal benefits.

Cisgenic – more natural, more acceptable

A significant number of those who accept the naturalness argument against GM technology, hold cisgenic modification to be less problematic than transgenic modification, as the former uses only genetic material from within the species. This means that the variety could have been produced through systematic breeding. One example of this kind of GM plant is a fungus resistant strawberry plant where the plant's own fungus resistance was reinforced (Schaart, 2004). Although the technique was the same as with transgenic modification of plants, the fact that the production did not involve transferring genetic material across species borders, makes a difference for a majority of respondents (Schaart, 2004: 94), a result that is confirmed in later studies of other cisgenic varieties (Gaskell *et al.*, 2010). In a focus group study on cisgenic modification, Mielby *et al.* (2013) identifies five different naturalness arguments, expanding on Siipi's three. In addition to the history-based arguments, the ones based on substance are those that primarily support the claim that cisgenic crops are less unnatural than transgenic, and hence, more acceptable (Mielby *et al.*, 2013: 475). It is not a matter of human intervention as such, but about respecting some barriers set up by nature. If we accept that naturalness arguments are about degrees, this kind of modification is more acceptable. The implication is not that cisgenic plants should be allowed just because the technique is less offensive, but that there is less reason to object to them. Thus, less is required in terms of benefits for making them acceptable. The benefits of the strawberry – if it had been commercially produced – would have amounted to reduced loss and easier management for farmers, in addition to less use of fungicides. This last point has significance also for consumers who are concerned about pesticide residue in food products. Combined with the more acceptable degree of perceived naturalness, the arguments for accepting this variety is even stronger in theory compared to Golden Rice. But the benefits of the latter clearly outweigh those of the strawberry in significance.

Biofuels – a different issue

In a recently started project led by the Norwegian NIBIO research institute, funded by the Research Council of Norway, the aim is to contribute to low cost production of key cell wall degrading enzymes to be used for biofuel production. The so-called Bioboost project will develop GM tobacco plants to replace energy demanding and expensive fermenter-based systems, for turning cellulose from forest raw materials into biofuel. Tobacco is a non-food and non-feed crop with high biomass yield, and the production will be in closed greenhouse environments. If successful, the GM tobacco will be essential in providing an alternative to non-renewable energy, and to the use of food crops for production of biofuel. A GM tobacco would be no less unnatural than the GM soybean, but the production should still be regarded as less problematic. First, this is a non-food product going into the production of fuel used in industry and transportation, which are already clearly man-made activities. This means that naturalness is less relevant for the assessment. Second, it is the solution to a complex of generally accepted problems connected to destructive and irreversible human interventions in nature. The implication is that the GM tobacco is meant to be part of a replacement of practices that are regarded as negative due to unnaturalness. Third, the production is clearly beneficial, making up for any objection based on perceived unnaturalness or risk.

Conclusions

The naturalness objection to GM plants is not likely to disappear due to increased scientific knowledge in the public. Many people are not primarily concerned with the risks, but see no reason to accept food products that they consider more unnatural than conventional ones. The suggestion in this paper is that the degree of unnaturalness is significant for how one should evaluate GM plants, but the benefits of the product are probably even more important. The benefits must, however, be for those who are the intended users of the products. It is not sufficient that industry or farmers benefit.

References

Borlaug, N. (2000). Ending world hunger. The promise of biotechnology and the threat of antiscience zealotry. Plant Physiology 124: 487-490.

Gaskell, G., Stares, S., Allansdottir, A., Allum, N., Castro, P., Esmer, Y., Fischler, C., Jackson, J., Kronberger, N., Hampel, J., Mejlgaard, N., Quintanilha, A., Rammer, A., Revuelta, G., Stoneman, P., Torgersen, H. and Wagner, W. (2010). Europeans and biotechnology in 2010. Winds of change? European Commission, Brussels, Belgium, 176 pp.

Hansen, J., Holm, L., Frewer, L., Robinson, P. and Sandøe, P. (2003). Beyond the knowledge deficit: recent research into lay and expert attitudes to food risks. Appetite 41: 11-121.

International Service for the Acquisition of Agri-biotech Applications (ISAAA) (2014). Pocket K No. 16: global status of commercialized biotech/gm crops in 2014. Available at: http://tinyurl.com/jvemn33.

Kronberger, N., Wagner, W. and Nagata, M. (2014). How natural is 'more natural'? The role of method, type of transfer, and familiarity for public perceptions of cisgenic and transgenic modification. Science Communication 36: 106-130.

Mielby, H., Sandøe, P. and Lassen, J. (2013). Multiple aspects of unnaturalness: are cisgenic crops perceived as being more natural and more acceptable than transgenic crops? Agriculture and Human Values 30: 471-480.

Moen, O.M. (2012). Cosmetic surgery. Think 11: 73-79.

Myskja, B.K. (2006). The moral difference between intragenic and transgenic modification of plants. Journal of Agricultural and Environmental Ethics 19: 225-238.

Schaart, J.G. (2004). Towards consumer-friendly cisgenic strawberries which are less susceptible to Botrytis cinerea. PhD thesis, Wageningen University, Wageningen, the Netherlands.
Siipi, H. (2008). Dimensions of naturalness. Ethics and the Environment 13: 71-103.

Section 21. Food and environmental ethics

81. Limits to growth in the context of nature and land use

M. Thorseth
Department of Philosophy and Religious Studies, Norwegian University of Science and Technology, NTNU, 7491 Trondheim, Norway; may.thorseth@ntnu.no

Abstract

Nature is no longer abundant, and there is a need to put limits to growth. The current living population has to make commitments to ourselves at an inter-generational as well as an intra-generational level. This commitment, also expressed in the Brundtland report, has strong support worldwide. However, there is no univocal agreement how this commitment should be implemented. Limits to growth may be based on a defence of either environmental rights and/or a survivalist view, holding that nature, including ecosystem goods and services, are considered commons rather than private property. However, it is also possible to base limits to growth in a liberal and individualist view. As I shall argue, in a liberal perspective dating back to Locke it is possible to find strong arguments in favour of limitations to growth. Although Locke basically is identified with a possessive and individualist position that gives primacy to individuals and their rights to property, there are other aspects of Locke's writings that undermine this position. This becomes clear when looking at his views on the harm principle, the common goods, and future generations. The upshot of this consideration of his view on private property supports the position that an individual's right to exploit nature is limited. This interpretation of Locke's view of property is, however, conditioned by an extension of the concept of 'nature' to be composed of ecosystems and providing ecosystem goods and services. A profound thought in Locke is that land produces abundantly. The failure to protect the ecosystems that provide this abundance harms the property of others as well as the viability of the society. Additionally, Locke's writing also supports the view that the government should protect the abundance of nature. At this point, we can see a similarity between Locke's liberal view and a survivalist position that defends limits (as opposed to a Promethean view) in the discourses on sustainability. The defence of commons appears to converge in these two approaches that otherwise are associated with divergent arguments in favour of limits. In this paper, I want to compare these different approached to limits to growth, in view of a possible overlapping consensus between them.

Keywords: survivalism, commons, Locke, possessive individualism

Limits to growth: survivalist view

The survivalist view acknowledges that there is a limited stock of natural resources on the earth. This view is embedded in what Dryzek labels 'the discourse of ecological limits' (Dryzek, 2005: 28). The environmental problem results from the anticipation that survival of humankind is incompatible with unconstrained human procreation and consumption. As an example, sufficient fertile soil and responsible land use require constraints on how these resources are governed. As long as, for example, soil is considered a private property and a commodity for sail and profit, there seems to be poor incentives for concerns beyond private survival. The problem is well grasped in the influential essay 'The Tragedy of the Commons' by Garrett Hardin (1968). The commons is a good that belongs to all, like the village commons for peasants in the village. By putting in an extra cow on the village common, each individual will increase his or her own benefit while simultaneously putting stress upon the commons. The result of this logic is that when each peasant acts in rational self-interest by putting more cows on the commons, they simultaneously destroy the village commons. According to a survivalist view, the same logic is at play for all kinds of commons, like fish, soil, trees as well as sewage and air pollution. Each individual decision to increase private benefit points in an opposite direction with respect to public interests. For each decision maker, the question is whether to catch more fish, cut down more trees or to drive one's own

EurSafe 2016 – DOI 10.3921/978-90-8686-834-6_81, © Wageningen Academic Publishers 2016

car instead of traveling collectively or biking instead. The tragedy of the commons consists in there being a conflict between private benefits and public interests unless we act upon there being limits to growth.

Limits to growth: liberal, individual, Lockean view

Common goods may, however, also be of pivotal importance from a liberal point of view, not because of its collective character, but rather because it is required for realising individual interests. This is how group rights may be defended from a liberal stance (Kymlicka, 1995; Taylor, 1989). If some common good is necessary for obtaining an individual good, then it should be protected. Fertile soil and responsible land use are commons that are necessary for obtaining individual goods, and as such they are commons that are preconditions for individual gains, as well. Thus, on a liberal account, it is possible to argue in favour of limits and constraints to actions driven by pure rational self-interests. This is analogous to a liberal defence of group rights (Taylor, 1989). In comparing commons in the environmental debate to group rights in the debate on multiculturalism, we argue that limits to growth might be based on a liberal defence of limits to growth. Contrary to a survivalist defence of limits, a liberal, Lockean defence would accept limits to the extent that limits are necessary for the protection of individual goods. The main difference between a survivalist and a liberal account of limits to growth is that the former argues from a collectivist view of commons, whereas the Lockean one is based on protection of commons as a means to realise individual goods. Broadly speaking we can envisage the division line between a communitarian and a liberal defence of commons.

Food, soil and responsible land use are goods that include both access to safe food, but also sufficient food for all, although the reach of 'all' varies, from 'all' within a limited range to 'all' within current generations on a global scale, or even comprising both current and future generations. In this sense, responsible land use is an irreducible common good. If we consider this good from a communitarian perspective, it goes without saying that entitlement to it is by nature something collective and social, not least because the survival of the collective is a prerequisite for the individual. Safe and sufficient food would simply not exist as a purely individual commodity. From a communitarian position, it is an easy enterprise to appeal to a collective concern for the protection of responsible land use. On a similar ground, the survivalist view agreement to protect limited commons is considered a shared responsibility.

Although strongly diverging from the survivalist approach, the value of responsible land use as a collective and common good could also be recognised from a more liberal stance, although it would rest with a different kind of arguments to be developed elsewhere. Safe and sufficient food for all would still be important, but it would no longer be looked upon as irreducibly collective commons. On the liberal ground, it could be argued, as Locke does, that individuals are entitled to the goods they appropriate. In this rather extreme version, there need not be any mechanisms for compensating those who would not be in a position to appropriate land. At Locke's time, this particular problem had not yet entered the board, as natural supplies were still affluent. There are two different contexts for the argument in this paper: one is the context of affluence (cf. at Locke's time), whereas the other is the context of scarcity, i.e. present time. I shall return later on to interpretations of the possible existence of common goods in Locke. Across the different contexts there is perhaps a more profound question about entitlement to goods as such: on a liberal, Lockean account the entitlement is established through appropriation. This way of reasoning requires, however, that there is a possibility of ownership, a question that does not only require affluence. One important scenario is overlooked. There might be affluence, but still no good reason for appropriation, neither by individuals nor groups. As an example recreational areas are accessible to all, e.g. the Norwegian 'almenning' (English: commons) serves as an example. A reason for not allowing individuals to appropriate all recreational areas is because such areas should be available to all, as opposed to being individually possessed. The argument for protection runs across any particular context, independent of the affluence/scarcity divide.

When discussing collective (survivalist) and individual (liberal) arguments for protection of common goods, it is important to keep the two dimensions of individual/collective and scarce/affluent apart, at least at an analytical level. The reason why those who can afford to buy food should be entitled to eat is not only because they are capable of buying it. Rather, most people would agree that even poor people have a right to eat even if they can't afford to buy food (in order to survive). The argument in favour of food for poor people is, however, hard to base on a possessive individualist position á la Locke. Rather, most people would be prepared to base their argument on some theory of justice, e.g. Rawls' appeal to the difference principle, holding that inequalities are legitimate only as long as they serve those worst off better than any alternative (Rawls, 1985). Food for survival is a common good that many people would be prepared to consider a commons rather than a commodity at a very basic level. In the case of scarcity of commons, there is a need to implement limits, independently of the cause of scarcity: either due to an imbalance between the amount of food and people to be fed *or* scarcity for some people due to an unjust distribution of the resource. These are two different contexts of which only the first one is relevant to the argument of this paper. Responsible land use policies thus have to take as a vantage point that there is a need for limits to growth.

Commons vs commodities

At this point we need to make some provisos, basically having to do with the profound difference between commons and commodities. Briefly, the division line concerns how certain goods and services are valued. Certain goods, such as food and water supplies, are goods that should be available to all, not only to the extent that it can be traded as a commodity. The scarcity of food supplies is caused by varying conditions: uneven access to fertile soil; displacement of poor people; bad land use policies; uneven distribution, just to mention a few. Across these, partly overlapping, partly divergent causes these contexts still have to cope with the fact of scarcity and a need for limits. Another proviso to be mentioned is, of course, the role of technological innovation. We already know that improved technologies may contribute substantially to increasing food production. Given these provisos, there is still a need for limits to growth, which is based on an ethical argument rather than empirical circumstances. The ethical argument may be based on refusion of the conflation of commons with commodities (Hardin, 1968), as a basis for limits to growth. There is, however, also a divergent argument for limits to growth that seems not to presuppose an ideological difference between commodities and commons. This is the liberal, Lockean argument based on the existence of common goods whose protection is primarily based on the harm principle. The rationale for limits is then framed as a limitation on individuals' rights to exploit nature.

Provisos for limits based on a liberal, Lockean and individual view

The harm principle is the basis for limitations to exploiting nature in Locke. One proviso for this interpretation is that 'nature' is extended to comprise ecosystem goods and services, i.e. goods that nature provides abundantly. A second proviso for limits is then that failure to protect ecosystems that provide this abundance harms the property of others. Additional to these two there is a third proviso for limits to be found in Locke, which is that government should protect the abundance of nature. The main difference from the survivalist view to be emphasised here is that a Lockean defence of limits still seems to be based on the need for protection of individual rights, rather than protection of commons as basically different from private property.

Survivalism vs Locke

Survivalism addresses the need for increased interdisciplinary and international collaboration regarding survival and sustainability issues. Dryzek (2005) links survivalism to the concept of 'carrying capacity' as discussed by Hardin (1968), i.e. there is a limit for survival of the ecosystems with respect to population growth. Briefly, there is a limit to growth that cannot be transgressed. This is because some urgent goods are commons. These commons are collective goods in the sense that individual advantage (one more cow) will cause stress and costs upon the commons (exceeds the carrying capacity of the village commons). As we have seen, the limits are defined by the fact that private benefit and public interests point in opposite directions. Thus, there is a need for limits to growth as well as for dumping of sewage (Dryzek, 2005: 29). Compared to a Lockean liberal position the striking dissimilarities to survivalism far surpass the similarities: survivalists care about commons and limits whereas Locke rather is concerned with possessive individualism (Locke, 1690; Macpherson, 1973). Still, there are some principles and values that might appeal to some common ground for legitimating limits to growth. Despite Locke's possessive individualism there is also in his works acknowledgment of the harm principle, of some common good and even concern for future generations, or offspring (Bailey and Thorseth, submitted for publication 2016). Limits to growth would be legitimate to the extent that it is required for protection of some common good. As an example of such a legitimising strategy, we could mention environmental rights as described in the South African constitution, chapter two section 24, which affirms every citizen's right to a non-harmful environment (Cullett, 1995). The role of government is here to protect the environment in order to prevent harm, and thus also put limitations to human actions. To the extent that the basis for such limitations is also present in Lockean liberalism, there certainly is some common ground between his position and a survivalist position.

Conclusions

One aim of this paper is to establish a common ground between otherwise divergent arguments for limits to growth found in a survivalist and a Lockean liberal view, respectively. The argument holds true if we succeed in deriving a plausible defence of the following in both positions: a need for limits to growth, acknowledgment of commons, legitimate limitations to human action and governmental protection of common goods. The paper argues that the two different defences of limits converge in an overlapping consensus, i.e. an agreement on legitimating limits and constraints on individual actions, but for different reasons. Whereas the survivalist view is based on the tragedy of the commons, against individual appropriation of commons like food supplies, Locke's liberal view argues from the harm principle, which originates from concerns for individual property.

Having established a common ground between the two positions, it still has to be explored where such a common ground further leaves us with respect to responsible land use. Implementation of limits to growth is subject to economic, social, technological, as well as psychological dimensions of governance. It is hard to imagine how a liberal appeal to individual ownership could possibly converge in an overlapping consensus on policies based on a survivalist appeal to the profoundly collective dimension of goods, i.e. commons. However, as a beginning, to find some common ground is a non-trivial gain pointing in the right direction. An optimistic prospect for the outcome of this paper is to diminish the cleavage between a survivalist view and possessive liberalism. Across diverging motivations, there is perhaps still an overlapping consensus that converges in acknowledgment of limits in respect of untradeable commons like soil and food, and thus leaves us with some hope for increasing consensus on political governance of the commons in the future.

References

Cullett, P. (1995). Definition of an environmental right in a human rights context. Netherlands Quarterly of Human Rights 13: 25-40.

Dryzek, J. (2005). The politics of the earth. Environmental Discourses. Oxford University Press, Oxford, UK, 261 pp.

Hardin, G. (1968). The tragedy of the commons. Science 162(3859): 1243-1248.

Kymlicka, W. (1995). Multicultural citizenship. A liberal theory of minority rights. Clarendon Press, Oxford, UK, 296 pp.

Locke, J. (1690). Second treatise of government. Project Gutenberg eBook. Available at: http://tinyurl.com/6d62k9o.
Macpherson, C.B. (1973). Democratic theory. Essays in retrieval. Cambridge University Press, Cambridge, UK, 255 pp.
Rawls, J. (1985). Justice as fairness. Political not metaphysical. Philosophy and Public Affairs 14: 223-251.
Taylor, C. (1989). Cross-purposes: the liberal-communitarian debate. In: Rosenblum, N.L. (ed.) Liberalism and the moral life. Harvard University Press, Cambridge, MA, USA, pp. 159-182.

82. Territorial impact of greenhouses in Portugal

D. Lopes
Faculty of Law of the University of Coimbra, Pátio da Universidade, 3004-528 Coimbra, Portugal; dulcel@fd.uc.pt

Abstract

Although greenhouses have been around since the Roman age, their configuration, impact and potentialities have changed dramatically over the years. Such changes are motivated by the structure and function of agricultural areas, but at the same time raises concerns in what regards sustainability and the environmental aspects of food production. In this article we discuss whether or not Portuguese legislation and practices provide appropriate answers to problems brought on by the installation of greenhouse clusters, specifically the modification of the use of land, the impact on the landscape, and the environmental consequences. We conclude that a very clear definition of the conditions for the installation and functioning of greenhouses is lacking and should be adopted. Also, planning instruments and impact assessment analysis should be adjusted to the specificity of greenhouse clusters, especially by defining areas devoted to their location.

Keywords: greenhouses, planning, protected areas, environmental impacts

Introduction

Although greenhouses have been around since Roman times (Casey, 2009), their configuration, impact and potentiality have changed dramatically over the years. One such change deals, in particular, with the type of greenhouse constructed. A clear diversity and complexity of the greenhouse structure exist: some are very rudimentary, fragile, temporary and removable while others are very intricate, robust and stable.

Another transformation deals with the territory greenhouses cover. Moving away from usually limited areas of land, greenhouses are nowadays assembled in enormous agro-industrial compounds or clusters that include wide infrastructures and warehouses that cover large areas.

Finally, in the transition from greenhouses that harboured traditional soil cultures, recent greenhouses include technically advanced agro-activities such as hydroponics in which the soil is no longer the essential element but just another factor of a multipart equation.

These changes in the manner agricultural uses are being developed have brought about welcomed modifications to the structure and function of agricultural areas as well as to of food production. However, at the same time, this raises concerns regarding sustainability and the environmental aspects of food production.

We focus here on these concerns to bring to the table the Portuguese experience in which the rapid increase in the number of greenhouses is a product of both the very inviting biophysical conditions and very lax government regulation.

Results and discussion

In order to further our research, a descriptive method is used, based on the interpretation of legal provisions and the analysis of their results. We discuss whether or not Portuguese legislation and practices provide appropriate answers to problems brought on by the installation of greenhouse clusters.

EurSafe 2016 – DOI 10.3921/978-90-8686-834-6_82, © Wageningen Academic Publishers 2016

Specifically, we examine: (1) the modification of the use of land with the abandonment of traditional agricultural uses; (2) the impact on the landscape in agricultural and protected areas; and (3) the environmental impacts, mainly in the management of water, the deposit of waste and the dismantlement of the greenhouses.

Greenhouses: from a legal point of view what are they after all?

Article 2.°, paragraph (a) of Decree-Law 555/99, 16th September (with the latest modifications introduced by Decree-Law № 214-G/2015, 2nd October) takes a traditional approach regarding the meaning of edification. Under that legal concept, edification is defined as 'activity or the result of a construction, reconstruction, enlargement, modification and conservation of immovable property destined for human use, as well as any other construction that is incorporated in the soil with a permanent character'. This raises the question of what exactly, after all, a 'construction incorporated in the soil having a permanent character', is since municipalities only control such edifications.

The essence of the concept permanence appears to rely upon the technical features of the particular construction at hand, which must be portrayed and presented in the building project to the municipality. However, since greenhouse characteristics are so diverse and variable (greenhouse to greenhouse) (Arteca, 2015), it is hard for public entities – at least in Portugal – to apply a strict criterion of permanence.

Thus, even if most greenhouses nowadays have what appears to be a more permanent structure, and do not follow a simple model that uses light materials, such as wood and plastic (FAO, 2002) (Figure 1), but instead have a more rigid and stable format (Figure 2), most Portuguese municipalities consider them not to be included in the strict definition of edification, unless it is evident that the greenhouse corresponds to an immovable construction similar to, for instance, a sitting house.

Figure 1. Non permanent greenhouse.

Figure 2. Permanent greenhouse.

This conceptual exclusion would not be so problematic if the instalment of these 'heavy' greenhouses were included in a broader notion of urban operations.

These include, according to article 2, paragraph (j) of the above-mentioned Decree-Laws, the material operations of use of the soil (such as in bungalows or mobile construction sites), as long as they were for purposes of a non-exclusive agricultural nature. Given, however, that the agricultural activity taking place in a greenhouse is also under this functional criterion, most greenhouses continue to be excluded from any urban control.

Naturally, the proper functioning of greenhouses demands that basic conditions of the proposed terrain are laid down and that certain infrastructures are available, such as, for example, water reservoirs. These basic conditions correspond to urban uses and operations that might be, individually, subject to municipal authorisation. However, the main use of the soil in greenhouse compounds continues to be excluded from municipal jurisdiction, not allowing for a full consideration of urban and environmental impacts, mainly in rural landscapes.

Therefore, an activity that should be controlled by municipalities falls outside their interventional scope. And since Municipalities are the entities that have major powers in regulating the occupation of the land and in controlling illegal activities in Portugal, the loss of that control means a *de facto* 'deregulation' of the conditions for the installation of greenhouses.

This situation is unsatisfactory since no public entity authorises the installation of greenhouses or controls their concrete location and impact on other private or public interests. But this needn't be necessarily the case.

According to our understanding, the concept of permanence of a construction should be interpreted according to the specific features and goals of the field of urban planning law. Therefore, it should not be required that a construction have an immovable structure with, for instance, cement foundations; it should be sufficient for it to have a stable structure, which means that it must be fixed to the soil in a firm and secure manner, in the sense that it is not easily removed – at least not without the help of heavy mechanical means (Oliveira *et al.*, 2016). In addition, the Portuguese field of technical sciences (Meneses, 1990), has tended towards this semantic notion of permanence regarding greenhouses since the lifetime of a greenhouse is limited, but should, nevertheless, be supported by stable structures in order to make them more robust and resistant.

By adopting this viewpoint, greenhouses would be subject to the due administrative legal controls, ensuring their impacts are appropriately taken into consideration and allowing for the sustainable development of these new forms of agriculture.

However, without a legal clarification of concepts or the approval of a specific legal regime for greenhouses, the control of this use of the soil will most probably continue to fall between the fingers of municipalities that, in some cases, either willingly or reluctantly, as in other cases, accept the instalment of greenhouses with almost zero control or supervision.

Planning instruments: are they adjusted to greenhouse clusters?

Generally, when positioning, most planning instruments do not take into account, nor regulate specifically, the installation of greenhouses in municipal areas.

This is due to two main circumstances: the fact that in the past most municipal plans did not pay sufficient attention to the rural parts of the territory (Oliveira, 2011); and the fact that, even if they had rural concerns, greenhouses weren't thought of as a foremost subject for regulation during the first wave of planning laws in Portugal during the eighties and nineties.

Of course, if greenhouses were actually considered edifications, they would be subject to the legal parameters of construction established for rural areas; but since they still continue to fall outside the application of that concept, as referred to above, it is even more necessary that planning instruments do not ignore but explicitly regulate such activity.

An example can be found within the governmental plan applicable to a protected area: the special plan for the natural park of the southwestern Alentejo and Vicentin coast, adopted by the Resolution of the Council of Ministers no. 11-B/2011, 4th February 2011.

According to this plan, which has clear environmental purposes, the installation of greenhouses is forbidden, except in the intervention area of the irrigation perimeter of the river Mira [article 8.º, paragraph (e)]. This perimeter, included partly within the natural park, is regulated by special provisions that try to balance the environmental values at hand with the indisputable agricultural potential of the terrain.

The rules applicable to greenhouses are established in article 46.º, paragraph o), under which – among other prescriptions – the maximum height of the greenhouse is 6 m, the maximum length is 400 m and the maximum area of adjoining greenhouses is 5 acres. Also, greenhouses should have an autonomous system for water management and should only consume up to half of the area of the agricultural terrain. As a whole, the area occupied by greenhouses should not exceed 30% of the total area of occupation of the defined irrigation perimeter, criteria that seems to entail a very close and tight control of the physical occupation of those areas and of their accumulated impact.

At first sight, these rules might appear sufficient in preventing the uncontrolled dispersion of greenhouses in protected areas, but this hasn't been the case. Frequently, large areas of greenhouses are observed in spaces that are most in need of protection, such as cliffs, or in areas that should have had very careful landscape insertion studies (Figure 3).

This problem is due to the fact that – absent the level of municipal control mentioned above – the state authorities responsible for the protection of environmental and agricultural interests do not intervene as official authorities. They only provide an oversight of the protected areas and of the irrigation perimeter, without stringently enforcing the rules of the special plan they are responsible for and meant to safeguard.

Therefore, private parties interested in occupying the land are faced with a framework characterized by very weak preventive control that is due to the absence of municipal authority and a very relaxed supervisory jurisdiction due also to the distancing of central government authorities.

So, it is not surprising that Portugal is perceived as a natural contender for agriculture based on greenhouses. This is due not only to its favourable climate and biophysical conditions, but also to the loose administrative criteria and procedures applied to enforcing the applicable rules.

It has reached the point in which relevant voices are pushing for better regulation of the greenhouse setting in Portugal. Good examples of this is are an interview with Paula Ferro in the Diário Online Algarve (4 December, 2015) and a public petition entitled 'Impeçam a destruição do Sudoeste Alentejano e Costa Vicentina'(http://tinyurl.com/hsbc6av).

Resistance in Portugal to major landscape change is nothing new and stakeholders' perceptions towards greenhouses aren't necessarily objective, as rightfully pointed out by Rogge *et al.,* 2011: 334. It follows then that, absent due levels of public planning private participation and administrative control, these critiques that keenly touch the heart of the problem will simply continue without any results.

Figure 3. Greenhouses near a cliff.

In effect, it would be desirable to establish planning devices such as landscape planning that would define criteria for allocating greenhouses with a minimum of visual nuisance, as proposed by Rogge *et al.*, 2008: 76; however, here again questions of competence arise and remain unanswered: who would be responsible for these plans and under which criteria and legal framework would they fall?

Environmental impacts: are they duly taken into consideration?

In the absence of plans and projects that are evaluated by public entities and subject to authorisation procedures (or similar forms of administrative control), the environmental impacts of greenhouse installations will not be taken into due consideration.

The legislation of the environmental impact assessment is apparently in denial of the greenhouse reality since it continues to include this specific type of agriculture in the scope of 'agriculture development projects that include irrigation and draining infrastructure', which only mandates an assessment whenever the area equals or is higher than 2,000 acres or 700 acres in sensitive areas (Annex II to Decree-Law no. 151-B/2013, 31st October). This issue needs to be addressed.

These quantitative limits are, from our point of view, very high when considering the extraordinary environmental impacts greenhouses might have. These include degradation of the soil and possible chemical pollution and habitat loss, besides the above mentioned intrusion in urban and rural landscapes. Just to mention one example in the south-western part of Portugal, greenhouses were viewed as threats to habitat 3170 (temporary Mediterranean ponds) due to the abandonment of traditional forms of agriculture and the construction of agricultural infrastructures, draining and revolving the soil (Canha and Cruz, 2010).

Therefore, in what concerns the environmental area, before greenhouse installations are authorised, they should be subject to specific regulation, more adjusted to their onerous nature, especially by ensuring the effective consideration of accumulated impacts of greenhouse compounds. In addition, effective regulation and administrative control over the actual environmental impacts of greenhouses activity and dismantlement are lacking, since their specific features, in comparison to traditional forms of agriculture, are not being seriously discussed and tackled.

Conclusions

Given the area covered by greenhouses in Portugal (in 2014, 4.1% of the horticultural area (INE, 2015)) and the perspective of more yet to come, it is important that good practices are soon introduced into Portuguese legislation. The fear of another Almería (Wollosin, 2008), now located in the southern part of Portugal, is not without substance.

The preceding arguments lead us to believe that a very clear definition of the conditions for the installation and functioning of greenhouses is imperative. Problems as diverse as the modification of the use of land, the impact on the landscape and the environmental impacts of greenhouses are all not being undertaken in a sufficient manner by existing legislation and administrative regulations and practices.

Since a heavy greenhouse should clearly be considered an edification, it should thus be subject to similar criteria, both in terms of urban parameters and of authorisation procedures. Indeed, public interests can only be safeguarded adequately by involving public authorities – mainly municipalities, due to their specific knowledge and close proximity to the land – in the procedure of installing greenhouse compounds.

Also, planning instruments and impact assessment analysis should be adjusted to the specificity of greenhouse clusters, especially by defining areas devoted to their location. This could be accomplished, for instance, by the development of agro-industrial parks predominantly destined for greenhouse horticulture, and where the environmental and landscape impacts could be adequately managed and minimized.

References

Arteca, RN. (2015). Introduction to horticultural science. Cengage Learning, Boston, MA, USA.

Canha, P., Cruz, C.P. (2010). Plano de gestão de charcos temporários mediterrânicos no concelho de odemira. Available at: http://tinyurl.com/z7fbmy9.

Casey, W. (2009). Firsts: origins of everyday things that changed the world. Alpha Books, New York, NY, USA.

FAO (2002). El cultivo protegido en clima mediterráneo, estudio fao, producción y protección vegetal, 90. FAO, Rome, Italy.

INE (2014). Estatísticas Agrícolas de 2014. Instituto Nacional de Estatística, I.P. Lisbon, Portugal.

Meneses, J.P. (1990) Portuguese standards applied to greenhouse structural design. Acta Horticulturae 281.

Rogge, E., Nevens, F., Gulinck, H. (2008). Reducing the visual impact of 'greenhouse parks' in rural landscapes. Landscape and Urban Planning 87.

Rogge, E., Dessein, J., Gulinck, H. (2011). Stakeholders perception of attitudes towards major landscape changes held by the public: the case of greenhouse clusters in Flanders. Land Use Policy 28.

Wolosin, R.T (2008). El milagro de almería, españa: a political ecology of landscape change and greenhouse agriculture. Thesis paper. Available at: http://tinyurl.com/hy2joso.

Oliveira, F.P. (2011). Novas tendências do direito do urbanismo. De um urbanismo de expansão e de segregação a um urbanismo de contenção, de reabilitação urbana e de coesão social. Coimbra, Portugal.

Oliveira, F.P., Neves, M.J., Castanheira, L. (in press). Dulce, regime jurídico da urbanização e edificação comentado. 4th edition. Coimbra, Portugal.

83. The value of purity

H. Siipi
Turku Institute for Advanced Studies, Philosophy Unit, Assistentinkatu 7, University of Turku, Turku, Finland; helsii@utu.fi

Abstract

What does the purity of food mean? Is it desirable? The questions are answered by two kinds of conceptual analyses. First, the meaning of the term 'pure' is approached through its possible contrasts: dirtiness, unclarity, and richness and redundancy. As a contrast to dirtiness, purity refers to the lack of elements that would spoil the food in question. These elements can harmful to health (such as mercury), but also edible elements can challenge purity of food (e.g. pure hot chocolate does not contain tomato). As a contrast to unclarity, purity refers to accordance with conventional categories. When purity is contrasted with richness and redundancy, it means to lack of unnecessary elements. Second, the meaning of 'pure' is approached by comparing it to closely related terms 'natural' and 'authentic'. All three terms sometimes refer to the lack of certain elements. Moreover, both naturalness and purity may mean nutritious suitability and familiarity. Authenticity and purity sometimes refer to correspondence between claimed identity and actual properties.

Keywords: conceptual analysis, contrasts, related terms

Introduction

Claims about the purity of food are common. We are constantly told that the products offered are instances of 'pure olive oil' or 'pure rye bread', for example. A chocolate bar may be presented as a source of 'pure energy' and restaurants and food companies have names such as 'Pure Hero Salad Bar', 'Pure Food Co' and 'Just Pure Food Snacks'. But what does the purity of food mean? Is it desirable and is pure food better that food that is not pure?

The presented questions are answered by two types of conceptual analysis. The meanings of the term 'pure' are first approached through three of its contrasts: dirtiness, unclarity, and richness and redundancy. After that, the meanings of the term 'pure' are compared to meanings of two closely related terms, 'authentic' and 'natural'.

First contrast of pure: dirty

An obvious contrast to 'pure' is 'dirty'. But what is dirt and what distinguishes dirty objects from the pure ones? A very few substances are dirt in all contexts. A piece of clay, for example, is dirt when found on the kitchen table. The very same piece of clay is not a piece of dirt in a pottery class where cups are made of it. Furthermore, almost any substance can in some context be dirt. Fine champagne, for example, turns into dirt when spilled to a suit. Whether something is dirt is relational and dependent on the context (Douglas, 2003; Lagerspetz, 2002). Generally speaking, something is dirt when it is in a place where its presence spoils something.

I follow this general description of dirtiness and understand the purity of food to mean that it does not contain elements which spoil it. Accordingly, food that is not pure is contaminated by elements that should not be present in it. But which elements spoil and which do not spoil food? How to tell elements that belong apart from the ones that spoil?

EurSafe 2016 – DOI 10.3921/978-90-8686-834-6_83, © Wageningen Academic Publishers 2016

According to Mary Douglas (2003) the idea of purity combines views regarding hygiene and conventions. Hygiene is, of course, important in the food context since it contributes to food safety. Therefore, non-edible or harmful substances are one instance of spoiling elements. Pure food does not contain poisonous fungi, soil, mercury or other substances that are dangerous if eaten (beyond very minimal quantities).

The term 'pure' is often used in this sense also to designate the lack of elements that are considered as non-edible only by some people. The reasons for seeing something as non-edible include moral commitments, religious views, and health issues. 'Pure vegan food', for example, does not contain any elements originating from animals; 'pure halal food' is free of non-halal elements; and 'pure gluten free dish' does not contain wheat, barley, rye or oats. Similarly, a person who believes additives to be harmful may describe a food item 'pure' as long as it does not contain them.

Purity of food is desirable in the described sense. We do not want non-edible or harmful elements in our food, since they are risky for our health. Neither do we want our food to contain elements that are against our religious views, moral beliefs or life-style choices as they might – even though not directly be harmful to our physical health – threaten our psychological integrity and felt identity (as vegans or Muslims, for example) (Chackal, 2016).

However, dirt is not always risky to one's health, psychological integrity or felt identity. A stain of champagne in a suit does not, at least usually, form any hazard to these. Analogously, the purity of food does not merely mean absence of non-edible elements. Purity of food may also be compromised by elements that are considered perfectly edible. For example, a pure hot chocolate does not contain tomato and blueberries are not the ingredients of a pure *insalata caprese*. Blueberries in an *insalata caprese* are analogous to champagne stain in a suit – something which spoils the dish. This kind of purity of food is not an issue of hygiene. Rather it is a question of conventions (Douglas, 2003). Presence of a certain element compromises the purity of a food item because – according to the conventions regarding the food in question – that element spoils it.

Second contrast of purity: unclarity

Dirt is not usually seen to compromise the identity of an entity. A suit that has a stain of champagne in it is still a suit (although a dirty one). Analogously, a salad containing few blueberries might still be an *insalata caprese* provided that it in other ways follows the conventions regarding the dish in question. However, if the salad contained proportionally more blueberries than tomatoes, it would no-longer be appropriate to call that salad an *insalate caprese*.

As a contrast to unclarity, purity of food refers to lack of elements that would confuse its identity. A salad consisting of blueberry, tomato, mozzarella and basil is not true to the essence of *insalata caprese*. Rather it, because of the presence of a large amount of blueberries, lacks a clear identity. As a contrast to unclarity, to be pure is to be true to one's essence or to one's self. Purity in the described sense is based on the idea of clear categories, entities having essences and non-belonging elements confusing and mixing those categories and essences. (Bergin 2009; Schösler *et al.*, 2013) Being pure means being a characteristic and typical instance – that is only following mere ('pure') conventions (Douglas, 2003).

Is purity as accordance with conventions desirable? From the consumer point of view, this kind of purity manifests itself as a sufficient correspondence between the claimed identity (e.g. '*insalata caprese*') and the actual properties (ingredients, a way of preparing) of a food item. Consumers make their food choices on the basis of the names and identities given to food products. When the claimed identities and actual properties are compatible consumer gets what he expects. When they are incompatible, consumer may feel disappointed or even betrayed. Food journalist Mats-Eric Nilsson (2007) claims this kind of betrayal to be common. According to him, food industry intentionally names products in ways that do not conform to their actual properties. As examples he mentions 'smoked ham' that has not been produced by smoking and 'pistachio ice cream' which does not contain any pistachio. In similar lines, Michael Pollan (2009) guides us to 'avoid foods that are pretending to be something that they are not'. Both Nilsson and Pollan, thus, seem to highly value purity as a contrast to unclarity.

On the other hand, mixing categories or confusing essences is not problematic as such. Moreover, failing to question conventions may sometimes be very problematic. As Lisa Bergin (2009) and Kim Q. Hall (2014) point out, the 'metaphysics of purity' and purity metaphors have been widely used for sexist and racist purposes. Mixed-race peoples and transgendered folks, for example, have met hostility exactly because they have not fitted to conventional categories. Thus, we should be cautious about following conventions and metaphysics of purity also in other contexts:

> The metaphysics of purity ... assumes there are clear and stable boundaries that distinguish between what is and isn't food, what is and isn't edible. This is similar to assumption that there are clear and stable boundaries between genders that must be legible to others at all times. (Hall, 2014)

Bergin (2009) and Hall (2014) emphasize that instead of trying to make the reality follow the conventions, we should accept that categories are constantly mixed. This is, of course, not to say that faking a food product to be something should be accepted. Rather Bergin and Hall's criticisms highlight that purity as a contrast to unclarity is not self-evidently valuable or desirable as such. There is usually nothing problematic in mixing categories when it is done without an intention to deceive (as the example of cross-kitchen nicely shows).

Third contrast of purity: richness and redundancy

The term 'pure' can also be understood as a contrast to richness and redundancy. Purity is then taken to mean the lack of unnecessary elements. 'Pure' in this sense means minimal and simple – something that contains only the necessary and nothing else (Schösler *et al.*, 2013). A pure *insalata caprese*, for example, contains solely tomatoes, basil, mozzarella, black pepper and olive oil. This understanding of purity seems to be common in food marketing: expressions 'pure rye bread' and 'pure yoghurt' indicate that the products contain nothing else than the very necessary ingredients of rye bread and yoghurt.

Nilsson (2009) seems to have this kind of purity in mind when he reveals to his readers the great amount of ingredients and additives of certain processed foods. Even though Nilsson is quite uneasy about his finding, there is not necessarily anything problematic in non-purity in this sense. There is nothing problematic, as such, in eating and desiring to eat foods that do not qualify as pure due to containing numerous ingredients.

Purity as a contrast to richness and redundancy is similar to purity as a contrast to dirtiness and purity as a contrast unclarity in defining purity through lack of something. The three senses of purity, however, differ with respect to their normativity. As a contrast to dirtiness purity is necessarily normative in using the language of spoiling. As a contrast to unclarity purity is not valuable as such, but may nevertheless offer normative guidance for accurate naming of food products. As contrast to richness and redundancy purity lacks normative elements. Being simple, as such, is not *prima facie* better than rich and redundant, or vice versa. Thus, it can be concluded that the three senses of purity are conceptually and normatively distinct. Something can be pure on one sense of the term without being pure in another sense. Furthermore, one cannot infer from purity being desirable in one sense that it is also desirable in other senses.

Pure and closely related terms: natural and authentic

In addition to claims about purity, also claims about authenticity and naturalness are common in food marketing and food related discussions. How are the three terms connected to each other? All three terms are ambiguous (for discussion on naturalness see e.g. Sagoff, 2001; Siipi, 2013; for discussion on authenticity see e.g. Gilmore and Pine, 2007; Siipi, 2014). Some of their meanings overlap. However, terms 'authentic' and 'natural' also have meanings that are distinct from the meanings or 'pure'. For example, the meaning of naturalness as a contrast to the supernatural (Sagoff, 2001) and the meaning of authenticity as being typical to certain area (Sims, 2009) are not connected to the above described meanings of purity.

All three terms can be understood to mean absence of certain elements. In one of its most commonly used senses the term 'natural' refers to the lack of (certain) human influences (Sagoff, 2001). Naturalness then means that which has not been (or has only minimally been) modified by human beings. Natural food, according to this line of thinking, has only minimally been processed and does not contain synthetic or artificial additives. The term 'authentic' is sometimes used in a similar sense (Sagoff, 2001; Siipi, 2013). A claim about authenticity of vanilla, for example, means that vanilla comes from a plant and is not an artificial synthetic product. Sometimes authenticity may also refer to peasant production process. Claims about the purity of food are quite synonymous to these senses of 'natural' and 'authentic, when synthetic or artificial ingredients are seen to spoil the food.

The term pure has meanings that it either shares with naturalness or authenticity (but not with both of them). As noted above, the terms 'pure' can be understood to refer to a sufficient correspondence between a claimed identity and the actual properties of a food item. It seems the term 'authentic' is often used in the same sense. For example, a claim about a dish being 'authentic *insalata caprese*' can be taken to mean that it contains all and only necessary elements of *insalata caprese*.

Sometimes the naturalness of food is understood as nutritious suitability. For example, a statement that 'cow's milk is natural food for calves but not for human beings' can be taken to mean that cow's milk is very good nutrition for calves but not for humans. The term 'natural' is constantly used in this sense in disagreements about best dietary choices (Siipi, 2013.) This meaning of 'natural' is closely connected to the purity of food when purity is understood (as a contrast to dirt) as a lack of harmful or risky elements.

Finally, the term 'natural' is often used of familiar entities. We tend to call customary and conventional entities natural, whereas something odd and foreign is described as being unnatural. (Sagoff, 2001; Siipi, 2013.) Since the purity of food may, as presented above, be taken to mean that it follows and is according to conventions, also these two senses of naturalness and purity are intimately connected.

Conclusions

The term 'pure' is ambiguous. It offers good reasons for food choices when 'pure food' is understood as a contrast to food that contains harmful elements or elements that are considered non-edible for religious or moral reasons. However, purity is not merely a question of good hygiene. It is also about following conventions. As a contrast to unclarity or unconventionality the desirability of purity is quite questionable. On the one hand, a sufficient correspondence between a claimed identity and the actual properties of food products is valuable for informed food choices. On the other hand, there is nothing wrong with breaking the conventions and mixing categories as such, only when it includes an intention to deceive. As a contrast to richness and redundancy, purity does not offer unanimous reasons for food choices. Because of their ambiguity, claims about purity (just like claims about authenticity and naturalness) should be met with a critical attitude. That something is 'pure' may not be sufficient for it being desirable.

References

Bergin, L. (2009). Latina feminist metaphysics and genetically engineered foods. Journal of Agricultural and Environmental Ethics 22: 257-271.

Chackal, T. (2016). Autonomy and the politics of food choice: from individuals to communities. Journal of Agricultural and Environmental Ethics 29: 123-141.

Douglas, M. (2003). Collected works: volume II:purity and danger: an analysis of concepts of pollution and taboo. Routledge, London, UK, 193 pp.

Gilmore, J.H. and Pine, B.J. (2007). What consumers really want: authenticity. Harvard Business School Press, Boston, MA, USA, 299 pp.

Hall, K.M. (2014). Towards a queer crip feminist politics of food. philoSOPHIA 4: 177-196.

Lagerspetz, O. (2002). Puhtaus ja lika tosiasian ja käytännön rajalla. In: Pihlström, S., Rolin, K. and Ruokonen, F. (eds.) Käytäntö. Yliopistopaino, Helsinki, Finland, pp. 92-100.

Nilsson, M.-E. (2008). Petos lautasella. WSOY, Juva, 309 pp.

Pollan, M. (2009). Food rules: an eaters manual. Penguin Books, London, UK, 140 pp.

Sagoff, M. (2001). Genetic engineering and the concept of the natural. Philosophy and Public Policy Quarterly 21: 2-10.

Schösler, H., De Boer, J. and Boersema, J.J. (2013). The organic food philosophy: a qualitative exploration of the practices, values and beliefs of Dutch organic consumers within cultural-historical frame. Journal of Agricultural and Environmental Ethics 26: 439-460.

Sims, R. (2009). Food, place and authenticity: local food and the sustainable tourism experience. Journal of Sustainable Tourism 17: 321-336.

Siipi, H. (2013). Is natural food healthy? Journal of Agricultural and Environmental Ethics 26: 797-812.

Section 22. Responsible research and innovation

84. The role of responsible innovation in the technology assessment of smart farming technologies in Europe

V. Blok[1,2] and T.B. Long[1]*
[1]Management Studies Group, Wageningen University, Droevendaalsesteeg 4, 6708 PB Wageningen, the Netherlands; [2]Philosophy Group, Wageningen University, Droevendaalsesteeg 4, 6708 PB Wageningen, the Netherlands; vincent.blok@wur.nl

Abstract

Smart farming technologies like global positioning systems, sensors, nanotechnology, etc. provide several opportunities to address the grand challenges of our time (e.g. climate change, poverty alleviation) but raises ethical issues as well. For instance, it is expected that herd sizes in livestock farming will expand, raising issues in terms of the instrumental use of animals, the increasing industrialisation of farming, animal welfare issues, etc. Due to the existence of these ethical issues, technology developers who launch smart farming innovations find themselves in an important and potentially influential position. They have an important influence on the form and function of Smart Farming Technologies, which raises the question how they deal with the ethical issues involved. In this paper, we explore the opportunity the emerging concept of responsible innovation provides for the evaluation and management of ethical issues within smart farming technology development.

Keywords: responsible innovation, technology assessment, smart farming

Introduction

Agriculture has been undergoing a period of industrialisation for several decades. Technological advances now mean a new period of agricultural development is beginning, with industrialisation evolving into a period of increasing digitalisation. Smart farming involves the use of Global Positioning Systems (GPS), sensors, nanotechnology, etc., which promise to provide greater control and enhanced management capabilities. Whilst smart farming has been the focus of research and innovation for 20 years, it is only in the last few years that smart farming technologies are starting to be applied within real world contexts (Bos and Munnichs, 2016).

Smart farming technologies provide the potential to save on inputs (such as fertilizers, pesticides, veterinary medicines or feed), increase revenue, lower environmental risks, reduce labour inputs and provide higher quality produce (Kempenaar, 2010). The enhanced management capabilities of smart farming will also help farmers to deal with fluctuating food prices, greater weather variability, as well as be more reactive to demands from the market, such as quality requirements or reduced environmental impacts. It is for these reasons that smart farming offers several ways to address grand challenges of our time (European Commission, 2015).

One area where smart farming technologies could play a substantial role is in combating climate change as well as challenges related to food security. The concept of Climate Smart Agriculture (CSA) provides a synthesised approach to achieve sustainable agricultural development for food security whilst under climate change impacts (FAO, 2013). Specifically, CSA aims for three objectives. These include sustainable increases in agricultural productivity and incomes, the adaptation and building of resilience to climate change impacts and the reduction of GHG emissions where possible. This creates three distinct 'pillars', and potentially a 'triple-win' (FAO, 2014).

EurSafe 2016 – DOI 10.3921/978-90-8686-834-6_84, © Wageningen Academic Publishers 2016

As smart farming technologies could contribute towards CSA, their successful diffusion and adoption is critical. However, smart farming technologies also raise ethical questions (Bos and Munnichs, 2016). For example, smart farming technologies will change the role and position of the farmer in agricultural processes. As farmers become more concerned with data management rather than with animal husbandry, animal welfare issues could arise. Such concerns are already raised in the increasingly used practice of robotic milking (Holloway *et al.*, 2013; Wathes *et al.*, 2008), which has changed farming routines, reducing farmer to cattle contact and making farmers (big) data managers. Further, digitalisation does not mean a move away from the industrialisation of farming, but is likely to exacerbate such trends, allowing farms to become larger, changing supply chain relationships, and the location and nature of production more generally (Gray and Boehlje, 2007).

Smart farming technologies raise issues for farmers and society, but also technology developers. Technology developers are the actors involved in the design, development and diffusion of smart farming technologies into agri-food systems. They impact the form and function of these technological developments, and as such, are implicated in the ethical and societal outcomes that result from the technologies used. As such, technology developers find themselves in a position that can fundamentally alter farming practices, with corresponding ethical responsibilities. This raises questions as to whether or how technology developers deal with such ethical issues. In addition, the influence that technology developers have over how smart farming technologies are developed and diffused could impact adoption rates. Correspondingly, responsible innovation in this context could act to enhance or drive an increase in the adoption of smart farming technologies by increasing societal embeddedness.

Responsible innovation as a concept and practice has emerged due to the failure of innovations to embed themselves in society, leading to reduce levels of diffusion (von Schomberg, 2013). Responsible innovation seeks to make innovation and associated processes more inclusive and democratic in an effort to provide innovation outcomes that are more (ethically) acceptable, sustainable and socially desirable (Stilgoe *et al.*, 2013; von Schomberg, 2013). As such, responsible innovation may offer a method for the identification, evaluation and management of ethical and societal issues within smart farming technology development and innovation.

Indeed, responsible innovation may not only provide a systematic way for the identification and management of ethical issues, but could offer the potential for enhanced opportunity recognition and improved understanding of market and user demands, and enhanced social legitimacy (Husted and De Jesus Salazar, 2006). How smart farming technology developers access the opportunities presented by responsible innovation is particularly important because these actors often have limited financial and capital resources and must generate profits like any other business (Schaltegger and Wagner, 2011). Many processes consistent with responsible innovation use precious resources (Orlitzky *et al.*, 2011), meaning there could be tensions between the potential opportunities and the challenge to be profitable. Indeed, research notes that the prioritization of economic motives can compromise social and environmental improvements (Schaltegger and Wagner, 2011). In a cynical way, one can argue that responsibility only exists if there is 'a clear and directly foreseeable return on investment' (Waldman and Siegel, 2008).

This research seeks to understand how smart farming technology developers currently understand and manage socio-ethical issues in the innovation process, and to what extent responsible innovation can enhance such processes. A review of responsible innovation, including existing frameworks, will be undertaken to provide a base of knowledge from which to launch the planned empirical enquiry.

Literature review: responsible innovation

Responsible innovation emerged as a concept and practice due to the ethical questions and potential unintended consequences emerging from the innovation process (Owen *et al.*, 2013). Unforeseen and negative consequences are considered probable, not just possible, in many cases, increasing the need for methods for the management and governance of innovation processes (Groves, 2006; Hacking, 1983). Although it has long been recognised that control and management of the innovation process cannot be left to the market, it is increasingly recognised that the contemporary approach of retrospective regulation is limitations in its effectiveness (Owen *et al.*, 2013). Indeed, as science and innovation become increasingly globalised, complex and dynamic, new control measures are needed (Owen *et al.*, 2013).

Responsible innovation emerged due to such issues and denotes a lack of early, ethical reflection and inclusive deliberation as a problem in current innovation processes (Owen *et al.*, 2013). Responsible innovation attempts to instigate an innovation process that specifically addresses a societal challenge which is consistent with societal demands (Stilgoe *et al.*, 2013). Responsible innovation is defined by von Schomberg (2011) as:

> A transparent, interactive process by which societal actors and innovators become mutually responsive to each other with a view to the (ethical) acceptability, sustainability and societal desirability of the innovation process and its marketable products (in order to allow a proper embedding of scientific and technological advances in our society).

Recent research on responsible innovation has focused on the development and articulation of frameworks. These seek to provide a structured and categorised approach to responsible innovation, incorporating political and ethical elements into the innovation process, whilst also enabling dilemmas and tensions to be considered and incorporated.

These efforts have resulted in responsible innovations often being articulated through four dimensions (Owen *et al.*, 2013; Stilgoe *et al.*, 2013). These include:

- Anticipation, aimed at prompting innovators to ask 'what if ...' questions, being open to myriad possibilities and thinking systematically about possible impacts and futures within a context that recognizes the unpredictable and uncertain processes that govern innovation.
- Reflexivity, which concerns the moral boundaries and roles of innovators, seeking self-critique of assumptions and commitments as well as reflection on how issues are being framed.
- Inclusion, which requires a wider set of societal actors to engage in dialogue and engagement processes, and
- Responsiveness, which seeks to ensure that innovation processes have the capacity and leadership to respond to the questions and concerns raised through the first three dimensions.

These frameworks allow key components of responsible innovation to be highlighted and considered, however how responsible innovation is applied and practiced is less understood. Why responsible innovation is needed and the positive outcomes that could accrue from its wider use, are well covered in the literature (Scholten and Blok, 2015; Van den Hoven *et al.*, 2012; Von Schomberg, 2013). Science and policy contexts, in particular, have been widely explored in terms of the need and application of responsible innovations. However, what responsible innovation means for commercial or private sector contexts, or how it can be characterised, is less understood (Scholten and Blok, 2015).

Empirical investigation into responsible innovation for smart farming technologies

To provide full answers to the research question, this paper will conduct an empirical investigation. This investigation will explore the way technology developers consider the ethical and societal issues linked to their innovations, as well as how responsible innovation could provide a more structured and organised method to embed smart farming technology into society and address related ethical issues.

We posit that the development and diffusion of smart farming technologies is heavily influenced by the technology developers (Giarratana, 2004). The role they play in the innovation process means that they often have high levels of influence over the direction and character of the innovation, and as such, their actions and decisions will have ethical implications. This research will explore how these technology developers consider the ethical issues linked to smart farming technologies.

As smart farming technologies, as well as the application of responsible innovation in commercial contexts, are a relatively new phenomenon, with little previous research conducted, the research takes exploratory approach. This utilises semi-structured interviews as the data collection method. This is felt appropriate as it enables rich and in-depth data to be gathered.

Semi-structured interviews are conducted with smart farming technology providers. The individuals will be required to be senior decision-makers (entrepreneurs), with responsibility for how the technology is developed and diffused. A non-probabilistic purposive sampling strategy. Participants are primarily identified through internet searches and co-nomination and then approached for interviews. All questions were designed to be neutral and non-leading and sought to explore firstly, how technology developers understand and manage the socio-ethical issues in the innovation process, and second if they thought the framework of responsible innovation could enhance such process.

The semi-structured interviews are recorded and transcribed, and as such, the data takes the form text containing the thoughts, opinions, and knowledge of the respondents. The analysis involves noting frequent and typical elements key to the answering of the research questions. When completed, this process provides a range of codes and categories that highlight key themes. Initial coding is organised using the Nvivo software package.

Initial results will be presented at the conference, and will provide:

1. An exploration of if and how smart farming technology developers consider and deal with these ethical issues through the innovation process.
2. And to what extent responsible innovation could provide a systematic and improved method for the evaluation and management of ethical issues involved.

The results will be relevant to the developers of smart farming technologies, their users, and policymakers.

References

Bos, J. and Munnichs, G. (2016). Digitalisering van dieren: verkenning precision livestock farming. Rathenau Instituut, Den Haag, the Netherlands.

European Commission (2015). Societal challenges. Horizon 2020. European Commission, Brussels, Belgium.

Food and Agriculture Organization of the United Nations (FAO) (2013). Climate-smart agriculture – sourcebook. FAO, Rome, Italy.

Food and Agriculture Organization of the United Nations (FAO) (2014). FAO success stories on climate-smart agriculture. FOA, Rome, Italy.

Giarratana, M.S. (2004). The birth of a new industry: entry by start-ups and the drivers of firm growth: the case of encryption software. Research Policy 33: 787-806.

Gray, A.W. and Boehlje, M.D. (2007). The industrialization of agriculture: implications for future policy. Department of Agricultural Economics, Working Paper 07-10, Purdue University, West Lafayette, IN, USA.

Groves, C. (2006). Technological futures and non-reciprocal responsibility. International Journal of the Humanities 4: 57-61.

Hacking, I. (1983). Culpable ignorance of interference effects. Center for Philosophy and Public Policy, University of Maryland, College Park, MD, USA.

Holloway, L., Bear, C. and Wilkinson, K. (2013). Robotic milking technologies and renegotiating situated ethical relationships on UK dairy farms. Agriculture and Human Values 31: 185-199.

Husted, B.W. and De Jesus Salazar, J. (2006). Taking friedman seriously: maximizing profits and social performance*. Journal of Management Studies 43: 75-91.

Kempenaar, C. (2010). Precisielandbouw: slim combineren van nieuwe technologieen. Syscope Magazine 26: 22-27.

Orlitzky, M., Siegel, D.S. and Waldman, D.A. (2011). Strategic corporate social responsibility and environmental sustainability. Business & Society 50: 6-27.

Owen, R., Stilgoe, J., Macnaghten, P., Gorman, M., Fisher, E. and Guston, D. (2013). A framework for responsible innovation, responsible innovation. John Wiley & Sons, Ltd, New York, NY, USA, pp. 27-50.

Schaltegger, S. and Wagner, M. (2011). Sustainable entrepreneurship and sustainability innovation: categories and interactions. Business Strategy and the Environment 20: 222-237.

Scholten, V. and Blok, V. (2015). Responsible innovation in the private sector. Journal on Chain and Network Science 15(2): 101-105.

Stilgoe, J., Owen, R. and Macnaghten, P. (2013). Developing a framework for responsible innovation. Research Policy 42: 1568-1580.

Van den Hoven, J., Lokhorst, G.-J. and Van de Poel, I. (2012). Engineering and the problem of moral overload. Science and Engineering Ethics 18: 143-155.

Von Schomberg, R. (2011). Prospects for technology assessment in a framework of responsible research and innovation. In: Dusseldorp, M. and Beecroft, R. (eds.) Technikfolgen abschätzen lehren. VS Verlag für Sozialwissenschaften, Wiesbaden, Germany.

Von Schomberg, R. (2013). A vision of responsible research and innovation. In: Owen, R., Bessant, J. and Heintz, M. (eds.) Responsible innovation. Wiley, New York, NY, USA, pp. 51-74.

Waldman, D.A. and Siegel, D. (2008). Defining the socially responsible leader. The Leadership Quarterly 19: 117-131.

Wathes, C.M., Kristensen, H.H., Aerts, J.M. and Berckmans, D. (2008). Is precision livestock farming an engineer's daydream or nightmare, an animal's friend or foe, and a farmer's panacea or pitfall? Computers and Electronics in Agriculture 64: 2-10.

85. What have we learnt? If glyphosate were to become a late lesson we have now ignored the early warnings

M. Silva[1*] *and J.L. Garcia*[2]
[1] *CBQF – Centro de Biotecnologia e Química Fina, Laboratório Associado, Escola Superior de Biotecnologia, Universidade Católica Portuguesa/Porto, Apartado 2511, 4202-401 Porto, Portugal;* [2] *Institute of Social Sciences – University of Lisbon (ICS-UL), Av. Professor Aníbal de Bettencourt 9, 1600-189 Lisbon, Portugal; msilva@esb.ucp.pt*

Abstract

The current European Union reapproval procedure for glyphosate is developed as a case study testing whether the 12 precautionary lessons identified by European Environment Agency research as needed to avoid unnecessary harm to society have been incorporated into current regulatory approaches. The working hypothesis that those lessons have not been learnt to a significant degree could not be falsified.

Keywords: precaution, European Union, assessment, herbicide, policies

Introduction

Glyphosate is the active ingredient in the world's best-selling herbicides and is currently available in more than 300 different formulations in Europe alone (Glyphosate Task Force). The European Commission (EC) had been expected to reauthorize glyphosate friction-free until the International Agency for Research on Cancer of the World Health Organization (IARC) classified this herbicide as probably carcinogenic to humans (IARC, 2015).

However, when the European Food Safety Authority, as mandated by law, later published its own conclusions, it stated the herbicide was 'unlikely to pose a carcinogenic hazard to humans' (EFSA, 2015). The EC has already published a proposal to reapprove glyphosate based on EFSA's advice, to be voted on by Member States (Giannopolitis, 2016).

This kind of controversy can be expected but how careful should the political decision be in the face of scientific incoherence? In 2001 the European Environment Agency demonstrated a low precaution level was not in society's interest and derived twelve lessons that should be learnt so history doesn't repeat itself and harm is avoided (EEA, 2001).

The current glyphosate reauthorization procedure will be checked below against each of these lessons (paraphrased in the title of each section below) to determine whether they have been incorporated at all into European thinking, with the underlying hypothesis that no such thing happened – not to a significant degree. If, on the other hand, the learning effectively took place, then it should not be possible to find glaring examples to the contrary.

Lesson 1: deal with unknowns

It is not within the reach of scientists to immediately and fully answer questions as they arise, let alone identify beforehand all questions that need asking. Are there ways of safeguarding against current unknowns that could translate into future risks?

EurSafe 2016 – DOI 10.3921/978-90-8686-834-6_85, © Wageningen Academic Publishers 2016

Glyphosate has been shown to exhibit synergistic behaviour when in the presence of coadjuvants (other chemicals, considered inert by the regulatory authorities) included in the herbicide formulation). Richard *et al.* (2005) showed over ten years ago that Roundup (a commercial glyphosate-based herbicide) was consistently more toxic than glyphosate alone. An Argentinian group (Chaufan *et al.*, 2014), working with agriculturally realistic concentrations, found toxic effects in a glyphosate formulation where none could be found with glyphosate alone.

In its draft renewal text (DG SANTE, 2016) the EC recognises the issue when it mandates that one particular group of coadjuvants (POE-tallowamine, or POE-15) not be used in glyphosate-based herbicides (GBH). However, problems have been shown to occur with other chemicals also present in GBH (Mesnage *et al.*, 2013), and various combinations have not been tested at all (Mesnage *et al.*, 2014), which indicates major data gaps. This illustrates glyphosate's clear potential for future surprises while at the same time hinting at regulatory unwillingness to consider it.

Lesson 2: look for early warnings

Early warnings may take the shape of solitary publications in lesser journals, but they must not be disregarded, particularly if there is no monitoring in place for early detection of unexpected impacts. Such is the case for glyphosate as an endocrine disruptor since the EC has yet to define the necessary official testing criteria (Horel, 2015).

The inability to pinpoint the origin of the endocrine disrupting effects (whether glyphosate itself or coadjuvants) in papers by the Séralini group was enough for the Renewal Assessment Report (RAR) to disqualify the data (Anonymous, 2013). Where research used pure glyphosate, however, such as in Thongprakaisang's work, the disruption detected was reviewed apologetically. In the final listing of peer-reviewed scientific studies on *in vitro* endocrine disrupting effects, 11 out of the 12 publications mentioned are classified as 'not reliable' and 'not relevant' for one reason or another. The remaining paper was considered 'reliable/relevant with restrictions' with the added consideration that 'no conclusion can be drawn that the observed effects are the result of glyphosate exposure.' All 12 articles had detected some sort of negative impact in glyphosate or GBH exposure.

Such a blanket repudiation of peer-reviewed research points to a deliberate prejudgment of any evidence harming an authorisation. The introduction to the RAR states the work overload prevented reviewers from analysing the primary literature, relying instead on the industry's own assessment. This conveys a tainted overtone and is not compatible with a serious determination to find warnings where they might exist.

Lesson 3: react to detected weaknesses

One aspect of the glyphosate controversy is its teratogenicity. A 2012 review (Antoniou *et al.*, 2012) points to such effects after glyphosate or GBH exposure. It also describes how a previous (1998) German assessment report equivocally stated that the regulatory (industry commissioned) studies showed glyphosate did not cause teratogenicity while mentioning increased bone and soft organ anomalies. The studies themselves are not public, but the summaries could be analysed and, still according to the same review, show teratogenic effects at various dosages.

German assessors apparently glossed over the inconvenient (albeit limited) data and no regulatory action ensued. In addition, the 2012 review was ignored during the present renewal procedure. That is the exact opposite of what Lesson 3 should have taught, and shows it remains unlearnt.

Lesson 4: bridge knowledge areas

European regulation 1107/2009 on pesticides requires the EC to address endocrine disruptors and determines such problematic chemicals not be accepted under any dosage. EC's Environment unit (DG ENVI), tasked with defining the appropriate criteria, set up a diverse working group in 2010 comprised of over 40 elements, including representatives from Member States, academic institutions, other DGs, civil society and the industry itself.

Industry, however, lobbied hard until DG SANCO (Health and Consumers, historically more industry-friendly) replaced DG ENVI in 2014 (Horel, 2015). Part of this takeover included the set up by DG SANCO's EFSA, in 2012, of a nine-member endocrine task force. Irrespective of other details, the first composition of this group seems noteworthy. It is particularly relevant only two scientists were appointed, with none of those being a mammalian endocrinologist. Other members included several officials from two countries opposing DG ENVI and the majority in the group had no endocrinology background at all, administrative or otherwise (PAN, 2015).

The contrast between the two DGs couldn't be more suggestive of a learning inversion as regards Lesson 4. The inability, or unwillingness, to be inclusive prevailed in the European process.

Lesson 5: account for real life

It was around the epidemiological studies of glyphosate's cancer causing properties that the current German Renewal Assessment Report, most visibly clashed with the 2B classification (probable human carcinogen) from IARC. The two assessments differed in a number of crucial aspects, as described by Portier *et al.* (2016). Many would be worth exploring but a particular one, the discordant weight attributed to the studies available, is remarkable.

The RAR dismissed on technicalities all case-control studies as not reliable (as opposed to IARC), but valued data from the Agricultural Health Study. This large cohort study (which was also noted by IARC) elicited no cause-effect relationship for glyphosate and non-Hodgkin lymphoma (NHL, a type of blood cancer showing up in the case-control studies as being connected to glyphosate exposure). One problem, however, is the time dimension. In Olsson *et al.* (1988), for example, the average latency period for NHL after exposure to solvents averaged 21 years. In the AHS, on the other hand, the time between exposure and assessment averaged a meagre 6.7 years (Portier *et al.*, 2016). This puts the AHS data in the possible false negative range, largely reducing its validity at this stage: cancers may be yet to arrive. Real life takes time, and the RAR failed to recognise it.

Lesson 6: look into the wider costs and benefits

Pesticides, although defended as a whole by the agrochemical industry, are generally recognised as unsustainable to the point of justifying the introduction of genetically modified food crops as a better option (Klümper and Qaim, 2014). The question is not asked at the assessment and decision-making levels, however, if or when each pesticide's prospective benefits outweigh any potential impacts. Moreover, even if that kind of justification were sought, the analysis would be incomplete. In case there are better options, that attain the same objectives with less or no harm, officials would be hard pressed to approve any pesticide at all. Whether such alternatives do exist for glyphosate is the topic for the next Lesson.

Lesson 7: look for the best alternative

No official focused effort has been carried out to look for alternatives that could pave the way for a glyphosate free future. Nevertheless, the Julius Kühn-Institut, a German federal research agency, did look into the matter (Kehlenbeck *et al.*, 2015) and found that, for the German reality, agricultural and non-professional users alike can adjust to lower glyphosate use, concluding 'glyphosate should not be regarded as a standard measure in arable crop production systems'. This information has not percolated into any EC decision thus far.

Lesson 8: welcome non-experts too

This lesson is a reminder that non-experts and the public, in general, should be enticed to participate and actively heard in technical discussions having a bearing on society, mostly because there is knowledge and perspective to be gained. In glyphosate's case, a public hearing was held by EFSA in 2014. It was, unfortunately, 'entirely unfit for purpose' according to an environmental group that participated (John, 2014). In addition EFSA, according to the same source, eliminated all comments not entirely focused on glyphosate, even though it could have chosen to consider formulations such as Roundup which constitute the vehicle that channel glyphosate into the world. This approach does not bode well for the role of non-experts in European decision making.

Lesson 9: consider the people's feeling

Regarding glyphosate public values can be gleaned through various ways. In 2015 a petition hosted by Avaaz and delivered to DG SANCO contained 1.4 million signatures requesting glyphosate be banned (Neslen, 2015). More recently a Yougov poll of 7,000 Europeans in the five largest Member States evidenced an expressive two third majority in favour of glyphosate's demise (Neslen, 2016). The decision by the European Commission to go ahead with glyphosate's reapproval regardless of public views argues strongly in favour of this Lesson having been ignored.

Lesson 10: stop undue influence

The German RAR was put out by the Bundesinstitut für Risikobewertung (BfR), an official agency reporting to the Federal Ministry of Food and Agriculture. The BfR's 13 member Commission on Plant Protection Products and their Residues includes three experts from either Bayer or BASF, among others multinationals (BfR, 2016). In addition, the BfR report was not signed by the experts who wrote it which, while not corrupt *per se*, is certainly not transparent and opens the door to undue influence.

Lesson 11: the administration could be the problem

The glyphosate RAR summarises industry toxicological studies, but the full documents are not available because of commercial confidentiality (Antoniou *et al.*, 2012). This obstructs evidence cross-examination and has been criticized for years, to the point where DG SANCO has announced it is considering introducing increased transparency (Matthews, 2016). However, until the EC actually changes the rules, the underlying tension will not be solved.

Lesson 12: do not stall

A major example of how Lesson 12 has yet to be learnt is to be found within the European endocrine disruptors proceedings, which intimately intertwine with the glyphosate reapproval timeline.

The 1107/2009 mandate for the definition of endocrine disruptor criteria put the deadline at December 2013. A draft proposal (disparaged by industry) was ready by July 2013 but the EC then made a U-turn and chose to start an impact assessment instead (as demanded by industry), knowing full well this would introduce a three-year delay in the process (Horel, 2015). In March 2014 Sweden (later followed by the European Council and the Parliament) took the EC to court for failure to act and in December 2015 the General Court of the European Union ruled the EC had been illegally stalling (Jacobsen, 2015).

Conclusions

The glyphosate exercise described seems to negatively illustrate every one of the 12 precautionary lessons the European Union should have learnt and, at some level, didn't. As described elsewhere for various other instances (EEA, 2013), scientific data is available, but no early protection measures take place. As the saying goes, to ignore history is to repeat it. Glyphosate is poised to become a late lesson, which would translate into decades of avoidable human suffering and death. A late early warning is hereby served.

References

Anonymous (2013). Renewal assessment report. Glyphosate 3, annex B.6.1.

Antoniou, M., Habib, M., Howard, C., Jennings, R., Leifert, C., Nodari, R. and Fagan, J. (2012). Teratogenic effects of glyphosate-based herbicides: divergence of regulatory decisions from scientific evidence. Journal of Environmental and Analytical Toxicology S4(006): 1-13.

Bundesinstitut für Risikobewertung (2016). Mitglieder der BfR-kommission für pflanzenschutzmittel und ihre rückstände. Available at: http://tinyurl.com/z6nzg39.

Chaufan, G., Coalova, I. and Molina, M. (2014). Glyphosate commercial formulation causes cytotoxicity, oxidative effects, and apoptosis on human cells differences with its active ingredient. International Journal of Toxicology 33(1): 29-38.

DG SANTE (2016). Annexes to the Commission implementing Regulation (EU) renewing the approval of the active substance glyphosate in accordance with regulation (EC) no 1107/2009. European Commission, Brussels, Belgium. Available at: http://tinyurl.com/hzpopvh.

European Environment Agency (2001). Late lessons from early warnings: the precautionary principle 1896-2000. Office for official publications of the European Communities.

European Environment Agency (2013). Late lessons from early warnings: science, precaution, innovation. Office for official publications of the European Communities.

European Food Safety Authority (2015). Conclusion on the peer review of the pesticide risk assessment of the active substance glyphosate. EFSA Journal 13(11).

Giannopolitis, C. (2016). Possible Europe-wide withdrawal of glyphosate herbicide formulations by member states. Available at: http://tinyurl.com/zh6g3gh.

Glyphosate Task Force (2016). How is glyphosate used? Available at: http://tinyurl.com/jxjemb8.

Horel, S. (2015). A toxic affair – how the chemical lobby blocked action on hormone-disrupting chemicals. Corporate Europe Observatory.

International Agency for Research on Cancer (2015). IARC Monographs. Vol. 112, evaluation of five organophosphate insecticides and herbicides. Available at: http://tinyurl.com/hutfqt3.

Jacobsen, H. (2015). Sweden wins case against Commission on endocrine disruptors. Available at: http://tinyurl.com/ht9nr8c.

John, B. (2014). EFSA's public consultation on glyphosate is 'entirely unfit for purpose'. Available at: http://tinyurl.com/gok5z46.

Kehlenbeck, H., Saltzmann, J., Schwarz, J., Zwerger, P., Nordmeyer, H., Roßberg, D. and Freier, B. (2015). Folgenabschaetzung fuer die landwirtschaft zum teilweisen oder vollstaendigen verzicht auf die anwendung von glyphosathaltigen herbiziden in Deutschland.

Klümper, W. and Qaim, M. (2014). A meta-analysis of the impacts of genetically modified crops. PLoS One 9(11): e111629.

Matthews, J. (2016). EU health commissioner moves towards disclosure of pesticide industry studies. Available at: http://tinyurl.com/hk732cw.

Mesnage, R., Bernay, B. and Séralini, G. (2013). Ethoxylated adjuvants of glyphosate-based herbicides are active principles of human cell toxicity. Toxicology 313(2): 122-128.

Mesnage, R., Defarge, N., Vendômois, J. and Séralini, G. (2014). Major pesticides are more toxic to human cells than their declared active principles. BioMed Research International: 1-8.

Neslen, A. (2015). EU watchdog opens the door to new licence for controversial weedkiller. The Guardian. Available at: http://tinyurl.com/jrkozmh.

Neslen, A. (2016). Two-thirds of Europeans support a ban on glyphosate, says Yougov poll. The Guardian. Available at: http://tinyurl.com/zkbqpfw.

Olsson, H. and Brandt, L. (1988). Risk of non-Hodgkin's lymphoma among men occupationally exposed to organic solvents. Scandinavian Journal of Work, Environment and Health 14(4): 246-251.

Pesticide Action Network (2015). How EU Commission Health service DG SANTE joined forces with economic DG's on the criteria for endocrine disruption and crushed not only their enemy, lead DG Environment, but also the entire new EU policy on endocrine disruption.

Portier, C., Armstrong, B., Baguley, B., Baur, X., Belyaev, I., Bellé, R. and Zhou, S. (2016). Differences in the carcinogenic evaluation of glyphosate between the International Agency for Research on Cancer (IARC) and the European Food Safety Authority (EFSA). Journal of Epidemiology and Community Health, 207005.

Richard, S., Moslemi, S., Sipahutar, H., Benachour, N. and Seralini, G. (2005). Differential effects of glyphosate and roundup on human placental cells and aromatase. Environmental Health Perspectives 113(6): 716-720.

86. Agricultural technologies as living machines: toward a biomimetic conceptualization of technology

V. Blok[1,2] and B. Gremmen[2]*
[1]Management Studies Group, Wageningen University, Hollandseweg 1, 6707 KN Wageningen, the Netherlands; [2]Philosophy Group, Wageningen University, Hollandseweg 1, 6707 KN Wageningen, the Netherlands; vincent.blok@wur.nl

Abstract

In this paper, we raise the question whether we can conceptualize precision livestock farming (PLF) in biomimetic terms as 'living machines', which are no longer characterized by the exploitation, domination, instrumentalisation and commodification of nature, but instead, are embedded in and in accordance with the natural environment. The concept of biomimicry will be used to conceptualize PLF technologies as ecological innovations which are embedded in and in accordance with the natural environment. Such a biomimetic approach of PLF could take advantage of its potential advantages like mitigating climate change, while at the same time avoiding the ethical issues related to the industrialization of the agricultural sector by traditional PLF technologies.

Keywords: precision livestock farming, biomimicry, agricultural technology

Introduction

A relatively new phenomenon in the agricultural sector is the application of precision livestock farming (PLF). By the integration of smart technology and the internet of things in which computers, sensoring devices, global positioning systems but also robots and even animals communicate with one another and function autonomously in an integrated farm management system – farmers can reduce farm inputs (fertilizers and pesticides) and increase yields, while reducing emissions to the environment (Bos *et al.*, 2016). However, there are a number of ethical issues related to the industrialization of the agricultural sector by smart farming technologies, and for this reason society is reluctant to accept PLF and calls for the human and natural scale of agricultural practices.

In this paper, we raise the question whether we can conceptualize PLF in biomimetic terms as 'living machines', which are no longer characterized by the exploitation, domination, instrumentalisation and commodification of nature, but instead, are embedded in and in accordance with the natural environment. The concept of biomimicry will be used to conceptualize PLF technologies as ecological innovations which are embedded in and in accordance with the natural environment. Such a biomimetic approach of PLF could take advantage of its potential advantages like mitigating climate change, while at the same time avoiding the ethical issues related to the industrialization of the agricultural sector by traditional PLF technologies.

Precision livestock farming

PLF can be defined as 'the management of livestock production using the principles and technology of process engineering ... PLF treats livestock production as a set of interlinked processes, which act together in a complex network' (Wathes *et al.*, 2008). The introduction of this type of integrated farm management systems enables farmers to control the production process by monitoring and controlling animal growth, behaviour and health, the production of milk and eggs, the physical environment of livestock buildings, and greenhouse gas emissions and other pollution to the environment. Furthermore,

EurSafe 2016 – DOI 10.3921/978-90-8686-834-6_86, © Wageningen Academic Publishers 2016

the exchange of information about oestrus detection, health, milk quality, etc., enables supply chain actors to optimize coordination and efficiency throughout the supply chain.

Although PLF provides economic, social and environmental opportunities for the agricultural sector to feed the world and to mitigate climate change, they also raise ethical issues associated with the increased corporatization and industrialization of the agricultural sector (Bos *et al.*, 2013; Harfeld, 2010). Therefore, society is reluctant to accept PLF and calls for the human and natural scale of agricultural practices. This reluctance to accept the industrialization of farming practices can be understood as a resistance against the conceptualization of the natural environment as a commodity for human needs, in which nature's own strategies and principles of operation are neglected – natural animal growth and behaviour for instance – and instead, nature is challenged to supply efficiently agri-food products and animals as commodities in an instrumental economic exchange among chain actors. A subsequent call for farming practices which are better embedded in the natural environment, like multi-functional agriculture, organic farming, etc. can be recognized in current western societies, even if these practices are disadvantageous to feed the word and to mitigate climate change.

The concept of biomimicry

According to the German philosopher Peter Sloterdijk, the twenty-first century is characterized by 'a paradigm shift in the basic idea of technology' (Sloterdijk *et al.*, 2006: 329). Emerging technologies like biotechnology and nanotechnology are increasingly biomimetic, embedded in and in accordance with the natural environment: 'It appears that we are for the first time on the threshold of a form of technology, which will be sufficiently developed to enable us to radically imitate nature' (Sloterdijk *et al.*, 2006: 329). One of the founding mothers of the concept of biomimicry, Jane Benyus, defines biomimicry or biomimetics as 'a new science that studies nature's models and then imitates or takes inspiration from these designs and processes to solve human problems' (Benyus, 2002: 1). Biomimicry provides a new and ecosystem-friendly approach to nature, which is no longer characterized by the domination and exploitation of nature, but by learning and by the exploration and cooperation with nature (cf. Sloterdijk, 2001: 228). Benyus, for instance, argues that the first industrial revolution is characterized by the domination and exploitation of nature, whereas the second – biomimetical – industrial revolution is characterized by learning from, and exploring nature (cf. Blok *et al.*, 2016). The pretention of biomimetic or homeo-technologies is that they act and perform in accordance with or are similar to the operation principles of nature and for this reason, can claim to be sustainable *and* embedded in the environment (Benyus, 2002; Todd, 2013).

The evolution of nature is seen here as an operation principle in which plants, animals and microbes were able to perfect themselves and learned to fly, swim, navigate, capture the sun's energy, etc. 'In short, living things have done everything we want to do, without guzzling fossil fuel, polluting the planet, or mortgaging their future. What better models could there be?' (Benyus, 2002: 2; cf. Todd, 2013). According to Benyus, nature functions here as a model – photosynthesis, self-assembly, self-sustaining eco-systems, but also eyes and ears for instance – which design and manufacturing processes are imitated by biomimetics to solve technological problems. The objective of these innovations inspired by nature is no longer to exploit the natural environment but to explore and learn from nature in our effort to respectfully imitate nature. Benyus provides the example of biomimicry in farming practices: 'In a biomimetic world, we would manufacture the way animals and plants do ... Our farms, modelled on prairies, would be self-fertilizing and pest-resistant. To find new drugs or crops, we would consult animals and insects that have used plants for millions of years to keep themselves healthy and nourished. ... In each case, nature would provide the models: solar cells copied from leaves, steely fibres woven spider-style, ... perennial grains inspired by tallgrass ...' (Benyus, 2002: 2-3). According to Benyus, the main lesson we can learn from nature is the 'hand-in-glove harmony' of natural systems, in which 'organisms are adapted to their places and to each other', which should inspire our future nature-based technological innovations.

According to Benyus, there are nine laws, strategies and principles of nature behind the maintenance of the eco-systems of planet earth:

- nature runs on sunlight;
- nature uses only the energy it needs;
- nature fits form to function;
- nature recycles everything;
- nature rewards cooperation;
- nature banks on diversity;
- nature demands local expertise;
- nature curbs excesses from within;
- nature taps the power of limit.

Although the nature of these 9 laws or strategies is not clear and calls for a philosophical reflection on the concept of nature which is presupposed in biomimetic practices, we call attention to another problem which remains unsolved in Benyus' concept of biomimetics; the dualism between nature and (human) technology which is presupposed in biomimetic practices.

The dualism between nature and technology[1]

This presupposition becomes clear in Benyus concept of nature as a model, measure, and mentor. If biomimicry studies the design principles of nature and then imitates these designs, if it uses these principles to judge the 'rightness' of technological designs, if it learns from nature's models and applies it to human technologies, a dualism between nature as that which is pre-given and copied by biomimicry and technology as its copy, a dualism between nature as normative standard and human technology which can be judged based on this standard, a dualism between nature as endless pool of learning opportunities and human technologies which are developed based on this learning experience is presupposed. The concept of biomimicry only makes sense in case there is a difference between that which is mimicked (nature) and that which is the product of this mimicking process.

This dualism between nature and (human) technology also becomes clear in Benyus' focus on engineering, and in this respect, on 'human' problems which should be solved by human technology: Biomimicry is the conscious emulation of life's genius' (Benyus, 2002). This emulation 'aims to mimic life, not to reproduce it' (Bensaude-Vincent, 2002). Contrary to the primacy of human technology in the exploitation of nature in the industrial age, in this dualist biomimetic approach of technology development, primacy is given to nature's principles as a model, as normative standard and as an opportunity to learn from these natural principles. The ideal of biomimicry is not to bridge this dualism by reproducing life, but by 'doing it nature's way in our technology, we mimic nature's principles and strategies in our technological design: In a biomimetic world, we would manufacture the way animals and plants do, using the sun and simple compounds to produce totally biodegradable fibres, ceramics, plastics, and chemicals' (Benyus, 2002: 2). Doing it nature's way primarily means that technological developments imitate the design principles of nature, not necessarily to incorporate nature's design principles in technological designs. Also, philosophers of biomimicry like Henry Dicks and Freya Mathews presuppose a dualism between nature and human technology: technology should be designed like the operating system of nature and even collaborate with nature, not necessarily become natural itself (Dicks, 2015; Mathews, 2011).

[1] Parts of this subsection are published in Blok, 2016.

There are at least two reasons to question this dualism. According to philosophers like Sloterdijk and Plumwood, the exploitation of planet earth is rooted in anthropocentric humanism, i.e. in the 'standpoint of mastery' of the human will to master and exploit the natural world as a commodity for human needs (cf. Blok, 2014). According to Plumwood, this dualism also gave rise to the idea that human agency can solve the environmental crisis we face today by engineering and technology (Plumwood, 2002; cf. Van der Hout, 2015: 438), which in fact consists in the exploitation of nature (Sloterdijk *et al.*, 2006: 330). In other words, it is questionable whether biomimetic technologies can contribute to climate smart solutions, as long as they are embedded in dichotomies like nature-human, nature-culture, nature-technology.

A second reason to question the supposition of the dichotomy between nature and human technological agency is a 'paradigm shift in the basic idea of technology' in the twentieth century (Sloterdijk *et al.*, 2006: 329). Recent developments in technology and science like biotechnology, nanotechnology, and synthetic biology show that they are not purely natural or purely technological, but hybrid forms of technology which are similar to nature. Sloterdijk calls these new forms of technology 'homeo-technologies', derived from the Greek *homoios,* i.e. 'similar', or 'alike.' In other words, new and emerging technologies show that a dualist perspective on biomimicry is no longer appropriate to conceptualize current developments in science and technology and call for a monist approach of biomimicry.

Because the concept of biomimicry presupposes a dualism between nature and human technology, it may be the case that Sloterdijk prefers to speak about homeotechnology instead. Although he conceptualizes homeotechnology as imitation of nature in his work, a shift can be seen from the imitation of nature to the incorporation of nature's design principles in technology and *vice versa*, like in biotechnology. Myers even goes so far to say that his concept of bio-design goes further than biomimicry, because it specifically refers to 'the incorporation of living organisms as essential components, enhancing the function of the finished work. It goes beyond mimicry to integration, dissolving boundaries and synthesizing new hybrid typologies' (Myers, 2012: 8-9). In this respect, we can characterize these technologies as biomachines or living machines. Sloterdijk believes that the integration of the biosphere and the technosphere under the direction and guidance by human cognition can guarantee a sustainable future (cf. Van Hout, 2015: 428).

The characterization of these technologies as biomachines or living machines raises the question what principles of nature should be incorporated in biomimetic technologies. While Benyus introduced nine principles of nature, Sloterdijk has completely different principles in mind, such as replication, selection and transgenesis. In bionics, which concerns the supplementing or duplicating of neurophysiological characteristics of the human body by the integration of electronic devices and mechanical parts, again other principles seem to be at stake. Nature is seen here as a 'system that uses information to achieve heightened regulation and control', and is understood 'in terms of the concepts of feedback, information, control, regulation, teleonomy' (Dicks, 2015: 10). This conceptualization of nature raises the question whether the proposed principles are in fact principles of nature or principles of technology which are applied to nature.

Freya Mathews was one of the first authors who tried to lay a philosophical ground for the nine design principles of Benyus. The first principle of nature which is identified by Mathews is the principle of conativity: 'It asserts that all living beings and living systems act in accordance with a will or impulse to maintain and increase their own existence' (Mathews, 2011: 368). She associates this principle of conativity with the concept of *auto-poiesis* and the 'will wherewith everything strives to persevere in its own existence', and takes it as a defining characteristic of all living things (Mathews, 2011: 368). The idea that nature has to be understood as *auto-poiesis* is confirmed by others, like Dicks (2015: 6), Maturana and Varela (1980: 85-87) and Capra (1997: 95-97). According to Maturana and Varela, the notion of *autopoiesis* 'is necessary and sufficient to characterize the organization of living systems' (Maturana *et al.*, 1980: 82).

It is questionable however whether *autopoiesis* can be considered a principle of nature. Not only because nature is directly understood then in technological terms like a biomachine (Sloterdijk, 2006) or living machine (Maturana *et al.*, 1980). If nature is understood as a regulation and control system like in bionics, the 'paradigm of the artificial machine' is applied to nature (Dicks, 2015: 10). This technological preoccupation in the concept of nature is however already embedded in the concept of *auto-poiesis* itself. The conceptualization of nature as *poiesis* already implies that nature is understood in technological terms, namely in terms of its productivity or makeability (Heidegger, 1999: 88, 133-134; cf. Blok *et al.*, 2016); while the technological production of artefacts is characterized by the *poiesis* by an external agent, nature is characterized by *auto-poiesis* or self-making.

Conclusions

In this paper, we raised the question whether we can conceptualize PLF in biomimetic terms as 'living machines', which are no longer characterized by the exploitation, domination, instrumentalisation and commodification of nature, but instead, are embedded in and in accordance with the natural environment. We can conclude, first, that the concept of biomimicry is in fact governed by a technological concept of nature, and second, that we have to reject this technological model of nature in favour of a natural model of nature; one can only claim a more eco-system friendly type of technology and innovation, in contrast with the exploitation of nature as a commodity for human needs, if it mimics or incorporates a natural principle and is embedded in nature, rather than a technological principle (Blok and Gremmen, 2016). Future research has to identify such a natural principle which can subsequently be incorporated in PLF practices.

References

Bensaude-Vincent, B., Arribart, H., Bouligand, Y. and Sanchez, C. (2002). Chemists and the school of nature. New Journal of Chemistry 26: 1-5.

Benyus, J.M. (2002). Biomimicry. Innovation inspired by nature. Harper Perennial, New York, NY, USA.

Blok, V. (2014). Reconnecting with nature in the age of technology. The Heidegger and radical environmentalism debate revisited. Environmental Philosophy 11(2): 307-332.

Blok, V. (2016). Earthing technology: global warming and the biomimetics of technology in the anthropocene. Working Paper.

Blok, V. and Gremmen, B. (2016). Ecological innovation: biomimicry as a new way of thinking and acting ecologically. Journal of Agricultural and Environmental Ethics 29: 203-217.

Bos, J. and Gremmen, B. (2013). Does PLF turn animals into objects? Precision Livestock Farming 2013. EC-PLF, Leuven, Belgium.

Bos, J. and Munnichs, G. (2016). Digitalisering van dieren. Verkenning Precision Livestock Farming. Rathenau, Den Haag, the Netherlands.

Capra, F. (1997). The web of life. A new scientific understanding of living systems. Anchor Books, New York, NY, USA.

Dicks, H. (in press). The philosophy of biomimicry. Philosophy and Technology.

Heidegger, M. (1999). Contributions to philosophy. Indiana University Press, IN, USA.

Mathews, F. (2011). Towards a deeper philosophy of biomimicry. Organization and Environment 24 (4): 364-387.

Maturana, H.R. and Varela, F.J. (1980). Autopoiesis and cognition. Reidel Publishing Company, Dordrecht/Boston, the Netherlands, USA.

Myers, W. (2012). Biodesign. Nature, science, creativity. Thames & Hudson, London, UK.

Plumwood, V., (2002). Environmental culture: the ecological crisis of reason. Routledge, New York, NY, USA.

Sloterdijk, P. (2001). Nicht gerettet. Versuche nach Heidegger. Suhrkamp, Frankfurt am Main, Germany.

Sloterdijk, P. and Heinrichs, H.-J. (2006). Die sonne und der tod. Dialogische untersuchungen. Suhrkamp, Frankfurt am Main, Germany.

Todd, J. (2013). John Todd living machines lecture. Available at: http://tinyurl.com/he7o7qk.

Van der Hout, S. (2014). The homeotechnological turn. Sloterdijk's response to the ecological crisis. Environmental Values 23: 423-442.

Wathes, H.H., Kristensen, J.M. and Berckmans, D. (2008). Is precision livestock farming an engineer's daydream or nightmare, and animal's friend or foe, and a farmer's panacea or pitfall? Computers and Electronics in Agriculture 64: 2-10.

Posters

87. Antimicrobial residues in milk: a food policy problem in an ethical framework

M. Vieira[1,2], *L. Batista*[1] *and C. Prudêncio*[1,2*]
[1]*Ciências Químicas e Biomoléculas, Centro de Investigação em Saúde e Ambiente, Escola Superior de Tecnologia da Saúde, Instituto Politécnico do Porto, Rua Valente Perfeito, 322, 4400-330 V. N. Gaia, Portugal;* [2]*i3S, Instituto de Investigação e Inovação em Saúde, Rua Alfredo Allen, 535, 4200-135 Porto, Portugal; cps@estsp.ipp.pt*

Abstract

Antimicrobial residues in food is a public health concern and a major ethical problem. Intensification of large-scale animal production is at the top of the causes for contamination of water and soil, but other questions are raised. On the other hand, the presence of antimicrobial residues in foodstuffs constitutes a potential risk to the human health and undesirable effects on consumers. In fact, the presence of these compounds in food may have direct consequences in human health, as allergies and alteration of the commensal flora, and indirect consequences enabling the acquisition and development of resistance to antibiotics. In the present study, a new chromatographic method was developed and proved to be appropriate to simultaneous quantification of five antimicrobial agents selected from a pool of twenty. Milk samples were run in order to determine the presence of antimicrobial residues, after the application of a solid phase extraction-based clean-up method and the presence of one antimicrobial agent (sulfamethoxazole), used in animal production, was detected. The detection and quantification of antimicrobial agents on food from animal origin is an important factor in food quality control. The presence of this kind of contaminants appears to influence the appearance and spread of resistant strains of bacteria, by exposure to these drugs. Data obtained suggests the presence of antimicrobial agents in milk samples analysed, according to the maximum residue limits (MRLs) established for EU members. In this context the ethical issues underlying this problem will be discussed.

Keywords: antimicrobial quantification, SPE-HPLC-DAD, European MRLs

Introduction

Antibiotics are widely used in veterinary medicine as therapeutic for bacterial infection, disease prophylaxis and as growth promoters (food additives) (Andreotti and Nicodemo, 2004). These drugs are used to avoid epidemic propagation of infection diseases, prevention of animals' zoonosis transference to humans and possible foodborne diseases, as well as improving the efficiency of animal production and food stuff, like meat, milk or eggs, as well as for security for human consumption (Arias and Carrilho, 2012). These drugs may be administrated orally as feed additives or directly by injection and this use, especially when they are not used properly, may result in drug residues present in animal origin food, for example in milk (Schenck and Callery, 1998).

No matter the proved contribution in animal performance, antibiotics with grow promoting effect are now being seen as risk factors to human health due to high consume of animal products like milk and meat. Part of the resistant bacteria end up spreading and creating resistances to human organism defences (Branco *et al.*, 2011). In 2006, European Union forbade the use of antibiotics as grow promoters in production of meat at the same time that dramatically changed the demands for aviary meat importation (Souza *et al.*, 2010). But antibiotics are equally important tools for treatment, maintain a healthy condition and animals' well-being in cattle raising. Nowadays, these drugs are being used as therapy according to objectives of use, dose, and duration of treatment (Spinosa *et al.*, 2006).

EurSafe 2016 – DOI 10.3921/978-90-8686-834-6_87, © Wageningen Academic Publishers 2016

Taking into account the principles of Principlism enunciated by Beauchamp and Childress (Beauchamp and Childress, 2001) the ethical issues underlying this problem will be discussed.

Materials and methods

Standard antimicrobial solutions

The developed method performs the simultaneous analysis of five antimicrobial agents, in acidic medium. The selection relayed on the antibiotic class and peak resolution observed. The antimicrobials chosen represent the ß-lactam (penicillin and cephalosporin); sulphonamides, quinolones and chloramphenicol.

A stock solution for each antimicrobial studied (amoxicillin (AMOX), ciprofloxacin (CIP), sulfamethoxazole (SULFO), chloramphenicol (CLOR), cloxacillin (CLOX), Sigma-Aldrich, Co., USA) was prepared with a concentration of 0.8 mg/ml, dissolving the weighted powder in mobile phase (80% TFA 0.02% pH=2.2; 20% CH_3CN). TFA (trifluoroacetic acid), CH_3CN (acetonitrile). Serial diluted standards were prepared. Final concentrations ranged from 10 to 200 µg/ml. A mixture of the five antimicrobial agents was prepared in a mixture containing all the five. Standard dilutions were prepared as described above, ranging the same concentrations.

Samples collection and treatment

In the present study 50 ml from 18 different commercial brands and kinds (fat, medium fat, low fat, dehydrated, without lactose) of milk were used and analysed. Samples were randomly chosen from different commercial stores in Porto. Samples were identified and kept refrigerated at -80 °C until laboratory analysis. At the time of the analysis, samples were unfrozen at environment temperature and homogenized in vortex for 1 minute; 2 ml were taken into a falcon tube, 3 ml of trichloroacetic (TCA, Sigma-Aldrich, Co., USA) in water at 20% were added and samples were centrifuged at 8,000 rpm during 10 minutes. Supernatant was recovered and stored at -80 °C, until solid phase extraction (SPE) clean-up and analysis.

Procedure for sample analysis and identification

In order to establish a SPE clean-up method and evaluate the recovery percentage, the 18 samples were split into four groups as referred in Table 1. In A group samples were directly analysed in HPLC-DAD, as suggested by van Bruijnsvoort *et al.* (2004); In B group 200 µl of SULFA standard solution to a final concentration of 160 µg/ml. were added and afterwards analysed in HPLC-DAD; in group C samples were cleaned-up using SPE technique and then analysed in HPLC-DAD; finally the D group 200 µl of SULFA standard solution were added before SPE clean-up and HPLC-DAD analysis.

Table 1. Sample distribution over 4 groups of analysis.

Number of samples	Group	Procedure[1]
18	A	HPLC-DAD analysis
18	B	AB + HPLC-DAD analysis
18	C	SPE-HPLC-DAD analysis
18	D	AB + SPE-HPLC-DAD analysis

[1] AB: antimicrobial agent; SPE: sample clean-up by solid phase extraction.

Solid phase extraction

SPE extraction apparatus used was a 12 sample container coupled to a vacuum bomb Laboport Knf Neureberger. SPE-C18 with 500 mg and 3 ml volume cartridges were used (Lichrolut RP-18 Standard PP tubes).

A sample method of clean-up and pre-concentration was developed, for SPE in acid environment, in order to preserve antimicrobial agents' structure. The extract was kept at 4 °C until HPLC analysis. 1 ml was filtered (Pall Life Sciences® GHP Acrodisc 13 mm Ø Syringe filter 0.45 μm GHP membrane; GHPolypro 47 mm Ø 0.45 μm hydrophilic Polypropylene Membrane filters) and transferred to 1 ml vials before injection.

High-performance liquid chromatography – diode array detector (HPLC-DAD)

The equipment used was a Hitachi® HPLC LaChrom Elite system, constituted for a quaternary bomb HTA L-2130, a vacuum degassing of solvent, in line, incorporated, auto sampler L-2200. The used detector was DAD L-2455, LaChrom Elite serie and column hoven L-2300, LaChrom Elite serie. Data collection and treatment was made using Software EZChrom Elite, Lachrom Elite serie.

A Purospher® STAR C18 chromatographic collumn, with 5 μm particle size and 4.6 × 250 mm internal dimensions, at 40 °C was used. Samples were kept at 4 °C in the auto sampler tray and gradient eluent conditions were applied, with a two solvent mobile phase: acetonitrile and water with trifluoroacetic acid (TFA) 0.02% (pH=2.2), with a flow rate of 1 ml/min, for 25 minutes. Runs were performed in triplicated with 20 μl volume. Detection was made with a DAD system, ranging from 190 to 400 nm.

Results

Calibration of a mixture of five antibiotics

Four concentration standard mixture solutions were injected (n=3) and calibration curves were drawn. Table 2 and Figure 1 present the results obtained. Selectivity was determined by analysis of an antimicrobial agents' free matrix. Linearity was assessed by the determination of the calibration curve for each antimicrobial agent, in the mixture of five antimicrobial agents, at four different concentrations. It was verified that alterations on the mobile phase's pH and concentration of the organic eluent influenced the time of retention for all the antimicrobial agents tested so pH was adjusted to 2.2 and maintained at all stages of treatment, to ensure stability.

Table 2. Limits of Detection (LD) and Quantification (LQ) for the five antimicrobial agents studied.

	AMOX	CIP	SULFO	CLOR	CLOX
LD (μg/ml)	0.162569	0.044277	0.020993	0.00243	0.135142
LQ (μg/ml)	0.492633	0.134173	0.063616	0.007363	0.409523

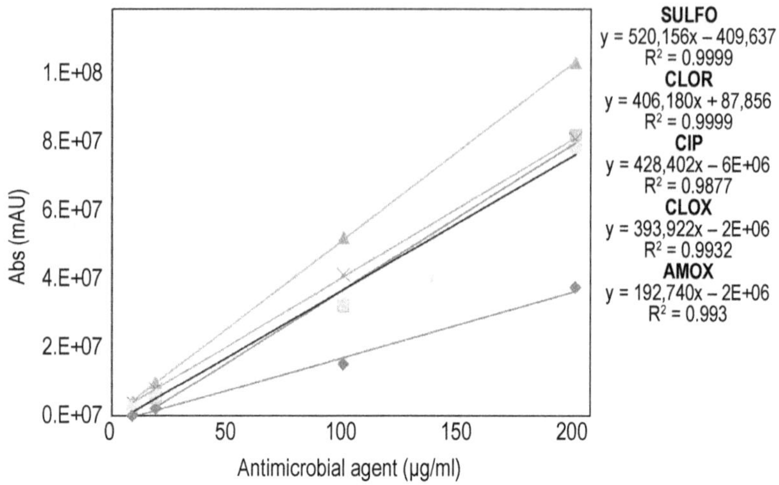

Figure 1. Calibration curves for a five antimicrobial agents' mixture. (SULFO: sulfamethoxazole; CLOR: chloramphenicol; CIP: ciprofloxacin; CLOX: cloxacillin; AMOX: amoxacillin).

Quantification of antibiotics in milk samples

Clean-up of milk samples, with SPE improved the chromatographic analysis, allowing the evaluation of antibiotic residues. No antimicrobial residues were, however, detected, for the antimicrobial agents tested, with exception to SULFA. Samples spiked with SULFA were analysed and the results indicated the presence of this antimicrobial agent, in two samples, although respecting the MRLs established for the EU members (Table 3).

The mean concentration for SULFA was 1.25±0.15 µg/ml. Recovery was of 121.81±13.60%.

Discussion and conclusions

Antimicrobial residues in animal origin food are a public health issue. The presence of these compounds may have consequences in human health, including allergies, alteration of the commensal flora and the emergence of antibiotic resistance and its dissemination in the environment.

Table 3. SULFA quantification and recovery percentage, for spiked samples according to clean-up method B.

Sample	SULFA µg/ml	Recovery% SULFA
1B	1.36	131.13
2B	1.19	115.06
3B	1.33	128.37
4B	0.99	98.45
5B	1.41	136.07

In the present study, the chromatographic method developed proved to be appropriate to simultaneous quantification of five antimicrobial agents, representing five antimicrobial groups. The clean-up step has proved to be an indispensable step to obtain cleaner matrices and improved the detection and quantification of antimicrobial agents. Limit of Detection and Limit of Quantification determined for the presented chromatographic method is below the European MRL's for the antimicrobial agents studied (Commission Regulation (EU) No 37/2010), thus proving to be an efficient method for quality control on foodstuff.

These low concentrations now detected may be responsible for inducing undesirable effects that may be viewed under the maleficent bioethics principle, as discussed below.

The recovery percentage determined for SULFA indicates that the SPE sample clean-up is efficient and allows reliable measurements on milk samples. This recovery was determined and evaluated as 121.81±13.60%.

The discussion of results obtained in this work in the light of the main Bioethical Principles underlying Principlism highlight the conflict between beneficence and non-maleficence. In fact, the usefulness of antibiotics use in veterinary, may be viewed under the Beneficence principle to the animals extended to humans, but this is limited by the non-maleficence principle '*primum non nocere*' since this practice leads to the presence of antibiotic residues, namely in milk, which in turn may lead to unconscious and harmful human ingestion of antibiotics. On the other hand, human elimination of antibiotics especially as urine or faeces, contributes to its environment dissemination. Thus, in this particular case an action under the beneficence principle from an initial point of view may finish in turn as a maleficent perspective to human consumers that without knowledge may be exposed to antibiotic residues leading in turn to antibiotic resistance development. Furthermore, the development of methods such the present proposal that enable the detection of lower concentrations of antibiotics may contribute to achieve better control and also a food policy of higher quality, thus diminishing the maleficent effects on humans and environment. More investment on policies and research on these issues are needed to address this problem.

References

Andreotti, R. and Nicodemo, F. (2004). Uso de antimicrobianos na produção de bovinos e desenvolvimento de resistência. Embrapa Gado de Corte: 1-50.

Arias, M.V.B. and Carrilho, C.M.D.d.M. (2012). Antimicrobial resistance in animals and in human being. There is reason for concern? Semina: Ciências Agrárias, Londrina 22(2): 775-790.

Beauchamp, T.L. and Childress, J.F. (2001). Principles of biomedical ethics. Oxford University Press, New York, NY, USA.

Branco, P.A.C., Soares, R., Vieites, F., Cabral, N. and Tavares, E. (2011). Efects of essential oils as growth promoters on performance of weaned pigs. Archives of Zootech. 60(231): 699-706.

Schenck, F.J. and Callery, P.S. (1998). Chromatographic methods of analysis of antibiotics in milk. Journal of Chromatography A 812(1-2): 99-109.

Souza, A.F.L., Araújo, D.N., Astolphi, L.L.J., Dias, M.B.L., Ambiel, C.A. and Santos, S.L. (2010). Probiótico e antibiótico como promotores de crescimento para frangos de corte. Colloquium Agrariae 6(2): 33-39.

Spinosa, H., Górniak, S. and Bernardi, M. (2006). Considerações gerais sobre o uso de antimicrobianos – farmacologia aplicada à medicina veterinária. Guanabara Koogan, Rio de Janeiro, Brasil.

Van Bruijnsvoort, M., Ottink, S.J.M., Jonker, K.M. and De Boer, E. (2004). Determination of streptomycin and dihydrostreptomycin in milk and honey by liquid chromatography with tandem mass spectrometry. Journal of Chromatography A 1058(1-2): 137-142.

88. Goat production in the county of Terras de Bouro: forest fire prevention, organic goat production and parasitism

F. Gandra[1], M. Vila-Viçosa[2,3], H. Cortes[2,3], J. Araújo[1,4] and T. Mateus[1,5,6,7]*
[1]Agrarian High School of Ponte de Lima, Polytechnic Institute of Viana do Castelo, Portugal, Refoios do Lima, 4990-706, Ponte de Lima, Portugal; [2]Laboratório de Parasitologia Victor Caeiro, DMV, ECT, IIFA, Universidade de Évora, Portugal [3]ICAAM – Instituto de Ciências Agrárias e Ambientais Mediterrânicas, Portugal; [4]Mountain Research Center (CIMO), Agrarian High School of Ponte de Lima, Polytechnic Institute of Viana do Castelo, Portugal, Refoios do Lima, 4990-706, Ponte de Lima, Portugal; [5]Veterinary Medicine Department, Vasco da Gama University School, Coimbra, Portugal; [6]Animal and Veterinary Research Centre, Vila Real, Portugal; [7]Institute of Public Health of the University of Porto, Porto, Portugal; pedropi@esa.ipvc.pt

Abstract

In Terras de Bouro, in the northwest of Portugal, goat production is a relevant activity for social, economic and environmental development. Often animals and shepherds are seen along the landscapes of Gerês mountains and grazing plays an essential role in the landscape mosaic. Many of the grazing animals are goats, many of which are kept in organic production systems, as encouraged by local organizations. Goat production produces meat but also has an essential role in natural control of vegetation. The dry matter content and the flammability of the natural vegetation increases in the summer, which consequently enhances the risk of fire. Goats grazing influence the vegetal biodiversity, reduce the scrubland and contribute to reducing the fire risk. However, human management is important to reduce the negative impact to the ecosystem. The aim of this study is to estimate the environmental contamination with parasitic forms in goat faeces, in Terras de Bouro region, because the animals walk long distances during grazing and contaminate the environment with these parasitic forms. The parasitic forms remain infective in the environment and may also infect other domestic animals, wild animals and even humans. A coprological survey in goats (n=59) (4 flocks: 2 organic; 2 conventional units) was performed using the FLOTAC method. It is a very sensitive method and allows the detection of parasitic forms in faeces. Parasitic forms were present in all samples and a high diversity of parasitism was observed: *Eimeria* spp. (59/59), Strongylida (51/59), *Dicrocoelium dentriticum* (45/59), *Nematodirus* spp. (44/59), *Skrjabinema* spp. (26/59), *Moniezia* spp. (15/59) and *Trichuris* spp. (14/59). Fifty seven samples had concomitant infections. When it is necessary to control the animal parasitism, pharmaceutical compounds (most often) are used, which can also contaminate the environment. This selective pressure has led to the rise of parasitic resistance to antiparasitic compounds. The present preliminary results are part of an on-going project on the study and promotion on use of goats in forest areas, as a factor in the sustainability of human activities in rural areas, and having an important role avoiding fires in forest areas.

Keywords: forest prevention of fire, grazing goats, animal health

Introduction

Goat production is a relevant activity for social, economic and environmental development in Terras de Bouro, where goats are raised in the natural vegetation that exists in the lands of the mountains of Gerês. As the goat production is an important activity it is necessary to study and to evaluate the animal health and welfare. In this paper, we review the role of goats in development and environmental protection and present a study of parasitism with potential consequences for environmental contamination with parasitic forms caused by animal faeces.

Goat production and local development

Gerês mountains are known not only for their beautiful landscapes but also for the permanent presence of wild and domestic animals (Figure 1). The domestic animals found in the mountains are goats, cattle, horses and sheep. The goats are in larger groups and characterize the mountain landscapes.

Goat production in this area has aroused the interest of several scientific bodies, such as universities. There are several scientific studies about goat production in the Gerês mountains in recent years addressing goat kid growth and meat quality (Araújo *et al.*, 2011, 2012; Pires *et al.*, 2011).

Due to the importance of this activity in the region, in 2005, the Terras de Bouro municipality started a project called 'Territory vs Sustainability' and with the support of six goat farmers began a process of conversion of flocks to organic production (APBTB, 2011). In this phase they created the Biological Producers Association of Terras de Bouro (APBTB). In 2014, continuing the promotion of goat production and agriculture in Terras de Bouro, the municipality and the Association for the Development of The High Lands of Homem, Cávado and Ave (ATAHCA) organized the first fair of Gerês organic goat kids, to promote goat production and gastronomy, showing the importance of this activity for the region (CMTB, 2014).

The most abundant goats in Gerês mountains are of the Bravia breed, a traditional Portuguese breed (Figure 2). The animals of indigenous breeds are better adapted to the climate and topography characteristics of the regions where they are native. The Bravia goats are animals with a very selective feeding behaviour (they are browsers). They are constantly moving around as they forage, seeking the most nutritious parts of plants (between grass, shrubs and trees) (Correia, 2004). Therefore, they are useful to reduce the forest understory, reducing the combustible and inflammable materials.

Goat production in Terras de Bouro also contributes to the conservation of genetic resources in Portugal as it relies almost exclusively on goats of the indigenous Bravia breed. This breed is classified as a C degree of risk of extinction (lower extinction threat level), but with some extinction danger (MAM, 2015).

Figure 1. Some animals of local breeds in the Gerês mountains.

Figure 2. Goats in the Gerês mountains.

Goat production and forest fire prevention

Over the last years, in many places livestock production using pasture in forest areas have diminished or even been abandoned. As a result, the negative impacts of the forest fires increased (DRAEDM, 2006). In Terras de Bouro, more than 1000 ha of forest burned in 2015 (ICNF, 2015).

The causes of forest fires in Portugal are multiple. Mostly they are caused by people (accident, negligence or intentionally). Fires from natural causes are rare. Approximately 1% are caused by natural causes, 23% intentionally, 37% by negligently and 39% by unknown causes (PROCIV, 2009). In the summer, the dry matter content and the flammability of the natural vegetation increases and can burn easily in case of fire. Therefore, reducing the amount of combustible materials in the forest is important to reduce the negative impacts of fires and reduce the burned area.

The manual or mechanical forest cleaning is important to reduce the fuel load. However, such cleaning is typically non-selective. Complete destruction of plants may occur as well as extinction of some vegetal species. Moreover, the mechanical cleaning contributes to greenhouse gas emission, because they typically use fossil fuels. (Mancilla-Leytón *et al.*, 2013; Lovreglio *et al.*, 2014) refer that the costs of forest cleaning with animals represent approximately 23% of the manual or mechanical cleaning costs. In addition to being a cheaper way to clean the forest, and the farmers can also produce and sell goat meat to increase their profits (Lovreglio *et al.*, 2014; Mancilla-Leytón and Martín Vicente 2011). Also, in the mountains, many fields are difficult to access, but the goats morphological characteristics allow access even in the fields where machines cannot reach (Mancilla-Leytón *et al.*, 2013).

However, the use of goats to clean forest lands has been criticized, arguing that the goats destroy vegetation and reduce vegetal biodiversity. Indeed, some studies have been conducted to determine how goat foraging affects the forest. In Andalusia, Spain, in Doñana National Park (a forest with similar characteristics to the north of Portugal), studies have been conducted using goats in forest cleaning. Mancilla-Leytón *et al.* (2014) concluded that in non-grazed lands the vegetal biodiversity was greater when compared to the grazed lands for goats. They also found that in grazed lands a significant reduction of the vegetal volume had occurred (28.5%) as well as a significant reduction of flammability (18.7%) compared to the non-grazed lands. They observed that in non-grazed lands there was a significant increase of the plant volume (28.1%) and consequently a significant increase of flammability (22.6%).

Goats eat different vegetal species during the day and during the year. In March, the goats eat more variety of plants (about six species) and in July and December they ingest less (about three species). The quantity of ingested plants of each species is different possibly due to the palatability and plant toxins, which thus affect the goats feeding behaviour. The dispersal of seeds is one of the functions of the goats during grazing. Goats return to the ground (via faeces) 30% of the ingested seeds. This means that animals can absorb some nutrients of the seeds and still disseminate some in the environment, thus contributing to the maintenance of vegetal biodiversity. Some seeds even increase their germination power after passing through the goats' digestive system (Lovreglio *et al.,* 2014; Mancilla-Leytón *et al.,* 2014). These authors also noticed that the plants that the goats eat more are plants of the more flammable species, which strengthens the role in reducing the risk of forest fires.

Based on these results, it is argued that the secret to an effective and sustainable reduction of combustible material is a good animal management of the lands (using scientific knowledge) and adopting a low animal density in pasture (about 2.7 animals/hectare/year) (Lovreglio *et al.,* 2014; Mancilla-Leytón *et al.,* 2014).

Goat production and parasitism

In the mountains of Terras de Bouro, goats are still raised in territories with little human activity, where flocks cross large areas to reach places where they have access to natural food. During these journeys, the delivery of parasitic forms from goat faeces to the environment is a permanent threat. These parasitic forms remain infective in the environment and may also infect other domestic or wild animals and even humans, hence it is important to estimate the environmental contamination with parasites from freely roaming goats.

For this, a coprological survey in goats (n=59) (4 flocks: 2 organic; 2 conventional units), was performed using the FLOTAC method. The flocks under study are located in the northwest of Portugal, in Terras de Bouro, in Gerês mountains, in the National Park of Peneda-Gerês. The identification and quantification of parasitic forms was performed using FLOTAC method. FLOTAC is a Faecal Egg Count (FEC) technique that can be used for a broad range of parasite identification and collection (Cringoli *et al.*, 2010). FLOTAC method is based on the centrifugation and flotation on a solution, with a density higher them 1.18 of the diluted sample, which sensibility can be of one egg/g of faeces (Cringoli *et al.*, 2010).

Parasitic forms were present in all samples and a high diversity of parasitism was observed: *Eimeria* spp. (59/59), Strongylida (51/59), *Dicrocoelium dentriticum* (45/59), *Nematodirus* spp. (44/59), *Skrjabinema* (26/59), *Moniezia* spp. (15/59) and *Trichuris* spp. (14/59). Fifty seven samples had concomitant infections.

Organic animal production has specific rules. According to the regulation number 889/2008 of the European Commission, in organic animal production the use of synthetic pharmaceutical compounds is limited. (CE, 2011). To comply with organic production rules, organic farmers need to use different approaches to control parasites and prevent parasitic disease development, such as moving animals between pastures (Burke *et al.*, 2009; Hoste and Torres-Acosta, 2011), increasing the amount of dietary protein (it has been proven that a diet rich in protein reduces the parasitic load in goats; (Hoste and Torres-Acosta, 2011) as well as genetically select for resistance to some parasitic infections (Hoste and

Torres-Acosta, 2011; Torres-Acosta and Hoste, 2008). Provision of certain plants and minerals (e.g. copper) may alos reduce the parasitic load (Hoste and Torres-Acosta, 2011; Piluzza *et al.*, 2015).

Conclusions

Forest fires have many negative impacts. It is necessary to educate the population and make a more effective forest management.

Goat production in Terras de Bouro is an essential activity to the region development.

The production of indigenous breed animals contributes to the conservation of the local genetic heritage. Animal production allows us to use the genetic heritage in the management of natural resources.

Forest cleaning using goats allows resource management, cost reduction and reducing of the forest fire risk. It is known that goat grazing has a negative impact in the biodiversity reduction, however, the benefits are greatest.

When farmers produce goats, it is necessary to promote animal health and welfare. It is important to know the parasitism of the animals to estimate the environmental contamination that may occur.

References

Araújo, J.P., Pacheco, L.F., Cerqueira, J.O., Fernandes, A.C., Ramos, S. and Costa, H.R. (2011). Crescimento de cabritos de raça Bravia em modo de produção biológico no concelho de Terras de Bouro. In Livro de Resumos do 3. Colóquio Nacional de Horticultura Biológica / 1. Colóquio Nacional de Produção Animal Biológica, Braga, Portugal, 49 pp.

Araújo, J.P., Fernandes, E., Cerqueira, J., Barros, M. and Pires, P., (2012). Caraterísticas da carcaça e conteúdo mineral da carne de cabrito de raça Bravia em modo de produção biológico. IV Congreso Internacional de Agroecoloxía e Agricultura Ecolóxica. Universidade de Vigo. Iniciativas agroecoloxicas innovadoras para a transformacion dos espazos rurais. Vigo, Spain.

Associação de Produtores Biológicos de Terras de Bouro (2011). Available at: http://tinyurl.com/zr8pwdo.

Burke, J., Miller, J. and Terrill, T. (2009). Impact of rotational grazing on management of gastrointestinal nematodes in weaned lambs. Veterinary Parasitology 163(1-2): 67-72.

Câmara Municipal de Terras de Bouro (2014). Primeira feira do cabrito biológico da Serra do Gerês. Available at: http://tinyurl.com/hoj8a32.

Comissão Europeia (2011). Regulamento (CE) No. 889/2008 da Comissão de 5 de Setembro de 2008. Brussels, Belgium.

Correia, T. (2004). Estudo da variabilidade e relações genéticas em raças caprinas autóctones mediante microssatélites. Available at: http://hdl.handle.net/10198/2294.

Cringoli, G., Rinaldi, L., Maurelli, M.P. and Utzinger, J. (2010). FLOTAC: new multivalent techniques for qualitative and quantitative copromicroscopic diagnosis of parasites in animals and humans. Nature Protocols 5: 503-551.

Direcção Regional de Agricultura de Entre Douro e Minho (2006). Plano regional de ordenamento florestal do baixo Minho. Available at: http://tinyurl.com/z4qjo7h.

Hoste, H. and Torres-Acosta, J. (2011). Non chemical control of helminths in ruminants: adapting solutions for changing worms in a changing world. Veterinary Parasitology 180(1-2): 144-154.

Instituto de Conservação da Natureza e Florestas (2015). Relatório provisório de incêndios florestais – 2015. Instituto da Conservação da Natureza e das Florestas – Departamento de Gestão de Áreas Públicas e de Proteção Florestal. Available at: http://tinyurl.com/zalbjx4.

Lovreglio, R, Meddour-Sahar, O. and Leone, V. (2014). Goat grazing as a wildfire prevention tool: a basic review. iForest – Biogeosciences and Forestry 7(4): 260-268.

Mancilla-Leytón, J. and Martín Vicente, A. (2011). Goat grazing silviculture for the prevention of forest fires in Doñana Natural Park. In: Bernués, A., Boutonnet, J.P., Casasús, I., Chentouf, M., Gabiña, D., Joy, M., López-Francos, A., Morand-Fehr, P. and E Pacheco, F. (eds.) Economic, social and environmental sustainability in sheep and goat production systems, pp. 169-172.

Mancilla-Leytón, J., Martín Vicente, A., Parejo-Farnés, C., Fernández-Ales, R. and Leiva, M. (2014). A vegetation management experiment: goats grazing shrublands in Doñana Natural Park. Russian Journal of Ecology 45(5): 384-390.

Mancilla-Leytón, J., Pino Mejías, R. and Martín Vicente, A. (2013). Do goats preserve the forest? Evaluating the effects of grazing goats on combustible Mediterranean scrub. Applied Vegetation Science 16(1): 63-73.

Ministério da Agricultura e do Mar (2015). Portaria 55/2015 de 27 de Fevereiro. In Diário da República, 1ª Série, No. 41.

Piluzza, G., Virdis, S., Serralutzu, F. and Bullitta, S. (2015). Uses of plants, animal and mineral substances in mediterranean ethno-veterinary practices for the care of small ruminants. Journal of Ethnopharmacology 168: 87-99.

Pires, P., Coelho, R., Fernandes, E., Cunha, C., Cerqueira, J., Ramos, S. and Araújo, J.P. (2011). Perfil dos ácidos gordos da carne de cabrito de raça Bravia em modo de produção biológico em Terras de Bouro. In: Livro de Resumos do 3. Colóquio Nacional de Horticultura Biológica / 1. Colóquio Nacional de Produção Animal Biológica, Braga, Portugal, 133 pp.

Protecção Cívil (2009). Causas dos incêndios florestais em Portugal. Available at: http://tinyurl.com/z5jgvdd.

Torres-Acosta, J. and Hoste, H. (2008). Alternative or improved methods to limit gastro-intestinal parasitism in grazing sheep and goats. Small Ruminant Research, Special Issue: Sheep and Goat Farming: Grazing Systems of Production and Development 77(2-3): 159-173.

89. Influence of quinoa (*Chenopodium quinoa* Willd.) varieties on chemical composition

C. Encina-Zelada[1,2,3]*, J. Pereda*[2]*, J.A. Teixeira*[3]*, L. Gómez-Pando*[4]*, M. Ibañez*[4]*, V. Cadavez*[1] *and U. Gonzales-Barron*[1]

[1]*CIMO Mountain Research Centre, School of Agriculture, Polytechnic Institute of Braganza, Portugal;* [2]*Faculty of Food Industries, National Agricultural University La Molina, Lima, Peru;* [3]*Department of Biological Engineering, School of Engineering, University of Minho, Portugal;* [4]*Faculty of Agronomy, National Agricultural University La Molina, Lima, Peru; cencina@ipb.pt, cencina@lamolina.com.pe*

Abstract

Quinoa has been considered by the Food and Agriculture Organization (FAO) as one of the crops destined to offer food security because of both the superior nutritional quality of its grains, and its ability to grow in very difficult environments, showing tolerance to frost, hail, snow, heat, salinity and drought. The objective of this study was to assess whether variety had any effect on the proximate composition of quinoa. One hundred and twenty (120) accession samples of Peruvian quinoa grains from five varieties of different grain colours (yellow, beige, grey, orange and black) were employed. Proximate analyses were performed by AOAC official methods. Analyses of variance evidenced that moisture (8.11-13.4%), protein (8.33-11.7% db), fat (4.79-9.46% db) and ashes (2.51-4.62% db) differed strongly (P<0.001) among varieties; although fat (coefficient of variation, CV=0.11) and ashes (CV=0.13) presented a higher variability than moisture (CV=0.08) and protein (CV=0.07). Quinoa of black (11.21%, SD=0.24), yellow (11.24%, SD=0.12) and orange (11.29%, SD=0.25) grain varieties presented higher moisture levels than quinoa of grey (10.24%, SD=0.24) and beige (10.24%, SD=0.09) grains. Similarly, protein contents (db) were higher for orange (10.15%, SD=0.25), yellow (10.17%, SD=0.12) and black (10.40%, SD=0.23) quinoa, while total fat (db) was higher for yellow (6.96%, SD=0.11) and orange (7.16%, SD=0.24) grain varieties. Beige quinoa was the studied variety having the lowest (P<0.05) protein (9.75%, SD=0.09), fat (5.97%, SD=0.22) and ashes (2.95%, SD=0.04) contents. Although a high variability in chemical composition was found within the same variety; yellow, orange and black quinoa stood out as the varieties having the best nutritional quality in terms of total fat and protein; while beige and grey quinoa were comparable, yet both with a significantly lower fat and protein content. The plasticity of quinoa can transform this culture into an important protein source especially for regions where the harsh environmental conditions put limitations to the growth of other sources of protein.

Keywords: proximate composition, grain quality, varieties

Introduction

According to scientific research, quinoa (*Chenopodium quinoa* Willd.) originates around the Titicaca Lake, and is grown in the Andean countries. For 7,000 years, indigenous people have maintained, controlled, protected and preserved different quinoa's varieties, in different ecological zones in natural germplasm banks. Quinoa seeds are small, round and flat, and can be found in yellow, red, grey, pink, orange, red wine and beige colour. Quinoa is a plant that produces grains even if cultivated up to 4,500 m.a.s.l. and with higher nutritive value than traditional cereals, although most varieties differ in morphology, phenology and the chemical composition of the tissues (Vega-Gálvez *et al.*, 2010).

Research interest in quinoa has been renewed due its high nutritional value (i.e. good amino acid profile and fatty acid profile and good antioxidant capacity) (Wrigley *et al.*, 2004). Quinoa seeds contain carbohydrates (~77.6%), lipids (~6.5%), protein (~12.9%), and a balanced amino acid spectrum of

EurSafe 2016 – DOI 10.3921/978-90-8686-834-6_89, © Wageningen Academic Publishers 2016

high lysine and methionine contents. It is rich in dietary fibre, and rich in mineral nutrients (~3.0%), especially calcium and sodium (Ando *et al.*, 2002). Due to these nutritional qualities, there has been growing interest, in a number of countries, especially in Europe, in initiating introduction of quinoa and research work (Hirose *et al.*, 2010). Thus, the objective of this study was to assess whether variety had any significant effect on the proximate composition of quinoa.

Methodology

Proximal analyses

Quinoa crops (*Chenopodium quinoa* Willd.) harvested in Peru (from the National Agricultural University La Molina and Regional Development Centre – Highland), in different seasons between 2010 and 2012, were utilised. Quinoa grains samples amounted to 120 accessions which were of orange (n=8), beige (n=56) and yellow (n=36), grey (n=9) and black (n=9) colour. Moisture, protein, crude fat and ash contents were determined using the reference methods 925.10, 920.87 (conversion factor of 6.25), 923.05 (FOSS Soxhtec) and 923.03, respectively, described in AOAC (2000). Determinations, done in triplicate, were averaged and converted to dry basis (g/100 g).

Statistical analysis

An analysis of variance was performed for each of the proximate components, using a generalised linear model with quinoa variety as fixed effect. Varieties were then contrasted using Tukey's test of comparison of means at a confidence level of 0.95. Statistical analyses were conducted using the function *lm()* in the R software version 3.2.2 (R Development Core Team).

Results and discussion

Analyses of variance evidenced that moisture (in the range 8.11-13.4%), protein (8.33-11.7% db), fat (4.79-9.46% db) and ashes (2.51-4.62% db) differed strongly ($P<0.001$) among varieties; although fat (coefficient of variation, CV=0.11) and ashes (CV=0.13) presented a higher variability than moisture (CV=0.08) and protein (CV=0.07).

Table 1 shows the mean and standard error of the proximate analysis; this is the contents of moisture, protein, fat and ashes (g/100 g) in quinoa grains of different colour varieties. Quinoa of black (11.21%), yellow (11.24%) and orange (11.29%) grain varieties presented significantly higher moisture levels than quinoa of grey (10.24%) and beige (10.24%) grains. The content of moisture could be different between samples of quinoa. For example, Repo-Carrasco-Valencia *et al.*, (2010) evaluated ten quinoa varieties with contents between 8.26% to 11.51%, all values inside the range that we obtained in this study; and Gallego-Villa *et al.*, (2014) found a moisture content of 10.8%.

Similarly, protein contents (db) were higher for orange (10.15%), yellow (10.17%) and black (10.40%) quinoa, while grey (9.83%) and beige (9.75%) quinoa had statistically lower contents. These protein levels are lower than those found by Vega-Gálvez *et al.* (2010) who showed that protein content in quinoa grains can range from 15.6% to 17.9% (db), with an average of 15%. Earlier on, De Bruin (1963) had quantified protein content of four genotypes of quinoa, reporting a range of 12.9-15.1%. Repo-Carrasco-Valencia *et al.* (2010) indicated that, in ten quinoa varieties, the protein content was in a range between 11.32 and 14.72% (db). While Sharma *et al.* (2015) found 16.5% (dry wt), Gallego-Villa *et al.* (2014) found a quinoa protein of 14.35% (db).

Table 1. Mean and standard error (in brackets) of moisture, protein, fat and ashes (g/100 g) contents in quinoa grains of different colour varieties.[1]

Colour variety	Moisture	Protein	Fat	Ashes
Grey	10.25^{a} (0.2401)	9.83^{ab} (0.2330)	5.98^{a} (0.2233)	3.05^{ab} (0.0875)
Beige	10.25^{a} (0.096)	9.75^{a} (0.0930)	6.52^{a} (0.0895)	2.95^{a} (0.0351)
Black	11.21^{b} (0.2401)	10.40^{b} (0.2329)	6.69^{ab} (0.2233)	3.22^{b} (0.0875)
Yellow	11.24^{b} (0.1203)	10.17^{b} (0.2471)	6.96^{b} (0.1117)	3.63^{c} (0.0437)
Orange	11.29^{b} (0.2550)	10.15^{ab} (0.1165)	7.16^{b} (0.2369)	3.66^{c} (0.0928)

[1] Different superscript letters denote difference in means at α=0.05.

Protein content in quinoa is generally higher than in common cereals. In addition, it has a higher content of essential amino acids (Drzewiecki *et al.*, 2003; Gorinstein *et al.*, 2002). In contrast to most common grains, the proteins in quinoa are composed mainly of globulins and albumins, and contain very little or no storage prolamin, which are the main storage proteins in cereals, and also the toxic proteins in celiac disease (Drzewiecki *et al.*, 2003).

An important feature of the composition of the pseudocereals is their fat content. Lipid content in quinoa is known to be 2-3 times higher than in buckwheat and other common cereals such as wheat (Alvarez-Jubete *et al.*, 2009). Quinoa lipids are characterised by a high degree of unsaturation, which is desirable from a nutritional point of view. Linoleic acid is the most abundant fatty acid (50% of the total fatty acids in quinoa, and approximately) followed by oleic acid (25% in quinoa) and palmitic acid (Alvarez-Jubete *et al.*, 2009). Amongst the pseudocereal seeds, high α-linolenic acid (C18:3 n-3) content is found in quinoa seeds, with values ranging from 3.8% (Ruales and Nair, 1993) to 8.3% (Alvarez-Jubete *et al.*, 2009). From our work, fat content was significantly higher for yellow (6.96%) and orange (7.16%) grain varieties, whilst beige quinoa was the studied variety having the lowest fat (5.97%) content. Other authors have reported different fat contents, for instance Repo-Carrasco-Valencia *et al.* (2010) reported fat content in ten quinoa varieties in a range between 4.5-7.7% (db). Sharma *et al.* (2015) obtained a fat content of 6.3% (db), and Gallego-Villa *et al.* (2014) a fat content of 7.8% (db). In general, our levels are in agreement with those found by Vega-Gálvez *et al.* (2010), who reported a fat content range between 1.8 to 9.5%, with an average of 5.0-7.2%, which is higher than that of maize (3-4%).

According to Vega-Gálvez *et al.* (2010), quinoa has a high content of calcium (32.9 mg/100 g), magnesium (206.8 mg/100 g), zinc (1.8 mg/100 g) and iron (5.5 mg/100 g). A good way to get an insight into the minerals' level is through the determination of the ash contents. The highest levels were obtained from orange (3.66%), yellow (3.63%) and black (3.22%) quinoa varieties, while both grey (3.05%) and beige (2.95) had significantly lower contents. Other authors mention a different content of ashes. According to Repo-Carrasco-Valencia *et al.* (2010) in ten quinoa varieties, the ash contents were in a range between 2.5-3.2% (db), while for Sharma *et al.* (2015) they were as high as 3.8% (db).

This short communication has underscored the high nutritional value of quinoa. In conjunction with these nutritional qualities, it is known that quinoa can resist to drought, poor soils and high salinity, and grows well in temperatures between -8 to 38 °C. There are hundreds of quinoa ecotypes, varying in morphology, phenology, and in its ability to adapt to multiple agro-environmental conditions. This great genetic variability of quinoa allows selection to develop new cultivars for a wide range of worldwide environmental conditions (Jacobsen *et al.*, 2003). Thus, quinoa offers to communities living in harsh environments options to improve their livelihoods, generate income, and enjoy better nutrition and health (Zurita-Silva *et al.*, 2014). In fact, the recognition of both the nutritional quality and its great adaptability to diverse environmental conditions led the United Nations to declare the year of 2013 as the International Year of Quinoa, in the view that this crop can contribute to global food security. This is true, especially in areas where the population has no access to adequate sources of protein, or where there are environmental constraints to food crop production in the aftermath of climate change. In developing countries like China, India, Kenya, Morocco, and Brazil, where the diet is based on rice and wheat, quinoa is presently investigated as the best alternative to challenge the lack of good nutrition and low protein diets (FAO, 2011; FAO 2013).

Conclusions

Contents of moisture, protein, fat and ashes differed strongly ($P<0.001$) among quinoa colour varieties. Although a high variability in chemical composition was found within the same variety, yellow, orange and black quinoa stood out as the varieties having the best nutritional quality in terms of total fat and protein; while beige and grey quinoa were comparable, yet both with a significantly lower nutritional quality (i.e. fat and protein content). Yellow quinoa presented both the highest fat (~7% db) and protein (~10.2% db) contents. Along with the high adaptability to multiple agro-environmental conditions (resistance to drought and frost, poor soils and high salinity), and its high nutritional profile, quinoa can become an important alternative to traditional cereals, contributing to global food security problems.

References

Alvarez-Jubete, L., Arendt, E.K. and Gallagher, E. (2009). Nutritive value and chemical composition of pseudocereals as gluten-free ingredients. International Journal of Food Science and Nutrition 60(1), Suppl. 4: 240-257.

Ando, H., Chen, Y., Tang, H., Shimizu, M., Watanabe, K. and Mitsunaga, T. (2002). Food components in fractions of quinoa seed. Food Science and Technology Research 8(1): 80-84.

AOAC (2000). Official methods of analysis of AOAC International. In: Horwitz, W. and Latimer, G.W (eds.) AOAC International, Galthersbrug, MD, USA, 2200 pp.

De Bruin, A. (1963). Investigation of the food value of quinoa and cañihua seed. Journal of Food Science 29: 872-876.

Drzewiecki, J., Delgado-Licon, E., Haruenkit, R., Pawelzik, E., Martin-Belloso, O., Park, Y.S., Jung, S.T., Trakhtenberg, S. and Gorlnstein, S. (2003). Identification and differences of total proteins and their soluble fractions in some pseudocereals based on electrophoretic patterns. Journal of Agricultural and Food Chemistry 51(26): 7798-7804.

FAO (2011). Quinoa an ancient crop to contribute to world food security. Regional Office for Latin America and the Caribbean. Available at: http://tinyurl.com/3nnsgm4.

FAO (2013). Home-international year of quinoa 2013. Available at: http://www.fao.org/quinoa-2013/en/.

Gallego-Villa, D.Y., Russo, L., Kerbab, K., Landi, M. and Rastrelli, L. (2014). Chemical and nutritional characterization of Chenopodium pallidicaule (cañihua) and *Chenopodium quinoa* (quinoa) sedes. Emirates Journal of Food and Agriculture 26(7): 609-615.

Gorinstein, S., Pawelzik, E., Delgado-Licon, E., Haruenkit, R., Weisz, M. and Trakhtenberg, S. (2002). Characterisation of pseudocereal and cereal proteins by protein and amino acid analysis. Journal of the Science of Food and Agriculture 82: 886-891.

Hirose, Y., Fujita, T., Ishii, T. and Ueno, N. (2010). Antioxidative properties and flavonoid composition of *Chenopodium quinoa* seeds cultivated in Japan. Food Chemistry 119: 1300-1306.

Jacobsen, S.E., Mujica, A. and Jensen, C.R. (2003). The resistance of quinoa (*Chenopodium quinoa* Willd.) to adverse abiotic factors. Food Reviews International 19: 99-109.

Repo-Carrasco-Valencia, R., Hellstrom, J.K., Pihlava, J.M. and Mattila, P.H. (2010). Flavonoids and other phenolic compounds in Andean indigenous grains: Quinoa (*Chenopodium quinoa*), kaniwa (*Chenopodium pallidicaule*) and kiwicha (*Amaranthus caudatus*). Food Chemistry 120: 128-133.

Ruales, J. and Nair, B.M. (1993). Content of fat, vitamins and minerals in quinoa (*Chenopodium quinoa*, Willd) seeds. Food Chemistry 48: 131-136.

Sharma, V., Chandra, S., Dwivedi, P. and Parturkar, M. (2015). Quinoa (*Chenopodium quinoa* Willd.): a nutritional healthy grain. International Journal of Advanced Research 3(9): 725-736.

Vega-Gálvez, A., Miranda, M., Vergara, J., Uribe, E., Puente, L. and Martínez, E. (2010). Nutrition facts and functional potential of quinoa (*Chenopodium quinoa* willd.), an ancient Andean grain: a review. Journal of Science and Food Agriculture 90: 2541-2547.

Wrigley, C., Corke, H. and Walker, C.E. (2004). Encyclopedia of grain science. Elsevier Academic Press, Waltham, MA, USA, 1700 pp.

Zurita-Silva, A., Fuentes, F., Zamora, P., Jacobsen, S.E. and Schwember, A. (2014). Breeding quinoa (*Chenopodium quinoa* Willd.): potential and perspectives. Molecular Breeding 34: 13-30.

90. Precision livestock farming for reproductive performance optimization: a survey

S.I. Lopes[1,3], R. Bexiga[2,4], J.P. Araújo[1,5], J.L. Cerqueira[1,6], C. Abreu[1,7], C. Paredes[1] and J.M. Alonso[1]*
[1]Instituto Politécnico de Viana do Castelo, 4900-347 Viana do Castelo, Portugal; [2]Faculdade de Medicina Veterinária – Universidade de Lisboa, 1300-477 Lisboa, Portugal; [3]IT – Instituto de Telecomunicações, 3810-193 Aveiro, Portugal; [4]CIISA, Centre for Interdisciplinary Research in Animal Health, 1300-477 Lisboa, Portugal; [5]CIMO, Mountain Research Center, ESA-IPVC, 4990-706 Ponte de Lima, Portugal; [6]CECAV, Animal and Veterinary Research Center, 5001-801 Vila Real, Portugal; [7]CMEMS-UMinho, Universidade do Minho, 4800-058 Guimarães, Portugal; sil@estg.ipvc.pt

Abstract

Precision Livestock Farming (PLF) can be seen as the application of state-of-the-art methodologies and technologies in favour of optimizing animal production processes. Adopting PLF includes automatic, continuous and nearly real-time monitoring of animal production, welfare and health condition, which results in valuable information that can be used to support the farmer in decision-making and to improve the production performance. A major challenge in livestock farming is related to the improvement of cattle reproductive performance. Reproductive inefficiency has a negative impact on the technical and economical sustainability of farms and on their ecological footprint, since those unproductive animals (i.e. less than a birth/year for cattle) continue to emit methane into the atmosphere. This paper surveys the state-of-the-art technologies and methods used to optimize the reproductive efficiency of cattle, and introduces the guidelines for the specification of a PLF monitoring system focused on the improvement of the reproductive performance of cattle at pasture. In this sense, several devices will be discussed using the following main criteria: (a) method used; (b) technology involved; (c) invasiveness; (d) telemetry range; and (e) application (oestrus/mounting detection and calving prediction). Based on these criteria, a discussion will be undertaken to have in mind animal welfare and ethical implications.

Keywords: cattle fertility, mounting detection, oestrus detection, calving prediction

Introduction

Feeding the world is one of humanities' growing concerns. With a growing population predicted to surpass 9 billion people by 2050, producing sufficient food to feed every human without constraints poses a serious challenge for future generations. This challenge is made more difficult for several reasons. The needs of the population are not growing uniformly: the expectations of a fast-growing middle class in India and China mean that certain items at the table are a lot more in demand than others, including meat and dairy products. Cattle farming, for either milk or meat production will necessarily continue to increase its efficiency, with higher productions and lower greenhouse gas emissions per animal being essential to accomplish food security and environmental sustainability. It is estimated that livestock supply chains account for 14.5% of all human-induced greenhouse gas (GHG) emissions, with cattle accounting for approximately 61% of GHG produced by livestock (Gerber *et al.*, 2013).

With an increase in farm size and in the cost of labour, many technological developments have been introduced on a large scale to improve production on farms. Initially, technologies were used for animal identification, recording of milk yield and allocation of feed to individual animals. Subsequent developments focused on reproductive aids, and the most recent developments have brought the possibility to detect disease or under-performance. PLF technology clearly has great potential to create added value for many stakeholders, especially as a management tool for farmers, making it possible to

EurSafe 2016 – DOI 10.3921/978-90-8686-834-6_90, © Wageningen Academic Publishers 2016

improve animal welfare, health, efficiency and the environmental impact (Berckmans, 2014). While PLF will not be able to necessarily resolve all welfare-related questions, it will allow interested parties to detect and act upon time periods when animals are kept under sub-optimal conditions (Banhazi *et al.*, 2012). PLF has the potential to monitor, manage and control many issues of livestock production, simultaneously and automatically (Wathes *et al.*, 2008).

Reproductive inefficiency has a negative impact on the technical and economical sustainability of farms and on their ecological footprint. One of the main drivers of increasing efficiency of production is related to the optimization of the reproductive performance of cattle (Capper, 2011). In Europe, this problem has been recognized by the EC under the CAP 2014-2020, in which it is determined that the suckler-cow premium must be adjusted taking into account minimum levels of fertility (80%). To improve cattle fertility, two decisive moments in the reproductive process should be detected: (1) the time when natural mating occurs and (2) the time when the animal starts calving. Knowledge of these two key moments allows the farmer to concentrate births, adapt the food availability to animals with different nutritional needs, and maximize the attention given to animals during calving.

Cattle fertility impact on production and environment

Central to this increase in production efficiency is the dilution of the maintenance effect (Capper, 2011), i.e. the decrease in the relative proportion of the amount of resources used for the animals' maintenance when compared with the resources used for milk or meat production. The increase in efficiency of cattle production has been mostly driven by improvements in nutrition, genetics, reproduction and animal health. Reproductive improvements have the potential to decrease the relative proportion of unproductive animals in a herd, as the amount of replacements needed to guarantee herd size would be lower. It is estimated that improving fertility levels to values considered ideal, could lead to 24% reductions in methane emissions (Garnsworthy, 2004).

Either for dairy or beef production, it is generally accepted that the reproductive objectives include a calving interval for the herd of 365 days, leading to one live calf per year (Rae, 2006). In the case of beef herds, a short duration of the breeding season is also desirable to optimize feed resources and to generate groups of animals of similar size that can be marketed together. A breeding goal for beef herds is that 90 to 95% of cows are bred within a 60 day period (Rae, 2006). Reducing the calving period from 80 to 60 days has been modelled to lead to reductions of 3.2% in land use, water use and greenhouse gas emissions (White *et al.*, 2015).

PLF for reproductive performance optimization

Sensors that play a role in cattle reproduction take advantage of certain physiological responses that occur in these animals and that are relevant to their reproductive function. The cow presents cyclic periods of sexual receptivity (oestrus), in which there are behavioural changes that are a result of a particular hormonal profile, leading to ovulation approximately 24 hours after the onset of oestrus (Hartigan, 2004). These behavioural changes last between 8 and 30 hours, and include increased activity (walking substantially more than their cohorts), grouping of sexually active animals, reduction in the time spent eating, resting, ruminating, and often a decrease in milk yield. The most reliable sign of oestrus is standing of the cow when mounted by another animal, for approximately 5 seconds or more. A transparent mucoid vulval discharge may be observable, the hair of the tail-head may be lifted and there may be skin abrasions due to the animal being mounted by other animals. The body temperature also fluctuates around oestrus, decreasing around 0.5 °C the day before oestrus, increasing during oestrus and decreasing at the time of ovulation (Noakes *et al.*, 2001).

The use of sensors to detect sexual receptivity, and thus to predict the timing of ovulation, is central to the generalized use of artificial insemination in dairy cattle. The most common way to detect oestrus in these animals is through the use of pedometers. These measure the number of steps each animal takes, and allow for comparison with previously recorded activity for the same animal, leading to detection of oestrus, when a certain factor of increase is measured (Mottram, 2015). It has been observed that the use of activity meters for automated detection of oestrus, may represent an improvement when compared with visual observation of oestrus signs (Michaelis *et al.*, 2014), which is especially important with the current economic constraints on farmers.

Other commercially available sensors to detect oestrus include pressure-sensitive mount detectors, collars with inertial measurement units (IMU), thermometers to measure body and/or milk temperature (Mottram, 2015), thermal imaging systems for skin temperature (Talukder *et al.*, 2014), and measurement of milk progesterone, with cows showing a sharp decrease in progesterone concentrations before oestrus (Mottram, 2015).

Another event that needs to be closely monitored to obtain an optimal reproductive performance is calving. Calving itself may lead to catastrophic results for the cow when dystocia occurs, defined as calving difficulty resulting from prolonged spontaneous calving or prolonged or severe assisted extraction (Mee, 2004). Following calving, there can also be perinatal mortality of the calf, and a series of complications may follow, including prolapse of the uterus, or inflammation of several layers of the uterus (metritis, endometritis).

Timely assistance at calving is therefore necessary to ensure the survival of cows and calves alike. There are some calving detection sensors on the market, which resort to the detection of different changes that occur around or during calving. Rectal and vaginal temperatures have been measured to be 0.4 to 0.6 °C and 0.6-0.7 °C lower on the day of calving than when measured 48 hours before, for both beef and dairy cattle (Saint-Dizier and Chastand-Maillard, 2015). On the calving day the animals tend to isolate themselves from the rest of the herd, have increased activity and lie down and stand up more often. There are also increased movements by the tail close to parturition, with increased rising of the tail head as early as 5 days before parturition (Saint-Dizier and Chastand-Maillard, 2015). Monitoring of feeding, drinking and rumination activity has also shown that there are decreases in all these activities around parturition, even though these are yet to be translated into commercially available sensors. On the market, there are inclinometers and accelerometers (i.e. IMUs), to detect tail rising and behavioural changes, abdominal belts to detect uterine contractions, intravaginal sensors that detect a decrease in temperature and the expulsion of foetal membranes, and devices attached to the vagina to detect calf expulsion (Saint-Dizier and Chastand-Maillard, 2015). Based on the methodologies previously introduced, we compiled in Table 1 several commercially available PLF devices that are being used for reproductive performance optimization of cattle.

Discussion

The increasing attention to farm animal welfare has led to many different programs designed to ensure a certain level of animal welfare in food production. Evidence shows that the physical environment alone is not a good predictor of animal welfare. Genetic factors, human factors, including management practices and animal care skills, can be major determinants of farm animal welfare (Hemsworth *et al.*, 2002). Given this complexity, it is widely accepted that valid assessment of farm animal welfare requires actual on-farm assessment methods. Measures used in on-farm welfare assessment systems are often classified into resource-based measures (e.g. housing systems, space allowances, animal management practises) and animal-based measures (e.g. low incidence of disease or injury, normal behaviour) (Main

Table 1. Commercial Precision Livestock Farming (PLF) devices used for reproductive performance optimization.

Method	Technology Involved	Device	Invasiveness	Telemetry range	Application
Pressure based	Pressure detector	HeatWatch II	outer body	Medium (<400 m)	mounting detection oestrus detection
		AccuBreed	outer body	High (<800 m)	mounting detection
Temperature based	Temperature monitoring	Bella AG (active bolus)	inner body (rumen)	low (<10 m)	oestrus detection
		TempTrack	inner body (rumen)	medium (<100 m)	oestrus detection calving prediction
		Gyuonkei	inner body (vagina)	low (<20 m)	oestrus detection calving prediction
		Vel'Phone	inner body (vagina)	low (<10 m)	calving prediction
		Anemon	inner body (vagina)	low (<10 m)	oestrus detection
	pH and temperature monitoring	SMAXTEC (active bolus)	inner body (rumen)	low (<50 m)	oestrus detection
	Themal imaging	FLIR A310	external	low (<10 m)	oestrus detection calving prediction
Motion based	IMU (activity level and animal behaviour patterns)	AfiAct II (Pedometer Plus)	outer body (leg)	low (<50 m)	oestrus detection
	IMU (activity level)	IceCube	outer body (leg)	low (<10 m)	oestrus detection
	IMU (tail activity level)	Moocall	outer body (tail)	high (cellular-based)	oestrus detection
	IMU (activity level and animal behaviour patterns)	Track-a-Cow	outer body (leg)	high (<700 m)	oestrus detection
	IMU (activity level and animal behaviour patterns)	CowScout	outer body (neck or leg)	high (<1 km)	oestrus detection
	IMU (activity level and animal behaviour patterns)	SmartBow EarTag	outer body (ear)	low (<50 m)	oestrus detection
	IMU (activity level)	SensOor EarTag	outer body (ear)	low (<50 m)	oestrus detection
	IMU (activity level and animal behaviour patterns)	HR Tag	outer body (neck)	medium (<500 m)	oestrus detection calving prediction
Multi-metric based	IMU (activity level and animal behaviour patterns) Skin temperature	Alanya Data Centre	outer body (neck)	medium (<500 m)	mounting detection oestrus detection calving predition

et al., 2003). Animal-based measures provide more direct assessment of the state of the animals (Barnett and Hemsworth, 2009) and nowadays PLF has an important role in the welfare assessment.

With regard to the invasiveness criterion, inner-body devices, cf. Table 1 – such as the intra-vaginal devices – are highly invasive, can cause internal wounding and may interfere negatively with the reproductive cycle. These types of devices are typically used in housed production systems. Outer-body devices, cf. Table 1 – such as neck, leg, ear or tail attached devices – are potentially invasive, can cause external wounding and may influence the animals' behaviour (Rushen *et al.*, 2012). Notwithstanding the above, these types of devices have been extensively used on commercial dairy farms with rare situations of injury occurrence, if managed correctly. Other factors regarding the physical nature of the attached devices should also be considered – i.e. device size, antenna size, and battery – due to the animals' natural ability to remove the attached devices (Rushen *et al.*, 2012). Outer-body devices have increased potential in pastured production systems since radio communications from outer-body are more effective allowing telemetry systems to have large communication ranges. On the other hand, external devices – such as thermal imaging systems – are non-invasive and can be used for long-term monitoring of animals in housed production systems.

Despite the possibility of a device causing injury not being completely ruled out, the advantages of PLF devices greatly outweigh the possible disadvantages. In fact, such devices are able to improve the reproductive efficiency of cattle herds, with all the inherent economic and environmental advantages, but can also bring significant welfare improvements through increased predictability of calving time.

References

Banhazi, T.M., Lehr, H., Black, J.L., Crabtree, H., Schofield, P., Tscharke, M. and Berckmans, D. (2012). Precision livestock farming: an international review of scientific and commercial aspects. International Journal of Agricultural and Biological Engineering 5(3): 1-9.

Barnett, J.L. and Hemsworth, P.H. (2009). Welfare monitoring schemes: using research to safeguard welfare of animals on the farm. Journal of Applied Animal Welfare Science 12: 114-131.

Berckmans, D. (2014). Precision livestock farming technologies for welfare management in intensive livestock systems. Scientific and Technical Review of the Office International des Epizooties 33(1): 189-196.

Capper, J.L., Cady, R.A. and Bauman, D.E. (2009). The environmental impact of dairy production: 1944 compared with 2007. Journal of Animal Science 87: 2160-2167.

Capper, J.L. (2011). The environmental impact of beef production in the United States: 1977 compared with 2007. Journal of Animal Science 89: 4249-4261.

Garnsworthy, P.C. (2004). The environmental impact of fertility in dairy cows: a modelling approach to predict methane and ammonia emissions. Animal Feed Science and Technology 112: 211-223.

Gerber, P.J., Steinfeld, H., Henderson, B., Mottet, A., Opio, C., Dijkman, J., Falcucci, A. and Tempio, G. (2013). Tackling climate change through livestock – a global assessment of emissions and mitigation opportunities. Food and Agriculture Organization of the United Nations (FAO), Rome, Italy.

Hartigan, P.J. (2004). Reproductive physiology in cattle. In: Andrews, A.H, Blowey, R.W., Boyd, H. and Eddy, R.G. (eds.) Bovine medicine – diseases and husbandry of cattle, pp. 471-505.

Hemsworth, P.H., Coleman, G.J., Barnett, J.L., Borg, S. and Dowling, S. (2002). The effects of cognitive behavioral intervention on the attitude and behavior of stockpersons and the behavior and productivity of commercial dairy cows. Journal of Animal Science 80: 68-78.

Main, D.C.J., Kent, J.P., Wemelsfelder, F., Ofner, E. and Tuittens, F.A.M. (2003). Applications for methods of on-farm assessment. Animal Welfare Journal 12: 445-456.

Mee, J.F. (2004). Managing the dairy cow at calving time. Veterinary Clinics: Food Animal Practice Journal 20: 521-546.

Michaelis, I., Burfeind, O. and Heuwieser, W. (2014). Evaluation of oestrus detection in dairy cattle comparing an automated activity monitoring system to visual observation. Reproduction in Domestic Animals Journal 49: 621-628.

Mottram, T. (In press). Animal board invited review: precision livestock farming for dairy cows with a focus on oestrus detection. Animal.

Noakes, D.E. (2001). Endogenous and exogenous control of ovarian cyclicity. In: Noakes, D.E., Parkinson, T.J. and England, G.C.W. (eds.) Veterinary Reproduction and Obstetrics, pp. 3-53.

Rae, D.O. (2006). Assessing performance of cow-calf operations using epidemiology. Veterinary Clinics: Food Animal Practice Journal 22: 53-74.

Rushen, J., Chapinal, N. and De Passille, A.M. (2012). Automated monitoring of behavioural-based animal welfare indicators. Animal Welfare – The UFAW Journal 21(3): 339.

Saint-Dizer, M. and Chstand-Maillard, S. (2015). Methods and on-farm devices to predict calving time in cattle. Veterinary Journal 205: 349-356.

Talukder, S., Kerrisk, K.L., Ingenhoff, L., Thomson, P.C., Garcia, S.C. and Celi, P. (2014). Infrared technology for estrus detection and as a predictor of time of ovulation in dairy cows in a pasture-based system. Theriogenology 81: 925-935.

Wathes, C.M., Kristensen, H.H., Aerts, J.M. and Berckmans, D. (2008). Is precision livestock farming an engineer's daydream or nightmare, an animal's friend or foe, and a farmer's panacea or pitfall? Computers and Electronics in Agriculture 64: 2-10.

White, R.R., Brady, M., Capper, J.L., McNamara, J.P. and Johnson, K.A. (2015). Cow-calf reproductive, genetic, and nutritional management to improve the sustainability of whole beef production systems. Journal of Animal Science 93: 3197-3211.

91. Stakeholder's perceptions about food safety

J. Rodrigues[1], J. García-Díez[2], J. Castro[3] and A. Esteves[2]*
[1]DRAPN – Direção Regional de Agricultura e Pescas do Norte, 5370-347, Mirandela, Portugal; [2]CECAV – Centro de Ciência Animal e Veterinária. Universidade de Trás-os-Montes e Alto Douro, 5001-801, Vila Real, Portugal; [3]IPB – Instituto Politécnico de Bragança. Escola Superior Agrária. Departamento de Ambiente e Recursos Naturais. 5301-855, Bragança, Portugal; alexe@utad.pt

Abstract

The objective of food public policies (FPP) in Portugal is to ensure the safety of foodstuffs and the public health. Currently, food safety is controlled by several official authorities that are responsible for the National Food Control Program of Foodstuffs and Food Operators (NFCP). This program is aimed to control all foodstuffs and food operators. The evaluation of FPP depends on the efficiency of the public administration to assist consumers and organizations to improve their food safety control programs, good manufacturing practices and compliance with the food policy for the benefit of citizens. Thus, the present work was aimed to verify the perceptions of different stakeholders about the FPP and the NFCP. The risk control and compliance of food policy were the priorities of the FPP indicated by the private livestock associations and official inspectors. Other stakeholders presented similar perceptions regarding the risk control; however, they attributed a similar importance to the compliance of the food policy, the guarantee of the public health of consumers and the guarantee of animal and plant health. As expected, the guarantee of the public health of consumers was perceived as the most important characteristic of the FPP by consumers. The perception of food public policies differs among the stakeholders and their professional activity. All of them perceived the risk control as the most important objective of the FPP. However, the perception of the compliance with food policy and the implementation of food controls was mainly revealed by consumers and official inspectors. In contrast, the low perception attributed to the control of animal and plant health indicated a low knowledge about the role of animal and plant health in the food safety.

Keywords: public health, food public policies, risk

Introduction

In Portugal, the control of foodstuffs and food business operators (FO), that include food industries and food establishments, are carried out by several official authorities according to the guidelines defined in the National Food Control Program of foodstuffs and food operators (NFCP).

This program, defined by the food public policies of the government, includes all kind of food inspections that must be carried out in FO (number of FO inspected, frequency, etc.) as well the foodstuffs characteristics that must be surveyed (microbial and chemical analysis, traceability, labelling, etc.) The objective of the NFCP is aimed to ensure food safety, compliance with food policy and public health (Finn *et al.*, 1992).

The food public policies (FPP) could be defined as the legislative framework and other government actions that are carried out to ensure the food safety and public health. Thus, the main perception of the FPP by the general population is the guarantee of the public health by control of food safety and compliance with food law.

However, the development of FPP is difficult since it is necessary an agreement between two groups: the FO and their economical and commercial interest and the defence of the food safety and public health by the government (Martinez *et al.*, 2007). Thus, the objective of this work was aimed to evaluate the perception of the different stakeholders about the FPP and NFCP in Portugal.

Material and methods

The study of the FPP perception by stakeholders was carried out by a web-based voluntary questionnaire. Information and objectives about the study were explained, and all the respondents must previously agree to complete the questionnaire. The stakeholders' overall perception about food public policies was assessed by 12 questions divided into 4 categories as follows: (1) demographic and professional characterization; (2) general perceptions of NFCP; (3) NFCP compliance by food operators; and (4) difficulties and barriers to the application of the NFCP.

According to the professional category declared by respondents, they were grouped in 5 stakeholders. Stakeholder 1 includes all respondents that work at private livestock associations; stakeholder 2 includes farmers; stakeholder 3 includes all respondents that belong to livestock public agencies; stakeholder 4 includes the consumers; and stakeholder 5 includes the official inspectors of livestock and foodstuffs.

Analytic hierarchy process (AHP), a multi-criteria decision-making method (Saaty, 1980), is used to explain the perceptions of stakeholders by pairwise comparisons about the FPP and the NFCP objectives, importance and compliance level. This approach showed the perception matrices of pairwise comparisons obtained from five stakeholders: stakeholder 1 (livestock' organizations – 21 answers), stakeholder 2 (livestock farmers – 31 answers), stakeholder 3 (livestock public agencies – 15 answers), stakeholder 4 (consumers – 58 answers) and stakeholder 5 (food inspectors – 22 answers). The comparisons are processed mathematically, and importance and fulfilment level of perceptions are derived for each stakeholder. Like probabilities, importance and fulfilment level perceptions results are absolute numbers between zero and one, without units or dimensions. The procedure guarantees a consistent result in comparisons. The AHP model also included the stakeholders' responses about the difficulties and barriers, as well as practices and actions needed, for a better performing of NFCP. Those responses were then weighted in function of the power assigned to each stakeholder's the final decision.

Results and discussion

A total of 147 respondents completed the questionnaire. The sample set consisted of 67 farmers (45.57%), 58 consumers (39.46%) and 22 official inspectors (14.97%). By stakeholders, respondents were classified as follow: stakeholder 1 (21-14.29%), stakeholder 2 (31-21.1%), stakeholder 3 (15-10.20%), stakeholder 4 (39.46%) and stakeholder 5 (22-14.96%).

The stakeholders' perceptions about the NFCP are presented in Figure 1. The risk control and compliance of food policy were the priorities of the FPP indicated by the private livestock associations. However, they attributed the lowest importance regarding the animal and plant health. The importance of risk control was also considered by farmers and livestock public agencies; however, they attributed a similar importance to the compliance with food policy, guarantee of consumers' health and guarantee of animal and plant health.

The guarantee of consumers' health was perceived as the most important characteristics of the FPP by consumers. Although with less importance, this stakeholder attributed similar importance to risk control, implementation of food controls and compliance of food policy as same as indicated by the livestock public agencies. In contrast, consumers presented a lower perception of the FPP regarding the animal and plant health. The perception of the FPP by the official inspectors revealed that risk control

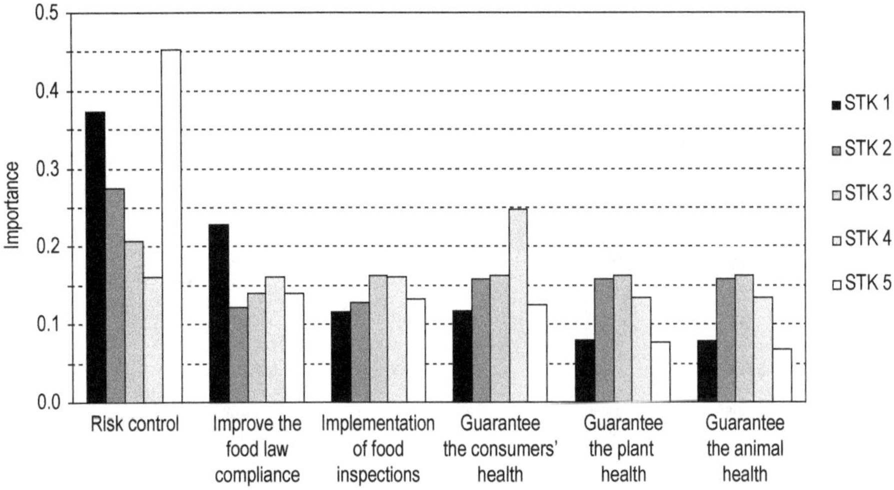

Figure 1. Stakeholders' perception of the National Food Control Program.

should be the most important feature. Moreover, this stakeholder attributed less priority to the others characteristic under study than the other stakeholders.

The importance of the risk control given by the livestock public agencies and official inspectors could be associated to a misinterpretation with the concept 'hazard analysis' since risk assessment (RA) includes this stage. Hazard analysis consists in the identification of unacceptable biological, chemical or physical contamination, rendering the food unfit for consumption whereas risk assessment consists in the methodology to assess the risk from specific hazards through different exposure pathways.

The similar perception about the FPP by the livestock public agencies and could be explained by the fact that they work in the same area (Dosman *et al.*, 2006). The perception attributed by the consumers was expectable since the FPP should be aimed to the guarantee of the consumers' health (Miles *et al.*, 2004). In addition, the perception attributed by consumers regarding the compliance of food policy and food inspection may be explained as the previous steps necessary to achieve to guarantee an appropriate consumer defence.

However, the absence of knowledge of consumers about the food safety at the primary production may explain the low importance attributed to these characteristics (Redmon *et al.*, 2004). The perception of the official inspectors about the FPP is influenced by their professional activity. Since they are government' inspectors, the perception given to risk control and food policy compliance is expectable. However, the greater perception given by the livestock public agencies and consumers regarding the official controls than official inspectors could be explained by the fact that the work developed by these stakeholders are subject to various types of official controls. The barriers observed in the implementation of the NFCP are presented in Figure 2. The lack of communication and variety of public agencies involving in the NFCP were recognized by almost to 60% of the stakeholders as the most important barriers to the implementation of the NFCP.

The lack of training and education has also been referred as an important constrains in the application of the NFCP. Thus, lack of food safety knowledge was indicated by 50% of food inspectors whereas

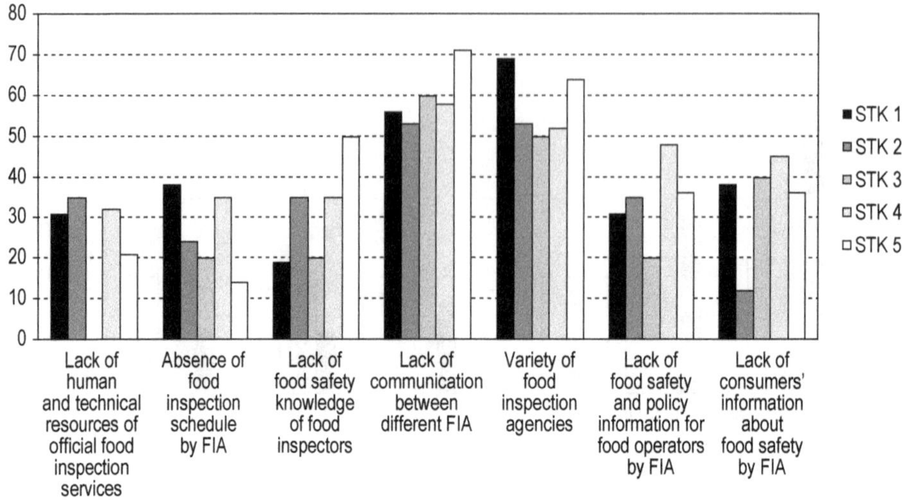

Figure 2. Barriers observed by stakeholders in the National Food Control Program. Results expressed in %.

almost 42% of consumers indicated that lack of food safety and food policy compliance information were important barriers of the NFCP.

The problematic of food inspection by different FIA have been described in the literature (Holley *et al.*, 2010; Varzakas *et al.*, 2006) as the main barrier of the NFCP due to the variety of inspection criteria.

Although the lack of information for FO and consumers was indicated by less than 50% of respondents, this information is necessary to guarantee the food safety and the public health (Holley *et al.*, 2010; Soares *et al.*, 2013; Wilcock *et al.*, 2011). Nowadays, information about food safety and food policy is easily obtained from the Internet. However, the use of Internet in the primary production is scarce given the low use of new technologies (García-Díez *et al.*, 2015). Since the food safety in the primary production is satisfactory (García-Díez, 2012), the current work showed that food training should be adapted to farmers (García-Díez *et al.*, 2015). The lack of information indicated by consumers could be explained by the fact that this stakeholder is only in contact with one part of the intervenient in the food chain, the retail sector. Thus, the results could be explained due to the absence of knowledge about the food safety measures and controls carried out at the animal and plant production (Ménard and Valceschini, 2005).

Regarding the alternative actions (Figure 3), all the stakeholders indicated that it is necessary more training and education to improve the NFCP. For farmers and the private livestock associations, education programs aimed at consumers may improve the information about the quality of its products in a commercial perspective, to increase sales (Botonaki *et al.*, 2006). The importance of the food law compliance could be related to the specific inspections that this sector is usually subjected. Also, the implementation of measures to increase the efficiency of food controls as indicated by food inspectors may be associated to facilitate their work and decrease the potential differences among FIA.

The increase and coordination of food inspections of FO by FIA have been referred in the literature to improve the food safety (Powell *et al.*, 2011). Also, the low importance in the increase of human and technical resources of FIA could be associated to the large number of inspections reported by

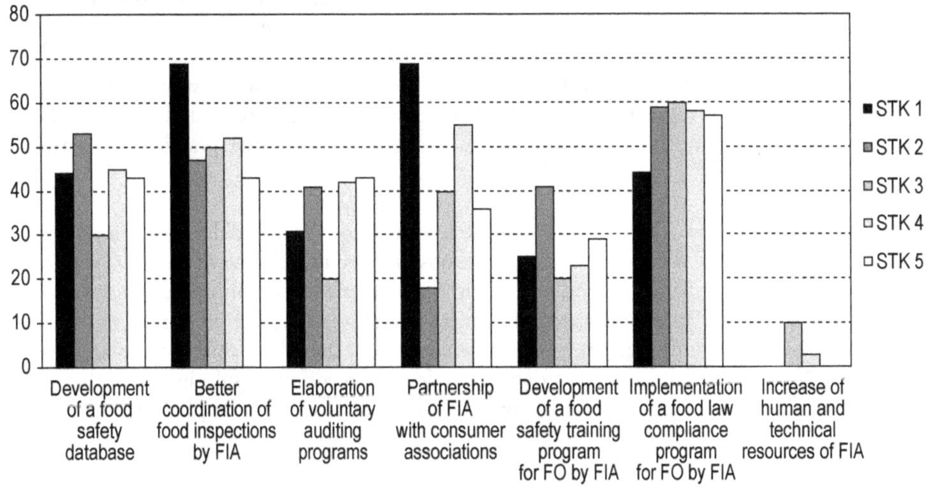

Figure 3. Alternative actions observed by stakeholders to improve the compliance of the National Food Control Program. Results expressed in %.

communication media. Also, this result could be explained because this factor was not indicated as a barrier to the implementation of the NFCP.

The more importance of an implementation of a food law training program for FO by FIA than implementation of food safety education program could be explained by economical purposes. Food safety implies an important economical budget for any FO. Thus, in case of non-compliance of food policy, the application of fines by FIA decrease the budget destined for food safety.

Currently, in Portugal, there are several food inspection databases specific for each FIA. Thus, there are databases for inspections to establishments, animal health surveillance, international food trade or for food sampling for microbiological and chemical analysis. The level of access is limited for each FIA and restricted for general public. Since the development of a food safety database may improve the efficiency of the different FIA (also according to the food public policies) may explain the importance indicated by the stakeholders (Rocha, 2010).

In the performance sensitivity graph for the model (Figure 4) data for each of the seven practices and actions needed to improve the NFCP is shown with a different colour. On the x-axis, the five NFCP stakeholders are represented whereas the overall scores of the seven practices and actions needed are represented in the y-axis.

It should be noted that the implementation of measures to increase the efficiency of food controls (i.e. coordination of inspections) presented a great advantage compared to other alternative practices for all stockholders. Thus, these actions can be very suitable to improve the NFCP performance as they are perceived by all stockholders that operate under low levels of coordination. In Figure 4, the bars situated on the x-axis show the relative importance given for each stakeholder as decision makers (whose values are shown on the left-hand-side y-axis). As the length of a bar increased, the influence/prominence of the corresponding stakeholder increased. In spite of the attribution of more influence to each stakeholder (i.e. modification of the hierarchical predominance of a specific stakeholder), we can observe that ranks and priorities of alternative actions do not present evident changes (Figure 4).

In these cases, we can consider that our decision model is solid, and we can be sure about our choice. Since research about AHP regarding food safety is scarce, the current work could be used by the FIA in the processes of evaluation and comparison of decisions about alternative practices and actions on food safety policies.

In conclusion, the perception of the NFCP is variable according to the stakeholder and their professional activity. All of them perceived the risk control as the most important objective of the NFCP. However, the perception of the compliance of food policy and the implementation of food control was mainly revealed by consumers and official inspectors. In contrast, the low perception attributed to the control of animal and plant health indicated the lower knowledge about the role of animal and plant health in the food safety. Despite small differences among stakeholders' perceptions, results indicated that a better coordination of food inspections by FIA, the implementation of a food law compliance programme for

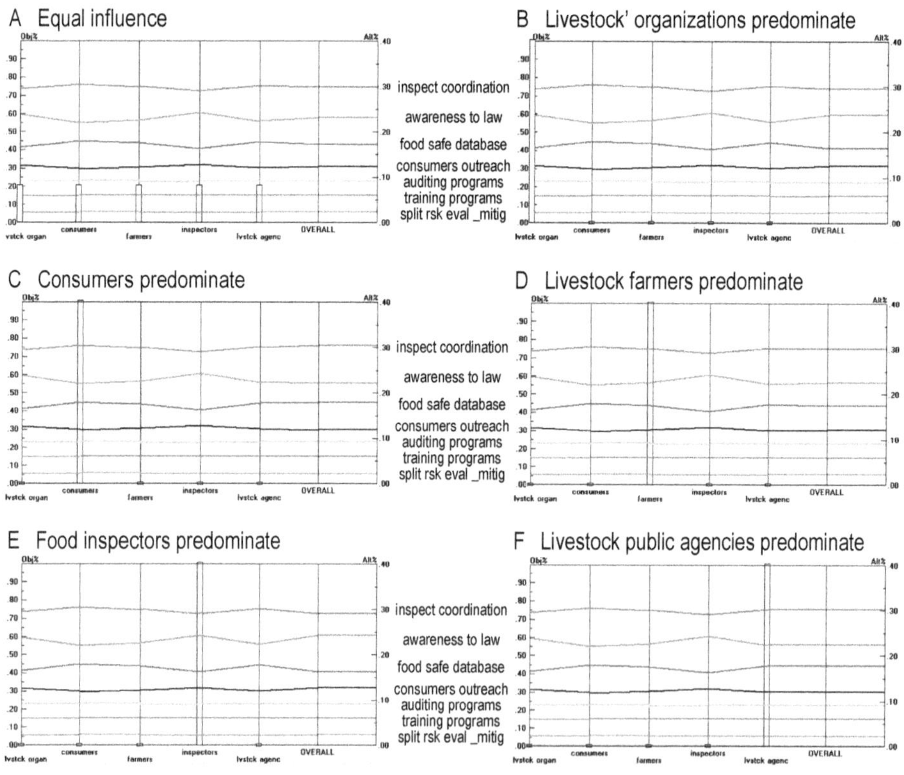

Figure 4. Performance Sensitivity graph (Expert Choice software output): the five NFCP stakeholders on the x-axis, the seven practices and actions needed on right-hand-side y-axis, varying the influence of stakeholders in the decision taking (A = equal influence; B = livestock' organizations predominate; C = consumers predominate; D = livestock farmers predominate; E = food inspectors predominate; and F = livestock public agencies predominate).

FO by FIA, and the development of a food safety database are the right actions and practices to improve the NFCP performance in a robust decision taking.

References

Botonaki, A., Polymeros, K., Tsakiridou, E. and Mattas, K. (2006). The role of food quality certification on consumers' food choices. British Food Journal 108: 77-90.

Dosman, D.M., Adamowicz, W.L. and Hrudey, S.E. (2001). Socioeconomic determinants of health – and food safety – related risk perceptions. Risk Analysis 21(2): 307-318.

Finn, A. and Louviere, J.J. (1992). Determining the appropriate response to evidence of public concern: the case of food safety. Journal of Public Policy & Marketing: 12-25.

García-Díez, J. (2012). Evaluación de la seguridad alimentaria en explotaciones de vacuno lechero de pequeña y mediana dimensión en los municipios de vila real y sabrosa (Portugal) a través de la aplicación de prácticas correctas y medidas de bioseguridad. Revista Electrónica de Veterinaria 13: 1-14.

García-Díez, J., Saraiva, C. and Coelho, A.C. (2015). Cattle farmers' perceptions about livestock policy. Large Animal Review 21(3): 115-123.

Holley, R.A. (2010). Smarter inspection will improve food safety in Canada. Canadian Medical Association Journal 182(5): 471-473.

Martinez, M.G., Fearne, A., Caswell, J.A. and Henson, S. (2007). Co-regulation as a possible model for food safety governance: opportunities for public – private partnerships. Food Policy 32(3): 299-314.

Ménard, C. and Valceschini, E. (2005). New institutions for governing the agri-food industry. European Review of Agricultural Economics 32(3): 421-440.

Miles, S., Brennan, M., Kuznesof, S., Ness, M., Ritson, C. and Frewer, L.J. (2004). Public worry about specific food safety issues. British Food Journal 106(1): 9-22.

Powell, D.A., Erdozain, S., Dodd, C., Costa, R., Morley, K. and Chapman, B.J. (2013). Audits and inspections are never enough: a critique to enhance food safety. Food Control 30: 686-691.

Redmond, E.C. and Griffith, C.J. (2004). Consumer perceptions of food safety risk, control and responsibility. Appetite 43(3): 309-313.

Rocha, J.A. (2010). Gestão do processo político e políticas públicas. Escolar Editora, Lisbon, Portugal.

Saaty, T.L. (1980). The analytic hierarchy process. McGraw Hill, New York, NY, USA.

Soares, K., Garcia-Diez, J., Esteves, A., Oliveira, I. and Saraiva, C. (2013). Evaluation of food safety training on hygienic conditions in food establishments. Food Control 34: 613-618.

Varzakas, T.H., Tsigarida, E.T., Apostolopoulos, C., Kalogridou-Vassiliadou, D. and Jukes, D.J. (2006). The role of the Hellenic Food Safety Authority in Greece – implementation strategies. Food Control 17(12): 957-965.

Wilcock, A., Ball, B. and Fajumo, A. (2011). Effective implementation of food safety initiatives: managers', food safety coordinators' and production workers' perspectives. Food Control 22: 27-33.

92. Adoption of proteomics in traditional meat products: the case of *Khliaa Ezir*

*H.R. Boudechicha, M. Sellama, K. Hafid, A. Boudjellal and M. Gagaoua**
Equipe Maquav, INATAA, Université Frères Mentouri, Constantine 1, Route de Ain El-Bey, 25000 Constantine, Algeria; gmber2001@yahoo.fr

Abstract

Khliaa Ezir is a typical Algerian meat product, prepared according to a traditional process based on empirical observations and traditional recipes, consisting of steps of trimming, marinating, cooking, ripening and ageing in earthenware jar (*Ezir*). Understanding the biological mechanisms behind the development of the sensory properties of this cultural product is of great interest for eventual product marketing. The aim of the present study was first a preliminary proteomic characterization of myofibrillar and sarcoplasmic protein extracts of the product during preparation, ripening and ageing in earthenware jar. A comparison of the proteome of the two extracts at different days and steps was carried out with 1-DE electrophoresis, showing quantitative difference in several bands. Some bands might be potential markers of the proteolysis that takes place during the preparation steps, especially during marinating and ripening. This first and preliminary proteomic data from both myofibrillar and sarcoplasmic protein extracts of *Khliaa Ezir* are consistent with a differential release of specific proteins as a function of preparation step.

Keywords: *Khliaa Ezir*, Algerian meat product, preliminary proteomic characterization

Introduction

Traditional meat products cover a wide class of food items and their characterization is of great importance. Numerous traditional ethnic meat products are commonly prepared in Algeria since the highest antiquity. They have been widely consumed for their enhanced quality, flavour, and storage properties. The inimitable colour, aroma, taste, and simulated shape of Algerian ethnic meat products enjoy a positive reputation and are regarded as the essence of human experience and wisdom. Among them, *Khliaa Ezir*, which is a traditional cured meat mostly produced in the north-east of Algeria using fresh (beef, lamb, goat or camel) meat. The particularity of its traditional process is the ripening step in an earthenware jar able to be preserved for more than one year (Boudechicha *et al.*, 2015). *Khliaa Ezir* preparation and processing, cooking habits and its organoleptic properties exhibit the deep connotation of Algerian foods of delectable, healthy, and cultural heritage.

During the processing of cured meat products which involve both (1) seasoning/curing; and (2) cooking/ripening steps, complex chemical and biochemical changes occur on the main components of raw meat (proteins and lipids) leading to a numerous modifications on colour, flavour and texture (Toldra, 2006). Cured meat products manufactured from beef are particularly a great source of proteins and free amino acids (Garrido *et al.*, 2012). Many of these free amino acids come from proteolysis, which is at the central of myriad biochemical reactions occurring in cured meat preparation, contributing to the typical flavour and texture of the final product (Lorenzo and Franco, 2012).

To understand the type and extent of proteolysis, several proteomic techniques have been successfully applied for the identification of generated peptides and their sequencing. This approach provides new insights to achieve better quality of cured meat products and reveal key biochemical mechanisms controlling the process. This may play a great role on the preservation of this meat product both by its

EurSafe 2016 – DOI 10.3921/978-90-8686-834-6_92, © Wageningen Academic Publishers 2016

cultural and historical dimension. In addition, the nutritional value of the product would be well known, hence providing a full *Khliaa Ezir* data sheet for consumers and butchers. Thus, the aim of the present study was a preliminary proteomic characterization of myofibrillar and sarcoplasmic protein extracts of *Khliaa Ezir* during preparation, ripening and ageing in earthenware jar.

Materials and methods

Preparation of Khliaa Ezir and sampling

Khliaa Ezir is prepared from fresh boneless beef meat using the traditional preparation diagram (Figure 1). Briefly, fresh selected cuts of *Semimembranosus* muscle (5-8 cm length, 4-6 cm thick) obtained from a local butcher were marinated for 7 days in a mixture of spices (salt, coriander, caraway, and smashed garlic) before cooking at an average temperature of 80 °C on water. After that, the cooked meat was immersed in a mixture of melted bovine fat and olive oil and preserved in an earthenware jar for numerous months. In this study, nine preparations of *Khiaa Ezir* were conducted in the laboratory following the traditional diagram given in Figure 1. Aliquots of each preparation were sampled at different times of the process: on fresh meat (day 0), marinade time (days 1, 3, 5 and 7), after cooking (day 8), during ripening and storage (days 2, 6, 10, 15 and 21).

Extraction of sarcoplasmic and myofibrillar muscle proteins

The sarcoplasmic and myofibrillar proteins were extracted according to the procedure described by Diaz *et al.* (1997). Briefly, each aliquot of *Khliaa Ezir* (from fresh meat to 21 day of storage on the

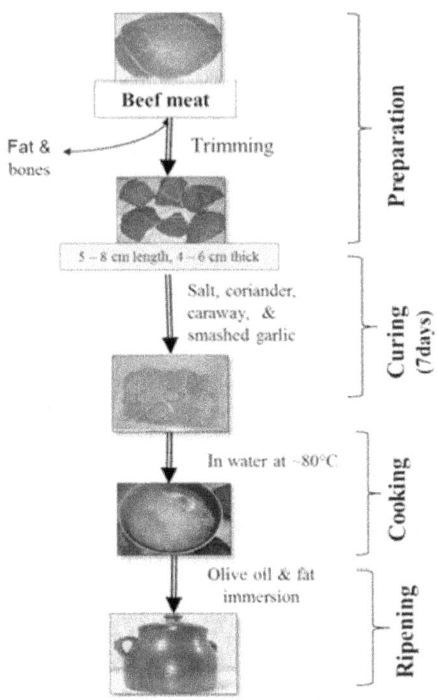

Figure 1. The traditional diagram used for the preparation of Khliaa Ezir *(Boudechicha* et al.*, 2015).*

earthenware jar) at the desired time was used for protein extraction by homogenization on 1:10 (w/v) of 40 mM sodium phosphate buffer containing 0.02% NaN_3 using a Polytron homogenizer (×20,000 rpm) (Polytron ® PT- MR 2100, Kinematica AG, Switzerland). The homogenate was then filtered through clean cheesecloth before centrifugation at 4,000 rpm for 15 min at 4 °C (Bench top centrifuges NF 400R). The supernatant was collected as the sarcoplasmic protein fraction and stored in adequate tubes at -20 °C until use. The resulting pellet was re-suspended in 10 ml of the same buffer and centrifuged twice. The final pellet was once again re-suspended in 10 ml of buffer, filtered through Whatman paper and collected as the myofibrillar protein fraction and stored at -20 °C until use. The protein concentration of both sarcoplasmic and myofibrillar protein extracts was determined using the Bradford protein assay kit obtained from Bio-Rad. A calibration curve was constructed using bovine serum albumin as the standard (Bradford, 1976).

1DE-electrophoretical analysis (glycine SDS-PAGE)

The extracts of the sarcoplasmic and myofibrillar proteins were mixed in a ratio of 1:1 (v/v) with 312.5 mM Tris buffer (pH 6.8), 0.4 mM EDTA, 7.5% (w/v) SDS, 150 mM dithiothreitol, 25% glycerol and 0.05% bromophenol blue. The mixture was then heated at 95 °C for 5 min. The protein extracts (10 µg for each well) were analysed using one dimensional SDS-PAGE electrophoresis on 3.75% stacking and 12% resolving gels according to Laemmli, (1970). A mixture of proteins with a known molecular weight ranging from 10 to 250 kDa (#161-0374) obtained from Bio-Rad Laboratories, Hercules, CA, was used. After separation, the gel was stained overnight as described by Gagaoua *et al.* (2015) with a staining solution of 4.9 mM Coomassie Brilliant Blue G-250, 50% (v/v) ethanol, and 7.5% (v/v) acetic acid. Protein patterns were then visualized after destaining the gel until a clear background was achieved. The molecular weight of the protein bands was calculated using the Un-Scan-It Gel 6.1 analysis program (Silk Scientific, Orem, UT).

Results and discussion

During the processing of *Khliaa Ezir*, the proteins undergo various modifications which depend on numerous factors. The separated sarcoplasmic and myofibrillar are depicted on Figure 2. The overall banding patterns between replicates were similar and protein bands were identified based on their molecular weights.

Myofibrillar proteins

Significant change on the protein profile of the myofibrillar proteins of *Khliaa Ezir* during preparation was observed (Figure 2a). For example, there is a progressive reduction in intensity of the two common myofibrillar and structural proteins, myosin heavy chain (MHC), and actin, appearing approximately at 200 kDa and 47 kDa, respectively. The significant decrease in abundance of these proteins seems to clearly occur after the cooking phase. In accordance, several research papers reported the effect of cooking/ripening on meat products (Diaz *et al.,* 1997; Ruiz *et al.,* 2007). On another hand, Thorainsdottir *et al.* (2002) has shown that myosin is more sensitive to processing than actin. Since thermal treatment of meat can have an impact on the primary structure of the meat proteins, which is closely associated with both the nutritional and sensory qualities of cooked meat, characterizing cooking-induced protein modifications in meat is one of the fundamental steps to understand these effects. This work is now under investigation as a complementary work with both MS identification and characterization of the other changed proteins.

Otherwise, other high MW bands, such as the 100 kDa also decreased in intensity before its disappearance after cooking. At the same time, a progressive disappearance of other proteins such as the 61, 35 and 30

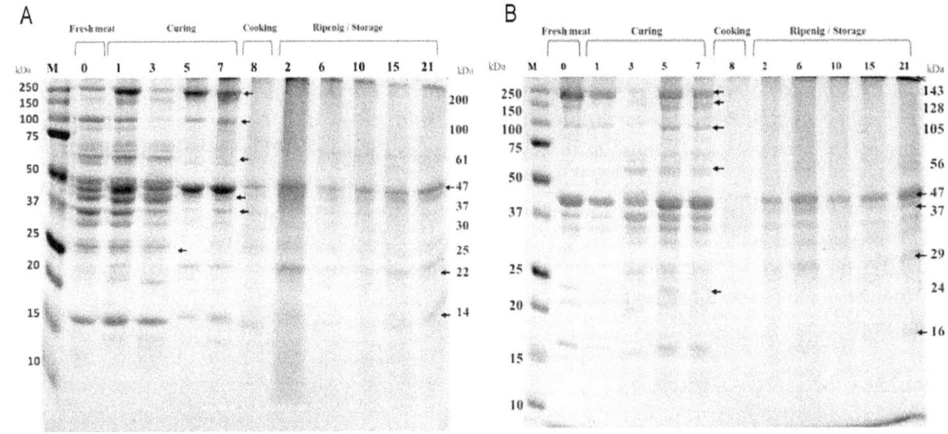

Figure 2. SDS-PAGE profile of (A) myofibrillar and (B) sarcoplasmic proteins at different preparation steps of Khliaa Ezir *(the numbers indicate days of preparation and storage).*

kDa are also observed during the last period of curing. The intensity of the band at 14 kDa decreases throughout the curing time and disappear completely after cooking. Several studies conducted on numerous traditional products showed similar changes to those described in myofibrillar proteins during meat product processing. For example, Larrea *et al.* (2006) observed that in dry cured ham the bands at 97, 22 and 14 kDa decrease progressively during the curing time, probably due to a specific sensitivity to salt. Garcia *et al.* (1997) studied the changes in proteins during the ripening of Spanish dried beef *Cecina* and also observed changes in the myofibrillar proteins: disappearance of MHC and troponin C after the smoking step (a thermal treatment).

Sarcoplasmic proteins

The electrophoretic profile of sarcoplasmic proteins is shown in Figure 2b. As for the myofibrillar protein pattern, numerous sarcoplasmic proteins were found to be changed and affected by the preparation steps of *Khliaa Ezir*. We can suppose that these proteins may undergo both truncation and aggregation after cooking as recently proposed by Wen *et al.* (2015) and Yu *et al.* (2015). Numerous proteomic studies reported that the mainly affected proteins are namely glycolytic enzymes such as lactate dehydrogenase A chain isoform 1, beta-enolase, fructose bisphosphate aldolase A isoform 1, pyruvate kinase isozymes M1/M2 isoform 4, glyceraldehyde-3-phosphate dehydrogenase and triosephosphate isomerase 1 (for a review see Picard *et al.,* 2016). In this study, the intensity of a band with a mobility very close to heavy myosin chain and an estimated MW of 143 kDa decreased throughout the preparation steps leading to its disappearance after cooking. The appearance of bands at about 105, 56 and 37 kDa has been observed during the last period of curing. The formation of compounds with similar MW has been sometimes observed during ripening of fermented sausages (Garcia de Fernando and Fox, 1991). The appearance of these proteins could result from the treatment effect leading to the formation of protein dimmers and/or aggregation of numerous proteins together. A complete mass spectrometry is under process for a better characterization of the phenomenon really occurring during *Khliaa Ezir* preparation. The intensity of band 47 kDa increased throughout the curing time and became very intense by days 5 and 7. After cooking, several bands disappeared; only this later band which seems to be highly dominant and present with a slight intensity.

According to the related literature, proteolytic processes that occur during the curing and ripening of cured meat products are mainly due to endogenous proteolytic activity (Toldra *et al.,* 1997). To this regard, Zhao *et al.* (2005) observed that the activity of these enzymes was dependent on the pH, temperature and salt content. In our study, the behaviour of the proteins is due to curing conditions (time, temperature and salt content), but this phenomenon is now under investigation with accurate techniques. The appearance of compounds of MW between 100 and 30 kDa observed throughout the marinating time could result from the breakdown of myosin heavy chain and other high molecular weight proteins. The denaturation of proteins seems to be affected by the length of curing time (unpublished data). These findings are in agreement with those reported by Martín *et al.* (1998) and Perez-Palacios *et al.* (2010) who found that salt content and the curing time affect strongly the proteolytic activity. The effect may be more pronounced on the endogenous enzymes such as cathepsins, dipeptidyl peptidases and aminopeptidases, which release large peptide fragments from proteins.

During the processing of *Khliaa Ezir*, the temperature applied during thermal treatment (80 °C) can have enhanced protein denaturation. As can be seen from Figure 2, the myofibrillar and sarcoplasmic proteins of MW greater than 47 kDa disappeared after cooking. The only exception was the band at 47 kDa which still present throughout the ripening step. Earlier studies conducted on the effects of heating on protein denaturation revealed similar changes (Di Luccia *et al.,* 2005). Tornberg, (2005) reported that between 65 and 87 °C many changes are taking place in the myofibrillar and sarcoplasmic proteins, which causes their aggregation, decreases the solubility and disappearance of the bands. Meat protein aggregation induced by cooking can lead for example to a compact structure (Promeyrat *et al.,* 2010). Furthermore, the proteins with MW less than 50 kDa still present after cooking in some ripened meat products process. This is consistent with our study suggesting that actin, the myofibrillar protein previously shown to be relatively heat stable below 100 °C (Huang *et al.,* 2011), was still detected. Finally, the identification of numerous proteins above and below their theoretical MW probably relates to *post-mortem* actomyosin complex formation and proteolysis (the work still in progress).

Conclusions

This preliminary proteomic investigation has shown the complex pattern of protein changes occurring during the traditional preparation of *Khliaa Ezir*, a cured and cooked meat product of Algeria. In summary, cooking combined with the marinating step seems to affect considerably the textural properties of the product, by significant changes on the structural proteins. Complimentary studies on the characterization and identification of the affected proteins by accurate proteomic and peptidomic techniques are now in progress for a well characterization of the mechanisms behind the sensory qualities development.

References

Boudechicha, H.R., Gagaoua, M., Hafid, K., Becila, S., Boudjellal, A. and Astruc, T. (2015). Khliaa Ezir, a traditional cured meat product of Algeria: preparation and characterization. In 61st International Congress of Meat Science and Technology, Clermont-Ferrand, France.

Bradford, M.M. (1976). A rapid and sensitive method for the quantitation of microgram quantities of protein utilizing the principle of protein-dye binding. Analytical Biochemistry 72: 248-254.

Diaz, O., Fernandez, M., Garcia de Fernando, G.D., De la Hoz, L. and Ordóñez, J.A. (1997). Proteolysis in dry fermented sausages: the effect of selected exogenous proteases. Meat Science 46: 115-128.

Di Luccia, A., Picariello, G., Cacace, G., Scaloni, A., Faccia, M., Liuzzi, V., Alviti, G. and Spagna Musso, S. (2005). Proteomic analysis of water soluble and myofibrillar protein changes occurring in dry-cured hams. Meat Science 69: 479-491.

Gagaoua, M., Hoggas, N. and Hafid, K. (2015). Three phase partitioning of zingibain, a milk-clotting enzyme from Zingiber officinale Roscoe rhizomes. International Journal of Biological Macromolecules 73: 245-252.

Garcia de Fernando, G.D. and Fox, P.F. (1991). Study of proteolysis during the processing of a dry fermented pork sausage. Meat Science 30: 367-383.

Garcıa, I., Diez, V. and Zumalaca'rregui, J.M. (1997). Changes in nitrogen fractions and free amino acids during ripening of Spanish dried beef: 'Cecina'. Journal of Muscle Foods 9: 257-266.

Garrido, R., Domínguez, R., Lorenzo, J.M., Franco, I. and Carballo, J. (2012). Effect of the length of salting time on the proteolytic changes in dry-cured lacóon during ripening and on the sensory characteristics of the final product. Food Control 25: 789-796.

Huang, F., Huang, M., Xu, X. and Zhou, G. (2011). Influence of heat on protein degradation, ultrastructure and eating quality indicators of pork. Journal of the Science of Food and Agriculture 91(3): 443-448.

Lorenzo, J.M. and Franco, D. (2012). Fat effect on physico-chemical, microbial and textural changes through the manufactured of dry-cured foal sausage Lipolysis, proteolysis and sensory properties. Meat Science 92: 704-714.

Laemmli, H.K. (1970). Cleavage of structural proteins during the assembly of the head of the bacteriophage T4. Nature 227: 680-685.

Larrea, V., Hernando, I., Quiles, A., Lluch, M.A. and Perez-Munuera, I. (2006). Changes in proteins during Teruel dry-cured ham processing. Meat Science 74: 586-593.

Perez-Palacios, T., Ruiz, J., Barat, J.M., Aristoy, M.C. and Antequera, T. (2010). Influence of pre-cure freezing of Iberian ham on proteolytic changes throughout the ripening process. Meat Science 85: 121-126.

Picard, B., Gagaoua, M. and Hollung, K. (2016). Chapter 13: proteomics as a tool to explain/predict meat and fish quality. In: Purslow, P. (ed.) Meat quality aspects: from genes to ethics. Elsevier, New York, NY, USA.

Promeyrat, A., Gatellier, P., Lebret, B., Kajak-Siemaszko, K. and Aubry, L. (2010). Evaluation of protein aggregation in cooked meat. Food Chemistry 121(2): 412-417.

Ruiz A.G., Mariscal, C. and Soriano, A. (2007). Influence of hunting-season stage and ripening conditions on nitrogen fractions and degradation of myofibrillar proteins in venison (Cervus elaphus) chorizo sausages. Meat Science 76: 74-85.

Thorainsdottir, K.A., Arasonm, S., Geirsdotir, M. and Bogason, S.G. (2002). Changes in myofibrilar proteins during processing of salted cod (Gadus morhua) determined by electrophoresis and differential scanning calorimetry. Food Chemistry 77: 327-335.

Toldra, F., Flores, M. and Sanz, Y. (1997). Dry-cured ham flavour: enzymatic generation and process influence. Food Chemistry 59: 523-530.

Toldra, F. (2006). The role of muscle enzymes in dry-cured meat products with different drying conditions. Trend in Food Science Technology 17: 164-168.

Tornberg, E. (2005). Effects of heat on meat proteins – implications on structure and quality of meat products. In 50th International Congress of Meat Science and Technology, (ICoMST), 8-13 August 2004, Helsinki, Finland.

Wen, S., Zhou, G., Li, L., Xu, X., Yu, X., Bai, Y. and Li, C. (2015). Effect of cooking on *in vitro* digestion of pork proteins: a peptidomic perspective. Journal of Agricultural and Food Chemistry 63: 250-261.

Yu, T.Y., Morton, J.D., Clerens, S. and Dyer, J.M. (2015). Proteomic investigation of protein profile changes and amino acid residue level modification in cooked lamb meat: the effect of boiling. Journal of Agricultural and Food Chemistry 63: 9112-9123.

Zhao, G.M., Zhou, G.H., Tian, W., Xu, X.L., Wang, Y.L. and Luo, X. (2005). Changes of alanyl aminopeptidase activity and free amino acid contents in biceps femoris during processing of Jinhua ham. Meat Science 71: 612-619.

Workshop

93. Utopian thinking and serious game design: two powerful tools to fight food waste

F. Vieira[1] and M. Alves de Barros[2]
[1]Faculdade de Letras da Universidade do Porto; Via Panorâmica s/ n.° 4150-564 Porto, Portugal; [2]Universidade Federal de Campina Grande, Departamento de Sistemas e Computação, Rua Aprígio Veloso, 882, Bloco CN, Bairro Universitário, 58429-900 Campina Grande-Paraíba, Brazil; vieira.mfatima@gmail.com; mbarros@computacao.ufcg.edu.br

Abstract

This paper offers part of the information delivered to the delegates attending the workshop facilitated by Fátima Vieira (on site, at the EurSafe 2016 Conference) and Marcelo Alves de Barros (via Skype). It aims at familiarising the reader with the principles of utopian thinking and the basics of alternate reality games design. In the first part, it offers theoretical knowledge about the concept of utopia, how it has worked as a driving force for change over the centuries, and the kind of (prospective, inclusive, critical, holistic and creative) thinking it entails. By resorting to the idea put forward by Frédéric Lenoir that as Modernity has made the individual the centre of everything the only way to solve the problems of the ultra-modernity we live in lies in the possibility of changing the way the human being thinks, the authors propound there are two ways of internalizing new perspectives and behaviours: by reading books and by playing serious games. The second part of the paper highlights the advantages of alternate reality games, the 'porousness' of the activities they promote in the virtual and the real worlds, and the way they engage a crowd in the adventure of looking for a solution to a real problem. At the end of the paper, the reader will find the challenge set to the EurSafe Conference delegates participating in the workshop: to contribute to the collaborative design of a new game that, by resorting to the dynamics of utopian thinking, succeeds in appealing to a wide community of players committed to the unswerving purpose of finding new possibilities to fight food waste.

Keywords: Thomas More, future, gaming

Utopian thinking

This year we are commemorating the 500th anniversary of the publication of Thomas More's 'Utopia'. This is no doubt an important event, as it marks the birth of a new literary genre that created space for a new way of thinking. We cannot forget that 'Utopia' was written and published during the Renaissance and that Thomas More was a prominent humanist: 'Utopia' conveys, above all, a ground-breaking discovery – the human being is not simply supposed to accept his or her fate, but expected to reason with it in order to build the future. This discovery implied a broadening of horizons to which the geographical expansion of the 15th and 16th centuries has much contributed. In actual fact, these discovery periods forced Europeans to contact with new cultures and new forms of social, economic and political organisation. Moreover, by borrowing the narrative techniques from travel literature, Thomas More created a genre which entailed the travel of a protagonist unto an unknown place, the visit to its institutions, and a return back home to spread the word about the newly discovered possibilities. 'Possibilities' is, in all truth, the keyword to understand what utopia is really about; it entails searching, experimenting, creating, and innovating.

Nevertheless, Utopia is also the name of the island described by Thomas More, and the title of his book. When he started to write his masterpiece, More first called the island *Nusquama*, which in Latin means 'nowhere', 'in no place', 'in no occasion'. However, he soon found out that this would not suit the

EurSafe 2016 – DOI 10.3921/978-90-8686-834-6_93, © Wageningen Academic Publishers 2016

extension of the humanistic discovery, and as the search for new possibilities entailed a never-ending process, he created a neologism by adding the suffix *ouk,* which means *not* (then reduced to *u*) to the Greek word for place (*topos*). Utopia thus means a place which is not a place, and is marked by a tension, a perpetual movement of affirmation and denial. Right at the end of his book, More introduced another neologism which further complicated this tension: *Eutopia*, which means 'place of happiness'. As in English these two words are pronounced in exactly the same way, the search for utopia is also to be seen as a search for happiness, but done with the awareness that this place of happiness does not exist; still, the Morean message seems to be that it is important to go on looking for it. Right from the very beginning, utopia has always been seen as a process, this is, a dynamic and never-ending search.

Although humanistic thought implied a major advancement for the way human beings were to think about their role in the world – no longer passive, but positive, constructive, and creative – it was nonetheless marked by a Christian worldview based on the idea of the Fall. As human beings were then seen as sinful, the only effective way of ensuring that his or her sins would not affect society would be by investing in a form of social organisation which would, on the one hand, be governed by laws that would repress instincts and behaviours that were prejudicial to others, and on the other hand stimulate beneficial comportments.

Over the years, this conception of the human being has changed, and it has thus entailed a different conception of the utopian strategy. The very positive view of the infinite possibilities of human perfection that characterised the Enlightenment led to the publication of optimistic utopias, visions of societies that worked well because they were a reflection of the individuals that they had built. In the 19th century, socialist utopias implied an investment in the possibilities created by the Marxist social, economic and political organisation; the abolition of property would solve every problem and would originate a 'new' social human being.

The onset of the 20th century, with its two World Wars, brought pessimism about the future back into Western thinking. For most of the century – except for a period of optimistic revival in the late 60s and 70s, dystopian thought dominated – and we have had to acknowledge its pervasiveness until the current day. This does not mean, though, that utopian thinking has ended. It is still to be found mostly in ecological and feminist utopian thinking; and it persists in other fields, although often hiding behind other labels: social betterment, innovative thinking, or critical thinking. Contemporary utopian thinking is in fact marked by these ideas, and these are part of its strategy.

As Frédéric Lenoir has contented, since Modernity has made the individual the centre of everything, the only way to solve the problems of the ultra-modernity we live in lies in the possibility of changing the way the human being thinks (Lenoir, 2012). This may be done as it used to be done in the past – as More did himself, i.e. through literature. Literature is in fact one of the most effective ways of affecting people's thinking about the world because: (a) it takes you to places and allows you to come into contact with people and social groups you would otherwise never encounter; (b) it is a simulator of reality, as it provides you with opportunities to live the characters' experiences, thus preparing you for life; (c) it provides you with the opportunity to experience a variety of feelings: empathy, understanding, disgust, indignation, revolt, and rebellion; (d) it inspires you when you are confronted with individuals living difficult situations, and it facilitates the acceptance of your own failure when you witness the failure of fictive characters; and (e) it enables you to understand the Other, this is, to appreciate different cultures and different ways of life, to identify the fundamental questions and to understand the consequences of your actions. Furthermore, utopian literature adds other experiences to this process as it develops prospective thinking, inclusive thinking, critical thinking, holistic thinking and creative thinking. Utopian literature is thus a powerful tool for changing the world.

As the 2004 report 'Reading at risk: a survey of literary reading in American' has evinced, there is a direct relation between people's reading habits and the development of a civic awareness and readiness to participate in the cultural life of a society. Gioia Dana, the then Chair of the National Endowment for the Arts, the organisation that promoted this study, expressed his concern for the results of the survey which revealed that, for the first time in the history of modern American society, less than half of the population has reading habits. These numbers are very alarming because, as Gioia highlighted, the consequences will be visible in the long-term, and will affect not only literature, but all the arts and social activities, such as political engagement, philanthropy and voluntary work (Gioia, 2004: VII).

Serious games

Reading is thus important, not only because of the enjoyment of the aesthetics it involves, but also because of the empathy it promotes with the characters' lives. However, there are other ways of promoting empathy; such is the case of 'serious games,' which may well be seen, regarding the identification process they imply, as being complementary to the process of reading.

Interestingly, in his 'Utopia', Thomas More recognises the important value of games for the character building of citizens. The Utopians read or are read texts aloud and are lectured every day, early in the morning, before lunch and before dinner, but in the evening, before going to bed, they play two games that are very similar, in their principles, to chess:

> The one is the battle of numbers, wherein one number stealeth away another. The other is wherein vices fight with virtues, as it were in battle array or a set field. In the which game is very properly shewed both the strife and discord that vices have among themselves, and again their unity and concord against virtues; with what power and strength they assail them openly; by what wiles and subtlety they assault them secretly; with what help and aid the virtues resist and overcome the puissance of the vices; by what craft they frustrate their purposes; and, finally, by what sleight or means the one getteth the victory.'

What Thomas More is talking about here is in fact what we would currently call a process of internalisation of knowledge, following Nonaka's and Takeuchi's description of knowledge conversion (1995). The Utopians receive knowledge (as to the consequences of vices, how they attack and how virtues are to be fought back), apply it by playing the game and internalise it so that it becomes part of their own knowledge. The 'spiral' of knowledge creation then moves on to another level, which implies that through 'socialisation', 'externalisation' and 'combination' the spread of good values is ensured.

This process of knowledge conversion is possible via game playing because there is an obvious pervasiveness between the realms of the game and that of real life. The concept introduced by Johan Huizinga in 'Homo ludens: a study of the play-element in culture' (first published in 1938) will be useful here to understand the importance of the play element of culture and society. As Huizinga propounded, social life includes play in a variety of situations, from the card-table and the stage to the temple or the law court, where a series of rites and rituals are to be observed; these are temporary worlds within the real world, with rules of their own and devoted to peculiar performances (Huizinga, 10) – and they are 'magic circles' in that sense.

The so-called 'serious games' that are on offer nowadays in the digital world have been inspired by this idea of the magic circle. Katie Salen and Eric Zimmerman, in 'Rules of Play: Game Design Fundamentals' (2004), have highlighted how the porousness of this magic circle is instrumental to the effectiveness of games (both digital and non-digital, although they focus especially on the former) for affecting real life; this happens because games, considered as 'culture', are 'extremely open systems: ... as a cultural system the focus is on the way the game exchanges meaning with culture at large' (Salen and Zimmerman, 2004). The players enter the magic circle in order to experience what it has to offer in terms of entertainment, dreams, narrative, and catharsis, among other aspects; and when they return to the real world they carry with them both experience and meaning. Consequently, the magic circle is a place for learning and for cultural transformation, and it is twofold: the real world influences the virtual world (via the game designers) and the virtual world has an impact on the real world (via the players). That is why Salen and Zimmerman stand for the idea, in the concluding chapter of their book, that 'game designers create meaningful play' (*ibid.*: 13, 4). This is of course more visible when the player is trusted in the games with missions which, even if he or she is not aware of it, are to be fulfilled in the virtual worlds but also, in part, in the real world.

This pervasiveness of the effects of the game in both the virtual and the real worlds is one of the main characteristics of the so-called 'alternate reality games'. These games identify problems and transform them into projects of public utility by engaging a high number of players (a crowd) in an individual or collective transformation that is only to happen if the activities take place in both the real and the virtual worlds. These games normally use the real world as a platform for a shared narrative that lasts over several weeks or even months, and that challenges the community of players to work collaboratively in order to solve a real problem – something that no one would ever be able to solve alone. They normally imply transmedial storytelling (web sites, emails, blogs, MP3s and DVDs, web cams, text messages, instant messages, networked game consoles, and hand-held GPS devices, among other possibilities) and unfold in 'real-time'. As the title of the 2011 Jane MacGonigal book reveals, 'Reality is Broken: Why Games Make Us Better and How They Can Change the World,' these games are really meant to transform society via the transformation of the individual. And here, as in the case with literature, empathy – with the community that experiences a problem and is committed to solving it – is instrumental to awaken, in each person, the responsibility to participate in the collective epic challenge, and the awareness of his or her capability to find the best solution for the problem.

In a pervasive experience of knowledge conversion, literature and serious games of alternate reality can be used together to change the world, mainly when a lot of readers and players become an engaged crowdsourcing with utopian missions.

The workshop

Run by Fátima Vieira (on site) and Marcelo Alves de Barros (via Skype), this workshop will welcome up to 30 participants to be divided into five groups. Each group will count on three consultants – high school and university students who have been working on ways of solving the problem that has been identified for the workshop – the need to fight food waste –, and who are, themselves, very familiar with the world of digital games.

The first part of the workshop will be mainly informative, and will be devoted to a description and explanation about the way utopian thinking works. The five types of thinking it involves – prospective, inclusive, critical, holistic and creative thinking – will be explored through a variety of examples. The second part of the workshop will aim at providing insights into the way serious games have been used to solve problems and to collectively engage communities of players in a collaborative strategy. In the third (and main) part of the workshop the participants will be invited to play a few games and to contribute to the collaborative design of a new game that will hopefully result in a joint project of the Federal University of Campina Grande, Brazil, and the University of Porto, Portugal – a 'serious game' of alternate reality which, by resorting to the dynamics of utopian thinking, effectively succeeds in appealing to a wide community of players committed to the unswerving purpose of finding new possibilities to fight food waste. Moreover, critical thinking and innovative thinking, aiming at the betterment of society – the very basis of utopian thinking – will be called for; the game designers (the participants in the workshop) will definitely need it, as will the future players, who will be expected to internalise not only new knowledge about the consequences of food waste but also the awareness that it is their responsibility to take action and look for new 'possibilities'. In this way, we will not be very far from Thomas More's 'Utopia'; the only difference will be that now we will be expected to work together and engage together in this quest for alternatives. We will thus be stronger, will we not?

References

Gioia, D. (2004). Preface. Reading at risk: a survey of literary reading in America. Research Division Report 46. National Endowment for the Arts, Washington, DC, USA.

Huizinga, J. (1949). Homo ludens; a study of the play-element in culture. Routledge, London, UK.

More, T. (1985). Utopia. Dent, Everyman [1516].

Lenoir, F. (2012). La Guérison du Monde. Fayard, Paris, France.

MacGonigal, J. (2011). Reality is broken: why games make us better and how they can change the world. Jonathan Cape, London, UK.

Nonaka, I. and Hirotaka T. (1995). The knowledge-creating company: how Japanese companies create the dynamics of innovation. Oxford University Press, Oxford, UK.

Salen, K. and Zimmerman, E. (2004). Rules of play: game design fundamentals. MIT. MA, USA.

Author index

www.ingramcontent.com/pod-product-compliance
Lightning Source LLC
Chambersburg PA
CBHW061150240926
51846CB00006B/969
9789086862887